IET HISTORY OF TECHNOLOGY SERIES 34

The Struggle for Unity

Colour television, the formative years

Other volumes in this series:

Volume 4 **The history of electric wires and cables** R.M. Black
Volume 6 **Technical history of the beginnings of radar** S.S. Swords
Volume 7 **British television: the formative years** R.W. Burns
Volume 9 **Vintage telephones of the world** P.J. Povey and R. Earl
Volume 10 **The GEC research laboratories 1919–1984** R.J. Clayton and J. Algar
Volume 12 **A history of the world semiconductor industry** P.R. Morris
Volume 13 **Wireless: the crucial decade 1924–34** G. Bussey
Volume 14 **A scientists war – the diary of Sir Clifford Paterson 1939–45** R.J. Clayton and J. Algar (Editors)
Volume 16 **Curiosity perfectly satisfied: Faraday's travels in Europe 1813–1815** B. Bowers and L. Symonds (Editors)
Volume 17 **Michael Faraday's 'Chemical Notes, Hints, Suggestions and Objects of Pursuit' of 1822** R.D. Tweney and D. Gooding (Editors)
Volume 18 **Lord Kelvin: his influence on electrical measurements and units** P. Tunbridge
Volume 19 **History of international broadcasting, volume 1** J. Wood
Volume 20 **The early history of radio: from Faraday to Marconi** G.R.M. Garratt
Volume 21 **Exhibiting electricity** K.G. Beauchamp
Volume 22 **Television: an international history of the formative years** R.W. Burns
Volume 23 **History of international broadcasting, volume 2** J. Wood
Volume 24 **Life and times of Alan Dower Blumlein** R.W. Burns
Volume 26 **A history of telegraphy: its technology and application** K.G. Beauchamp
Volume 27 **Restoring Baird's image** D.F. McLean
Volume 28 **John Logie Baird: television pioneer** R.W. Burns
Volume 29 **Sir Charles Wheatstone, 2nd edition** B. Bowers
Volume 30 **Radio man: the remarkable rise and fall of C.O. Stanley** M. Frankland
Volume 31 **Electric railways, 1880–1990** M.C. Duffy
Volume 32 **Communications: an international history of the formative years** R.W. Burns
Volume 33 **Spacecraft technology: the early years** M. Williamson

The Struggle for Unity

Colour television, the formative years

Russell W. Burns

The Institution of Engineering and Technology

Published by The Institution of Engineering and Technology, London, United Kingdom

First published 2008

The Institution of Engineering and Technology
Michael Faraday House
Six Hills Way, Stevenage
Herts, SG1 2AY, United Kingdom

www.theiet.org

British Library Cataloguing in Publication Data
A catalogue record for this product is available from the British Library

ISBN 978-0-86341-824-2

Typeset in India by Newgen Imaging Systems (P) Ltd, Chennai
Printed in the UK by Athenaeum Press Ltd, Gateshead, Tyne & Wear

Contents

Preface vii

Acknowledgements ix

List of figures xi

List of tables xviii

Chronology relating to colour television xix

List of abbreviations xxiii

Introduction xxv

1 A brief history of colour photography and colour cinematography 1

2 Low-definition colour television 19

3 RCA and pre-war colour television 43

4 EMI and high-definition television 59

5 J.L. Baird and colour television 71

6 CBS, RCA and colour television 101

7 The 1949–50 FCC colour television hearings 125

7.1 Flicker, motion continuity, and allied effects 136
7.2 Brightness-contrast 138
7.3 Superposition of colour images 139
7.4 Colour fidelity 139
7.5 Resolution 140
7.6 Picture texture (structural) 141
7.7 Susceptibility to interference 141

7.8 Adaptability and convertibility 141
7.9 Equipment considerations 142
7.10 Conclusions (FCC) 142

8 RCA's resolve **147**

9 The work of the NTSC **165**

10 Colour television broadcasting in the USA, 1953–1966 **185**

11 British developments, 1949–1962 **197**

12 Camera tubes for colour television to *c*. 1967 **221**

13 The development of display tubes to *c*. 1967 **243**

14 An attempt at unity **267**

15 Epilogue **303**

Appendix A: The NTSC signal specifications **313**
A.1 General specification 313
A.2 The complete colour picture signal 314

Bibliography **317**

Index **319**

Preface

Curiously, although colour television broadcasts are now ubiquitous throughout the civilized world, no scholarly history of colour television appears to have been written in the English language. In the present book the author has attempted to consider the development of colour television and the relative merits of the several systems from a general, technical and political viewpoint.

The history commences in 1928–9, when rudimentary colour television was demonstrated for the first time ever by J.L. Baird and, soon afterwards, by engineers of the Bell Telephone Laboratories. Nineteen sixty-six has been chosen as a convenient end-date for this history because by this date the detailed investigations and deliberations, from 1949 to 1950, of the Federal Communications Commission, the National Television Systems Committee and the Condon Committee in the United States of America, and of the European Broadcasting Union and the Comité Consultatif International Telecommunications in Europe had restricted the viable colour television systems to just the NTSC system and its variants, the PAL and SECAM systems, and these systems were in general use from *c*. 1966.

The development of colour television presented in this book is based predominantly on written primary source material, that is on committee reports, committee minutes, memoranda, letters and published learned society papers, together with editorials, articles and reports in the technical and non-technical press.

Great care has been taken to ensure that an unbiased, accurate and balanced history has been written. Numerous references are given at the end of each chapter and a bibliography has been provided to enable readers to pursue further particular aspects of the history of colour television.

Some of the topics discussed were once contentious – for example, the controversy relating to CBS and RCA and the use of sequential/simultaneous scanning, and the disputed choice of the NTSC or PAL or SECAM system for European television. To maintain impartiality of treatment, the author has quoted extensively from the relevant primary source material. This approach preserves any nuances of opinion that may exist in an official ruling and thereby prevents any inadvertent distortion that may result from a paraphrasing of the ruling. Thus, a reading of the FCC's actual conclusions, following the 1949–50 Hearings on colour television, should show clearly why the FCC ruled in favour of CBS and not RCA.

Among the many topics discussed in the book, mention may be made of the following: compatibility and non-compatibility; mechanical and all-electronic systems; field-sequential, line-sequential, and dot-sequential scanning; bandwidth constraints and band-sharing techniques; the CBS-RCA conflict; the relative merits of the different systems; the attempt to achieve unity of purpose in Europe; standards; and the development of colour cameras and display tubes.

Since the author's writings on the history of technology, and especially the history of television, are quite extensive, selected extracts from these have been used in the present book. In particular the sections on Baird's contributions to colour television are based on the author's biography, *John Logie Baird: television pioneer* (IET, London, 2000).

Both colour photography and colour cinematography preceded colour television. Since some systems of colour television were based on the principles of these subjects, a brief history of early colour photography and colour cinematography has been given in Chapter 1 to form a background to the subject of colour television. Again, brief historical details of low-definition and high-definition black-and-white television have been included to enable a coherent account of the history of colour television to be presented.

Acknowledgements

Many persons and organizations have assisted me in my researches on the history of colour television and I am most grateful for all the contributions and help which I have received.

In the first place I must acknowledge my sincere gratitude to Radio Rentals Ltd for generously awarding me a travelling scholarship to visit the United States of America. I am especially grateful to Mr S.L. McCrearie, the service director of this company, for all his kindness and cooperation in making the necessary arrangements for the award and for introducing me to some very useful personal contacts. Without Radio Rentals Ltd's bounteousness this book would not have been possible, since much of the material on which the book is based was given to me by various persons in the USA.

Among these persons, I should like to thank most warmly Mr J. Dixon, Mr J. Linthicum, and Miss K. Fagen of the Federal Communications Commission, Washington, for giving me a mass of documentation relating to the FCC's Hearings on colour television. Mr E. Schamel, of the National Records Centre, Suitland, and Mr J. Finiston and Mr Hess of the National Archives, Washington, were very helpful in locating primary source material on colour television in their centre's voluminous archives.

During my stay in the USA I spent some time at Bell Laboratories, Murray Hill, and am pleased to say that the staff who looked after me – Mr G. Schindler, Mr M.D. Fagen, Mr W. Bullock and Miss R. Stumm – could not have been more supportive. I am very appreciative of their efforts. The same remark applies to Mr A. Pinsky, Mr Russinoff, and Mr J. Stranix of the Radio Corporation of America, Princeton, and also to Mrs Stepno of the same organization in Crystal City. It was always a pleasure to experience the kind, courteous and helpful attitude of the staff of the RCA and Bell Laboratories. Miss M. Lyons, and Mrs E. Romano of the American Telephone and Telegraph Company, New York, made me most welcome and provided me with photographs relating to Bell Telephone Laboratories' researches in the field of television. Mr W. Howard, of the National Broadcasting Company, New York, received me kindly and gave me some useful papers. To all these persons I am most obliged.

In addition I must acknowledge my indebtedness to the librarians of the Library of Congress, Washington; the US Patent Office, Crystal City; the Television Information Office, New York; the Museum of Broadcasting, New York; and the Broadcast Pioneers Library, New York.

In the United Kingdom my chief debt is to the British Broadcasting Corporation, and the Independent Television Commission. Both of these bodies have splendid archive collections – at the BBC's Written Archives Centre, Caversham, and at the ITC's offices, London. The staff at these centres were always most helpful and it was ever a pleasure to research in conditions where knowledge, experience and expertise were so freely available. Particular mention must be made of Mrs J. Kavanagh and Mrs T. Hayes, of the BBC; and of Mr P. Troake, of the ITC. They provided excellent services, which I gratefully acknowledge. Additionally, the staff at the Post Office Records Office, INTELLECT and the Alexandra Palace Television Group were ever responsive to my requests for information, and I thank Mrs J. Farrugia, Mr R.E. Norman, Dr J. Lewis, and Mr N. Glover for all their assistance.

I am especially grateful to the late Mr R.M. Herbert, Dr G.E. Winbolt, AT&T Bell Laboratories, Syndication International, the Royal Television Society and Radio Rentals Ltd, who generously provided me with the photographs that are reproduced in this book.

Figures

Figure 1.1	Sectional diagram of a chronoscope for still colour photographs (*c.* 1990). The three black-and-white positive transparencies – each with its appropriate red, green or blue filter – are placed at A, B and C respectively. With the aid of the half-silvered mirrors D and E, an observer views the three images in colour and in register through the eyepiece F [Source: *The First Colour Motion Pictures*, HMSO, London, 1969]	3
Figure 1.2	A contemporary engraving of Lumière's projector of 1896 [Source: *The Origins of the Motion Picture*, HMSO, London, 1964]	9
Figure 1.3	Diagram of a stencil-making machine. An enlarged image of one frame of a film A is projected onto the ground glass at B. By means of a rod C, an operator traces the outline of a part of the image. Depression of C closes an electric circuit and activates the cutting tool [Source: *The First Colour Motion Pictures*, HMSO, London, 1969]	10
Figure 1.4	Diagram of the filter disc of the Kinemacolor projector. The shaded portion of the green sector represents a double thickness of gelatine [Source: *ibid.*]	12
Figure 1.5	Sectional diagram of the beam splitter of an early Technicolor camera (*c.* 1920) [Source: *ibid.*]	14
Figure 1.6	Diagram of the Kodacolor process [Source: *Photographic Journal*, September 1929, pp. 404–5]	16
Figure 2.1	Baird *c.* 1915 as a young mains engineer at the Rutherglen substation of the Clyde Valley Electrical Power Company [Source: Royal Television Society]	21
Figure 2.2	Baird in his Frith Street laboratory [Source: *ibid.*]	24
Figure 2.3	Baird and 'Stookie Bill' being televised in the early days of 'floodlight' television [Source: *ibid.*]	25
Figure 2.4	Diagrams showing the principles of floodlight and spotlight scanning [Source: Author's collection]	26
Figure 2.5	Schematic diagrams of Baird's colour television system. The system used sequential three-colour scanning. (a) of the patent shows the transmitter and (b) the receiver. Three-spiral	

	Nipkow discs were utilized – each of the spirals being associated with a primary-colour filter [Source: British patent application 321 390]	28
Figure 2.6	Schematic diagram of the circuits for Bell Telephone Laboratories' 1927 television demonstration [Source: AT&T Bell Laboratories]	33
Figure 2.7	(a) The schematic layout of the lamps and filters used in the Bell Telephone Laboratories' apparatus for colour television. (b) With the exception of the photoelectric cabinet on the left, the apparatus was identical with that used for the original demonstration of monochrome television [Source: *ibid.*]	35
Figure 2.8	(a) Sectional view of television-sending apparatus for transmitting from coloured picture film. A, light source; D, scanning disc; M, synchronous motor; C, photoelectric cell cabinet containing three sets of colour-sensitive cells with appropriate filters; F, motion-picture film; L, projection lens; S, white screen. (b) Diagram of the scanning apparatus. A, light source; C, condensing lens; F, film; S, slot, transverse to direction of film motion; D, scanning disc, provided with radial slots R; L, projection lens; T, position of colour filters used for screen projection; P_1, P_2, P_3, three photoelectric cells; M, mirrors. (c) View of the slotted scanning disc, film and method of scanning [Source: *Journal of the Optical Society of America*, 1931, vol. 21, p. 3]	37
Figure 2.9	Schematic diagram of BTL's three-zone television system: (a) receiving-end disc with spiral holes provided with prisms; (b) sending-end disc with circle of holes with prisms; (c) general arrangement of apparatus [Source: *Journal of the Optical Society of America*, 1931, vol. 21, pp. 8–19]	39
Figure 3.1	Electron lens – optical lens analogy: diagram showing the analogy between the focusing of a beam of electrons by the use of electron lenses, and the focusing of a beam of light by optical lenses [Source: *Journal of the IEE*, 1933, vol. 73, p. 443]	44
Figure 3.2	Schematic diagram of Zworykin's 1924 colour television system, showing the reciver (a) and the transmitter (b), which incorporated a Paget plate (c) [Source: British patent application 255 057, 3 July 1926]	48
Figure 3.3	Details of Zworykin's 1929 cathode-ray display tube (the kinescope) [Source: *Radio Engineering*, 1929, vol. 9, pp. 38–41]	49
Figure 3.4	Sketch of the iconoscope camera tube – the cathode C, control grid G, first anode A, second anode P_a and mosaic elements P_c, are indicated [Source: *Journal of the IEE*, 1933, vol. 73, p. 441]	52

Figure 3.5 Block diagram showing the television system that incorporated an iconoscope and a kinescope [Source: *Proceedings of the IRE*, 1934, **22**(11), pp. 1241–5] 53
Figure 4.1 Diagram of the emitron television camera tube [Source: J.D. McGee] 62
Figure 4.2 The Marconi-EMI Television Ltd video waveform [Source: Author's collection] 64
Figure 5.1 John Logie Baird, E.O. Anderson and dummy model in Baird's private laboratory [Source: Syndication International] 72
Figure 5.2 The 20-facet mirror drums of Baird's colour television camera and projector used at the Dominion Theatre Demonstrations in December 1937 and February 1938 [Source: R.M. Herbert] 73
Figure 5.3 Baird's 120-line colour television mirror drum camera, which was utilised for the Dominion Theatre demonstrations [Source: Dr G.E. Winbolt] 76
Figure 5.4 Baird's colour projector (a) and the associated power supply (b) for the 1937–8 Dominion Theatre demonstrations [Source: R.M. Herbert] 77
Figure 5.5 The original cathode-ray-tube large-screen projection equipment used in the Tatler cinema, London, on the occasion of the Trooping of the Colour in 1938 [Source: Dr G.E. Winbolt] 79
Figure 5.6 J.L. Baird, E.O. Anderson and model (Paddy Naismith) during work on colour television [Source: Syndication International] 84
Figure 5.7 Chromaticity diagram to illustrate two-colour television [Source: Author's collection] 86
Figure 5.8 The principle of operation of the anagylph method of stereoscopic viewing [Source: *Electronic Engineering*, February 1942, p. 620] 88
Figure 5.9 The layout of the apparatus that Baird used to show stereoscopic colour television [Source: *Electrician*, December 1941, p. 359] 89
Figure 5.10 The Thomas system of colour cinematography and colour television [Source: *Tele-Tech*, February 1947] 91
Figure 5.11 The optical arrangement of Baird's three-colour television system [Source: British patent application 555 167] 92
Figure 5.12 Diagrams illustrating Baird's line-sequential colour television system [Source: British patent application 562 334] 93
Figure 5.13 Diagrams showing the principles of Baird's two-gun and three-gun telechrome tubes [Source: British patent number 562 168] 95
Figure 5.14 Baird with the world's first multi-electron gun, colour television receiver (1944) [Source: Radio Rentals Ltd] 96

Figure 6.1	Block diagrams showing the principles of CBS's colour television systems [Source: FCC Docket 8736]	102
Figure 6.2	Diagram of a CBS colour filter disc [Source: *Proceedings of the IRE*, 1942, **30**(4), p. 180]	104
Figure 6.3	Comparison of CBS's black-and-white and colour television systems [Source: Federal Communications Commission]	113
Figure 6.4	Schematic diagram of RCA's colour television system (1946) [Source: *Journal of the Television Society*, 1947, vol. 5, part 1, p. 32]	116
Figure 7.1a	Scanning patterns for CBS's field-sequential colour system, line-interlaced [Source: Condon Committee Report]	127
Figure 7.1b	Scanning patterns for CBS's field-sequential colour system, dot-interlaced [Source: *ibid.*]	129
Figure 7.2a	CTI 30-frame continuous film pick-up or recorder [Source: FCC Docket No. 7844–8704]	130
Figure 7.2b	Schematic diagram of CTI's colour receiver [Source: *ibid.*]	130
Figure 7.3	Scanning sequence patterns, showing pattern with 'line crawl' (a), and desirable patterns (b) and (c) [Source: *ibid.*]	132
Figure 7.4	Block diagram of RCA's colour television transmitter [Source: RCA paper on 'A six-megacycle compatible high-definition color television system', undated (but *c.* September 1946, unsigned)]	133
Figure 7.5	Scanning and interlace pattern of RCA's system [Source: *ibid.*]	134
Figure 7.6	Block diagram of RCA's colour television receiver using bypassed highs [Source: *ibid.*]	135
Figure 9.1	The NTSC waveform specification [Source: Author's collection]	168
Figure 9.2	The experimental set-up for investigating the transmission of mixed-high television signals [Source: *ibid.*]	171
Figure 9.3	The line-frequency harmonics of a typical monochrome video frequency spectrum [Source: *Sylvania Technologist*, 1952, vol. 5, p. 41]	173
Figure 9.4a	The components of the equipment for generating NTSC colour video signals [Source: *Proceedings of the IRE*, 1955, **43**(11), pp. 1575, 1578]	176
Figure 9.4b	The components of the equipment for receiving NTSC colour video signals [Source: *ibid.*]	177
Figure 10.1	The Bell System television network [Source: Author's collection]	187
Figure 11.1	Classification of luminance/chrominance systems [Source: *Journal of the Television Society*, 1955, **7**(12), p. 493]	202
Figure 11.2	Map of England and Wales showing the areas where cross-talk could be experienced [Source: BREMA]	203

Figure 11.3 Complete waveform for the proposed British adaptation of the NTSC standards [Source: *Journal of the Television Society*, 1954, **7**(6), pp. 241–7] 205
Figure 11.4 The essential components of the systems (excluding the picture-signal generators and displays) investigated by MWT [Source: *ibid.*] 207
Figure 11.5 Characteristics of the available signals that could be demonstrated by the systems of Figure 11.4 [Source: *ibid.*] 208
Figure 11.6 Cartoon [Source: Sprod in *Punch*, 1 March 1961] 215
Figure 12.1 Diagram of the super-emitron [Source: *Electronics*, A.C.B. Lovell, p. 178] 222
Figure 12.2 Diagram of the orthicon [Source: *ibid.*, p. 190] 225
Figure 12.3 Diagram of the c.p.s. camera tube [Source: *Proceedings of the IRE*, 1950, **38**(6), p. 601] 226
Figure 12.4 Schematic diagram of EMI's chromacoder [Source: *Journal of the Television Society*, 1955, **7**(7), p. 307] 229
Figure 12.5 Diagram of the image orthicon [Source: *Journal of the Television Society*, 1963, **10**(8), p. 257] 231
Figure 12.6 Image orthicon colour camera optics [Source: *Journal of the Television Society*, 1961, **9**(10), p. 422] 232
Figure 12.7a Cross-sectional diagram of the vidicon tube [Source: *Transactions on Electronic Devices*, 1976, **ED-23**(7), p. 744] 234
Figure 12.7b Diagram of a three-tube colour camera employing three lead oxide vidicons [Source: *ibid.*] 236
Figure 12.8 Target structure for an early tri-colour vidicon target having 870 internal red-green-blue filter strips and a target output lead for each primary colour [Source: *Transactions on Electronic Devices*, 1976, **ED-23**(7), p. 746] 237
Figure 13.1 A multicolour phosphor screen: (a) shows a multiple-layer screen with colour depending on beam velocity; and (b) shows how saturation in a two-component screen makes colour dependent on current density [Source: *Proceedings of the IRE*, 1951, **39**(10), p. 1181] 244
Figure 13.2a A line-screen colour kinescope using one electron beam. To assure correct colours, beam scanning must be highly accurate. No automatic registry means are shown [Source: *ibid.*] 245
Figure 13.2b RCA's 16 in. (40.6 cm) diameter line-screen kinescope [Source: *ibid.*] 245
Figure 13.3a Greer's three-electron gun display tube [Source: FCC Docket 8736] 247
Figure 13.3b Greer's ideas for the production and manufacture of the screen, and for depositing the phosphors upon it [Source: *ibid.*] 248

Figure 13.4	Two forms of a non-planar direction-sensitive colour screen [Source: *Proceedings of the IRE*, 1951, **39**(10), p. 1185]	250
Figure 13.5	Flechsig's proposal to use colour phosphor strips, shadowed by a wire grid and (a) three electron beams or (b) one beam deflected at the electron gun [Source: *ibid*., p. 1184]	251
Figure 13.6	Selected figures from the colour tube patent of A.C. Schroeder, showing three beams deflected by a single deflection yoke and the relative arrangement of the mask apertures and the phosphor dots that produces a nested phosphor dot screen [Source: US patent application 2 446 791]	253
Figure 13.7a	The principle of the shadow mask tube [Source: *Journal of the Royal Television Society*, 1967–68, **11**(12), p. 279]	255
Figure 13.7b	Cross-sectional diagram of RCA's shadow mask tri-colour kinescope [Source: *Proceedings of the IRE*, 1954, **42**(1), p. 315]	255
Figure 13.8	Beam control, at the phosphor screen, for changing colour: (a) simple line screen colour switching; (b) deflection switching of colour with line screen; (c) deflection switching without requiring registry [Source: *Proceedings of the IRE*, 1951, **39**(10), p. 1182]	257
Figure 13.9a	The chromoscope tube arranged for direct viewing [Source: *Electronic Engineering*, June 1948, p. 191]	258
Figure 13.9b	RCA's three-colour, three-gun, grid-controlled colour kinescope [Source: *Proceedings of the IRE*, 1951, **39**(10), p. 1214]	259
Figure 13.10a	Illustration showing the principle of operation of the single-gun Lawrence tube, the chromatron [Source: FCC Docket 8736]	260
Figure 13.10b	Schematic diagram of the three-gun chromatron [Source: *Journal of the Royal Television Society*, 1967–68, **11**(12), p. 279]	261
Figure 13.11	Rosenthal's proposed subtractive colour display scheme [Source: *Journal of the Royal Television Society*, 1954, **7**(6), pp. 241–7]	263
Figure 14.1	Modulation and demodulation processes of the colour signals in the NTSC system [Source: *EBU Review, Part A – Technical*, 1965, no. 93, pp. 194–204]	269
Figure 14.2	The NTSC transmission characteristic [Source: *Journal of the Royal Television Society*, 1954, **7**(6), pp. 236–240]	270
Figure 14.3	(a) Block diagram of the TSC system (1956); (b) block diagram of the H. de France system (1957–58) [Source: *ibid.*]	270
Figure 14.4	Block diagrams of the SECAM system (a) 1959–60 and 1960; and (b) of the FAM system (1960) [Source: *ibid.*]	272

Figure 14.5 Block diagrams of the SECAM-NTSC system (1961–62), first version (a), and the SECAM-NTSC system, second version (b) [Source: *ibid.*] 273
Figure 14.6 Block diagrams of the PAL system (1962), and of the PAL system (1965) [Source: *ibid.*] 274
Figure 14.7 Block diagram of the ART system (1963) [Source: *ibid.*] 275
Figure 14.8 BBC Television Centre: general layout of original colour-production studio control suits and apparatus areas [Source: *IEE Reviews*, August 1970, vol. 117, p. 1479] 293
Figure 14.9 General layout of Type-1 and Type-2 colour mobile control rooms (c.m.c.r.s) [Source: *ibid.*, p. 1480] 294

Tables

Table 2.1	Baird's patents on the subject of colour television	29
Table 6.1	Possible parameters for various colour television systems	103
Table 6.2	The RCA colour television systems (1940)	107
Table 7.1	Summary of performance characteristics (Condon Committee) System	137
Table 8.1	The annual production of black-and-white television sets	155
Table 10.1	Progress of colour television in the United States	193
Table 11.1	Possible luminance/chrominance systems, based on 1955 European monochrome standards	204
Table 12.1	A comparison between the c.p.s. emitron, the emitron, the super-emitron and the orthicon	228
Table 12.2	Photoconductive materials investigated by Miller and Strange	233
Table 12.3	Comparison of image orthicon and vidicon colour cameras	235
Table 14.1	The activities of the EBU Ad Hoc Group on Colour Television	296

Chronology relating to colour television

May 1861	J.C. Maxwell produced an image in colour by a natural colour process
1869	L. Du Hauron suggested reproducing colours photographically by means of a mosaic of minute red, green and blue filters applied on a photographic plate
1895	Joly produced colour pictures by ruling fine lines, coloured red, green and blue, on a glass plate
1899	E.R. Turner applied Maxwell's ideas to cinematography (the Lee–Turner three-colour process)
1906	First trial, by G.A. Smith, of two-colour cinematography
1907	The Lumière brothers used a random mosaic screen
1908	L. Dufay invented the dioptochrome plate – a regular mosaic screen
1908	First demonstration, by G.A. Smith, of two-colour cinematography
26 January 1926	First demonstration of television, by J.L. Baird
1928	Introduction of Kodacolor by the Kodak Company
3 July 1928	First demonstration, by Baird, of colour television using red, green and blue colour signals transmitted sequentially
27 June 1929	Demonstration, by Bell Telephone Laboratories, of colour television using red, green and blue colour signals transmitted simultaneously
1930	H.E. Ives described a method of colour television using Kodacolor motion picture film
1932–4	RCA proposed to fabricate a kinescope in which the screen would be coated with successive very narrow strips of red-, green- and blue-sensitive phosphors. Some work on a two-colour filter wheel colour television system.

15 June 1936	FCC Hearings to determine its long-term policies on the future allocation of frequency channels
2 November 1936	The world's first, regular, high-definition television service inaugurated at Alexandra Palace
2 January 1937	Adoption by the BBC of the 405-line television standard
4 February 1938	Public demonstration, by Baird, of two-colour images at the Dominion Theatre, London
27 July 1939	Demonstration, by Baird in his private laboratory, of colour television using a projection c.r.t. (cathode-ray tube – see 'List of abbreviations' on Page xxvii) and a rotating filter disc
6 February 1940	RCA's three-colour, additive, sequentially scanned television system demonstrated to the FCC
1940	Demonstration, by Baird in his private laboratory, of 600-line, two-colour television
31 July 1940	First meeting of NTSC
27 August 1940	CBS broadcasts colour television images using Kodachrome film, a three-colour filter disc and an image dissector tube
9 January 1941	First public showing, by CBS, of direct pick-up colour television
March 1941	RCA and NBC commenced testing of sequentially scanned 343- and 441-line, all-electronic, colour television systems
20–4 March 1941	FCC Hearings on NTSC report on television standards. Panel 1 of NTSC mentioned it had considered the colour television systems of CBS (4), GE (1), RCA (4) and Baird (3).
December 1941	Demonstration, by Baird in his private laboratory, of 600-line two-colour, stereoscopic, colour television
13 January 1944	Demonstration, by Baird in his private laboratory, of two-colour, stereoscopic television using the telechrome tube
30 April 1941	New FCC rules on television standards adopted
25 May 1945	FCC approves 480–920 MHz frequency band for experimental television
October 1945	CBS commences colour television broadcasting on an experimental basis
December 1945	RCA demonstrates three-colour television using colour filter drum

October 1946	RCA's all-electronic, three-colour, three-channel, television system shown to representatives of the press, industry and government. Use of 16 mm colour motion film
9 December 1946	FCC Hearing on CBS's petition on colour television
18 March 1947	FCC issued its colour television report and denied the CBS petition
September 1948	FCC again investigates colour television to determine progress
20 May 1949	Letter from Senator E.C. Johnson to Dr E.U. Condon requests a factual appraisal of colour television by technical experts not associated with the radio and television industry
26 September 1949 to 26 May 1950	FCC Hearing on colour television. Systems of CTI, CBS, and RCA investigated.
1 September 1950	FCC issues its 'First Report on Color Television'. Any representations to be submitted by 29 September 1950
11 October 1950	FCC issues its 'Second Report on Color Television' and denies RCA's petition re colour television
17 October 1950	RCA, NBC, and RCA Victor file suit in the Federal District Court, Chicago, against the US and the FCC
20 November 1950	Ad Hoc Committee of the NTSC appointed
22 December 1950	Court judgement not in RCA's favour but temporary injunction granted pending appeal
25 January 1951	RCA appeal heard by the Northern District Court of Illinois. RCA permitted to appeal to the Supreme Court. Papers served on 30 January 1951.
19 April 1951	Report of Ad Hoc Committee circulated
29 May 1951	Supreme Court ruling held that the FCC had not acted 'arbitrarily and capriciously . . . [or] against the public interest, and contrary to law'
26 June 1951	FCC announcement: all regular television stations permitted to broadcast colour television programmes in accordance with its standards
19 November 1951	Office of Defense Mobilization asks CBS to cease manufacture of colour television receivers
24 March 1953	First meeting of the US House of Representatives Committee on Interstate and Foreign Commerce

25 June 1953	RCA petition to the FCC to adopt its (RCA's) colour television standards
17 December 1953	FCC approves signal specification of RCA's all-electronic, compatible, colour television system. Oppositions filed.
17 December 1953	NTSC's signal specification approved by FCC
1 January 1954	First major outside broadcast in colour televised by NBC
7 October 1954	First broadcast of compatible colour television from Alexandra Palace, London
5 November 1956	Full-scale experimental colour television transmissions from the 60 kW (e.r.p.) transmitter at Crystal Palace
July 1962	Report of the Pilkington Committee issued
November 1962	EBU's Ad Hoc Group on colour television established
14–26 February 1964	Meeting, in London, of the CCIR Study Group XI on colour television
9 April 1964	First Eurovision colour television demonstration
February 1965	Report of the EBU Ad Hoc Group on colour television published
24 March 1965	Meeting, in Vienna, of the CCIR Study Group XI on colour television
22 June 1965	XIth Plenary Assembly of the CCIR commences in Oslo. No unanimity of view on the adoption of a single system of colour television for Europe. PAL and SECAM to be used.

Abbreviations

a.c.	alternating current
ARD	Arbeitsgemeinschaft der offentlich-rechtlichen Rundfunkanstalten der Bundesrepublik Deutschland
ART	Additional Reference Transmission
AT&T	American Telephone and Telegraph Company
BBC	British Broadcasting Corporation
BIT	Baird International Television Ltd
BREMA	British Radio Equipment Manufacturers Association
BTDC	Baird Television Development Company
BTL	Baird Television Ltd
BTL	Bell Telephone Laboratories
CBS	Columbia Broadcasting System
CCIR	Comité Consultatif International des Radiocommunications
CEA	Cinematograph Exhibitors Association
CFT	Compagnie Française de Thomson-Houston
CSF	Compagnie Générale de Telegraphe Sans Fil
CTI	Color Television Incorporated
c.r.o.	cathode ray oscillograph
c.r.t.	cathode-ray tube
d.c.	direct current
DSIR	Department of Scientific and Industrial Research
EBU	European Broadcasting Union
EEA	Electronic Equipment Association
EMI	Electric and Musical Industries Ltd
ENR	Emissora Nacional de Radiodifusão
e.r.p.	effective radiated power
FAM	Frequency and Amplitude Modulation
FCC	Federal Communications Commission
FRC	Federal Radio Commission
G-BPC	Gaumont – British Picture Corporation
GE	General Electric (US)
GPO	General Post Office (UK)
HF	High frequency
HMV	His Master's Voice

IBTO	International Broadcasting and Television Organization
IFRB	International Frequency Registration Board
IRT	Institut fur Rundfunktechnik
ITA	Independent Television Authority
IRE	Institute of Radio Engineers
ITU	International Telecommunications Union
JTAC	Joint Technical Advisory Committee
M-EMI	Marconi-EMI Television Company Ltd
MWT	Marconi (or Marconi's) Wireless Telegraph Company Ltd
NBC	National Broadcasting Company
NHK	Nippon Hoso Kyokai
NPA	National Production Authority
NRK	Norsk Rikskringkasting
NTSC	National Television Systems Committee
OB	Outside Broadcast
ODM	Office of Defense Mobilization
OIRT	Organization Internationale de Radiodiffusion et Television
ORF	Osterreichischer Rundfunk
PAL	Phase Alternation by Line
PMG	Postmaster General
PTT	Post, Telegraph and Telephone
QAM	Quadrature Amplitude Modulation
ORTF	Office de la Radiodiffusion Television Française
RAI	Radiotelevisione Italiana
RCA	Radio Corporation of America
RECMF	Radio & Electronic Component Manufacturers' Federation
RMA	Radio Manufacturers Association
RTF	Radio Television Française
RTPB	Radio Technical Planning Board
SB	Simultaneous broadcast
SECAM	SEQuential A Memoire
TAC	Television Advisory Committee
TSC	Technical Subcommittee (of the TAC)
TSC	Two sub-carrier
UHF	ultra-high frequency
VHF	very high frequency
WDR	West Deutscher Rundfunk
WEM	Westinghouse Electric and Manufacturing Company

Introduction

On 3 July 1928 J.L. Baird demonstrated a rudimentary system of colour television based on the additive colour principle, that is the principle of synthesizing a coloured image of an object/scene from the three separate red, green and blue images of the same object/scene. One year later engineers of the Bell Telephone Laboratories (BTL) also demonstrated colour television using the same additive colour principle. The essential difference between the two systems lay in the method whereby the three images were generated and transmitted. Baird presented his primary-colour images sequentially to the viewer, whereas BTL displayed the three images simultaneously. Both methods had been tried by the pioneers of colour cinematography.

Following BTL's early success, the Laboratories, for several years until *c*.1935, endeavoured to adapt the Kodacolor film process of cinematography to colour television, but their work did not become commercially acceptable.

In 1928–9 neither Baird nor BTL envisaged that their demonstrations of low-definition colour television would lead to colour television broadcasting services. Rather, they were exploring the possibilities that television seemed to offer. The subject was in its infancy and there was much to be investigated – for example, long-distance television, news by television, stereoscopic television, phonovision, noctovision, large-screen television, two-way television, and, of course, the means to increase the definition of the images that could be synthesized. There was a view, particularly held by the British Broadcasting Corporation (BBC) of London, that the small 'head and shoulders' images of low-definition television were unsuitable for a television service, and that only when tennis from Wimbledon, cricket from Lords, and national events could be televised and reproduced in adequate size on a domestic receiver could a service be attractive and viable.

Baird, from the time of his first television patent (dated 26 July 1923), held the opinion that cinema television was an important application of the new medium. He wrote at about that time, 'Personally I look forward with confidence to the time when we shall not only speak with, but also see those with whom we carry on a telephone or wireless telephone conversation, and the distribution of a cinematograph film will be superseded by the direct transmission from a central cinema.' He gave a demonstration of large-screen television, on 28 July 1930, at the London Coliseum, and televised the Derby on 3 June 1931 and reproduced the images on a large screen at the Metropole cinema, London.

In 1933, after he had been ousted from his position of technical director of Baird Television Ltd's research and development programme, he set up a private laboratory at his home in London and commenced work on cinema television with the object of showing the images in colour. A demonstration of black-and-white cinema television, using a 6 ft x 4 ft (1.8 m x 1.2 m) screen, was presented at the Dominion Theatre, on 4 January 1937, and on 4 February 1938 Baird showed colour cinema television at the same theatre. Further demonstrations were given and, after the commencement of hostilities in September 1939, he continued to experiment with various systems of colour television, including 600-line colour television and stereoscopic colour television. Of these systems, one was an adaptation of Thomas colour, which had been proposed for colour cinematography.

Elsewhere, the Columbia Broadcasting System, on 4 September 1940, exhibited its three-colour, frame-sequential colour television system. A rotating colour filter disc was utilized as in an early method of colour cinematography. The Second World War interrupted this work, but after the cessation of hostilities CBS returned to its former interest. CBS's proposal for colour television conflicted with the view of the Radio Corporation of America that colour television should be based on the simultaneous presentation of the three primary images of an object/scene, as in modern colour cinematography. Several Federal Communications Commission (FCC) Hearings were held to resolve the issue and eventually, on 17 December 1953, the FCC approved the colour television scheme of the National Television Systems Committee (NTSC). On New Year's Day 1954, the first major outside broadcast in colour was televised by the National Broadcasting Company.

In war-torn Europe colour television, for most countries, was not considered an urgent priority of the postwar years. The BBC carried out much investigative work on colour television from 1953 and on 7 October 1954 broadcast its first compatible colour television programme from Alexandra Palace, London. Later, on 5 November 1956, full-scale experimental colour television transmissions were radiated from the 60 kW (e.r.p.) transmitter at Crystal Palace. However, it was not until November 1962 that the European Broadcasting Union (EBU) established an Ad Hoc Group on colour television. By this time some difficulties of the NTSC system had been exposed and to overcome them variants of the NTSC system, such as the PAL system and the SECAM system, were advanced. In February 1965, after many meetings and demonstrations, the EBU's Ad Hoc Group published its report on colour television. A few months later the Comité Consultatif International des Radiocommunications, at a meeting held in Oslo, concluded that it was not possible to obtain unanimity of view, regarding a common European colour television standard, among the various countries represented. The PAL and SECAM systems would prevail in Europe, and the NTSC, PAL and SECAM systems elsewhere. Public colour television services on approved standards could now, at last, be realized.

Colour television and colour cinematography have some common factors. Additive colour mixing, two-colour, and three-colour systems, sequential and simultaneous displays, rotating filter discs, the use, or proposed use, of screen plates, and of the Kodacolor and Thomas colour methods, are all part of the histories of colour television and colour cinematography. For this reason, and to show how the latter subject influenced the former, a brief description of early colour photography and colour cinematography follows in Chapter 1.

Chapter 1
A brief history of colour photography and colour cinematography [1,2,3]

In 1839 L.K.M. Daguerre in France and W.H. Fox in England publicised the first practical techniques for creating permanent images by the agency of light. The important factor in their work was that they had each discovered and published a way of developing a latent image so that it became visible on paper or on a plate.

The problem that had faced artists and scientists using the camera obscura during the early years of the nineteenth century had been how to fix the image that they had obtained by the action of light, without having to trace it onto translucent paper. Clearly, a light-sensitive chemical was required which was capable of being developed and fixed. Berzelius in his *Text Book of Chemistry*, published in 1808, had listed more than one hundred substances that had their chemical or physical structure altered by light, and indeed the influence of light on silver nitrate had been reported by an Italian physician named Angelo Sala in 1614. Not surprisingly, several workers in Britain, Europe and America experimented with these substances in the late eighteenth and early nineteenth centuries in the hope of obtaining permanent pictures.

This activity was probably spurred on by the demand for inexpensive naturalistic pictures, particularly portraits, which existed towards the end of the eighteenth century. A simple way of reproducing pictures was by means of silhouettes, made by tracing the outline of a projected image of the face and filling it in with black paint. G.L. Chrétien in 1786 invented a machine, the physiontrace, in which the projected image of a head was traced by a stylus, and by a pantograph arrangement an engraving tool could cut a copper plate that could be inked and printed. Aloys Senefelder invented lithography in 1798, but although it was introduced in Paris in 1802 it was not until 1813 that it became a success and a fashionable hobby.

Thomas Wedgwood, son of the famous Josiah Wedgwood, and Humphry Davy achieved some fame with their use of silver nitrate and silver chloride in 1802 [4], and Nicephone Niepce in 1822 succeeded in making a heliographic copy of an engraving on a glass plate coated with bitumen. The first permanent camera picture was taken by Niepce in 1826 using his asphalt process on a pewter plate. The exposure was inordinately long, about eight hours on a bright summer's day, and hence the shadow and intermediate tone effects recorded on the plate represented a distortion of the scene viewed at a given instant of time [5].

L.J.M. Daguerre, a painter, was also experimenting with silver salts at this time. In December 1839 he and Niepce formed a partnership. Subsequently, a full description of their inventions and methods was presented at a joint meeting of the Académie des Sciences and of the Académie des Beaux Arts, by François Arago, on 19 August 1839.

Meanwhile, Fox Talbot had been working with light-sensitive substances from 1835 and had used paper coated with silver chloride [6]. He publicized his process in 1839, after the first announcement of the daguerreotype, but before the official description of it, to the two *académies*. Two years later he patented the calotype process, which used silver iodide as a sensitive material together with silver nitrate and potassium iodide. The sensitivity of Talbot's paper was further increased by treatment with gallic acid, the sensitizing properties of which had been discovered in 1837 by J.B. Reade.

Another important advance was made in 1851, when F.S. Archer introduced the wet collodion process, in which silver salts in a film of collodion [7] were coated on glass. The plates were exposed while still wet and gave very clear glass negatives. Talbot believed that his patents covered the collodion process, but, following a lawsuit in 1854 (*Talbot v. Laroche*), a favourable verdict was given to Laroche and the use of Archer's process was thenceforth free for general utilization, as its inventor had intended. The British patent of the daguerreotype expired in 1853 and, as Talbot did not renew the calotype patents, all forms of photography could be used in Britain, without restrictions, by amateurs and professionals.

Prior to 1861 photographs in colour could be reproduced only by the hand painting of daguerreotypes. But, in 1855, one of the nineteenth century's greatest physicists, James Clerk Maxwell, suggested that Young's three-colour theory could be applied to the art. Maxwell's idea was to photograph an object and obtain three negatives, each of which contained an image of the given object, such that one was exposed through a red filter, one through a green filter and one through a blue filter. In this way the three black-and-white negatives would record information about the hue of the object. Consequently, if three positive black-and-white transparencies were made from the negatives and these were projected in register onto a screen by means of three lanterns, each fitted with the appropriate filter in front of the projection lens, a coloured image would result.

Maxwell showed the correctness of his idea at a meeting of the Royal Institution in May 1861, when, for the first time ever, an image was produced in colour by a natural colour process [8].

In retrospect, Maxwell's achievement was quite astonishing because the light-sensitive materials available to photographers at that time were sensitive to only a small region of the visible spectrum, namely, the blue and ultraviolet region. And yet the method depended on the utilization of photographic emulsions responsive to the green and red portions of the spectrum. This apparent paradox was solved in 1960, when R.M. Evans [9] repeated Maxwell's experiments and found that, although Maxwell's plates were quite insensitive to red light, the red filter that he used allowed some ultraviolet radiation to be transmitted and the red part of the object reflected some ultraviolet rays from the light source. Nevertheless, very long exposure times were required to obtain negatives having a satisfactory density.

A major advance in the progress of colour photography occurred when, in 1873, H.W. Vogel found that the addition of small quantities of certain dyestuffs to photographic emulsions made the plates sensitive to a wider range of colours [10]. Vogel's outstanding discovery opened the way forward and culminated in 1906 in the manufacture of the first panchromatic emulsions, sensitive to almost the whole of the visible spectrum. Before 1906, but certainly by 1890, sufficient development work had been carried out on improving the spectral sensitivity of emulsions to enable Maxwell's process to be in common usage. The three black-and-white positive transparencies could be viewed either as in Maxwell's experiment or by means of a chromoscope – a portable device incorporating filters that permitted the three slides to be seen, in register, through an eyepiece (see Figure 1.1).

In 1869 L. and A.D. du Hauron published several suggestions for reproducing colours photographically by means of a mosaic of minute red, green and blue filters applied on the surface of a photographic plate [11]. Their notions were possibly inspired by the work of the *pointillistes* – a school of French painters who painted tiny dots of colour side by side to give the effect of continuous areas of colour and tone.

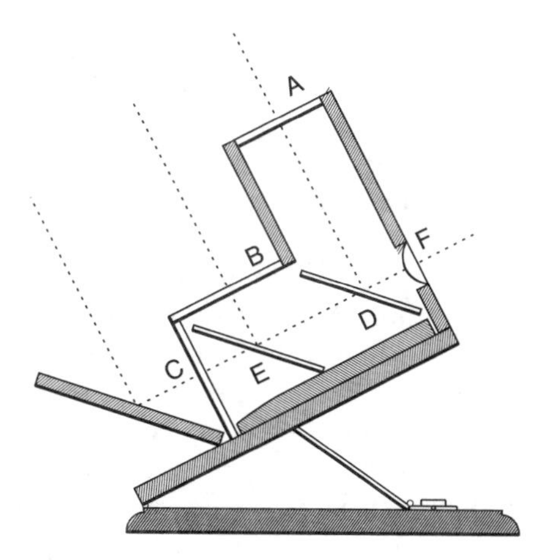

Figure 1.1 *Sectional diagram of a chronoscope for still colour photographs (c. 1990). The three black-and-white positive transparencies – each with its appropriate red, green or blue filter – are placed at A, B and C respectively. With the aid of the half-silvered mirrors D and E, an observer views the three images in colour and in register through the eyepiece F [Source:* The First Colour Motion Pictures, *HMSO, London, 1969]*

Until *c*. 1890 the ideas of the du Haurons could not be implemented satisfactorily, but by the early 1890s the preparation of suitable plates had progressed to the stage where the mosaic screen method of additive colour photography could be explored. Two important versions of it were:

1. the random mosaic screen, based on the use of starch grains dyed red, green and blue, of the autochrome plate of the Lumière brothers (1907); and
2. the regular mosaic screen, whose surface consisted of a field of red on which there were parallel lines of blue and above them at right angles parallel lines of green, of the dioptochrome plate due to Louis Dufay (1908).

The random mosaic screen process gave rise to difficulties due to the clumping of dyed grains of the same colour, which caused an increase in the appearance of mottle. Furthermore, both types of plate suffered from two serious limitations due to the absorption of light by the filters, and to the loss of image definition caused by the mosaic structure. Nevertheless, these plates, and others based on similar principles, were successfully employed for many years until subtractive methods of colour photography gradually displaced them.

An alternative to the above methods was suggested by Professor Joly. In *c*.1895, he successfully produced colour pictures by mechanically ruling fine lines, coloured by red, green, and blue dyes, across a sheet of glass [12]. This was placed in contact with a photographic plate, which was then exposed in a camera and subsequently processed to produce a black-and-white transparency. By viewing the transparency and the screen by transmitted light, the original photographed view was rendered in colour.

The earliest proposal to use regular mosaic screens in a colour television system was advanced by V.K. Zworykin in 1925. He suggested using screens of the Paget type positioned in front of the transmitter and receiver cathode-ray tubes. The Paget plate, introduced in 1913, had a chequerboard-patterned screen on which the squares were coloured red, green and blue – the number of blue squares being twice those of the other colours.

Interestingly, the first proposals for 'seeing by electricity' and cinematography were made at almost the same time: Figuier described his telectroscope in June 1877 and Wordsworth Donisthorpe, an English barrister, filed a patent for a moving picture camera on 8 November 1876 [13]. But while motion pictures took only a few years to develop – they were first shown in 1893 – the gestation period of television was to be almost half a century. Hence, when public television broadcasting services commenced in the late 1920s, much experience and expertise on the production of films had been gained which could be applied to the production of television programmes.

Donisthorpe's invention had for its object:

> ... to facilitate the taking of a succession of photographic pictures at equal intervals of time, in order to record the changes taking place in or the movement of the objects being photographed, and also by means of a succession of pictures so taken of any moving object to give to the eye a presentation of the object in continuous movement as it appeared when being photographed. [13]

At that time magazine dry-plate cameras were being introduced and Donisthorpe described a version in which plates were to be changed rapidly, the exposure taking place while they were stationary. Positives were to be printed on a long roll of paper and viewed in rapid succession. No clear idea as to how the necessary intermittent viewing was to be implemented was given in the patent.

Subsequently in a letter to *Nature* dated 24 January 1878, following Edison's 1877 invention of the phonograph, Donisthorpe wrote:

> By combining the phonograph with the kinesigraph [Donisthorpe's invention] . . . the life-size photographs shall move and gesticulate . . . the words and gestures corresponding as in real life.
>
> Each picture as it passes the eye is instantaneously lighted up by an electric spark. Thus, the picture is made to appear stationary while the people or things in it appear to move as in nature.
>
> I think it will be admitted that by this means a drama acted by daylight or magnesium light may be recorded and reacted on the screen or sheet of a magic lantern, and with the assistance of the phonograph the dialogues may be repeated in the very voices of the actors . . . [14]

Donisthorpe did not achieve success at this early date but in the late 1880s he collaborated with W.C. Crofts in designing another motion-picture camera and projector, which were patented on 15 August 1889. Only a few frames, taken at eight to ten frames per second, in Trafalgar Square, London, in 1890, survive but they indicate that Donisthorpe and Croft's approach seemed to work well. Of particular interest is Donisthorpe's attempt to obtain financial backing, a matter which was to be experienced by some of the television pioneers. He approached Sir George Newnes, who:

> submitted the matter to two 'experts' selected, by Sir George Newnes, to pronounce on its merits. One I [Donisthorpe] afterwards learnt was an artist, a painter who was as ignorant of the physical sciences as Noah's grandmother, and the other was, I believe, a magic-lantern maker.
>
> I need hardly [say] that both these 'experts' reported adversely. They agreed that the idea was wild, visionary and ridiculous, and that the only result of attempting to photograph motion would be an indescribable blur.
>
> What could Sir George Newnes do in the face of such 'expert' testimony? [15]

Meanwhile, the English photographer E.J. Muybridge had acquired some considerable fame in analysing motion by sequence photography [16,17]. At a demonstration on 15 June 1878 the press and other visitors saw 12 photographs, made in about 0.5 second, of a horse in movement. Since Muybridge used the wet-plate process, the results were little more than silhouettes, but nevertheless the successive movements of the horse were recorded with considerable accuracy. (It is perhaps pertinent to note that the earliest televised images, by J.L. Baird, C.F. Jenkins and D. von Mihaly, were silhouettes.)

The following year the number of cameras was increased to 24 and by the end of the year Muybridge had obtained hundreds of photographs of animals, birds and athletes. His financial sponsor, L. Stamford, a former Governor of California and the president of the Central Pacific Railroad, had now spent, it is believed, approximately

$42,000 since he commissioned Muybridge in 1872 to photograph a record-breaking trotter of Stamford's called Occident.

The success of Muybridge's sequence photographs spread throughout America and Europe and the images were reproduced in many of the leading magazines. Articles in the press in 1879 suggested that the photographs should be mounted in a zoetrope so that the motion of the horse could be synthesized. One London magazine implemented such a system and exhibited it in its office window. Realistic reproductions of the original movements were created and soon afterwards Muybridge designed a projector, based on a similar principle, which he first called the zoogyroscope, but later, in 1881, renamed the zoopraxiscope [18].

Muybridge's zoogyroscope was finished in the autumn of 1879 and was first described in the *Alta California* newspaper on 5 May 1880: 'Mr Muybridge has laid the foundation of a new method of entertaining the people, and we predict that the instantaneous photographic magic lantern zoetrope will make the rounds of the civilized world.'

During 1881 and 1882 he toured Europe, lectured to learned societies and, in the spring of 1882, demonstrated his zoopraxiscope to the Royal Society, the Royal Academy, the Society of Arts and the Royal Institution of London. Everywhere, he met with marked interest and enthusiasm.

The publication of Muybridge's work stimulated others to investigate sequence photography. E.J. Marey [19], of France, in the 1880s pioneered the science of chronophotography (the analysis of movement) using a single-sequence camera; and O. Anschutz, a Prussian photographer, in 1887 devised the electrotachyscope, a type of viewing apparatus.

Marey's 1882 camera was in the shape of a gun, with an eccentric cam rather than a Maltese cross to achieve the intermittent motion, and took 12 exposures, the duration of each being approximately 1/720 of a second, on a circular glass plate. The glass plate was inherently unsuited for recording long sequences of action, but in 1885 a development occurred that transformed the taking and reproduction of serial images. In that year, George Eastman, an American manufacturer of dry plates, with W.H. Walker, designed a roll holder to replace the conventional camera plate holder. The new device enabled a paper roll coated with a gelatine emulsion to be used. Four years later, Eastman introduced a thin, tough, flexible, transparent celluloid film to supersede the paper film in the roll holders. This was the essential invention that made cinematography possible [20].

In 1888 Marey began experimenting with moving film and evolved what is deemed to be the first successful cine camera. At first rolls of paper were employed, but when celluloid film became available Marey worked with the new material. The roll of film passed from one spool to another and was stopped in its motion, for the exposure, by an electromagnet mechanism. The images were about 9 cm by 9 cm in size and as the rolls of film were 4 metres in length only approximately 40 exposures could be recorded. Each exposure lasted 1/1,000 of a second and the camera was capable of taking up to 50 pictures per second.

Marey was primarily a scientist and was uninterested in pursuing his investigations for general entertainment. He had endeavoured to devise means to study animal motion and in this he had been eminently successful.

Anschutz's sequence negatives were printed as 90 mm by 120 mm transparencies and fixed around the circumference of a large steel disc [21]. As the disc rotated a Geissler discharge tube produced a brief intense flash of 1/1,000 of a second duration every time a transparency passed in front of it and behind the viewing aperture. An electric switch mounted on the disc ensured that the flashes were synchronised with the appearance of the transparencies. These followed each other at about 1/30 of a second and, because of the persistence of vision, created the illusion of a continuously moving picture.

One hundred of the machines were manufactured by the Siemens and Halske Company and were widely distributed throughout Europe and America. In December 1892 the electrotachyscope arrived in London under the name 'an electrical wonder', and was operated by a penny-in-the-slot mechanism. According to the *Amateur Photographer*, the impression gained from viewing the transparencies was so lifelike that 'we think this new wonder will become a very good thing'. But not everyone was convinced. In London's Strand one lady was overheard to say to another, 'It's a show of moving figures. It's awfully stupid.' [22]

Among others who in the 1880s participated in the advancement of the display of moving images, mention should be made of William Friese-Green, an English professional photographer, and Louis-Aimé-Augustin Le Prince, a Frenchman who worked in New York as the manager of a chain of panoramas. They both secured patents for their inventions, in 1889 and 1888, respectively [23,24].

However, it was the 'Wizard of Menlo Park', Thomas Alvar Edison [25], who provided the idea and the means which made cinematography practical. Once again it was Muybridge who supplied the stimulation. He had lectured in New Jersey on 25 February 1888 and on the 27th visited Edison to discuss with him the possibility of combining the zoopraxiscope with the phonograph [26]. Edison was intrigued with the notion and set his assistant W.K.L. Dickson, a Scotsman who had emigrated to the USA in 1879, to work on the project.

In advancing their ideas Edison and Dickson were influenced by the work of Anschutz on his electrical tachyscope, and in a patent dated 20 May 1889 Edison and Dickson used the same arrangement of continuous movement and momentary light flashes in their viewing device, the kinetoscope.

Little success seems to have been achieved by the two inventors until after Edison's meeting, in August 1889, with Marey at the Paris Exposition. Marey showed him his new roll film camera and the results that had been obtained with it. On his return, Edison applied for a caveat that covered the use of a roll of film and mentioned the crucial innovation of perforations along the edge of the film. By their means, and a toothed sprocket wheel driven by an intermittent escapement, a positive drive could be obtained that would ensure the correct registration of successive frames.

An experimental kinetoscope viewer was shown to the National Federation of Women's Clubs on a visit to Edison's laboratory on 20 May 1891. Limited production began in the summer of 1892 and full manufacture in 1893. By the beginning of April 1894 a batch of ten kinetoscopes had been sent to a firm, Holland brothers, at 1155 Broadway, New York, where the first kinetoscope parlour opened on 14 April 1894 [27]. The viewer came to Europe in the same year and was on exhibition in August in the Boulevard Poissonière in Paris, and in October at 112 Oxford Street, London. By the end of the year kinetoscope parlours had been opened all over Europe and North America. More than 1,000 machines were sold before their popularity declined.

The coin-in-the-slot kinetoscope was a viewing cabinet that permitted one person at a time to see a film of around 50 ft (15.2 m) in length. Each frame of the continuously moved film was seen momentarily by light transmitted by a slit in a revolving shutter placed between the film and an electric lamp. The picture rate of 46 frames per second restricted the viewing time to about 15 seconds. The subjects shown included wrestling, Highland dancing, a trapeze act and the strong man Eugene Sandow.

Strangely, Edison, who acquired 1,093 patents during his life, failed to see the potential of the kinetoscope [28,29], even though it was initially successful in Europe and America, and he regarded it as a novelty, which in time would pass out of fashion. He failed to patent it in Great Britain, to the advantage of Robert W. Paul [30], a London scientific-instrument maker, who soon began making copies of the machine. When Edison retaliated by attempting to deny Paul the use of films for his kinetoscope, he gave Paul an incentive to construct his own moving-film camera and associated projector. Paul abandoned Edison's use of continuously moving film and devised a mechanism based on the utilization of two seven-star Maltese crosses to transport the film intermittently past the projector lens.

The projector was in use in 1895 and received its first public demonstration on 28 February 1896 before the Royal Institution. In the following month Paul began showing films at the Alhambra, Leicester Square, London. The nightly presentations were popular and continued for the succeeding four years.

Almost simultaneously with these events the Lumière brothers [31], Auguste and Louis – who had patented in France, on 13 February 1895, a combined camera, printer and projector (see Figure 1.2) – had given on 20 February 1896 the first showing of their *cinématographe* in London at the Marlborough Hall, Regent Street [32]. The Empire Theatre of Varieties in Leicester Square soon contracted to include the *cinématographe* in its shows, and it featured in their performances from 9 March 1896. In the USA the machine was exhibited at Koster's and Bial's Music Hall on 34th Street, New York, and subsequently at halls throughout the country. The Lumière agents took the *cinématographe* all over the world during 1896.

And in Germany Max and Emil Skladnowsky projected films with their own projector, touring Europe in 1896.

Great interest was shown by the general public in cinema entertainment and this led to a rapid growth in the motion-picture industry. By the end of the nineteenth century many different cameras and projectors, utilizing a variety of film sizes, were available and cinemas had opened in many large towns and cities [33].

Figure 1.2 A contemporary engraving of Lumière's projector of 1896 [Source: The Origins of the Motion Picture*, HMSO, London, 1964]*

As with the first coloured photographs, the first coloured cine films were hand-painted. Later, machines were developed so that the various dyes could be applied to the film automatically – but only after hand-cut stencils had been made (see Figure 1.3). Then, in 1899, E.R. Turner applied Maxwell's ideas to motion-picture film projection. He received financial backing from F.M. Lee and their process became known as the Lee and Turner process. A patent (no. 6202) was issued in 1899 [34].

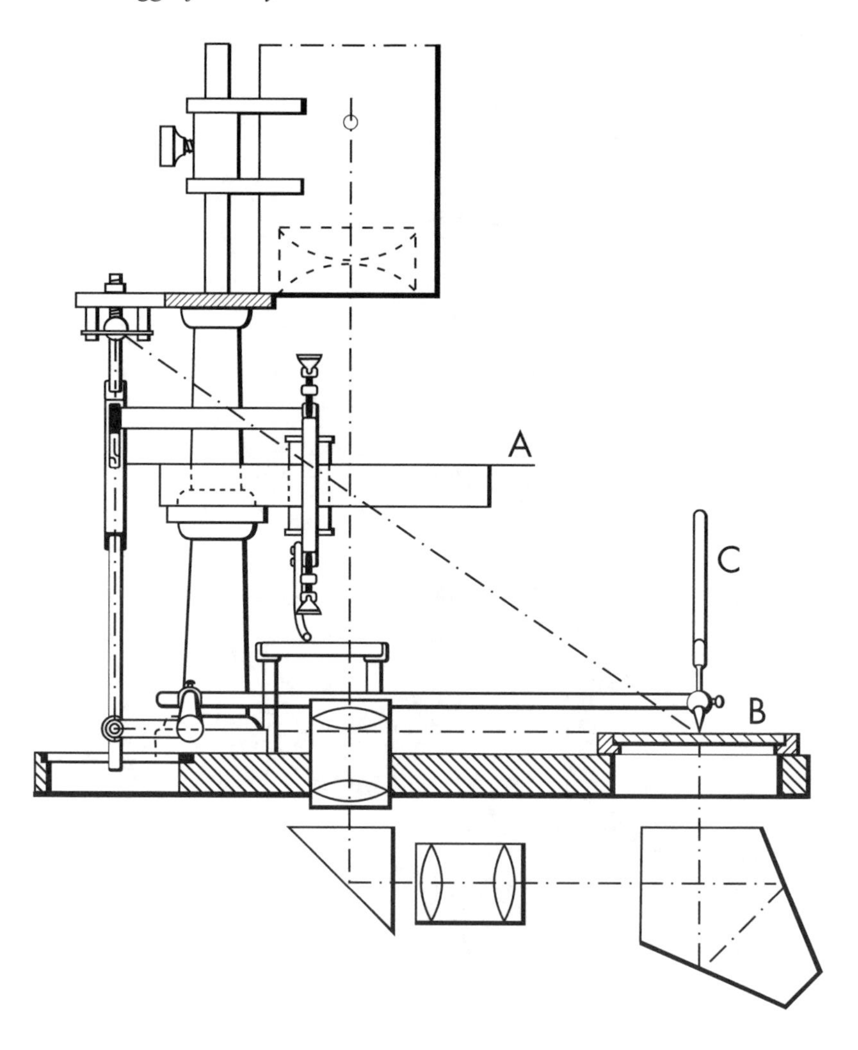

Figure 1.3 *Diagram of a stencil-making machine. An enlarged image of one frame of a film A is projected onto the ground glass at B. By means of a rod C, an operator traces the outline of a part of the image. Depression of C closes an electric circuit and activates the cutting tool [Source:* The First Colour Motion Pictures, *HMSO, London, 1969]*

The patent described the use in a cine camera of a rotating shutter, which consisted of three opaque sectors alternating with three colour filters, red, green and blue. By synchronizing the shutter with the film transport mechanism, successive frames of the film could be exposed through the filters. They had different optical densities to enable the correct exposure times to be obtained for each of the primary colours.

After processing, three consecutive frames of the film could be projected onto a screen simultaneously by a special projector fitted with three lenses, placed very close together, and a synchronously driven three-sector shutter bearing concentric bands of colour filters. Each frame of the film had to be projected three times.

In 1901 Lee withdrew his backing from the project and Turner approached Charles Urban [35] (1867–1942), an energetic American who was the managing director of the British Warwick Trading Company. Urban agreed to finance Turner's project and a camera and projector were manufactured for the Warwick Trading Company by A. Darling of Brighton. The work was completed in April 1902. Shortly afterwards Turner died of a heart attack.

The projector was not an unqualified success:

> As soon as the handle of the projecting machine was worked, the three pictures refused to remain in register and no knowledge that any of us could bring to bear upon the matter could even begin to cure the trouble . . . The difficulty [was] mainly due to the fact that the cinematograph pictures [were] small to begin with (about the size of a postage stamp), and they [had] to be enormously magnified in exhibiting . . . The slightest defect in registration [was] pitilessly magnified. [36]

Following Turner's death, Urban acquired the patent rights of the Lee and Turner method and sought to further its development. He did not have the necessary scientific or technical expertise to initiate a suitable programme of improvement but, fortunately, one of his associates, G.A. Smith (1865–1959), had an inventive turn of mind. Smith had been a photographer in Brighton before he embarked in 1897 on his life's work in the field of cinematography. He had produced his first films for Urban's company in 1898, and, in 1900, had signed a two-year exclusive agreement with them for the distribution of Urban's films.

Urban now financed Smith to work on progressing the Lee and Turner method, but his discouraging attempts led him to conclude the three-colour process was impracticable. About 1902, Smith began to devise a simplified form of the apparatus. He found that by using the colours red and green only he could produce pleasing and acceptable results. His first trial of the two-colour approach took place in July 1906; a few months later Smith applied for patent protection of his method and apparatus [37].

The inventor described his ideas as follows:

> An animated picture of a coloured scene is taken with a bioscope camera in the usual way, except that a revolving shutter is used fitted with properly adjusted red and green colour screens [i.e. filters]. A negative is thus obtained in which the reds and yellows are recorded in one picture and the greens and yellows (with some blue) in the second, and so alternately throughout the length of the bioscope film.
>
> A positive picture is made from the above negative and projected by the ordinary projecting machine which, however, is fitted with a revolving shutter [see Figure 1.4], furnished with somewhat similar coloured glasses to the above, and so contrived that the red and green pictures are projected alternately through appropriate coloured glasses.
>
> If the speed of projection is approximately 30 pictures per second, the two colour records blend and present to the eye a satisfactory rendering of the subject in colours which appear to be natural.
>
> The novelty of my method lies in the use of two colours only, red and green, combined with the principle of persistence of vision. [38]

Figure 1.4 *Diagram of the filter disc of the Kinemacolor projector. The shaded portion of the green sector represents a double thickness of gelatine [Source:* ibid.*]*

The first demonstration of Kinemacolor, as the process was called, took place on 1 May 1908 on the occasion of the inauguration of Urbanora House, the well-appointed new premises, in Wardour Street, London, of the Charles Urban Trading Company. The audience, mostly newspapermen, was impressed. Among the plaudits that followed the occasion, the *Morning Post* described the demonstration as 'the great photographic event of the moment' [39]; the *Daily Telegraph* said it was 'a remarkable advance' [40]. Several months later, on 26 February 1909, the first public cinema show of a colour motion film that had not been hand- or machine-tinted was given at the Palace Theatre, London. Subsequently, screenings of the new Kinemacolor film were given daily for the next 18 months, and Urban began to recoup some of his investment.

He formed the Natural Color Kinematograph Company (NCKC) in March 1909; purchased Smith's interest in Kinemacolor for £5,000; retained the inventor for a further period of five years, at a fee of £500 per annum; and set about, with his characteristic drive and enthusiasm, to exploit the potential of the new medium. By May 1910 Kinemacolor films were being shown in seven theatres in England and Scotland. Initially, the films comprised mostly news events – the unveiling of the Queen Victoria Memorial (May 1911), the Coronation (June 1911), the Naval Review (June 1911) and the investiture of the Prince of Wales at Caernarfon (July 1911) – but soon dramas and comedy films began to be produced at NCKC's studios in Hove, England, and Nice, France [40].

These ventures led the *Bioscope* to comment, in October 1911:

Within the year – almost within the last six months – Mr Charles Urban's Kinemacolor process has come right to the front, and has become a formative influence upon the future of the business, the importance of which cannot be overestimated. 'Colour' has become

> the *sine qua non* of the picture theatre programme, and one cannot pass along the streets without seeing from the announcements of exhibitors that they are fully alive to this, and, if they have not a Kinemacolor licence, they are making a special feature of tinted or coloured films in order to cope with public demand. [40]

Urban's greatest success, prior to the outbreak of the First World war, was filming the Delhi Durbar, India, of 1911. It was a spectacular event which required colour film to record all the multifarious costumes, displays and peoples of the great occasion. Approximately 16,000 ft (4,880 m) of Kinemacolor film was exposed, which was sufficient for a two-and-a-half-hour cinema showing. Its public inauguration was given at the Scala, London, which Urban leased in 1911. Not one to do things by halves, Urban had a special stage setting of the Taj Mahal built, and music composed especially for the opening event. An 'augmented orchestra [of] 48 pieces, a chorus of 24, a fife and drum corps of 20, 3 Scotch bagpipes and beautiful electrical effects' added to the splendour of the film [40]. Urban wrote:

> It was a brilliant and a splendid financial success. We sent out five road shows in England, Ireland, Scotland and Wales, always, mind you, following where black and white pictures of the same subjects had been previously shown. In fifteen months we grossed more than £150,000 after the interests in the monochrome production were exhausted in three weeks. [40]

The *Morning Post* enthused, 'It is quite safe to say nothing so stirring, so varied, so beautiful, so stupendous, as these moving pictures, all in their natural colours, has ever been seen before.' [40]

Kinemacolor was an undoubted success. From April 1911 to March 1914 Kinemacolor's total receipts were *c.* £300,000, expenditure was *c.* £260,000 and profit *c.* £37,000. In addition the sale of patent rights in the USA, Italy, Japan, Russia and Finland, Canada, Holland and Belgium, Brazil, and Switzerland brought in £98,500; and licence fees for foreign exhibition rights yielded another £21,000. (None of these marketing strategies and tactics were lost on J.L. Baird and his supporters when television was demonstrated for the first time in January 1926.)

After the onset of hostilities in August 1914, and the departure of Urban to the United States of America to assist in the promotion of the British war effort by means of films, Kinemacolor declined until by 1917 it had become non-existent. By then about 700,000 ft of Kinemacolor film had been exposed: this was sufficient for 80 to 90 hours of continuous projection.

Kinemacolor film was not without its limitations. It was shot at twice the speed of black-and-white film, which required each frame to receive just half the exposure of a black-and-white film; and some of the incident light was absorbed in passing through the rotating colour filters. Very bright filming conditions, such as good sunlight, were necessary.

Colour fringing was the most serious defect of the colour process. Since 1/30 of a second elapsed between the exposure of two consecutive 'red' and 'green' frames, movement of the subject being filmed would lead to the image on, say, the 'green' frames being out of register with the images of the subject on the 'red' frames. (This was a problem with some of the sequential colour television systems that were demonstrated to the Federal Communications Commission in 1950.) An improvement could be achieved if the elapsed time could be reduced by a redesign of the intermittent mechanism of the Kinemacolor camera. In the Urban–Joy Process (Kinakrom), a faster pull-down mechanism, designed by H.W. Joy, was employed, but though the process was promoted in the USA it never progressed beyond the experimental stage.

If a camera has two lenses, one above the other spaced apart by, say, 18 mm, the 'red' and 'green' images may be recorded simultaneously. W. van D. Kelly of Prizma, Inc., worked on several camera designs based on this principle. A demonstration of the Prizma process given in New York on 8 February 1917 was not commercially successful. The method solves one problem but introduces another – that of parallax – when close-ups are being filmed.

The ideal solution to the problem is to photograph both the 'red' and 'green' frames simultaneously using an optical beam-splitting system based on mirrors or prisms. Such systems had been utilized in still colour cameras from the early 1900s, with good results; nevertheless, the method was not used successfully for colour cinematography until *c.* 1920. (This solution was later applied to some colour television systems.)

The Technicolor Motion Picture Corporation (founded in 1915) also adopted the two-colour principle, and from the commencement of its operations the Corporation based its endeavours on the simultaneous recording, and not the sequential recording, of the two colour images. (During the 1949–50 Federal Communications Commission Hearing on Color Television much evidence was heard on the relative merits of the two types of image acquisition.) Figure 1.5 illustrates, schematically, an early Technicolor

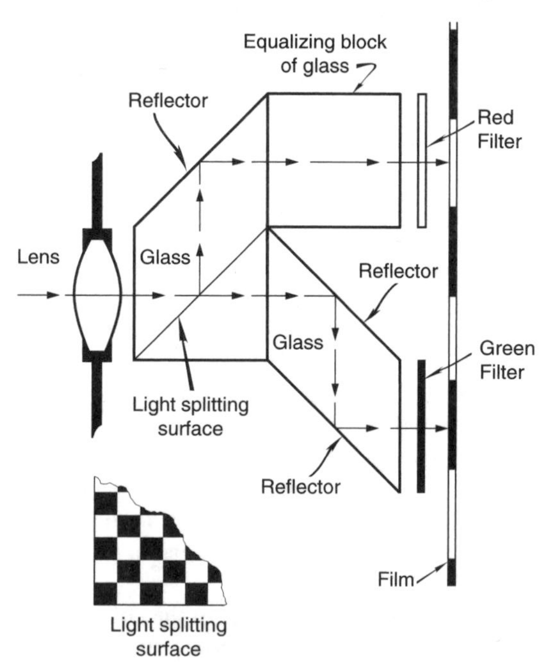

Figure 1.5 *Sectional diagram of the beam splitter of an early Technicolor camera (*c. *1920) [Source:* ibid.*]*

camera of *c*. 1920. The light-splitting device consists of a piece of glass on which a pattern of rectangular squares of silver, alternating with squares of clear glass, is formed. Light reflected from the mirror surfaces subsequently passes through a red filter onto the film. Similarly, the light that is transmitted by the clear spaces passes through a green filter onto the film. In use, the film in the camera moves forward two frames at a time.

The disadvantages of the early Technicolor camera arose from the loss of light, by absorption in the glass; and from the length of the optical path between the taking lens and the film. A longer-focal-length lens than was usual was needed.

The first Technicolor film, *The Gulf Between*, was produced in 1917 and used the additive two-colour principle. Though the film was acclaimed, the projection system had its defects. One writer noted, 'I concluded that the operator would have to be a cross between a university professor and an acrobat.' [40] After 1917 Technicolor adopted the subtractive method of colour photography.

Of the early television schemes, that of J. Adamian was the first to advance a means, based on the two-colour principle, that 'imitate[d] the natural colouring' of objects/scenes [41]. Possibly he was stimulated by Smith's efforts since the patent is dated 1 April 1908. Later, in 1925, J.L. Baird obtained two patents for two-colour television systems, and in 1944 he demonstrated a two-colour telechrome receiver tube.

By the late 1920s the various processes of natural-colour still and motion photography could be classified into three groups:

1. additive processes, as exemplified by the Kinemacolor and Gaumont processes;
2. screen plate processes; and
3. subtractive processes, as manifested by the Kodachrome and Technicolor processes.

According to an account published in the 1929 *Photographic Journal*,

> ... successive projection additive methods [suffered] from the defects of stereoparallax and fringing. The best results [were] given by the three-colour additive process using simultaneous projection, but its scope [was] limited in appeal, as it [required] special apparatus. Subtractive processes [gave] somewhat inferior results ... [and] no process of colour cinematography using the screen-plate principle [had] yet been described which showed signs of commercial practicability. [42]

Nonetheless, taken together these several types of colour film methods permitted the professional cinematographer to produce results that were of good quality. There were difficulties, however. The equipment tended to be cumbersome, costly and heavy, the film was expensive, and specialised knowledge and experience were needed to achieve pleasing images. Effectively, until 1928, the subject and practice of colour cinematography was for the expert only.

During the summer of 1928 (the year when colour television was first demonstrated, by Baird) the Kodak Company announced details of a colour film suitable for amateurs. This was Kodacolor; it was based on a screen film process developed and patented by R. Berthon in 1908. Sometime later, following Berthon's association with A. Keller-Dorian, who had an engraving company in Alsace, the Société Keller-Dorian-Berthon was established to further the development of the process for commercial purposes. Many problems relating to the process were solved but the difficulties of production seemed to inhibit a profitable commercial venture. However, when the process was brought to the attention of the Kodak Company, it recognized that some of these difficulties were similar to those it had encountered in the development of Cine Kodak amateur film. Consequently, the patents were purchased and experimental work was commenced at the Rochester works of the firm.

In the Kodacolor process the film was embossed on its support side (the side opposite the emulsion) with a large number of small, parallel, cylindrical lenses (559 per inch) extending lengthwise along the film (see Figure 1.6). The effect of each of these lenses, when a filter consisting of three bands (red, green and blue) of coloured material was positioned over the camera lens, was to form behind the cylindrical lenses, and hence in the emulsion, an image comprising three parallel microscopic coloured lines. The intensity and colour of each line depended upon the light flux passing through the given filter, the camera lens, and the minute cylindrical lens. Hence the image recorded in the panchromatic emulsion contained information not only about the brightness of the object or scene being photographed but also about its hue.

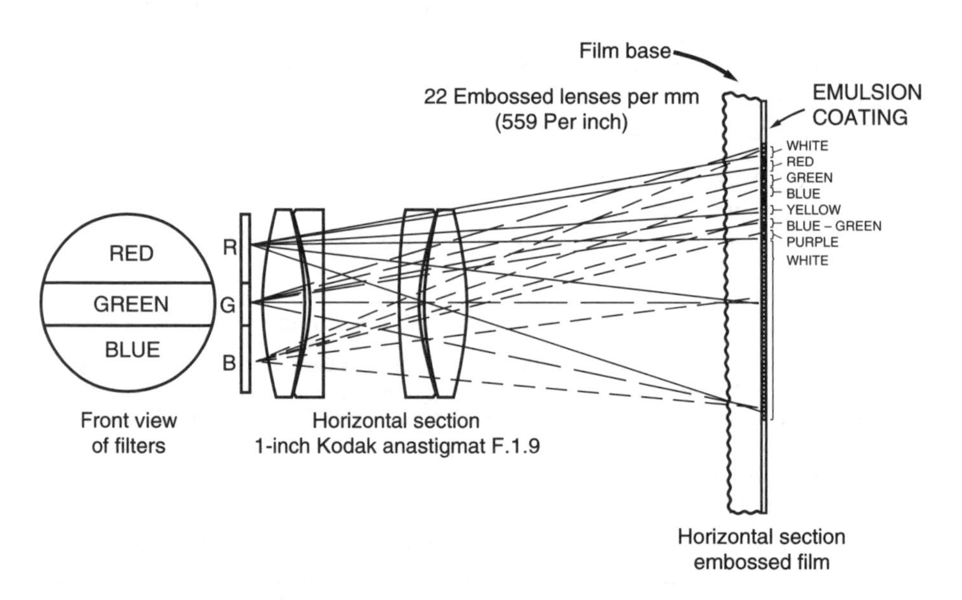

Figure 1.6 Diagram of the Kodacolor process [Source: The Photographic Journal, *September 1929, pp. 404–5]*

In projecting the processed film, a projector was utilized that was optically similar (*mutatis mutandis*) to the camera.

The advantage of Kodacolor derived from the simultaneous recording of the red, green and blue colour records, which eliminated the production of colour fringes when moving objects were photographed. Again, the panchromatic emulsion was relatively cheap, although it had to be very fine-grained to enable the resolving power necessary to record the images produced by the 22-minute cylindrical lenses per millimetre to be achieved. This limitation restricted the use of Kodacolor to cameras fitted with f/2 lenses, and even then the photographed scenes/objects had to be lit by sunlight.

The film appears to have given very satisfactory results:

> '... it gave the most excellent colour rendering [the commentator] had ever seen.'
> '... the outstanding feature in the colour rendering was the lack of crudity in the colours.'
> '... the quality of the colour seemed to be most excellent in anything that was rapidly moving.'
> '[Another observer] had never seen anything like the delicacy of colour shown in [the examples]. Some of them were absolutely charming.' [42]

A scheme to use Kodacolor in a telechromocinematography (i.e. colour television) system was suggested by Dr H.E. Ives in October 1930. This, and Baird's early work on colour television, are considered in Chapter 2.

References

1. Gernsheim H., Gernsheim, A. *The history of Photography*. London: Thames and Hudson; 1955
2. Trimble L.S. *Color in Motion Pictures and Television*. North Hollywood: Optical Standards; 1954
3. Cornwell-Clyne A. *Color Cinematography*. London: Chapman & Hall; 1951
4. Eder J.M. *History of Photography* (tr. E. Epstean). New York: Columbia University Press; 1945
5. Anon. 'Photography'. *Encyclopaedia Britannica XVIII*. 9th edn. pp. 484–90.
6. Arnold H.J.P. *William Henry Fox Talbot. Pioneer of photography and man of science*. London: Hutchinson Benham; 1977. pp. 97–127
7. *Ibid.* pp. 175–216
8. Maxwell J.C. 'On the theory of three primary colours', *Proceedings of the Royal Institute of Great Britain*. 1861;**3**:370–4
9. Evans R.M. Letter. *Journal of Photographic Science*. 1961, pp. 243–6
10. Jongh A.I. de. 'The origins of colour photography; scientific, technical and artistic interactions'. *History of Photography*. 1994;**18**(1):111–19
11. Hauron L. du, Hauron A.D. du. *Traite pratique de photographie des couleurs: système d' heliochrome Louis Ducos du Hauron*. Paris: Gauthier-Villars; 1878
12. Joly J. 'On a method of photography in natural colours'. *Transactions of the Royal Dublin Society,* 1898;**6**(5):127–38
13. Donisthorpe W. *Apparatus for taking and exhibiting photographs*. British patent application 4344. 9 November 1876
14. Donisthorpe W. 'Talking photographs', Letter. *Nature*. 1878;**18**:242
15. Coe B. *Muybridge and the Chronophotographers*. London: Museum of the Moving Image; 1992. p. 45
16. Haas R.B. *Muybridge – Man in Motion*. Berkley: University of California Press; 1976

17. Muybridge E. *Method and apparatus for photographing objects in motion.* US patent application 212 864. 4 March 1879
18. Muybridge E. *Animal Locomotion: an electro-photographic investigation of consecutive phases of animal movement.* Philadelphia: Lippincott; 1887
19. Marey E.J. *Le photographie du movement les methodes chronophotographiques sur plaques fixes et pellicules mobiles.* Paris: Carré; 1892
20. Musser C. *History of the American cinema. The emergence of cinema.* New York: Charles Scribner; 1990
21. Anschutz O. *Projektions Apparat fur stroboskopisch bewegte Bilder.* German patent application 85 791. 5 November 1894
22. Coe B. *Maybrige and the Chronophotographers.* London: Museum of the Moving Image; 1992. p. 36
23. Prince L-A-A Le. *Animated photographic pictures.* US patent application 376 247. 2 November 1886
24. Green W.F. *Photographic printing apparatus.* British patent application 4956. 29 March 1889
25. Clark R.W. *Edison, the Man who Made the Future.* London: Macdonald and Jane's; 1977. pp. 170–9
26. *Ibid.*, p. 172
27. *Ibid.*, p. 176
28. Edison T.A. *Apparatus for exhibiting photographs of moving objects.* US patent application 493 426. 24 August 1891
29. Edison T.A. *Kinetographic camera.* US patent application 589 168. 24 August 1891
30. Thomas D.B. *The Origins of the Motion Picture.* London: HMSO, London; 1964. p. 30
31. Lumière A., Lumière L. '*Appareil servant a l' obtention et à la vision des epreuves chronophotographiques*'. French patent application 245 031. 13 February 1895
32. Muybridge E. *Animal Locomotion: an electro-photographic investigation of consecutive phases of animal movement.* Philadelphia: Lippincott; 1887. pp. 52–3
33. Sklar R. *Film: An international history of the medium.* London: Thames and Hudson; 1993
34. Lee F.M., Turner E.R. British patent application 6202. 1899
35. Thomas D.B. *The First Colour Motion Pictures.* London: HMSO; 1969. pp. 7–9
36. Smith G.A. 'Animated photographs in natural colours'. *Journal of the Royal Society of Arts.* 1908;**LVII**:70–6
37. Smith G.A. British patent application 26671. Nov 1906
38. Anon. Typewritten notes dated 1921. Science Museum (London), Urban Collection
39. Quoted in *Ibid.* p. 17
40. Anon. Typewritten notes dated 1921. Science Museum (London), Urban Collection, pp. 20–40
41. Adamian J. *Electrically controlled apparatus for seeing at a distance.* British patent application 7219. 1 April 1908
42. Clark T. 'Amateur kinematography in colour: the Kodacolor process'. *Photographic Journal.* September 1929, pp. 402–8

Chapter 2
Low-definition colour television

The history of television, from the earliest impractical notions to the commercial success of the new medium, comprises three distinct periods, namely: the period of speculation (*c*. 1877 to *c*. 1925); the period of low-definition television broadcasts (*c*. 1926 to *c*. 1933); and the period of high-definition television broadcasts (1936 to date).

During the first period a few suggestions were advanced for systems of colour television. Most were based on ideas put forward for colour photography but none were investigated experimentally and so are not considered further in this history.

The First World War gave an enormous impetus to the utilization of valves in signalling systems and so stimulated advances in technique that by 1918 triodes could be manufactured to cover a wide power range and were suitable for both transmitting and receiving purposes; their theory and operation, over the frequency bands used at that time, were both well understood.

Accordingly, when interest in television was revived a few years later by scientists and inventors in the UK, the USA, France, Germany and elsewhere, the basic components of a distant vision system were available. The principles of scanning an object or image by means of apertured discs, lensed discs and mirror wheels had been expounded during the era of speculation; methods and apparatuses for synchronizing two non-mechanically coupled scanners had been suggested by many workers and demonstrated in facsimile-transmission systems in the nineteenth century; much development work had taken place on photoelectric cells; the means for amplifying the weak currents obtained from these cells seemed to be available; and now, at the end of the 1910–20 decade, the subject of radio communications had evolved to the state where commercial broadcasting could be seriously contemplated. And, if sound signals could be propagated by radio, then surely vision signals, too, could be transmitted.

Undoubtedly the growth of commercial radiotelephony and domestic broadcasting influenced the progress of television. The time was opportune for further attempts at the practical implementation of a television system. Whereas from 1911 to 1920 only a few isolated attempts had been made to investigate, on an experimental basis, the subject of 'distant vision', from early in the 1920–30 period determined efforts to advance television were made. Initially, these endeavours were mainly those of individuals working in isolation from others. J.L. Baird of the UK, C.F. Jenkins of the USA, E. Belin of France and D. von Mihaly, a Hungarian working in Germany, were four of the principal early investigators in this period. For a short time in 1923, V.K. Zworykin pursued some personal work on an all-electronic television camera while at the Westinghouse Electric and Manufacturing Company, USA, but the only determined effort by an industrial/government organization was that which was initiated at the British Admiralty Research Laboratory, Teddington, UK, in 1923 [1].

From 1925 this situation changed. Bell Telephone Laboratories, of the American Telephone and Telegraph Company began an ambitious programme of work that led to an impressive demonstration in April 1927 of well-engineered apparatus for the transmission and reception of television images by landline and radio links [2]. Later in the USA, General Electric, Westinghouse Electric and Manufacturing Company and the Radio Corporation of America, in addition to a number of smaller companies, also began to be associated with television projects; while in Germany both Fernseh A.G. and Telefunken were active in this field by the end of the decade [3]. Leading companies in the UK adopted a rather reserved position on television matters until 1930, and before then only the Baird companies (Television Ltd, Baird Television Development Company and Baird International Television Ltd) vigorously engaged in the pursuit of 'distant vision' research and development [4]. The Marconi Wireless Telegraph Company, and the Gramophone Company started their television activities in 1930.

Of the work of Jenkins, Belin, von Mihaly and Baird, that of Baird from 1923 to 26 January 1926, when, for the first time anywhere a crude form of television was demonstrated, was remarkable. Baird's life and work during this period is unique in the annals of British twentieth-century electrical engineering. Here was a 35-year-old man with no extended experience of research and development work, no workshop or laboratory facilities, no scientific apparatus of any sort, no employment and no external source of funding, no access to acknowledged expertise or experience, and only one friend in Hastings to give encouragement, seeking in a small room in a suburban house the solution to a problem that had defied the efforts of inventors and scientists around the world for approximately 50 years, and which Dr C.V. Drysdale, the superintendent of the Admiralty Research Laboratory, described in 1926 as 'extremely difficult' [5]. That Baird succeeded is to his everlasting credit. It was a most astonishing and outstanding accomplishment that has not been paralleled since.

John Logie Baird was born on 13 August 1888, in the town of Helensburgh, a small seaside resort situated approximately 22 miles northwest of Glasgow [6]. He was brought up in a comfortable middle-class professional household. His father was the minister of the local church and an intellectual of some merit.

Baird's acquired interest in science, while he was a schoolboy, seems to have been self-generated, for no form of science was taught at the school, Larchfield in Helensburgh, which he attended in his youth. Nevertheless, he engaged in various experiments and projects in his spare time. He installed electric lighting in his home, The Lodge, at a time when such an event could make news in the local press; he constructed a small telephone system so that he could easily contact his friends; he tried to make a flying machine; and he fabricated selenium cells [7].

This curiosity in science appears to have greatly influenced Baird's post-Larchfield education, for he rejected his father's request to enter the ministry and enrolled at the Royal Technical College, Glasgow, in 1906, to follow a course in electrical engineering [8]. Eight years later Baird was awarded an Associateship of the College. An examination of the course curriculum shows that the subject timetables for the second and third years of the mechanical engineering and electrical engineering programmes (following a common first year) were almost identical [9].

This fact had an important bearing on Baird's work on television, for he had a penchant for designing and inventing devices that had a mechanical basis rather than an electrical foundation. Baird displayed considerable ingenuity and innovativeness in the fields of optics and mechanics and produced many patents on aperture-disc, lens-disc and mirror-drum scanning mechanisms, but only a few on electronic devices or systems. Electronics was not Baird's forte. Neither electric telegraphy nor wireless telegraphy formed part of the course programme.

Baird was 26 when he left the Royal Technical College (RTC). He tried to enlist in August 1914 and when he was declared unfit for service entered Glasgow University as a BSc student. He stayed for six months but did not sit the examinations.

Subsequently, he obtained work as an assistant mains engineer with the Clyde Valley Electrical Power Company (CVEP) [10] (see Figure 2.1). This job entailed the supervision of the repair of any electrical failure in the Rutherglen area of Glasgow, whatever the weather, day or night.

Figure 2.1 *Baird* c. *1915 as a young mains engineer at the Rutherglen substation of the Clyde Valley Electrical Power Company [Source: Royal Television Society]*

Throughout his life Baird was subjected to colds, chills and influenza, which necessitated lengthy periods of convalescence. His studies at the RTC were constantly interrupted by long illnesses and his numerous absences from his employment as an assistant mains engineer because of illness militated against any promotion in the company. Because of this he disliked the job and eventually resigned in 1919.

Actually, Baird's departure from the CVEP Company was hastened by his entrepreneurial exploits during the period 1917–19. In 1917 boot polish was difficult to obtain. Baird seized the opportunity to enliven his existence by registering a company and employing girls to fill cardboard boxes with his own boot polish [11]. This venture seems to have escaped the notice of his employers, but the next did not.

Baird had always suffered, and always did suffer, from cold feet, and, on the principle of capitalizing on one's deficiencies, he devised an undersock – consisting of an ordinary sock sprinkled with borax [12]. He arranged its commercial exploitation with such a degree of business acumen and skill that, when he sold the enterprise twelve months later, he had made roughly £1,600 – a sum of money that would have taken him 12 years to earn as an engineer with CVEP.

By 1919 the future television pioneer appeared to be on the threshold of a lucrative commercial life. Unfortunately, continuous good health was not a blessing that had been bestowed on Baird and, during the winter of 1919–20, he suffered a cold, which entailed an absence of six weeks from his venture. He decided to sell out and, following glowing accounts from a friend of the possibilities that seemed to exist in the Caribbean, travelled to the West Indies. His stay there was short and unprofitable and he returned to London in September 1920 [13].

Again he set about establishing a trading business and for the next two years dealt in honey, fertilisers, and coir fibre. The business was successful, but again another illness caused him to remain in bed for several weeks, 'the business meanwhile going to bits', and when his cold did not improve he sold his undertaking [14].

Later, in 1922, on his return to good health, another trading enterprise, based on soap, was started. The concern flourished so much that Baird imported large quantities of soap from France and Belgium and formed a limited company. But another illness compelled him to sell out and convalesce. He went to Hastings, where a friend from childhood lived [15], to convalesce.

Television development was not initially in Baird's mind when he settled there, for in some autobiographical notes he related how, when his health improved, he attempted to invent a pair of boots having pneumatic soles [16] and also a glass safety razor. The author, in his biography of Baird, *John Logie Baird: television pioneer* (Institution of Electrical Engineers, London, 2001), has postulated [17] that his interest in television was stirred by his reading an article: 'A development in the problem of television' by N. Langer in the *Wireless World and Radio Review* issue of 11 November 1922. Langer's paper was optimistic in tone and endeavoured to indicate the lines along which progress could be made.

Baird commenced his investigation at an advantageous time for, in addition to the ideas that had been put forward by others, the technology existed for narrowband television broadcasting He commenced his experiments in various rooms, which he rented when staying in Hastings and elsewhere. Baird did not have any particularly original suggestions to put forward at the outset of his investigations and modelled his schemes on the ideas of others. His earliest vision apparatuses were based on proposals that had been advanced by Nipkow and others in the late nineteenth and early twentieth centuries.

Whatever the source of Baird's inspiration, the solution to the television problem seemed to him to be comparatively simple. Two optical exploring devices rotating in synchronism, a light-sensitive cell and a controlled varying light source capable of rapid variations in light flux were all that were required, 'and these appeared to be already, to use a Patent-Office term, known to the art'. Baird, however, appreciated the difficult nature of the problem: 'The only ominous cloud on the horizon', he wrote, 'was that, in spite of the apparent simplicity of the task, no one had produced television' [18].

Baird's approach to the television problem necessarily, because of his impecunious state, had to be entirely different from those of his contemporaries. Still, undaunted by the formidable difficulties that faced him, he commenced his experiments in 1923 by collecting bits and pieces of scrap material and assembling them into a scanning system. The constraints imposed by his finances severely limited the type of investigations he could carry out, but nonetheless he pursued his objective with dogged determination, ingenuity and resourcefulness [19]. A Nipkow disc could be made from a cardboard hat box, the apertures could be formed using a knitting needle or pair of scissors, electric motors could be obtained cheaply from scrap metal merchants and bull's-eye lenses could be bought at low cost from a cycle shop (see Figure 2.2).

Results were achieved: images of simple objects such as letters of the alphabet were transmitted with a certain degree of clarity. 'The hand appeared only as a blurred outline, the human face only as a white oval with dark patches for the eyes and mouth. The mouth, however, can be clearly discerned opening and closing and it is possible to detect a wink' (Baird, April 1925) [20].

Baird at this time was experiencing difficulties with the poor transient response of his selenium cell and was not able to reproduce halftones. A number of experimenters and inventors had previously encountered or appreciated this property of the cell and had advanced various solutions to overcome its defect, but these were unsuitable for television purposes. Baird's solution had a simplicity characteristic of his early work. He added to the cell's output current a current proportional to the first derivative of the output current and obtained a much-improved response. Success followed. In October 1925 the dummy's head ('Stookie Bill'), which Baird employed as an object, showed up on the screen not as a mere smudge of black and white, but as a real image with details and gradations of light and shade.

A public demonstration to 40 members of the Royal Institution was given on 26 January 1926 at 22 Frith Street [21] (see Figure 2.2).

Following his successful but crude demonstration of television, Baird wished to capitalize on his invention and develop and extend it to the stage where it would

Figure 2.2 Baird in his Frith Street laboratory [Source: ibid.*]*

form the basis of a public television service. This objective required a substantial sum of money. Previously, in 1923, Baird had tried to obtain some support from a major manufacturer of communication equipment, but without any favourable outcome. An advertisement in *The Times* for assistance led to a poor response; and though some financial sponsorship was given by a Mr W.E.L. Day [22] this was insufficient for Baird's needs. Baird's invention had no immediate application to warfare, or to safety, and he received no patronage or encouragement from the one body that could assist him, the British Broadcasting Corporation. Baird's financial resources from 1923 to 1927 were minimal, and so, necessarily, companies had to be formed that would attract investment from the general public. But the public had to be persuaded that the prospects of television were favourable.

The appointment, late in 1925, of a business partner, O.G. Hutchinson, enabled Baird to concentrate on laboratory work (see Figure 2.3). As a consequence, he was able to devote all his energy and inventive skill to the achievement of 'firsts' without being encumbered by the need to attend to business matters: he was able to demonstrate a number of applications of his basic scheme before any other person or industrial organization succeeded in doing so. These demonstrations attracted favourable comments from several notable scientists, many newspapers (including the *New York Herald Tribune* [23]) and two Postmaster Generals.

Figure 2.3 Baird and 'Stookie Bill' being televised in the early days of 'floodlight' television [Source: ibid.*]*

Unfortunately, Hutchinson's business and publicity methods were justifiably regarded with some suspicion and concern in some quarters, notably the BBC, and these caused Baird to suffer some criticism.

There is little doubt that Hutchinson engaged in gross exaggeration to advance Baird television. Nevertheless, it has to be said that he brought the accomplishments of Baird to the attention of the populace. Hence, when the shares of the Baird Television Development Company and Baird International Television Ltd became available for purchase in 1927 and 1928, respectively, there was no shortage of buyers. As one writer noted, 'Television was born at a time when company promoting was running rather wild. Speculators, greedy for fat profits and quick returns, were ready to gamble on very slim chances.' [24]

Baird had to devote a great deal of time and labour to amass a collection of patents [25], which would place his companies in a favourable position commercially, and until *c.* 1930 he engaged in this task almost single-handedly. Not surprisingly, Baird had little time for writing scientific papers and engaging in extensive field trials. He tried to anticipate, and be the first to implement, every likely development and application of the new art. Spotlight scanning (see Figure 2.4), daylight television, noctovision, news by television, stereoscopic television, long-distance television, phonovision, two-way television, zone television, large-screen television and colour television were all demonstrated, in a rudimentary way by Baird during a hectic four-year period of activity from 1926 to 1930 [6]. Additionally, these demonstrations of 'firsts' were required to stimulate and maintain the interest of members of the public. Their support was vital to the

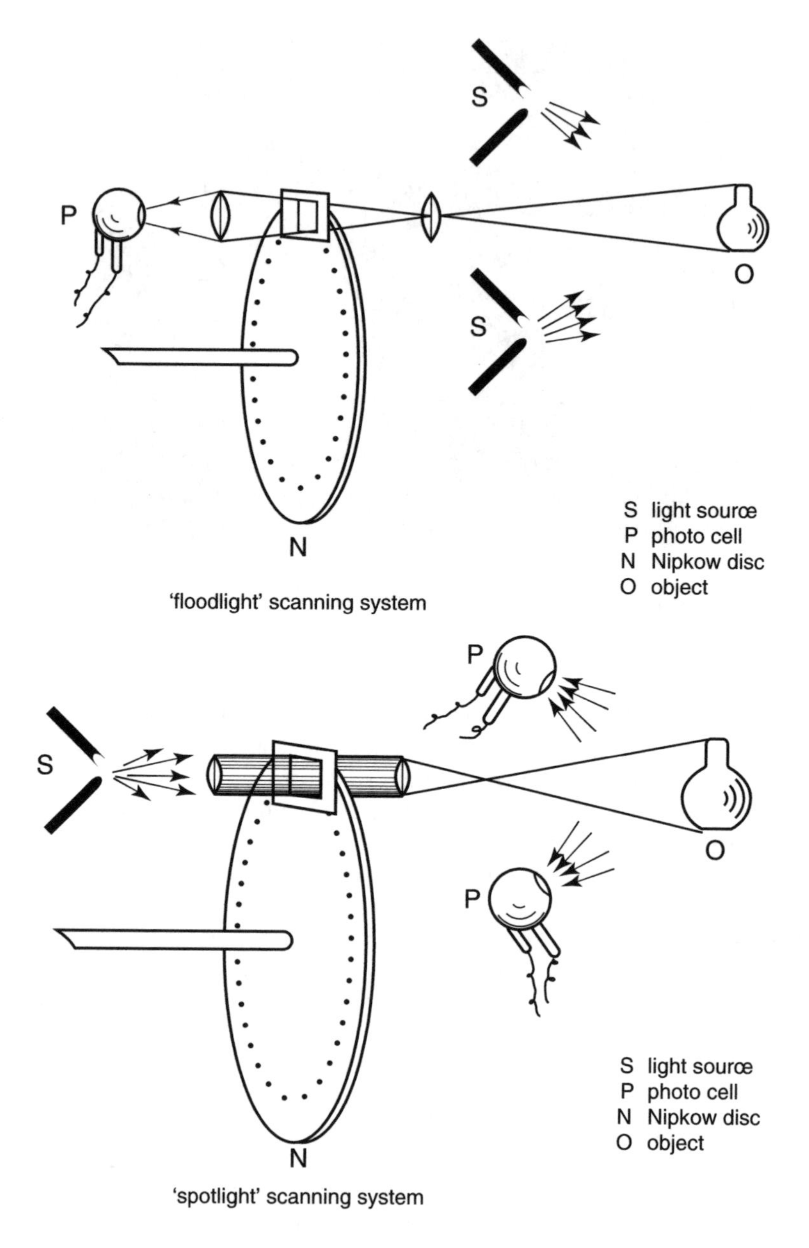

Figure 2.4 Diagrams showing the principles of floodlight and spotlight scanning [Source: Author's collection]

wellbeing of the Baird companies, and to the initiation of a television broadcasting service.

Baird's plans for television were ambitious and extensive, as were those of Marconi for marine wireless communications. By the late 1920s Baird had accumulated a considerable number of patents and hoped to establish his system in many countries, including, of course, the United Kingdom. He wished to establish a monopolistic position in this country, but the rather pushing methods employed by Hutchinson caused antagonism with the broadcasting monopoly. The lack of enthusiasm shown by the BBC towards Baird's low-definition system was a source of much concern and frustration to Baird and his supporters, and resulted in delays in the execution of their desires [26].

The monopolistic position of the Corporation during this period was obviously a considerable obstacle to Baird's aspirations. Essentially, the BBC was not interested in participating in the advancement of television on the basis of a system that could not reproduce images of, say, a Test match at Lord's or tennis at Wimbledon. The BBC considered that low-definition television was inappropriate to its services. As a consequence, the BBC's policy towards Baird's work was necessarily negative in outlook and did not conduce to the rapid advancement of Baird's ambitions.

Patronage and encouragement are important factors in the early development of an invention. Marconi initially was fortunate in this respect. In America and elsewhere facilities for television broadcasting were given by broadcasting stations in the late 1920s, but in Britain the chief engineer of the BBC opposed the use of the BBC's stations for this purpose [27]. This opposition led a former Postmaster General to write:

> It is understood that the Chief Engineer of the BBC holds the view that progress cannot be made along the lines so far pursued. He may be right. On the other hand he may be wrong. In any case the road to further experiment ought not to be closed. If the apparatus proves to be valueless, it will find no patrons and the question will solve itself. [28]

During Hutchinson's protracted discussions and correspondence with the BBC and the GPO, Baird continued to improve and adapt his basic system of television. He demonstrated colour television, for the first time anywhere in the world, on 3 July 1928. The *Morning Post* for 7 July 1928 contained an account of the progress that Baird had made:

> One of the party went onto the roof of the building where a transmitting televisor had been set up, and the rest of the party went with Mr Baird into a room where there was a receiving apparatus. The receiver gave an image, about half as large again as an average cigarette card, but the detail was perfect.
>
> When the sitter opened his mouth his teeth were clearly visible, and so were his eyelids and the white of [his] eyes and other small details about his face. He was a dark-eyed, dark-haired man, and appeared in his natural colours against a dark background. He picked up a deep red coloured cloth and wound it round his head, winked and put out his tongue. The red of the cloth stood out vividly against the pink of his face, while his tongue showed up as a lighter pink.
>
> He changed the red cloth for a blue one, and then, dropping that, put on a policeman's helmet, the badge in the centre standing out clearly against the dark blue background.

> Colour television proved so attractive that the sitter was kept for a long time adjusting his/her postures and expressions at the request of the spectators. A cigarette showed up white with a pink spot on the end when it was lit. The fingernails on a hand held out were just visible and the glitter of a ring showed on one of the fingers.

Although Baird's first demonstration of colour television did not take place until July 1928, the inventor had been reflecting on the problem for a number of years. His British patents nos. 266 564 and 267 378 were applied for on 1 September 1925 and they both contained references to the transmission of images in their natural colours.

For the 1928 demonstration Baird used transmitting and receiving Nipkow discs, each with three spirals and associated filters (one for each of the three primary colours), a single transmission channel and effectively three receiver light sources (giving red, blue and green outputs). These sources were switched sequentially by a commutator so that only one lamp was excited at a time (see Figure 2.5).

Another person who saw the display was Baird's supporter, Dr Alexander Russell. He contributed an article on the progress of television to the 18 August 1928 issue of *Nature* [29]. His report was naturally more technical than the *Morning Post* article:

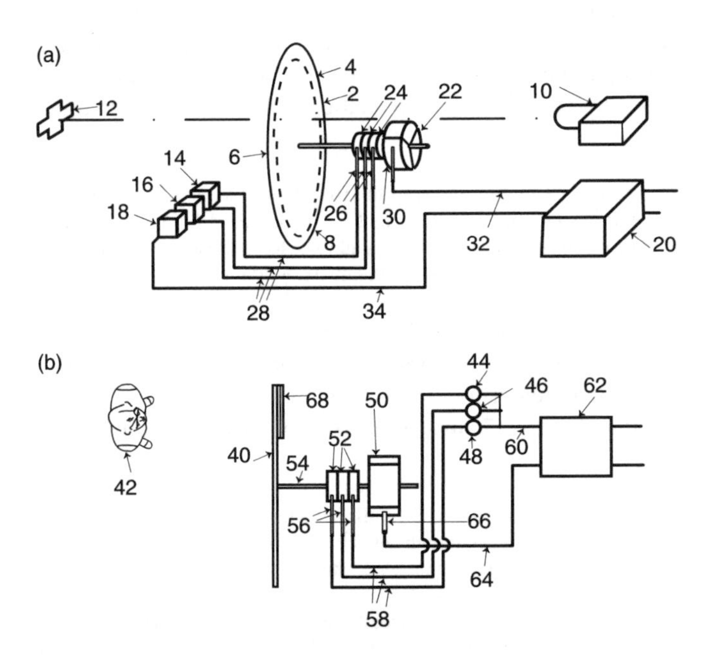

Figure 2.5 Schematic diagrams of Baird's colour television system. The system used sequential three-colour scanning. (a) of the patent shows the transmitter and (b) the receiver. Three-spiral Nipkow discs were utilized – each of the spirals being associated with a primary-colour filter [Source: British patent application 321 390]

> The process consisted of first exploring the object, the image of which is to be transmitted, with a spot of red light, next with a spot of green light, and finally with a spot of blue light. At the receiving station a similar process is employed, red, blue and green images being presented in rapid succession to the eye. The apparatus used at the transmitter consists of a disc perforated with three successive spiral curves of holes. The holes in the first spiral are covered with red filters, in the second with green filters and in the third with blue. Light is projected through these holes and an image of the moving holes is projected onto the object. The disc revolves at 10 revolutions per second and so thirty complete images are transmitted every second – ten blue, ten red, and ten green.
>
> At the receiving station a similar disc revolves synchronously with the transmitting disc, and behind this disc, in line with the eye of the observer, are two glow discharge lamps. One of these lamps is a neon tube and the other a tube containing mercury vapour and helium. By means of a commutator the mercury vapour and helium tube is placed in circuit for two-thirds of a revolution and the neon tube for the remaining third. The red light from the neon is accentuated by placing red filters over the view holes for the red image. Similarly, the view holes corresponding to the green and blue images are covered by suitable filters. The blue and green lights both come from the mercury helium tube, which emits rays rich in both colours.
>
> The coloured images we saw which were obtained in this way were quite vivid. Delphiniums and carnations appeared in their natural colours and a basket of strawberries showed the red fruit very clearly.

Baird's successful demonstration, on 3 July 1928, of rudimentary colour television, by the three-colour process, was noteworthy. His 1928 patents on the subject of colour television show that he had considered various arrangements prior to this date. These are listed in Table 2.1.

Several technical problems had had to be overcome by Baird and his staff before the demonstration. These concerned the photosensitive cells utilized at the transmitter; the light sources used in the receiver; and the equalization of the spectral responses of the system. The first and third of these stemmed from the unavailability of a photoelectric cell having a uniform response. J.C. Wilson, a member of Baird's staff,

Table 2.1 Baird's patents on the subject of colour television

Patent	**Transmitter**		**Receiver**	
	Scanner	**Light source**	**Scanner**	**Light source**
314,951 (4.1.1928)	3-spiral disc	single lamp (white)	3-spiral disc	single lamp (white)
321,389	3-spiral disc	single lamp (white)	3-spiral disc	3 lamps (coloured)
322,776 (9.6.1928)	3-spiral disc	3 lamps (coloured)		
319,307 (20.6.1928)	3-spiral disc	single lamp (white)	1-spiral disc	3 lamps (coloured)
322,823 (11.7.1928)	Types of glow discharge lamps for colour television			

described this difficulty and the solution adopted in a lecture given in 1934 to the Royal Society of Arts [30]:

> Ordinary potassium coatings, suitably sensitized, do not respond to light of wavelength longer than about 585 nm, while monatomic potassium layers on silver, although having a lower critical frequency, are not sufficiently responsive to blue light; for use with incandescent tungsten, which is very deficient in blue rays, the high sensitivity of ordinary potassium cells to blue is very desirable, and a mixture of the two forms of cell is necessary in practice.

Two positive column gas-discharge tubes, one containing neon and the other filled with mercury vapour with a 'little' helium, provided the red, green and blue light components necessary for the reconstitution of the receiver's image. An attempt was made to construct a special glow-discharge lamp by mixing neon, helium and mercury vapour but this was unsuccessful because the proportions of the essential spectral components depended upon variable conditions in the lamp (patent application 322,776). Fortunately the two tubes used had a composite line spectrum, which included important isolated spectral lines in the blue-violet, green, and red-orange regions of the spectrum and these could be separated by means of filters attached to the receiver scanner. In this patent Baird gave details of the filters that he employed to equalize the overall spectral response of his colour television system.

Colour	**Bandpass filter**	**Density filter**
Red	Wratten no. 25	–
Green	Wratten no. 63	0.3010
Blue	Wratten no. 49a	0.6021

Stereoscopic television was demonstrated by Baird in August 1928; it was also demonstrated together with colour television at the British Association meeting in Glasgow in September.

Eventually, the efforts of Hutchinson and Moseley, a great friend of Baird, to obtain transmission facilities were successful and on 30 September 1929 the Baird experimental television service, transmitted by the BBC, was inaugurated [31]. The broadcasts lasted until 22 August 1932, when they were taken over by the BBC and formed the BBC's low-definition television service.

Of all the demonstrations of television given in the 1920s, none surpassed in technical excellence those mounted by the Bell Telephone Laboratories (BTL), of the American Telephone and Telegraph Company (AT&T) [2]. The Laboratories were formed in 1925, when the engineering department of Western Electric was reorganized and became Bell Telephone Laboratories, with a total staff of approximately 3,600.

The results that were obtained by BTL in 1927 are particularly important historically because they were the best that could be expected with the technology as it existed at that time. With its vast resources in finance and equipment, and in staff expertise and experience, BTL was uniquely able to demonstrate what could be engineered in the field of television. Subsequently, colour television and two-way television systems were implemented in 1929 and 1930 respectively, all at great cost.

From 1925 to 1930 (inclusive) AT&T approved the expenditure of $308,100 on low-definition television – a sum far in excess of anything available to any of the 'lone' television workers of whom Baird, Jenkins, Mihaly, Belin and Karolus were in the vanguard.

After 1930 the Laboratories continued their work on television, but without achieving successes of the type that were being manifested contemporaneously by RCA of the USA and EMI of the UK, despite the allocation of $592,400 to the work from 1931 to 1935 (inclusive) [32]. Thereafter, television research and development declined and ceased, sometime in 1940, to be part of BTL's interests.

AT&T commenced its experimental study of the television programme when 'it began to be evident that scientific knowledge was advancing to the point where television was shortly to be within the realm of the possible' [33]. The company was of the opinion that television would have a real place in worldwide communications and that it would be closely associated with telephony. It was certainly well placed to advance television, not only because of the extensive facilities of the newly formed BTL but also because of the experience acquired in the R&D work that had made transcontinental and transoceanic telephony and telephotography possible.

In January 1925 development work under the direction of Dr H.E. Ives [34] had been completed on a system for sending images over telephone lines, and R&D resources and expertise existed for a new scientific venture [35].

Ives was an outstanding experimental physicist. He was born in Philadelphia on 31 July 1882. His father, Dr F.E. Ives, was a noted scientist who during his life invented the first halftone process and the halftone engraving process, and who originated several of the early methods of colour photography. Following graduation from the University of Pennsylvania in 1905, H.E. Ives subsequently obtained a PhD from Johns Hopkins University in 1908. Afterwards he was an assistant physicist of the Bureau of Standards from 1908 to 1909; a physicist of the Electric Lamp Association of Cleveland from 1909 to 1912; a physicist with the United Gas Improvement Company from 1912 to 1919; and finally a physicist of the Western Electric Company and Bell Telephone Laboratories from 1919 until his retirement in 1947.

Ives was eminently well qualified to lead a television project team. His experience and erudition at that time (1925) had been founded on work and investigations on colour photography, phosphorescence, illumination, colour measurement, intermittent vision, photometry, photoengraving, photoelectricity and picture transmission. His standing in his chosen fields had been recognized by the award of three Longstreth Medals (in 1906, 1915 and 1918) by the Franklin Institute of Philadelphia. Later he was to receive three more medals: in 1927 the John Scott Medal, in 1937 the Frederick Ives Medal of the Optical Society of America, and, after the Second World War, the Medal of Merit, the highest civilian award of the US government, for his war work. Ives's publication list comprises *c.* 250 papers and *c.* 100 patents.

Ives felt that the television problem could be examined by utilizing a mechanically linked transmitter and receiver, each incorporating a Nipkow disc scanner operating on a 50-lines-per-picture, 15-pictures-per-second standard. A photographic transparency, later to be superseded by a motion-picture film, would be used at the sending end, together with a photoelectric cell and a carbon arc lamp. His plan was based on the transmission of light through the 'object' rather than on the reflection of light from an opaque body. At the receiving end Ives proposed the use of a crater-type gaseous glow lamp. A sum of $15,000 was approved for the project.

By 7 April 1927 the system was ready to be demonstrated. The demonstration, using a wire link, consisted of the transmission of images from Washington, DC, to the auditorium of the Bell Telephone Laboratories in New York, a distance of over 250 miles (402 km). During the radio demonstration, images were sent from the Laboratories' experimental station 3XN at Whippany, New Jersey, to New York City, a distance of 22 miles (35.4 km). Reception was by means of two forms of receiver. One receiver produced a small image of approximately 2.0 inches × 2.5 inches (5.08 mm × 6.35 mm), which was suitable for viewing by one person. The other receiver gave a large image of nearly 24 inches × 30 inches (60.9 cm × 76.2 cm) for viewing by an audience of considerable size [36].

Ives and his colleagues used a Nipkow disc with 50 apertures for scanning purposes. They arrived at this figure by taking as a criterion of acceptable image quality the standard of reproduction of the halftone engraving process in which it was known that the human face can be satisfactorily reproduced by a 50-line screen. Thus, assuming equal definition in both scanning directions, 2,500 elements per picture had to be transmitted at a rate of 16 pictures per second. The frequency range needed to transmit this number of elements per second was calculated to be 20 kHz.

A spotlight scanning method [37] was adopted to illuminate the subject, the beam of light being obtained from a 40 A Sperry arc (see Figure 2.6). Three photoelectric cells of the potassium hydride, gas-filled type were specially constructed and utilized to receive the reflected light from the subject. At that time they were probably the largest cells that had ever been made and presented an aperture of 120 square inches (774 cm^2).

For reception a disc similar to that at the sending end was used together with a neon glow lamp. The disc had a diameter of 36 in. (91.4 mm) and synthesized the 2.0 in. × 2.5 in. (5.1 × 6.3 cm) image. Another form of receiving apparatus comprised a single, long, neon-filled tube bent back and forth to give a series of 50 parallel sections of tubing. The tube had one interior electrode and 2,500 exterior electrodes cemented along its rear wall. A high-frequency voltage applied to the interior electrode and one of the exterior electrodes caused the tube to glow in the region of that particular electrode. The high-frequency modulated voltage was switched to the electrodes in sequence from 2,500 bars on a distributor with a brush rotating synchronously with the disc at the transmitting end. Consequently, a spot of light moved rapidly and repeatedly across the grid in a series of parallel lines, one after the other, and in synchronism with the scanning beam. With a constant exciting voltage the grid appeared uniformly illuminated, but when the high-frequency voltage was modulated by the vision signals an image of the distant subject was created. To transmit the vision, sound and synchronizing signals three carrier waves were employed: 1,575 kHz for the image signals, 1,450 kHz for the sound signals and 185 kHz for the synchronization control signal [38].

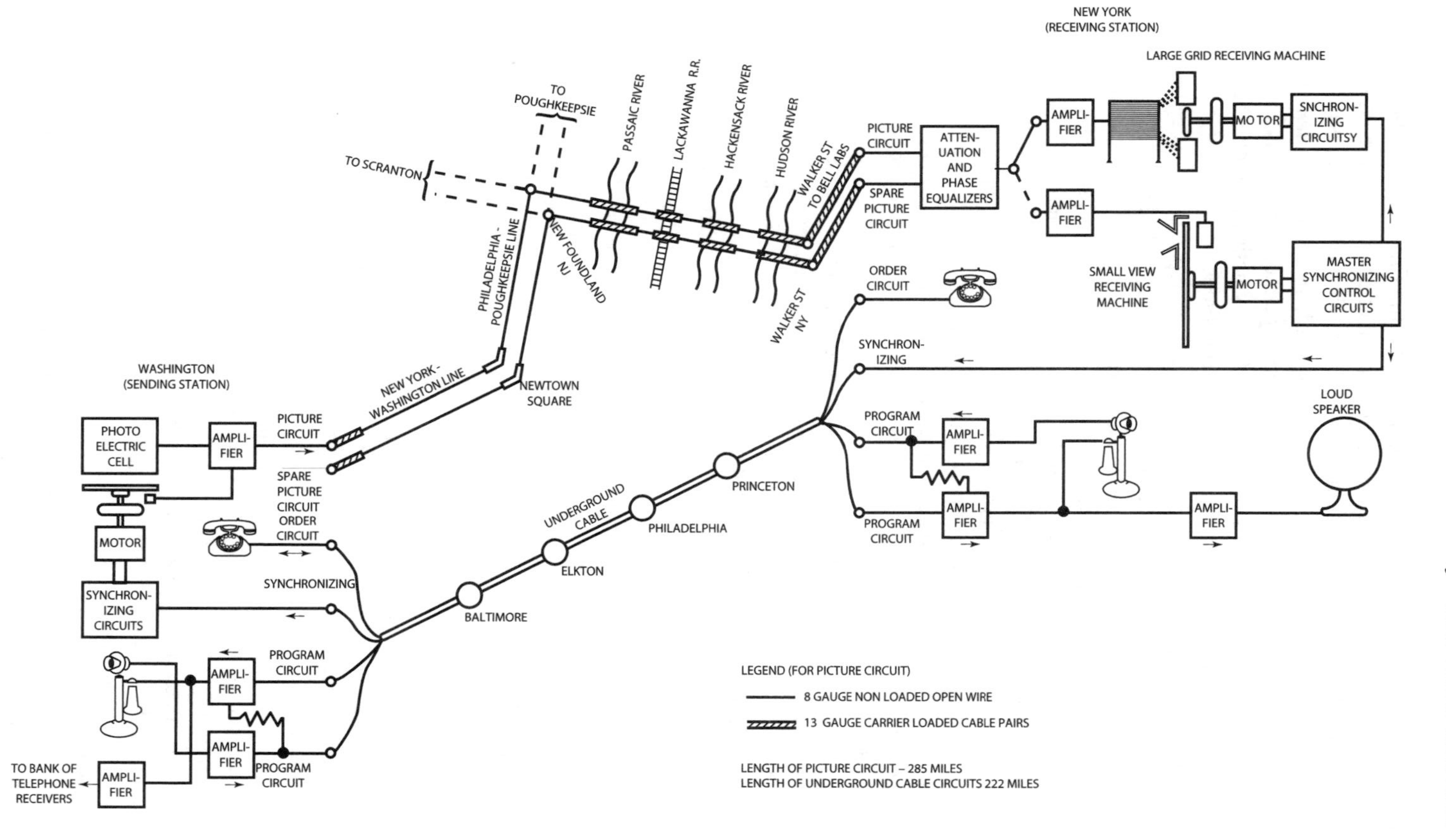

Figure 2.6 *Schematic diagram of the circuits for Bell Telephone Laboratories' 1927 television demonstration [Source: AT&T Bell Laboratories]*

The April 1927 demonstrations were undoubtedly the finest that had been given anywhere, even though no especially novel features had been incorporated into the various systems. They established standards from which further progress could be measured. Moreover, the publication in October 1927, in the *Bell System Technical Journal* (vol. 6, pp. 551–653), of five detailed papers on the factors that led to Ives's group success enabled other workers to ponder on whether their own ideas and practices were likely to lead to similar favourable outcomes.

F.B. Jewett, president of BTL, believed that Ives's group should proceed as vigorously as possible with the preliminary work concerned with the development of public-address television apparatus, without there being any definite commitment [39]. In addition, Jewett felt the Laboratories should carry on, as adequately as possible, whatever fundamental work would be necessary to safeguard the company's position and advance the art along lines that would be likely to be of interest to the company. This mandate gave Ives ample scope to investigate a quite wide range of television problems. He seized the opportunities made available to him and during the next three years daylight television, large-screen television, television recording, colour television, and two-way television were all subjected to the group's scrutiny and engineering prowess [40].

When Ives was admitted in 1905 to Johns Hopkins University, as a PhD student, it was to study colour photography under the supervision of R.W. Wood, then the leading authority on optics in the United States. Ives's choice of subject had possibly been influenced by his illustrious father's substantial contributions to the art of photography and the science of optics. Dr F.E. Ives had invented a trichromatic camera, various processes of colour photography and a 'device for optically reproducing objects in both full modelling and natural colors', *inter alia*.

H.E. Ives's first two papers (both published in the *Physical Review*) relate to improvements in methods of colour photography. The earlier paper (1906) pertains to Wood's device and the later paper (1907) to Lippman's scheme. The papers were written before Ives had completed his doctoral thesis (1908), which had the title 'An experimental study of the Lippman colour photograph'.

With such a background it was perhaps inevitable that Ives and his coworkers would want to attempt to create coloured televised images. Indeed, Ives and his group had already transmitted by wire a colour still photograph of Rudolph Valentino, and so some knowledge of the likely difficulties to be anticipated in transmitting televised colour images had been gained. On that occasion the three-colour additive process was used – the black-and-white colour separation transparencies having been produced by M. Hofstetter of the Powers Photo Engraving Company of New York. They showed Valentino dressed for the part of Monsieur Beaucaire in the film of that name. On 15 July 1924 Ives sent the three separations from Chicago to New York, where Hofstetter made a coloured lantern slide transparency from them by the Uvachrome process.

On 27 June 1929 colour television [41] was shown by Bell Telephone Laboratories to an invited gathering of scientists and journalists. Ives employed three signal channels so that the three colour signals could be sent simultaneously from the transmitter to the receiver. An advantage of this arrangement was that the same scanning discs and motors, synchronizing equipment and light sources, and the same type of circuit and method of amplification were used as in the monochrome scheme. The only new features were the form and disposition of the specially devised photocells at the sending end and the type and grouping of the neon and argon lamps at the receiving end (see Figure 2.7).

A neon glow lamp gave the desired red light but for the sources of green and blue light 'nothing nearly efficient as the neon lamp was available'. Two argon lamps, one with a green filter and one with a blue filter, were finally adopted for the demonstration; however, various expedients were needed to increase their effective

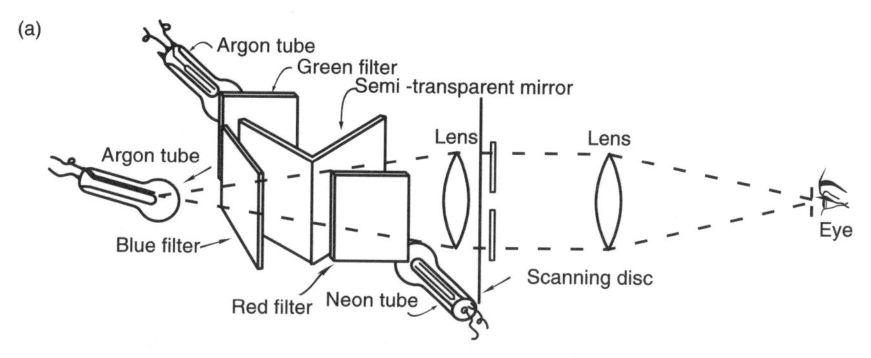

Figure 2.7 (a) The schematic layout of the lamps and filters used in the Bell Telephone Laboratories' apparatus for colour television. (b) With the exception of the photoelectric cabinet on the left, the apparatus was identical with that used for the original demonstration of monochrome television [Source: ibid.*]*

luminous intensity. Special lamps with long, narrow and hollow cathodes cooled by water were utilized and these were observed end on so that the thin glowing layer of gas was greatly foreshortened and the apparent brightness thereby increased.

To render the correct tone of coloured objects it was essential to obtain photoelectric cells that would be sensitive throughout the visible spectrum. A.T. Olpin and G.R. Stilwell constructed a new kind of cell that used sodium in place of potassium. Its active surface was sensitized by a complicated process involving sulphur vapour and oxygen instead of by a glow discharge of hydrogen as with the former class of cell.

An account of the demonstration, in which the transmission was over lines, was published in *Telephony* on 6 July 1929. The display [42]:

> ...opened with the American flag fluttering on a screen about the size of a postage stamp. The observer saw it through a peep hole in a darkened room. The colours reproduced perfectly. Then the Union Jack was flashed on the screen and was easily recognized by its coloured bars.
>
> The man at the transmitter picked up a piece of watermelon, and there could be no mistake in identifying what he was eating. The red of the melon, the black seeds and the green rind were true to nature, as were the red of his lips, the natural colour of his skin and his black hair.

For Ives, the coloured images were quite striking in appearance, notwithstanding their small size and low brightness. He felt that the addition of colour contributed notably to the naturalness of the images, but observed that, since colour television was intrinsically more complicated and costly than monochrome television, it was likely some considerable time would elapse before it was practically implemented.

From the inception of his work on television, in 1925, H.E. Ives had appreciated the advantages that would be gained if motion-picture film were used at the sending end of a television link. In his January 1925 memorandum to Arnold, Ives had referred to measurements he had made of the brightness of the image of a sunlit landscape as projected onto a photoelectric cell by means of a wide-aperture lens. The measurements showed that the magnitude of the light flux which could be concentrated on the cell was about 1/500 of that employed in picture transmission. Hence, the degree of sensitivity of the photocell and the degree of amplification necessary in the proposed television system would be far greater than those that had been acceptable in connection with picture transmission. Because of this Ives advocated using transmitted light (from a film) rather than reflected light (from a scene) to ease the solution of the television problem. He noted [43]:

> It may be pointed out that the use of a moving picture film as the original moving object is the equivalent to a very great amplification of the original illumination brought about by the photochemical amplification process involved in the production and subsequent development of the photographic latent image.

Ives did not employ cine film in his group's impressive 1927 demonstrations of small- and large-screen television, but at a meeting of the Optical Society of America held on 30 October 1930 he described a method [44] of accomplishing the transmission of images in colour, from Kodacolor motion picture film, using the receiving apparatus described above.

Figure 2.8a shows how images could be generated using film, in which the colours are incorporated by dyes (e.g. Technicolor), and the three-colour transmitting and

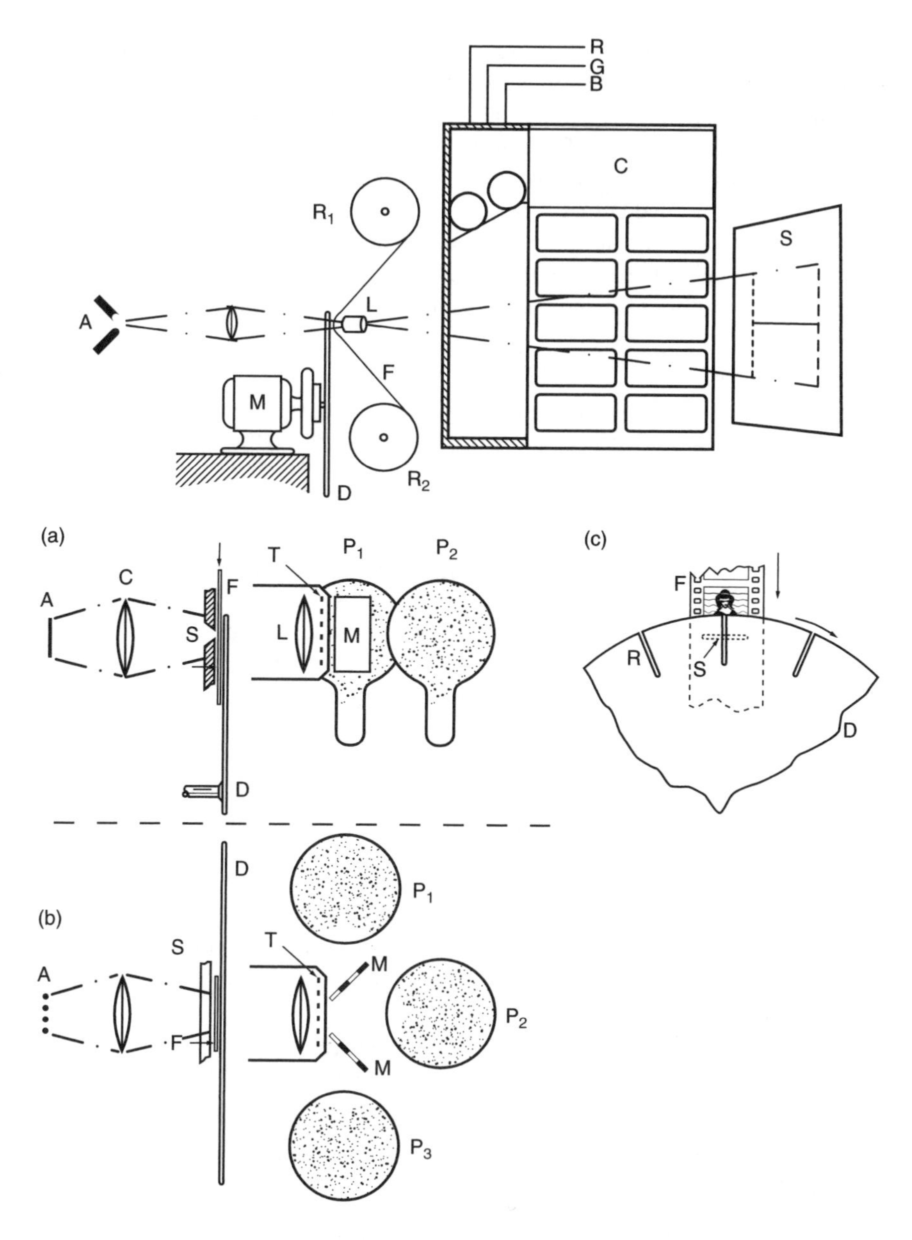

Figure 2.8 *(a) Sectional view of television-sending apparatus for transmitting from coloured picture film. A, light source; D, scanning disc; M, synchronous motor; C, photoelectric cell cabinet containing three sets of colour-sensitive cells with appropriate filters; F, motion-picture film; L, projection lens; S, white screen. (b) Diagram of the scanning apparatus. A, light source; C, condensing lens; F, film; S, slot, transverse to direction of film motion; D, scanning disc, provided with radial slots R; L, projection lens; T, position of colour filters used for screen projection; P_1, P_2, P_3, three photoelectric cells; M, mirrors. (c) View of the slotted scanning disc, film and method of scanning [Source:* Journal of the Optical Society of America, *1931, vol. 21, p. 3]*

receiving apparatus recently (1929) developed in the Bell Telephone Laboratories. The light source A, the scanning disc D, the synchronous motor M, the photocell cabinet with its filters C, the motion picture film F, the projection lens L, and the white screen S onto which the film image would be projected are indicated. 'This method', said Ives, '... while completely practical, suffers under the disadvantage that it requires an original colored film of a sort which is both expensive and time-consuming to produce. Should television transmission from film become popular it is probable that the chief demand would be for films which would be shown but once, and for showings within a few hours at most of the event.'[44] Kodacolor met Ives's requirements.

Figure 2.8b illustrates the configuration of the apparatus:

> After passing through the film and disc the light is projected as if to a screen by the lens L, in front of which is placed, in the regular projector, the set of red, green and blue filters T.... [Thence] the light is diverted into three photoelectric cells, P_1, P_2, and P_3 by the mirror M. These cells are all similar, and need not be color sensitive. [44]

Work on this method of colour television continued until *c.* 1936 but regrettably no eyewitness reports on the results obtained appear to be available.

Following the completion of the colour and two-way-link television demonstration, Ives undertook an important appraisal of the progress that had been made by his group and attempted to define the course of action that had to be implemented for the future advancement of television, particularly high-definition television. His prognosis was gloomy in outlook [45]. He concluded, 'The existing situation is that if a many-element television image is called for today, it is not available, and one of the chief obstacles is the difficulty of generating, transmitting, and receiving signals over wide frequency bands.' A partial solution was to employ multiple scanning and multiple channel (zone) transmission.

Ives's multi-channel experimental set-up [45] is illustrated in Figure 2.9. This used scanning discs with prisms over their apertures, so that, at the sending end, the beams of light from the successive holes were diverted to different photoelectric cells. At the receiving end, the prisms enabled beams of light from the three lamps to be deflected in a common direction.

Ives found that his three-channel apparatus yielded results strictly in agreement with the theory underlying its conception and observed that the 13,000-element image was a marked advance over the single-channel 4,000 image:

> Even so, the experience of running a collection of motion picture films of all types is disappointing, in that the number of subjects rendered adequately by even this number of image elements is small. 'Close-ups' and scenes showing a great deal of action, are reproduced with considerable satisfaction, but scenes containing a number of full length figures, where the nature of the story is such that the facial expression should be watched, are very far from satisfactory. On the whole the general opinion ... is that an enormously greater number of elements is required for a television image for general news or entertainment purposes. [45]

This point had been appreciated by a few workers for several years. Swinton had outlined an all-electronic system of television in 1911 [46], and V.K. Zworykin had patented a version of such a system in 1923. When, in 1926, Farnsworth started his

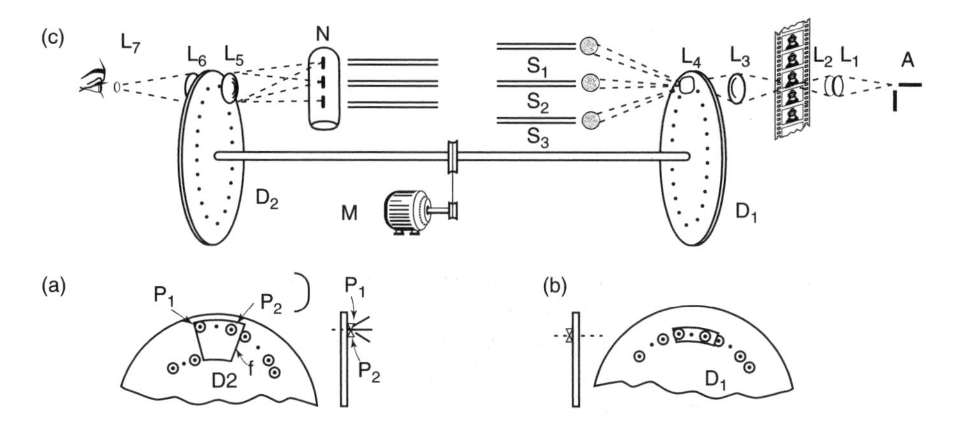

Figure 2.9 Schematic diagram of BTL's three-zone television system: (a) receiving-end disc with spiral holes provided with prisms; (b) sending-end disc with circle of holes with prisms; (c) general arrangement of apparatus [Source: Journal of the Optical Society of America, *1931, vol. 21, pp. 8–19]*

work on television it was on the basis of an all-electronic scheme. In Great Britain, a similar approach would be adopted in 1931 by Electric and Musical Industries Ltd (EMI). Subsequently, Zworykin, Farnsworth and EMI satisfactorily demonstrated all-electronic television before the end of 1935.

References

1. Beatty R.T. 'Report on television', parts 1 and 2, 34 pp. AIR 2/S24132, PRO; December 1925
2. Burns R.W. *The Contributions of the Bell Telephone Laboratories to the Early Developments of Television*. London: Mansell; 1991. pp. 181–213
3. Burns R.W. *Television: an international history of the formative years*. London: Institution of Electrical Engineers; 1998. pp. 242–82
4. Burns R.W. *British Television: the formative years*. London: Institution of Electrical Engineers; 1986
5. Lefroy H.P. Minute, AIR 2/S24132, Public Record Office; 9 June 1926
6. Burns R.W. *John Logie Baird: television pioneer*. London: Institution of Electrical Engineers; 2000
7. *Ibid*. pp. 8–10
8. Baird J.L. *Sermons, Soap and Television*. Royal Television Society, London, 1988
9. Burns R.W. *John Logie Baird: television pioneer*. London: Institution of Electrical Engineers; 2000. Note 2, p. 28
10. Baird J.L. *Sermons, Soap and Television*. Royal Television Society, London, 1988. pp. 28–30
11. Watkins, H. 'How television began – Part 13', *London Home magazine*, *c*. 1962

12. Baird J.L. *Sermons, Soap and Television.* Royal Television Society, London, 1988. pp. 30–2
13. Burns R.W. *John Logie Baird: television pioneer.* London: Institution of Electrical Engineers; 2000; pp. 20–3
14. Baird J.L. *Sermons, Soap and Television.* Royal Television Society, London, 1988. 147 pp.
15. *Ibid.* p. 42
16. *Ibid.*
17. Burns R.W. *John Logie Baird: television pioneer.* London: Institution of Electrical Engineers; 2000. pp. 31–3
18. Baird J.L. 'Television'. *Journal of Scientific Instruments.* 1927;**4**:138–43
19. Anon. 'Television. Amateur scientist's invention. Secret plans. Outline of objects transmitted'. *Daily News.* 15 January 1924
20. Editor. 'Note by the Editor'. *Discovery.* April 1925, p. 143
21. Burns R.W. *John Logie Baird: television pioneer.* London: Institution of Electrical Engineers; 2000. pp. 90–1
22. Burns R.W. *John Logie Baird: television pioneer.* London: Institution of Electrical Engineers; 2000. Chapters 2 and 3
23. Anon. Report, *New York Herald Tribune*; 11 February 1928
24. Moseley, S.A. *The Private Diaries of Sydney Moseley.* London: Max Parish; 1960. p. 292
25. Burns R.W. *John Logie Baird: television pioneer.* London: Institution of Electrical Engineers; 2000. p. 166, pp. 397–403
26. Burns R.W. *British Television: the formative years.* London: Institution of Electrical Engineers; 1986, Chapter 4, 'The BBC view', pp. 94–108
27. Burns R.W. *British Television: the formative years.* London: Institution of Electrical Engineers; 1986, Chapter 7, 'Television and the BBC', pp.155–76
28. Samuel Rt Hon. Sir H. Letter to the Postmaster General, 14 January 1929, Post Office minute 4004/33
29. Russell, A. Report, *Nature.* 18 August 1928
30. Wilson J.C. 'Trichromatic reproduction in television'. *Journal of the Royal Society of Arts.* 1934;**82**:841–63
31. Burns R.W. *British Television: the formative years.* London: Institution of Electrical Engineers; 1986; 'The start of the experimental service, 1929', pp. 132–47; and chapter 7, 'The low definition experimental service, 1929–1931', pp. 148–75
32. Burns R.W. *The contributions of the Bell Telephone Laboratories to the Early Developments of Television.* London: Mansell; 1991. p. 209
33. 'Remarks by Frank B. Jewett at the television demonstration'. *Bell Laboratory Record.* May 1927, p. 298
34. Findley P.B. 'Biography of Herbert E. Ives'. Internal report, date unknown. AT&T Archives
35. Ives H.E. 'Television: 20th anniversary'. *Bell Laboratory Record.* 1947;**25**:190–3
36. Ives H.E. 'Television'. *Bell System Technical Journal.* October 1927, pp. 551–9
37. Gray F., Horton J.W., Mathes, R.C. 'The production and utilization of television signals'. *Bell System Technical Journal.* October 1927, pp. 560–81

38. Nelson E.L. 'Radio transmission for television'. *Bell System Technical Journal.* October 1927, pp. 633–53
39. H.S.R. 'Television'. Memorandum for file, 18 March 1966, Case Book No. 1538, Case File 20348 and Case File 33089, Vol. A, pp. 1–9, AT&T Archives
40. Burns R.W. *The Contributions of the Bell Telephone Laboratories to the Early Developments of Television.* London: Mansell; 1991. p. 193
41. Ives H.E. 'Television in colour'. *Bell Laboratory Record.* July 1929, pp. 439–44
42. Anon. 'Television in colour successfully shown'. *Telephony* 1929;**97**:23–5
43. Ives H.E. 'Television'. Memorandum for file, 10 July 1925, Case File, 10 July 1925, Case File 33089, Vol. A, pp. 1–2, AT&T Archives
44. Ives H.E. 'Television in full color from motion picture film'. *Journal of the Optical Society of America.* 1931;**21**:2–7
45. Ives H.E. 'A multi-channel television apparatus'. *Journal of the Optical Society of America*, 1931;**21**:8–19
46. Campbell Swinton A.A. 'Presidential Address'. *Journal of the Roentgen Society.* 1912;**8**:1–5

Chapter 3
RCA and pre-war colour television

During Bell Telephone Laboratories' programme of work on television, Ives's group had undertaken some investigations (in 1926) on the appropriateness of utilizing a cathode-ray tube in a television receiver. Images of simple objects, such as a letter A, a bent wire and so on, had been received on a modified cathode-ray tube. The received picture signals had been impressed on an extra grid in the tube and controlled the intensity of the electron beam incident on the fluorescent screen. In one type of tube, the grid was close to the hot filament and the picture signals acted on the beam before it reached the accelerating field. In a second type of tube, the cathode beam passed through two parallel wire gauzes just before striking the screen and the image signals were applied across these two gauzes. Both types of tube reproduced images of the simple objects just mentioned, but they did not reproduce more complex objects, nor did they show halftones in a satisfactory way [1].

Baird never utilized cathode-ray tubes in the 1920s. His view on their application was given in an article published in January 1928. 'The use of the cathode ray', he wrote, 'is beset with the greatest difficulties and so far, no practical success has been met with in its application' [2]. Dr R.T. Beatty, of the Admiralty Research Laboratory, held a similar opinion. 'Although this method sounds attractive the practical difficulties are so great that little progress has been made by experimenters on these lines.' These difficulties were delineated in a report [3] that he wrote in December 1925.

1. Since the angular displacement of the spot is not proportional to the voltage applied, and owing to the curvature of the fluorescent screen, considerable distortion of the picture takes place. This is increased by the fact that the deflection voltages are sinusoidal instead of being linear functions of the time.
2. The modulations produced in the anode voltage for the purpose of increasing the luminosity cause radial displacements of the spot, whereby serious confusion of the picture is produced.
3. The successive pictures will not register unless the amplitudes and frequencies of the simple harmonic motion deflection voltages are regulated with great accuracy. [3]

During the 1920s three methods of focusing the electron beam in a cathode-ray tube (c.r.t.) were investigated, namely: gas focusing, magnetic-field focusing and electric-field focusing [4]. The last method resembles the focusing of a light beam by a lens system (see Figure 3.1).

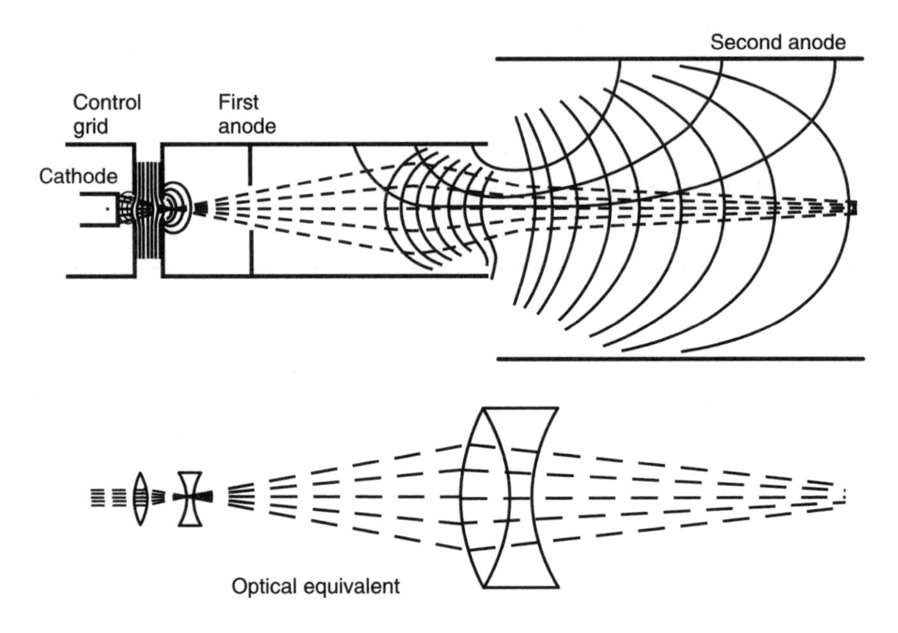

Figure 3.1 *Electron lens – optical lens analogy: diagram showing the analogy between the focusing of a beam of electrons by the use of electron lenses, and the focusing of a beam of light by optical lenses [Source:* Journal of the IEE*, 1933, vol. 73, p. 443]*

Among the early investigators of electrostatic focusing there were W. Rogowski and W. Grosser, who applied, in December 1923 for a patent [5] for a c.r.t. based on this type of focusing; A. Dauvillier [6], who studied the potential of such a method for television purposes and filed a patent in February 1925; P.T. Farnsworth [7], who, from 1926, endeavoured to evolve an all-electronic television system based on hard vacuum tubes; F. Holweck and P.E.L. Chevallier of the Laboratoire des Établissements Edouard Belin, Paris, whose experimentation [8] led to a patent application dated 4 March 1927; R.H. George [9] of Purdue University, whose work led to an important paper in 1929; and V.K. Zworykin [8] of the Westinghouse Electric and Manufacturing Company, who, from 1923, had sought to develop an all-electronic television system.

Vladimir Kosma Zworykin has been described as the 'father of modern television', so a few words on his background seem appropriate. He was born on 30 July 1889 at Murom, Russia, which is 220 miles east of Moscow. After leaving school, Zworykin enrolled at the University of St Petersburg determined to become a physicist. This did not accord with his father's wishes because he felt that Russia's rising new industries offered a richer future in engineering than in physics, and Zworykin was persuaded to transfer to the Imperial Institute of Technology. Here he remained for six years from 1906 till 1912. The move greatly influenced his subsequent career.

During the above period Professor B. Rosing, of the Institute, was working on distant vision (an early name for television). The subject soon attracted the attention of Rosing's young student, Zworykin, and he began to assist Rosing in his private laboratory, which was 'a little cubby hole in the basement of the artillery school', which was situated across the street from the Institute. Though much of the apparatus had to be manufactured by them, including the photocells and all the glass vessels needed for their work, Zworykin found the research work much to his liking. The period from 1910 to 1912 he later described as 'the glorious three years'.

Following his graduation in 1912 Zworykin's love for physics took him to Paris, where he engaged in work on X-rays, for his doctorate, under the guidance of Paul Langevin.

When the First World War broke out, Zworykin returned to Russia and was drafted immediately. He was sent, a few months later, to the Grodno fortress near the Polish frontier, but after a year and a half was transferred to the Officers Radio School. There he was commissioned and began teaching soldiers how to operate and repair electrical equipment. A further period of time was spent in the Russian Marconi factory, which was constructing radio equipment for the Russian Army and which was situated on the outskirts of St Petersburg. He was attached to the factory as an inspector of radio equipment.

In 1917 the Russian Revolution commenced and Zworykin, fearing that it would disrupt his scientific career, decided to leave his native land. At first he could not obtain the necessary permission and the United States refused him a visa. For months he wandered around Russia to avoid arrest during the chaos of the civil war between the Reds and the Whites. Then, when an Allied expedition landed in Archangel, in September 1918, to aid Russia's northern defences against the Germans, Zworykin decided to make his way to that town. Pleading his case with an American official, Zworykin told him about the work he could do in advancing television. He was given a visa and arrived in the USA on 31 December 1918.

His first requirement, of course, was to seek employment. After several unsuccessful applications, a locomotive company called Baldwin recommended him to the head of research at Westinghouse's factory in East Pittsburg, Pennsylvania. Zworykin applied for a post and was engaged. His initial task was to assemble vacuum tubes on a production line. 'The assembly was cumbersome and took a lot of time. There [were] a tremendous number of rejects – about 70 per cent. I did this for about three months and almost went crazy,' Zworykin has written. However, when an explosion occurred that burnt his right hand and resulted in his spending some time in hospital – and filing a claim for damages – his fortunes changed. Because Westinghouse felt responsible for the accident the company allowed him to work in his beloved field of endeavour – television – in order to 'humour him'.

Zworykin subsequently spent a year developing a high vacuum cathode-ray tube, but in 1920 he left Westinghouse following a dispute over some patents relating to the WD11 valve. He obtained employment with the C&C Development Company in Kansas City but soon received a most attractive proposal from Westinghouse. It appears that Zworykin had previously much impressed O.S. Schairer, the manager of the Westinghouse patent department: he persuaded the new manager of the research laboratory, S.M. Kintner, to extend an invitation to Zworykin to return to the company. According to Zworykin, he was offered a three-year contract at about three times his former salary. Under the terms of the contract Zworykin would retain the rights to his prior inventions with Westinghouse, although the firm would hold the rights to purchase his patents at a later date. Zworykin accepted the offer and recommenced his employment with Westinghouse in February/March 1923.

On arrival he was asked by Kintner to suggest a suitable research project and was allowed to engage in the field that had held his attention about a decade previously. Zworykin worked rapidly and applied for a patent on his television system on 29 December 1923, but it was not granted until 20 December 1938. The patent which was based on Zworykin's electronic camera tube, a later version of which became known as the 'iconoscope', was the subject of a number of interference actions. [4] There is no evidence that any effort was made in 1923 to reduce the patent to practice (to use a Patent Office expression), and until *c*. June 1924 it appears that Zworykin worked on other projects.

Zworykin began his practical work in that year using a modified Western Electric type 224A gas-filled cathode-ray tube, and in 1925 demonstrated his work to Kintner. He was 'very impressed by [its] performance', but to further the work more effort, space and financial resources were needed. It was decided to show the system to H.P. Davis, the general manager of the company.

In the late summer/late autumn of 1925 Davis, Kintner and Schairer witnessed a demonstration of Zworykin's electronic television scheme. Schairer was 'deeply impressed' but, unfortunately for Zworykin, Davis was not. Zworykin has described the consequences of the test: 'Davis asked me a few questions, mostly how much time I [had] spent building the installation, and departed saying something to Kintner which I did not hear. Later I found out that he had told him to put this "guy" to work on something useful.'

The immediate effect of the 1925 demonstration was the relocation of the television project to Dr F .Conrad, who had participated in the engineering of Westinghouse's KDKA transmitter station, and who was highly regarded by Davis. Conrad's approach to his new task was to return to more conventional methods of scanning an object. He devised a system of radio movies in which 35 mm motion-picture film was scanned to provide a source of video signals. Conrad's efforts were rewarding and, on 25 August 1929, KDKA began broadcasting radio movies on a daily basis. A 60-lines-per-picture, 16-frames-per-second standard was adopted.

Meanwhile, Zworykin had been working on the recording and reproduction of sound on cine film. This work led to a new recording camera, and the loss of two of Zworykin's associates, who received handsome offers from a Hollywood film company. Zworykin was proffered a similar proposal but decided to remain in research: he obtained permission to transfer his efforts to the field of facsimile transmission. A notable feature of this project was his development, with Dr E.D. Wilson, of the gas-filled caesium photocell, which was many times more sensitive than the commonly used potassium hydride cell. The cell aided Zworykin in the implementation of a new type of high-speed facsimile machine that enabled reproductions to be made rapidly on special paper without photographic development.

These endeavours, and the publication of papers, enhanced Zworykin's standing in Westinghouse. He was given more independence in the choice of the problems on which his group worked and, obviously, this choice concentrated more and more on television.

In the late 1920s the major electrical-engineering organizations in the USA that were carrying out R&D work on television were the American Telephone and Telegraph Company (AT&T), the General Electric Company (GE), the Westinghouse Electric and Manufacturing Company (WEM), and the Radio Corporation of America (RCA). When RCA was formed in 1919 the ownership of the company was: GE 30 per cent; WEM 20 per cent; AT&T 10 per cent; United Fruit and Company 4 per cent; others 36 per cent [10]. GE and WEM had the exclusive right to manufacture receivers, and AT&T retained the exclusive right to manufacture, lease and sell transmitters. RCA could operate point-to-point radio communications, though not exclusively, and market receivers. In 1928 all radio and television research carried out by General Electric and the Westinghouse Electric Manufacturing Company was on behalf of RCA. Both firms had engineers working independently on the elucidation and engineering of the principles of television. However, since RCA was primarily a sales organization, the influence of D. Sarnoff, vice-president and general manager of RCA, on developments was especially strong.

Of the senior executives in GE, WEM and RCA, Sarnoff was particularly forward-looking. His thoughts on television were expressed in an article, 'Forging an electric eye to scan the world', published on 18 November 1928 in the *New York Times* [11]. He wrote, 'Within three to five years I believe that we shall be well launched in the dawning age of sight transmission by radio.' He predicted that television broadcasting would be classified as radiomovies (following the policy adopted by WEM), and as radio television, in which vision signals would be generated directly – as was being done by, for example, Baird.

Sarnoff's use of the expression *electric eye* is intriguing. In the summer of 1928 he had visited France, where it was known that television experimentation with cathode-ray tubes had been conducted by E. Belin and Dr F. Holweck, by Dr A. Dauvillier, and by G. Valensi. It appears that, while in Paris, Sarnoff, who closely followed television developments, saw something relating to television that provoked an interest in electronic television. He knew that Zworykin was a proponent of cathode-ray television (see Figure 3.2) and instructed him to visit Europe to determine 'the status of ideas applicable to cathode-ray television there'. The visit was to be, possibly, the most valuable ever to be undertaken by a television pioneer, since RCA's television efforts were to be greatly advanced by the knowledge and hardware which Zworykin acquired, and by the appointment in July 1929 of G.N. Ogloblinsky (the former chief engineer of the Laboratoire des Établissements Edouard Belin).

Zworykin sailed for Europe on 17 November 1928 with a schedule to visit England, France and Germany. Of these visits, that which Zworykin made to the Laboratoire des Établissements Edouard Belin was of prime importance. Here he met Belin, Dr F. Holweck, G.N. Ogloblinsky and P.E.L. Chevallier, and was shown one of Belin's and Holweck's latest continuously pumped, all-metal cathode-ray display tubes [12]. Zworykin immediately recognized the superiority of Holweck's and Chevallier's cathode-ray tube (c.r.t.), compared with the Western Electric Type 224, which employed gas focusing, and the likelihood that their approach to the design of

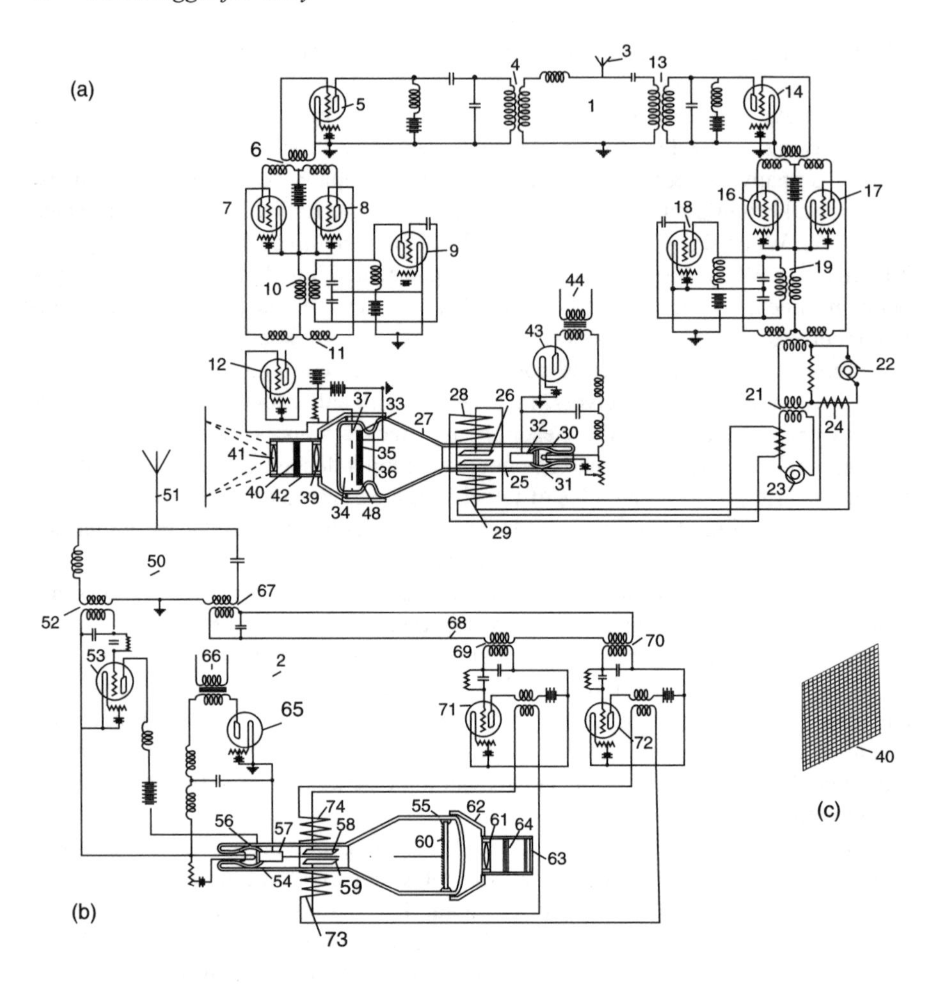

Figure 3.2 Schematic diagram of Zworykin's 1924 colour television system, showing the receiver (a) and the transmitter (b), which incorporated a Paget plate (c) [Source: British patent application 255 057, 3 July 1926]

the c.r.t. would have a great impact on the evolution of an all-electronic television system. Some modifications to the construction of the tube would be necessary before a practical version suitable for the domestic market could be manufactured, but, nevertheless, so convinced was Zworykin that the Holweck and Chevallier tube held the promise of providing the essential foundation for his cherished idea that he sought to purchase the tube from the Laboratoire des Établissements Edouard Belin. He was successful and on 24 December 1928 arrived back in the USA with the latest Holweck and Chevallier metallic, demountable cathode-ray tube, and a Holweck rotary vacuum pump. On this visit Zworykin's biographer has written, 'Zworykin's trip to Paris changed the course of television history.' [13] The consequence of his journey would be the engineering of an electron gun, with electrostatic focusing, which would be at the heart of all camera and display tubes built and sold by the RCA and some of its licensees.

On his return to New York, Zworykin reported to S.M. Kintner, now the vice-president of Westinghouse. He suggested that since he (Zworykin) had gone to Europe on behalf of RCA rather than the Westinghouse Electric Manufacturing Company he should go to New York and discuss his work with Sarnoff [14]. The meeting took place in January 1929. 'Sarnoff quickly grasped the potentialities of my proposals,' Zworykin later wrote, 'and gave me every encouragement from then on to realize my ideas.' [15]

Good progress was made by Zworykin and his team and on 18 November 1929 he was able to present a lecture on his new 'kinescope' or picture tube (see Figure 3.3) to a meeting of the Institute of Radio Engineers (IRE) [16]. The production of moving images, obtained from a film scanner-photocell unit, and their reception and display, using a kinescope, were described. Conrad of WEM had been working, from 1925, on a system of television known as radio-movies, in which 35 mm motion picture film was scanned, using vibrating mirrors, to provide a source of video signals. As the

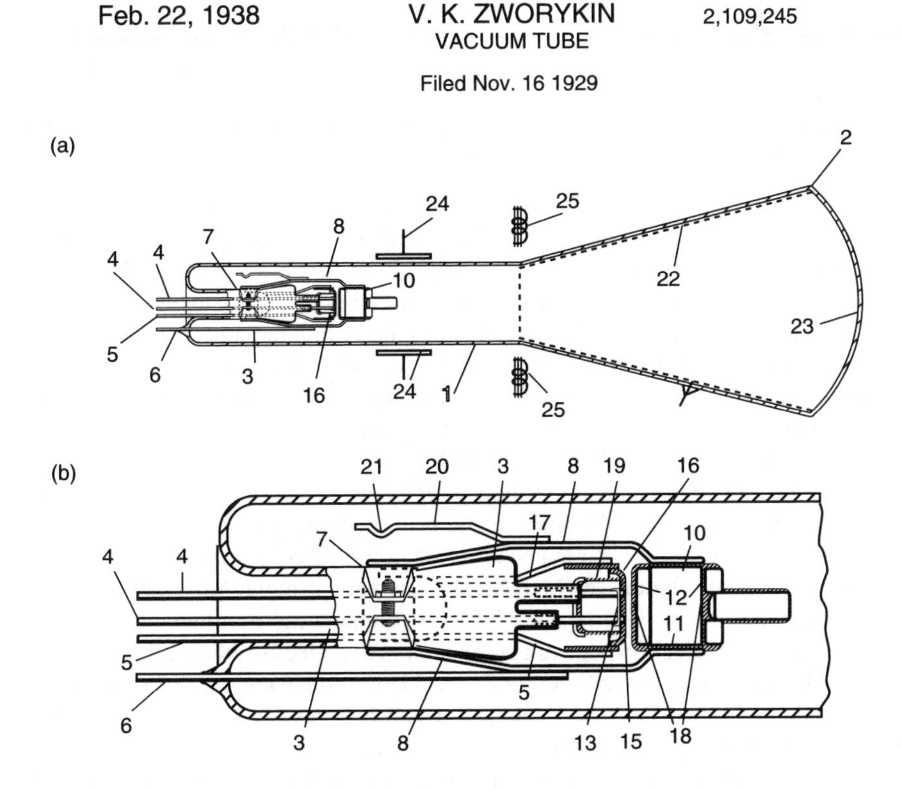

Figure 3.3 Details of Zworykin's 1929 cathode-ray display tube (the kinescope) [Source: Radio Engineering, *1929, vol. 9, pp. 38–41]*

principles of mechanical scanning were well understood it was natural that, lacking a suitable camera tube, Zworykin should have chosen Conrad's method of generating images for the tests with the kinescope.

Zworykin's efforts to develop a vacuum picture receiving tube were an important step forward in the slow march towards an all-electronic television system. Ives was not impressed, though. A demonstration of the kinescope had not been given at the IRE meeting and the account of Dr Zworykin's work was 'chiefly talk'. 'This method of reception', wrote Ives, 'is old in the art and of very little promise. The images are quite small and faint and all the talk about this development promising display of television to large audiences is quite wild.' [17]

Zworykin's and his team's work, in 1929, showed that the non-mechanical method of image synthesis had the following desirable attributes:

1. the image was visible to a large number of people at once, rather than to only one or two as with some mechanical scanners, and no enlarging lenses were required;
2. there were no moving parts and hence no noise;
3. the picture was brilliant enough to be seen in a moderately lit room;
4. the framing of the picture was automatic;
5. the use of a fluorescent screen aided the persistence of the eye's vision;
6. the motor previously used, together with its power amplifier, was redundant, and the power consumed by the kinescope was no greater than that used in ordinary vacuum tubes;
7. 'the inertia-less electron beam was easily deflected and could be synchronised at speeds far greater than those required for television'.

Towards the end of November 1929 Zworykin's team had constructed six television receivers. He was permitted to use Westinghouse's KDKA sound transmitter for three nights a week from 2–3 a.m. to broadcast carrier waves, modulated by vision signals and synchronizing pulses, derived from the cine-film projector equipment, and sent to KDKA by landline [18]. At other times the station radiated, on a daily basis, television signals generated by Conrad's film projector. The received signals could be displayed using Zworykin's kinescope. With this the images had a size up 3 in. by 4 in. (7.6 cm × 10.2 cm) and were 'surprisingly sharp and distinct', notwithstanding the 60 lines per picture standard utilized [19].

There can be no doubt that the short period (from February 1929 to November 1929) it took Zworykin and his team to engineer the kinescope resulted from RCA's acquisition of the Chevallier c.r.t., and the work of R.H. George of Purdue University. He had presented his findings, in March 1929, at a meeting of the American Institution of Electrical Engineers. George's hard vacuum cathode-ray oscillograph featured a special hot cathode electron gun, and made use of a new electrostatic method of focusing the electron beam. His solution to the problem of devising a satisfactory means of producing and focusing a high-intensity beam over a wide range of accelerating voltages constituted the chief contribution of his investigation.

Chevallier applied, in France, for a patent [20] to protect his c.r.t. invention on 25 October 1929: Zworykin's c.r.t. US patent application [21] was dated 16 November 1929. Both patents were for c.r.t.s of the hard vacuum, hot cathode, electrostatically focused type, with grid control of the fluorescent spot intensity and post-deflection acceleration.

The prior disclosures by George and Chevallier did not enable Zworykin, and RCA, to establish an undisputed patent claim to the use of grid control, electrostatic focusing and post-deflection acceleration. Since these characteristics were essential to the proper functioning of the kinescope it was necessary for RCA to purchase the rights of George's and Chevallier's patents.

Zworykin's hard vacuum c.r.t. (the kinescope) was well suited to adaptation as a camera tube – an essential component in any all-electronic black-and-white or colour television system. Basically, the fluorescent screen had to be replaced by a mosaic target plate, and an additional collecting electrode incorporated into the tube. At first, however, when Zworykin and Ogloblinsky began their work on the iconoscope they used the demountable, metallic tube which Zworykin had obtained in Paris. This had the advantage, compared with a glass tube, that it could be dismantled, the electrode configuration/mosaic modified and the tube reassembled for testing without the need to construct new tubes.

In his 1923 iconoscope patent Zworykin specified potassium hydride as the photoelectric substance. Now, in 1929, following the 1928 discovery by Koller [22], the new, highly sensitive caesium-silver oxide photoelectric surface could be deposited onto his target plate and fabricated into a mosaic that exhibited charge storage.

With a single scanning cell, as used in the Nipkow disc-photocell arrangement and also in Farnsworth's image dissector tube, the cell must respond to light changes in 1/(Nn) of a second where N is the frame rate and n is the number of scanned elements. For N equal to ten frames per second and n equal to 10,000 elements (corresponding to a square image having a 100-line definition), the cell must react to a change in light flux in less than ten-millionths of a second. But if a scanned mosaic of cells is used each cell has 100,000-millionths of a second in which to react, provided that each cell is associated with a charge storage element. Theoretically, the maximum increase in sensitivity with charge storage is n, although in practice the early 1930s iconoscopes only had an efficiency of about 5 per cent of n.

This most important principle is implicit in a patent [23], dated 21 May 1926, of H.J. Round, of the Marconi Wireless Telegraph Company, but it is not described in the body of Zworykin's 1923 patent application. The charge storage principle is a property not only of the iconoscope but also of the camera tubes that followed it, including the image iconoscope, the orthicon, and the image orthicon. From *c.* 1950 until the introduction of the Plumbicon tube, the image orthicon was the standard colour television camera tube.

In January 1930 Zworykin and his group moved to RCA Victor at Camden, New Jersey [24]. His main task was to engineer a reliable, sensitive, high-definition electronic camera tube. Initially, Zworykin and Ogloblinsky tried to fabricate a two-sided camera tube, but when 'the technical difficulties associated with the making of [the] targets with the high degree of perfection required [became] very great' they decided in May 1931 to experiment with single-sided target camera tubes [25] (see Figure 3.4).

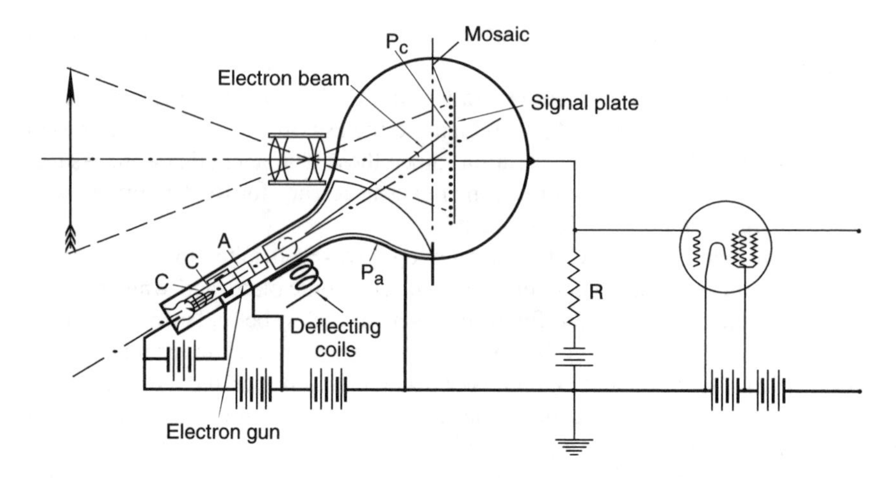

Figure 3.4 Sketch of the iconoscope camera tube – the cathode C, control grid G, first anode A, second anode P_a *and mosaic elements* P_c*, are indicated [Source:* Journal of the IEE, *1933, vol. 73, p. 441]*

In such tubes the plate consisted of a very thin (between 25 μm and 75 μm thick) mica sheet, onto one side of which a mosaic of minute silver globules, photosensitized and insulated from each other, was deposited; and onto the other side of which a metal film (known as the signal plate) was deposited. Each globule was capacitatively coupled to the signal plate and hence to the input stage of the video amplifier. When an optical image was projected onto the mosaic, each photosensitive element accumulated charge by emitting photoelectrons. The information contained in the optical image was stored on the mosaic in the form of a charge image. As the electron beam uniformly scanned the mosaic in a series of parallel lines the charge associated with each element was brought to its equilibrium state ready to start charging again. The change in charge in each element induced a similar change in charge in the signal plate and, consequently, a current pulse in the signal lead. The train of electrical pulses so generated constituted the picture signal.

Although promising results were obtained just one month after Zworykin's group transferred its efforts to the fabrication of single-sided plates much experimentation was required before the preparation of the mosaic was perfected. As with many successful undertakings, an adventitious factor aided the group. On one occasion S. Essig accidentally left one of the silvered mica sheets in an oven too long and found, after examining the sheet, that the silver surface had broken up into a myriad of minute, insulated silver globules. This, of course, was what was needed: by October 1931 good results were being obtained with the new processing method. According to H. Iams, of RCA, the first satisfactory tube (no. 16) was fabricated on 9 November 1931. One of the early experimental tubes was employed by the National Broadcasting Company for three years but general adoption of the iconoscope had to wait until 1933, when it could be manufactured to have uniform and consistent properties.

The Radio Corporation of America was one of the first to recognize the necessity for conducting investigations in the 43–80 MHz band and during the 1930 to 1940 decade it carried out a number of field tests, in 1931–2, 1933, 1934 and 1936–9. These tests were conducted with the same thoroughness and engineering excellence as the Bell Telephone Laboratories' 1927 trials. An important feature of RCA's enquiries was the use of a systems approach, an approach that is concerned with the formulation and understanding of the individual units that make up the whole system and with the interconnection of the units to form an integrated system.

With progress being made in the development of an electronic camera tube, it was to be expected that RCA's 1933 field test would embrace an appreciation of this device. In the earlier 1931–2 tests the major limitation to adequate television performance had been the studio scanning apparatus, since the lighting in the studio had been of too low an intensity to give a satisfactory signal-to-noise ratio; only when a motion-picture film was being scanned could a reasonable ratio be achieved. Fortunately, the sensitivity of the iconoscope was sufficient to allow for a further increase in the number of lines scanned per picture, in addition to permitting outdoor, as well as studio, scenes to be televised.

RCA's second experimental television system (see Figure 3.5) employed a picture standard of 240 lines per picture, sequentially scanned at 24 pictures per second, with

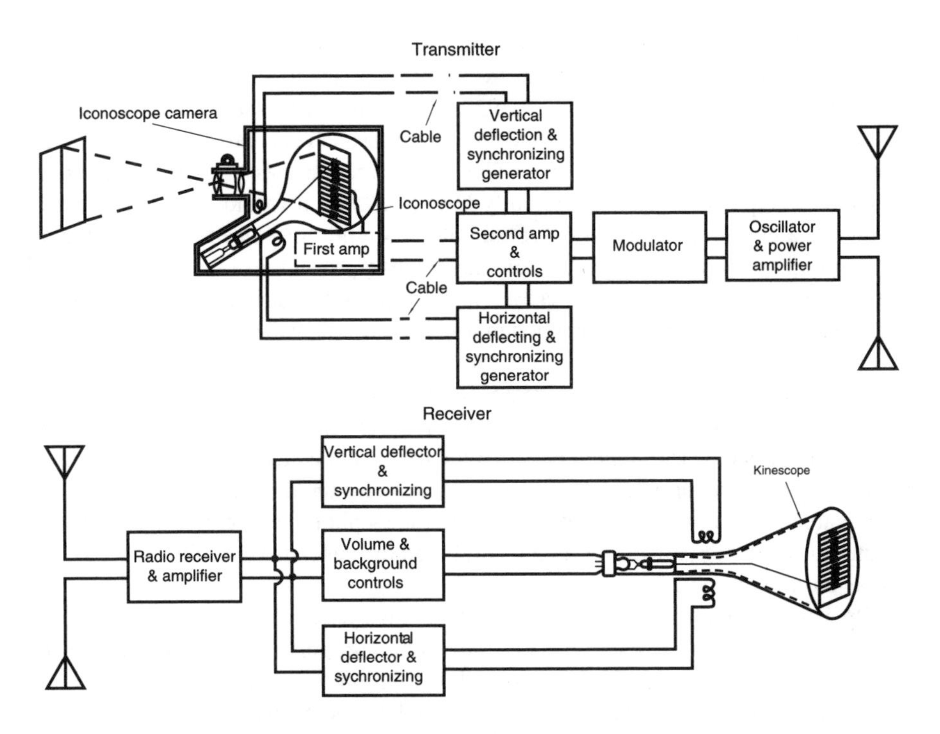

Figure 3.5 Block diagram showing the television system that incorporated an iconoscope and a kinescope [Source: Proceedings of the IRE, *1934,* ***22****(11), pp. 1241–5]*

an aspect ratio of 4:3 (width to height). The testing programme was comprehensive in concept and execution: line and radio links; film, studio and outside broadcasts; and relay working – all were included.

Among the group's findings Engstrom noted, in a paper [26] published in November 1934:

> The use of the iconoscope permitted transmission of greater detail, outdoor pick-up, and wider areas of coverage in the studio. Experience indicated that it provided a new degree of flexibility in pick-up performance, thereby removing one of the major technical obstacles to television . . . The iconoscope type pick-up permitted a freedom in subject material and conditions roughly equivalent to motion picture camera requirements.
>
> The choice of 240 lines was not considered optimum, but all that could be satisfactorily handled in view of the status of development . . . The increase of image detail widened very considerably the scope of the material that could be used satisfactorily for programmes. Experience with this system indicated that even with 240 lines, for critical observers and for much of the programme material, more image detail was desired. The desire was for both a greater number of lines and for a better utilization of the detail capabilities of the system and lines chosen for the tests.
>
> As in the [1931–2] New York tests, much valuable experience was obtained in constructing and placing in operation a complete television system having standards of performance abreast of research status. Estimates of useful field strengths were formulated. The need for a high power television transmitter was indicated.

Apart from undertaking much research-and-development work on black-and-white television in the early 1930s, RCA also began to consider the problems associated with colour television. In 1932 RCA proposed to fabricate a kinescope in which the screen would be coated with successive very narrow strips of red-, green- and blue-sensitive phosphors. The idea advanced was that the strips would be scanned sequentially at a sufficiently high rate, which would have the effect of producing a coloured image as though the three slightly displaced coloured images were in register on the kinescope's screen. However, the control technique required to ensure that the electron beam scanned just one phosphor strip during a given scan, without impinging on an adjacent phosphor strip, was not available and the suggestion was not tried.

The following year, in January 1933, RCA's research staff at Camden developed a two-colour mechanically scanned television system in which a motor driven two-colour filter wheel was mounted between the television camera and the scene/object being televised, and similarly at the receiver. Among the defects of the system, the appreciable infrared sensitivity of the camera tube's photoelectric mosaic screen and the storage characteristic of the screen combined to cause the two colours to mix and hence mar the reproduced image.

By January 1934 the same R&D staff had constructed and tested an iconoscope having a lower infrared sensitivity but the storage property of the tube still caused impure colours to be manifested. A possible solution to this problem, namely, the use of two (or three) camera tubes – one for each of the primary colours used – was put forward at a demonstration in the RCA laboratories in 1934. No records exist to indicate whether this idea was implemented and subsequently interest in colour television seems to have declined.

In September 1939 the television research group carried out a detailed study of the various colour methods that were being applied in still and motion-picture photography. Extended facilities for the investigation of colour television were introduced during January 1940, and Dr E.W. Herold was appointed chief coordinator for all research undertaken in this field. A three-tube camera and optical unit (of the type used in early trichromatic still cameras) were designed and built to permit either sequential or simultaneous scanning. Corresponding projection-tube receivers and rotating filter wheels were also produced. On 6 February 1940 RCA's three-colour, additive, sequentially scanned television system was demonstrated to the Federal Communication Commission, at RCA's Camden site. Slides, studio and outdoor scenes were shown in natural colour [27].

The demonstration highlighted the need for further endeavour, and so, a few months later, facilities were established at RCA's Harrison plant to test the properties of camera and receiver tubes suitable for colour television. (At this time RCA's work was still based on the use of rotating trichromatic filters.) Of especial importance to this effort was the need to fabricate receiver display screens having short persistence colour phosphors, and improved colour responses.

Much thought and industry were committed to RCA's mechanical-type system, but its inherent limitations persuaded RCA's executives in the autumn of 1940 to abandon further work and to concentrate instead on an all-electronic colour television system. Among the potential advantages of the latter were: (1) compatibility with monochrome television, (2) higher brightness levels, (3) lack of colour break-up with eye movement, and (4) much larger display tubes. Furthermore, RCA's management felt the promotion and marketability of such a colour television receiver would be enhanced relative to their present receiver. The economic and scientific arguments for the change appeared compelling.

In March 1941, RCA and the National Broadcasting Company commenced comprehensive testing of sequentially scanned, 343 and 441 lines per picture, 60 pictures per second, colour television systems using the RCA-NBC transmitter in the Empire State Building, New York. The signals were received at distances of up to 50 miles (80 km) away. Both tele-cine and studio scenes were televised, and both colour and black-and-white receivers were employed. The latter used the green signal only for the display.

A problem soon arose concerning the conflicting requirements of the need for a large number of lines per picture, N, and high frame rate, R. In any television system the bandwidth is dependent on RN^2. Thus, if the bandwidth is constrained to a maximum figure, either (1) N can be made large to give good picture definition, in which case the flicker effect would be objectionable; or (2) R can be high to prevent flicker, but the picture resolution would be unsatisfactory. In the tests it was found that the flicker of the black-and-white images was intolerable (since the green field rate was just 20 Hz). Increasing the rate by a factor of three restored the flicker to its usual unobtrusive level, but led to an absence of a picture on a normal black-and-white receiver.

Further consideration of RCA's early work on colour television is given in Chapters 6 and 7.

From 1931 RCA's only competitor, apart from Farnsworth (whose image dissector tube did not possess the property of charge storage), in the all-electronic television field was Electric and Musical Industries (EMI) of the United Kingdom. The company's R&D effort led to the inauguration of the world's first, high-definition, public television service on 2 November 1936, and is considered in the next chapter. Additionally the question of standards for high-definition black-and-white television – which had an influence on those for colour television – is discussed.

References

1. Gray F. 'The cathode-ray tube as a television receiver'. Memorandum to H.E. Ives, 16 November 1926, Case File 33089, pp. 1–2, AT&T Archives, Warren, New Jersey
2. Dunlap O. 'Baird discusses his magic'. *New York Times*. 25 October 1931, section IX, p. 10:1
3. Burns R.W. 'Early Admiralty interest in television'. *Institution of Electrical Engineers Conference Publication of the 11th IEE Weekend Meeting on the History of Electrical Engineering*; 1983. pp. 1–17
4. Burns R.W. *Television: an international history of the formative years*. London: Institution of Electrical Engineers; 1998. pp. 381–2
5. Rogowski W., Grosser W. DRP patent application 431 220, 4 February 1927
6. Dauvillier A. *Procédé et dispositifs permettant de realizer la télévision*. French patent application 592 162. 29 November 1923. And first addition 29 653 on 14 February 1924
7. Everson G. *The Story of Television, the Life of Philo T. Farnsworth*. New York: Norton; 1949
8. Abramson A. *Zworykin, Pioneer of Television*. Chicago: University of Illinois Press; 1995. pp. 71–5
9. George R.H. 'A new type of hot cathode oscillograph and its application to the automatic recording of lightning and switching surges'. *Transactions of the AIEE*, July 1929, p. 884
10. Burns R.W. *Television: an international history of the formative years*. London: Institution of Electrical Engineers; 1998. pp. 404
11. Sarnoff, D. 'Forging an electric eye to scan the world', *New York Times*, 18 November 1928, X, 3:1
12. Abramson A. *Zworykin, Pioneer of Television*. Chicago: University of Illinois Press; 1995. p.72
13. Abramson A. *Zworykin, Pioneer of Television*. Chicago: University of Illinois Press; 1995. Chapter 6, 'The kinescope', pp. 62–86
14. 'Interview with Dr Zworykin on 3 May 1965', transcription of tape, pp. 1–25, Science Museum, UK
15. Zworykin V.K. 'The early days: some recollections'. *Television Quarterly*. 1962;**1**(4): 69–72

16. Zworykin V.K. 'Television with cathode ray tube for receiver'. *Radio Engineering*. 1929;**9**: 38–41
17. Ives H.E. Memorandum to H.P. Charlesworth, 16 December 1929, Case File 33089, p. 1, AT&T Archives
18. Abramson A. *Zworykin, Pioneer of Television*. Chicago: University of Illinois Press; 1995. p. 83
19. Zworykin V.K. 'Television with cathode-ray tube for receiver', 9 September 1929, a report, Westinghouse Electric and Manufacturing Company, archives
20. Chevallier P.E.L. *Kinescope*. US patent application 2 021 252. 20 October 1930
21. Zworykin V.K. *Vacuum tube*. US patent application 2 109 245. 16 November 1929
22. Koller L.R. 'S1 photocathode', *Physical Review* 1930;**36**:1639
23. Round H.J. *Improvements in or relating to picture and like telegraphy*. British patent application 276 084. 21 May 1926. Ref. 7, pp. 105–6
24. Abramson A. *Zworykin, Pioneer of Television*. Chicago: University of Illinois Press; 1995. Chapter 7, 'The iconoscope', pp. 87 – 113
25. Abramson A. *Zworykin, Pioneer of Television*. Chicago: University of Illinois Press; 1995. p. 105
26. Engstrom E.W. 'An experimental television system', *Proceedings of the IRE*. Nov 1934;**22**(11). pp. 1241–5
27. Bucher E.E. 'Television and David Sarnoff'. Unpublished manuscript. David Sarnoff Museum, Princeton, New Jersey, Chapter 1

Chapter 4

EMI and high-definition television

On 2 November 1936 the world's, first, public, regular, high-definition television service was inaugurated at Alexandra Palace, London [1]. The decision to establish the service was made by Parliament following the submission to it of the Report of the Television Committee. This Committee had been constituted in May 1934 'to consider the development of television and to advise the Postmaster General on the relative merits of the several systems and on the conditions – technical, financial, and general – under which any public service of television should be provided' [2].

The Committee, chaired by Lord Selsdon, a former Postmaster General, worked with commendable speed and tendered their recommendations, a total of 17, to the Postmaster General, the Rt Hon. Sir Kingsley Wood, on 14 January 1935. During their work the committee had examined 38 witnesses, had received numerous written statements from various sources regarding television and had visited Germany and the USA to investigate television developments in those countries.

The principal conclusion and recommendation of the committee was that 'high-definition television had reached such a standard of development as to justify the first steps being taken towards the early establishment of a public television service of this type'.

Marconi-EMI (M-EMI) Television Company Ltd (a company jointly owned by Marconi's Wireless Telegraph Company and Electric and Musical Industries Ltd) and Baird Television Company were the two companies who were invited to submit tenders for studio and transmitting equipment and for a short period from 2 November 1936 to 13 February 1937 both companies transmitted television programmes, on an alternate basis from the London television station. Subsequently, until the commencement of hostilities in September 1939, only the M-EMI equipment was in operation.

Electric and Musical Industries Ltd (EMI) was formed in 1931 to acquire the ownership of the Gramophone Company Limited (sometimes known as HMV after the company's His Master's Voice records) and the Columbia Graphophone Company Limited. The research groups of the two companies were merged [3]. Previously HMV (in which RCA had a financial interest) had demonstrated in January 1931 a mechanical multi-zone television system, operating on 150 lines per picture, based on the use of cine film as the source of the televised images.

Following the merger, EMI's directors agreed that HMV's television development effort should continue to be an item of the new company's R&D programme. One of the first questions that had to be tackled was whether this work should proceed on mechanical or electronic lines. Mechanical scanners had the advantage that they had been made successfully, whereas electronic scanners were still in the development phase. On the other hand, electronic scanning had many potential advantages for high-definition television.

The challenge that faced EMI, and was being faced by Farnsworth and by Zworykin's group at RCA, was immense [4]. Photoelectricity, vacuum techniques, electron optics, the physics of the solid state and of secondary electronic emission, and wideband electronics and radio communications, were all in a rudimentary state of development. Many fundamental investigations would have to be undertaken before a high-definition television system could be engineered.

EMI's great success was undoubtedly due to the very powerful team that I. Shoenberg, EMI's director of research, had assembled. Professor J.D. McGee, in his 1971 Memorial Lecture on 'The life and work of Sir Isaac Shoenberg, EMI's Director of Research, 1880–1963' noted:

> Isaac Shoenberg was born of a Jewish family in 1880 in Pinsk, a small city in north west Russia. His great ambition in life was to be a mathematician, and it was this ambition that brought him later to England. But because of the difficulties facing a young Jew in Russia at that time, he had to settle for a degree in engineering, reading mathematics, mechanical engineering and electricity at the Polytechnical Institute of Kiev University. However, he was to retain his interest in mathematics throughout his life and, about 1911, he was awarded a Gold Medal by his old university for mathematical work.
>
> After leaving university, he worked for a year or two in a chemical engineering company, but very soon he joined the Russian Marconi Company and so – in the first decade of [the 20th] century – became involved in wireless telegraphy. From 1905 to 1914 he was Chief Engineer of the Russian Wireless and Telegraph Company of St Petersburg. He was responsible for the research, design and installation of the earliest radio stations in Russia.
>
> In 1914 he resigned his good job and emigrated to England where, in the autumn of the same year, he was admitted to the Royal College of Science, Imperial College, to work under either Whitehead or Forsythe for a higher degree in mathematics. [4]

Unfortunately, the outbreak of war brought this plan to a premature end and Shoenberg had to look for employment. 'He found [a post] with the Marconi Wireless and Telegraph Company at the princely salary of £2 a week.' His abilities and potential were soon noted and he became joint general manager and head of the Patent Department.

McGee, in his lecture, has described how Shoenberg joined the EMI organization:

> Shoenberg had no formal training as a musician, but he had an intense natural love of music; this made him a keen connoisseur of recorded music and the technique of recording. This in turn resulted in a friendship with Sir Louis Sterling, and an invitation to join the Columbia Graphophone Company and to put into practice his ideas on recording. He was clearly successful in this and soon the Columbia Company was competing effectively with the formerly almost unchallenged, prestigious firm The Gramophone Company or HMV. So began the association of Sterling with Shoenberg which was, to my mind, crucial in the field of television in the following decade. The shrewd business man and financier, Sterling, completely trusted Shoenberg, an engineer, scientist or applied physicist with a large dash of the visionary.

Progress, however, was swift and on 11 November 1932 Shoenberg invited the BBC's chief engineer, N. Ashbridge, to a private demonstration both in the transmission and in the reception of television [5].

Ashbridge visited EMI's R&D Department at Hayes on the 30 November and was shown apparatus for the transmission of films using four times as many lines per picture and twice as many pictures per second as Baird's current equipment. He was impressed and thought the demonstrations represented by far the best wireless television he had ever seen and felt they were probably as good as or better than anything that had been produced anywhere else in the world [6].

The actual demonstration consisted of the transmission of a number of silent films, over a distance of approximately two miles, by means of an ultra-short-wave transmitter using a wavelength of 6 m and a power of about 250 W.

The emergence of a competitor in the form of EMI caused J.L. Baird and his associates much unease and led to a rather bitter controversy between them and the BBC. This has been described and discussed in copious detail in the author's book on British television, and is not repeated here.

Following demonstrations [7,8,9] of the Baird Television Ltd and EMI systems on 18 and 19 April 1933, a conference was held on 21 April 1933, at the General Post Office, between some BBC and Post Office representatives. It was agreed [10] that:

1. the EMI results were vastly superior to those achieved by the Baird Company;
2. the results were incomplete because of the different transmission methods (line and wireless) used in the two cases;
3. further tests by wireless in a town area were essential to determine the range of reception and the effect of interference;
4. whatever system of synchronization was adopted in the first instance for a public service might be liable to standardise the type of receiving equipment;
5. a test of one system could not be a reliable judgement on the results achieved by the other.

Meanwhile, from *c.* May 1932 W. Tedham and J.D. McGee (both of EMI) had been undertaking many investigations on the chemistry of the preparation of electronic camera mosaic signal plates and on the physics of the mechanisms operating at the surfaces of the plates [11]. Both single- and double-sided signal plate camera tubes had been constructed, and patents obtained.

In addition further staff had been recruited. By June 1934 the Research Department totalled [11] 114 persons including 32 graduates, 9 of whom had PhDs, despite the fact that PhDs were not particularly common in the early 1930s, and 10 had been recruited direct from Oxford and Cambridge Universities. The department also included A.D. Blumlein, the greatest British electronics engineer of the twentieth century.

Funding for the television project was amply sufficient. Shoenberg succeeded in persuading the EMI Board to invest about £100,000 per year in EMI's R&D work on television. The expertise and funding of EMI's Research Department represented an ominous situation for Baird Television Ltd, which, initially, could not match EMI's staff and financial resources.

The enlargement of the R&D department soon led to favourable results. When camera tube no. 14 (of the type shown in Figure 4.1) was constructed it was so much better than previous tubes in picture quality and sensitivity that a very experimental

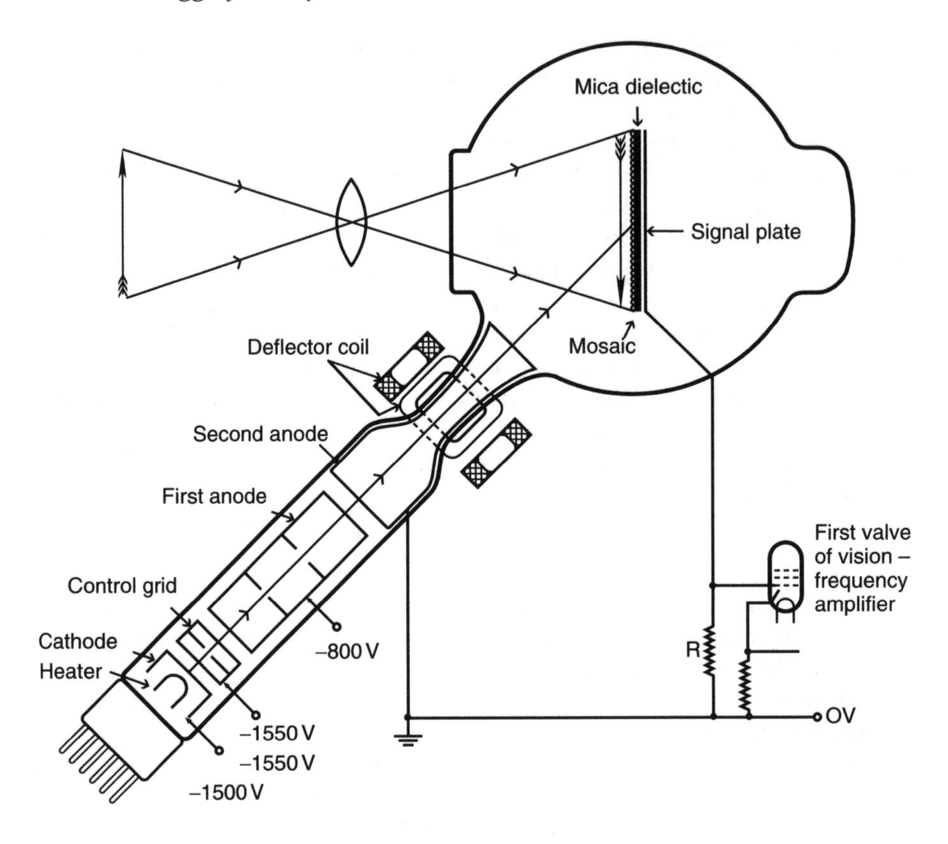

Figure 4.1 Diagram of the emitron television camera tube [Source: J.D. McGee]

electronic camera directed through the window of the laboratory, on 5 April 1934, enabled a daylight outside broadcast picture to be obtained [12].

Many visits to EMI were made by important persons during the formative period of EMI's system. Ashbridge was given a demonstration in January 1934. He was very impressed. 'The important point about this demonstration is, however, that it was far and away a greater achievement than anything I have seen in connection with television. There is no getting away from the fact that EMI have made enormous strides.' [13]

Shoenberg told Ashbridge the policy of the EMI was to develop television energetically since they believed there was a great commercial future for the firm, which was first in the field with something practicable. Two months later the M-EMI Television Company Limited was formed 'to supply apparatus and transmitting stations' [14,15].

In an attempt to settle the rival claims of EMI and Baird Television Ltd, Reith, the Director General of the BBC, on 15 March 1934 wrote to Kingsley Wood, the Postmaster General, and proposed a conference 'between some of your people and some of ours to discuss the future arrangements for the handling of television' [16].

Reith thought there were three aspects to discuss: the political ('using the term in a policy sense and for want of a better one'), the financial and the technical. Kingsley Wood agreed [17] .

The informal meeting [18] was held at the General Post Office on 5 April 1934. A number of general questions were examined by the BBC and GPO representatives, including: (1) the method of financing a public television service; (2) the use of such a service for news items and plays; (3) the relative merits of some of the systems available including those of the EMI, and Baird companies; (4) the arrangements necessary to prevent one group of manufacturers obtaining a monopoly on the supply of receiving sets; and (5) the possible use of film television to serve a chain of cinemas.

With two rival companies campaigning for the creation of a television service – EMI for a new BBC station and Baird Television for a station of its own – it was agreed by the conference that a committee should be appointed to advise the Postmaster General (PMG) on questions concerning television. The PMG concurred and the Television Committee was constituted. Lord Selsdon, a former Postmaster General. was appointed chairman of the committee.

Following acceptance by Parliament, in January 1935, of this committee's report [2], and the recommendations that:

1. 'a start should be made by the establishment of a service in London with two television systems operating alternately from one transmitting station;
2. 'Baird Television Limited and M-EMI Television Company Limited, should be given an opportunity to supply, subject to conditions, the necessary apparatus for the operation of their respective systems at the London station; and
3. 'the Postmaster General should forthwith appoint an advisory committee to plan and guide the initiation and the early development of the television service',

the two contracting companies submitted their proposals to the newly formed Television Advisory Committee (TAC) [19,20] in February 1935. They were based on standards of 240 lines/picture, 25 pictures/s, sequentially scanned (Baird Television Ltd); and 405 lines/picture, 50 frames/s, interlaced 2:1 to give 25 pictures/s (M-EMI Television Company Ltd) (see Figure 4.2).

Installation of M-EMI's and BTL's television equipment at Alexandra Palace, the site of the new London television station, was carried out during the period December 1935 to August 1936. Extensive building operations were necessary to accommodate the heavy machinery and sound and vision transmitters, and various rooms had to be converted for use as studios, dressing rooms, control rooms and offices.

The opening ceremony of the London television station on 2 November 1936 was a most modest affair. Although the station provided the world's first, high-definition, regular, public television broadcasting service, the inaugural programme [21], arranged by the BBC and approved by the Television Advisory Committee (TAC), lasted hardly a quarter of an hour. Subsequently, the two rival systems operated on an alternate basis.

The experiences gained by the BBC of the operation of both television systems under service conditions from 2 November to 9 December were described in an

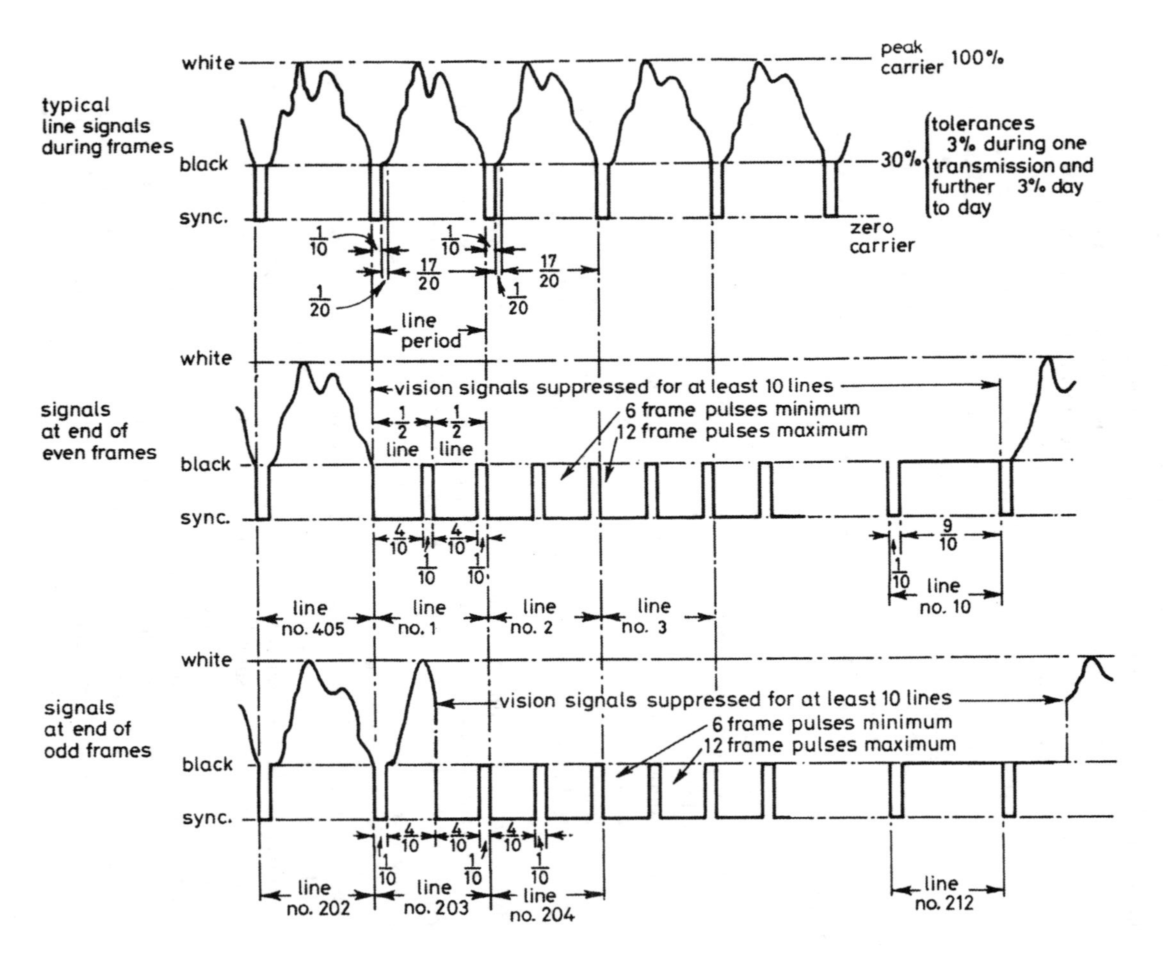

Figure 4.2 *The Marconi-EMI Television Ltd video waveform [Source: Author's collection]*

important report [22], written by Gerald Cock, the BBC's director of television, which was submitted to the TAC.

At its meeting on 16 December 1936 the TAC recommended to the Postmaster General the adoption of M-EMI's 1935 transmission standards as the standards for the London television station. They were good enough to last until 1986. And, when discussions were held in the UK in the 1950s on standards for colour television, several persons argued for a 405-line colour television service. The discontinuance of the BTL system would date from 2 January 1937 [23].

In the USA the adoption of nationwide television standards did not proceed as smoothly as in the UK and consensus was achieved only after a prolonged period of enquiry and discussion. Movement for an industry-wide consensus on standardization arose in 1935 after a demonstration by RCA to the members of the Radio Manufacturers Association (RMA) of the company's 343-line system. In that year the RMA [24] asked its engineering department to determine when it would be appropriate to adopt national television standards. This activity came to the notice of the Federal Communications Commission's chief engineer, T.A.M. Craven, who proposed to the RMA that various branches of the industry should come to an agreement on television performance standards among themselves prior to any Commission action.

Craven's view stimulated two responses. First, the FCC announced hearings, to commence on 15 June 1936, to determine its long-term policies on the future allocation of frequency channels; and, second, the RMA set up two television committees, one on standards and the other on frequency allocations, to prepare a joint report to the hearings.

At the hearings the RMA allocations committee recommended the FCC should establish seven 6 MHz television channels between 42 and 90 MHz and that experimental authorizations be permitted above 120 MHz. The 1936 RMA standards committee proposed [25] 441 lines per picture, 2:1 interlacing, to give 30 pictures per second (i.e. 60 frames per second), double-sideband negative picture modulation, 2.5 MHz video bandwidth and frequency modulation for the sound signal; the video and audio signal carriers to be spaced *c.* 3.25 MHz apart with the audio signal carrier being at the higher frequency.

In 1937 the FCC reacted to the findings of its 1936 hearings and established 19 6 MHz channels interspersed with other services in the spectrum from 44 MHz to 294 MHz [24]. On the issue of line standards, the FCC felt television engineering did not possess the necessary stability needed for governmental sanction.

Further work by the RMA standards committee led, in July 1938, to several additional recommendations to its 1936 standards. d.c. transmission of the brightness of the televised scene was specified: black in the picture had to be represented by a definite carrier level, as in British practice; the radiated electromagnetic wave had to be horizontally polarized; details of the line and frame synchronizing pulses were delineated; and the frame-synchronizing pulse had to be serrated and include equalizing pulses, again as in British practice [25]. In addition, the 1936 statement on double-sideband transmission was replaced by one on vestigial-sideband transmission, and the frequency channel needed to accommodate it was rearranged, thereby increasing the video bandwidth from 2.5 MHz to 4.0 MHz. This increase permitted an increase in the number of lines per picture from 441 to 560 (since the number of lines is proportional to the square root of the bandwidth), but this important variation was not made until March 1941.

These statements were circulated [26] to the RMA's members and, in the absence of dissent, forwarded in September 1938 to the FCC. However, the FCC again concluded that 'television [was] not ready for standardization or commercial use by the general public', because of the prospect of rapid obsolescence of equipment, and it therefore decided to continue its policy of not acting on the RMA recommendations.

Meanwhile, during the somewhat protracted RMA, FCC and industry discussions on standards, RCA had been conducting television field tests from its Empire State Building experimental station in New York. By the beginning of 1939 these had advanced to the point where the introduction of television to the public, in the area served by television stations, had become practicable. Furthermore, the business slump of 1937 which had impeded progress for the first half year of 1938 had declined and the second half of the year had witnessed a substantial improvement in all branches of the radio industry. Sarnoff, now president of RCA, seemed keen to commence a regular programme service and on 20 October 1938 announced that in April 1939 the RCA manufacturing company would place television receiving sets on the market to coincide with the opening of the New York World's Fair [27].

Actually, Sarnoff's 1938 television policy was being forced be several factors. The first concerned the threat to RCA's manufacturing base by other manufacturers; and the second related to the challenges to RCA's television system, which were being made by a number of rival organizations including Philco, DuMont, CBS and Don Lee.

RCA's plans soon received a setback when on 9 April 1939 it was reported that the FCC 'won't be stampeded blindly into launching [a] television industry'. In a *New York Times* article [28] headed 'FCC stops gold rush', Craven was quoted as saying:

> We want to know about the business end of television. There have been many misgivings on the amount of royalties to be received. Much to the astonishment of the FCC [television] committee, the RMA committee didn't give any consideration to the commission's problems. If the FCC committee accepts the standards offered by the RMA it means almost a monopoly. The standards they propose would put television on a par with the movies in about 1906.
>
> If the television development means a limited amount of channels, who, considered on a broad public basis, is entitled to them – the existing broadcasting industry, the moving picture industry or the newspapers? They are all vitally interested.
>
> Despite great pressure exerted upon the commission's committee to launch the television industry and still beyond that, pressure brought to bear for the adoption of certain standards, the FCC's watchword is "caution" in the public interest.
>
> What's the necessity of going so fast in this important matter of television?

Of the major industrial companies, RCA, GE and Farnsworth supported the RMA's position but, in 1939, Philco, Zenith and DuMont (and soon CBS) strongly dissented. The last group asserted that the proposed standards were not 'sufficiently flexible to permit certain future technical improvements without unduly jeopardizing the initial investment of the public in receivers'. This seemed to be the FCC's opinion also.

Actually, the FCC neither approved nor disapproved of the RMA's standards and accepted that they represented a consensus of expert engineering views. The commission did not believe the standards were objectionable but simply wished to be free to prescribe a better performance for the transmitters which it would licence in the future [29]. It was concerned, too, that its future pronouncements on standards should enable a 'radical reduction' in the price of sets to be accomplished. 'Unless the television receiver of the future is to be within the pocket book capabilities of the average American citizen', it argued, 'television as a broadcasting service to the general public cannot thrive as a sound business enterprise for any extended period.'

For the FCC the position was simple: 'Considered from the broadcast standpoint . . . television [was] now barely emerging from the first or technical research stage of development.' Therefore the commission counselled 'patience, caution and understanding'. The aim was to develop 'a new and important industry logically and on sound economic principles'.

The FCC's reservations did not deter many aspiring telecasters and, by the middle of October 1939, 19 licences had been issued in the New York, Philadelphia, Chicago, Washington, Fort Wayne, Cincinnati, Schenectady, Los Angeles and San Francisco areas. A further 23 applications were being processed [30].

However, there was a cloud on the horizon – in the form of colour television – which threatened to change completely the black-and-white standards of the RMA. Chairman J.L. Fly, of the FCC, alluded to amendments to come when, in October 1939, he opined: 'I feel that color in television offers the possibility of a vastly improved system, which gives not merely color but improvements in terms of definition, contrast and other related factors.' Obviously, the issue of television standards in the USA was going to drag on for some considerable time. Nevertheless, the FCC on 29 February 1940, adopted rules that permitted limited commercial operation of experimental television stations.

The FCC's reluctance to act decisively had an adverse effect on the sale of receivers. By August 1939 only about 800 sets had been sold throughout the USA, and another 5,000 were occupying space on dealers' shelves [31]. Three factors militated against a receiver boom:

1. the high cost of sets;
2. the limited programming schedules of the television broadcasting companies;
3. apprehension among the public that their considerable financial outlays could be wasted as a consequence of changing standards and possibly set obsolescence.

Moreover, the FCC's disinclination to set standards had a serious impact on the funding of television programmes. Unlike the position in the UK, where programmes were provided by just one organization, the British Broadcasting Corporation, the programmes in the US required commercial sponsors. But sponsors, for example advertisers, were not willing to invest huge sums to produce attractive programmes that would be seen by the few members of the public who had bought receivers. And the public was not willing to spend large sums on television sets that did not offer attractive programmes and that might after a short time be incapable of receiving a commercial broadcast. There appeared to be a deadlock.

Eventually, following: further FCC hearings; accusations of RCA's 'forcing the pace'; a public outcry against the FCC, which was denounced for endeavouring to impose an 'alien theory of merchandising' on the United States by attempting to protect consumer interests beyond 'acceptable bounds'; and an allegation by a senator that the FCC had exceeded its authority by interfering with the freedom of public and private enterprise, the impasse was broken at a meeting between Dr W.R.G. Baker, director of engineering for the RMA, and J.L. Fly, chairman of the FCC [32].

Cognisant of the success of Lord Selsdon's Television Advisory Committee, the RMA decided to establish a committee comprising representatives of all pertinent companies and organizations to draft acceptable standards. Both RMA and non-RMA members would be invited to participate in the discussions. And so the National Television Standards Committee (NTSC), sponsored by the RMA and supported by the FCC, came into being. The following companies were each invited to send one representative to the NTSC [33]: Bell Telephone Laboratories, Columbia Broadcasting System, DuMont Labs Inc, Farnsworth Television and Radio Corporation, General Electric Company, Hazeltine Service Corporation, John V.L. Hogan, Hughes Tool Company, IRE, Philco Corporation, Radio Corporation of America, Stromberg Carlson Telephone Manufacturing Company, Television Productions and the Zenith Radio Corporation. Later, the NTSC had a membership of 168 divided into nine panels each responsible for investigating a particular aspect of standardization. The nine panels were:

1. system analysis, chaired by Dr P.C. Goldmark of CBS;
2. subjective aspects, chaired by Dr A.N. Goldsmith, of the IRE;
3. television spectra, chaired by J.E. Brown, Zenith Radio Corporation;
4. transmitter power, chaired by Dr E.W. Engstrom of RCA;
5. transmitter characteristics, chaired by P.T. Farnsworth of Farnsworth Television and Radio Corporation;
6. transmitter-receiver coordination, chaired by I.J. Kaer of GE;
7. picture resolution, chaired by D.E. Harnett of Hazeltine Service Corporation;
8. synchronization, chaired by T.T. Goldsmith of DuMont;
9. radiation polarization, chaired by D.B. Smith of Philco.

The appointment of Dr Baker as the first chairman of the NTSC was announced on 17 July 1940 and the first meeting was held on 31 July that year. By operating according to parliamentary procedures it was anticipated that an industry-wide technical consensus could be achieved. The panels began their work in September.

Of the issues that divided the industry there were three that were of major importance and required firm resolution:

1. the conflict between fixed and flexible standards;
2. the method of synchronization; and
3. the line standard if a fixed standard were decreed.

These matters had to be clearly delineated by the NTSC. However, just before the nine panels commenced their work, CBS stated on 29 August 1940 [34] that it had perfected a system of colour television that would be ready by 1 January 1941. This was a new factor in the television debate and was one that possibly could challenge all previous notions on television standards.

Meanwhile, J.L. Baird, too, had been experimenting with various systems of colour television. His endeavours are described in the following chapter.

References

1. Burns R.W. *British Television: the formative years*. London: Institution of Electrical Engineers; 1986. Chapters 13, 14, and 15
2. Report of the Television Committee, Cmd. 4793, HMSO, January 1935
3. Burns R.W. *Television: an international history of the formative years*. London: Institution of Electrical Engineers; 1998, pp. 431–5
4. McGee J.D. '1971 Shoenberg Memorial Lecture'. *Royal Television Society Journal*. 1971;**13**(9)
5. Director General (BBC). Memorandum to the Chief Engineer, 1 January 1933, BBC file T16/65
6. Burns R.W. *British Television: the formative years*. London: Institution of Electrical Engineers; 1986. pp. 240–301
7. Phillips, F.W. Letter to Sir J.F.W. Reith, 10 April 1933, BBC file T16/42
8. Simon L. Memorandum on Baird and EMI demonstrations, 27 April 1933. Minute 4004/33, Post Office Records Office
9. Notes on a meeting held at the GPO, 21 April 1933, BBC file T16/42
10. Burns R.W. *Television: an international history of the formative years*. London: Institution of Electrical Engineers; 1998, pp. 431–5. pp. 456–9
11. Anon. Laboratory staff research department, 12 June 1934, EMI Archives
12. McGee J.D. 'The early development of the television camera'. Unpublished manuscript. Institution of Electrical Engineers Library. p. 16
13. Ashbridge N. Report on television, 17 January 1934, BBC file T16/65
14. White H.A. Letter to Sir J.F.W. Reith, 23 March 1934, BBC file T16/65
15. Notes of a meeting of the Television Committee held on 27 June 1934. Evidence of Messrs Clark and Shoenberg on behalf of EMI and the M-EMI Television Co. Ltd, minute 33/4682. Post Office Records Office
16. Reith Sir J.F.W. Letter to Sir Kingsley Wood, 15 March 1934, BBC file T16/42
17. Kingsley Wood Sir H. Letter to Sir J.F.W. Reith, 20 March 1934, BBC file T16/42
18. Notes on 'Conference at General Post Office, 5 April 1934'. Minute Post 33/4682, Post Office Records Office
19. Electric and Musical Industries. Response to questionnaire of the Technical Subcommittee on 'Proposed vision transmitter', Minute Post 33/5533
20. Baird Television Ltd. Response to questionnaire of the Technical Subcommittee on 'Proposed vision transmitter', Minute Post 33/5533
21. Television Advisory Committee. Minutes of the 32nd meeting, 15 October 1936, Post Office bundle 5536
22. Cock G. 'Report on Baird and M-EMI systems at Alexandra Palace'. TAC paper no. 33, 9 December 1936

23. Draft of public announcement. Post Office bundle 5536
24. Fink D.G. 'Perspectives on television: the role played by the two NTSCs in preparing television service for the American public'. *Proceedings of the IEEE*. September 1976;**64**(9):1322–31
25. *Ibid*. p.1326
26. Murray A.F. 'RMA completes television standards'. *Electronics*. July 1938, p. 28–9, 55
27. Sarnoff D. 'Statement on television'. *New York Times*. 3 January 1939, p. 31:3
28. Anon. 'FCC stops the gold rush'. *New York Times*. 9 April 1939, section X, p. 12:3
29. Anon. 'Public verdict awaited'. *New York Times*. 28 May 1939, section X, p. 10:2
30. Anon. 'Televiews on the air'. *New York Times*. 13 October 1939, section IX, p. 12:1
31. Anon. 'What's television doing now?' *Business Week*. 12 August 1939, p. 24
32. Fink D.G. 'Perspectives on television: the role played by the two NTSCs in preparing television service for the American public'. *Proceedings of the IEEE*. September 1976;**64**(9):1325
33. Anon. 'Television Committee organizes'. *Electronics*. August 1940, p. 34
34. Goldmark P.C., Dyer J.N., Piore E.R, Hollywood, J.M. 'Color television, part 1', *Proceedings of the IRE*. April 1942; **30**:162–82

Chapter 5
J.L. Baird and colour television

Although John Baird took the Television Advisory Committee's 1937 decision, following the adoption of M-EMI's television standards, as a personal misfortune he was not involved in the engineering, installation, commissioning and maintenance of any of Baird Television Ltd's equipment at Alexandra Palace. The responsibility for these tasks had been devolved to Captain A.G.D. West, the technical director, and his staff. Moreover, Baird had not been a party to the discussions that had taken place, from 1934, between the TAC and BTL: these had been handled by West and Major A.G. Church (a director of BTL). Baird was nominally the managing director of Baird Television Ltd but effectively he had been eased out of his position of technical control after the Gaumont-British Picture Corporation – which was controlled by I. Ostrer – became associated with BTL early in 1932. As S.A. Moseley, Baird's closest friend, noted:

> The new [1933] Board [with Greer and Clayton as Ostrer's appointments] was fast losing patience with Baird and his dilatory methods. Baird the visionary was still occupied in development, whereas the practical men of the Board wanted results. They wished to sell receivers, whereas Baird was still reaching out. This difference in conception led to hostility which developed and reached a climax. [1]

And so from *c*. July 1933 Baird reverted – essentially – to personal research and development in his own personal laboratory rather than company R&D. Thus Baird's distress in January 1937 was due to an injury to his pride.

But Baird had been a sufferer in adversity for many years and did not intend giving up his lifelong interest. If the BBC would not allow him to transmit from Alexandra Palace, 'It seemed to [Baird] that now being out of the BBC we [his company] should concentrate on television for the cinema, and should work hand-in-glove with Gaumont-British, installing screens in their cinemas and working towards the establishment of a broadcasting company independent of the BBC for the supply of television programmes to cinemas . . .' [2]

From the outset of his life's work on television Baird had been interested in cinema television. In June 1931, he had televised, albeit crudely, the Derby from Epsom, and in June 1932 he had shown images of the Derby on a large screen in the Metropole Cinema. The *Daily Herald* reporter who witnessed the event was enthusiastic about the prospect that appeared to be unfolding.

> With five thousand people in the Metropole Cinema, Victoria, S.W., fifteen miles from Epsom, I watched the finish of the Derby, while thousands on the Downs saw nothing of yesterday's great race. It was the most thrilling demonstration of the possibilities of

television yet witnessed. It made history .... As we sat in the darkened theatre distance was annihilated.' [3]

Baird was not alone in concluding that a viable market existed for cinema television. Various schemes were advanced in the 1930s in the USA, Germany, and the UK [4].

In July 1933 Baird purchased a property at 3 Crescent Wood Road, Sydenham, which was approximately one mile from the laboratories, workshops and offices of Baird Television Ltd, which were situated at the Crystal Palace. At his new abode Baird established his own private laboratory (see Figure 5.1), where he carried out research work, with his personal staff, independently of the general work which was being undertaken by BTL. Initially, he had one assistant, P.V. Reveley [5], who had been the technical assistant who had set up and operated Baird's mirror drum equipment at the Metropole Cinema.

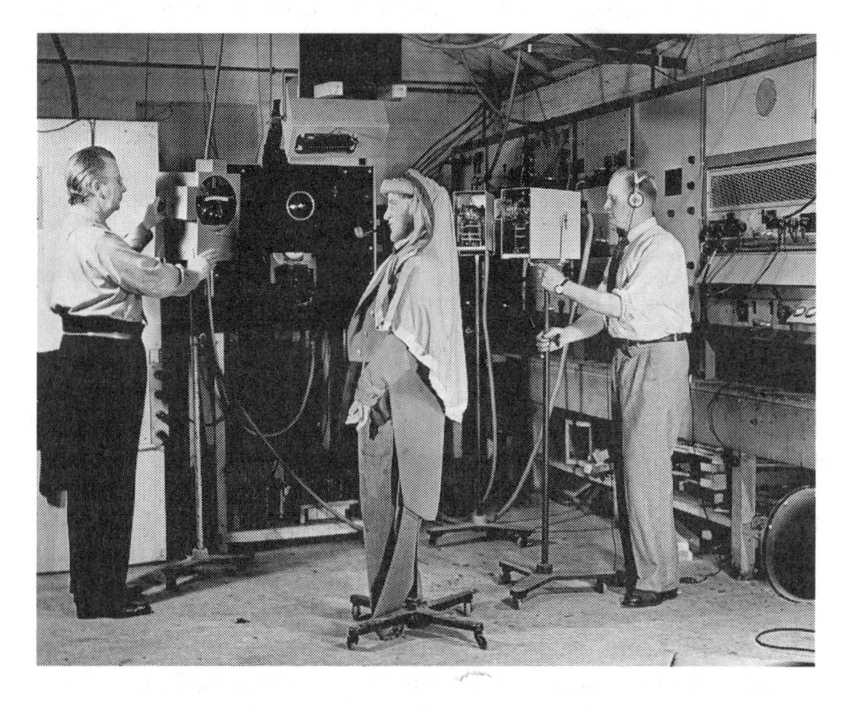

Figure 5.1 John Logie Baird, E.O. Anderson and dummy model in Baird's private laboratory [Source: Syndication International]

Special parts that had to be machined were manufactured by B.J. Lynes, who had his own small prototype and model-making business located in 'mews premises somewhere in the Euston Road area'. It was Lynes who made the scanning apparatus, to J.C. Wilson's design, which was installed in the basement studio of the BBC's Broadcasting House. From time to time Wilson, at Baird's request, would visit Crescent Wood Road to discuss with JLB and Reveley the design of a new part.

'I [Reveley] liked to define things by making sketches, but JLB was always anxious to push on and discouraged anything he thought might lead to loss of time.' 'Results, Mr Reveley, not reports' [5] Baird would say.

For his new scanner Baird abandoned the three-zone television system that he had employed for the Metropole Cinema demonstration and decided to use a television standard based on 120 lines per picture, 16.67 pictures per second – each picture comprising six interlaced frames, each of 20 lines scanned 100 times per second – the aspect ratio being 5.5 (high) to 4 (wide). Baird's choice of such a complex standard suggests that he was looking ahead to the time when he would be able to demonstrate colour television. In a colour television system of the frame-sequential type each picture comprises three frames, R, G and B, corresponding to the images of a scene as viewed (separately) in red, green and blue light. If now a colour picture is formed from two interlaced colour pictures (to reduce flicker as in black-and-white television) it is evident that six frames are required to constitute the final image.

For his purpose Baird used two mirror drum scanners (see Figure 5.2), driven at 6,000 r.p.m. by 100 Hz synchronous motors. The scanners were designed by Reveley and manufactured by Lynes from duralumin forgings, and contained glass mirrors that were toughened before silvering. The mean facet pitch diameter and the facet width were 12 in. and 4 in. respectively and the mirror drums had to be enclosed to reduce air turbulence to an acceptable level. Similarly the air-cooled drive motors had to be 'more spaciously enclosed to reduce noise. [The] containment panels were of [the] cavity wall type to permit [the] removal of heat by [a] through flow of water.' [5]

Figure 5.2 *The 20-facet mirror drums of Baird's colour television camera and projector used at the Dominion Theatre Demonstrations in December 1937 and February 1938 [Source: R.M. Herbert]*

At the studio end of the transmission link one of the scanners was employed in a flying spotlight configuration – the light source being a Zeiss 10 A automatic arc lamp. Banks of either caesium or rubidium photosensitive cells provided an adequate video signal. The receiving apparatus comprised a mirror drum scanner, another Zeiss 10 A arc lamp, and a Kerr cell. Pulse generation for synchronizing the sending-end and receiving-end scanners, and for interlacing the six frames of each picture, was achieved by using 'a rotating disc cum optical sensor pulse machine'. During tests of the system at Crescent Wood Road, Baird and Reveley used a cable link, the sending and receiving units being separated by the width of the coach yard. From the outset, a Cossor electrostatically deflected, intensity modulated cathode-ray tube was employed as a 'sending control room outgoing picture monitor. It had a green colour fluorescent screen about five inches in diameter.'

Baird first demonstrated his new system at the Dominion Theatre, London, on 4 January 1937. According to *To-days Cinema*:

> Gaumont-British were overwhelmed with requests for tickets for the opening show yesterday, and of significance is the fact that the bulk of these came from members of the trade anxious to watch the public reaction to the new entertainment.
>
> The actual program consisted of a stage prologue, and a fanfare of trumpets, after which George Lansbury, MP, made the first appearance on the special [6 ft × 4 ft (1.8 m × 1.2 m)] screen. A televised variety program followed, including Billy Bennet, Sutherland Felce (compere), and Haver and Lee, the latter team conducting a cross-talk act, one member appearing on the screen and the other on the stage.
>
> It was explained that the temporary studio [on the roof of the theatre] was due to the destruction [by fire] of the Baird station in the Crystal Palace. [6]

The *Daily Film Renter* noted that the experiment was received with 'moderate enthusiasm by a packed house' [7], and *To-days Cinema* reported that 'the performance had considerable novelty attraction' [6]. But not everyone was impressed:

> While the pictures transmitted were, though rather dim, perfectly visible, television on this scale has yet to attain clearness and precision of the pictures obtained by the smaller receiving sets. One may have been distracted by the incongruously amplified voices of those appearing on the screen, but the pictures certainly seemed to be disfigured by some flicker and rain, as in the early silent films, that is now unusual in ordinary transmissions of television. [8]

In addition to Baird, the Scophony Company [9] too felt that a substantial market existed in the UK for cinema television apparatus. For several years it had been developing the highly original inventions of G.W. Walton and in December 1933 S. Rowson, chairman of the company and a pioneer of the British film business, observed:

> From an entertainment point of view, scenes of current events can be transmitted by wireless or possibly by a wire, for immediate reproduction, or they may be filmed for later exhibition as desired. One master film could be sent out simultaneously to a chain of receiving stations in cinemas and shown directly on the screen. [10]

Later, in September 1936, Sir Maurice Bonham Carter, the new chairman, said, 'We expect to be able, before the close of the year, to have medium screen receivers installed for public demonstration in London, and by the middle of next year to have our cinema apparatus installed for public use.' [11]

Eleven months after the Dominion Theatre demonstration, Baird Television Ltd, in association with the Gaumont-British Picture Corporation (G-BPC), gave its first large-screen reproduction of a BBC television programme in the Palais-de-Luxe cinema, Bromley. The demonstration, on 7 December 1937, which occurred at a distance of 30 miles from the BBC's London Station, was a great success, noted the *Bromley and West Kent Mercury*:

> The programme was at all times perfectly clear, both from the front of the theatre and from the back. The focus was excellent and there was never sufficient interference to disturb the enjoyment of the audience.' [12]

Marsland Gander of the *Daily Telegraph* described the event as 'easily the most impressive I have seen' [13].

The pictures at the Palais-de-Luxe were seen by Isidore Ostrer, president of the Gaumont-British Picture Corporation (G-BPC), and Mark Ostrer, managing director, and they decided immediately to give demonstrations to the public. They intended to equip 15 of the 300 cinemas in the London area. Though the legal position concerning the BBC's monopoly of television broadcasting seemed to G-BPC to be obscure, nevertheless it considered the issue should be put to the test. Pending the possible establishment of a new central transmitter for this purpose, G-BPC proposed to reproduce in its cinemas short excerpts from the BBC's programmes, which would be added to the news film.

Baird's intention was 'to obtain the backing of the entire industry in an application to the Government for permission to broadcast special television programmes to cinemas' [14]. He had been refused a licence by the Postmaster General but, of course, he had had much experience in mustering support for his cause in the past when such rebuffs had been given. Now, with G-BPC's powerful help, he would seek the aid of the entertainment industry to persuade the government to accede to his request for a licence. Baird had convinced the Ostrer brothers of the need for cinema television and, with their financial backing, probably felt that he was on secure ground.

'Working all day and most of the night', Baird gave another demonstration at the Dominion Theatre on 4 February 1938. (This followed a press preview on 12 December 1937.)

At the public demonstration [15] images 12 ft × 9 ft (3.65 m × 2.74 m) in colour were shown, the television signals being received by radio, using a wavelength of 8.3 m, from the Crystal Palace, about 10 miles (16.1 km) away. The demonstration took place during the ordinary programme and appeared to be a complete surprise to those members of the audience who had not received a specific invitation to be present. It seems that the only people who were especially invited to the show were some BBC and GPO representatives and E.V. Appleton, who was a friend of John Baird.

The transmitting apparatus (see Figure 5.3), consisted of an 8 in. diameter mirror drum, provided with 20 mirrors inclined at different angles, revolving at 6,000 r.p.m.. These mirrors reflected the scene to be transmitted, through a lens, onto a rotating disc provided with 12 concentric slots positioned at different distances from the disc's axis. Each of the slots was covered by a colour filter, blue-green and red being used alternately. The disc rotated at 500 r.p.m. and by this means the fields given by the 20-line mirror drum were interlaced six times to give a 120-line picture repeated twice for each revolution of the disc.

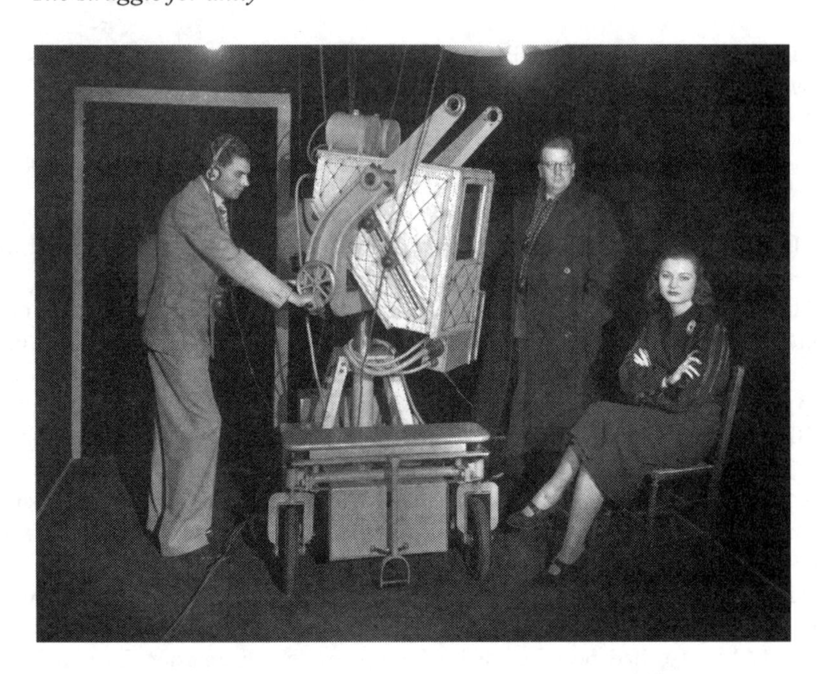

Figure 5.3 Baird's 120-line colour television mirror drum camera, which was utilised for the Dominion Theatre demonstrations [Source: Dr G.E. Winbolt]

At the receiving end a similar system of drum and disc was used with the exception that the mirror drum was 12 in. in diameter. Light from a 120 A high-intensity arc lamp was transmitted through a Kerr cell and then concentrated onto the moving aperture in the disc. Baird claimed that the projected picture could be seen from all parts of the Dominion Theatre, which had a seating capacity of 3,000.

The advantages claimed for Baird's 'multi-mesh' system were that it:

1. had a 'high frame frequency, with an accompanying high l.f. [low frequency] cut-off';
2. produced a 'reduction in flicker and a considerably increased light efficiency'; and
3. employed an 'exceedingly simple optical system of very high efficiency'.

Colonel A.G. Lee of the Post Office's Telecommunications Department was present at the demonstration and recorded his impressions:

> The scenes televised consisted of half-length views of announcers and mannequins, one at a time, the Union Jack and a side view of the head and shoulders colour portrait of the King. The picture was reasonably bright, but probably not bright enough for public performances, and the colours seemed satisfactory for a first demonstration, although in my view completely spoilt from a technical aspect by some 30 or 40 bright green vertical bars having a very appreciable width. The bars moved slowly across the picture and were always present and the definition was much inferior to that obtainable on the 405-line television from Alexandra Palace. The number of lines used was not stated, but it is estimated that not

> more than about 120 could have been used from the definition attained. This low definition was particularly noticeable in the reproduction of the King's portrait, the vertical edge of the King's forehead . . . weaving in and out very markedly, although there was no general movement of the picture thereon. [16]

Lee observed that, while the audience applauded the demonstration (and particularly the portrait of the King), they did not seem particularly enthusiastic: possibly the bright green bars and low definition disappointed them. Still, from a technical point of view, Lee found the show interesting and confirmed that 'considerable progress' had been made in colour television, but that there remained a lot of work to be done on the subject 'before the public [would] treat it as anything more than a mere novelty'. Those members of the public who saw television for the first time on the above occasion would probably have formed a poor impression of the possibilities of ordinary television now available to the London population, Lee concluded.

Another demonstration was given on Thursday morning, 17 February, before members of the Television Society, the press and a number of distinguished guests. Lord Selsdon was present on this occasion and later he wrote:

> Needless to say it [the system of colour television] is in a very crude state at present and, as he [Baird] can only use 8.33 frames/second of each colour, the resultant picture is badly framed by bars, the number of lines for each colour is of course admitting a very low order of definition, and a pretty strong arc light has to be used in order to get any illumination. Still with all its defects, he is producing a [12 ft × 9ft] colour picture. [17]

Figure 5.4 shows the projector equipment used at the demonstrations.

(a)

(b)

Figure 5.4 Baird's colour projector (a) and the associated power supply (b) for the 1937–8 Dominion Theatre demonstrations [Source: R.M. Herbert]

Ostrer was, of course, invited to the 4 February demonstration. Afterwards he and John Baird had a long talk on the possible future of cinema television.

It was decided to form a new company, which would progress the development and installation of television equipment in cinemas and would effectively control Baird Television Ltd. Cinema Television was registered with a nominal capital of £250,000. J.L. Baird became president at a salary of £4,000, Sir Harry Greer was appointed chairman and Clayton became managing director. Baird Television Ltd seemed to be on the verge of a potentially prosperous future, but the European situation was ominous and caused grave concern [18].

During the uneasy political position of 1938 the procrastination of the General Post Office in regard to cinema television led the leader writer of the *Daily Film Renter* (24 March 1938), to observe:

> Surely it would not be too much to expect from the Minister to say definitely, one way or the other, that cinemas may or may not transmit television programmes to their patrons. The television industry, as well as exhibitors, would welcome such a ruling, which would have a beneficial effect for all parties, including the BBC. To impede the progress of this new science, at a time when this country is still ahead of all its competitors, would be a great pity, but in refusing to define the policy of the Government the PMG is actually bringing this about. [19]

However, while the Postmaster General was not able or willing to state his policy, the Gaumont–British Picture Corporation Ltd was quite clear about its company policy. At the meeting held to adopt the directors' report and audited accounts for the year ended 31 March 1938, Isidore Ostrer mentioned, 'In regard to our television interest in Baird, owing to the cooperation between Gaumont-British and the Baird Company, the Baird Company are in a position to install immediately – and in fact have already installed one large screen set – in theatres and cinemas. The Derby was shown in one of our West End theatres and the audience saw the Derby finish more clearly and probably more satisfactorily than the great majority on the course.

'The importance of this revolutionary advance in television – where we are in a position to present to you the Test matches, the tennis matches, in fact almost everything of real interest whilst it is going on – is obvious. You can see the Test matches from your armchair; it is largely due to our vision and cooperation with the Baird company that we are in this unique position, and further that Baird leads the world at that moment in cinema reception of events.

'What I am leading up to is this: we have asked the BBC for permission to show television in our cinemas, other cinemas and theatres to have the same right.' [20]

The Derby transmission that Ostrer mentioned in his address was shown on 1 June 1938 in the Tatler, a Gaumont-British newsreel house in Charing Cross Road, London, before an invited audience (see Figure 5.5).

The equipment used included 'a giant cathode-ray tube' which projected an image of its screen onto a cinema screen comparable in size with that used for ordinary cinema work [21].

Figure 5.5 The original cathode-ray-tube large-screen projection equipment used in the Tatler cinema, London, on the occasion of the Trooping of the Colour in 1938 [Source: Dr G.E. Winbolt]

Reaction to the broadcast was mixed:

> Technically, big screen television is still behind the film; but yesterday's entertainment had a thrill which cannot be duplicated on a film of a news event; the thrill of the unexpected. . . . Yesterday's show once again demonstrated that big screen television can play a part in cinema entertainment if it is used at the time for special news events such as the Derby or a boxing match, in which the main thrill depends on not knowing in advance what is going to happen. . . .
>
> The show confirms my conviction expressed on the occasion of the last demonstration, that the time will come soon when every cinema will have to have television as a normal part of its entertainment; and it also confirms my view that television will not be a rival to films but an ally, to be used only for a special events. [22]

Not all comments were euphoric. The *Daily Film Renter* noted:

> This is bluntly exactly what the transmission was like. At moments, had it not been for the commentator, I shouldn't have known whether I was looking at this year's Derby or last year's winter sports! That was just now and again when the screen was a mass of blurred black and white, in which I could pick out here and there what appeared to be the top of a roundabout or the figure of a horse in the distance. At other moments I clearly saw the owners strolling about the paddock – the horses being paraded – and the serried rows of people on the grandstand. When, however, the commentator told me I was looking at the King and Queen walking up to the Royal Enclosure – I just had to take his word for it. [23]

This mixture of exclamatory reporting on the one hand and rather brutal, realistic coverage on the other, had been, from the time in 1925 when Baird gave his first public demonstration, a characteristic of JLB's and BTL's previous demonstrations, There is no doubt that Baird's publicity methods put the name of JLB firmly before the public. John Logie Baird was the inventor of the television in the general public's eyes; they knew of the struggles he had experienced in the early days of television and probably had much sympathy for him; he embodied the layman's concept of a rather absent-minded inventor, and his shy, kind and courteous manner endeared him to reporters. Above all, he was the one person in a company who could be identified with the invention of television. EMI had no such personality; neither Shoenberg nor Blumlein was known to the general public; and, while G.W. Walton's name was often mentioned in connection with Scophony's experiments and inventions, no aura surrounded this particular inventor. Not surprisingly, therefore, though Baird's 'Derby show' attracted considerable interest in the press, the Scophony cinema television performance seemed less popular with the news medium.

The sequel to the Tatler presentation was threefold:

1. the actual Baird equipment used was put on view at the official Gaumont-British Equipment stand at the Cinematograph Exhibitors Association (CEA) Exhibition at Folkestone;
2. similar apparatus remained installed in the theatre to give private trade shows of cinema television;
3. Gaumont-British Equipment invited immediate inquiries concerning the new medium.

Today's Cinema in an editorial headed 'To what use?' commented on point 3 mentioned above:

> We may not like the idea of television, but we cannot afford to be indifferent to it.... Therefore showmen will want to have a look at big screen television for themselves. ... They should ask themselves: 'To what use can we put this thing?' They must realize that there are many difficulties to be surmounted before television can be converted from a potential opposition to the film industry into a useful adjunct and ally. Many technical advances have to be made; there is the problem of negotiating rights before rediffusion; there is the undoubted fact that the enormous novelty interest which television now possesses is not a permanent asset, and that a greatly improved television programme service is just one of the many necessary developments. [24]

Apart from these points the important and complex issues of copyright and licensing greatly exercised the minds of the TAC, the BBC and the cinema trade for many months. (See the author's book *John Logie Baird: television pioneer*' (Institution of Electrical Engineers, London, 2001).)

On 18 January 1939 the TAC [25] received a deputation from G-BPC (Isidore Ostrer, Colonel Micklem and Captain West). At the meeting the BBC's representative made an announcement that surprised the film trade and led to television being shown in cinemas before fee-paying audiences. It was the breakthrough that the trade wanted.

For the BBC Mr Graves rather surprisingly said they 'were sympathetic to the idea of the reception of television [outside broadcasts] in cinemas'. Unfortunately, the minutes of the 57th meeting of the TAC give no explanation for this remarkable *volte face*.

Selsdon, Phillips, Brown and Smith had been either sympathetic to, or realistic in regard to, cinema television, but from January 1936 Ashbridge and Carpendale, and later Graves, had all argued against the introduction of television broadcasts in cinemas. The BBC had obtained counsel's opinion on the matter and Sir John Reith and the Board of Governors of the Corporation had made their apprehensions known to Lord Selsdon.

A number of factors may have influenced the Corporation's change of attitude. First, counsel's advice regarding the non-enforceability of any possible legal action taken by the Corporation to prevent the reproduction of televised sporting events in cinemas; second, the apparent inevitability of cinema television and the prospect of adverse criticism from the press and the House of Commons in the event that the BBC did not agree to Gaumont-British's request; third, a realization that cinema television could be beneficial to television broadcasting generally on the lines suggested by Ostrer; fourth, a feeling that cinema television would not come about for some considerable time because of the ominous state of the international situation.

The first occasion at which television was shown in cinemas before a paying audience occurred on 23 February 1939, when the BBC televised the lightweight title fight between Eric Boon and Arthur Danahar [26]. Both the Baird and the Scophony companies took advantage of the BBC's change of attitude and made their large-screen apparatuses available for the purpose.

BTL's equipment, which was described as the outcome of several years' development work on cathode-ray tubes, was installed at the Marble Arch Pavilion and at the Tatler Theatre. Each twin projector unit – one was a reserve – contained cathode-ray tubes, which operated at 45 kV. The size of the image on the screen of each tube was 5.5 in. by 4.4 in. (14 cm × 11 cm) and this was projected by 14 in. (35.6 cm) f/1.8 and 10 in. (25.4 cm) f/1.6 Taylor, Taylor and Hobson lenses at the above-named cinemas respectively. The units were mounted in the centre of the stalls of the two cinemas so that no seat was obstructed by the apparatus, and the 45 kV EHT supply was obtained from equipment housed in protective cages sited beneath the stages of the cinemas.

Much effort had been expended on the development of the projection cathode-ray tube, which was of a novel design and became known as the 'teapot' tube because of its shape. The fluorescent screen was deposited on a metal plate and the optical image formed on it by the electron beam was projected from the same side as the scanning beam. This arrangement had several advantages: [27]

1. an increased illumination, with anode voltage for a given beam current, compared with that obtained with a glass screen;
2. an increased illumination by a factor of two because of the lack of absorption of the emitted light in the screen material;
3. a greater uniformity of field illumination since the emitted light from the screen not did have to traverse the possibly variable thickness of the fluorescent material;
4. the possibility of controlling the temperature rise of the screen; and
5. the possibility of accurately shaping the image to conform to the field of focus of the projection lens.

At the Marble Arch Pavilion (1,400 seats) and at the Tatler Theatre (700 seats) 15 ft × 12 ft (4.57 m × 3.66 m) and 12 ft 6 in. × 10 ft (3.8 m × 3.05 m) silver screens were used. A contemporary account of the demonstrations mentioned that these screens gave an even distribution of light values throughout the theatre and that the projection units gave a highlight illumination of the screen at the order of 20 lux. This was considered to be adequate and no dimming of the exit lights or of the general theatre lighting was required.

On 27 July 1939 John Baird showed in his private laboratory at Sydenham his modified large-screen colour television system. The transmitter mirror drum now had 34 facets and produced – with the same angular velocities for the drum and disc as previously used, namely 6,000 r.p.m. and 500 r.p.m. respectively – an image of 102 lines per picture, triple-interlaced. Again, the picture frequency was 16.66 per second. The most interesting aspect of the modified overall system was the use of BTL's recently developed 'teapot' projection cathode-ray tube in the receiver in place of the rotating mirror drum and slotted disc. To achieve a coloured image Baird used a rotating disc having 12 circular filters, alternately red and blue-green, situated in front of the screen of the c.r.t. The test signals were transmitted from the Crystal Palace to Baird's home in Sydenham. No description of the quality of the images obtained appears to be available, but writers who reported on the system described it as 'a very great advance in television technique' [28].

In March 1939 Isidore Ostrer, encouraged by the successful cinema shows of the Boon–Danahar fight, announced plans for the furtherance of cinema television:

> Everyone in the trade now recognizes the power of television as a competitor. This is television year (1939), and the strides made lately have been enormous. Everyone in the film world visualises that home television will become as popular as radio and can see that the effect on cinema attendance within the area covered by television might well be disastrous. Television is at present confined to the area supplied by the Alexandra Palace service, but in that region is situated no less than a quarter of the cinema money capacity of the country.
>
> While the trade now admits the great danger of home television, I foresaw this 10 years ago, and that is why I insisted that Gaumont should take an interest in the radio-television fields. My insistence also extended to the development of cinema television, for only in doing this can we counteract the appeal of the home television. [29]

Notwithstanding BTL's dismal financial record, Ostrer felt BTL was now in a position to make profits from its cinema television expertise. He considered his Corporation had 'got in on the ground floor and consequently [stood] in the best position to reap the benefit of the big developments which must now follow the successful transmission of the Boon-Danahar fight . . .' In London alone the Corporation had 80 cinemas, each of which would require two cinema television sets (one for emergencies), at about £1,000 each, and so for Ostrer it was 'self-evident' that the first cinemas to be equipped with television would have 'a great advantage' over those without the apparatus.

Sir Harry Greer, too, was enthusiastic and stated in his chairman's address to BTL, 'I would like to stress the vast importance of cinema television in this country. Every cinema in the world must ultimately have its large television screen . . .' [30]. This was certainly optimistic forecasting at a time of considerable international unease.

John Baird has recorded:

> We had orders pending to fit the Gaumont-British cinemas with large screens, and our home receivers were considerably the best on the market and were in great demand. Orders were pouring in. Our stores were stocked with receivers and we had a staff of nearly 500 men. Television was coming into its own! [31]

Of course, for Baird, cinema television had to be *colour* cinema television. Just as black-and-white film photography and cinematography were slowly – very slowly – giving way to colour film photography and cinematography, so some visionaries felt that black-and-white television would be replaced by colour television. And Baird was certainly a visionary.

There seems little doubt that 1939 held the promise of being a turning point in the Baird company's fortunes. After many years spent in developing television, at a cost of *c.* £1,500,000, substantial orders had been taken for receiving sets and, if the government's proposed additional television transmitting stations had been completed at Birmingham and Manchester, the company's sales would have been further increased. BTL had shown a first-class exhibit at the Radio Show in August and, in addition, the development of BTL's large-screen projector had reached a point when an order for the provision of 55 large-screen sets at an average price of £2,700 each had been placed with the company. When hostilities commenced between the United Kingdom and Germany in September 1939 five of these sets had been installed in London cinema theatres [32].

On Friday, 1 September 1939, the London station at Alexandra Palace ceased to transmit television programmes. Essentially the arguments against the continuation of the television broadcasts were:

1. the frequency bands that had been allocated for television were required for defence purposes;
2. the technicians who were capable of providing the service were needed for war work;
3. the government had to conserve money and would not be justified in expending resources for a limited audience; and
4. manufacturers of television sets were heavily engaged in supplying communications and radar equipment for the war effort.

As a consequence of the cessation, orders that had been placed with Baird Television Ltd for home receiving sets were cancelled, sets that had been delivered to dealers were returned and the orders for cinema and theatre large-screen projectors were abrogated. All receiver development ceased at BTL's Sydenham works and the company was effectively reduced to bankruptcy. The board had to advise the debenture holders of the position [33].

Gaumont-British held *c.* £300,000 worth of bonds in BTL, the total issued capital of the company being £1,087,500 [34]. Eventually BTL was put into liquidation – the bondholders acquiring the company's assets in payment of their bonds. The receiver and manager for Baird Television Limited was appointed early in November

1939 – the motion for his appointment being made before Mr Justice Crossman by the plaintiffs of the debenture holders' company, the grounds being that the company was unable to carry on its business.

John Baird's contract and salary of £4,000 per annum were immediately terminated on the appointment of the receiver [35]. He 'sent in [his] name to the authorities and expected to be approached with some form of Government work, but no such offer materialised'. Since he was in the 'middle of some extremely interesting and, [he] believed, important work on colour television', he decided to continue this at his own expense in his private laboratory at 3 Crescent Wood Road, Sydenham. Here he had two assistants, E.G.O. Anderson and W. Oxbrow (see Figure 5.6).

Figure 5.6 J.L. Baird, E.O. Anderson and model (Paddy Naismith) during work on colour television [Source: Syndication International]

Baird pressed on with his television work and the task of developing a 600-line colour television system [36]. Mechanical scanning could not be used, so he employed a cathode-ray tube, of the type that had been employed in the projection units at the Marble Arch Pavilion, as a flying spot scanner. Again the two-colour process was adopted with the two-segment (blue-green and red) colour filter rotating in front of the screen of the c.r.t. The number of lines per frame was 200, three frames being interlaced to produce the picture, and the frame rate was 50 Hz. In operation the person being televised was scanned sequentially by blue-green and red light beams and the reflected light allowed to fall onto three large photoelectric cells. The receiver also utilized a similar cathode-ray tube and rotating filter. A 2.5 ft × 2.0 ft (0.76 m × 0.61 m) picture was obtained by enlargement of the image produced on the screen of the c.r.t. and was the largest ever produced – at that time – for domestic reception.

Few technical details were published by the technical press about Baird's latest system, but fortunately the inventor's patent [37] of 7 September 1940 enables the principle of scanning to be easily understood. The patent gives a clue to Baird's choice of a 600-line standard since the object of the invention was to provide a system for reproducing coloured or monochrome 600-line pictures and for receiving pictures 'using the normal number of lines at present in common use'.

Now a 405-line picture comprises two interlaced 202.5-line frames. Hence, as Baird observed:

> . . . for [the above] purpose a field consists of a primary scan of 200-lines interlaced three times to give a 600-line picture at a picture frequency of 16.6 per second and a colour picture frequency of 8.3 per second when reproduced on a receiver with a suitable interlacing circuit. On a receiver without a suitable interlacing device, satisfactory 200-line pictures are produced. [37]

With this method the same colours do not overlap in alternate complete 600-line pictures; instead, in the first picture a red 200-line frame is interlaced with a blue-green frame and a red frame, while in the second picture a blue-green frame is interlaced with a red frame and a blue-green frame, i.e. the red lines of the first picture are covered by the blue-green lines of the second picture and vice versa.

Baird described his 600-line colour images as having a 'very fine quality'. An indication of the quality was given in the April 1941 issue of *Electronics Engineering and Short Wave World*, which contained a Dufaycolor colour photograph of the colour television image. From this it is easy to appreciate Baird's remark. The photograph is of Paddy Naismith, one of the visitors to the press demonstration. Since 1941 Dufaycolor film gave results that were much inferior to present-day colour films, it may be conjectured that Baird's colour images were even better than portrayed by the print.

On the demonstration *Wireless World* commented:

> The demonstration can only be described as a very considerable success. The cover picture was of more than adequate brilliance, both pleasing and restful to watch. The various tone values were reproduced with a degree of truth comparable with the Technicolor films which we are now used to seeing at the cinema. A notable point in connection with viewing the colour pictures is an apparent stereoscopic effect which makes the picture stand out to a remarkable degree. The effect was quite apparent when still pictures were used as the subject, but became even more so when their place was taken by a girl with red hair, the tones and sheen of which were reproduced perfectly. [38]

Interestingly, although Baird's use of just two colours was probably based on a necessity for simplicity rather than a thorough scientific investigation, the use of two colours instead of three was vindicated by work reported in 1945 and later by several researchers [39].

The principles of colour mixing, based on the science of colorimetry [40], can be explained, and quantitatively expressed, by means of the chromaticity diagram. On this diagram three monochromatic radiations, with wavelengths of 0.700 μm (red),

0.5461 μm (green) and 0.4358 μm (blue) (as recommended by the Commission International de l'Eclairage (CIE)), represent the three primary colours. Use of the diagram enables any hue to be specified as a point on the diagram and to be defined in terms of a mixture of the standardised primary colours. The location of a point determines the hue and saturation – but not the brightness – of a colour. Fully saturated hues (with their wavelengths) are shown as points on the CIE colour triangle: as a particular hue becomes desaturated, so its location moves towards the centre of the triangle, which defines pure white. All realisable colours can be described by reference to points within the colour triangle.

In 1945, Willmer and Wright [41] reported their finding that any colour in a small area can be matched by mixing just two, and not three, primary colours. A few years later, Middleton and Holmes [42] carried out an experiment in which small pieces were cut from coloured sheets and then observers were asked to match the pieces to various large coloured sheets. The investigators plotted their results on the chromaticity diagram (see Figure 5.7). On this the circles show the chromaticities of the original sheets and the dots the chromaticities of the sheets that gave the best colour match. As the sizes of the pieces decreased so the positions of the dots moved towards the orange–cyan straight line shown. Middleton and Holmes (1949) found

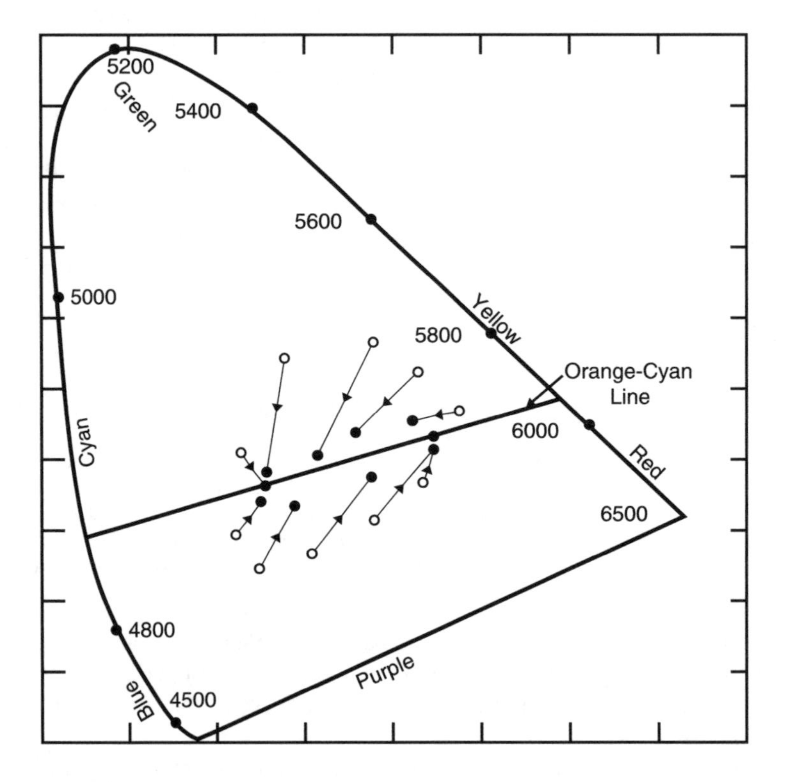

Figure 5.7 Chromaticity diagram to illustrate two-colour television [Source: Author's collection]

that just two primaries were required to define this line; these were an orange-red hue and a greenish-blue hue. These results were in accord with those of Hartridge [43] who had carried out an extensive investigation on colour acuity, in 1947.

Additional confirmation was provided from the field of colour photography. Studies of two-colour photography had shown that only the orange–cyan combination could lead to the adequate reproduction of the hues of the sky, skin and foliage. Moreover, all the evidence pointed to the fact that the eye appears to have the greatest visual acuity for colour differences when they lie on the orange–cyan line.

Without a salary or any income Baird's financial situation in 1941 became somewhat parlous. He had to maintain his house and laboratory at 3 Crescent Wood Road; support his wife and family who, at the start of the war, had evacuated from London to a rented house in Bude, Cornwall; pay the weekly wages (£12) of his staff; purchase the various materials and apparatus used in his R&D work; render hotel accounts for himself when bomb damage made his house uninhabitable; and be financially responsible for numerous miscellaneous matters. He decided to approach Sir Edward Willshaw, chairman of the Cable and Wireless company, and seek some assistance:

> ... he came to see me to say he had no funds and would have to shut down his laboratory at Sydenham and dismiss the few rather exceptional men who had worked with him over a long period of years and who would be irreplaceable if he started up again. He was very depressed and disconsolate and, after going into the matter carefully and in detail, I told him that we would take him on as a consultant at a fee which would cover the expenses of his laboratory and staff and so enable him to keep his men together and to go on with his experiments. [44]

As a consequence of Cable and Wireless Ltd's magnanimity John Baird was appointed a consulting technical adviser to the company at a salary of £1,000 per annum, for a period of three years, the appointment dating from 1 November 1941 [45]. A press notice was issued.

Willshaw told Baird that his 'appointment would be unconditional and there would be no agreement, and that [Cable and Wireless] would not tie his hands in any way'. The company clearly was showing considerable benevolence in assuaging Baird's difficulties at this time:

> We simply wanted to help him during the war period, but [said] we would like him to devote part of his time to the development of television in its application to telegraphy, and if anything came of it, or of television, he might feel – although under no obligation whatever – that, as we had been kind and generous to him, he might give us the first opportunity of benefiting from his experiments. [44]

On 18 December 1941 Baird demonstrated to some members of the press apparatus to show coloured images in relief [46]. His wartime demonstrations were civilized, relaxed events with Baird handing out ham sandwiches and a press release while the apparatus was being set up. Sometimes one of his voluntary helpers, who was an accomplished pianist, would provide an impromptu recital on Mrs Margaret Baird's Steinway grand piano [47].

Baird had previously experimented with the production of stereoscopic images in 1928 but at that time a stereoscope had to be used for viewing these: now his latest invention dispensed with this optical device and made use of anaglyphs.

The effect of a relief in a reproduced scene can be recreated, wrote Baird [48], by observing simultaneously with a stereoscope two plane views taken from positions corresponding to the separation of the observers eyes. Figure 5.8 illustrates the principle. Two views of a truncated cone are projected onto a screen at a, b, c, d and a′, b′, c′, d′. When seen through a stereoscope the views merge to give a solid figure image A, B, C, D.

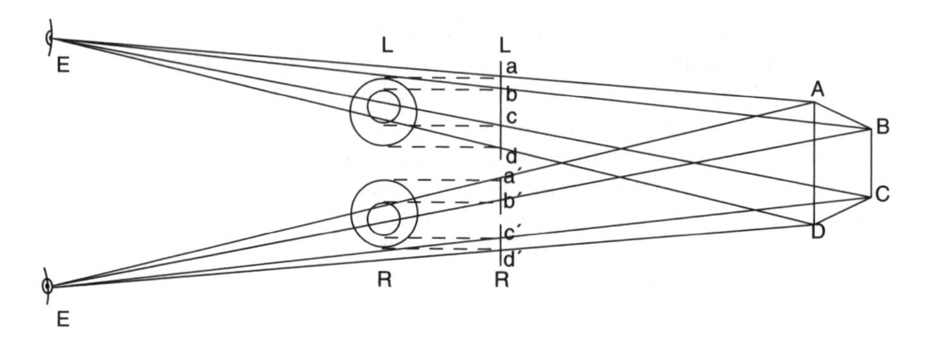

Figure 5.8 The principle of operation of the anagylph method of stereoscopic viewing [Source: Electronic Engineering, *February 1942, p. 620]*

In the method that incorporates anaglyphs, it is possible to combine the two separate views into one picture, provided they are capable of separation, the left and right eyes viewing only the left and right eye scenes respectively. This can be achieved by generating and transmitting two appropriate scene images using red and blue filters respectively and reproducing the images slightly displaced on a screen. If these are seen through glasses fitted with a red filter for, say, the right eye and a blue filter for the left eye, then each eye sees only the picture corresponding to its viewpoint and the scene appears to have relief.

Another way of achieving the effect is to use light at the transmitter that has been polarized into two components at right angles to each other and to discriminate between the images, subsequently reproduced at the receiver, by means of a viewer fitted with suitable polarizing glasses.

At his December 1941 private demonstration Baird showed a combined stereoscopic and colour television system which dispensed with the need for a stereoscope or special glasses. Schematic diagrams illustrating the principle of operation of the transmitter and receiver are given in Figure 5.9. In the apparatus demonstrated the frame frequency was 150 Hz and the scanning was arranged so that five interlaced 100-line frames gave a 500-line picture, successive 100-line frames being coloured green, red and blue.

At the transmitter the primary scanning beam of light, having passed through one of the colour filters and projection lens, was divided by a system of two pairs of parallel mirrors into two secondary beams spaced apart by a distance equal to the average separation of an observer's eyes. By means of the rotating shutter disc, the scene was scanned alternately by each secondary beam.

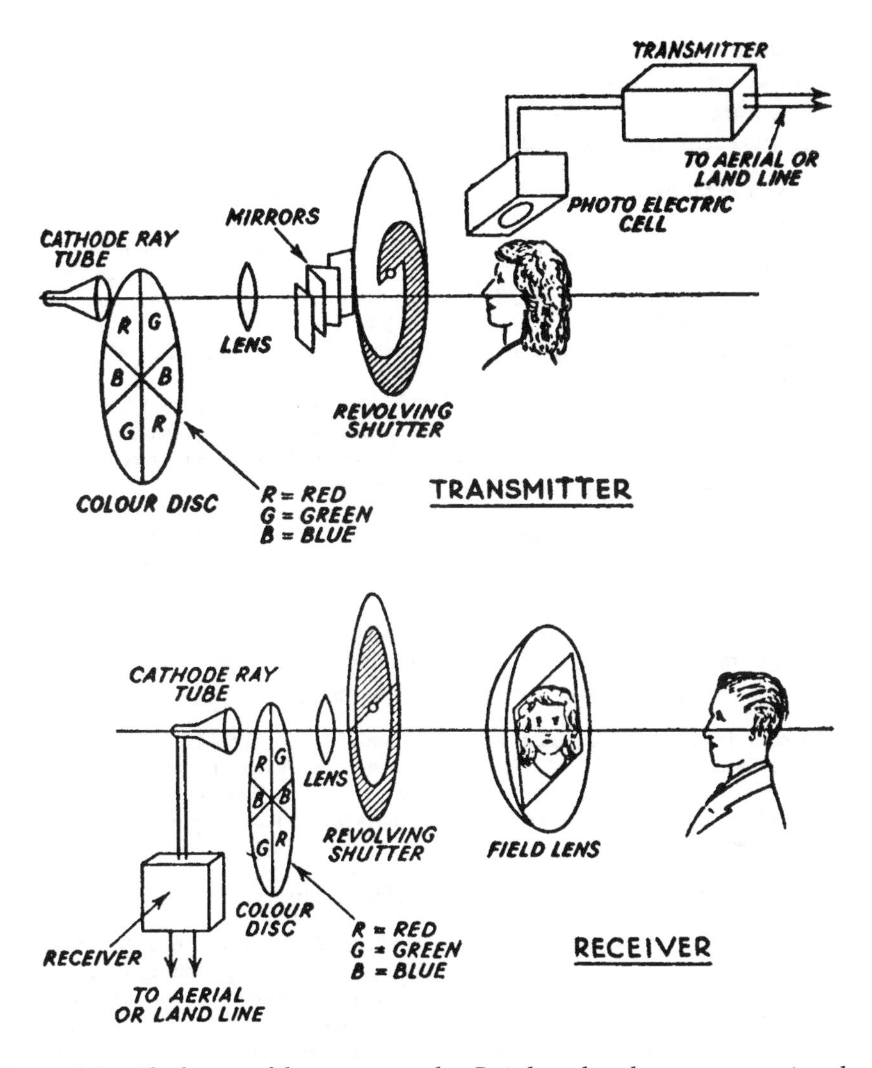

Figure 5.9 The layout of the apparatus that Baird used to show stereoscopic colour television [Source: Electrician, *December 1941, p. 359]*

The receiver included an identical colour disc to that of the transmitter and a rotating shutter, both synchronized to the corresponding discs at the sending end of the system. Hence each eye of the viewer alternately observed red, green and blue images. The shutters used differed at the transmitting and receiving ends, as shown in Figure 5.9. At the transmitter it was found desirable to maintain constant the total illumination of the scene, in order to avoid flicker, and the shutter was accordingly provided with a spiral boundary between its opaque and transparent regions. This was not necessary with the receiver shutter and it was found sufficient to provide a plain mask as indicated in the figure.

Both the transmitter and receiver utilized 'teapot' projection cathode-ray tubes, as previously employed at the Marble Arch Pavilion cinema in 1939.

The performance of Baird's stereoscopic television system was described in the February 1942 issue of *Wireless World*:

> As in some of Mr Baird's earlier 30-line demonstrations, it [was] necessary for the viewer to sit directly in front of an image forming lens and in this latest development, although the picture quality is vastly superior, the tolerance in the matter of movement of the head is smaller since the left eye must not know what the right eye is seeing, and vice versa. That is not to say that any special skill or endurance is called for in finding and holding the viewpoint which makes the picture 'come to life', and one can readily accede to the inventor's suggestion that the system might be usefully employed in a 'seeing telephone' system.
>
> If the colour reproduction lacked the ability to differentiate the subtler shades, it dealt faithfully with the bolder colours. The stereoscopic effects were an unqualified success and when the person being televised reached towards the 'camera' his arm at the receiving end seemed to project out of the lens towards the viewer. [49]

Fortunately, the truth of this assessment can be tested, for the prescient editor of *Electronic Engineering* had arranged for a colour photograph to be taken of the c.r.t. screen images.

Baird's anaglyph colour/stereoscopic television system was seen on 22 April 1942 by four members of the BBC's staff, namely, Birkinshaw, Campbell, Rendall and Schuster. On the stereoscopic-television demonstration Schuster wrote:

> [It] seemed to be a separate process in one medium which could not be combined with colour. Its effectiveness appeared to vary with each individual's sight. It seemed to create a greater sense of depth between the centre of the picture and the background. The greatest objection to it is that one has to wear peculiar glasses in which one side is coloured blue and the other red. [50]

On the colour television demonstration, Schuster observed:

> The transmission was, of course, by closed circuit and shown on an 8 in. × 6 in. tube. The two-colour process gave good blues and oranges, less good red and, as far as one could judge, poor green. The human face came through reasonably well and much more effectively than the still picture. The definition was poor and movement, if not very slow, produced colour blurring. The picture appeared adequately though not really well lit.' [50]

Baird's efforts were certainly most creditable – and especially so considering the difficulties under which he worked. In an interview he said, 'It is costly, but there is one compensation. I have so many expenses in connection with my own work that I have nothing to pay in income tax. I have no subsidy from the Government. . . . I had a very anxious time when blast from nearby bombs blew in my door. I was [seeking protection] under the dining room table most of the time, but my thoughts were all with my apparatus.'

Baird's thoughts were never far from the subject of television and, shortly after he had achieved some success with colour television and stereoscopic television using rotating filter discs, Baird pondered on the problem of devising a system that would dispense with mechanical components. He endeavoured to adapt a system of colour photography known as Thomascolour [51], which had been devised by

Richard Thomas to enable motion pictures in colour to be shown from black and white cine film. In 1923 Thomas and his father William had obtained US patents on a sequential colour system but the problems associated with flicker, loss of light and colour fringing led to its abandonment in 1930. From that time work was continued on a simultaneous system of colour photography which resulted in 11 patents. The method reached the stage of commercial exploitation in 1946, after more than 16 years had been spent on development work.

The system was based on the colour additive principle. Essentially, three images of a given scene were recorded simultaneously on each frame of a 35 mm black-and-white film (see Figure 5.10), the three images being dependent on the red, green and blue colour content of the scene. The images were of course negative images, but, by processing, black and white positive images could be obtained. In projection the three images per frame were combined simultaneously by a special projector lens to give a picture in natural colours. Again red, green, and blue filters were employed. The method was parallax free.

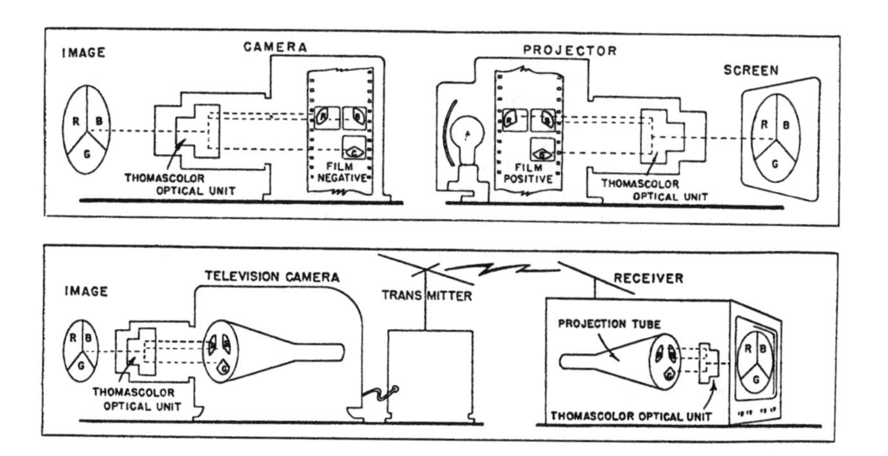

Figure 5.10 The Thomas system of colour cinematography and colour television [Source: Tele-Tech, *February 1947]*

The chief characteristics of Thomascolour were its fidelity, its economy and its speed. According to one enthusiastic report:

> Pictures of a good fidelity . . . are made available with unusual speed and economy. The entire process from exposure to projection can be done in 45 minutes, compared with days for the subtractive, dye-coloured films. The film and processing costs are those of low-cost black and white film. [51]

After the 1914–1918 war, many additive colour film systems were devised based on the use of two or three small images contained in the space of a normal 35mm frame. The disadvantages of these systems were the loss of definition, and the difficulty of registering the images when the film shrank. All additive colour systems lead to a loss of light in projection, theoritically two-thirds, but at least half. For this reason the subtractive colour film processes were investigated, notwithstanding the many advantages of the additive principle, notably the ease of processing.

With the Thomascolour film method each image frame corresponded in size with that associated with a 16 mm film. Hence, as the definition of 35 mm and 16 mm film images are equivalent to 1,080-line and 510-line televised images, it might be expected that the Thomascolour system would have an inferior resolution performance compared with a conventional 35 mm film image.

Regarding this point C. Haverline (vice president of the Thomascolour Company) stated at the Federal Communications Commission Hearing on Color Television standards:

> It appears that ... [with black-and-white television], dividing the picture into three separate images will reduce the mathematical definition by 50 per cent. However, based upon our experience with motion pictures where a 35 mm frame is divided into three images and re-combined onto the screen, we are certain that no loss of apparent definition will result. [52]

In Baird's colour television method [53], patented on 13 May 1942, and shown in Figure 5.11, the three images A, B, C, corresponding to the three primary colours red, green and blue are reproduced side by side in sequence on the fluorescent screen of the receiving cathode-ray tube and superimposed by means of an optical system D, E, F, G on a screen H. Each of the three lenses D, E, F is separated from the screen of the cathode-ray tube by a distance equal to its focal length and similarly the distance between G and H is made the same as the focal length of G. The three lenses D, E, F have their optical centres on perpendiculars through the centres of the three images respectively. By this arrangement, the rays of light issuing from each of these lenses are rendered parallel: they are brought to a focus on the screen H and therefore the

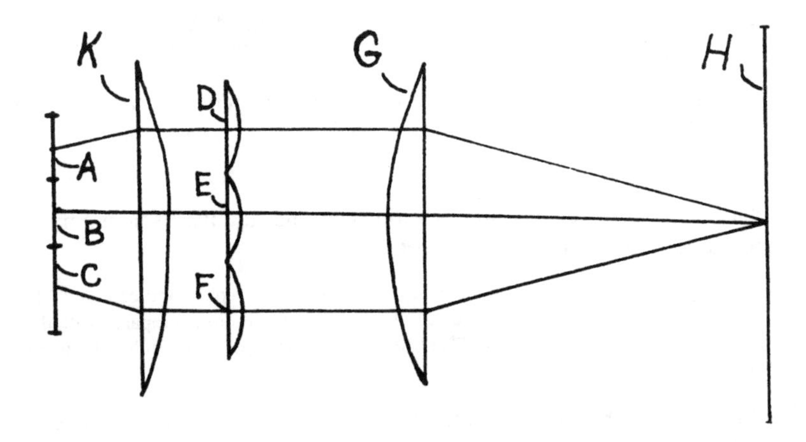

Figure 5.11 The optical arrangement of Baird's three-colour television system [Source: British patent application 555 167]

three images are combined. This optical projection scheme was not original but it had not been used previously in a television system. No account of a demonstration of Baird's system appears to be extant.

In additive colour television systems the methods of analysing, in colour, a scene or image can be classified according to the type of scanning used, namely:

1. scanning each picture frame successively in the three primary colours;
2. scanning each picture line successively in the three primary colours; and
3. scanning each picture point successively in the three primary colours.

Of these Baird and CBS (see Chapter 6) had demonstrated variants of the first method. The method could produce good colour images but gave rise to colour flicker. In an effort to reduce this effect Baird, in October 1942, patented a method based on type 2 above [54].

Figure 5.12 illustrates Baird's ideas in diagrammatic form. He wrote:

> ... I arrange that the lines of a television image which follow in sequence are produced in different colours by passing the light from the scanning spot in sequence through fixed

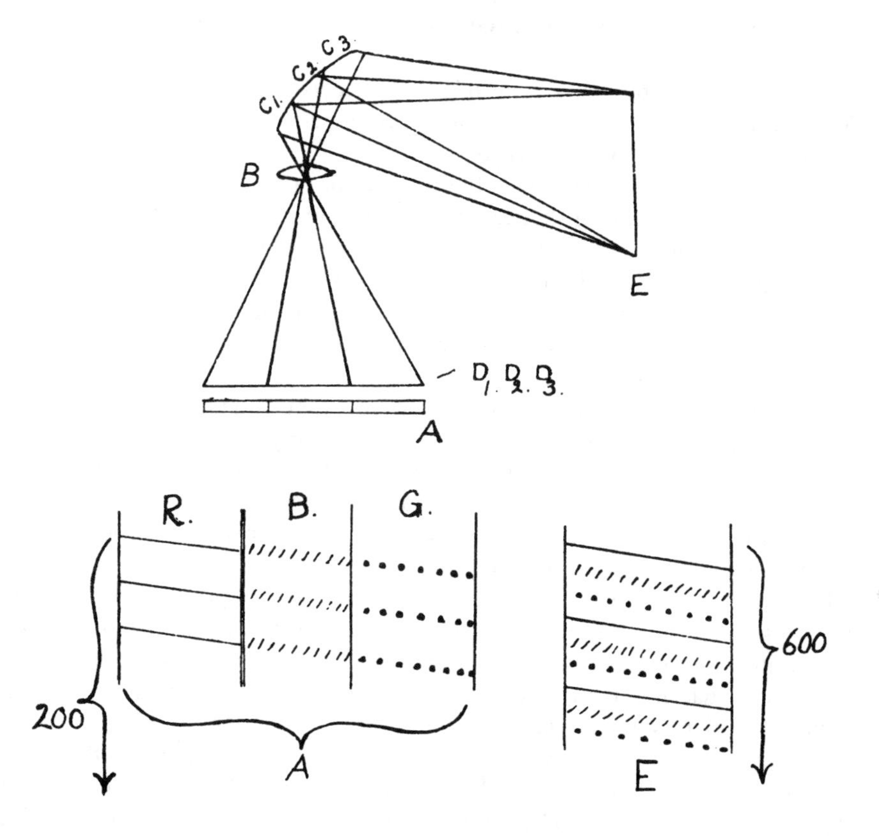

Figure 5.12 Diagrams illustrating Baird's line-sequential colour television system [Source: British patent application 562 334]

> colour filters, the passage through each filter giving colour to one line of the picture. These coloured lines [are then] projected on the field in the desired positions by means of suitable optical devices. [54]

One form of device comprised three mirrors C1, C2, and C3 slightly inclined to each other so that the three coloured parts were interlaced on the surface E to give a picture having three times the number of lines of frame A.

The idea was certainly most ingenious even though it meant coloured filters were required. It was thought the flicker effect would be reduced by displacing the successive frames of each set of three in such a way that each individual line was scanned in each of the basic colours in an appropriate sequence.

No evidence exists to show whether Baird actually put his ideas into practice. It would seem that the problem of interlacing three 200-line scans to form a picture having a 600-line definition would be formidable. Taking a picture size of 24 × 18 in. (61 x 46 cm) and assuming horizontal scanning the separation between two consecutive lines would be 0.015 in. (0.38 mm). Consequently, the positioning of the three interlacing frames would have to be carried out to a smaller tolerance than this – and be independent of vibration and temperature effects. Probably the successful accomplishment of the task eluded Baird and hence accounts for the lack of reports on the method.

A sequential line-by-line system of colour television was developed by Color Television Incorporated. It was one of the three principal systems examined by the 1949 FCC Hearing on Color Television.

The ideal colour receiver is clearly one that does not utilize either mechanical components (such as moving shutters or discs) or colour filters or other devices that introduce absorption losses. Baird pondered on this problem and subsequently on 25 July 1942 patented his highly original inventions which he called *telechrome tubes* [55]. These employed either double or triple, separate and independent, electron guns and multiple fluorescent screens depending upon whether two- or three-colour reproduction was required.

Baird conceived that if two 'teapot' tubes were combined and a fluorescent screen, coated on both sides with suitable phosphors, say orange-red and blue-green, was positioned so that the two electron beams independently scanned each surface a two-colour tube could be obtained. Figure 5.13 shows the arrangement envisaged by Baird. If now the two orange-red and blue-green screen images were properly superimposed a picture in colour would be obtained [56].

The practical implementation of the invention obviously posed some problems, since a large glass bulb was needed to accommodate the two electron guns and large screen. Baird's ingenious solution was to modify a mercury arc rectifier bulb. These bulbs were at that time fabricated in considerable numbers by the Hackbridge and Hewittic Rectifier Company, *inter alia*, and were used in the power supplies that provided the direct current for various railway systems, including the Southern Railway. Fortunately for Baird, a Mr Arthur Johnson, a retired glass technologist – who had previously been employed by Baird Television Ltd – was a past employee of the Hackbridge and Hewittic company and was familiar with the mercury arc bulbs. He undertook to convert a bulb to Baird's design.

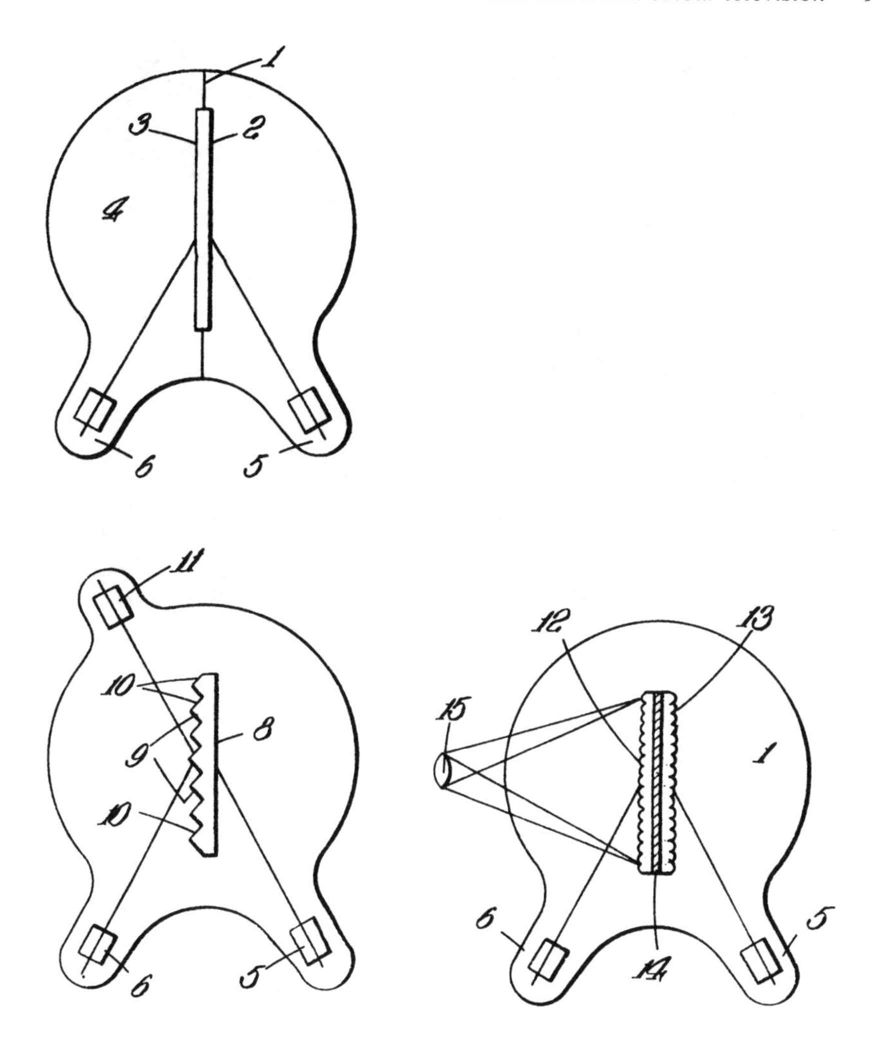

Figure 5.13 Diagrams showing the principles of Baird's two-gun and three-gun telechrome tubes [Source: British patent number 562 168]

Figure 5.14 shows Baird and his telechrome tube. The 10 in. (25.4 cm) clear mica screen was coated on one side with an orange-red phosphor and on the other side with a blue-green phosphor. Baird also considered a three-colour version of his telechrome tube in which one side of the mica screen was ridged to form two fluorescent screens. One of Baird's two-colour telechrome tubes has survived and is at the National Museum of Photography, Film and Television, Bradford. In this tube one of the electron guns has been positioned so that its axis is perpendicular to its associated surface. Baird, in his patent stated that this arrangement simplified 'the set-up of the apparatus'.

Figure 5.14 Baird with the world's first multi-electron gun, colour television receiver (1944) [Source: Radio Rentals Ltd]

On 13 January 1944 Baird invited G.B. Banks, of the Marconi Wireless Telegraph Company Ltd, and Messrs Higgett and Smale, of Cable and Wireless Ltd, to a demonstration of two-colour television and stereoscopy. Their report [57] makes interesting reading. As with many demonstrations given by Baird to technically competent observers, and the press, friends and the general public, reactions to Baird's work ranged from the harsh realism of the former to the effusive optimism of the latter. Baird's visitors observed:

> [Baird] seems to have given up the 3-colour scheme we saw on a previous occasion and now claims that 2-colour is sufficiently good. This [is] apparently because he thinks it simplifies the problem of adapting colour to the standard BBC transmission (pre-war) which is likely to be continued, at least for a period, after the war. We can confirm the improvement of 2-colour over monochrome and the demonstration leaves no doubt of the value of colour television, but we consider that the 3-colour system will have to be adopted eventually if true colour representation is to be attained.
>
> We also consider the modification to the existing transmitter and receivers necessary for colour television are so great that this simplification is of negligible value. Secondly, Mr Baird thinks the 2-colour system provides a temporary solution for stereoscopic television.
>
> He has discontinued the system of stereoscopy previously demonstrated because of insuperable difficulties of viewing .... We regard stereoscopic television as something in the nature of a stunt ...

On Baird's 2-colour television system the visitors reported:

> The colour and detail of the picture was very crude owing to imperfections of the coloured coatings. Nevertheless, the difficulties are not insurmountable and the system has definite possibilities... [W]e think that Mr Baird, despite difficult working conditions, has done

> a real service in demonstrating the value of colour television. Apart from the coated mica screen, we do not think any new invention has been demonstrated, and we consider that development on a scale far beyond the capabilities of Mr Baird's present organization is necessary for successful results. [57]

The first press demonstration of the two-colour tube was held on 16 August 1944 and a *News Chronicle* reporter subsequently wrote:

> By means of a new cathode-ray tube, Mr Baird has caused a new medium of perfect colour to blend with his stereoscopic television. As I stood by the camera I saw my own photograph flash onto the television screen across the room. The image was in colour as natural as any colour film I have seen. The light wood grain of my pipe stood out clearly, a bead of perspiration on my forehead was highlighted, and the book in my hand was pictured so plainly that the coloured title of it could be read. I moved my hand backwards and forwards, tilted a cigarette packet away from the camera and saw the effect of distance on the screen was quite natural. [58]

A number of reports and articles mention Baird's new colour television system but these are woefully inadequate and do not give any technical details of the phosphors used, or the operating voltages of the tube, or of the steps taken to superimpose the two images, or of the light flux produced. However, the magazine *Electronics* in October 1944 recorded:

> The tubes give a very bright picture due to the absence of colour filters and the fact that special powders are used giving only the desired colours, which are seen additively. The tubes give excellent stereoscopic television images when used with a stereoscopic transmitter, the blue-green and orange-red images forming a stereoscopic pair and being viewed through colour glasses.

References

1. Baird M. *Television Baird*. Cape Town: HAUM; 1973. pp. 128–9
2. Baird J.L. *Sermous, soap and television*. London: Royal Television Society; 1988. p. 140
3. Anon. Report, *Daily Herald*. 2 June 1932
4. Burns R.W. *Television: an international history of the formative years*. London: Peter Peregrinus; 1998. pp. 308–29
5. Reveley P.V. 'Some memories of John Logie Baird', 8pp. Personal collection
6. Anon. 'Trade flocks to see television. First public presentation of new system'. *Today's Cinema*. 5 January 1937
7. Anon. 'World television premier. Initial public demonstration'. *Daily Film Renter*. 5 January 1937
8. Press extract. 'Television in the cinema'. January 1937, Post Office bundle 5536
9. Singleton T. *The Story of Scophony*. London: Royal Television Society; 1988
10. Anon. 'Television may be "on the phone" soon. A new idea in transmission'. *Evening News*. 29 December 1937
11. Anon. 'The development of television. Scophony's lead in large-size pictures'. 22 September 1936

12. Anon. 'Television in a cinema'. *Bromley and West Kent Mercury*. 10 December 1937
13. Gander L.M. 'Television at 15 cinemas'. *Daily Telegraph*. 8 December 1937
14. *Ibid.*
15. Herbert R.M. 'Seeing by Wireless' Croydon: PW Publishing Ltd; 1997. p. 22
16. Lee Colonel A.G. Report on 'A demonstration of colour television by wireless given at the Dominion Theatre, London, 4th February 1938'. Post 33/5271
17. Anon. Report on 'Baird colour television'. *Television and Short Wave World*. March 1938, pp. 151–2
18. Moseley S.A. *John Baird* London: Odhams Press; 1952. pp. 221–2
19. Anon. Report, *Daily Film Renter*. 24 March 1938
20. Directors' Report, Gaumont-British Picture Corporation Ltd, 1938
21. Cricks R.H. 'A remarkable television feat. The Derby on an eight-foot screen'. *Kinematograph Weekly*. 9 June 1938
22. Anon. Report, *Today's Cinema*. 2 June 1938
23. Anon. Report, *Daily Film Renter*. 2 June 1938
24. Editorial. 'To what use?'. *Today's Cinema*. 3 June 1938
25. Minutes of the 57th meeting of the Television Advisory Committee, 18 January 1939, Post Office bundle 5536
26. Anon. 'The progress of television. A significant experiment. The cinemas look ahead'. *Observer*. 26 February 1939
27. Russell G.H. 'A survey of large-screen television'. *EBU Bulletin*. 1954;**24**(5): 123–44
28. Maybank N.W. 'Colour television. Baird experimental system described'. *Wireless World*. 17 August 1939. pp. 145–6
29. Anon. 'Gaumont and Baird. Reason for the association. Interview with Mr I Ostrer'. *Financial Times*. 31 March 1939
30. Anon. 'Baird Television Ltd. Success of large-screen projections in cinemas'. *Times*. 3 April 1939
31. Baird, J.L. *Sermons, Soap and Television*. Royal Television Society, London, 1988. p. 145
32. West A.G.D. 'Development of theatre television in England'. *Journal of the BKS*. 1948, pp. 183–200
33. Greer H. Letter to the Rt Hon. Oliver Stanley, president of the Board of Trade, 17 November 1939, Post 5474/33
34. Anon. 'Television financing'. Press extract, Post 5474/33
35. Baird M. *Television Baird*. Cape Town: HAUM; 1973. p. 143
36. Report. 'Baird high-definition colour television'. *Journal of the Television Society*. 1941;**3**(6):171–2
37. Baird J.L. *Improvements in television apparatus*. British patent application 545 078. 7 September 1940
38. Report. *Wireless World*. *c*. April 1941
39. Anon. 'Subjective colour tests'. *Wireless World*. January 1960, pp. 33–4
40. Wintringham W.T. 'Color television and colorimetry'. *Proceedings of the IRE*; October 1951, pp. 1135–72

41. Wilmer E.N., Wright W.D. 'Colour sensitivity of the fovea centralis'. *Nature*. 1945;**156**:119
42. Middleton W.E.K., Holmes, M.C. 'The apparent colours of surfaces of small subtense – a preliminary report'. *Journal of the Optical Society of America*. 1949; **39**:582
43. Hartridge, H. 'The visual perception of fine detail'. *Philosophical Transactions of The Royal Society*. 1947;**232**:519
44. Willshaw Sir E. Letter to S.A. Moseley. 7 August 1951. pp. 245–6
45. Minute 2080. Cable and Wireless archives. 28 October 1941
46. Anon. 'Television in depth'. *Scotsman*. 19 December 1941
47. Herbert R.M. *Seeing by Wireless*. Croydon: PW Publishing Ltd; 1997. pp. 24–6
48. Baird J.L. 'Stereophonic television'. *Electronic Engineering*. February 1942, pp. 620–1
49. Anon. 'Stereoscopic colour television'. *Wireless World*. February 1942, pp. 31–2
50. Schuster L.F. Memorandum to the Director General (BBC). 28 April 1942, BBC file T100
51. Anon. 'Thomascolour for television'. *Journal of the Royal Television Society*. 1947;**5**(3):107–8
52. Haverline C. Statement to the FCC. 13 December 1946, FCC Docket No. 7896, Exhibit No. 61. p. 5
53. Baird J.L. *Improvements in colour television apparatus*. British patent application 555 167. 13 May 1942
54. Baird J.L. *Improvements in colour television*. British patent application 562 334. 10 October 1942
55. Baird J.L. *Improvements in colour television*. British patent application 562 168. 25 July 1942
56. Anon. 'J.L. Baird's telechrome'. *Journal of the Royal Television Society*. 1944;**4**(3):58–9
57. 'Report on a visit to Mr J.L. Baird's laboratory, 13th January 1944', by G.B. Banks of MWT Co. Ltd, and Mr Higgett and Mr J.A. Smale of Cable and Wireless Ltd.
58. Anon. 'Baird gives television colour and depth'. *News Chronicle*. 17 August 1944

Chapter 6
CBS, RCA and colour television

On 27 August 1940, CBS, using its experimental television station in New York, broadcast colour television pictures for the first time. A private test of CBS's system was given the following day to the FCC's chairman, J.L. Fly, and some of his staff. Fly seems to have been impressed and stated 'that if we can start television off as a colour proposition, instead of a black and white show, it will have a greater acceptance with the public' [1]. CBS claimed that colour television was capable of being accommodated in a 6 MHz channel and that 'existing receivers need not suffer radical changes to adapt them to three colours instead of mere black and white' [1]. The FCC representatives were quite excited by the test, although they were concerned that the method could only work with film. But soon CBS was able to adopt a new camera tube, the orthicon, to its colour system, thereby providing a direct pick-up capability.

The driving force behind CBS's colour television work was Dr P.C. Goldmark [2], who was born in Budapest, Hungary, on 2 December 1906. His interest in television dates from his student days in Vienna when he applied for Austria's first television patent. J.L. Baird invited him to London but could not give him any employment. Goldmark then wrote to eight British radio companies seeking a suitable post. Only one company – Pye Radio Ltd – responded and Goldmark was appointed in 1932. He demonstrated a version of his television system, which used a cathode-ray tube, to Prince George, later the Duke of Kent, when he visited the factory in 1932. Pye's efforts to inspire the Prince with the romance of the new technology did not succeed, and, after inspecting Goldmark's prototype, he commented that 'it would not replace cricket'.

Goldmark's departure from the Cambridge company after just 18 months was possibly stimulated by Pye's opinion that television 'probably would never be useful for the home'.

Subsequently the young Hungarian, in 1936, joined the Columbia Broadcasting System. He was naturalised in 1937, and advanced in the company to become the vice-president of CBS in 1950. Among his achievements, other than his systems of colour television, was his invention, in 1948, of the long playing, 33⅓ r.p.m. record. He also developed the scanning system that enabled the United States' Lunar Orbiter spacecraft to relay photographs from the moon to Earth in 1966.

An impressive demonstration of CBS's colour television (see Figure 6.1), using colour film and slides, was mounted on 4 September 1940 [3]. A 16 mm Kodachrome film was run at 60 pictures per second, using a continuous motion projector, past a standard Farnsworth image dissector. A six-segment colour filter disc, of 7.5 in.

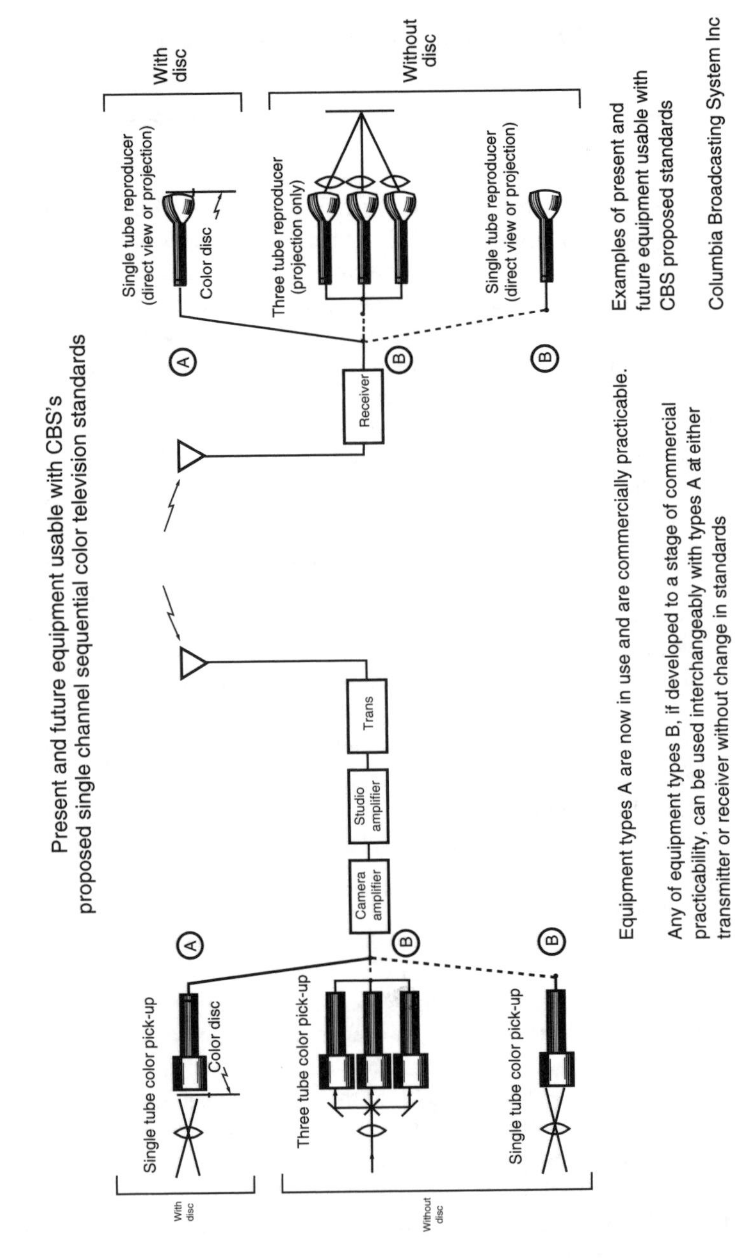

Figure 6.1 Block diagrams showing the principles of CBS's colour television systems [Source: FCC Docket 8736]

(19 cm) diameter, containing two sets of red, green and blue filters, placed before the image dissector tube, enabled the primary colour components of the film images to be televised. The disc was driven at 1,200 r.p.m. (i.e. six filters, RGB RGB were placed in front of the tube every 1/20 s) [4]. Thus, the frame rate was 120 frames per second per primary colour in contrast to the 60 frames per second that was currently being discussed by the NTSC.

At the receiver another colour filter disc was used with a 9 in. (23 cm) monochrome (white light) cathode-ray picture tube: the size of the image was 5.5 in. (height) × 7.25 in. (width) (14 × 18 cm).

Essentially, Goldmark's system was a frame sequential colour system similar to that which had been demonstrated previously by Baird. The number of lines employed was 343, and not 441, to allow use to be made of the higher frame rate. Consequently, the resolution of the colour images was correspondingly poorer. 'Disturbing flicker was evident in bright saturated colors and color break-up appeared at sharp edges. Otherwise it was a convincing demonstration', noted one observer [3].

The standard used by Goldmark was specified only after a detailed examination of the various ways of combining different interlace ratios, colour frames, scanning lines and frame and picture rates. In a paper [5],[6] published in the *Proceedings of the Institute of Radio Engineers* he tabulated five possibilities. These are given in Table 6.1.

Since the FCC, after the Second World War, spent much time considering several colour schemes, it is appropriate to outline Goldmark's comments on the reasons why he chose system 3.

System 1 was an adaptation of the black-and-white standards but, though the monochromatic definition remained unchanged, the colour picture flicker was intolerable at low illumination levels due to the low colour-frame frequency (20 Hz). 'The effect, known as "color break-up", which is purely a physiological one and increases with frequency with decreasing color frame frequency, was objectionable.'

Table 6.1 Possible parameters for various colour television systems

System	**1**	**2**	**3**	**4**	**5**
Colour fields (c)	60	120	120	180	120
Colour frames	20	40	40	60	40
Frames (f)	30	120	60	45	30
Colour pictures	10	40	20	15	10
Interlace ratio (c/f)	2:1	1:1	2:1	4:1	4:1
Lines per frame[1]	441	240	315	350	441
Lines per frame[2]	525	260	375	450	525
Colour break-up conditions	US	S	S	S	S
Interline flicker conditions	US	S	S	D	US
Picture flicker conditions	US	S	S	S	S

[1,2] corresponding to 441 and 525 black and white lines respectively
S = satisfactory; US = unsatisfactory and D = doubtful

System 2 used sequential scanning to eliminate interline flicker. However, the loss of definition was excessive. 'It became evident that in order to increase definition interlacing had to be used.' System 3 was free from flicker with screen brilliances 'up to 2 apparent foot-candles' and displayed no interline flicker. System 4 was a compromise between systems 3 and 5 with regard to the colour picture rate. Interline flicker was considered 'excessive' though the system flicker 'even at the highest brilliances [was] eliminated' [5]. System 5 used the same horizontal scanning frequency as for monochromatic television and resulted in excellent resolution, satisfactory flicker conditions but excessive interline flicker.

From his study, Goldmark selected system 3, with system 4 as a close second.

Good colour rendition of the displayed image was accomplished by careful attention to the spectral characteristics of the sending-end and receiving-end filters, the colour properties of the phosphors used in the camera and display tubes, and the colours of the illuminating sources. It was found that the infrared radiation from these sources contaminated all the colours as it passed freely through the red, blue and green filters because the camera tubes were very sensitive to the radiation. This effect had been noted by Baird in his work. He solved the problem, as did CBS, by utilizing an infrared filter.

Both receiver filter discs and receiver filter drums were tried. Figure 6.2 shows the shape of the disc suitable for a transmitting or receiving tube with a short decay

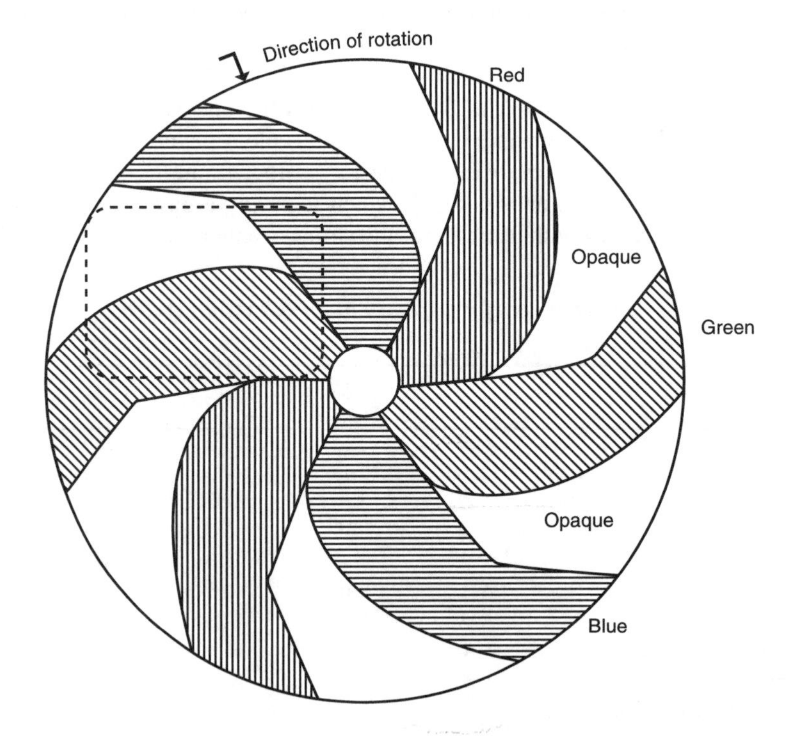

Figure 6.2 Diagram of a CBS colour filter disc [Source: Proceedings of the IRE, *1942,* ***30****(4), p. 180]*

of storage times. The curve was an envelope of the position of a scanning line as traced onto the filter which moved with the line. 'The required filter shape for a given mechanical arrangement [was] obtained by developing curves which [made] allowance for positive and negative tolerances to take care of fluctuations in the disk position, viewing angle, and screen decay.' [5]

The field-sequential method of television has both advantages and disadvantages. Since only one colour frame is present at any given time, the method is free from the mis-registration problem of a simultaneous television system, and the uniformity of colour gradation over the image area is good. On the other hand, for large image areas, the need for a rotating filter wheel poses problems – apart from size; namely, the need for dynamic balance, perfect synchronization (though this is much simplified if the mains frequency is utilized), and a considerable power input to the driving motor. Overall, the field-sequential system is a simple, low-cost solution to the colour television problem for small screen receivers.

The limitations of the system arise from the aversion of the human eye to excessive discontinuities of light, and of motion. If the eye were tolerant of these the field-sequential system and the black-and-white system could have the same standards. The intolerance is noticeably increased if the discontinuity in position is accompanied by a change of colour. Colour fringing and colour break-up may then be observed: both are ameliorated by an increase of frame rate.

Goldmark's attention to detail ensured good results. G. Cock, who had been the BBC's former director of television and was now (late 1940) its North American representative, enthused about a colour television demonstration that he witnessed:

> If television broadcasting had been launched in colour, it would now be a universal medium of entertainment and instruction, occupying a position comparable to that of present-day radio.
>
> It is an understatement to say that Columbia's colour television is astounding. It is a miracle! These are not hysterical remarks. I had naturally expected some distortion in reproduction. But I was exceedingly surprised to observe the precision with which the receiving set's screen reproduced the coloured images of the experimental films. Moreover, these colours registered with exactitude. The result was faithful – up to the limitations of the film's own colour sensitivity. Not only vivid colours but the delicate tints of nature were superbly shown.
>
> Black-and-white television fades out of the picture in comparison with Dr Goldmark's colour television. Apparent definition is greatly increased. One gains the impression that the colour images have the additional dimension of depth. Flatness – bugaboo of all photography – vanishes. The effect is stereoscopic and crystalline in clarity. I am now looking forward with greatest interest to what can be done with direct pick-up television cameras ... [7]

Cock did not have long to wait. On 9 January 1941 CBS gave the first public showing of direct pick-up colour television. (A private show had been given on 12 November 1940.) The picture signals were generated by an orthicon television camera in the CBS television laboratories on the fifth floor at 485 Madison Avenue, New York, transmitted by coaxial cable under 52nd Street, and reproduced on two different models of colour receivers in studio 21 of the new CBS studio building at 49 East 52nd Street [8].

Three receivers were used in the demonstration: a standard black-and-white receiver adjusted to receive colour transmissions in black and white, a standard black-and-white receiver that had been adapted for colour, and a compact table model colour receiver especially designed and constructed in the CBS laboratories. According to the CBS press release, the demonstration showed 'substantial progress along new fronts such as: direct pick-up itself; synchronization of color disks; phasing of color disks; and new lighting methods for color television'.

The introduction of CBS's colour system did not please everyone. RCA, GE, Farnsworth, Philco and DuMont had each invested heavily in all-electronic monochrome television and were highly critical of the reintroduction of mechanical devices into television and the lowering of picture definition. Zenith and Stromberg-Carlson, on the other hand, were just entering the television field and defended the utilization of colour. CBS itself seemed to be in no hurry to commence television broadcasting. It had bought an experimental television transmitter in 1938 but had never really pressed its television broadcasting facilities. The CBS radio system was quite prosperous and, since CBS was not a manufacturing organization, unlike RCA, and had no aspirations in manufacturing, it could afford to argue for a delay in the start of commercial television until it had perfected its colour system. Some support for CBS's work was tacitly given by the NTSC when it recommended its black-and-white 525-line standards to the FCC and additionally recommended field testing of the CBS system prior to any standardization. The NTSC stressed 'the potential importance of color to the television art'.

On 31 October 1940 the FCC issued a press release [9] announcing that it would hold a conference in January 1941 'for the purpose of receiving a progress report on [the NTSC's] study of the television situation'. Another FCC press release, dated 16 November 1940, mentioned that 'some 20 experimental authorizations had been authorized for television experimentation and that a total of $8,000,000 had been budgeted for that purpose' [10].

The NTSC's report was received [11] by the FCC at its one-day hearing on 27 January 1941, and, between 20 and 24 March, the FCC conducted hearings on the report. (Earlier, on 27 February 1941, the Commission had issued a public notice, which delineated alternative standards for black-and-white television.) It was clear that a general industry consensus had been accomplished on television standards; the only significant dissenting voice was that of DuMont Laboratories, which continued to press its case for flexible standards. Much thought and effort had been expended by the NTSC's nine panels. More than 60 meetings had been held, 20 tests/demonstrations had been conducted, and approximately 5,000 man-hours of work had been devoted to studies of their briefs [12].

Panel 1 made a thorough investigation of colour television [13]. During their enquiry the members analysed five (4 CBS and 1 GE) American and three (Baird) British colour systems; and attended three demonstrations of colour television: one of the three-colour film pick-up at the CBS laboratories on 25 September 1940; one of the two-colour slide pick-up at the GE Schenectady laboratories on 20 November 1940; and one of three-colour direct pick-up at the CBS laboratories on 11 December 1940. All of these systems were designed to operate within a 6 MHz channel bandwidth.

Table 6.2 The RCA colour television systems (1940)

System	1	2	3	4
No. of channels	1	3	1	1
No. of colours	2	3	3	3
Additive/subtractive	A	A	A	A
No. of lines	120	RMA standard	441 interlaced	441 interlaced
No. of colour pictures/s	24	RMA	60	40
All-electronic screening	Yes	Yes	Yes	Yes
Rotating colour filters	tr. & rec.	not in rec.	rec.	tr. & rec.
Transmitter			3 Os	
Receiver			1 K	
Operational use	1 month in lab	dem. to FCC Feb. 1940	some parts tested	several days in lab

K = kinetoscope, O = orthicon

In addition Panel 1 received a paper from RCA on 'Television systems investigated experimentally by RCA' (dated 12 September 1940), and considered and discussed briefly the colour systems of Philco, DuMont, and Kolorama.

Of these the CBS systems have been considered previously. The GE two-colour system operated on the 441 lines, 30 frames, 60 fields standard and used dichromatic filter discs synchronised to the mains. Odd lines were always one colour, and even lines another. The parameters of the four RCA colour systems are summarized in Table 6.2.

RCA stated that these systems had the following advantages (1) and disadvantages (2) [14].

System 1

1. Adds color to other transmitted intelligence. Compared to three-color systems, requires less bandwidth for same resolution and flicker frequency.
2. Fidelity of color reproduction [is] not as good as obtainable with three-color systems. Compares to black and white systems, requires twice the bandwidth for same resolution and flicker frequency.... Number of lines and flicker frequency too low. Low sensitivity of pick-up device and low brilliance of reproduced picture, due to use of color filters.'

System 2

1. Same resolution and flicker frequency as RMA system. Good brilliance in reproduced picture. No moving parts in receiver.
2. Three wideband channels required. Accurate registration of three superimposed images in receiver is required. Receiver is large. Sensitivity of pick-up camera is reduced.

System 3

1. Adds color intelligence without sacrificing resolution or flicker standards.
2. Wide frequency band required. Inefficient use of light at kinetoscope, and loss of sensitivity in pick-up. Vulnerable to 60 cycle interference ... Mechanically moving parts in receiver.

System 4

1. Adds color to other transmitted intelligence
2. For same resolution, wider frequency band (approximately twice as wide) is required as for black and white transmissions. Low flicker frequency results in noticeable flicker at high light levels. Inefficient use of light from kinetoscope, and loss of sensitivity in pick-up. Vulnerable to 60 cycle interference... mechanically moving parts in receiver. [14]

None of these systems was analysed by the FCC because they needed much more bandwidth than the RMA standards provided.

On RCA's aspirations for colour television, their letter of 3 October 1940 to the FCC declared:

> We believe in television in natural colors just as we believe in television in monochrome. We believe this because we know that color adds so much picture information and much to viewers satisfaction. We have worked with color systems and intend to continue this work. We particularly feel that investigation should look toward color television at the higher radio frequencies for broadcasting where the channel width may not need to be limited to six megacycles. We feel also that investigations should be made that will give definite answers to television in color when using six megacycle channels. [15]

The need for more bandwidth was also referred to by Dr E.W. Engstrom, the chairman of Panel 4 (transmitter power). He averred that 'it has not been demonstrated to my satisfaction that 343 line 60 frame color television has been proved to be adequate.... If it turns out that it is not adequate then of course the way to make the performance proper is to have more bandwidth. I know of no other solution.' [16] (In 1946 CBS would introduce their 16 MHz bandwidth colour television system.)

Prior to the arrival of colour television on the television scene the NTSC had had to consider the issue of either fixed standards as propounded by RCA and others or continuously flexible standards as advanced by DuMont [17]. Now there was another possibility, discontinuous flexible standards as advocated by CBS, namely, receivers would be required to receive two different standards, one for monochrome television and one for colour television.

In its report to the FCC, the NTSC recommended the following television standards [18]:

1. television channel: 6 MHz wide, with the picture carrier 4.5 MHz lower in frequency than the sound carrier;
2. number of lines: 441;
3. picture frequency: 30 pictures per second with 60 frames per second;
4. aspect ratio: 4 (horizontal): 3 (vertical);
5. vision modulation: amplitude modulation for both picture and synchronizing signals, with a decrease in light intensity causing an increase in radiated power;

6. black level: to be represented by a definite carrier level, at 75% of the peak carrier amplitude;
7. sound modulation: f.m. with a pre-emphasis of $100\mu s$ and a maximum deviation of ± 75 kHz;
8. polarization: horizontal;
9. synchronizing signals: specified.

Later, at its final meeting on 8 March 1941, the NTSC revised its recommendation on the number of lines from 441 to 525 and it recommended the alternative use of f.m. or a.m. for the synchronizing signals [19].

The new rules [20] of the FCC (adopted on 30 April 1941 and published in its report of 3 May 1941) specified that each commercial television station was to broadcast a regular programme schedule for a minimum of 15 hours per week, of which two hours had to be between 14.00 and 23.00 hours daily, except Sunday, with at least a one-hour programme transmitted on five weekdays between 19.30 and 22.30 hours. Of the various channels available for television, numbers one to seven inclusive were each permitted to provide a sponsored service, and channels eight to eighteen inclusive were restricted to experimental and television relay development and applications. Above 300 MHz further channels could be utilized for relay and experimental purposes [21]. The operation of commercial television stations was authorized from 1 July 1941.

On colour television, the FCC opined:

> The three color television demonstrated by the Columbia Broadcasting System during the past few months has lifted television broadcasting into a new realm in entertainment possibilities. Color television has been known for years but additional research and development was necessary to bring it out of the laboratory for field tests. The three color system demonstrated insures a place for some scheme of color transmission in the development of television broadcasting. [20]

And so after approximately five years of rancour and frustration in the industry rules and standards were available for commercial television broadcasting. But then, on 27 May 1941, less than one month after the FCC had issued its new rules, President Franklin D. Roosevelt declared a state of unlimited national emergency. No raw materials or production capacity could be directed to television. Indeed, RCA's television receiver manufacturing line had ceased to operate for more than a year, and as with other national companies its production capacity was engaged in work for national defence. Nevertheless, notwithstanding this constraint, the official opening of American television on a full commercial basis was held on 1 July 1941. Sponsored programmes were now permitted but the only station with paid programmes was NBC. Its inaugural broadcasts were supported by the Bulova Watch company, Sun Oil, Lever Bros and Proctor and Gamble, and included a news broadcast, a television show and a quiz show. Very few saw the television transmission; one source mentions a figure of *c.* 5,000 sets in the New York area. Another source gives the number of publicly owned television sets between 1940 and 1942 as just 3,000.

For the remainder of 1941 television broadcasting was carried on by only seven of the 22 licensed stations across the USA [22]. The bombing of Pearl Harbor by the Japanese on 7 December 1941 finally led to a cessation of domestic television progress and expansion for the war years. After the entry of the United States into the war, a government order dated 24 February 1942 ruled out the manufacture of television equipment for civilian use. At the same time the minimum number of hours of transmission required of a licensed station was reduced from fifteen to four hours per week.

By September 1943 the war was going well for the Allies. Rommel and the Africa Korps had been driven out of North Africa and Sicily; Italy had been invaded; the U-boat had effectively been defeated; and German night bombers no longer posed a threat to the United Kingdom. On the Russian front, German forces were reeling under the onslaught of Soviet troops; and in the Far East the Americans and their allies were retaking the territories that had been conquered by the Japanese. In the UK plans were being prepared for the invasion of Europe. Also, some thought was being given to the problems that would have to be resolved when the war was concluded. Churchill expounded his views on these in a memorandum dated 19 October 1943. The memorandum began, 'It is the duty of His Majesty's Government to prepare for the tasks which will fall upon us at the end of the war.'

One of these tasks concerned the reopening of the London television station at Alexandra Palace and the development of television in the early postwar years. In good British fashion, a committee, the Television Committee, was appointed in September 1943 to consider this matter. Its terms of reference were [23]:

> To prepare plans for the reinstatement and development of the television service after the war with special consideration of –
>
> (a) the preparation of a plan for the provision of a service to at any rate the larger centres of population within a reasonable period after the war;
> (b) the provision to be made for research and development;
> (c) the guidance to be given to manufacturers, with a view especially to the development of the export trade.

The committee comprised the Rt Hon. the Lord Hankey (chairman) and senior members of the General Post Office, the British Broadcasting Corporation, the Department of Scientific and Industrial Research, a government defence establishment and the Treasury.

Reference to the appointment of the Committee was made in the House of Commons on 18 January 1944, although the first meeting of the Committee had been held on 26 October 1943. Subsequently, a further 30 meetings were convened at which witnesses from GEC Ltd, M-EMI Television Ltd, Scophony Ltd, STC Ltd, the Board of Trade, the BBC, the British Film Producers Association, the Ministry of Education, the Radio Industry Council, the REP Joint Committee and J.L. Baird were examined. Baird was the only private individual invited to give evidence [24].

In its report, dated 29 December 1944, the Hankey Committee concluded that 'the television service should be restarted in London on the basis of the system in operation before the war . . .'; 'the aim should be to make the television service self-supporting as soon as possible . . .'; 'the aim should be to produce an improved television system having a standard of definition approaching that of the cinema [para. 25 mentioned 'of the order of 1,000 lines'] and possibly incorporating colour and stereoscopic effects'; 'the desirability of adopting common international standards should be kept constantly in mind'; and 'the successful development of television at home [would be] of prime importance to the development of an export trade'.

Television in the UK recommenced on 7 June 1946; the standard was still the 405-line standard of the prewar years, but no regular colour television service was broadcast until 1 July 1967, when BBC2 began Europe's first colour service. Essentially the factors that determined the re-establishment of the prewar standard were:

1. the 1936 London television station at Alexandra Palace, with its studios and offices, television and control equipment, transmitters and antennas, had survived the numerous bombing attacks on the capital and could be brought to a state where television broadcasting could be quickly re-started;
2. unlike the position in the USA, neither the British Broadcasting Corporation nor any major UK industrial company or research undertaking had carried out research on colour television;
3. the introduction of a higher standard for black-and-white television, and/or the introduction of acceptable colour television, would require several years research and development by a wealthy organization;
4. the death of J.L. Baird in 1946 removed the UK's principal proponent of colour television from the television scene;
5. the ravages of the war on British industry, and the immense cost of the war effort, necessitated an expeditious return to profitable, free-enterprise working based on the export of manufactured goods and services.

In the United States, further consideration of the colour television problem was given at the FCC Hearing of 1944–45 [25]. This proceeding was instituted on 15 August 1944 because 'all recognize that a complete review of present allocations of bands of frequencies in the radio spectrum is necessary as a result of the important advances in the radio art which have been made during the war and the greatly increased demands for the use of radio'. The hearing lasted from 28 September to 2 November 1944.

Of particular note, in the context of colour television, was the opinion of Panel 6 [26] of the Radio Technical Planning Board 'that adequate standards for color television for a six megacycle channel cannot be established at this time'. This view was given without prejudice to continued experimentation in colour in the VHF band. Panel 6 averred that provision should be made for practical work in high-definition monochrome and colour television in the UHF (i.e. the band of frequencies from 300 MHz to 3,000 MHz), with 20 MHz channels but with no other constraints. CBS, too, pressed for a move to the ultra-high frequencies. 'As we have . . . expressed it, the ultimate in 6-megacycle television is equivalent to approximately 250,000 picture elements per image, while in 16-megacycle television the ultimate is the equivalent of 585,000 picture elements.' [28] In addition, the majority of the RTPB's Panel 6 favoured a field rate of 180 Hz to give a colour television flicker performance similar to that of monochrome television, and 525 lines per frame to provide equal resolution with black and white television. These figures implied a bandwidth of *c.* 15 MHz.

Dr Goldmark, for CBS, proposed the utilization of the UHF for 735-line monochrome television and 525-line colour television (see Figure 6.3). A similar stance was taken by Dr Engstrom of RCA:

> RCA is preparing to experiment under broadcast conditions in New York at 288mcs [MHz] and expects to start in the early spring, but this is only a beginning as tests are needed in the 300 to 1,000mcs region [i.e. in part of the UHF band]. In the 400 to 1,000 mcs region 'we just don't know what we can do about broadcasting television. We have indications that the propagation conditions will be severe, but what troubles us most are the uncertainties about this propagation and our lack of experience. Therefore, I endorse the RTPB recommendation that commercial television be assigned channels not to exceed 300mcs top frequency.' [20]

After much discussion and deliberation the FCC ordered that the frequency band from 480 MHz to 920 MHz be made available for experimental television:

> The Commission repeats the hope in its proposed report that all persons interested in the future of television will undertake comprehensive and adequate experimentation in the upper portion of the spectrum. The importance of an adequate program of experimentation in this portion of the spectrum cannot be over-emphasised, for it is obvious from the allocations [12 VHF channels each of 6 MHz bandwidth] which the Commission is making for television below 300 megacycles that in the present state of the art the development of the upper portion of the spectrum is necessary for the establishment of a truly nation-wide and competitive system. [30] [FCC Report dated 25 May 1945]

Members of the press and industry did not have long to wait to view wideband colour television for, from October 1945, CBS began colour television broadcasting on an experimental basis and mounted many demonstrations of their basic colour system. Signals were transmitted from the 71st floor of the Chrysler Building, New York – using a 485 MHz, 1 kW transmitter and directional antenna – to a receiver situated on the 9th floor of the CBS Building at 485 Madison Avenue, New York. Double-sideband operation was utilized for the demonstration, but vestigial sideband transmission was proposed for which the sidebands of the transmitter would extend from 480 MHz to 496 MHz (i.e. a 16 MHz channel would be employed) [31].

Televised images of coloured 16 mm films and 35 mm slides were generated with the aid of an image dissector camera tube, a tube ideally suited for this purpose since it lacked a storage property and hence did not carry over stored charge from one field to the next. Its disadvantage was its low sensitivity. This was not a problem when films and slides were used but for direct studio television very powerful lights were needed. Each picture image, in each of the primary colours, was scanned using a standard of 525 lines per picture, 20 pictures per second. The scanning sequence followed the pattern given below.

Scanning lines	1st field	2nd field	3rd field
Odd lines	red	green	blue
Even lines	blue	red	green
Time interval (s)	1/60	1/60	1/60

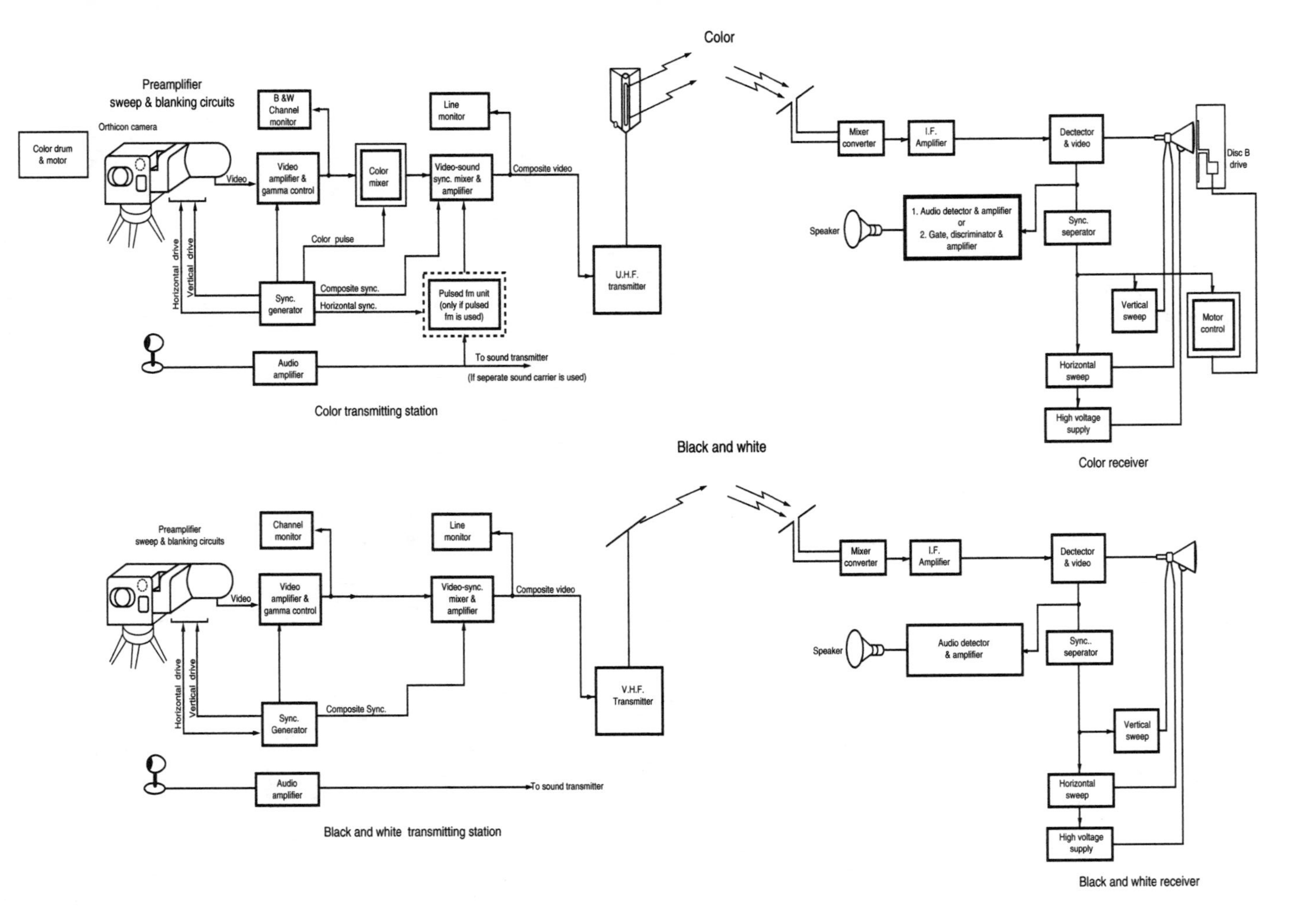

Figure 6.3 Comparison of CBS's black-and-white and colour television systems [Source: Federation Communications Commission]

(Each colour field had a duration of 1/120 s, each colour frame comprised two colour fields, and each complete colour picture required six fields.)

The receiver display used a 10 in. (25.4 cm) cathode-ray tube and a lens to magnify the screen image to a width of 12 in. (30.5 cm). A rotating six-segment disc, (substantially the same as that used in 1940), driven at 1,200 r.p.m. and automatically synchronised to the transmitter filter, was placed in front of the tube. One type of filter, developed by Kodak, was fabricated on an acetate base, coated with the appropriate coloured gelatine – Wratten numbers 61, 47, and 26 for the green, blue and red filters respectively – and protected by a layer of lacquer.

An appreciable improvement of the quality of the pictures, compared with those of 1940, were noted by several correspondents. Whereas the earlier images were of low definition (343 lines per picture) and suffered from 'occasional difficulties in the rendition of certain colors, particularly the dark shades', 'the reception of the 16 mm colored films on the CBS receiver [of the new 10 MHz bandwidth system] seemed satisfactory, with definition, contrast and brilliance comparable to that of black and white. The color itself was an improvement over anything previously seen. The basic colors were not overemphasized, and the results showed a good proportion of intermediate shades, tending to make the reception more realistic. Blurring of color, due to rapidly moving objects, was not evident. The 1,200 r.p.m. scanning disc caused a small amount of noise which was not noticeable with the sound channel on.' [32]

A major disadvantage of the CBS system, and of all colour systems that use filters in front of a receiver's screen, arises from the loss of light from the screen's phosphors. The CBS rotating filter produced an average light transmission loss of 86 per cent. Consequently, the phosphor image had to be exceedingly bright to produce a worthwhile picture. At the demonstration the 'light output was considerably under that generally deemed necessary for viewing in a subdued-light room, although the illumination was more than adequate for a darkened room'. This low output had one advantage: the absence of flicker.

On the performance of the CBS system, D.G. Fink, chairman of the RTPB subcommittee conducting colour television studies, said:

> The system . . . transmits excellent color pictures. The intrinsic definition of the picture is better than the 525-line black-and-white pictures by virtue of the more than proportionally wide band used and, of course, the apparent definition is greatly enhanced by the color contrasts presented in the image. From this standpoint the images are perhaps the best television pictures yet produced. [33]

With praise like that it seemed CBS's venture in colour television was assured. However, the future outlook was somewhat menacing.

RCA and, in general, the industry, its committees and receiver manufacturers were wholly opposed to sequentially scanned colour television systems based on rotating colour filter discs. The company's position followed work on such a system, which was demonstrated in early December 1945. It was seen by General Sarnoff [34], who asseverated that neither he nor his engineers cared much for their version of the home-entertainment receiver. He said an all-electronic method was under development, though he saw little hope of either method becoming commercially practicable for another five years.

The parameters of RCA's latest apparatus were [35]: r.f. carrier frequency, 9,500 MHz; bandwidth, 12 MHz; transmitter output, 50 mW; line standard, 525 lines per picture, 40 pictures per second; colour filter drum, 12 segments – red, green and blue; rotation speed, 600 r.p.m.; sound modulation, pulse width; display tube diameter, 12 in. (30.5 cm). Stereoscopic presentation was accomplished by means of Polaroid glasses rotating in conjunction with the colour drum.

On the demonstration Dr Engstrom was quoted as saying:

> RCA's color research had now advanced about as far as their black and white television was in 1930 [*sic*]. . . . RCA is vigorously carrying on improvements in color television. The time needed to make it commercial is at least five years. Color does add something to television.

RCA's new simultaneous all-electronic colour television system was shown to representatives of the press, industry and government in October and November 1946 [36,37]. A schematic diagram of the equipment is illustrated in Figure 6.4. At the sending-end, the scanning raster of the 5 in. (12.7cm) diameter cathode-ray tube was focused on a frame of the 16 mm colour motion film. After transmission through the film, the light was divided, by colour-selective mirrors, into red, green and blue components, which were then impressed onto three photoelectric cells. Three r.f. channels (one for each colour signal) having an overall bandwidth of 16 MHz were employed. The receiver display apparatus comprised three 3.5 in. (8.9 cm) diameter cathode-ray tubes, red, green and blue filters, and condensing lenses configured to produce a 15 in. × 20 in. (38 cm × 51 cm) image on a translucent screen.

Since each of the three-colour channels used the same standards as those then employed for black-and-white television transmission (525 lines per frame, 60 fields per second), the green channel was suitable for monochrome purposes. Thus, the system was characterized by compatibility of colour and monochrome standards, i.e. a colour receiver could display a monochrome picture when black-and-white signals were received; and a monochrome receiver could give a black-and-white picture when colour signals were received. This interchangeability property, which prevented any undesirable obsolescence of monochrome receivers, would be debated at length in further colour television FCC hearings [38].

A report of a demonstration mentioned the excellent register of the three-colour images, the absence of colour separation and colour fringing, and the wide colour range.

Interestingly, and perhaps pertinently, RCA's system configuration could be operated in a field-sequential mode, rather than in a simultaneous manner. To achieve this it was only necessary to key on and off in sequence the three photocells at the transmitter and the three cathode-ray tubes at the receiver. There is no evidence that CBS tried this method; it would have led to all-electronic field-sequential television, albeit at an increased cost to the receiver purchaser.

RCA's plan for colour television development was bold and comprised several stages: reproduction from colour slides (which was demonstrated 30 October 1946);

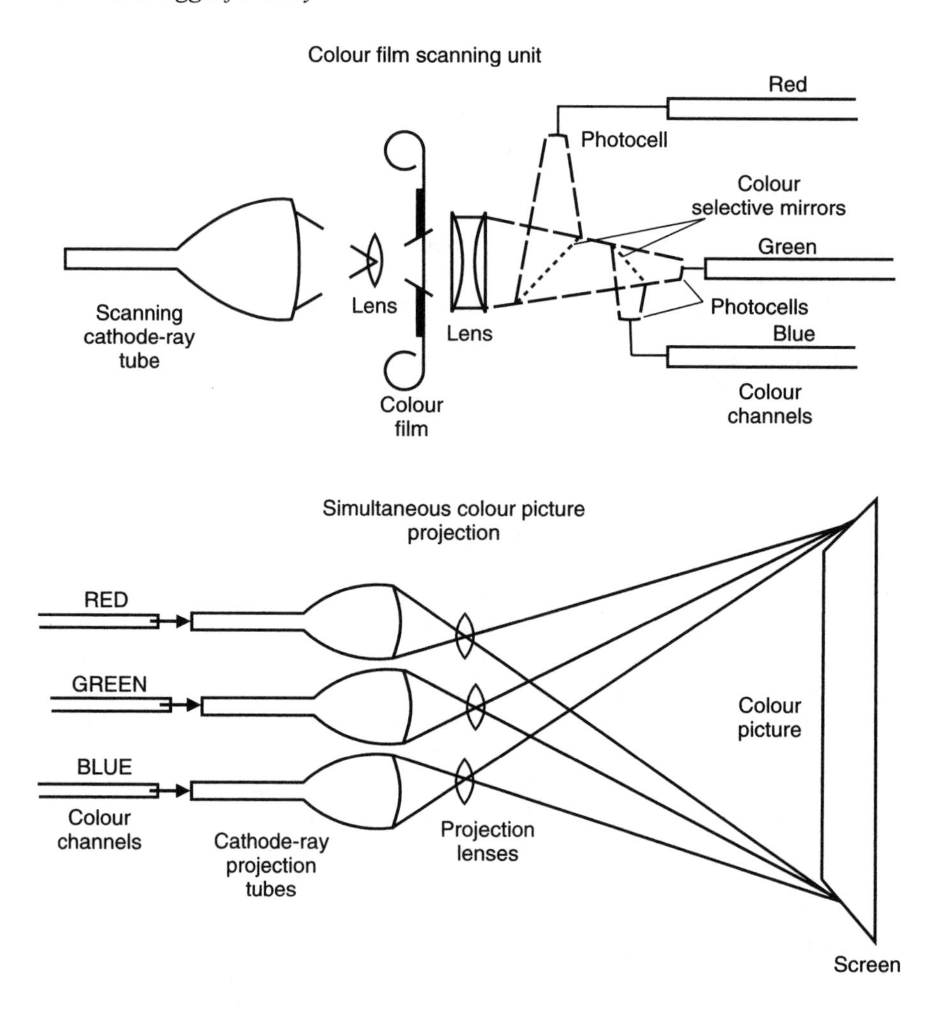

Figure 6.4 Schematic diagram of RCA's colour television system (1946) [Source: Journal of the Television Society, *1947, vol. 5, part 1, p. 32]*

reproduction from motion picture films (within three months); televised live action studio scenes (by the middle of 1947); and televised large-screen theatre-size pictures (in 1948). In referring to this plan Sarnoff felt that the difference between mechanical and electronic colour television was the difference between there being no colour television service and actual public colour television.

Presumably RCA's demonstration of all-electronic colour television was hastened by CBS's petition, on 27 September 1946, to the FCC for the commercialization of 16 MHz, UHF colour television based on 525 lines per picture, 48 frames per second, 144 fields per second. These standards were with the exception of the field rate in accord with the majority view of the industry concerning resolution. The field rate limited the brightness of the image to *c*. 25 ft lamberts which was appreciably less than the 100 ft lamberts of black-and-white television receivers.

CBS claimed it had:

> … developed, constructed, extensively operated, and successfully and repeatedly demonstrated a complete [UHF] color television system consisting of the following: (1) live and film color pick-up equipment, including control, monitoring, color mixing, and other accessories; (2) studio lighting for color pick-up; (3) [UHF] transmitting transmitter (for both sight and sound); (4) [UHF] transmitting antennas; (5) [UHF] receiving antennas; [and] (6) [UHF] color receivers.

Moreover, CBS had 'tested coaxial cable color transmission, [both] intra-city and between New York and Washington'; 'it [had] conducted field intensity surveys in the 700 mc and 490 megacycle regions'; and had 'conducted approximately 200 demonstrations of actual broadcasts of ultra-high frequency color television attended by a total of more than 2,700 representatives of government, industry, educational and cultural institutions, press, and public' [39].

CBS's petition requested the FCC: (a) to authorize the operation of commercial colour television stations in the 480–920 MHz band; (b) to modify its standards of good television engineering practice as proposed by CBS; and (c) to hold a public hearing on these issues. The FCC agreed and the hearing was scheduled for the 3 December 1946. It lasted until 13 February 1947.

In preparation for the hearing, the comparative advantages and disadvantages of the two systems of colour television were studied by experts of the Radio Technical Planning Board and the Radio Manufacturers Association [40]. Since the merits of the sequential and simultaneous methods were a feature of the FCC's hearings for the next six years it is necessary to deal with these methods in some detail.

Prima facie, though the two schemes seem entirely different, they do have several performance features in common, for example: the same fidelity of colour transmission for the same colour filters in the two cases; approximately the same bandwidth; and similar geographical coverage when operated in the same region of the electromagnetic spectrum (because, although the separate transmitters of the simultaneous method can radiate a higher power over the narrower band, three sets of valves are required, which, if combined in a single transmitter, would provide nearly the same power output over the wider bandwidth).

The two systems differ in their flicker, colour fringing and colour break-up characteristics. With the sequential method means must be adopted to prevent the colour of one field being carried over into the next field. Rapid decay phosphors are needed and these lead to a higher apparent flicker, which can only be diminished by an increase of the field rate to a figure appreciably above 100 Hz. (CBS initially used a rate of 120 Hz, but this was increased to 144 Hz.) In the simultaneous method the blending of the colours is inherent in the method, and so a lower field rate is acceptable. Furthermore, if phosphors having a slower decay rate are used (which is permissible) the flicker threshold frequency falls. Indeed, the flicker response of a simultaneous system operating at 60 fields per second may be superior to that of a sequential system working at 180 fields per second. For these conditions the bandwidths of the two systems are the same except that the simultaneous method must incorporate frequency stop bands between the channels to allow the receiver circuits to separate the three channels.

Colour break-up and colour fringing are not possible with the simultaneous system (since all the colours are presented continuously), but with the sequential method the effects can be caused by eye movement (due to a lack of register of the three images). Fortunately, it was found that many observers have a tolerance for colour break-up and colour fringing and these defects were not felt to be important factors in the debate on colour television.

Compatibility of colour standards was one of the major factors that eventually determined the outcome of CBS's system. Given the same scanning standard (525 lines per picture, 60 fields per second) for black-and-white television and colour television, a colour signal of the simultaneous system can be received by a black-and-white receiver. It is only necessary for an owner of such a set to purchase a frequency converter to receive the green image signal (which contains most of the black-and-white information content), to view the colour programmes transmitted, but the colour-sequential method was not compatible with the 1946 FCC black-and-white standard.

Brightness of the displayed images was another major factor that had to be examined. In the sequential method, only one electron beam contributes to the formation of the image on the screen of the cathode-ray tube; in the other method three beams are used. There is thus, apparently, a threefold advantage for the simultaneous method. However, because the colour phosphors of c.r.t.s do not have identical luminosities, a correcting red filter has to be utilized in the simultaneous system. This reduces the overall brightness of the displayed picture to *c*. 50 per cent of that without the filter. In the sequential system the use of a rotating filter effects a very substantial loss of brightness, *c*. 85 to 90 per cent in practice. Of course, if three c.r.t.s were employed in the sequential method – as mentioned above – the picture brightness would be one-third as bright as that of the simultaneous method.

Of the other factors that were thought to influence the selection of appropriate standards, the power mains frequency (60 Hz) and field frequency (120 Hz, or later 180 Hz) of the CBS system could lead to hum effects in sequential receivers unless precautions were taken in their design and construction. On the other hand, the use of three r.f. carriers and the effect of multipath effects on the three colour signals could be more serious in the RCA system than in the single r.f. carrier CBS system. Again, in general, the sequential transmitting and receiving equipment is simpler than that for the three-channel simultaneous equipment. Consideration of these points lead to an expensive receiver compared to a direct-viewed sequential receiver.

When the FCC's hearing commenced on 9 December 1946, the CBS president, Dr Stanton, said that the CBS system was ready for standardization and that colour television could progress further only under conditions of commercial broadcasting. He testified that CBS could build a 'substantial regular color television program schedule within a year'. But if the CBS petition were denied 'we would curtail . . . the work that we have carried on in the development of ultra-high frequency color television' [41]. In support of the CBS position the vice-president, A. Murphy gave nine points why they did not believe simultaneous standards could be satisfactory (see Note 1 at the end of the chapter).

Of course, by this time, RCA had exhibited their all-electronic simultaneous colour system and had no intention of approving a sequentially scanned system based on rotating filters. At the demonstration General Sarnoff had told newsmen that 'any claim color is here today is just pure bunk and nothing else'. Essentially, RCA's task was to delay FCC's approval of new television standards for black-and-white and colour television until RCA had progressed its system. Dr C.B. Joliffe, executive vice-president of RCA Laboratories, in his testimony, summarized RCA's position (see Note 2 at the end of the chapter) and recommended that:

1. the FCC deny the petition of the CBS and not authorize operation of commercial television stations in the frequency band 480 to 920 megacycles;
2. standards for color television not be established at this time; and
3. the radio industry, RTPB and RMA be encouraged to continue development and studies of colour television until such time as satisfactory standards, fully proven by adequate field tests, can be determined and recommended to the Commission.

Various witnesses were called upon to support/oppose the two opposing standpoints. Broadly, CBS received backing from Bendix Radio, Cowles Broadcasting, the Federal Telephone and Radio Corporation, Westinghouse Electric and Manufacturing Company, and the Zenith Radio Corporation. The opponents of the petition were [42], apart from RCA, the Allen B. DuMont Laboratories, Farnsworth Radio, the National Broadcasting Company (an RCA subsidiary), Philco, the Radio Technical Planning Board and the Television Broadcasters Association. Dr DuMont listed six reasons against the petition and made four proposals (see Note 3 at the end of the chapter).

On 18 March 1947 the FCC issued its colour television report [43] and denied the CBS petition. The Commission believed that 'there had not been adequate field testing of the CBS system for the Commission to proceed with confidence that the system [would] work adequately in practice'; and it was 'of the opinion that there may be other systems of transmitting color which offer the possibility of cheaper receivers and narrower bandwidths that have not yet been fully explored' [43].

As to compatibility, the [FCC] stated:

> The Commission is of the opinion that compatibility is an element to be considered, but that of greater importance, if a choice must be made, is the development of the best possible system, employing the narrowest possible bandwidth, and which makes possible receivers capable of good performance at a reasonable price.

Colour television was again investigated by the FCC in September 1948 to determine the progress that had been made in the development of black-and-white and colour television for operation in the UHF band. The testimony presented showed that progress was being made but the Joint Technical Advisory Committee, sponsored by the RMA and the IRE, reported that it was impractical at that time to establish commercial standards for colour television.

Meanwhile, the industry had resumed its production of black-and-white television sets. From a production figure of *c.* 5,000 receivers in 1946 the number increased to *c.* 178,500 in 1947. The need for colour systems to operate with the 'narrowest possible bandwidth' became obvious a year later, when the demand for channel capacity was greater than the number of channels available in the VHF band. By September 1948, the FCC had issued 107 construction permits or licences to television stations and had received 310 applications for permits. It was manifestly clear that this number could not be accommodated in the VHF band. As a consequence the FCC in 1949 published its 'freeze order' on the issuance of further television station licences pending further studies of the effects of co-channel interference. The need for 12 MHz for colour television was wholly unrealistic given the rapid growth of black-and-white television; advocates of colour television had to base their systems on 6 MHz channels.

In the same year CBS demonstrated its field-sequential system, in a closed-circuit configuration, at a meeting of the American Medical Association held in Atlantic City. The demonstration showed that a bandwidth of 6 MHz was practical provided care was taken to minimize the effects of limited resolution. Later, the CBS engineers used a new technique, known as 'crispening' to improve the horizontal resolution. CBS chose a standard of 405 lines per frame, and 144 fields per second, rather than the 1940 standard of 343 lines per frame, and 120 fields per second, since 144 fields per second had been found to be the minimum rate acceptable for freedom from flicker.

Note 1: Mr A. Murphy, of CBS, on why simultaneous standards cannot be satisfactory

1. No reasonably priced color receivers can be made for operation on simultaneous standards. The cheapest such color receiver would be in the 'Cadillac' class. Even in the luxury class, color receivers operating on simultaneous standards would be more expensive than receivers operating on sequential standards.
2. No satisfactory low cost black and white receiver can be made for operation on simultaneous standards. Receivers reproducing only partial images using the green component of UHF color programs could not provide satisfactory broadcast service. In fact, the picture would be inferior to present black and white picture. The same comment applies to any possible use of a converter utilizing only the green component to permit reception of low frequency black and white and UHF simultaneous transmissions. In addition to this deficiency I know of no converter which has proven practical.
3. Networking in color with simultaneous standards would be more difficult technically and economically than with sequential standards, and might be economically impossible. Accordingly, simultaneous standards would lead directly to a network service on a black and white basis.
4. Simultaneous standards are less flexible inherently than sequential standards in permitting advantage to be taken of improvement in the art.
5. Simultaneous standards, by comparison with sequential standards, would waste frequency space for ... technical reasons ... Such a waste might make a nation-wide competitive system impossible. There is an equally significant practical reason, in addition. To the extent simultaneous standards limit networking to black and white programs, the wide channels intended for color transmissions would be utilized for black and white programs which of themselves would require far less frequency space. The extent of the frequency waste is obvious.
6. There is no assurance whatsoever that simultaneous standards can be developed technically to a point adequate for a broadcast service.
7. Finally, there is no significant improvement in performance advanced in behalf of simultaneous standards which cannot be effected within sequential standards.

Note 2: Dr C.B. Jolliffe on RCA's position

1. We have today a well developed monochrome television service. All of the equipment – transmitting and receiving – has been engineered to the point where it is capable of excellent reproduction in the home of the best in current events, sports, drama and education. We urge the Commission to give full support to this proven service and continue to encourage its fullest use for the benefit of the public.
2. No steps should be taken under the guise of bringing color television to the public which, instead of advancing the art of television, confuse the public, the broadcaster and the equipment manufacturer and result in depriving the public of any television service now and for some time to come.
3. Further developments and improvements in television must and will be made. One of these developments will be a color television system which can become an integral part of television service. RCA has developed the basic elements of an electronic simultaneous color television system which can be introduced, when it is ready in the future, without obsoleting the present excellent electronic monochrome system.
4. Much work remains to be done before a determination can be made as to the proper standards for a system of color television which ultimately should be adopted. To adopt standards and authorize commercialization of any system of color television now will probably result in no television rather than in improved television.

Note 3: Dr A.B. DuMont on why the proposed CBS colour petition should be denied

I consider the proposed CBS system inferior because of the following reasons:

1. It does not provide wide enough coverage.
2. It does not provide for as bright or as large a picture as the present black and white standards and the standard proposed is totally inadequate in this respect.
3. It requires broadcasters to provide duplicate equipment for black and white and color transmissions increasing the cost of capital equipment and increasing the cost of producing programs.
4. It requires more expensive and complicated receivers – if provision is made for black and white and color.
5. It does not provide any practical method of allowing black and white receivers to be converted so they can receive color transmissions.
6. It is wasteful of frequency spectrum – 19 vs. 14.5 megacycles.

References

1. Anon. Report on 'Columbia colour television'. *Electronics and Television and Short Wave World*. October 1940, p. 465
2. Goldmark P.C. 'Maverick inventor: my turbulent years at CBS'. *Saturday Review Press*, New York. 1973
3. Anon. Report on 'Color television achieves realism'. *New York Times*. 5 September 1940, p. 1
4. Anon. 'Description of CBS colour television'. Submission to the NTSC, 17 September 1940
5. Goldmark P.C., Dyer J.N., Piore E.R., Hollywood J.M. 'Colour television – Part I'. *Proceedings of the IRE*. April 1942, pp. 162–82

6. Goldmark P.C., Dyer J.N., Piore E.R., and Hollywood J.M. 'Color television – Part II', *Proceedings of the IRE*. September 1943, **31**, p. 465.
7. Anon. Report, *Electronics and Television and Short Wave World*. November 1949, pp. 499–500
8. Anon. 'Salient facts about CBS's first public demonstration of direct pick-up of colour television'. CBS press release, 9 January 1941.
9. Federal Communications Commission. 'Regulation of television broadcasting (with emphasis on colour television) (July 1934 to July 1949)'. Third Hearing in Docket 5806 (on NTSC Status Report); Fourth Hearing in Docket 5806 (Tentative Commercial Rules and Standards) (1941); Commercial Standards adopted, April 30, 1941; Hearing in Docket 6651: 1944–45; Promulgation of Television Rules and Standards (Docket 6780); Hearing on CBS Petition: 1946–47 (Docket 7896)
10. *Ibid.* paras. 52–3
11. *Ibid.* para. 54
12. *Ibid.* para. 55
13. *Ibid.* paras. 57–8
14. Anon. Report on 'Television system investigated experimentally by RCA'. Submitted to Panel 1 of NTSC, 12 September 1940.
15. Federal Communications Commission. 'Regulation of television broadcasting (with emphasis on colour television) (July 1934 to July 1949)'. Third Hearing in Docket 5806 (on NTSC Status Report); Fourth Hearing in Docket 5806 (Tentative Commercial Rules and Standards) (1941); Commercial Standards adopted, April 30, 1941; Hearing in Docket 6651: 1944–45; Promulgation of Television Rules and Standards (Docket 6780); Hearing on CBS Petition: 1946–47 (Docket 7896) para. 58(b)
16. *Ibid.* para. 67
17. Burns R.W. *Television: an international history of the formative years*. London: Institution of Electrical Engineers; 1998. pp. 558–9
18. Anon. 'NTSC proposes television standards'. *Electronics*. February 1941, pp. 17–21, 60, 62, 64, 66
19. Federal Communications Commission. 'Regulation of television broadcasting (with emphasis on colour television) (July 1934 to July 1949)'. Third Hearing in Docket 5806 (on NTSC Status Report); Fourth Hearing in Docket 5806 (Tentative Commercial Rules and Standards) (1941); Commercial Standards adopted, April 30, 1941; Hearing in Docket 6651: 1944–45; Promulgation of Television Rules and Standards (Docket 6780); Hearing on CBS Petition: 1946–47 (Docket 7896) paras. 75–6
20. *Ibid.* paras. 92–3
21. Anon. 'Groundwork laid for commercial television'. *Electronics*. April 1941, pp. 18–19, 70
22. Fink D.G. *Television Standards and Practice*. New York: McGraw-Hill; 1943
23. 'Report of the Television Committee, 1943' (the Hankey Committee). Privy Council, 29 December 1944
24. Burns R.W. *John Logie Baird: television pioneer*. London: Institution of Electrical Engineers; 2000. pp. 404–7

25. Federal Communications Commission. 'Regulation of television broadcasting (with emphasis on colour television) (July 1934 to July 1949)'. Third Hearing in Docket 5806 (on NTSC Status Report); Fourth Hearing in Docket 5806 (Tentative Commercial Rules and Standards) (1941); Commercial Standards adopted, April 30, 1941; Hearing in Docket 6651: 1944–45; Promulgation of Television Rules and Standards (Docket 6780); Hearing on CBS Petition: 1946–47 (Docket 7896) para. 109
26. *Ibid.* para. 112
27. *Ibid.* para. 113
28. *Ibid.* para. 114
29. *Ibid.* para. 116
30. *Ibid.* para. 137
31. Nygreen A.C. 'Report on CBS 490-Mc colour television' *FM and Television.* February 1946, pp. 21–7
32. *Ibid.* p.27
33. Fink D.G. 'Colour television on UHF'. *Electronics.* April 1946, pp. 109–15
34. Anon. Report, 'RCA demonstrates colour television', *Journal of the Television Society*, Part 10. 1946;**4**:250
35. Anon. 'RCA colour television status'. *Electronic Industries.* March 1946, pp. 102, 136–38
36. Anon. 'RCA reveals first electronic colour television'. *Electronic Industries.* 1946; **5**:58–9
37. Kell R.D., Fredenhall G.I., Schroeder A.C., and Webb R.C. 'An experimental colour television system'. *RCA Review.* June 1946, p. 141
38. Federal Communications Commission. 'Regulation of television broadcasting (with emphasis on colour television) (July 1934 to July 1949)'. Third Hearing in Docket 5806 (on NTSC Status Report); Fourth Hearing in Docket 5806 (Tentative Commercial Rules and Standards) (1941); Commercial Standards adopted, April 30, 1941; Hearing in Docket 6651: 1944–45; Promulgation of Television Rules and Standards (Docket 6780); Hearing on CBS Petition: 1946–47 (Docket 7896). Paras. 152–4
39. *Ibid.* para. 154
40. Fink D.G. 'Two systems of colour television' *Electronics.* January 1947, pp. 72–7
41. Federal Communications Commission. 'Regulation of television broadcasting (with emphasis on colour television) (July 1934 to July 1949)'. Third Hearing in Docket 5806 (on NTSC Status Report); Fourth Hearing in Docket 5806 (Tentative Commercial Rules and Standards) (1941); Commercial Standards adopted, April 30, 1941; Hearing in Docket 6651: 1944–45; Promulgation of Television Rules and Standards (Docket 6780); Hearing on CBS Petition: 1946–47 (Docket 7896). Para. 157
42. *Ibid.* paras. 156–82
43. *Ibid.* paras. 183–5

Chapter 7
The 1949–50 FCC colour television hearings

The move for a reconsideration of the colour television issue was made by Senator E.C. Johnson, chairman of the United States Senate Committee on Interstage and Foreign Commerce, when, on 20 May 1949, he sent a letter [1] to Dr E.U. Condon, the director of the national Bureau of Standards. Johnson's opening paragraphs set the scene:

> The question of the present-day commercial use of color television has been a matter of raging controversy within the radio world for many months. There is a woeful lack of authentic and dependable information on this subject.
>
> Hundreds of applicants for television licenses, as well as those now operating television stations, are vitally affected by its settlement. The capital investment involved in the installation of a television station runs into a tremendous sum. The operational costs of such a station are extremely high also. All these expenses must be recovered through advertising. Those who are experienced in advertising believe that if color television were available now, attractive local advertising revenues could be obtained due to the strong consumer demand for it.
>
> The Federal Communications Commission has declined to authorize commercial licensing of color television. It seems reluctant to indicate when and if it will act with respect to authorizing commercial licensing of color. As we understand it, the Commission must first fix minimum standards for color television before licensing can be undertaken, but it refuses to attempt to do so on the premise that color television has not been developed sufficiently for standards to be determined. [1]

Johnson stressed that his objective and that of his committee was to encourage the development of colour television and press for a nationwide competitive television service in the public interest. The Interstage and Foreign Commerce Committee saw television as a great new industry, not only in providing new jobs and a new source of wealth but as the greatest medium of entertainment and diffusion of knowledge yet known to man. Accordingly, the Committee wished to learn whether the time had now approached when minimum standards could be fixed. For this purpose a factual appraisal, by a small group of technical experts unassociated with the radio and television industry, of the several methods of colour television was necessary. Johnson requested Dr Condon to form such a group, and said:

> I am particularly concerned with resolving once and for all the charges that have been made that the advance of color television has been held up by the Commission for reasons difficult for us to understand, and I feel certain that a committee headed by so eminent a scientist as you will help resolve these doubts and questions which have been tossed about. [1]

Condon accepted the request, and subsequently his small committee met to consider the issues of colour television on 3, 17 to 19 August; 7 to 10 October; 21 to 22

November 1949; 18 to 20 January; 1, 20 and 23 February; 11 and 14 March; 26 April; 22 May; and 5, 6 July 1950. During these meetings, demonstrations of colour television were attended by two or more members of the committee as follows: CTI system, 20, 23 February; 14 March 1950: CBS system, 6 to 10 October; 21 to 22 November 1949; 20 January; 1, 23 February; 26 April 1950: RCA system, 6 to 10 October; 21 to 22 November 1949; 19 23 January 1950: Hazeltine demonstration, 2 May 1950. All these meetings and demonstrations gave the committee an unrivalled view of the progress that was being accomplished in the new-media field. The committee approved unanimously its final report at its meeting on 5 and 6 July 1950 [2].

Meanwhile, the FCC, six days after Johnson sent his letter to Condon, announced that, at a hearing to be convened to consider the expansion of the commercial television service, evidence would be taken concerning the possibility of instituting a public television service. It issued its comprehensive 'Notice of Further Proposed Rule Making' on 11 July 1949 [3]. The FCC's objective was an extensive revision of its 'Rules and Regulations and Standards of Good Engineering Practice' as they related to the frequency spacing of broadcasting stations, service areas of stations and allocation principles. In particular, an extensive revision of the allocation table of the 12 VHF channels and 42 UHF channels was proposed. Since the UHF band of 480 MHz to 920 MHz was being used for experimental colour television investigations the FCC invited interested persons to submit comments concerning the utilization of this band [4].

Subsequently papers were filed by the Joint Technical Advisory Committee (JATC); the Radio Manufacturers Association (RMA); the Radio Corporation of America (RCA); the Columbia Broadcasting System, Inc. (CBS); Color Television, Incorporated (CTI); Charles Willard Geer; Leon Rubenstein; Philco Corporation; and Allen B. DuMont Laboratories, Inc. [5].

The hearing on the issues concerning colour television was held before the Federal Communications Commission from 26 September 1949 to 26 May 1950, a total of 62 hearing days, during which evidence and testimony covering 9,717 pages of transcript were taken. Every aspect of the subject was scrutinised, witnesses were examined and demonstrations of the systems being advanced by the CBS, RCA, and CTI were held. In all, 53 persons testified and 265 exhibits were offered [6]. There were a number of important matters for the FCC to consider: whether the systems offered by CBS, CTI and RCA had appropriate standards of performance for a broadcast television service; whether compatibility between monochrome and colour sets should be an issue in its decision-making; whether the ban on commercialization should be lifted; and whether new standards relating to colour television should be formulated.

One of the witnesses examined was Dr D.G. Fink, chairman of the Joint Technical Advisory Committee. He opined, 'I see no way you [the FCC] can proceed, except to invite proponents of all systems to get together for demonstrations . . . as soon as it can be done, and to make comparisons. . . . We certainly agree that it is improper to adopt a system until you know you can render the service to the public, because it visits trouble on everybody.' The FCC heeded this advice and from 6 October 1949 to 17 May 1950 it conducted 12 practical examinations (CBS, 5; CTI, 3; and RCA, 4) of the proponents' apparatuses [7].

Of the three systems, CBS's was similar to that which it had exhibited previously to the FCC (see Chapter 6) with the exception that now CBS was using 405 lines per picture, 144 frames per second interlaced to give 24 pictures per second, rather than its former (black-and-white) standard of 525 lines per picture, 60 frames per second interlaced to produce 30 pictures per second. In addition CBS showed how (1) by employing an extra synchronizing pulse, the problem of ensuring correct colour phasing between the camera filter disc and the display tube filter disc could be automatically accomplished, and (2) by using a technique (not unique to the CBS system), known as 'crispening' the vertical edges of objects could appear more sharply defined [8].

Figure 7.1a illustrates the manner in which the colour television image is scanned. Since six fields are necessary to complete the whole scanning sequence, the picture repetition rate is 144/6, i.e. 24 per second.

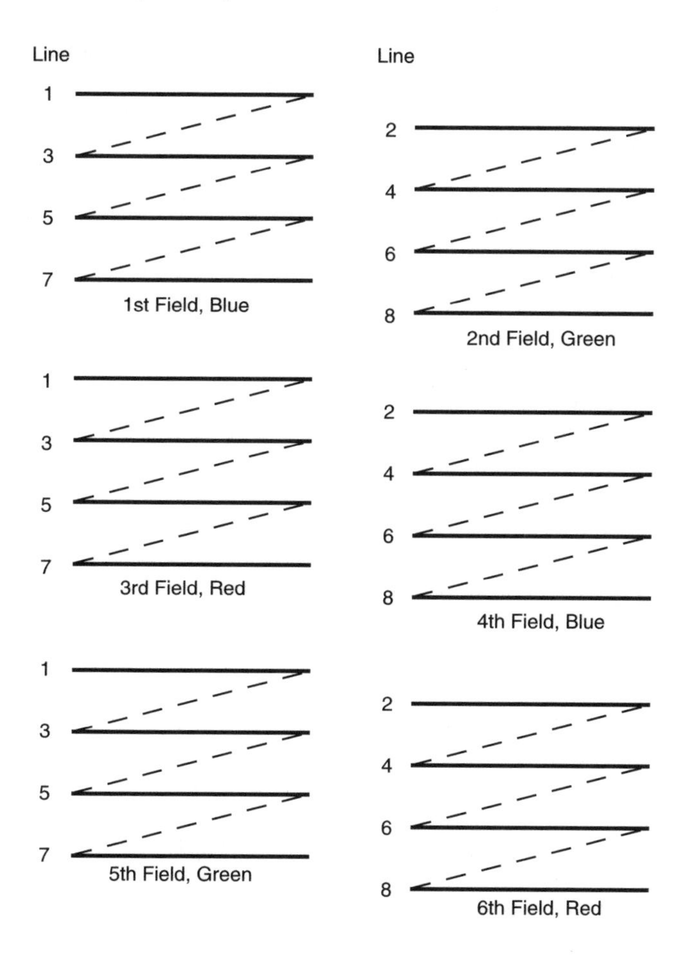

Figure 7.1a Scanning patterns for CBS's field-sequential colour system line-interlaced [Source: Condon Committee Report]

In a field-sequential system flicker is usually more pronounced than in a line-sequential or dot-sequential colour system because the eye is more sensitive to large-area flicker than to small-area flicker. To ameliorate the effect it is necessary to increase the field scanning rate. Observation has shown that, for equal flicker-brightness performance under all conditions, the field scanning rate of a colour television system should be approximately three times that of a black-and-white television system. CBS chose a figure of 144/60 = 2.4 rather than 3 to maintain as much geometric resolution as possible within the prescribed 6 MHz channel width. However, the comparable flicker rates for the CBS system and the black-and-white system are 48 Hz and 60 Hz, i.e. twice the complete colour picture and complete black-and-white picture rates respectively. From the Ferry-Porter flicker law, the difference of 12 Hz allows the black-and-white image to be *c.* 9 times as bright as the colour image for equal visibility of flicker.

Continuity of motion, like flicker, is affected in the CBS system by the composition of the colours transmitted. If the object in motion is predominantly of one primary colour, then the received image of the object is displayed for just one-third of the time possible for a multicoloured object, and the motion may appear jerky. When two or three colour components are present the synthesis of the received image is more continuous and the discontinuity is not so obvious.

In the dot-interlaced version (Figure 7.1b), a given dot in the image is scanned in each of the primary colours only after the completion of 12 scanning fields; hence the picture rate is 144/12, i.e. 12 per second. To achieve this type of scanning, the electrical output of the camera was switched on and off at a frequency of *c.* 9 MHz. Thus the camera was effectively connected to the video stages of the transmitter during the scanning of a particular dot, disconnected during the scanning of the adjacent blank space, reconnected for the scanning of the next dot, and so on. Apart from this switching the elements of the dot-interlaced system were the same as those for the field-interlaced system.

CTI's television scheme appears to have caused the FCC some interpretation problem. 'It is difficult', noted the FCC in its report, 'to make an adequate description of the CTI system because it was frequently changed during the course of the hearing, technical witnesses for CTI were not in complete agreement, and some of the more complicated points were never clearly expounded by CTI.' [9]

Color Television, Inc., of San Francisco, was formed on 31 May 1946 for the sole purpose of developing colour television. The vice-president and chief engineer was G.E. Sleeper and it was his patents that led to CTI's initiative [10]. His additive, tri-colour method employed a 525 lines per picture, 30 picture per second with a 2:1 interlace. The diagrams in Figures 7.2(a) and 7.2(b) illustrate schematically the camera and display apparatuses for film working. Briefly, the transmitter used a single camera tube of the orthicon type and an optical system – comprising three lenses and three filters – which focused upon the photoelectric target plate of the orthicon three separated primary-colour images of the scene/slide/film being televised. At the receiver three separated primary-colour images were displayed on the receiver's cathode-ray tube – the screen of which had three bands of red, green and blue phosphors – in

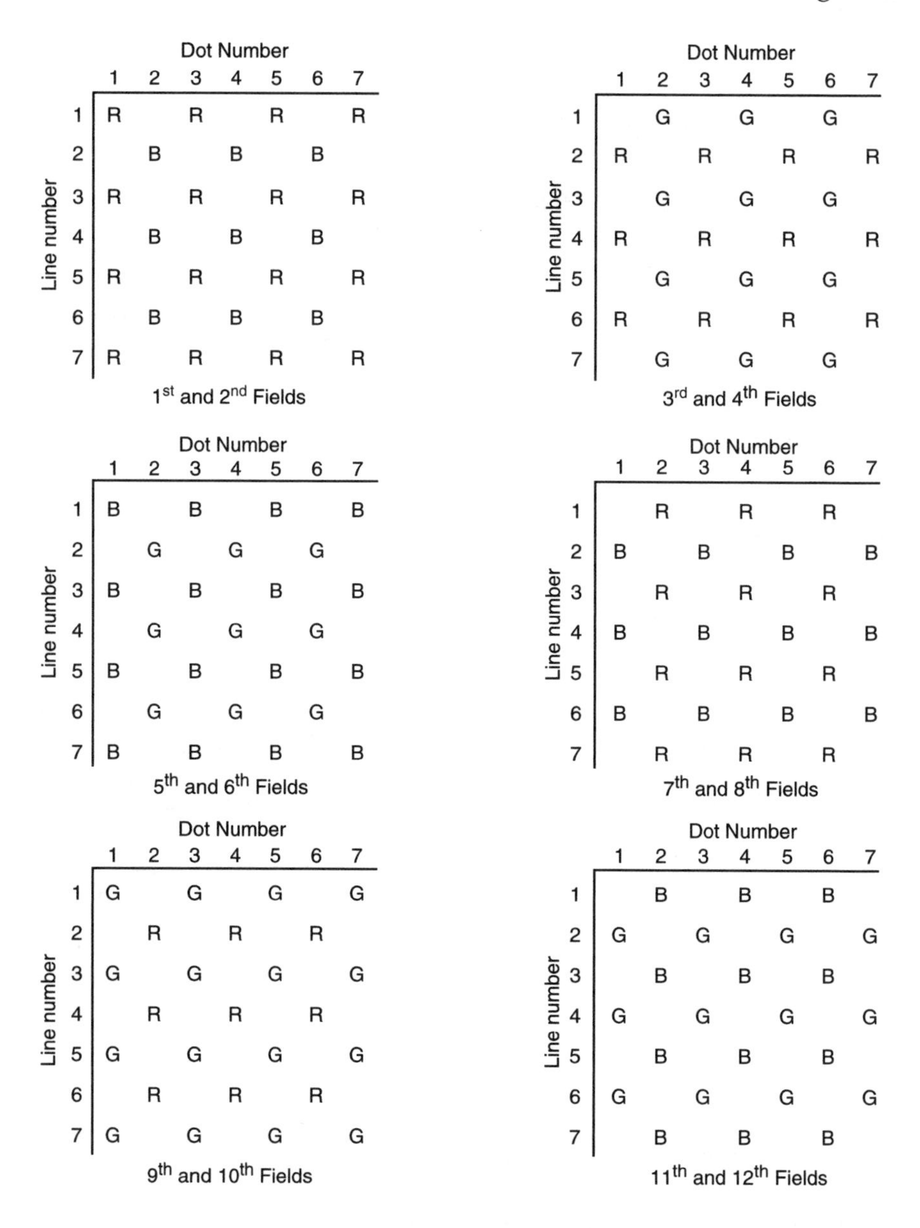

Figure 7.1b Scanning patterns for CBS's field-sequential colour system dot-interlaced [Source: ibid.*]*

positions corresponding to those on the target plate of the orthicon. These were combined in register by an optical system, similar to that of the camera, to give a picture in colour.

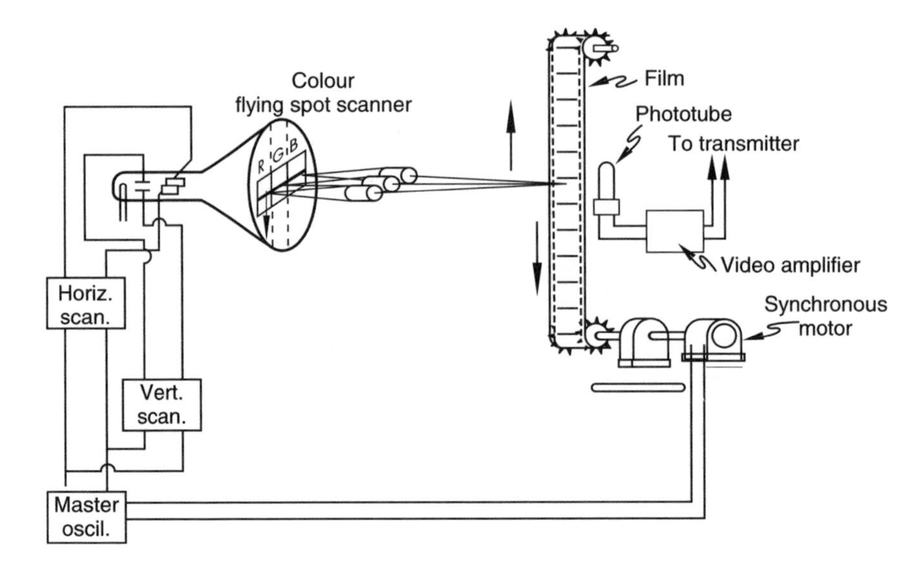

Figure 7.2a CTI 30-frame continuous film pick-up or recorder [Source: FCC Docket No. 7844–8704]

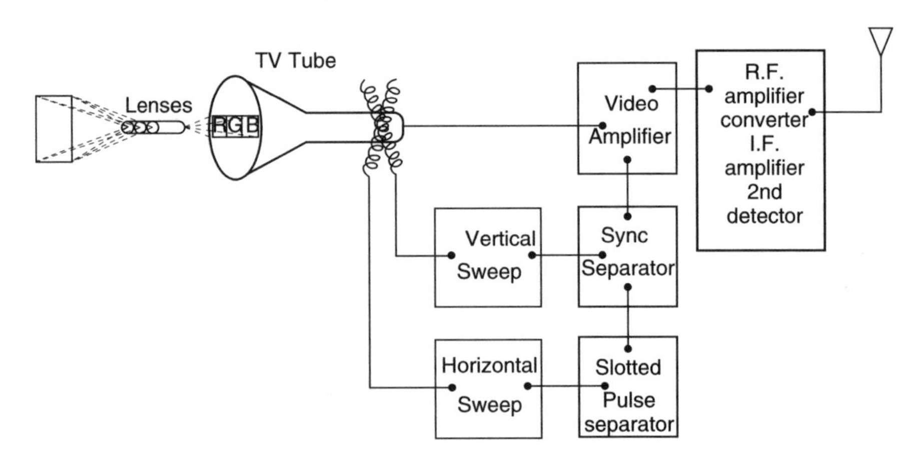

Figure 7.2b Schematic diagram of CTI's colour receiver [Source: ibid.*]*

Though the configuration of CTI's transmitter and receiver seems simple, the arrangement lent itself to problems of image registration. Mis-registration could manifest itself (1) from misalignment of the lenses of the camera optical system, which would produce colour images of different size, shape or orientation on the camera tube's mosaic plate; (2) from non-uniform motion, or non-alignment, of the camera tube's electron beam; (3) from non-congruency of the transmitted and received images; and (4) from poor superposition of the three received images by the receiver's projection lens system.

At the camera, the scanning operation commences at one edge of the analysed representation of the picture, the scanning proceeding from the red image to the green image to the blue image. Thus, line 1 is scanned in red, line 3 in green, line 5 in blue, line 7 in red, and so on until all the odd lines of a given field had been scanned. The electron beam then rapidly returns to the beginning of the next field to scan the even lines in a similar way, so that at the end of two fields all the lines have been scanned once. Since each line will have been scanned in just one colour (e.g. line 1 in red), in order for each line, say line 1, to be scanned in each colour (red, green and blue) it is necessary to repeat the above two-field scanning process (for fields 3 and 4) so that line 1 is now scanned in green, etc., and again to repeat for the next two fields (5 and 6) the scanning process so that line 1 is scanned in blue, etc. This procedure is shown in Figure 7.3(a).

These colour shifts from one line to the next were accomplished by counting circuits and synchronizing signals. By changing the shifts different scanning patterns could be achieved, e.g. patterns (a) and (c) in Figure 7.3. It is apparent that pattern (a) results in 'line crawl', which would be visually objectionable. To counter this CTI demonstrated several sequences of non-uniform colour shifts. One scanning sequence, known as the single shift, gave a pattern in which each picture line was scanned in only two of the three primary colours (see Figure 7.3(c)). Another sequence, which was preferred by CTI and demonstrated by them, was the double shift. In this, the order of scanning is changed from the normal pattern, where the odd lines are first scanned and then the even lines, so that the odd lines are scanned in three successive fields, followed by the even lines in three successive fields.

The picture repetition rate of the CTI scanning method was 60/6 = 10 per second.

On RCA's system, the FCC again found it 'difficult to make an accurate description ... because it involve[d] new and complex techniques, many of which were never clearly expounded during the hearing'. [11] The system had its origin in the simultaneous method first disclosed by RCA on 30 October 1946 and subsequently described at the FCC Hearing of 1946 (see Chapter 8).

An important feature of the new system was the means by which the signals of the simultaneous system were compressed into a 4 MHz band, suitable for use in a frequency channel of 6 MHz bandwidth, without any loss of image detail. This method permitted a high-definition colour picture to be synthesized at a colour receiver, while at the same time the transmitted signal was compatible with the current black-and-white television system (i.e. 525 lines per picture, 60 fields per second).

The compression of the simultaneous system was enabled by a combination of two processes; namely, use of the mixed-highs principle [12], and colour picture sampling and time-division multiplexing.

Figure 7.4 illustrates diagrammatically the sending-end equipment. It comprised three separate photosensitive camera tubes and an optical unit consisting of one lens and three dichroic mirrors, the individual mirrors of which reflected only one of the three primary colours. By configuring these elements as shown, each photosensitive tube received light of a specific colour, namely, red or green or blue. Since the three camera tubes were scanned from common line and field time-base generators, the images formed on the photosensitive surfaces of the three tubes were scanned

(a)

LINE FIELD →

	I	II	III	IV	V	VI	VII
1	RED		BLUE		GREEN		RED
2		RED		BLUE		GREEN	
3	GREEN		RED		BLUE		GREEN
4		GREEN		RED		BLUE	
5	BLUE		GREEN		RED		BLUE
6		BLUE		GREEN		RED	
7	RED		BLUE		GREEN		RED
8		RED		BLUE		GREEN	

(b)

LINE FIELD →

	I	II	III	IV	V	VI	VII
1	RED		GREEN		BLUE		RED
2		GREEN		RED		BLUE	
3	GREEN		BLUE		RED		GREEN
4		BLUE		GREEN		RED	
5	BLUE		RED		GREEN		BLUE
6		RED		BLUE		GREEN	
7	RED		GREEN		BLUE		RED
8		GREEN		RED		BLUE	

(c)

LINE FIELD →

	I	II	III	IV	V	VI	VII
1	RED		GREEN		RED		GREEN
2		BLUE		RED		BLUE	
3	GREEN		BLUE		GREEN		BLUE
4		RED		GREEN		RED	
5	BLUE		RED		BLUE		RED
6		GREEN		BLUE		GREEN	
7	RED		GREEN		RED		GREEN
8		BLUE		RED		BLUE	

Figure 7.3 Scanning sequence patterns, showing pattern with 'line crawl' (a), and desirable patterns (b) and (c) [Source: ibid.*]*

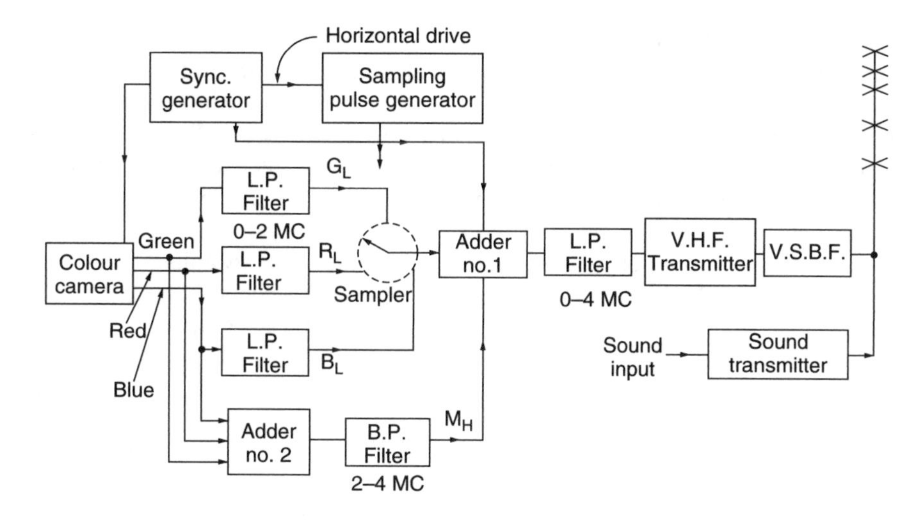

Figure 7.4 *Block diagram of RCA's colour television transmitter [Source: RCA paper on 'A six-megacycle compatible high-definition color television system, undated (but c. September 1946, unsigned)]*

simultaneously, thereby producing signals corresponding to the red, green, and blue contents of the scene being televised. All three camera tubes had frequency responses extending to 4 MHz.

From the camera the signals were sent to low pass filters, which eliminated the frequency components above 2 MHz, and then to an electronic switch operated at 11 MHz. Every 3.6 μs the switch sampled the appropriate input signal and so converted the three continuous colour input signals into a sequence of red, green and blue signal pulses. In addition, the three input signals were summed (adder no. 2) and the band of frequencies from 2 MHz to 4 MHz selected. The resulting signal was designated the mixed-highs signal and was combined with the output from the sampler. Thus the output from adder no. 1 was a signal that was partly the primary-colour signal (0–2 MHz) and partly the mixed-highs signal (2–4 MHz).

The mixed-highs principle arose from some work undertaken by RCA, which showed that the human eye is not sensitive to fine detail in colour. In a black-and-white television system, coarse detail in an image requires the transmission of signals up to *c.* 2 MHz, but for the display of fine detail additional signal frequencies in the band 2 MHz to 4 MHz must be received. However, because the physiology of the eye is such that it can distinguish colours only in coarse detail, RCA concluded that it would be appropriate to transmit the coarse detail of a picture in colour (i.e. from 0 to 2 MHz) and the fine detail in black-and-white (2 to 4 MHz).

The technique of using mixed highs to transmit fine detail in tones of grey was not applicable to line- or field-sequential television systems because these systems make no colour distinction between the dots along any given line of the image. Whatever detail is present in each line (of a particular colour) of the CTI and CBS scans must necessarily be provided in full.

RCA claimed that the above process led to a saving of bandwidth. Instead of requiring a bandwidth of 12 MHz (i.e. 3 × 4 MHz), the bandwidth of the new system was reduced to 8 MHz (i.e. 3 × 2 MHz, + 2 MHz). Since this bandwidth was still too large to be accommodated within the channel capacity of 6 MHz, RCA introduced a further bandwidth saving process known as dot interlacing. As its name implies dot interlacing is achieved by scanning each line in a series of dots rather than continuously [13], as illustrated in Figure 7.5.

Each line of a given field consists of dots in the three primary colours, arranged, in the sequence red, blue, green, so that the space between two dots of the same colour is equal to the width of a dot, i.e. the dots partially overlap each other.

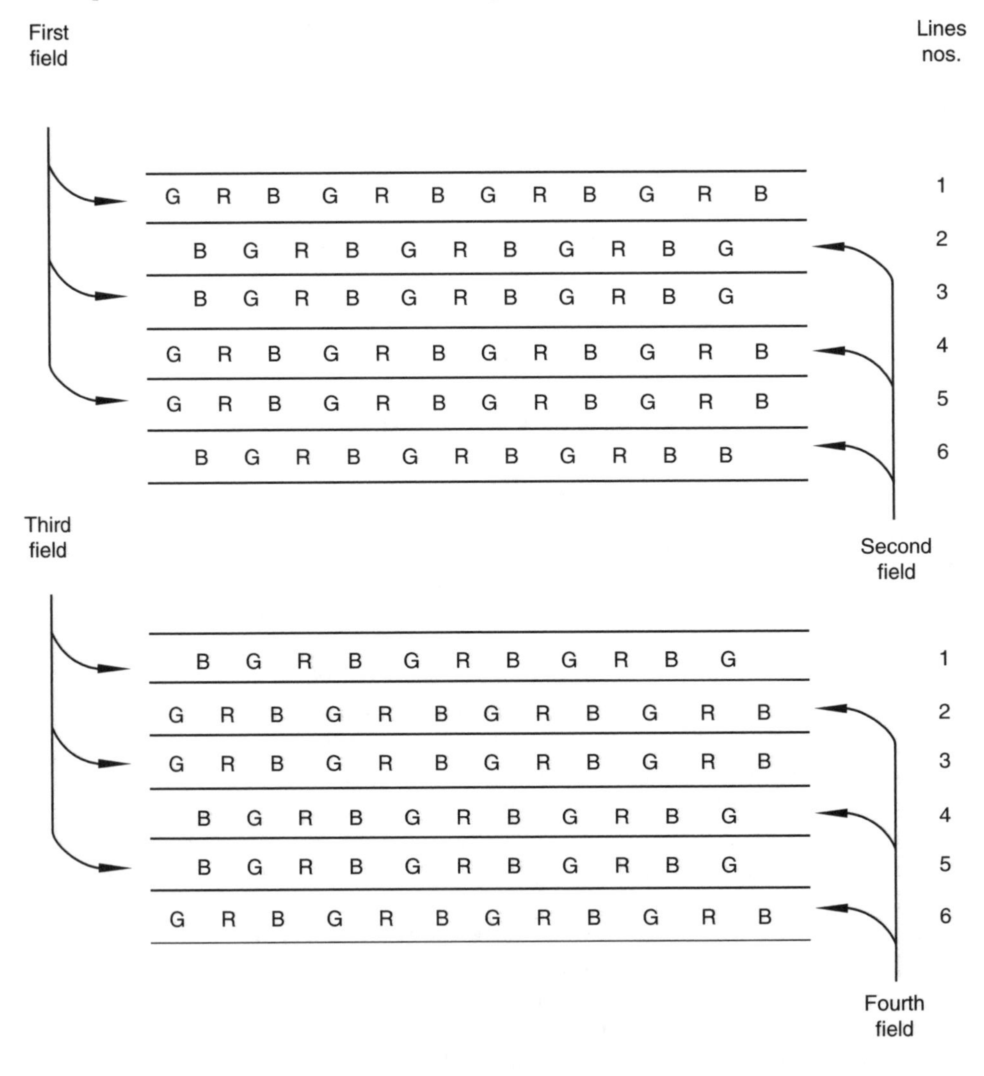

Figure 7.5 Scanning and interlace pattern of RCA's system [Source: ibid.*]*

On successive field scans of the same line the dot line pattern is changed such that, for example, a dot of a given colour (say blue) of line 1 of field 3 falls midway between the positions of the two dots of the other colours (green and red) of line 1 of field 1. Consequently at the end of four fields – two fields of odd lines and two fields of even lines – every primary colour dot will have been scanned. The colour picture rate is thus $60/4 = 15$ per second.

In the black and white 525-line system, *c.* 490 lines are actually visible on the screen, and each line has the equivalent of *c.* 420 dots along its length, giving a picture of *c.* 200,000 halftone dots. Each field comprises *c.* 100,000 dots. With the RCA colour dot interlace technique, all the dots of a given colour are defined in four consecutive fields (1/15 of a second), but because the image is blanked off for part of each field, this time is reduced to 1/20 of a second. Consequently 2 million green (say) dots are scanned per second, or 100,000 green dots during the complete duration of the colour picture period. Hence, nominally, the resolution in each colour is one half of that of a black-and-white picture.

Actually, the resolution is less than this since one colour dot overlaps an adjacent dot of another colour by *c.* 50 per cent. The result is a dilution of each colour by the other two colours and a combination of these colours to produce shades of grey. The overall effect is that fine details are reproduced in shades of grey rather than in their natural colours.

Figure 7.6 shows how the received signal, after standard signal processing, is sampled and filtered to recover the red, green and blue video signal components. The display unit consisted of three kinescopes, an optical unit and dichroic mirrors. Both direct-view receivers and projection receivers utilizing two- or three-colour tubes and dichroic mirrors were exhibited by RCA. Neither the two-tube dichroic receiver, for two-colour television, nor the projection receiver was demonstrated after the original demonstration in October 1949, and after this date, and until 6 April

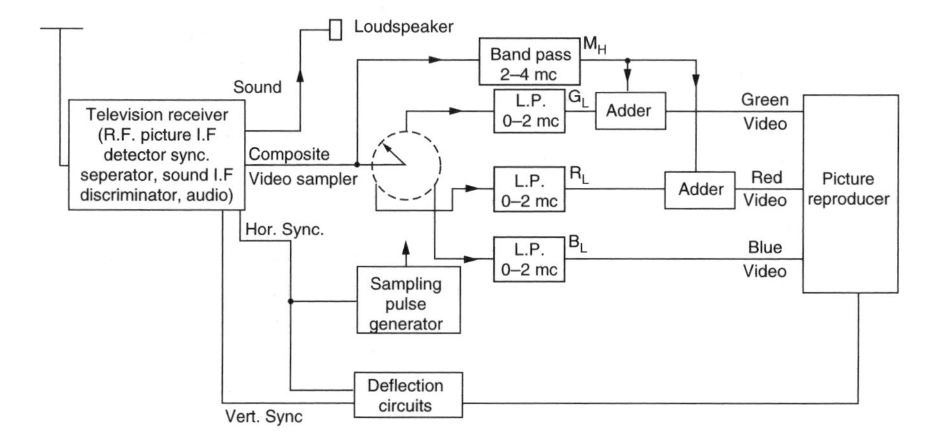

Figure 7.6 Block diagram of RCA's color television receiver using bypassed highs [Source: ibid.*]*

1950, all the receivers exhibited by RCA used three cathode-ray tubes. At the 6 April demonstration two models of a direct-view colour tube (known as a shadow mask tube) were displayed – one with three electron guns [14] and the other with a single electron gun [15]. (RCA's post-1950 superiority in the field of colour television was due to its development of the shadow mask colour television display tube. This was one of the truly outstanding inventions of the latter half of the twentieth century and is considered in Chapter 14.)

Necessarily, the FCC was greatly influenced by the demonstrations mounted by CTI, CBS and RCA. The Commission had to undertake a comparative evaluation of the various systems based on the factors that would determine the attractiveness of the systems to potential buyers when commercial colour television broadcasting commenced. Nine headings were selected, namely: large-area flicker, small-area flicker and continuity of motion; brightness-contrast; registration, colour break-up, and colour fringing; colour fidelity; resolution; picture texture (structural); susceptibility to interference; adaptability and convertibility; and equipment considerations [16]. These headings were similar to those adopted by the Condon Committee, namely: resolution; flicker-brightness; continuity of motion; effectiveness of channel utilization; colour fidelity; defects associated with superposition of primary colour images (improper registration, colour break-up, and colour fringing); and cost of colour receivers.

Because the FCC's conclusions caused some unease, particularly with RCA, the FCC's quoted comments on these factors are given below against the name of the demonstrating company. The Condon Committee's conclusions, when these are at variance with those of the FCC, are shown in Table 7.1.

7.1 Flicker, motion continuity, and allied effects [17]

In colour television display tubes, the screen of the tube is dark during the vertical flyback period, i.e. the period between two contiguous scanned frames. Since the appearance of motion is achieved by the changing succession of images on the screen, if the rate of change is too slow noticeable large area flicker occurs which is irritating to the viewer. This annoyance increases as the picture brightness increases. For a field-sequential system, if one colour, say green, predominates in a scene, the flicker frequency is effectively just one third of the field rate and the flicker becomes especially noticeable.

Small area flicker, such as interline flicker and interdot flicker, arises when adjacent colour lines or adjacent colour dots are scanned in different colour scanning fields. The effects are not manifested in motion pictures because an entire frame image is projected as a whole.

During the display of a television picture, continuity of motion is perceived by means of persistence of vision. Consequently, the repetition frequency must be sufficiently high to ensure smoothness of motion, absence of jerkiness, and non-blurring of moving objects.

Table 7.1 Summary of performance characteristics (Condon Committee) [30] System

Performance characteristic	Standard black and white	CTI colour	CBS colour, line-interlaced	CBS colour, dot-interlaced	RCA colour	Superior
Adaptability	–	Not needed	Adaptable	Adaptable	Not needed	CTI-RCA
Colour fidelity						
Large areas	–	Satisfactory	Excellent	Excellent	Satisfactory	CBS
Small areas and edges of objects	–	Fair	Excellent	Excellent	Fair	CBS
Compatibility						
Quality of image rendered on existing sets	–	Fair	Not compatible	Not compatible	Excellent	RCA
Continuity of motion						
Large objects	Excellent	Good	Good	Good	Good	
Small objects	Good	Fair	Good	Fair	Good	CBS (line)-RCA
Convertibility	–	Not easily[1]	Convertible[2]	Convertible[2]	Not easily[1]	CBS
Effectiveness of channel utilization	Good	Good	Satisfactory	Good	Excellent	RCA
Flicker-brightness relationship						
Large areas	Excellent	Excellent	Good	Good	Excellent	CTI-RCA
Small areas	Good	Fair	Good	Satisfactory	Good	CBS (line)-RCA
Interdot flicker	Absent	Absent	Absent	Fair	Fair	CTI CBS (line)
Interline flicker	Good	Poor	Good	Good	Good	CBS-RCA
Geometric resolution						
Number[3]	200,000	200,000	83,000	166,000	200,000	CIT-RCA
Vertical resolution	490 lines	490 lines	378 lines	378 lines	490 lines	RCA
Horizontal resolution	320 lines	320 lines	185 lines	370 lines	320 lines	CBS (dot)
Superimposition performance						
Registration	–	Fair	Excellent	Excellent	Fair	CBS
Colour break-up	–	Excellent	Satisfactory	Satisfactory	Excellent	CTI-RCA
Colour fringing	–	Excellent	Satisfactory	Satisfactory	Excellent	CTI-RCA

1 = not easily convertible at present; 2 = convertible, 12.5-inch tube diameter maximum; 3 = number of picture elements per colour picture

(a) The CTI system (frame rate 10 Hz)

'...no flicker was observed at any of the demonstrations ... but ... the pictures had very low illumination.' 'In all the demonstrations there was a most noticeable line crawl or jitter, which seriously marred the quality of the transmitted picture.... This line crawl or jitter was noticeable even with the very low level of picture illumination that CTI was able to produce on its projection receiver.' 'No problem was evident at the demonstrations as to continuity of motion ...'

(b) The CBS system (frame rate 24 Hz)

'...flicker is not objectionable on the CBS disc type receiver up to highlight brightnesses of from 20 to 30 foot lamberts.' 'No problem of small area flicker was observed at any of the demonstrations ...' 'As to continuity of motion, no problem was experienced at any of the demonstrations ...'

(c) The RCA system (frame rate 15 Hz)

'...no flicker [was] observed at any of the demonstrations ... but ... the pictures had very low illumination.' '... small area flicker in the form of dot motion or twinkle has been observed. How serious a problem this is cannot be entirely ascertained ... since the RCA demonstration produced only a dim picture.' 'As to continuity of motion, no problem was observed ...'

7.2 Brightness-contrast [18]

Under normal home viewing the brightness of the screen image must be adequate to give a sufficient contrast range (i.e. the range between the brightest and darkest portions of a picture, as determined by the electron scanning beam, and the amount of light which the surface of the tube reflects respectively). The FCC found from its observed demonstrations that a contrast ratio of 30:1 produced a satisfactory colour picture.

(a) The CTI system

'At none of the demonstrations did any of the CTI receivers possess sufficient brightness. The highest measured brightness for a CTI receiver was 4 foot lamberts.... The CTI pictures were so dim that all CTI demonstrations had to be conducted in a room that was virtually dark.'

(b) The CBS system

'[The FCC is] of the opinion that the color picture so produced is bright enough [up to a maximum of 22 foot lamberts] and has sufficient contrast range to be entirely adequate for use in the home under normal viewing conditions. Indeed, the CBS demonstrations were the only ones that were conducted in a lighted room and during one of the demonstrations a light from a 300[W] lamp was thrown directly on the face of the receiver without seriously affecting the quality of the picture.'

(c) The RCA system

> 'At none of the demonstrations ... did any of the RCA color receivers produce sufficient illumination for ordinary home use. Indeed, all the RCA demonstrations ... were conducted with virtually no illumination.'

7.3 Superposition of colour images [19]

Since a colour television picture (in 1950) was reproduced from several colour fields, problems of registration of the individual colour fields, colour break-up, and colour fringing could arise.

(a) The CTI system

> 'There is a severe registration problem at the camera and receiver.' 'The difficulty of securing accurate registration is illustrated by the fact that there was evidence of faulty registration at every one of the CTI demonstrations ...' 'Color break-up has not been observed ..., it does not appear to be a problem in a line sequential system. Color fringing has likewise not been observed so far as moving objects are concerned but line interlace causes color fringing on nearly [all] horizontal lines of other than primary colors.'

(b) The CBS system

> 'The CBS system is not troubled by other than minor registration problems either at the camera or receiver. These minor problems arise from power supply hum, stray fields, and vibration but they are easily cured.' 'At none of the demonstrations ... did the Commission observe any evidence of mis-registration.' 'Color break-up and color fringing were observed at the demonstrations on the disc receivers.'

(c) The RCA system

> 'Registration, both electrical and optical, is a severe problem both at the camera and receiver.... at all of the demonstrations ... there was evidence of faulty registration.' 'There should be no problem of color break-up or color fringing due to motion in the RCA system. However, faulty registration of pictures observed at the demonstrations produces an effect similar to color fringing.'

7.4 Colour fidelity [20]

(a) The CTI system

> 'At all of the demonstrations ..., CTI had difficulty with color fidelity because of faulty registration. There was also difficulty in accurately reproducing skin

tones. Moreover, the low levels of illumination at which the pictures were shown rendered difficult a judgment as to the quality of the color fidelity.'

(b) The CBS system

'The color fidelity of the CBS system as demonstrated on the disc receiver has been of a uniformly high quality.'

(c) The RCA system

'At all the demonstrations ... RCA had difficulty producing a color picture with adequate color fidelity. The difficulty undoubtedly arises from several factors which are part of the RCA dot sequential system.' (1) '...registration is most difficult to maintain and when mis-registration occurs there is color contamination and a loss in resolution'. (2) '...Color control is exceedingly difficult to maintain. A time error of only 1/11,000,000 of a second in the sampler, results in the wrong color being transmitted or received'. (3) '...the utilization of mixed highs, cross talk, and the fact that the colored dots are larger than a single picture element appear to prevent the production of color detail in small areas over the entire picture'. (4) '...the fact that the three primary color images are scanned on three separate surfaces at the camera, makes it exceedingly difficult to obtain uniform sensitivity for all colors over the whole picture area'.

7.5 Resolution [21]

The geometric resolution – the number of lines which the system can provide – is determined from readings on a test pattern. If a given solid angle of a scene viewed by a television camera is appropriate for a given resolution, then a decrease in the resolution (e.g. from 525 lines per frame to 405 lines per frame) necessitates the solid angle to be reduced in the same proportion to produce the same degree of satisfaction. The reduction may be implemented by the use of a camera lens of longer focal length.

(a) The CTI system

'... at none of the demonstrations ... did CTI produce a picture which could compare in resolution with black and white television.'

(b) The CBS system

'The CBS system produces 405 lines per picture as compared with 525 lines per picture for the present system.... There is a reduction in vertical resolution by 23% and horizontal resolution by 46% as compared with the present system'.

(c) The RCA system

'Although the RCA system produces lines at the same rate as the black and white system, its resolution even in theory is not equal to that of the present system for all types of scenes . . . the horizontal resolution ranges between 67% and 100% of the present system.'

7.6 Picture texture (structural) [22]

Under this heading the FCC evaluated the overall quality of the picture.

(a) The CTI system

The FCC found it difficult to make a final judgement since 'virtually all the pictures were shown on a projection receiver at low levels of illumination. . . . However, . . . the line structure is prominently apparent in areas of red and green primary colors and line crawl is visible over the entire picture.'

(b) The CBS system

'At all the demonstrations . . . the CBS picture compared favorably with the present system so far as contrast, sharpness of picture, and freedom from line structure are concerned.'

(c) The RCA system

'The RCA picture has a "soft" quality . . . This is probably due to difficulty in maintaining contrast, particularly in small areas. Picture texture was also marred . . . by the visibility of dot structure at distances at which lines begin to be unresolved.'

7.7 Susceptibility to interference [23]

Sources of interference, including automobile ignition systems, diathermy apparatus, and radiations from broadcast receivers and stations, can impair the quality of a televised image. The FCC concluded that the CTI, CBS, and RCA systems had about the same susceptibility to interference as the present system.

7.8 Adaptability and convertibility [24]

The FCC used (1) the word *adaptability* to cover the changes that would be required to enable receivers to receive a black-and-white picture from colour transmissions, and (2) the word *convertibility* to cover the changes that would be necessary to permit existing receivers to receive colour transmissions in colour.

(a) The CTI system

'No change whatsoever is required ... However, the picture so received is substantially inferior in quality to a regular black and white picture. This is due to prominent line structure, jitter, and line crawl ...'

(b) The CBS system

'Existing receivers are unable to receive a black and white picture ... without making some changes in the receivers to enable them to handle the different vertical and horizontal scanning rates. ... Existing receivers can be converted to enable them to receive CBS color transmissions in color by the addition of an adaptor ...'

(c) The RCA system

No change whatsoever is required ... However, the picture so received is somewhat inferior to present black and white pictures ...'

7.9 Equipment considerations [25]

(a) The CTI System

'Extensive changes are required in the camera and associated studio equipment. Based upon CTI's estimates the cost of converting a single existing studio camera chain is about $7,000.'

(b) The CBS System

'... CBS submitted evidence that an existing studio camera could be converted for $3,622 [*sic*] and that CBS had bought a new color camera chain from RCA for approximately $27,000 ...'

(c) The RCA System

'Extensive changes are required in the camera and associated equipment. RCA admitted that existing camera equipment cannot be converted. It estimated that the cost for new RCA color studio camera equipment would be $54,440 ...'

7.10 Conclusions (FCC) [26]

In assessing the merits of the three demonstrated systems, to determine whether they could be considered eligible for adoption for commercial purposes, the FCC established the following minimum criteria for acceptability: operation within a 6 MHz channel; capable of synthesizing a colour picture of high quality that was (1) not marred by defects such as mis-registration, line crawl, jitter or unduly prominent dot or other structure, (2) sufficiently bright to permit an adequate contrast range without objectionable flicker, and (3) had a suitable apparent definition. Moreover, the receiving apparatus had to be simple to operate and not to be unduly susceptible to interference; and the transmitting equipment had to be capable of sending colour programmes over intercity relay facilities already in existence (or about to be developed in the foreseeable future), and be such that its functioning would be within the competence of the type of trained personnel employed by a station owner.

One surprising conclusion of the FCC concerned its decision to abandon its compatibility requirement. No satisfactory compatible system had been demonstrated during the Hearings and the FCC felt that to insist on compatibility would put 'too high a price on colour'. In any case, compatibility was a temporary problem which would decrease progressively as time proceeded and new receivers were marketed and sold. Similarly, neither convertibility nor adaptability featured in the FCC's list of criteria.

Judged against the above benchmarks, the FCC's conclusions on the CTI, CBS and RCA systems were as follows.

(a) The CTI system [27]

The CTI system fell short of the FCC's criteria. The quality of the colour picture was not at all satisfactory; there was serious line crawl and the picture texture was unsatisfactory; there was great doubt as to whether CTI even qualified on what it claimed to be one of its most important advantages – compatibility, there was serious degradation in quality of the black and white picture which existing receivers could obtain from CTI colour transmissions; the equipment was unduly complex and at the receiver image registration was so critical that it was entirely unlikely that the average person could successfully operate it; and finally CTI had not offered sufficient evidence relating to susceptibility to interference on which the FCC could form an opinion.

(b) The RCA system [28]

The RCA system too fell short of the FCC's criteria. The colour fidelity was not satisfactory and there seemed to be no reasonable expectation of improvement because of mis-registration, mixed-highs, cross-talk between picture elements, and criticalness of colour control; the picture texture was not satisfactory, the picture had a 'soft' quality and it was difficult to determine how the defects could be eliminated. Additionally, the receiving apparatus was so bulky, so complicated and so difficult to operate that it was inconceivable that the public would purchase the sets in any quantity; the equipment utilized at the sending station was also exceedingly complex, there was no assurance that satisfactory commercial-type equipment could be constructed because at no demonstration could accurate registration be maintained throughout the demonstration, it was unlikely that the trained staff generally available at the average station could operate the equipment; and the system was much more susceptible to certain kinds of interference.

On the recently introduced direct view receivers, the FCC opined:

> As demonstrated, the tube had an inadequate number of dots, produced insufficient illumination, and had a serious moiré pattern in it. There is no assurance that the tube will not be unduly expensive ... Finally, even though the tube is developed, there is no assurance that the receivers will not continue to be unduly complex and difficult to operate. Since a time error of 1/11,000,000 of a second results in color contamination, it is difficult to see how color control can be simplified to a sufficient extent for home use.

(c) The CBS system [29]

In the FCC's view the CBS system produced a colour picture that was most satisfactory from the point of view of texture, colour fidelity and contrast; the receiver and station equipment were simple to operate; the system was subject to colour fringing or colour break-up under certain conditions but this was not felt to be a serious obstacle because many viewers after a while tended not to see these defects; the poorer geometrical resolution (compared to a black and white receiver) was more than outweighed by the addition of colour so far as apparent definition was concerned.

In arriving at a conclusion the FCC had to take into consideration the possibility of developments in tri-colour display tubes, horizontal interlacing, long-persistence phosphors and new colour systems. [31] Either the Commission could delay making a final decision until such time as one or more of these had been satisfactorily tested, or it could recommend that the CBS colour system should now be adopted. However, the number of lines scanned per second in CBS's field-sequential system was 29,160 ($405 \times 144 \times 0.5$), against 15,750 ($525 \times 60 \times 0.5$) for black-and-white television. The two systems were incompatible.

Which choice should be made depended, said the FCC, 'on whether a method exist[ed] for preventing the aggravation of the compatibility situation if a final decision [was] postponed. If there [was] no method to accomplish this, the Commission believe[d] that a final decision should not be delayed and that the CBS colour system should now be adopted'. [32]

The latter alternative would be consonant with the FCC's action in launching black-and-white television in 1941, and in the FCC's opinion the CBS colour system was as fully developed as was the black-and-white system in 1941. Nevertheless, if there was a means whereby the compatibility problem could be averted the FCC would be more confident in postponing a decision so that a definitive resolution could be enacted.

The solution was to have 'bracket standards' [33]. If instead of receivers being manufactured to a given standard, say, a horizontal synchronizing rate of 15,750 Hz and a vertical synchronizing rate of 60 Hz, they were designed to operate within limits of 15,000 to 32,000 Hz and 50 to 150 Hz respectively, then they would be capable of handling the different rates that the FCC might later specify for colour and monochrome television. It was envisaged that selection of one or the other standard would be by a manual or automatic switch.

The FCC took this course of action and issued a Notice of Proposed Rule Making simultaneously with its First Report on Color Television Issues (1 September 1950). Interested persons were given until 29 September 1950 to put forward comments. Then, if the bracket standards were adopted as final, the FCC would be in a position to postpone a decision relating to the colour phase of the 1949–50 proceedings since it would have the time to explore more fully the matters considered above. In this case, the FCC would issue a Second Notice of Proposed Rule Making proposing that colour standards be adopted on the basis of the CBS field-sequential system.

Alternatively, if the bracket standards could not be made final without a hearing the FCC would not 'feel free to postpone a decision, for every day that passes would aggravate the compatibility problem. In that event, a final decision would be issued adopting the CBS color standards'.

Overall, the FCC found that 'color is an important improvement in television broadcasting. It adds both apparent definition and realism to pictures. It opens up whole new fields for effective broadcasting . . .'

References

1. Johnson E.C. Letter to Dr E.U. Condon, 20 May 1949, reproduced as Annex A of the report on 'The present status of colour television'. *Proceedings of the IRE*. September 1950;**38**:980–1002
2. *Ibid*. p. 980
3. Federal Communications Commission. First report of Commission (adopted 1 September 1950), para. 6
4. *Ibid*. paras. 1 to 6
5. *Ibid*. para. 23
6. *Ibid*. para. 24
7. *Ibid*. para. 25
8. Anon. 'Description of the CBS transmitting and receiving facilities', FCC Docket No. 8736, Exhibit 301, 27 February 1950
9. Johnson E.C. Letter to Dr E.U. Condon, 20 May 1949, reproduced as Annex A of the report on 'The present status of colour television'. *Proceedings of the IRE*. September 1950;**38**:980–1002, para. 36
10. 'Affidavit of George E Slipper, Jr' to the FCC, 1949, FCC Docket No. 8736
11. Johnson E.C. letter to Dr E.U. Condon, 20 May 1949, reproduced as Annex A of the report on 'The present status of colour television'. *Proceedings of the IRE*. September 1950;**38**:980–1002, para. 45
12. Anon.: 'Synchronization for colour dot interlace in the RCA colour television system', *Journal of the British IRE*. April 1950;**10**:128–136
13. Federal Communications Commission. First report of Commission (adopted 1 September 1950), paras. 46–53

14. Law H.B. 'A three-gun shadow-mask colour kinescope' *Proceedings of the IRE.* October 1951, pp. 1186–94
15. Law R.R. 'A one-gun shadow-mask colour kinescope'. *Proceedings of the IRE.* October 1951, pp. 1194–201
16. Federal Communications Commission. First report of Commission (adopted September 1, 1950), para. 54
17. *Ibid.* paras. 55–71
18. *Ibid.* paras. 72–6
19. *Ibid.* paras. 77–84
20. *Ibid.* paras. 85–7
21. *Ibid.* paras. 88–91
22. *Ibid.* paras. 92–7
23. *Ibid.* paras. 98–101
24. *Ibid.* paras. 102–7
25. *Ibid.* paras. 108–19
26. *Ibid.* paras. 120–57
27. *Ibid.* paras. 127–31
28. *Ibid.* paras. 132–9
29. *Ibid.* paras. 140–3
30. Johnson E.C. Letter to Dr E.U. Condon, 20 May 1949, reproduced as Annex A of the report on 'The present status of colour television'. *Proceedings of the IRE.* September 1950;**38**:995
31. Federal Communications Commission. First report of Commission (adopted 1 September 1950), paras. 144–57
32. *Ibid.* para. 149
33. *Ibid.* para. 151

Chapter 8
RCA's resolve

The FCC's conclusions caused much dismay and disquiet among members of the Radio and Television Manufacturers Association generally. They met the RTMA's Television Committee in September to determine whether combined action against the FCC's findings could be taken but were persuaded that this was not possible since it would violate the antitrust laws. Consequently, member companies could only make representations on an individual basis, and any representations had to be submitted by 29 September 1950.

A trade journal summed up the views of some of the manufacturers when, on 18 September, it commented:

> With the FCC deadline only a few days off, many manufacturers threw up their hands in despair as they continued efforts to extract from technical and statistical experts at least a few general recommendations on circuitry and economic aspects of bracket receivers.
>
> At the weekend, unofficial manufacturer attitude seemed to boil down to rather general willingness to go along with the FCC as far as possible by building limited quantities of receivers capable of picking up CBS color in black-and-white. The technical problems are staggering, some contend, but they are willing to give it a try. [1]

A week later the journal reported,

> Television set makers covering a dominant share of total production told the FCC last week they couldn't possibly give a valid answer to the Commission's request for a September 29th promise to make receivers with bracket standards. Moreover, practically all of them said they couldn't start making bracket standard sets by the FCC's proposed November deadline. These views culminated a month of frantic engineering and production research in which the industry's best brains managed to make only scant progress toward the design of TV sets having continuously variable standards. [2]

RCA, too, was unhappy with the FCC's 'First Report of Commission (Color Television Issues)' of 1 September 1950 [3]. It considered the report was wrong and scientifically incorrect. In a response to the FCC's request for comments, RCA said it would continue its research, development and improvement of its compatible, all-electronic high-definition colour system notwithstanding the report. The company did not intend to adopt the stance of CBS, in the 1946 Hearing, when it stated that if its system were not adopted CBS would terminate its work on colour television.

> We expect [RCA observed] to continue to broadcast color signals under the RCA system and to use every scientific resource at our command to further the process of color. We cannot acquiesce in a decision with respect to difficult scientific questions where the professional judgement of practically the entire industry is united that the public is about to be saddled with an inferior system. We believe that such a decision would violate the Commission's

> obligations under the law and we deny the public, the broadcasters, and the manufacturers their rights under the law. [4]

RCA's trenchant comments also dealt with the CBS system.

> The Commission's Report clearly shows that the Commission itself is not satisfied with the CBS system and the CBS system is inadequate. Yet the Commission threatens to adopt the CBS system now unless the industry will yield to impossible and illegal conditions.
>
> The Commission's Report clearly recognized that compatibility, high resolution, large direct-view picture size, and other advantages of the RCA color television system [were] essential attributes of any color television system. Yet the Commission's report would outlaw the RCA system in favour of the CBS system which has shown none of these attributes. . . .
>
> We urge withdrawal from this unprecedented position, not only in respect of its future implications to American business but as well in the interest of 40,000,000 people who are now enjoying television. [5]

With that missive RCA's course was defined: it would strenuously oppose the FCC's conclusions as presented in the 'First Report of the Commission (Color Television Issues)'. The Corporation was, of course, aware of the opinions of the radio and television set manufacturers on bracket standards and of their view that such standards could not be adopted in the time period prescribed by the FCC. RCA's initial move, on 4 October 1950, was to petition the FCC to set aside the period from 5 December 1950 to 4 January 1951 to review the progress that RCA had made in the development of its colour television system, and then, until 30 June 1951, to hold comparative demonstrations of the systems of CTI, CBS and RCA. 'By June 29th 1951, we [RCA] will show that the laboratory apparatus which RCA has heretofore demonstrated has been brought to fruition in a commercial, fully compatible, all-electronic, high-definition system of color television available for immediate adoption of final standards.' [5]

The FCC was unmoved by all the criticism of its conclusions and denied the petition on 11 October 1950. In the 'Second Report on Color Television' the FCC dealt with some of the contentions that had been expressed. In particular, it repeated its conclusions that the RCA system was deficient in the following respects [6]:

1. the colour fidelity of the RCA picture was not satisfactory;
2. the texture of the colour picture was not satisfactory;
3. the receiving equipment was exceedingly complex;
4. the equipment utilized at the transmitting station was exceedingly complex;
5. the RCA colour system was much more susceptible to certain kinds of interference than the present monochrome system or the CBS system;
6. there was no adequate assurance in the FCC record that RCA colour pictures could be transmitted over the 2.7 MHz coaxial cable facilities; and
7. the RCA system had not met the requirements of successful field testing.

On the vexed question of compatibility, the FCC said it:

> ... was forced to conclude from the evidence in the record that no satisfactory compatible system [had been] demonstrated in [the] proceedings and the Commission [had] stated that in its opinion, based upon a study of the history of color development over the past ten years, from a technical point of view, compatibility, as represented by all compatible color systems which [had] been demonstrated to date, [was] too high a price to put on color. In

an effort to make these systems compatible, the result [had] been either an unsatisfactory system from the standpoint of color picture quality, or a complex system, or both. . . .

With no way of preventing the growth of incompatibility, the longer we wait before arriving at a final decision the greater the number of receivers in the hands of the public that will have to be adopted or converted if at a later date the CBS color system is adopted. [7]

The Commission stressed that it did not imply in its findings that there was:

. . . no further room for experimentation. Radio in general and television in particular [were] so new that extensive experimentation [was] necessary if the maximum potentialities of radio and television [were] to be realized. Many of the results of such experimentation can undoubtedly be added without affecting existing receivers. As to others some obsolescence of existing receivers may be involved if the changes [were] adopted. In the interest of stability this latter type of change [would] not be adopted unless the improvement [were] substantial in nature, when compared to the amount of dislocation involved. But when such an improvement does come along, the Commission cannot refuse to consider it merely because the owners of existing receivers might be compelled to spend additional money to continue receiving programs.

All of this led to the FCC's Order, released on 11 October 1950, which confirmed the standard of 525 lines per frame, 60 fields per second, interlaced 2:1 to give 30 frames per second for monochrome television; and adopted the CBS colour television standard of 405 lines per frame, interlaced 2:1 in successive fields of the same colour; the frame, field, colour frame, and colour field frequencies being 72 Hz, 144 Hz, 24 Hz and 48 Hz respectively. The Order also specified the colour sequence – red, blue and green in successive fields, and defined the trichromatic coefficients, based on the standardised colour triangle of the International Commission on Illumination.

Shortly afterwards, CBS commenced large scale public demonstrations of colour television in New York, Philadelphia, Boston, Chicago, Denver, New Orleans and several other cities.

Again, there was disappointment with the FCC's conclusions. Sarnoff, of RCA, issued a press statement on the same day as the FCC's Order in which he said:

We regard this decision as scientifically unsound and against the public interest. No incompatible system is good enough for the American public. The hundreds of millions of dollars that present set owners would have to pay to obtain a degraded picture with an incompatible system reduced today's order to an absurdity. [8]

Other key industry figures commented similarly, but it was the New York Times that suggested a solution in an editorial published in the 18th October 1950 issue:

In effect, [the FCC] is dictating to manufacturers which kind of television sets they are to make if images are to be received in color with converters and in black-and-white with adapters . . .

This coercion of an industry which has hitherto enjoyed a large measure of freedom is apparently without precedent. In the case of color television we have a usurpation of authority that needs correction . . .

The time for judicial interpretation of the phrase 'to encourage the large and more effective use of radio in the public interest' has arrived. [9]

The previous day RCA and two associate companies, NBC and the RCA Victor Distributing Company Inc., had filed suit [10] in the Federal District Court in Chicago against the USA and the FCC to 'enjoin and set aside an order of the Commission which promulgated standards for the transmission of colour television'. A special district court of three judges was convened to hear the case which was scheduled for the 14, 15 and 16 of November 1950. CBS and other parties were permitted to participate in the case. A temporary injunction was issued by the court while it considered the suit [11].

Unhappily for RCA, the two-to-one judgement (on 22 December 1950) was not in their favour. However, the court extended the temporary injunction pending an appeal. Nonetheless, RCA's determination to prevent what it considered to be the imposition of an inferior colour television system on the American public never waned. An appeal against the judgement was clearly necessary and so on 25 January 1951 the three plaintiffs and seven other intervenors requested leave of the Northern District Court of Illinois in Chicago to appeal the court's decision to the United States of America's Supreme Court [12]. The petitioners declared that:

1. the FCC Order of 10 October 1950 was based on insubstantial evidence;
2. the evidence put forward to sustain the Order was taken early in the colour hearings and had been superseded by improved RCA colour developments during the conduct of the hearing; and
3. the FCC had violated its statutory duty and abused its discretion by refusing to consider relevant matter put before it at the Commission's invitation and included in the record. [13]

The District Court signed an order on 25 January 1950 that permitted the plaintiffs to appeal the decision to the Supreme Court, and the papers were served on 30 January 1951 [14].

In court, the appellees (the USA and the FCC) and the appellee-intervenor (CBS) moved that the judgement of the District Court be affirmed and that the temporary restraining order issued by the court be dissolved. Naturally, the appellants (RCA and its two associate companies) replied and reiterated their views concerning the public interest, the incorrectness of the FCC's findings, the points of law at issue, the inadequacy of the FCC's record in the colour hearings, and the lack of grounds for the denial of a judicial review.

On 5 March 1951, the US Supreme Court agreed to review the findings of the lower court, and to hear the arguments of the opposing parties in three weeks' time. These arguments need not be dealt with further since they have been outlined above; suffice to say that the Supreme Court held that the basis of the complaint of RCA et al. was that the order had been entered 'arbitrarily, and capriciously, without the support of substantial evidence, against the public interest, and contrary to law' [15].

The Supreme Court was not persuaded by the arguments advanced by the appellees for the revocation of the decision of the District Court, and held that the FCC had not acted capriciously in its conclusions relating to the CBS system. In stating the majority opinion of the court, Mr Justice Black said (on 28 May 1951),

> The Commission's special familiarity with the problems involved is amply attested by the record. It had determined after hearing evidence on all sides that the CBS system will provide the public with color of good quality and that the television viewers should be

> given an opportunity to receive it if they so desire. This determination cannot be held to be capricious ... We cannot say the District Court misapprehended or misapplied the proper judicial standard in holding that the Commission's order was not arbitrary or against the public interest as a matter of law. The District Court's judgement sustaining the order of the Commission is affirmed. [15]

One month later (on 25 June 1951) the FCC announced that all regular television stations were permitted to broadcast colour programmes in accordance with its standards.

RCA's position was now clear. What it could not achieve by litigation it would attempt to achieve by public opinion founded on conclusions reached from 'public demonstrations of RCA's improved all-electronic system of color television' (Sarnoff) [15].

CBS was, of course, delighted by the Supreme Court's ruling and announced:

> The decision of the Supreme Court removes the last roadblock to the public's enjoyment of color television in the home. CBS will shortly expand its present color broadcast schedule in New York, and within a few months expects to be producing a substantial schedule of color programs. Many of these will be sent over existing circuits to stations in other cities of the Columbia television network. [16]

At this time (May 1951) CBS was without any manufacturing facilities. As it seemed that CBS's aspirations would be promoted by the acquisition of such facilities an approach was made to a Brooklyn electron tube manufacturer, Hytron Radio and Electronics Corporation, and its set-manufacturing subsidiary, Air King Products Company, which was then assembling sets for Sears Roebuck and Montgomery Ward. (On 4 June 1951 Air King Products had announced the forthcoming introduction of the first commercial CBS type colour television receiver. It would use a 10 in. (25.4 cm) cathode-ray tube and, with a lens, would display a 12.5 in. (31.8 cm) picture. The retail cost was stated to be $499.95, excluding excise tax, warranty, and installation charges.) P.C. Goldmark was sent to appraise the company.

> I shot out to Brooklyn to see Air King's set manufacturing and then to Newburyport, Massachusetts, to look over the firm's tube plant, and I came away with the feeling that the company knew how to make excellent tubes and TV sets at low cost. It was the fourth largest manufacturer of radio and television tubes in the country. [17]

Following the acquisition, CBS, at a 'gala premiere' held on 25 June 1951 in New York, transmitted sponsored colour television programmes from its studios to Philadelphia, Baltimore, Washington and Boston [18]. During the first two weeks the numbers of programme hours were 4.5 and 6 respectively; but by the week of 15 October the schedule had increased to 12 hours per week.

> Plans were under way [said CBS] for further expansion of the broadcasting of color programs; additional programs were sponsored; our sales force was engaged in vigorous efforts to interest other advertisers and there were several promising prospects; and in order to increase broadcasting by stations not owned by CBS, we had completed plans to purchase and pay for time on some ten affiliates along the eastern seaboard during which they would carry our color broadcasts. [19]

Several weeks later, on 28 September, the company, confident that its system was commercially viable, placed full page advertisements in the New York daily newspapers announcing the sale of a console receiver for $500 net of other charges [20]. By mid-October the first colour television receivers were being sold in retail outlets.

CBS's enthusiasm for colour television broadcasting was soon tempered by an unexpected US government directive. On 19 November 1951 the Office of Defense Mobilization asked the CBS to cease the manufacture of colour television receivers. The reason given to them was the need to conserve scarce materials which were required by the National Defense Program to prosecute the Korean War [21]. (This had commenced in 1950.)

Next day, the National Production Authority issued Order No. M-90 [22] confirming the ODM request. It mentioned the 'shortage of critical controlled materials' needed for the war effort and stated the penalties for any violation of the order. More specifically, Section 1 of the order defined its purpose:

> This order prohibits the manufacture of sets designed to receive color television, and items solely designed to permit or facilitate the reception of color television. The manufacture of color television for experimental, defense, industrial, and certain hospital and educational uses is permitted. [23]

Section 2 defined controlled materials as 'steel, copper, and aluminium in the forms and shapes indicated in Schedule I of CMP Regulation No. 1, as from time to time amended'; and Section 3 sought to remove any ambiguity or misunderstanding on the part of industrialists/manufacturers about the order's intent on prohibition of manufacture:

> Except as otherwise provided in this section, no person shall, after the effective date of this order, produce or assemble any television set designed to receive or capable of receiving color television; nor shall any person produce or assemble any product, attachment, or part designed solely to permit or facilitate, or capable only of permitting or facilitating, the reception of color television. [24]

CBS's reaction to the initial ODM request was hasty. Within minutes of receiving it CBS issued a public statement affirming its compliance. Such was the rapidity of the company's response that one correspondent noted that if the wire report from ODM had been delayed by just three minutes the CBS press release from New York would have preceded it [25]. Indeed, to the outsider, the almost simultaneity of cause and effect, together with the fact that only the CBS company had received the request, seemed to suggest some degree of collusion between CBS and the ODM. A sceptic might have argued that CBS had been given some intimation of RCA's energetic progress in colour television and felt it had to take steps to extricate itself from what would be an unequal forthcoming contest. Certainly, CBS could not compete against the known vast expertise, experience and resources of the Corporation. This *a priori* juxtaposition required Dr F. Stanton, the president of CBS, to confirm that CBS had not initiated any action that led to the issuance of Order No. M-90 [26]. The FCC, too, stated its non-involvement in the matter. (Actually, the apparently remarkable rapidity of CBS's response came about because the Director of Defense Mobilization had written to Dr Stanton on the issue mentioned in the order the day before it was announced.)

The NPA's order was illogical in that it was not drafted to conserve some essential materials: instead it banned the manufacture of a specific product. Presumably the essential materials used in the production of a colour television receiver could have been utilized – and therefore not conserved – in the making of another product. But the issue of the propriety and validity of M-90 was of no concern to CBS.

The order was amended on 24 June 1952 to allow, under certain conditions, the manufacture of colour television sets for domestic use provided authorization was obtained from the NPA.

Sarnoff, of RCA, alluded to his rival's difficulties at a conference held on 2 February 1952 between officials of the Office of Defense Mobilization and senior representatives of the radio and television industry. He asserted that RCA was not in any way associated with the order and, in a vigorous diatribe against CBS, harangued the company for its failure to implement promises it had made to the FCC:

> Now we come to a most important date – October 19, 1951, [Sarnoff] said. CBS was hopelessly on the hook by this date. It wasn't broadcasting its color as Mr Stanton had promised, the price of CBS color sets was about double what CBS witnesses had told the FCC it would be, the advertisers were not buying CBS colour television time, the CBS affiliated stations were accepting few color programs, expenses were mounting terrifically. [27]

On the question of CBS manufacturing receivers, and its application to the National Production Authority for 250,000 fractional horsepower motors for the year 1952, Sarnoff commented:

> Even according to the optimistic prediction of the CBS manufacturing subsidiary's President, David Cogan, CBS only intended to turn out color television sets at the rate of 80,000 a year. And it is doubtful whether CBS could have reached this figure during 1952 even if it were allotted all the material it asked for.

Sarnoff concluded by saying:

> If the NPA should decide to lift its ban, it is safe to conclude the CBS would be ingenious enough to stay off the hook; CBS failed to meet its commitments before NPA Order M-90 was issued. Even if Order M-90 is now lifted, I expect that CBS will protest that it cannot get sufficient materials to go ahead with its previously promised program for manufacturing color sets and broadcasting color programs. [27]

The NPA's Order M-90 effectively brought to an end the realistic hopes of the Columbia Broadcasting System to have its colour television system adopted by the public. After the 20 October order, only the Radio Corporation of America was in a position to apply the necessary resources required to accomplishing a practical solution to the colour television problem. And before this was resolved more than $20 million would be spent.

RCA began to implement its strategy to gain public support for its system during the summer of 1951. First, a two-day symposium on RCA's direct-viewing tube was held at Princeton on 19 and 20 June 1951 [28], to which more than 200 manufacturers were invited and given free samples of the tri-colour tube, together with kits of parts to enable them to construct a colour television receiver; and, second, demonstrations of colour television were given each day in New York during the week ending 9 July 1951. By this date the shadow mask tube had 585,000 phosphor dots, new red and blue phosphors, and a resolution and a brightness (25 ft lamberts) about equal to black-and-white tubes. The televised programmes, which covering a wide variety of subject material, were broadcast from the RCA–NBC television transmitter (WNBT) in the Empire State building and were seen by viewers on both colour and black-and-white sets [29].

Of these demonstrations, those on 11 July were seen by *c.* 150 leading members of the industry. Among the comments made, Dr Allen B du Mont noted, 'It was a lot better colour picture than RCA showed us in Washington last December. The picture was good enough, in fact, to start commercial operations immediately.' [30]

Dr W.R.G. Baker, a vice-president of General Electric, agreed: 'It was a most excellent picture. It was outstanding. I thought the color was really beautiful and the black-and-white was outstanding. All we've got to do now is get compatible standards adopted.' [30]

Of especial importance, for the networking of television programmes across the United States, were the successful transmissions on 16, 17, and 18 October 1951 of test colour programmes from New York to Los Angeles and back to New York – a distance of *c.* 8,000 miles.

These publicity ventures were extended, early in 1952, when a laboratory was established by RCA, in Long Island, New York, where colour television signals could be received, via a cable or via a receiving antenna, and which any manufacturer could use for testing or experimental purposes. Further public colour telecasts from NBC's station WNBT were scheduled, with FCC approval, for 9, 11 and 15 July 1952, and on 16 to 19, 24 to 26, and 30 September 1952, station WNBT radiated colour television signals based on the NTSC standard [31].

One month later, Dr Baker issued the following statement to the press:

> Highly satisfactory field tests of an all-electronic color television system have been going on for months in New York and other areas, with superb results. Many improvements in both transmission and reception have been made, and color pictures now can be transmitted over the existing video networks with very satisfactory results.
>
> As a result, the television industry may hope to be ready soon to demonstrate both its new technical achievement and the fact that its use will not make obsolete any present-day home video receiver. [32]

Of course, much of this would have been known to CBS. Dr Stanton and his senior colleagues must have realized that CBS, however reluctantly, had to change its stance and embrace the new developments. In CBS's annual statement issued in late December 1952 Dr Stanton acknowledged that compatibility was 'extremely desirable', and expressed CBS's hope 'that the industry committee and groups now working on such a compatible system [would] be successful in their efforts and [would] press forward to obtain approval of its standards in 1953. Because of the uncertainty and conflicting claims, the public is not receiving this important new development of color television ... All elements in the industry should strive promptly to resolve the question so that all can go forward in vigorous efforts under whichever system prevails.' [33]

For several reasons, Dr Stanton was being realistic in expressing his views:

1. RCA was the undoubted leader in television development in general and colour television in particular. It had achieved great progress in its work on all-electronic black-and-white television and had investigated several systems of colour television – all at huge cost to the Corporation.
2. CBS could not match RCA's financial resources, or the expertise and experience of its highly qualified technical staff.
3. The NTSC was nearing the end of its studies and tests of compatible colour television, and definite progress was being reported.
4. The majority of industry shared a common opinion regarding the need for compatibility, but CBS's system was incompatible.
5. Whereas, in 1946 the issue of compatibility was, perhaps, of no appreciable significance, the position now (1951) had changed markedly because of the rapid expansion of black and white television receiver manufacture (see Table 8.1).

The total number of receivers produced by the end of 1952 was *c.* 23 million, of which 21.5 million were estimated to be in use. Any move by the FCC to implement a non-compatible colour television standard, and thereby discriminate against black-and-white television viewers, would clearly be unpopular.

In an endeavour to seek the truth behind the apparent slowness of the implementation of colour television, the US House of Representatives Committee on Interstate and Foreign Commerce again initiated a move to clarify the situation. On 14 March 1953 it requested a few of those who had been involved with developments in this field to appear before the committee for questioning. At the first meeting, on 24 March 1953, the chairman, the Republican representative for New Jersey, C.A. Wolverton, noted: '. . . the purpose of these hearings is to get, if possible, a definite answer to the question when will color television become a reality in the homes of the American people' [35].

The first witness to be called was Dr E. Engstrom, director of research at RCA. He reiterated RCA's well-known pronouncements on compatibility, CBS's failure to

Table 8.1 The annual production of black-and-white television sets [34]

Year	Production
1940–42	*c.* 3,000
1946	*c.* 5,000
1947	178,500
1948	975,000
1949	*c.* 3,000,000
1950	7,500,000
1951	5,400,000
1952	6,000,000

fulfil the promises which it had made to the FCC and the public, the need for the FCC to authorize immediately commercial broadcasts of colour television, and the readiness of RCA 'today' 'to commence broadcasting compatible color programs which [could] be received in black-and-white sets now in the hands of the public without changing these sets at all and without any set owner being required to buy any new equipment to receive these broadcasts' [36].

Dr F. Stanton, of CBS followed Dr Engstrom. He presented a comprehensive review of the history of CBS's efforts in the field of colour television but concluded, 'First, I say reluctantly but realistically, that CBS has no plans, so long as present circumstances exist, to broadcast or manufacture under the approved field-sequential system' [37]. A complex set of factors compelled this decision:

1. the continued policy of the NPA with regard to its Order M-90;
2. the lack of industry support, although the president of the RTMA had testified to the FCC that the association would support whatever system the FCC decided to approve;
3. the abandonment of any expectation that the FCC's 1950 decision and the subsequent litigation would settle the issue of colour television, at least for a reasonable time;
4. the acknowledgement that the FCC's decision had not brought 'an end to the conflicting claims [of colour television] ... and to the public's confusion and bewilderment which resulted from such claims';
5. the appreciation that CBS's broadcasting and manufacturing endeavours, which had 'cost us enormous energies and millions of dollars', were not providing 'enough impetus and competitive incentive for other broadcasters and other manufacturers to begin to follow our lead'; and
6. 'the problem of incompatibility [which had] now grown to such proportions that in combination with other factors, it [became] quixotic and economically foolish for us single-handedly at this time to resume a large scale broadcasting and manufacturing program under the sequential system' [38].

Dr Stanton was commended for the frankness of his statement and for his efforts to encourage and stimulate the introduction of a new media form before the public.

Among the statements presented to the committee, that of the chairman of the NTSC, Dr Baker, was of some interest. He said the NTSC was 'unanimously convinced that under its supervision there [had] been prepared a set of standards capable of producing a superlative system of color television . . .' [39] The system was compatible with existing black-and-white receivers and actually produced a better monochrome picture, as well as a superior colour picture. Dr Baker put forward several recommendations including the need for the FCC to recognize formally the existence of the NTSC and to lend its active support to the NTSC programme [40]. There was an implication that the FCC was exercising a restrictive policy, was not being fully cooperative, and that the delay in the implementation of compatible colour television stemmed from the attitude of the FCC.

Of course, the chairman of the Federal Communications Commission (now P.A. Walker) could not let such a distortion of the facts (as the FCC saw them) pass without comment. Essentially, the gist of the reasons why the FCC had not approved a compatible colour television system was as follows [41]:

> We would be the first to recognize, Mr Chairman, that a color television system different from the one adopted by the Commission *may* now have been developed. It *may* even be a better system. It *may* be that the color quality is excellent. It *may* be that the required receiving and other equipment is practical for everyday use and for commercial production. It *may* be that the cost involved would not put it beyond the reach of the general public. It *may* be that it is not unduly susceptible to interference. Indeed, it *may* be a system that the Commission should approve as the basis for its color television standards. But *the fact* is that this has not yet been demonstrated. And *because* it has not yet been demonstrated – *because* these possibilities have not been established as realities – it also *may not* be all that its proponents hope for it. And, as the Chairman of the NTSC itself had made clear, a sound judgment cannot be made until data is available based upon essential field testing. It must be remembered that field testing of the revised NTSC standards began this month [March 1953]. [42]

Walker strongly defended the FCC's previous actions and policy statements. He reminded the Committee of the seven criteria of the FCC that had to be satisfied by a system, before it could be given approval by the Commission and stressed that it had a duty to protect the interest of the public.

On the question of the relationship between the NTSC and the FCC, Walker confirmed the FCC recognized that the NTSC:

1. [was] a group of representative manufacturers of electronics equipment organized by the Board of Directors of Radio Manufacturers' Association;
2. [had] stated [its] purpose with respect to color television [was] to assemble technical data on 'basic standards for the development of a commercially practicable system of color television and to undertake additional work as may be in the interest of providing more adequate television service to the American public';
3. [had] devoted a substantial amount of time, money, effort and engineering talent to the accomplishment of its stated purpose;
4. [had] standing to request the revision of the Commission's standards; and
5. [expected] any proposal which it advance[d] to the Commission [would] receive the most careful and objective consideration. [43]

However, notwithstanding all these points, the chairman of the FCC noted the NTSC was not 'entitled to any special recognition over other industry groups'; the Commission was not 'committed in advance to an endorsement of future NTSC findings and conclusions', and 'should not abdicate in any way to the NTSC the exercise of its judgment, critically and objectively, as to the public interest'. The principal issue for the FCC in making a ruling on the system to be adopted was 'the extent to which there [was] opposition to or competition with any system advanced for [its] consideration and the extent to which any such system [was] shown to achieve that degree of excellence which would warrant its acceptance as the basic color television standard' [44].

During its Hearing, the Committee on Interstate and Foreign Commerce witnessed, on 14 April 1953, a demonstration of compatible colour television at RCA's research centre in Princeton. The telecasts were produced in the NBC colour studios at the Colonial Theatre, New York, and were broadcast using the NBC experimental licence KE2XJV [45]. They included 20-minute programmes of vignettes starring

Nanette Fabray, Burton's performing birds, Dolores Gray in singing roles, and a colour version of *Kukla, Fran and Ollie* [46].

Wolverton, chairman of the Committee, was said to be 'astounded' by the quality of the colour. 'It's amazing,' he observed, 'color television has reached the stage of perfection where the public should have the benefits. It would seem justified to put it into production' [47]

The reaction of the press was similarly favourable to the new colour system. '[S]ome of the soft pastel shadings were breath-taking in their loveliness. For the first time in several years of assorted color television tests, the screen seemed free of an artificial and unreal quality,' wrote J Gould of the *New York Times*. He reported a comment of Wolverton that the Committee 'might make some sort of recommendation to the FCC with a view to hastening the advent of color television' [48].

The following day, CBS demonstrated its system, but, now, it was a 'poor second' to that of RCA. RCA's enormously costly research-and-development effort, from September 1949, had resulted in a system that was superior to the officially approved FCC system. CBS's colours 'seemed cold and harsh and the flesh tones were noticeably erratic', reported the *New York Times*. 'The images had little of either the depth or delicacy that marked the RCA tests.' [49]

Approximately one month later, on 19 May 1953, RCA demonstrated its system to members of the Federal Communications Commission, licensed radio manufacturers, and officials of the stations affiliated with the NBC television network [50].

These well-received demonstrations emboldened RCA, on 25 June 1953, to petition the FCC to adopt the signal standards that defined its colour television system. RCA supported its application with evidence – basically a history of its activities in the television field from *c*. 1930 – which amounted to 697 closely printed pages, fully illustrated with diagrams and photographs [51].

For the FCC, the position now was quite different to that which had prevailed during the 1949–50 FCC Hearing. The principal factors that would determine its course of action were:

1. the NTSC standards on colour television had been drafted and approved by representatives from the leading industries concerned with television, and had been field tested;
2. the standards utilized by RCA were those of the NTSC;
3. the RCA system was a compatible system and it satisfied the six points of the FCC for judging the merits of a colour television system;
4. the only rival, post-1950, contender, CBS, had effectively withdrawn from the colour television contest;
5. the RCA had expended *c*. $20.5 million [52] on colour television R&D, and had employed on the television problem staff and resources of the very highest quality;
6. the RCA system had been investigated by the House of Representatives Committee on Interstate and Foreign Commerce and had not been found wanting;
7. the demonstrations of colour television given by RCA in April 1953 had been much superior to those it had given in 1950 and had received very

favourable comment from the chairman of the Committee on Interstate and Foreign Commerce and others.

To support its application, RCA embarked on a programme of publicity and public relations, which included: (1) the first publicly announced colour television broadcast (on 30 August 1953) from the NBC studios; (2) a symposium (on 7 October 1953), attended by *c*. 250 representatives of the television industry; (3) a demonstration (held on 16 October 1953) given to the FCC and the press; and (4) a transmission (on 3 November 1953) of colour television programmes, using the AT&T's microwave and cable facilities, from New York to Los Angeles, to demonstrate the networking capability of the system [53].

Much favourable comment was engendered by these telecasts. The *Los Angeles Mirror* exclaimed, 'It's so beautiful, it knocks you right out of your seat ... There's no doubt about it, this is it ... I couldn't keep my eyes from the colorcast ... The public is going to love it ... A new and bigger boom in the television industry is just around the corner.' [54]

With all the evidence before it, the FCC on 17 December 1953 approved the signal specifications for the all-electronic compatible colour television system advanced by the RCA and the NTSC. [55] In the FCC's Report and Order the Commission noted that the petition for the replacement of the existing rules and standards had been supported by the RCA, the NBC, the NTSC, the Philco Corporation, Sylvania Electric Products Inc., the General Electric Co. and Motorola. Comments advocating the changes had been submitted by the CBS, the Hazeltine Corporation, the Admiral Corporation, the Westinghouse Radio Stations Inc. and Harry R. Lubcke. [56]

Oppositions to the adoption of the new rules were filed by Paramount Television Productions Inc., Chromatic Television Laboratories Inc., American Television Inc. and two individuals.

Under the heading 'Conclusions of the Commission's Report and Order' the FCC stated:

> The accomplishment of a compatible color television system within a 6 [MHz] bandwidth is a tribute to the skill and ingenuity of the electronics industry. The proposed color television signal specifications produce a reasonably satisfactory picture with a good overall picture quality. The quality of the picture is not appreciably marred by such defects as misregistration, line crawl, jitter or unduly prominent dot structure. The picture is sufficiently bright to permit a satisfactory contrast range under favourable ambient light and is capable of being viewed in the home without objectionable flicker. Color pictures can be transmitted satisfactorily over existing inter-city relay facilities and improvements in inter-city relay facilities may be reasonably anticipated. [57]

The use of the words 'satisfactory', 'sufficiently', and 'not appreciably marred' hinted that further developments would most likely lead to a diminution of the 'defects and imperfections of this new medium' (Commissioner Lee) [58].

Shortly after the FCC's announcement of its new rules, the NBC interrupted its regular programme with a colour slide reading 'Color News Bulletin' and an announcement: 'Attention, please. You are looking at the first color picture telecast since compatible standards for color television were approved.' [59] This was followed by televised views of the NBC's colour television studios in the Colonial

Theatre in New York, and a declaration that viewers could expect 'most exciting moments of tour entertainment in the weeks and months ahead'. Approximately 30 minutes later General Sarnoff was televised. He noted the 'great victory for RCA, but an even greater triumph for the public and the television industry. RCA developed the compatible color television system. We have fought and worked hard and long for its adoption for commercial use because we were confident from the beginning that it is the right system in the public interest' [60].

References

1. Anon. Report, *Broadcasting-Telecasting*, 25 September 1950, p. 60
2. Anon. Report, *Broadcasting-Telecasting*, 2 October 1950, p. 57
3. O'Rourke J.S. 'The role of the Radio Corporation of America in securing Federal Communication Commission approval of an all-electronic compatible system of television in color, 1932–1953: a descriptive study'. A thesis submitted in partial fulfilment of the requirements for the Master of Science degree at Temple University, 17 March 1970
4. Quoted in *ibid*. pp. 111–12
5. Quoted in *ibid*. pp. 112–13
6. O'Rourke J.S. 'The role of the Radio Corporation of America in securing Federal Communication Commission approval of an all-electronic compatible system of television in color, 1932–1953: a descriptive study'. A thesis submitted in partial fulfilment of the requirements for the Master of Science degree at Temple University, 17 March 1970, p. 114
7. Federal Communication Commission Docket 8975, 3, FCC 50–1224
8. Anon. Report, *New York Times*. 11 October 1950, p.9
9. Anon. Report, *New York Times*. 18 October 1950
10. *Radio Corporation of America, et al., v. United States of America, et al.*, Civ. A. No. 50-C-1459 United States District Court, Northern District of Illinois, Eastern Division, 20 Dec 1950. Federal Supplement, 95. Also see Ref. 3, pp.123–5
11. O'Rourke J.S. 'The role of the Radio Corporation of America in securing Federal Communication Commission approval of an all-electronic compatible system of television in color, 1932–1953: a descriptive study'. A thesis submitted in partial fulfilment of the requirements for the Master of Science degree at Temple University, 17 March 1970, p.124
12. *Ibid*. p. 125
13. *Radio Corporation of America, et al., v. United States of America, et al.*, Appeal from United States District Court for the Northern District of Illinois, Eastern Division, United States Reports, 1951, **341**, pp. 412–27
14. O'Rourke J.S. 'The role of the Radio Corporation of America in securing Federal Communication Commission approval of an all-electronic compatible system of television in color, 1932–1953: a descriptive study'. A thesis submitted in partial fulfilment of the requirements for the Master of Science degree at Temple University, 17 March 1970, pp. 127–35

15. Quoted in *ibid.* p. 133
16. Anon. Report, *Broadcasting-Telecasting*, 18 June 1951, p. 62
17. Goldmark P.C. 'Maverick inventor. My turbulent years at CBS'. *Saturday Review Press*, New York, 1973. pp.114–16
18. Anon. Report, *Broadcasting-Telecasting*, 26 June 1951, p. 34
19. Stenographic transcript of the Hearings before the Committee on Interstate and Foreign Commerce, House of Representatives, 25 March 1951, **2a**, 'Color television', p. 8
20. Anon. Report, *New York Times*, 28 September 1951, p. 18
21. Anon. Report, *Broadcasting-Telecasting*, 29 October 1951, pp. 23, 28
22. National Production Authority, Department of Commerce, National Production Authority Order M-90, Federal Register, **16**, No. 226, 21 November 1951, p.11773
23. Quoted in O'Rourke J.S. 'The role of the Radio Corporation of America in securing Federal Communication Commission approval of an all-electronic compatible system of television in color, 1932–1953: a descriptive study'. A thesis submitted in partial fulfilment of the requirements for the Master of Science degree at Temple University, 17 March 1970, p. 147
24. *Ibid.* p. 148
25. Anon. Report, *New York Times*, 20 October 1951, p. 1
26. O'Rourke J.S. 'The role of the Radio Corporation of America in securing Federal Communication Commission approval of an all-electronic compatible system of television in color, 1932–1953: a descriptive study'. A thesis submitted in partial fulfilment of the requirements for the Master of Science degree at Temple University, 17 March 1970, p. 152
27. *Ibid.* pp. 149–52
28. Bucher E.E. 'Television and David Sarnoff'. Typewritten manuscript at David Sarnoff Museum, Princeton, New Jersey, **XXII**, Chapter CVI, p. 2789
29. Baun J.R. De, Montfort RA., Walsh A. A. 'The NBC New York color television field test studio'. *RCA Review*. 1952;**13**(3):107–24
30. Quoted in O'Rourke J.S. 'The role of the Radio Corporation of America in securing Federal Communication Commission approval of an all-electronic compatible system of television in color, 1932–1953: a descriptive study'. A thesis submitted in partial fulfilment of the requirements for the Master of Science degree at Temple University, 17 March 1970, p. 141
31. Cahill J.T. *et al.* 'Petition of Radio Corporation of America and National Broadcasting Company Inc. for Approval of Color Standards for the RCA Color television System', 25 June 1953, RCA, Princeton, pp. 519–24
32. Anon. Report, *Broadcasting-Telecasting*, 1 September 1952, p. 58
33. Anon. Report, *Broadcasting-Telecasting*, 5 January 1953, p. 57
34. Stenographic transcript of Hearings before the Committee on Interstate and Foreign Commerce, House of Representatives, 2 March 1953, **1**, 'Color television', pp. 3–9
35. *Ibid.* p. 2

36. Quoted in O'Rourke J.S. 'The role of the Radio Corporation of America in securing Federal Communication Commission approval of an all-electronic compatible system of television in color, 1932–1953: a descriptive study'. A thesis submitted in partial fulfilment of the requirements for the Master of Science degree at Temple University, 17 March 1970, p. 166
37. Stenographic transcript of the Hearings before the Committee on Interstate and Foreign Commerce, House of Representatives, 25 March 1951, **2a**, 'Color television', pp. 1–20
38. *Ibid.* pp. 14–16
39. O'Rourke J.S. 'The role of the Radio Corporation of America in securing Federal Communication Commission approval of an all-electronic compatible system of television in color, 1932–1953: a descriptive study'. A thesis submitted in partial fulfilment of the requirements for the Master of Science degree at Temple University, 17 March 1970, p.169
40. Stenographic transcript of Hearings before the Committee on Interstate and Foreign Commerce, House of Representatives, 31 March 1953, **5**, 'Color Television', p. 14
41. *Ibid.* pp. 1–26
42. *Ibid.* p. 3
43. *Ibid.* p. 17
44. *Ibid.* p. 26
45. O'Rourke J.S. 'The role of the Radio Corporation of America in securing Federal Communication Commission approval of an all-electronic compatible system of television in color, 1932–1953: a descriptive study'. A thesis submitted in partial fulfilment of the requirements for the Master of Science degree at Temple University, 17 March 1970, p.170
46. Cahill J.T. *et al.* 'Petition of Radio Corporation of America and National Broadcasting Company Inc. for Approval of Color Standards for the RCA Color television System', 25 June 1953, RCA, Princeton, p. 534
47. O'Rourke J.S. 'The role of the Radio Corporation of America in securing Federal Communication Commission approval of an all-electronic compatible system of television in color, 1932–1953: a descriptive study'. A thesis submitted in partial fulfilment of the requirements for the Master of Science degree at Temple University, 17 March 1970, p. 172
48. Anon. Report, *New York Times*, 15 April 1953, p. 1
49. Anon. Report, *New York Times*, 16 April 1953, p. 43
50. O'Rourke J.S. 'The role of the Radio Corporation of America in securing Federal Communication Commission approval of an all-electronic compatible system of television in color, 1932–1953: a descriptive study'. A thesis submitted in partial fulfilment of the requirements for the Master of Science degree at Temple University, 17 March 1970, p.173
51. *Ibid.* pp. 173–74
52. *Ibid.* p. 175
53. *Ibid.* pp. 177–80
54. Anon. Report, *Los Angeles Mirror*, 4 November 1953, p. 9

55. Federal Communications Commission. Text of the FCC's decision. *Broadcasting-Telecasting*, 21 December 1953, pp. 58a–58h
56. *Ibid.* para. 2, p. 58a
57. *Ibid.* para. 40, p. 58c
58. O'Rourke J.S. 'The role of the Radio Corporation of America in securing Federal Communication Commission approval of an all-electronic compatible system of television in color, 1932–1953: a descriptive study'. A thesis submitted in partial fulfilment of the requirements for the Master of Science degree at Temple University, 17 March 1970, p.185
59. *Ibid.* pp. 186–7
60. Anon. Report, *Broadcasting-Telecasting*, 21 December 1953, p. 20

Chapter 9
The work of the NTSC

Mention has been made earlier to the establishment, in July 1940, of the National Television System Committee (NTSC) to formulate standards for black-and-white television that would be acceptable to the television industry. Working with much skill, the Committee achieved a consensus among the disparate interests of the industry and (on 8 March 1941) defined standards that were adopted by the Federal Communications Commission. From 1 July 1941 the FCC authorized a public television service. A few months later hostilities commenced between the USA and Japan and, though the NTSC was not disbanded, it was felt that its work on monochrome television had been completed.

Following the 1949–50 FCC Hearings on colour television, the subsequent opposition encountered by CBS in its petition and the conflict that arose between various members of the industry on the standards to be adopted, a decision was taken to reactivate the NTSC. The re-emergent body became known as the Second NTSC [1]. Again the Committee was organized on a panel basis but, because of the complexity of colour television, the second NTSC was much larger than the first NTSC. Panel membership was open to all interested and competent engineers. An important feature of the second committee was the link with the previous committee to ensure continuity of thought, methodology and organization. Dr W.R.G Baker and M.E. Kinzie, chairman and secretary respectively of the first NTSC, now served the second committee; of the eight (later ten) panel chairmen of the new committee, all but two had been actively associated with the first NTSC; and 31 engineers who had served in 1940–41 became members of the second NTSC.

There was much for the NTSC to investigate. This was reflected in the charge imposed by the board of directors of the Radio Technical Planning Board on the NTSC, namely, 'to assemble data on:

1. the allocations of channels in the ultra high frequency band;
2. the procedures which [would] enable [the] FCC to lift the "freeze" on very high frequency allocations; and
3. the basic standards for [the] development of a commercially practicable [compatible] system of color television and to undertake such additional work as may be in the interest of providing [a] more adequate television service to the American public.'

The extent of the task that faced the second NTSC may be gauged from the time – 32 months – needed to complete the many necessary investigations and tests, and from the 18 volumes of evidence totalling *c.* 4,100 pages that were produced by the committee and its panels and subcommittees. For comparison, the first NTSC completed its work in less than nine months and the 1940–41 record covered just *c.* 60,000 words: in 1940, 168 individuals served on 9 panels, but by 1953 315 persons were members of 10 panels and 55 sub-panels of the second NTSC.

Initially, the work of the second NTSC was, according to D.G. Fink [1], vice-chairman of the NTSC (1950–52), 'rather pointedly ignored in Washington'. This arose because several members of the FCC had misunderstood the purpose of some advice given, in 1950 to the Commission, by the television industry's technical experts. Fortunately, any false impressions were soon corrected and by 1952 members of the FCC's engineering staff were regularly attending the major NTSC field trials.

Within a few months of the formation of the second NTSC the investigations of compatible colour television, being undertaken by the General Electric Company, the Hazeltine Corporation, the Philco Corporation and the Radio Corporation of America, had progressed to the position where a systematic review of the subject was needed. Accordingly, the work of the eight panels of the NTSC was temporarily suspended and an Ad Hoc Committee was appointed (on 20 November 1950) to determine whether there was any unanimity of opinion among the investigators, and to recommend the future course of action. There was a view that industry agreement on any proposed specifications was almost essential to a favourable FCC consideration.

The committee comprised D.B. Smith (Philco), chairman, E.W. Engstrom (RCA), T.T. Goldsmith (DuMont), I.J. Kaar (General Electric), and A.V. Loughren (Hazeltine). R.M. Bowie (Sylvania) joined the committee later, and A.G. Jensen (Bell Laboratories) served as an observer. During its work, the Ad Hoc group attended demonstrations of several system proposals at the Hazeltine, Philco, DuMont and General Electric laboratories, and at the NBC-RCA colour test facilities in Washington. By 19 April 1951 the group's deliberations had reached the point where a number of recommendations could be made on further detailed investigations and a report was issued.

The report of the Ad Hoc Committee noted the similarities of the various systems and recommended an intensive and extensive programme to develop a television system that would result from studies of all the relevant scientific and technical advances. In its report, the Committee recommended [2]:

> That color be added to the existing broadcast service by utilizing the present black-and-white standards to transmit all the information necessary concerning brightness – i.e., all the information necessary for a good black-and-white picture (comparable to the present service) – and by adding the necessary chromatic information (to color the picture) on a sub-carrier transmitted simultaneously with the 'brightness' signal and contained within the video band. To detect this sub-carrier, reference or color sync information [should be] added to the present sync signal during an interval of time available for this purpose [as defined] in the present standards.

The following recommendations were also affirmed [3]:

1. the present FCC transmission standards for black-and-white television shall continue to be used for the transmission of compatible color television.

2. chromatic information shall be transmitted by means of a color sub-carrier modulated in amplitude and phase with respect to a reference sub-carrier of the same frequency. The color sub-carrier shall be transmitted simultaneously with the video signal and during only the video portion of the composite signal. Synchronizing signals to transmit information concerning the reference sub-carrier shall be transmitted only during the synchronizing and blanking intervals of the composite video signal.
3. to ensure practical invisibility of the color sub-carrier, its normal frequency, but not phase, shall be related to the horizontal scanning frequency in the following manner: The color sub-carrier frequency shall be an odd multiple of half the horizontal scanning frequency.
4. for standard operating conditions, the amplitude of the primary video signal and the amplitude and phase of the color sub-carrier shall be specified in terms of a 'proper' set of taking characteristics. (Definition: A 'proper' set of taking characteristics is defined as a set, each one of which is a linear combination of CIE distribution characteristics.)
5. the color sync signal shall be transmitted by means of a burst of the reference carrier superimposed on the back porch following each horizontal sync pulse in accordance with the detail shown ... [see Figure 9.1, notes 7, 8, 9 and 11]

The Ad Hoc Committee stated its conviction about the technical merits of the above system as follows:

It is the firm conviction of the members of this committee:

1. that the (above) system of color television ... will provide for the maximum utilization of the existing 6 [MHz] channels as assigned to television broadcast service by the FCC. By this is meant that this type of system will transmit the maximum amount of information useful to the viewer with regard to picture clarity, color fidelity, picture brightness, freedom from flicker, and other deleterious effects, of any system of color television now known to us, and
2. that the (above) system of color television ... will be compatible, in the sense that existing black-and-white receivers will be able to derive a black-and-white picture from a color signal in accordance with this system, with no change in the receiver and with picture quality comparable to present black-and-white. In other words, telecasting stations could transmit the new color signal with no significant loss of service, nor inconvenience, nor any added cost, to their existing audience.

The Ad Hoc Committee's course of action was approved by the RTMA in spring 1951, following which the NTSC constituted nine new Panels to advance its programme on a much expanded scale. (See Note 1 at the end of the chapter.)

It is now necessary to digress and consider a few essential features of a colour television system in order to appreciate fully the above recommendations, and others, made by the participating companies associated with the NTSC.

Any scheme for broadcasting pictures in colour from one place to another comprises equipment that can conveniently be divided into two, broadly independent, groups – the terminal equipment (that is, the camera at the sending end and the display tube at the receiving end) and the transmission equipment, which carries the signals from the camera to the display tube.

In the NTSC system [4] the colour camera incorporates three image-orthicon camera tubes (R, G and B) mounted so that the tubes view the scene from the same vantage point. Each tube is fitted with a colour filter (red or green or blue) arranged so that tube R receives an image of the scene being televised in red light, tube G an

1
Max. carrier voltage
Blanking level
Reference black level
Reference white level
Zero carrier
Picture
Horizontal blanking
Vertical blanking
Bottom of pictur(see notes 3 and 5)
Equalizing pulse interval
Vert. sync. pulse interval
Equalizing pulse interval
Horizontal sync. pulses
3H
3H
3H
3·02H
0.5
t_1
0·07v± 0·01v 0v
Top of picture
(0·075 ± 0·025)P
(0·75 ± 0·025)C
(0·125 ±0·025)C
S
P
C

2
0·5H
4
4
t_1+V
3
3
Horizontal dimensions not to scale in 1, 2 and 3

3
Detail between 3–3 in 2
hor SYNC.
Blanking level
Rear slope of Vert. blanking See Note 3
Reference white level
zero carrier
H
FRONT PORCH
Colour burst
Back porch
9 (See note 8)
(z) 0·18H max
1/10 Of max. blanking
5
5

4
Details between 4–4 in 2
Blanking level
0·04H max.
0·004H max.
Equalizing pulse
0·004H max.
VERTICAL SYNC. PULSE
0·004H max.
0·004H max.
9/10 S
1/10 S
0·07H ±0·01H
0·04H See note6
0·5H
H

5
Detail between 5–5 in 3
S
1/10 S
(x) 0·02H Min.
0·075H ±0·005H
0·125H Max
(y) 0·145H Min
0·90 S TO 1·1 S
8 Cycles min.
0·006H min

TIME →

[Source: Author's collection]

Notes to Figure 9.1:

1. H = Time from start of one line to start of next line.
2. V = Time from start of one field to start of next field.
3. Leading and trailing edges of vertical blanking should be complete in less than 0.1 H.
4. Leading and trailing slopes of horizontal blanking must be steep enough to preserve minimum and maximum values of (x+y) and (z) under all conditions of picture content.
5. Dimensions marked with asterix indicate that tolerances given are permitted only for long time variations and not for successive cycles.
6. Equalizing pulse area shall be between 0.45 and 0.5 of area of a horizontal sync. pulse.
7. Colour burst follows each horizontal pulse, but is omitted following the equalizing pulses and during the broad vertical pulses.
8. Colour bursts to be omitted during monochrome transmissions.
9. The burst frequency shall be 3.579545 mc. The tolerance on the frequency shall be ± 0.0003% with a maximum rate of change of frequency not to exceed 1/10 cycle per second per second.
10. The horizontal scanning frequency shall be 2/455 times the burst frequency.
11. The dimensions specified for the burst determine the times of starting and stopping the burst, but not its phase. The colour burst consists of amplitude modulation of a continuous sine wave.
12. Dimension ‘P’ represents the peak excursion of the luminance signal from blanking level, but does not include the chrominance signal. Dimension ‘S’ is the sync. amplitude above blanking level. Dimension ‘C’ is the peak carrier amplitude.

image in green light, and tube B an image in blue light: thus, the colour values of the televised scene/object are analysed into three primary colours.

The colour value of a scene/object is characterized by three quantities based on its luminance, its hue and its saturation. Luminance, or brightness when used with reference to a sensory response, is a measure of the darkness or lightness of a colour; hue specifies whether a colour is red, green or blue, or the like; and saturation is a measure of the mixture of a given hue with white light. For black-and-white television only the luminance component is required for the transmission and reception of a televised picture, but, in a colour television system, the luminance, hue and saturation values of each point of a scene/object must be sent and received. The hue and saturation values together represent the chromatic values of a colour.

From the camera the three electrical signals representing the three primary colours must be transmitted to the display unit. A simple arrangement would be to use three channels and modulate three separate carrier signals with the red, green and blue signals so that there is a one-to-one correspondence between the camera's output signals and the display unit's input signals. This configuration has two disadvantages. First, neither the red nor the green nor the blue signal is ideally appropriate for the operation of a black-and-white receiver, since a scene that is largely composed of one colour (for example, that associated with green) would lead to excessive large area flicker. Second, and more important, the use of three channels requires a large bandwidth and therefore limits the number of television transmissions that can be accommodated in a given region of the electromagnetic spectrum.

In the receiver, the picture reproducer – such as a shadow mask tube – synthesizes an image in colour from the red, green and blue signals that are fed to it. The shadow mask tube consists of several hundred thousand individual phosphor dots arranged in the form of triplets, each of which comprises one red phosphor, one green phosphor and one blue phosphor. When all the red dots are excited a red image is produced, and similarly for the green and blue dots. In viewing the display screen the three images are superimposed by a viewer's eye-brain combination to give an image in full colour.

Fortunately, the eye is much less sensitive to changes in hue and in saturation than to changes in brightness. As the sizes of coloured objects decrease, it is found that blues and yellows become indistinguishable from greys of equivalent brightness. In the size range where this occurs, browns are confused with crimsons, and blues with greens, but reds remain clearly distinct from blue-greens. Generally, it appears that hues having a pronounced blue lose blueness, while colours lacking this hue gain blueness: additionally, the hues become desaturated. As the dimensions of the objects become ever smaller the reds and blue-greens merge with greys of similar brightness; and for exceedingly small objects – coloured threads, for example – the individual colours are not recognized and only brightness can be perceived [5].

When the first colour television systems were demonstrated in the UK and the USA in the late 1920s no detailed knowledge on the necessary subjective sharpness, or resolution, of the images and colour existed. Indeed, at that period such knowledge would not have contributed to any improvement of the low-definition images. The earliest reference to visual acuity for coloured light is dated 1865, but it was not until 1940 that the relevant data were rediscovered by television engineers in papers on physiological optics. In 1940 A.N. Goldsmith applied for a patent [6] for a colour television system having a low resolution in the blue region of the spectrum. Five years later A.V. Bedford wrote a paper [7] on 'Mixed highs in color television' in which he described how by limiting the frequency range of the red, green and blue colour signals to below 2 MHz, and combining the three signals together above 2 MHz, a saving of bandwidth could be achieved.

The need for careful investigations on the subjective sharpness of additive colour television images led to an extensive series of tests by M.W. Baldwin [8] of Bell Telephone Laboratories. His results showed that in a colour image composed of red, green and blue colours, the highest resolution was required in the green image, rather less resolution in the red image, and substantially less in the blue image. These demonstrations were followed by work undertaken by Panels 2 and 11 of the NTSC. Panel 2 conducted its study by sending a detailed questionnaire to its members to elicit from them the most important facts relating to colour vision as known to science and medicine. Panel 11 (particularly A.V. Loughren and C.J. Hirsch [9] of the Hazeltine Electronic Corporation) endeavoured to determine the necessary bandwidths of the luminance and chrominance signals so as to accommodate the signals in a 6 MHz channel. Figure 9.2 shows the layout of Loughren's and Hirsch's experiment in which the frequency – defined as the crossover frequency – at which the change from colour to mixed highs takes place is 2 MHz. The crossover frequencies investigated were 0.1 MHz, 0.5 MHz, 1.0 MHz, 2.0 MHz and 4.0 MHz. These led to effective video bandwidths of 4.2 MHz, 5.0 MHz, 6.0 MHz, 8.0 MHz and 12.0 MHz.

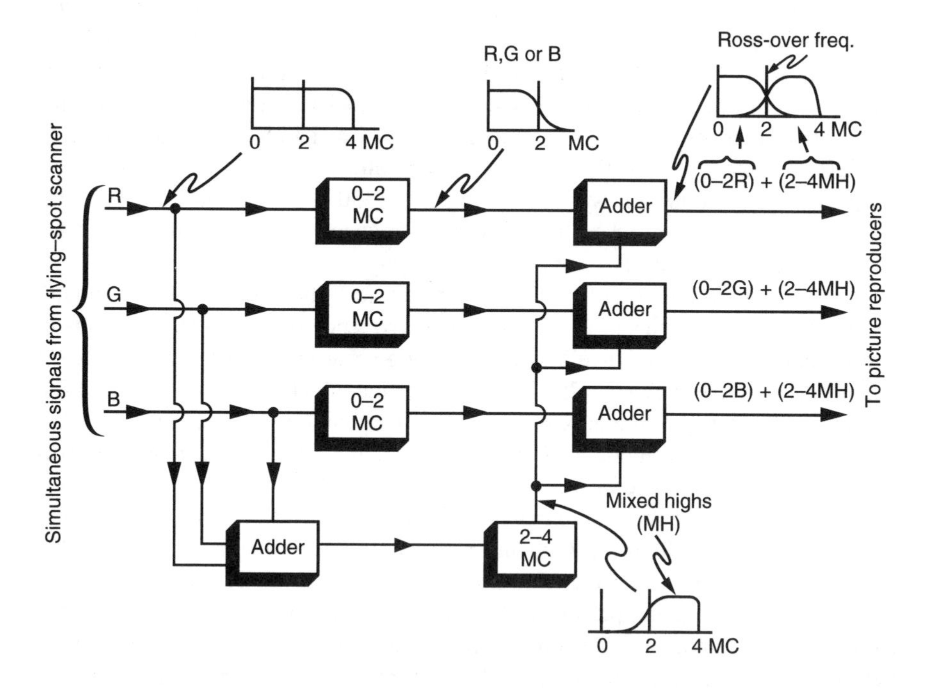

Figure 9.2 The experimental set-up for investigating the transmission of mixed-high television signals [Source: ibid.*]*

Loughren and Hirsch found from subjective comparisons made under critical conditions by a number of observers, at a normal viewing distance of four times picture height, that:

1. with a crossover frequency of 2.0 MHz (and a bandwidth of 8.0 MHz) the reproduction was 'virtually indistinguishable from the 12 MHz yardstick picture';
2. with crossover frequencies of 1.0 MHz or 0.5 MHz (and bandwidths of 6.0 MHz and 5.0 MHz) the reproduction was 'as good as the 12 MHz picture in all respects save that very small colored areas [were] partially de-saturated';
3. a crossover frequency of 0.1 MHz 'may not be fully satisfactory on some subjects'.

Overall, the authors concluded:

> The evidence ... supports completely the statement that, by means of the mixed-highs technique, color television pictures can be produced fully as sharp as the best monochrome pictures, with excellent color rendition, using a band not materially greater than that required for monochrome television.

In 1946 RCA, in its simultaneous colour television system, attempted to save some bandwidth by limiting the resolution of the blue image to approximately one-third that of the red and green images. However, their system was still not practicable since it used two sub-carriers (one for the blue signal and another for the red signal), and a bandwidth of 14.5 MHz.

The successful solution of the bandwidth problem came about in 1950, when it was realized that a given region of the spectrum can be shared by two independent modulated signals if their carrier frequencies are so related that the sideband components of one signal are interleaved with the sideband components of the other signal [10]. Interestingly, this very important principle had been stated as long ago as 1930 by F. Gray [11], of Bell Telephone Laboratories, but had lain unobserved in the patent literature until it was rediscovered in 1950. The relation is satisfied in a colour television system when the sub-carrier frequency is an odd multiple of half the line-scanning frequency.

Any periodic waveform can be shown, by Fourier's theorem, to comprise an infinite number of harmonics of the fundamental frequency of the waveform. Thus the line synchronizing pulse waveform consists of a great many harmonics of the line synchronizing frequency –15,750 Hz for a 525-line, 60 Hz television standard. The spacing between any two adjacent harmonics is, of course, 15,750 Hz in this case. When this waveform is modulated by the picture signals and the field-synchronizing pulses, the resulting waveform can still be represented by an infinite series of harmonics but the spectrum is more complex. Figure 9.3 shows the spectrum of such a waveform in the region of the 82nd and 83rd harmonics of the line frequency. Now, the single spectrum lines of the 82nd and 83rd harmonics of the previously unmodulated line-synchronizing pulses have groups of sidebands associated with each of the harmonic components; the components of each group being spaced apart by the field frequency. The scanning of any television picture always results in a spectrum of this nature, the differences between the spectrum of one picture and that of another

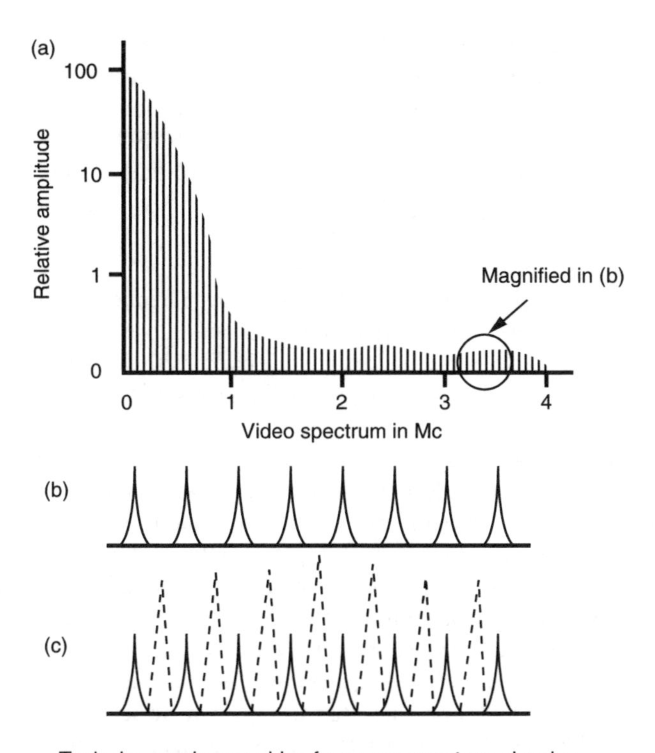

Figure 9.3 The line-frequency harmonics of a typical monochrome video frequency spectrum [Source: Sylvania Technologist, *1952, vol. 5, p. 41]*

being represented by changes in the relative amplitudes and phases of the various components.

Fortunately, the amplitudes of the sidebands of each group decrease rapidly and so each group occupies only a small fraction of the frequency space between successive harmonic groups. Consequently, there is space between these groups into which another like signal can be inserted. P. Mertz and F. Gray [12] have estimated that *c.* 46 per cent of the space between any pair of harmonics is not occupied. If the harmonic pattern of another modulated television signal is to be interleaved midway with a given harmonic pattern, the necessary condition is that the frequency of the inserted signal must be an odd multiple of half the line-scanning frequency.

The principle of frequency interleaving suggests how information concerning the luminance, hue and saturation of a televised scene/object could be accommodated within the existing (1950) black-and-white standard while maintaining the principle of compatibility and channel spacing. If the luminance, or brightness, signal – which contains all the information necessary for a good black-and-white, or monochrome, picture – occupies the band from 0 to 4.5 MHz of a 6 MHz channel, the necessary chromatic information can be modulated onto a sub-carrier transmitted simultaneously with the luminance signal and contained within the same video band.

The sub-carrier frequency chosen initially by the NTSC was 3.898125 MHz, which is the same frequency as the 495th harmonic of half the line frequency of 15.75 kHz. The choice of this frequency had the further advantage that the visibility of any effect due to the sub-carrier when receiving a colour television signal on a monochrome set was minimized (see Note 2).

Two advantages were obtained by this frequency interleaving. First, the standards for colour television were fully compatible with those of the existing black-and-white television service. Second, a greater bandwidth for both the luminance and the chrominance signals was practicable within the standard channel than would otherwise be possible.

To detect the sub-carrier in the television receiver, a reference signal was added to the 'back porch' of the waveform of the black and white standard.

The report of the Ad Hoc Committee was circulated (on 19 April 1951) to the industry, the FCC, and the technical press, together with a letter from the chairman of the NTSC. He stated the report outlined the broad framework of a new composite system, based on 'the best elements of the furthest advances in existing systems', which could be developed 'by individual coordinated effort on the part of our industry'. This required: the NTSC panels 'to be reorganized to permit a more direct attack on the problems to be resolved'; all companies 'to participate in the necessary equipment and field testing, by either establishing a system or by cooperating with a company that propose[d] to build a system and carry on field testing'; and the NTSC to 'coordinate the work of the competing companies'.

The report of the Ad Hoc Committee was accepted, and the first meeting of the reorganized NTSC (see Note 1) was held on 18 June 1951.

Among the many problems that had to be resolved, that of the encoding of the colour signal was of prime concern. Briefly, there are four principal ways of expressing a colour signal symbolically, namely: in terms of the three primary colour signals; in terms of the red, green and blue colour-difference signals; in terms of the red and blue colour-difference signals; and in terms of symmetrical components. The NTSC chose the third of these possibilities for reasons that will now be discussed.

In black-and-white television, the signal transmitted is closely related to the luminance, and it is this component of a colour television signal that must be transmitted if compatibility between black-and-white television and colour television is to be satisfied. If a scene is viewed by a three-tube colour camera, the electrical signal outputs from the camera can be represented by E_R, E_G, and E_B for each picture element. These signals may be weighted (by factors k_1, k_2 and k_3) and combined to give a signal (E_Y), which is representative of the sum of the red, green and blue components of the televised scene, and is the luminance, i.e. $E_Y = k_1E_R + k_2E_G + k_3E_B$. It is this signal that is utilized in monochrome television. Several different values of the k factors were suggested [13]. At the FCC hearings held in 1950, RCA used $k_1 = k_2 = k_3 = 1/3$. Sylvania proposed to the Ad Hoc Committee the values of $k_1 = k_3 = 0$ and $k_2 = 1$, which meant the green primary signal was the monochrome signal. Hazeltine and Philco advocated as values the relative luminous efficiencies of the three primary colours, though the systems were not the same in other respects. Panel 7 of the NTSC proposed $k_1 = 0.33$, $k_2 = 0.57$, and $k_3 = 0.11$ to accord with the phosphors available for use in television display tubes. (Later Panel 13 considered another set for which $k_1 = 0.30$, $k_2 = 0.59$ and $k_3 = 0.11$.) Systems employing these values are known as constant luminance systems: they require a bandwidth of 4 MHz – the same as for black-and-white television.

In the NTSC scheme, two of the colour components are transmitted as colour difference signals, namely $(E_R - E_Y)$ and $(E_B - E_Y)$; the third component, green, is determined from the equation that defines E_Y. Of course, the three colour components, R, G and B, could be sent separately, but extra bandwidth would be needed. Similarly, the R and B components could be transmitted without modification. However, the use of difference signals eases the utilization of the signals in colour television receivers (see Note 2). An additional important reason is that when $E_R = E_G = E_B$, they are also equal to E_Y (because $0.33 + 0.57 + 0.11 = 1.0$). The advantage of this choice is given in Note 2.

Each colour-difference signal has a bandwidth (initially chosen to be 1 MHz) and must be accommodated within the 6 MHz channel, which also accommodates the 4 MHz luminance signal. This is achieved by using a modulation process known as quadrature modulation in which a sub-carrier, C_1 (having a frequency of 3.898125 MHz), is generated and from it another signal, C_2, of the same frequency, but in quadrature (i.e. 90° apart in phase) with C_1, is derived. The two sub-carriers are modulated, separately, by the two colour-difference signals. Mathematical analysis of the modulation process shows that the amplitude of the modulated colour sub-carrier corresponds to the saturation of the colour signal, while the phase angle of the modulated sub-carrier corresponds to the hue of the colour signal. Following modulation, the luminance signal is added to the modulated sub-carrier and the resultant used to modulate the radio frequency carrier. Figures 9.4a and 9.4b show simplified schematic diagrams of the encoding and decoding processes employed in the transmitter and receiver respectively.

At the receiver the combined signal, after amplification, frequency changing and detection, is processed to separate the luminance and chrominance components; they are then applied to two quadrature demodulators, from which the two colour-difference signals are obtained. These signals and the luminance signal are applied to a decoding matrix and the three individual colour components recovered.

On 21 July 1953, the NTSC approved the final form of the compatible colour television signal specifications and petitioned the FCC for their adoption. During this period the ten panels had undertaken a vast amount of fundamental work:

> ... into the nature of human vision; explored the transmission systems best adapted to such vision; wrote and rewrote signal specifications; conducted field tests of the signals for color reception as well as compatible monochrome reception; studied the special problems of network connections for color; wrote tutorial papers; compiled definitions, and finally arrived at an unanimous agreement on 22 signal specifications which [would serve] as the basis of color telecasting in the United States. The record of this accomplishment fill[ed] 18 volumes, containing 4,104 pages. [1]

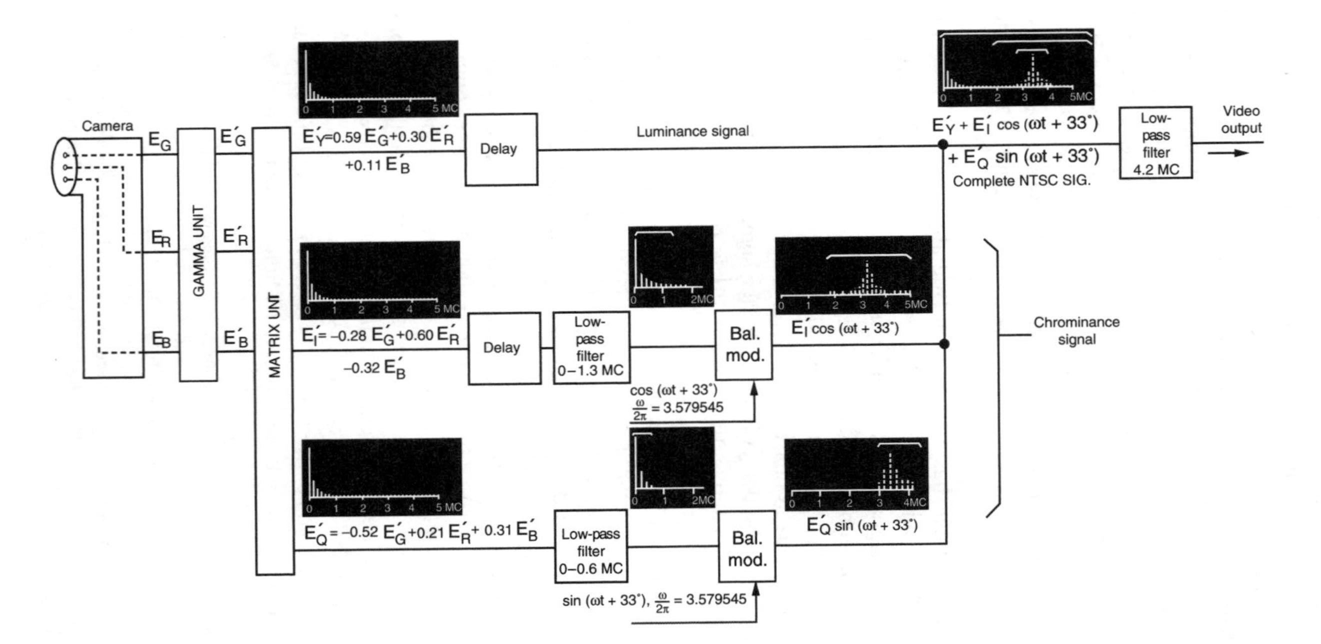

Figure 9.4a *The components of the equipment for generating NTSC colour video signals [Source:* Proceedings of the IRE, *1955,* ***43****(11), pp. 1575, 1578]*

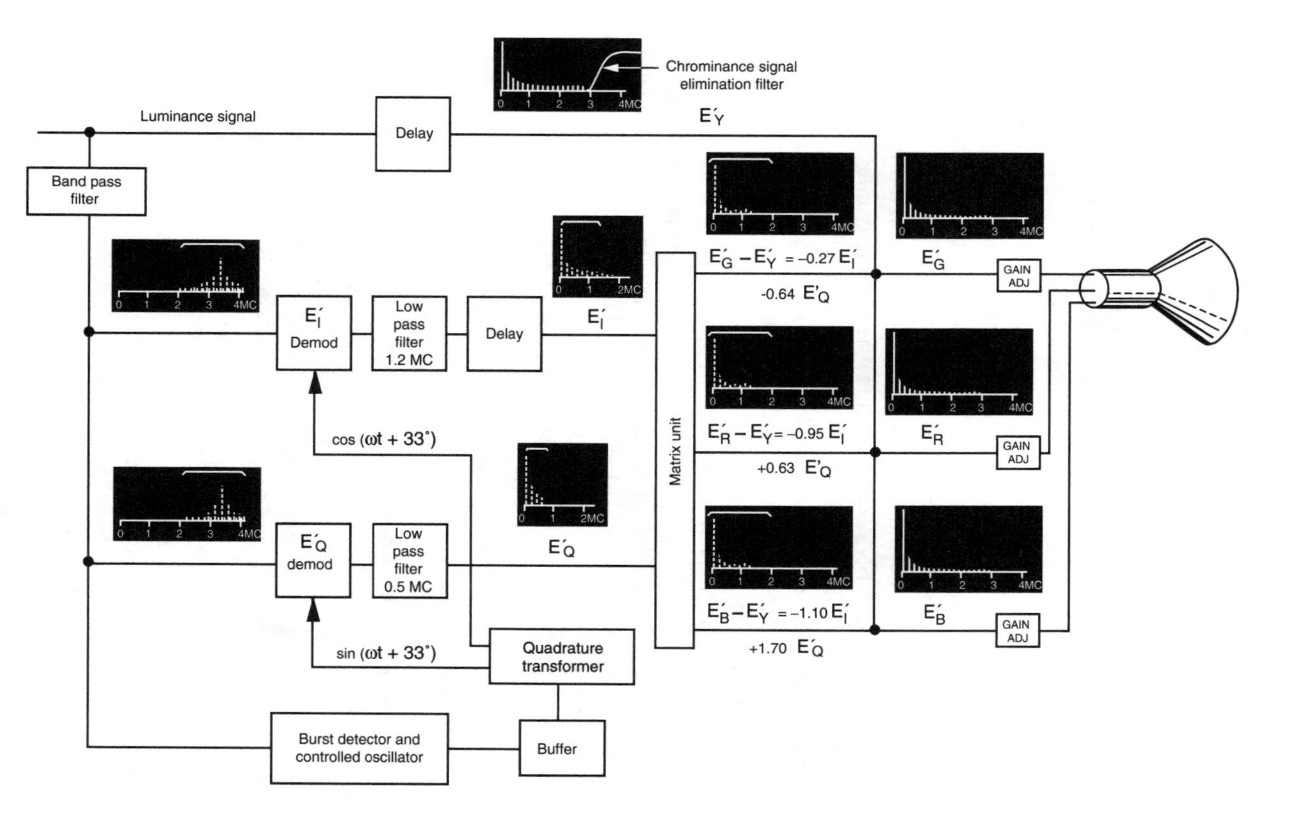

Figure 9.4b The components of the equipment for receiving NTSC colour video signals [Source: ibid.*]*

The consensus achieved from such a disparate group was an outstanding achievement and a splendid exemplar of the methodology established by the main committee.

A demonstration of the NTSC system was given on the 15 October 1953. It was witnessed by F.C. McLean (the BBC's deputy director of engineering) and was described by him as 'a great triumph of technical skill and very probably the answer [to the colour television problem] as far as the US [was] concerned' [14]. During the demonstration signals obtained from NBC's three-tube image orthicon cameras were transmitted: (1) direct by line; (2) by 540 miles of a centimetric radio link; (3) by 540 miles of 2.7 MHz cable; and (4) by an antenna. Also, CBS gave an outside broadcast (OB) presentation using 180-frames-per-second sequential cameras, the outputs of which were converted by a 'tube plus camera arrangement to an NTSC system'.

On the results achieved, McLean observed, 'The colours were good but a trifle garish. The sky was an unnatural blue, grass was rather a bright yellow green, and flesh tints were not quite faithful. On the whole however the effect was a very pleasing and good home entertainment.' CBS's demonstration was 'definitely inferior' to that of NBC but 'showed promise'. Because of its economic advantages 'we [the BBC] should think more about this'. 'Everybody that I talked to was most impressed. No one knows what will be the attitude of the Commission but everybody seems to think that they cannot do otherwise than approve.' [14]

The NTSC's signal specifications were approved by the FCC on 17 December 1953 as the technical transmission standards for commercial colour television broadcasting in the United States. They formed the foundation of all colour television standards (including the European PAL and SECAM systems) used around the world and so are of great historical importance. Of course, the system was not perfect – there was scope for improvement, and this was acknowledged by the FCC [15], as previously stated in Chapter 8

(The general specification and the complete colour picture signal specification are presented in Appendix A.)

The day after the FCC approved the NTSC's colour standards, RCA scheduled full-page newspaper advertisements headlined 'RCA wins fight for compatible color television'. No mention was made of the NTSC in the text – which was signed by D. Sarnoff and F.M. Folsom. Indeed, RCA claimed 'the FCC has approved the standards on which the RCA color television system operates'. There seemed to be an implicit assumption that the FCC had adopted standards that had been expounded and developed by RCA alone.

Naturally, its competitors were outraged. Philco counterattacked with a full-page advertisement, headlined 'Facts the American public should know about color television', which stated, 'These standards were developed by the leading scientists of the electronics industry. They are not the work of *any one company*.' To stress this fact, Philco listed the names of 42 firms, including RCA, that had contributed to the NTSC's efforts.

Soon afterwards, in February 1954, R.H. Hyde, chairman of the Federal Communications Commission, in his foreword to a book [1] titled *Color Television Standards. Selected papers and records of the National Television System Committee*, gave credit to the accomplishments of 'many minds'. He wrote, 'The solution incorporates the best thinking and striving by the different interests which contributed to the effort. As such, it is a tribute to modern engineering genius.' [1] The book, published in 1955, included documents, records, commentaries, diagrams and photographs prepared by the panels of the NTSC.

However, one member of the editorial committee, Dr E.W. Engstrom of RCA, did not believe the manuscript properly presented the contributions of RCA, and submitted a statement to support his contention. He observed:

> The standards approved by the FCC in December 1953, are in many basic respects similar to the signal specifications which RCA proposed to the FCC at the 1949–1950 hearings, but naturally reflect substantial improvements resulting from the knowledge gained during the research and field tests conducted since those hearings. For example, the signal specifications proposed by the Radio Corporation of America during the course of the FCC hearings embraced the following characteristics:
>
> (a) A signal component for the operation of black-and-white receivers. (Exhibit 209, Docket No. 8736, et al., Sept. 26, 1949, p.10, 'A Six-megacycle Compatible High-definition Color Television System.')
> (b) Color information carried on a sub-carrier. The sub-carrier phase or relative time position was the code used to convey hue information, while the strength of the sub-carrier conveyed the saturation information. (Exhibit 209, Docket No. 8736, et al., Sept. 26, 1949, pp.2, 3, and 4, 'A Six-megacycle Compatible High-definition Color Television System.')
> (c) The frequency of the sub-carrier was chosen to be an odd multiple of one-half the line scanning frequency. This was an application of the interlace principle, designed to reduce the visibility of the sub-carrier in the picture displayed on black-and-white receivers. It was this method of choosing the frequency of the sub-carrier that gave rise to the name 'dot interlace'. (Exhibit 314, Docket No. 8736, et al., February, 1950, pp. 2 and 3, Recent Developments in Color Television System.')
> (d) The 'mixed-highs' principle was used in transmitting the fine detail of the picture. (Exhibit 209, Docket No. 8736, et al., Sept. 26, 1949, pp. 1 and 2, 'A Six-megacycle Compatible High-definition Television System.')
> (e) The sub-carrier phase reference was provided by transmitting several cycles of a 3.6-mc sine wave during the horizontal blanking interval. (Exhibit 314, Docket No. 8736, et al., February, 1950, p.3, 'Recent Developments in Color Synchronization in the RCA Color Television System'). [16]

The following account [17] of the work of Ad Hoc Group is taken from the final report of the NTSC. (It is quoted verbatim because the information is not easily accessible or well known, and it enables the relative contributions of the companies concerned to be assessed accurately.)

> Hazeltine, using a wideband display as a reference, showed the following studies, among others:
>
> 1. The relationship between bandwidth and definition for the several [color] primaries and the effect of narrowing the bandwidth of color-difference signals.
> 2. The effect of varying the sampling angles in dot-sequential systems.
> 3. The effect of assigning different bandwidths to the two color difference components.
> 4. The effect of using 'shunted-monochrome' circuitry instead of narrow angle sampling.
> 5. The improvements obtained by a 'constant-luminance' type of system.
> 6. The effect of using 'oscillating color sequence'.

RCA demonstrated to the committee the system which they had previously demonstrated to the radio industry and others during December, 1950. This system was characterized by the following technical characteristics: The video signal consisted of the additive sum of the three primary colors from 0 to 4[MHz]. To this was added the signal produced by equal-angle (120-degree) three-phase sampling of the three primary colors band limited to 2 [MHz] or less. Color sync was provided by a burst of several cycles of the sampling frequency superimposed on the back porch of the horizontal sync signal.

The Allen B. DuMont Laboratories showed comparisons, using a wideband simultaneous display as a reference, of the following, among others:

1. Comparison of dot-sequential and field-sequential systems.
2. Studies of the effect of varying the bandwidth of the color-difference signals.

Philco, using a wideband simultaneous display as a reference, demonstrated their 'white-brightness' system. In this case the taking characteristics of the picture-generating equipment were adjusted to conform to the CIE distribution characteristics x, y, and z, from which output signals, corresponding to the CIE luminosity component Y and color-difference signals X-Y, and Z-Y, were obtained.

In the 'white-brightness' system, the Y, or brightness information, was transmitted as a standard black-and-white signal with a bandwidth of 3 [MHz]. The color-difference signals with bandwidths of 0.6 [MHz] were frequency multiplexed in phase quadrature on a 3.46 [MHz] color carrier. Studies were shown of the effect of interference at varying levels of color carrier. In addition, the effect of varying the bandwidth of the brightness component from 3 to 5 [MHz] (by the use of dot-interlace techniques) was shown.

A number of noise tests were shown in which the effect of noise in the several circuits, including the sync circuits, was demonstrated.

General Electric demonstrated studies of various forms of 'frequency-interlace' systems. These studies included the use of color signals of differing bandwidths and frequency multiplexed on different sub-carriers. In addition, demonstrations were given of the method of 'alternating highs' and of 'alternating lows'.

Sylvania presented a memorandum and several reports on the progress of its systems work.

Finally, the Committee had the benefit of information from many technical papers which had been published or presented at various meetings of the Institute of Radio Engineers, the American Institute of Electrical Engineers, the Radio Club of America, the Franklin Institute, etc.

Note 1: The National Television System Committee (at 20 June 1951) [18]

Chairman: W.R.G. Baker
Vice Chairmen: E.W. Engstrom, D.G. Fink (later A.V. Loughren), D.B. Smith
Secretary: M.E. Kinzic

Committee Member Affiliations

American Broadcasting Company
Admiral Corporation
Bendix Radio Division
Columbia Broadcasting System
CBS-Columbia, Inc.
Chromatic Television Laboratories
Color Television Inc.
Crosley Division AVCO
Allen B. DuMont Laboratories
Electronics McGraw-Hill
Emerson Radio and Phonograph
Federal Telecommunications Lab, Inc.
General Electric Company
General Teleradio Inc. (WOR)

Goldsmith, Dr Alfred N.
Hallicrafters Company
Hazeltine Corporation
Hogan Laboratories
Magnavox Company
Motorola, Inc.
Philco Corporation
Radio Corporation of America
Raytheon TV and Radio Corporation
Sentinel Radio Corporation
Sylvania Electric Products Inc.
Tele King Corporation
Tele-Tech Caldwell Clements Inc.
Zenith Radio Corporation

The titles, chairmen and scope of the panels that engaged in the detailed study of the various issues regarding colour television are as follows.

Panel 11 – Subjective aspects of color (Dr A.N. Goldsmith)

'It will be the responsibility of this panel to determine, on the subjective basis, the efficient distribution of information as between brightness and color. This determination will take into account the subjective importance of the various constituents of a color picture. In addition, the panel will give due consideration to the requirement for a satisfactory compatible black and white picture from such a color transmission.'

Panel 11-A – Color transcription (from 2 April 1952) (Dr A.N. Goldsmith)

'It will be the responsibility of this panel to study and report on methods of recording in color for television purposes, on methods of producing color release prints, and on methods and equipment for the utilization or transmission of color transcriptions.'

Panel 12 – Color System Analysis (D.G. Fink)

'The function of Panel 12 shall include the detailed analysis and interpretation of such new technology as may be evolved in connection with the NTSC signal and the preparation of tutorial papers concerning such new technology.'

Panel 13 – Color Video Standards (A.V. Loughren)

'It will be the responsibility of this panel to provide recommended standards relating to the complete video signal. As such, it will include the determination of both colorimetric and electronic specifications. The purview of this committee will include the following: (1) camera-taking characteristics; (2) gamma characteristics; (3) color carrier frequency and its phase relation with respect to horizontal synchronizing signals; (4) The color sequence to be used and whether or not it should be of oscillating type; (5) bandwidths of monochrome and color signals; (6) relative amplitudes of the monochrome signal and the color carrier; (7) determination of the maximum system amplitude demands at critical colors to enable the determination of picture-to-synchronizing ratios; (8) specification of radiated signal.'

Panel 14 – Color Synchronizing Standards (D.E. Harnett)

'It will be the responsibility of this panel to provide recommended standards for color synchronizing signals and the interrelation with normal deflection synchronizing signals. This committee will also consider the interaction between the proposed color standard and the existing black and white standards in this regard.'

Panel 15 – Receiver compatibility (D.E. Noble, later R. DeCola)
'It will be the responsibility of this panel to insure by actual observations during the field tests that the proposals of the panels as executed will result in satisfactory compatible black and white pictures and sound when viewed by available black and white receivers including the effects on broadcasting coverage.'

Panel 16 – Field testing (T.T. Goldsmith, later K. McIlwain)
'It will be the responsibility of this panel to insure by actual observation during field tests that the proposed standards will result in a signal which will satisfactorily operate color receivers and provide the public with service which, in color, is comparable in performance to that established by the monochrome standards.'

Panel 17 – Broadcast System (F. Marx, later R.E. Shelby)
'It will be the responsibility of this panel to insure by actual observation during field tests that the proposed standards will result in a signal which can be satisfactorily broadcast by present broadcast transmitters with only minor changes and can be transmitted from city to city by means of existing or presently contemplated inter-city program circuits. In addition, the panel will analyse the equipment required at the studio as well as the transmitter for the NTSC color signal as compared with the present black and white signal.'

Panel 18 – Coordination (D.B. Smith)
'The function of this panel will be liaison and co-ordination between the panels to reduce duplication of effort, and to resolve matters of scope and jurisdiction among the panels.'

Panel 19 – Definitions (R.M. Bowie)
'It will be the responsibility of this panel to produce a glossary of terms and working definitions as used in color television and color generally as required in the formulation of color television standards.'

Note 2: Miscellaneous aspects of the NTSC colour television standard

1. To minimize the possibility of interference between the luminance and chrominance signals the sub-carrier frequency must be located in a part of the frequency spectrum where the amplitudes of the harmonics of the luminance signal have decreased to acceptable values, i.e. as far away from the luminance carrier frequency as possible. On the other hand, the spectrum of the sub-carrier chrominance signal needs to extend to *c.* 0.6 MHz above the sub-carrier frequency. Since the video bandwidth available is *c.* 4.5 MHz it follows that the sub-carrier frequency must be no higher than *c.* 3.9 MHz.
2. Using the 1952 NTSC standard, the number of cycles of the sub-carrier signal per frame is 3,898,125 divided by 30, i.e. 129,937.5. Because a shift of half a

cycle in the time domain corresponds to a phase shift of 180°, during a given frame scan the colour sub-carrier signal must be exactly out of phase with that of the previous scan. The eye-brain averages the images of the two scans and so the sub-carrier is not visible on the screen of a black-and-white receiver. This process of averaging out the brightness variations can only work if the signal does not swing below the zero brightness axis as there can be no negative light. Because E_Y determines the background brightness and the colour signal can be represented by $\{E_Y + K_1(S_R \cos\omega t + K_2 S_B \sin\omega t)\}$, where S_R and S_B are the red and blue colour difference signals and the Ks are constants, appropriate selection of the Ks ensures that the colour sub-carrier in the complete colour picture signal does not cause a negative swing. Panel 13 chose $K_1 = 1/1.14$ and $K_2 = 1/1.78$ for initial field test purposes. [18]

3. The main reason for transmitting colour difference signals rather than the colour signals directly is that the reception of a monochrome (black-and-white) television signal by a colour television receiver causes the display screen to assume the colour that corresponds to zero sub-carrier, namely, white, which is required. This follows because, for white $E_R = E_G = E_B = E_Y$ (by choice) and the colour difference signals are zero. Panel 13 of the NTSC selected CIE illuminant C as zero sub-carrier white as it is close to the centre of the JETEC tolerance band for the colour of monochrome c.r.t.s. Furthermore, this white could be readily reproduced by means of a standard lamp which could serve as a reference both in a television studio, and a manufacturer's quality control department. [18]
4. In February 1953, the NTSC revised its standards [19] to: sub-carrier frequency, 3.579545 MHz; line synchronizing frequency, 15,734.26 Hz; field frequency, 59.94 Hz; harmonic, 455. These changes were introduced so that the spacing of the sound and vision carrier frequencies could be kept at the exact value of 4.5 MHz for which existing black-and-white television receivers were designed. However, it was thought that in some monochrome receivers the attenuation of the sound carrier would be insufficient to prevent an objectionable 0.9 MHz beat signal occurring between the sound carrier and the chrominance sub-carrier. Experiments showed that this beat signal was much less disagreeable if its frequency was an odd multiple of one half of the line frequency – because of the frequency-interleaving effect mentioned previously. Careful consideration led to the sound carrier being chosen as the 286th harmonic of the line frequency (f_L). Hence $f_L = 15{,}734.26$ Hz. Since the number of lines per frame remained unchanged, the field frequency became $f_L/525/2 = 59.94$ Hz. The sub-carrier frequency was given by $455 \times f_L/2 = 3{,}579{,}545$ Hz. Using these revised figures the beat frequency was $17 \times f_L/2$ which satisfied the experimental finding.

References

1. Fink D.G. *Color Television Standards*. New York: McGraw-Hill; 1955
2. *Ibid*. p. 23
3. *Ibid*. pp. 24–25
4. Hirsch C.J., Bailey W.F., Loughlin B.D. 'Principles of NTSC compatible television'. *Electronics*. February 1952, pp. 88–95
5. Brown G.H. 'The choice of axes and bandwidth for the chrominance signals in NTSC color television'. *Proceedings of the IRE*. January 1954;**42**:58–9
6. Goldsmith, A.N. US patent application 2 335 180. 23 November 1943
7. Bedford A.V. 'Mixed highs in color television' *Proceedings of the IRE*. September 1950;**38**:1003
8. Baldwin M.W. 'Subjective sharpness of additive color pictures' *Proceedings of the IRE*. October 1951;**39**:1173–6
9. Loughren A.V., Hirsch C. J. 'Comparative analysis of color television systems'. *Electronics*. February 1951, pp. 92–6
10. Dome R.B. 'Frequency-interlace color television'. *Electronics*. September 1950, pp. 70–5
11. Gray F. US patent application 1 769 920. 30 April 1929
12. Mertz P., Gray F. 'A theory of scanning and its relation to the characteristics of the transmitted signal in telephotography and television'. *Bell System Technical Journal*, July 1934;**18**:464
13. Bowie R.M., Tyson B.F. 'The National Television System Committee color transmission, Part 1'. *Sylvania Technologist*. January 1952;**V**(1):10–16
14. McLean F.C. Letter to DTS, 15 October 1953, BBC file R53/39
15. FCC. Report and Order, adopted 17 December 1953, para. 40, p. 669
16. Fink D.G. *Color Television Standards*. New York: McGraw-Hill, New York; 1955, pp. 22–3
17. Anon. 'National Television Systems Committee'. *Proceedings of the IRE*. January 1954;**42**:15–16
18. Bowie R.M., Tyson B.F. 'The National Television System Committee color transmission, Part 1'. *Sylvania Technologist*. January 1952;**V**(1):15
19. Abrahams I.C. 'Choice of chrominance sub-carrier frequency in the NTSC standards'. *Proceedings of the IRE*. January 1954;**42**:79–80

Chapter 10
Colour television broadcasting in the USA 1953–1966

On the day, 17 December 1953, the Federal Communications Commission approved the NTSC's compatible colour television standards, the National Broadcasting Company commenced to broadcast colour television programmes on its entire television network. This historic occasion was followed almost immediately by many more colour 'firsts' for the NBC. The first major outside broadcast in colour, the Tournament of the Roses Parade, in Pasadena, California, was televised by the Company on New Year's Day 1954 and, again, was transmitted over its network. In 1955 a full Broadway production of *Peter Pan* was broadcast in colour, which attracted an audience estimated at 65 million viewers; President Eisenhower's address at West Point was telecast using the NBC's new outside broadcast colour television units; and the first colour coverage of a Yankees versus Dodgers game was distributed by the NBC. Other colour productions scheduled for 1955 included *Our Town*, *Cyrano de Bergerac, The Caine Mutiny Court Martial*, the Sadler's Wells Ballet Company, and the American premiere of Laurence Oliver's *Richard III* [1].

The NBC, like the RCA, was keen to further the new transmission medium and together they introduced various measures, which, it was hoped, would attract not only customers but also advertisers and sponsors. Initially, NBC made programme time available free of charge to sponsors so they could gain experience of the new facility; and RCA offered to modify free of charge existing RCA transmitters to enable them to transmit colour television signals [2]. By October 1955, 111 stations of the NBC network alone were equipped to broadcast in the new medium; by *c.* 1958 the figure was *c.* 50 per cent of all the transmitting stations in the USA; and the number of homes that could receive the signals was 96 per cent of the total of the nation's homes with television. RCA also provided training courses for television servicing technicians [3], and six months after the FCC's order more than 30,000 personnel had attended special courses provided in 65 cities.

Microwave links and coaxial cable networks also had to be modified to handle the colour television signals. Although the bandwidths of the monochrome and colour signals were the same, the latter required apparatus in which the amplitude and delay characteristics – especially the phase versus frequency response in the region of the colour sub-carrier frequency – of the various networks were more carefully controlled than with black-and-white circuits. Since the hue and saturation of the reproduced colour depended upon the phase and amplitude of the resultant chrominance signal, unnecessary differential phase shifts and amplitude variations in the transmission path had to be minimized to reduce the errors of these colour properties.

With the 1954 modern wideband microwave relays there was little difficulty in meeting the required transmission tolerances, but with the older coaxial cable system – which had a bandwidth of just 3 MHz – a problem existed, since the bandwidth of the colour television channel was 6 MHz. The solution adopted was to heterodyne (automatically) the colour sub-carrier frequency from *c*. 3.3 MHz to 2.6 MHz prior to the transmission of the signal along the cable, and then to restore the frequency at the receiving terminal. Of course, some signal components were lost in the process resulting in degradation of the received images. For the newer wideband cables no special measures were needed to transmit the colour signals.

By November 1955 the Bell System had *c*. 47,000 channel miles of cable capable of propagating the programmes of the colour television providers (see Figure 10.1). This represented approximately 70 per cent of the total installed capacity of the television network. The cable colour television service was accessible to about 250 broadcasting stations in *c*. 150 cities; and of the total of 440 television stations nearly half were equipped to broadcast colour television signals so that *c*. 90 per cent of the homes in the United States could receive a colour television programme [4].

One of the major providers of colour programmes was the National Broadcasting Company. It rapidly established its colour television facilities and in April 1955 had four studios in operation. The first large colour studio was the 190 ft × 90 ft (58 m × 27 m), studio at Avenue M, Brooklyn. This was followed in 1955 by the Center Theatre, the Ambassador Theatre, and large studios on 67th Street, New York. The Ziegfield and Hudson Theatres were added in 1956. 'Color City', in Burbank, California, a $7 million project, went on air on 27 March 1955: it was the first studio facility (with the main studio 140 ft × 110 ft/42.7 m × 33.5 m in size) in the United States to be specially designed and constructed for colour television. At that time it was the best colour television studio complex in existence and was used for many of NBC's early colour television programmes. For the reproduction of colour film images, the Radio City 4G/J film studios were equipped with six RCA TK-26 colour film cameras: it became operational in 1954. The 5H film facilities, with six additional cameras, were added in 1955 [5].

CBS had two studios at this time: studio 72, at Broadway and 81st Street, New York; and a large, 110 ft × 110 ft (33.5 m × 33.5 m), modern studio at their Television City Building at Hollywood, California [6].

The lighting power available to programme producers in some of these studios was impressive: 1.25 MW, divided into 640 kW of dimmable, and 610 kW of non-dimmable lighting power in the NBC Brooklyn studio; and 750 kW of lighting power in the CBS Hollywood studio. An indication of the complexity of a lighting installation for a large colour television studio may be given by considering the Brooklyn studio. Here, the whole of the upper portion of the studio was a complex arrangement of lamps all supported on battens, which could be raised and lowered by electric winches, of which there were 126 – a number that would be increased to 180. Means to control the winches, either singly or in banks, were provided at the control desk so that whatever the needs of the producer, and whatever the shape of the lighting installation required,

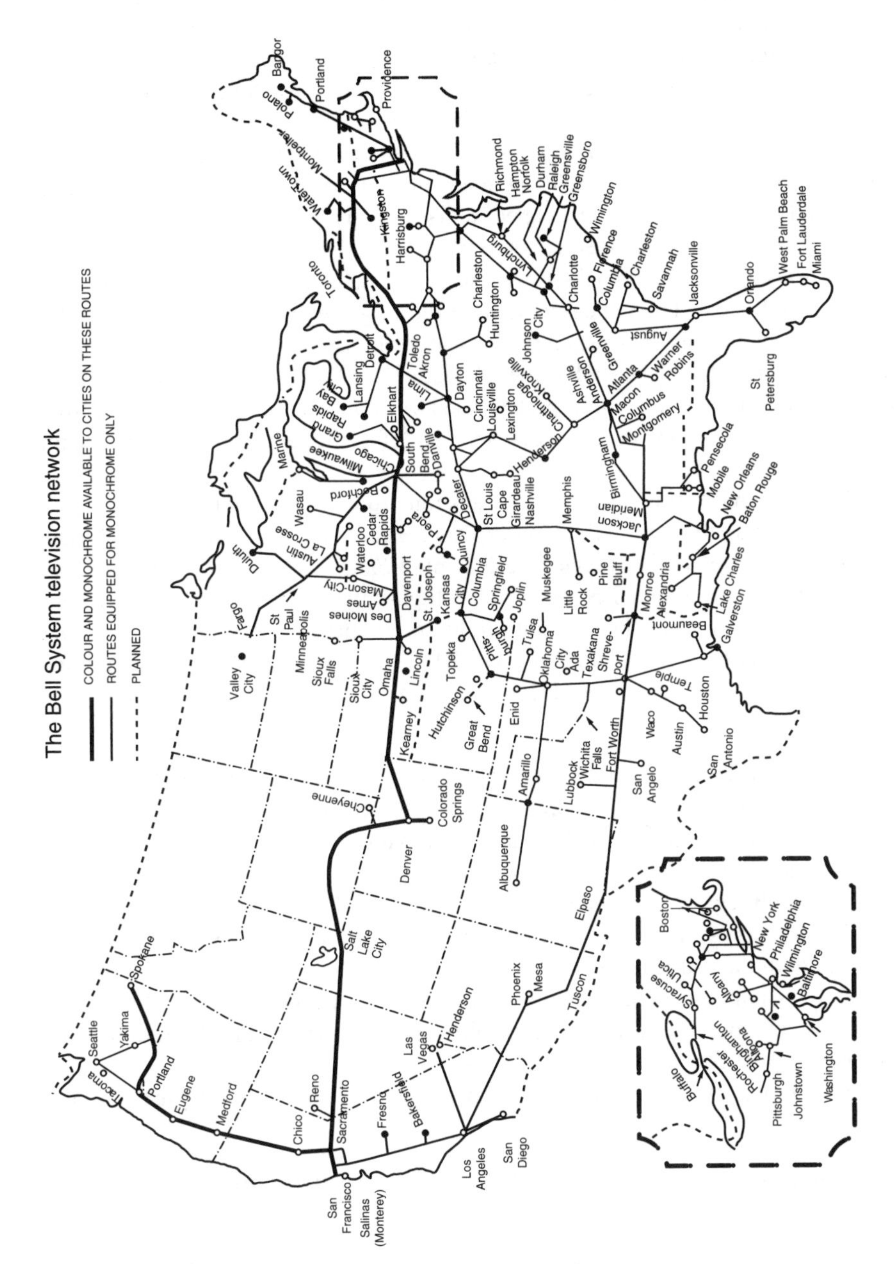

Figure 10.1 The Bell System television network in 1955 [Source: Author's collection]

the lighting supervisor could manoeuvre the lamps to give the planned illumination. Further control of the lighting pattern was provided by the use of dimmers. A major studio might have 100 dimmer circuits each enabling 10 preset positions to be selected, so that 10 complete lighting sequences could be pre-set before a production and subsequently selected by operating a single control [6].

One of the early visitors to experience the development of American colour television was D.C. Birkinshaw, of the BBC Television Service, who had been appointed to the first BBC post in television in 1932. In April 1955 with I. Atkins, the senior dramatic producer of the BBC Television Service, Birkinshaw viewed some studio facilities and subsequently wrote a paper on his findings [6]. He identified four problems associated with the management of a colour television studio: first, the relatively high cost as compared with black-and-white television; second, the need to limit the contrast range to 20:1; third, the requirement for a lighting level of 400 foot candles incident on the acting area corresponding to a power consumption of 180 W per square foot of acting area (for black-and-white television 100 foot candles sufficed, and for outside broadcasts pictures could be televised with an illumination of just 25 foot candles, or even less if absolutely necessary); and, fourth, the urgent requirement for a more sensitive camera.

Birkinshaw and Atkins saw several television productions, and their accompanying rehearsals, namely: a play *Reunion in Vienna*, and a musical play Merry Widow televised from the NBC Brooklyn studio; a play *Stage Door* and an experimental news programme from the CBS Studio 72; a light-entertainment production, *Kaleidoscope*, from the Colonial theatre; and a light-entertainment show, *Showers of Stars*, from the CBS studio at Television City. The BBC engineer was frankly disappointed with the appearance of the reproduced televised images and noted the following imperfections.

1. Faulty colour values. 'The whites and blacks . . . were always tinted; [f]lesh tones were far from faithful; sometimes they appeared to be too sallow, sometimes too vivid, sometimes too pallid, and they were far from consistent during a transmission'.
2. Lack of subtlety in the colours of the picture. 'There does not appear to be a sufficient variety of shades in the colour picture and in this respect it compare[d] poorly with the original scene'.
3. Lack of resolution. 'This was most noticeable. The pictures, viewed either by radio or by local monitor, were woolly and well below the definition to be expected from 525 lines'.
4. Poor lighting. 'Pictures were artistically inert due to the employment of flat lighting'.

Several factors contributed to these defects. Birkinshaw identified seven possible causes:

1. Poor registration in the display tube. He considered this to be its outstanding failing. There was a requirement for a tube that reproduced a white screen when fed with a white signal. 'Unbidden tinting, so noticeable in the whites and blacks,

must distort the values of any other colours which the tube is asked to transmit and must contribute to the faulty colour and lack of subtlety [noted] above.'

2. Incorrect effect of gamma, and rectification of chrominance information by the picture tube. Both of these technical defects distorted the colour values of the displayed image. Fortunately, their effects were partially opposed.
3. Restriction of luminance bandwidth. Complete implementation of the NTSC standards should ensure the invisibility of the chrominance information in the displayed image, 'but the difficulty of achieving this in practice has led American engineers to cut the luminance bandwidth at the receiver, starting at 3 MHz. This must impair the resolution'.
4. Registration of luminance and chrominance at the receiver screen. According to US standards the registration should be correct to 0.1 s, but in experiments in which Birkinshaw participated such precision was not maintained throughout the colour television broadcasting and receiving chain.
5. Instability of camera chain electronics. Colour values sometimes varied during a particular programme and often were not consistent from camera to camera. Birkinshaw ascribed this problem to 'the difficulty of adjusting, and keeping adjusted, the large amount of electronic circuitry associated with the camera equipment. This type of gear require[d] to be brought to a much higher state of electrical stability than [was] . . . manifested'.
6. Flat lighting. 'The pictures appeared to be flat, lifeless, and devoid of the attempt which [was] habitually made in black-and-white film and television technique to introduce a feeling of depth by means of back and side lighting.' The reason for this resulted from the use of the orthicon tube and its operation in a non-linear manner. When employed in black-and-white television studios, the orthicon was able to accept high contrast ratios, but the colour image orthicon was only able to accept low contrast ratios. In the UK, the BBC, from 1948, had utilized a proportion of c.p.s. (cathode potential stabilised) emitron cameras and this camera tube had no other mode of operation than the linear mode. A lighting technique had been evolved which gave the impression of depth while satisfying the characteristic of this tube. 'It follows', wrote Birkinshaw, 'that in the field of studio lighting, operators over here will be much better prepared when colour transmissions are eventually launched.'
7. Orthicon camera tube shading. From Birkinshaw's discussions with American engineers, it seemed the shading errors associated with the orthicon tube could not be cancelled out by the use of shading signals. There was a need for further circuit development since the shading errors were visible.

Birkinshaw's first-hand observations, and his interpretations based on many years of experience of British black-and-white television, led him to conclude:

> It is evident that we in this country [the UK] have some serious lessons to learn from these American experiences. It seems to be widely held that colour television in America has been launched too early. The author was told this several times by various authorities well qualified to express an opinion. . . . We must therefore be on our guard against premature launching of a colour service before the various problems [mentioned above] have been adequately surmounted . . . . The second conclusion is that much hangs on the receiver

> display tube. Nobody in any country engaged on these problems has yet made an adequate colour display tube and this is the most pressing development problem to be solved.... The third problem is that if it can be agreed that an important feature of television programming, and one without which it could not live, is the ability to do outside broadcasts, then it is difficult to see how colour television can be launched while they are virtually impracticable. This in turns waits for an improvement in camera sensitivity, a problem just as hard to solve as that of making a good display tube. [6]

Birkinshaw's comments were invaluable in providing guidelines for success when colour television broadcasting in the United Kingdom and other European countries eventually started.

However, there were additional matters that determined the progress of the new medium. One of these concerned the cost of colour receivers. Initially, the high prices of receivers, and the poor quality of reception, inhibited sales. A black-and-white receiver in the USA, in 1955, cost between $170 and $400, whereas a colour television receiver was priced between $795 and $890 [1]. The increase was due essentially to the price of $100 for a shadow mask tube, charged to set manufacturers by RCA, Sylvania, and one or two other tube manufacturers. This was five times the cost of a 21 in. black-and-white tube, and so, following the rule of thumb that the price of components had to be tripled to arrive at a retail price, the shadow mask tube represented *c.* $300 of that figure. Because of this the rest of the industry hesitated before engaging wholeheartedly in set fabrication. There was always a hope that a new and simpler colour display tube would be invented that would lead to substantially cheaper receivers. 'We are still working on tubes', C.B. Jolliffe, the RCA vice-president, reported in 1955, 'but the shadow mask tube is the best tube in sight at a reasonable price. There is nothing on the horizon.' [1]

This view was not quite correct. The Lawrence tube, or chromatron invented by E.O. Lawrence, who conceived the cyclotron, gave bright pictures; as did GE's post-acceleration tube which was demonstrated to the industry in September 1955. Both tubes were, at that time, inferior to the shadow mask tube – producing less sharp images – but there was always the prospect of further developments. Also Philco was working on a tube known as the 'apple' tube. 'It's still a horse race', the vice-president, D. Smith, said. 'No one knows which tube will win out. The initial tooling on a colour tube will cost several million dollars. You have to be sure you're right before you go ahead.' [7]

Apart from the cost of receivers and the imperfect quality of the displayed receiver images, several other factors hindered the early progress of American colour television, namely: the small number of hours per week of colour transmissions; the size of the displayed colour image; and broadcasting costs.

By the Spring of 1955 the total output of colour television programmes amounted to just one hour per week from NBC and the same from CBS, and an occasional transmission from Dumont. Such a paucity of provision did not encourage the general public to spend *c.* $800 on a colour set. This led to a 'Catch-22' situation. Colour television programmes were very expensive to produce (being *c.* 50 per cent more costly than black-and-white for a simple production, but *c.* 250 per cent higher for a 'spectacular' [8]); producers were dependent on sponsors and advertisers for financial backing; but without an adequate return, as measured by large scale viewer interest, the programme backers were not willing to provide the essential finance for the new service. Black-and-white television gave advertisers large audiences for their products, but the potential return, in 1955, from colour television advertising was most unattractive.

The initial advancement of colour television was further slowed by the relatively small screen size, *c.* 15 in. (38 cm), of the early receivers [8]. Birkinshaw observed a colour transmission on a 15 in. (38 cm) colour receiver and a 21 in. (53 cm) monochrome receiver side by side, and 'found it very difficult to take much notice of the colour picture in the presence of the black-and-white one' [8]. Picture size was important but the trend to manufacture larger colour tubes (e.g. 21 in. tubes) meant even higher receiver costs and did not conduce to an expansion in the number of viewers.

All of this was discouraging to the television industry and one by one the colour television set makers, except RCA, ceased production. CBS withdrew from the television scene leaving just NBC to carry on. Naturally, RCA wished to see a flourishing market for colour television sets [9]. It had spent a vast sum of money on research and development – one source [10] quotes a figure of $130 million from 1949 to 1963 – and so, with its broadcasting arm, NBC, had to persist in its endeavours to stimulate an awareness of the attractions of the new medium. Among the developments that originated in the NBC laboratories were special types of video effects, including chroma key, and the pioneering, with RCA engineers, of a colour heterodyne system for recording colour television images on magnetic tape. In 1957 NBC began the first delayed broadcast of colour shows from its Hollywood studios. One year later this operation was transferred to NBC's new Burbank studio, where eight colour television recorders were installed to serve the NBC network. Afterwards, the number was 16 at Burbank and 20 in the New York studios. These video recorders were provided with remote start-stop and mode selection facilities developed by NBC's engineers, and means were devised to record and edit video tape using a 16 mm magnetic sound recorder and a 16 mm kinescope recorder.

The prime determinants in the strategy to promote sales of receivers were the reduction of the purchase price of a colour set, and the increase in the number of hours per week of broadcast colour television. Two factors aided the achievement of the former objective [11]. The number of valves was reduced from 44 in the first models to 26 in 1956; and the original metal-coned display tube (21AXP22) was replaced by an all-glass colour display tube (21CYP22). (By 1964 RCA claimed the life expectancy of its colour tube was the same as that of a monochrome tube, approximately six years.)

The installation and servicing of receivers was also simplified. At first, the installation of a colour television receiver, which involved the measurement of signal strength and the adjustments associated with convergence and colour temperature, was a long and hence costly process. One report mentions that with the early television set chassis, about 30 adjustments were necessary and that the time required for this was four to five hours. In later sets the number of adjustments was a maximum of *c.* 15, but usually only minor alterations (three or four) were needed and the time required was reduced to less than half an hour. Together these improvements and modifications enabled the average receiver cost to fall from £300 in 1955 to £180 in 1956.

The uphill struggle to popularize colour television in the USA was pursued, until *c.* 1959, almost single-handedly by RCA and its subsidiary NBC. A hundred and thirty million dollars had been spent by the Corporation on colour television [12] but its conviction that success – as measured by sales of sets – would be achieved never wavered. In 1959 NBC expanded its colour studio facilities, continued its seven-days-per-week colour schedule, and increased the number of hours per week when colour television was shown. All of this cost NBC $4.75 million for a 26-week season.

Slowly, the position improved. Sales of black-and-white receivers, which had achieved a *c.* 85 per cent saturation of the potential market, began to decline. By 1959 this fall had been at an annual rate of $100,000,000 for several years. Also, NBC found that the addition of colour could increase a show's audience by as much as 80 per cent. These two factors possibly stimulated advertisers and sponsors to reconsider their previous hesitation and to engage to a greater extent in funding the new medium [13].

In 1960 RCA made its first profit – measured in seven figures – on colour television receiver sales since the introduction of colour television in 1954. Most of these sales were for the 'middle-of-the-line' models – those retailing in the $600 to $750 price range – and the average buyer was in the $8,000–9,000 income group. By the following year (1961) colour television products – receivers, tubes, videotape and other related equipment – achieved the status of a $100 million business [14], a feat of no mean proportions in seven years. For comparison, this figure had been attained by the petroleum industry only after *c.* 40 years of endeavour.

Nineteen sixty-one was the year of the breakthrough in colour television acceptability. One by one the television receiver manufacturers gave up their wait-and-see attitude and embraced the new medium. Before the end of 1962 nearly every major television manufacturer was actively fabricating and marketing colour receivers, resulting in an industry sales volume of *c.* $200 million. 'RCA's set sales for 1962 doubled over the year before; its profits from color manufacturing and services increased fivefold; and color sets and tubes became the *largest* single profit contributor of any products sold by the company.' [12]

This growth was linked to an expansion of the number of programmes broadcast. In 1958 there were 291 stations equipped to broadcast just *c.* 700 hours of colour television programmes [12]. Slowly, the NBC schedule of colour broadcasts increased from the initial four hours per month to *c.* 120 hours per month by May 1961. The company's listing for that month showed 4.25 hours of regular colour programmes – each of 30 minutes' duration – every weekday, plus an average of an hour each weeknight, two hours on Saturdays and 2.5 hours on Sundays. Additionally, 'specials' were broadcast from time to time, and there were several day-time colour broadcasts to allow dealers to demonstrate their goods [9]. Five years later the number of stations was 406 and *c.* 2,000 hours of programmes were transmitted in the year.

Table 10.1 Progress of colour television in the United States [10]

	Sets sold (1,000s)	**Broadcast hours average per week**	**Cost in £***
1955	30	5.0	300
1956	90	10.8	180
1957	90	13.4	180
1958	85	13.2	180
1959	90	14.1	180
1960	120	20.0	180
1961	160	31.7	180
1962	425	37.5	160
1963	725	42.0	150
1964	1,400	45.0	132
1965 estimated	2,250	77.0+	107

* cheapest colour television set

The correlation between the number of sets sold, the cost of the sets and the number of broadcast hours, is immediately evident from the sales of colour receivers for the period 1955 to 1965 given in Table 10.1.

The figures clearly showed that as the average cost of a set decreased and as the number of hours broadcast per week increased so the sales of sets increased. (For comparison, in 1955 the industry's forecast for colour receiver sales varied between 100,000 and 500,000; and for the subsequent years Sylvania optimistically estimated the following sales figures: 1957, 750,000; 1958, 1,500,000; 1959, 2,250,000; 1960, 3,000,000.)

The Advertising Research Foundation in New York estimated in August 1965 that approximately 7 per cent of all households in the United States had a colour television set. Their geographical distribution was, of course, not uniform. The Nielson Station Index (March 1965) found three west coast markets – Las Vegas, Santa Barbara and Fresno – as having the highest penetration at 14 per cent, though it was thought the figure for Metropolitan Los Angeles had increased to *c.* 20 per cent by the end of 1965 [10].

In 1964 the consumer market of 1.4 million colour TV receivers was shared between RCA with 40 per cent, Zenith with 25 per cent and another 25 per cent shared between Admiral, Motorola and Warwick Electronics (a Sears Roebuck subsidiary). The remaining 10 per cent comprised the output of all the other manufacturers.

Notwithstanding the welcome progress being made, some consumer resistance to colour television still existed ten years after the first colour television broadcasts. A 1964 Consumer Research Service survey on the attitudes of the populace to the non-purchase of a colour television set found various reasons were given (the total of more than 100 per cent is due to multiple choices) [10]:

- cost: too high: 59 per cent;
- colour TV not perfected: 18 per cent;

- too expensive to maintain and operate: 10 per cent;
- satisfied with present set: 5 per cent;
- poor reception in area: 5 per cent;
- lack of colour programmes: 5 per cent;
- just don't care for them: 4 per cent.

However, these points were being addressed. Colour television broadcasts, as noted previously, were increasing and considerable effort was being made to provide attractive and interesting schedules. In 1964 NBC broadcast 70 per cent of its peak time (19.30 to 23.00 hours) output in colour. Towards the end of the year NBC's colour TV output had consisted of: quiz shows, 35 per cent; variety, 32 per cent; drama, 10 per cent; films, 7 per cent; children's programmes, 2 per cent; and other programmes including sport and 'specials', 14 per cent. Of these, the majority were from prerecorded videotape; namely, videotape, 61 per cent; filmed material 26 per cent; and 'live' (including sport), 13 per cent [10].

Other programme makers were currently producing 50 per cent (CBS) and 40 per cent (ABC) of their peak output in colour.

Not everyone was happy with this *prima facie* desirable state of affairs. A *New York Times* television critic noted in his review of the autumn 1965 colour television programmes that they were suitable only for the 'under-developed segment of the population' [10]. Apart from the worth, or perhaps worthlessness, of the content of the productions, there was a view that the quality of reception was, at times, less than what could be attained:

> It is believed that in America there [was] a lack of agreement between stations and a degree of carelessness amongst studio staffs and technicians which [was] not usually experienced here [in the UK]. Both these faults affect[ed] the consistency of the colour information in the transmitted signal and cause[d] odd and uneven colours to be viewed on the sets. The unevenness [was] particularly noticeable when a viewer switche[d] from one station to another. [10]

As the number of hours of colour television broadcasts increased, so the number of advertisers also increased: 31 in 1962, 92 in 1964, and 130 (estimated) in 1966. The number of colour commercials in 1966 was approximately 20 per cent of the total of all broadcast commercials. No premium was charged by the network operators for transmitting colour commercials but it was anticipated such a charge would be made when audience participation had grown to a suitable figure. Some networks introduced a handling charge for colour advertisements. CBS levied $125 for each such commercial shown in the daytime and $250 when shown during peak evening time [10].

Elsewhere in the world, much development work had been progressed in the United Kingdom from *c.* 1954 by the BBC, MWT and EMI, and various demonstrations, field tests and regular experimental transmissions had been conducted. This work is described in the next chapter. In Japan, following experimental work and field trials undertaken from 1957 by Nippon Hoso Kyokai (NHK) and Nihon-TV, the first regular colour television broadcasts commenced on 10 September 1960. By the second week in November eight stations – four in Tokyo and four in Osaka – were providing regular daily schedules varying in length from 30 minutes to 2.5 hours in total. Approximately 1,000 sets had been installed by this time, mostly in bars and other public places. According to a report of the Television Society, the Japanese government predicted the growth of installed sets as: 12,000 in 1961; 30,000 in 1962; and 100,000 in 1964 – the year of the Olympic Games [10].

References

1. Bello F. 'Color TV: Who'll Buy a Triumph?' *Fortune*. November 1955, pp. 136–7, 201–2, 204, 206
2. Mayer C.G. 'Color television broadcasting in the USA'. *Electronic Engineering*. August 1954, pp. 342–7
3. Engstrom E.W. 'RCA an historical perspective', Part II: 'The years 1938–1958'; Part III: 'The years 1958–1962'. *RCA Engineer*. 1967
4. Mayer C.G. 'Color television in the USA'. *Electronic Engineering*. November 1955, pp. 488–91
5. Howard W. 'NBC Engineering – a fifty-year history', Part II. pp. 32–9, personal collection
6. Birkinshaw D.C. 'American colour television: the present state'. *Journal of the Television Society*. 1955;**7**(5):509–15
7. Bello F. 'Color TV: Who'll Buy a Triumph?' *Fortune*. November 1955, p. 137
8. Birkinshaw D.C. 'American colour television: the present state'. *Journal of the Television Society*. 1955;**7**(5):509
9. Darr J. 'Color television in the US'. *Wireless World*. July 1961, pp. 358–60
10. Damber D., Ran A. *Colour Television – a report*, Booklet 16. London: J. Walter Thompson; 1966. 37 pp.
11. Darr J. 'Color television in the US'. *Wireless World*. July 1961, p. 358
12. Engstrom E.W. 'RCA an historical perspective', 'Part III – The years 1958–1962'. *RCA Engineer*. 1967, p. 7
13. Anon. 'Colour television in the United States', BBC file R53/395/1
14. Anon. 'Color television's rosy hue', *Broadcasting*, 2 January 1961, p. 50

Chapter 11
British developments, 1949–1962

During the immediate postwar period, and, indeed, until the mid-1950s, little original research and development work on colour television was in progress in war-torn Europe. The need to make good the ravages wreaked by the munitions of war, the requirement for post-war financial stringency, and the necessity to convert industry from a wartime to a peacetime role, inhibited the establishment of what essentially would be a luxury service. Those European countries that aspired, ultimately, to providing a colour television broadcasting service were content to maintain a watching brief, augmented by appropriate experimental work, on the post-war R&D activities of, *inter alia*, CBS, RCA and CTI, and on the proceedings of the Federal Communications Commission, the Condon Committee and the National Television System Committee. All their efforts were widely reported and could be studied and investigated at low cost. Significantly, the main rivals to the NTSC system, the French SECAM system and the German PAL system, were not invented until 1959 and 1962 respectively.

Among the post-war European studies of colour television those of the BBC may be related because, in the 1950s, the BBC carried out more extensive, carefully conducted investigations of the relative merits of the several systems that were being advanced than were being undertaken elsewhere. This work resulted in 70 technical publications.

The BBC's experiments on colour television commenced in 1949. An internal memorandum, dated 26 October mentions, 'Experiments on the basic principles of colour television have been started in the Corporation's Research Department Laboratories, where a study is being made of all [sic] developments, at home and abroad, relating to colour television' [1].

Among these developments a sequential system comprising an iconoscope camera and a rotating colour filter disc was tried but it was felt to be impracticable because of the use of mechanical components and because of the high light levels that were required.

At about this time the BBC was giving much thought to the expansion of its television service. A 13.5-acre site had been purchased at White City, London, and a Television Centre Development Committee had been set up. In planning for the future growth of television, the Committee had to consider the need for more studios and control rooms but no knowledge existed in the BBC on the demands that colour television would make for space, lighting, ventilation and power. Fortunately, in 1950, CBS had a colour television broadcasting service in operation, using its sequential colour television scheme, and NBC was experimenting with RCA's colour television systems. Both of these organizations had acquired some experience, albeit very limited, on the new medium and so a small BBC working party was sent to the USA to view the CBS and NBC colour studios at Hollywood and Burbank. The party found that relative to black-and-white television studio apparatus, the colour television studio apparatus needed approximately 1.5 times as much floor space. This finding was used in the planning of the Television Centre [2]. The BBC believed there was a need to keep abreast of colour television developments even though these were unlikely to be a practical factor for many years [3]: its policy was predicated on the establishment of nationwide monochrome television before that of colour television [4].

In 1950, the BBC's Board of Governors reviewed its policy on colour television and, on the advice of its Scientific Advisory Committee, decreed that the Corporation should refrain from undertaking a major effort on the development of a colour television system, and that any preliminary work on colour television should be undertaken by private industry [5]. Possibly, the Governors were somewhat alarmed by the reports from the United States regarding the costs that RCA and CBS had borne in developing their colour television systems. As a consequence, letters [6] were sent, in June, to Electric and Musical Industries (EMI), Marconi's Wireless Telegraph Company (MWT) and Pye Ltd: the BBC expressed its desire to acquire, for experimental work, two or three single camera units, each employing a different method of working and each capable of giving colour television of studio and outdoor subjects. These firms were selected for various reasons: EMI, with MWT, had engineered the world's first public high-definition television system and had been very active in this field before the war; MWT were RCA licensees; and the Pye company, which had obtained the British rights of the CBS mechanical system, had developed colour television on the Goldmark principle and had demonstrated it at the 1949 Radiolympia [7], though only as a medical and industrial aid and not, as CBS intended, for domestic receivers. The response to the BBC's letter was initially disappointing [5]. All three firms felt unable to allocate scientific staff to the request because of the pressure of defence orders. A contributing factor was probably the fact that, as the BBC admitted, no firm in the United Kingdom had made a profit from the sale to the BBC of cameras, camera equipment, or transmitters [8]. The companies engaged in monochrome television work had borne the losses involved in these sales for prestige reasons because they were manufacturers selling equipment around the world.

By 1953 this position had changed. In June G. Barnes (the BBC's director of television) saw a demonstration of EMI's system for converting sequential colour television signals (of the CBS type) into simultaneous colour television signals. He wrote to EMI's research manager, G. Condliffe, and mentioned [9]:

> This seemed revolutionary, particularly the effect in an outside broadcast. I had hoped to put on one side all thinking about colour for at least a year, but I see now that once [the] United States has made up its mind which way it intends to jump, not only have we got to think how to broadcast colour (this is your headache rather than mine) but we have to get an experimental channel on BBC premises in order to teach ourselves how to get those effects in colour which we have been achieving so laboriously in monochrome over the last few years.

At this time the BBC's Research Department appears to have scaled down its previous (1949) objective of studying 'all developments, at home and abroad, relating to colour television'. The main subjects of investigation in this field were now just the colour and optics associated with the new medium [10]. A BBC memo dated 10 September 1953 noted: 'There was little chance of a new independent British system being produced; and it would take 6–9 months to reproduce [the] American equipment' [11]. The well-publicised expenditure of $25 million on colour television by RCA and its estimate of a further $15 million to bring it to a commercial state must have had a constraining effect on any proposal for an entirely new system [12]. Furthermore, the available reports of the American colour television scene, and the views of the BBC, did not conduce to an early start of a colour television service in the UK. Their technical staff were not convinced that the NTSC system was sufficiently satisfactory to 'warrant heavy expenditure on the part of both the broadcaster and the viewer'; the RCA colour camera, consisting of three image orthicon tubes, was 'elaborate, very large, and costly'; and the colour receiver was still the 'Achilles Heel of all colour systems'. In the opinion of a member of the BBC's Research Department who saw RCA's tri-colour picture tube demonstrated, in New York in April 1953, it was 'not yet a practical proposition for use in the home' [5].

Of the British companies likely to further colour television, EMI was working on a system known as the chromacoder; Pye were proposing to demonstrate the Lawrence receiver tube, but had undertaken little, if any, fundamental research; and the Marconi Company had set up at their Chelmsford works a version of the NTSC system suitable for British standards, and were working on ideas that might lead to an improved colour camera and colour receiver. The company also had 'at least one engineer apparently attached to the RCA research organisation'. Marconi's policy was to produce an improved system that could be manufactured and sold in world markets in competition with the products of American companies. However, the latter had a long start and Marconi's efforts were dependent on just a few engineers. Nonetheless, the head of the BBC's Research Department was able to report, following a visit to Marconi's works, that a 'great deal has been done in the past two years' [13].

All of this caused the BBC's director of television services (H. Bishop) some disquiet. Though EMI and MWT had been encouraged on several occasions by the BBC, and had been told the Corporation would be prepared to purchase 'equipment of promise', the slowness of their activities on colour television was a matter of concern. Bishop expressed his unease to the BBC's Board of Management:

> We, who started a public television service four years before the Americans, are now likely to be left well behind in colour television and may eventually have to adopt an American system because the Americans have so thoroughly explored the field. This position would be full of danger to the BBC if our future competitors were to find some way of getting into a more favourable position with America while we might wait hopefully for a British firm to produce a different system. The Americans might also be in a very favourable position on patents and this might be costly to us. [14]

In a further memorandum [12], dated 4 November 1953, Bishop opined it was most unlikely an entirely new British system independent of US patents, or differing materially from the NTSC system, could be evolved. On this system, Bishop felt it was not yet certain that the endeavours being made to adapt it for use in the UK's Bands I and III would be successful, or, if so, when. Nevertheless, to aid its aspirations the BBC invited L.H. Bedford and L.C. Jesty, of MWT, to the BBC's Research Department to discuss the question of collaboration [15]. This led in January 1954 to a BBC engineer being seconded to join the MWT R&D team, led by Jesty, at Great Baddow, which was developing equipment for a demonstration of colour television later that year.

In 1953 F.C. McLean (deputy chief engineer, BBC), as mentioned previously, visited the USA and attended the first demonstration of the NTSC system. From the various technical reports that he brought back with him, the Engineering Division agreed the NTSC system represented the correct approach and that experiments should be initiated at once [16]. By 1954 an NTSC type of signal, suitable for the British 405-line system, had been synthesized at the BBC's Kingswood Warren Research Department and had become the basis for further studies. Soon afterwards the Television Advisory Committee (TAC) requested its Technical Subcommittee to advise on the system of colour television to be standardised in the UK and to report within 12 months [17].

The Corporation would certainly have preferred to have delayed, if possible, the introduction of colour television so that greater progress could be made with its current developments. Much effort was being expended by the BBC on the establishment of the first chain of television stations to cover the UK, and on the expansion of the White City Television Centre, which were planned to be completed in 1955 and 1958 respectively. The Corporation averred that '...it seems improbable that the introduction of colour can be postponed as long as that, but an early and ill-considered introduction of colour before the system has been properly developed for British standards and before receivers [were] comparatively foolproof and somewhat cheaper, would seem to the Corporation to be inadvisable.' [18]

For the BBC, the process of introducing colour transmissions would involve two phases: namely, a period of experimental transmissions, and then general and advertised transmissions. The latter would begin with a weekly schedule of, say, half an hour per week on all BBC transmitters, but would steadily increase as resources allowed until in the long run the majority, if not all, of its programmes would be in colour. This was the way forward, said the BBC. And it noted, 'It need hardly be emphasised that the Corporation must be first in the field in a development of this importance, and will certainly proceed with the maximum speed that is justified by the technical factors.' [18]

At this stage in the narrative it is desirable to digress and consider some aspects of the NTSC system that, possibly, could be improved. (The following section may be omitted by non-technically minded readers.)

The work of the National Television Standards Committee was outstanding in its investigation of the colour television problem, and its solution based on the constant luminance principle, the channel-conserving principle of mixed 'highs', the channel-sharing principle of the interleaving of the luminance and the chrominance harmonic components, and the use of colour-difference signals and quadrature modulation was a splendid endorsement of NTSC's methodology to embrace the best ideas that American industry could offer at that time. Nevertheless, there seemed to be some scope for improvement. The Federal Communications Commission in its 1953 cautiously worded approval statement [19] referred to the NTSC's specifications as capable of producing a *reasonably satisfactory* picture, the quality of which was not *appreciably marred* by certain defects. The picture contrast was *satisfactory* under *favourable ambient light conditions*, and colour pictures could be transmitted *satisfactorily* over intercity relay facilities, but *improvements* in intercity facilities could be *reasonably expected* (author's italics). This suggested that further effort would be worthwhile. Confirmation came when reports began to appear concerning the variability of the hue of the received colour pictures. One wag commented that NTSC was an abbreviation for 'Not Twice the Same Colour'.

One cause of colour picture deterioration stemmed from the use of frequency interleaving of the luminance and the chrominance harmonic components since there was a possibility of cross-talk between the two signals. This feature of the NTSC signal led to several luminance/chrominance schemes [20] (see Figure 11.1) being proposed to overcome or reduce any limitations produced by this NTSC process. (For these figures, the red, green and blue colour information has been encoded into luminance and two chrominance signals, e.g. two colour-difference signals.)

Figure 11.1(a) shows the NTSC configuration and Figure 11.1(b), (e) and (f) illustrate variants of it. Two wideband arrangements are displayed in Figure 11.1(c) and (d).

Scheme (b), proposed by Dome of GE, and demonstrated (*c.* 1955) to the CCIR in another form by the Philips Laboratories of Eindhoven, uses two different colour subcarriers located inside the luminance channel to negate the difficulties of quadrature modulation. It is discussed later in this chapter.

Scheme (c) was proposed by P. Raibourn (vice-president of Paramount Pictures, USA) to the NTSC in 1952, in an attempt to simplify the problem of adapting an existing monochrome receiver to receive compatible colour transmissions in colour. The chrominance signals would be positioned in the adjacent channel to the luminance signal so that all that would be necessary to enable an existing black-and-white receiver to receive colour information would be an additional receiver and tri-colour tube [21]. Additionally, the method would eliminate cross-talk between the luminance and chrominance components but, of course, cross-talk could result from the chrominance signals of one channel interfering with the luminance signal of the adjacent channel. By careful geographical distribution of interfering transmitter stations, or by the use of directional antennas and/or different directions of polarization for the luminance and chrominance components some reduction of cross-talk could be achieved. This method was investigated by the British Radio & Electronic Equipment Manufacturers' Association (BREMA) and a map of England and Wales was produced by them (see Figure 11.2) illustrating the areas where cross-talk difficulties might arise if scheme (c) were implemented. (This point is discussed in Note 1 at the end of the chapter.)

The absence of an adjacent signal channel would eliminate cross-talk but a given transmitting station would require a much wider frequency band, as depicted in Figure 11.1(d). In this band the luminance and chrominance components would be

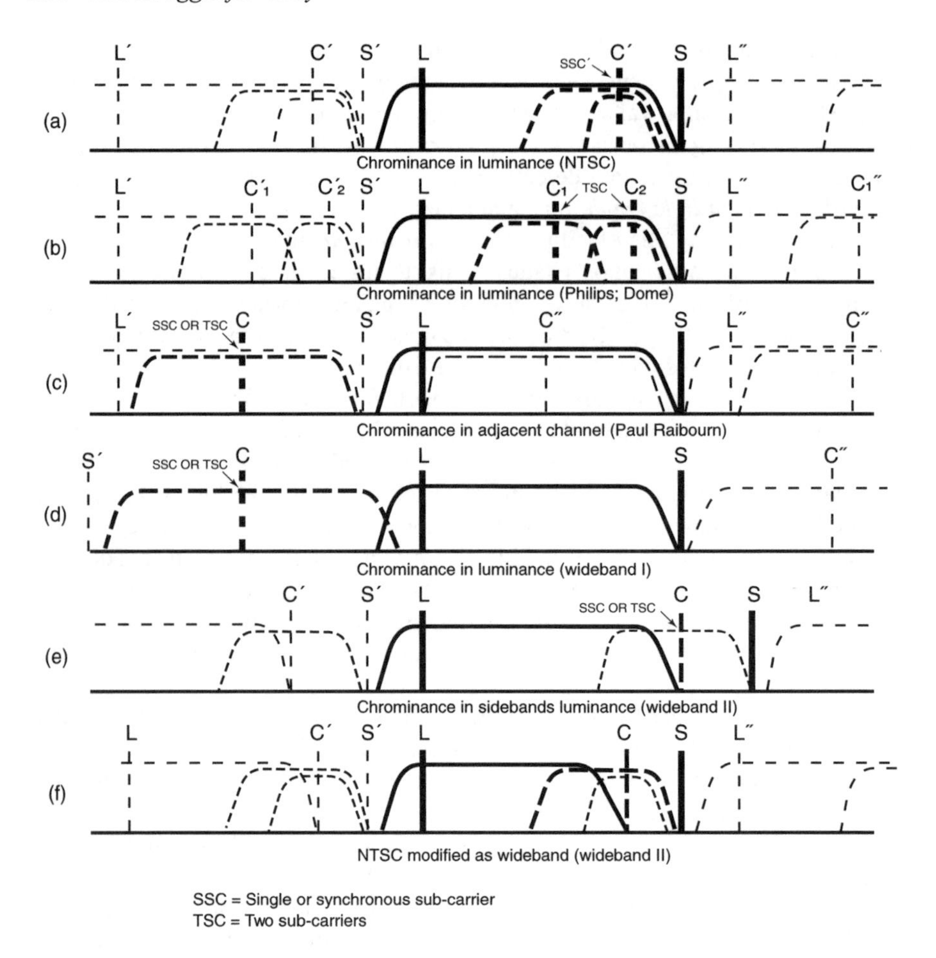

Figure 11.1 Classification of luminance/chrominance systems [Source: Journal of the Television Society*, 1955, 7(12), p. 493]*

transmitted separately and without any mutual interference. An advantage of the wideband scheme, and that of the Raibourn proposal, is that a wider chrominance bandwidth would be possible, thus reducing the distortion present on colour transitions.

An alternative to (d), which is more economical in channel bandwidth, is shown in Figure 11.1(e). The luminance bandwidth remains unaltered but the channel bandwidth is increased to permit the colour sub-carrier being positioned outside the luminance bandwidth. Since the principal harmonic chrominance components would be outside the luminance channel, the arrangement would lead to a reduction of cross-talk. The arrangement would necessitate a reallocation of the sound carrier and thereby the configuration would not be compatible with existing monochrome transmissions.

A modification of the NTSC system is illustrated in Figure 11.1(f). Here the luminance channel bandwidth has been restricted to reduce the cross-talk between the

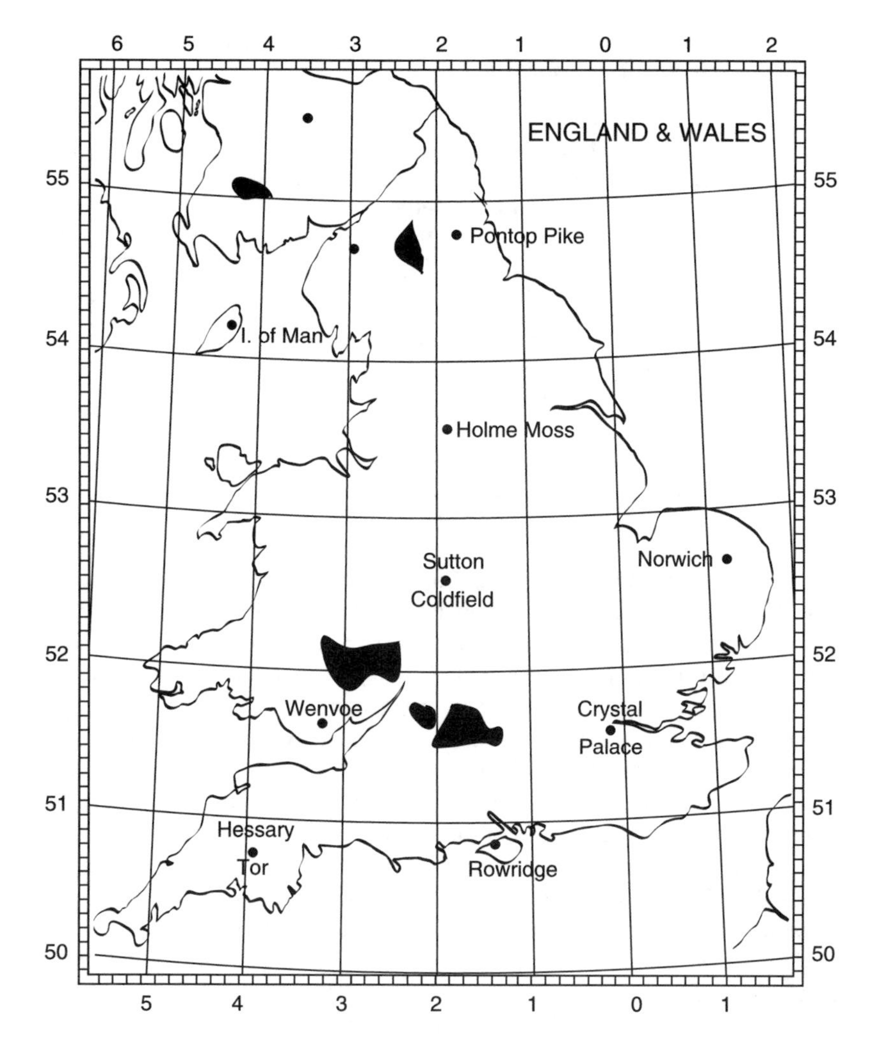

Figure 11.2 Map of England and Wales showing the areas where cross-talk could be experienced [Source: BREMA*]*

luminance and chrominance signals. By reducing the chrominance power transmitted, the method can lead to a reduction in colour dot structure on those monochrome receivers having a bandwidth sufficiently large to include the colour sub-carrier.

Table 11.1 summarises the channel and luminance bandwidths required for monochrome, NTSC and wideband colour television systems as a function of the number of lines per picture. The figures quoted were based on the assumption that the attempt to introduce chrominance signals into the luminance channel in order to maintain compatibility of transmission standards automatically reduced the effective bandwidth of the luminance channel by at least 25 per cent, due to cross-talk between

Table 11.1 Possible luminance/chrominance systems, based on 1955 European monochrome standards [22]

	Monochrome		NTSC		Wideband	
	Channel width MHz	Luminance bandwidth MHz	Channel width MHz	Luminance bandwidth MHz	Channel width MHz	Luminance bandwidth MHz
405 lines	5.0	3.0	5.0	2.25	7.5	3.0
525 lines (for reference)	6.0	4.0 (3.3)*	6.0	3.0 (2.5)*	–	–
625 lines	7.0	5.0	7.0	3.75	10.5	5.0
819 lines	14.0	10.4	14.0	7.8	21.0	10.4
625 lines (2:1 aspect ratio	–	–	–	–	14.5	7.5

* effective at 50 Hz field frequency

the chrominance and luminance signals. Fink, of the NTSC, admitted a figure of 12.5 per cent. The channel width of the wideband system was assumed to be 1.5 times the monochrome channel width that it replaced. A 625-line wideband system was considered because a suggestion had been made that all new television standards should take into account recent developments and should conform to the latest motion-picture practice to ensure a ready supply of films for future television programmes.

The work of the Marconi Company (described below) demonstrated an interesting paradox, namely, that a receiver having a poor frequency response (say, a bandwidth of just 2.2 MHz instead of the 3.0 MHz permitted) gave a better picture, when receiving a compatible colour television signal, than a receiver having a bandwidth of 3.0 MHz. This arose from the lack of cross-talk between the luminance and chrominance components in the frequency range (2.3–3.0) MHz.

On 11 May 1954 the BBC's patience with the companies – EMI, MWT and Pye – which it had encouraged on several occasions, from 1950, to develop colour television apparatus was finally rewarded. Following a comprehensive investigation of colour television, apparatus developed by MWT was demonstrated to the press, government bodies and scientific organizations, including the Television Advisory Committee, the General Post Office and the British Broadcasting Corporation. The Company's press release [20] mentioned:

> The following could be demonstrated: the colour camera; the colour tele-cine and slide scanner; the signal encoding equipment built to the British version of the NTSC standards; the effects of bandwidth alteration with the above; the 'wideband' colour system; the effect of bandwidth alteration with the above; a comparison of colour pictures produced without bandwidth compression; compatibility, with particular reference to the effect of various forms of bandwidth compression on this; receivers using tri-colour tubes; and receivers using three projection tubes.

Altogether, the demonstration was an exemplar of what could be done with very limited resources.

The picture-generating equipment consisted of a flying-spot scanner adaptable, by rotating the scanning tube through 90°, to scan either 35 mm colour transparencies or 16 mm colour films; and an experimental two-tube colour camera of a novel design [23]. Following encoding and decoding, the colour television signals were displayed on two pairs of monitors so that comparisons could be made of two different system configurations. One pair of colour monitors was based on Philips's 2 in. (5.1 cm) projection tubes, the red, green and blue images being optically superimposed with the aid of dichroic mirrors. Another pair of monitors used RCA 16 in. (40.6 cm) tri-colour tubes with modified toroidal deflection and convergence circuitry. Spot wobble was made available to cut out beating patterns between the colour dots on the tube screens and the scanning lines. This would otherwise have been objectionable with 405-line scanning due to this type of tube having almost exactly 400 sets of phosphor dots in the picture height.

Two methods of transmission with alternative choices of bandwidth were considered. The first of these methods was a straightforward adaptation of the NTSC system to the British 405-line television standard (see Figure 11.3). In this method the full bandwidth of 3 MHz available for the vision signal was utilized for the luminance signal, and the two colour-difference signals were quadrature modulated on a colour sub-carrier of 2.6578125 MHz within the luminance band. In addition a colour synchronizing signal, comprising nine cycles of the sub-carrier, was superimposed on the back porch of the video signal; and the peak white signal to synchronizing

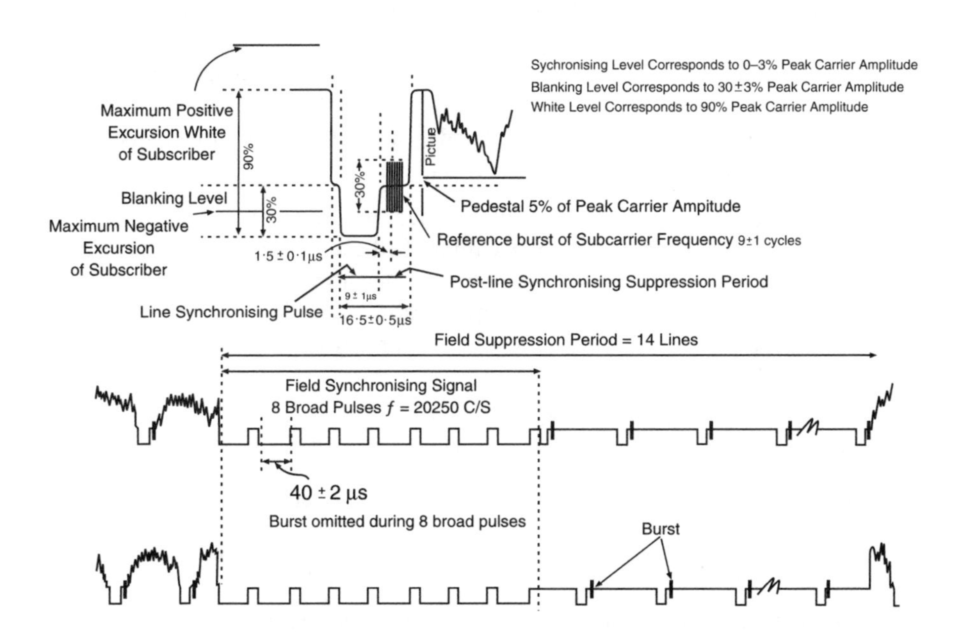

Figure 11.3 Complete waveform for the proposed British adaptation of the NTSC standards [Source: Journal of the Television Society, *1954,* 7*(6), pp. 241–7]*

signal ratio was reduced from 70/30 to 60/30 to accommodate partially the sub-carrier overswing.

In the second method, known as the wideband system, the chrominance information was transmitted outside the 3 MHz luminance channel – the chrominance signal occupying a bandwidth of *c.* 1.5 MHz for the two colour-difference signals of 1.0 MHz and 0.4 MHz bandwidth. (Non-technically minded readers may omit the following section.)

The essential components of the apparatus, excluding the picture signal generators and display devices are illustrated in Figure 11.4; and the various system frequency characteristics are given in Figure 11.5. Simultaneous 405-line red, green and blue video signals, each occupying a 3 MHz channel, were provided from the picture generator, and fed by line either direct, or after coding and decoding, to the colour and monochrome picture display tubes. The following basic schemes and signals were available from the Marconi equipment:

1. simultaneous 3 MHz channels for red, green and blue (Type I);
2. a band-saving and band-sharing (NTSC type) system decoded to red, green, and blue signals (Type IIA);
3. a band-saving (wideband) system decoded to red, green, and blue signals (Type IIB);
4. a coded signal of Type IIA, for display on laboratory monochrome monitors (Type IIIA);
5. a luminance signal of Type IIB, for display on laboratory monochrome monitors (Type IIIB);
6. a coded signal of Type IIA, modulated on a 45 MHz carrier for compatibility tests on standard production TV receivers (Type IVA); and
7. a luminance signal of Type IIB, modulated on a 45 MHz carrier for compatibility tests on standard production TV receivers (Type IVB)

(Types I to IIIB inclusive operated wholly at video frequencies.)

The following notes and comments refer to these systems: they have been extracted from the literature which the Publicity Department of the Marconi Company distributed at the comprehensive demonstrations – which were of compatible methods of colour television – which it gave at Marconi House, the Strand, London on 11 May 1954:

> **The British NTSC System (Signal IIA)**
>
> This [was] one of the many possible adaptations of the American NTSC colour signal to British 405-line standards. The colour video signal [was] designed to give three-colour presentation of the scene being scanned for video frequencies from DC to 0.4 MHz, two-colour presentation for intermediate video frequencies lying between 0.4 and 1 MHz, and monochrome presentation for the frequencies lying between 1 and 3 MHz, i.e., it [was] contained in a total video band of 3 MHz. This signal consist[ed] of:
>
> (a) A full bandwidth component carrying the luminance information corresponding to any picture element (E_Y).
> (b) Two-colour difference signals E_I and E_Q modulated on quadrature sub-carriers. The choice of the particular colour difference signals [was] governed by the most desirable two-colour presentation.

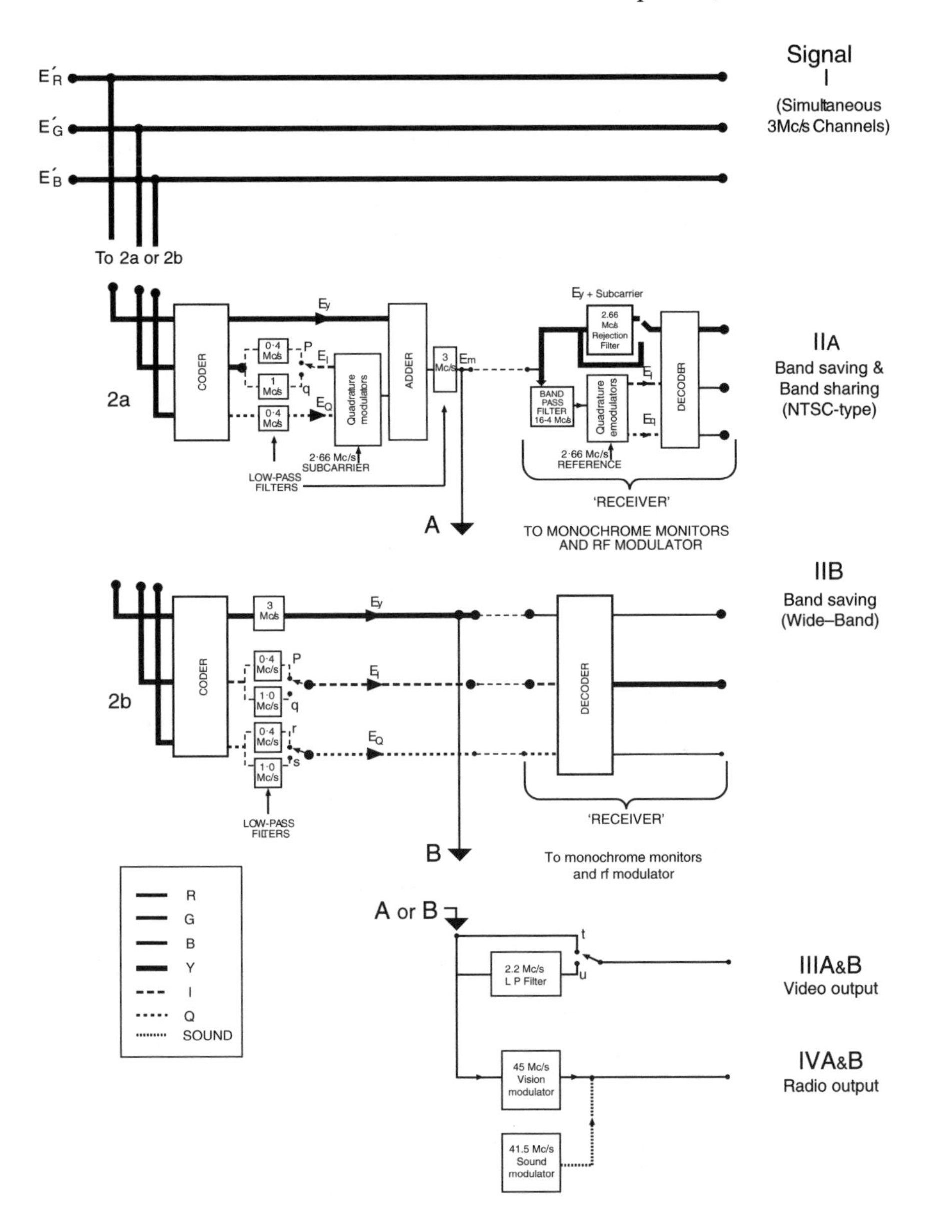

Figure 11.4 The essential components of the systems (excluding the picture-signal generators and displays) investigated by MWT [Source: ibid.*]*

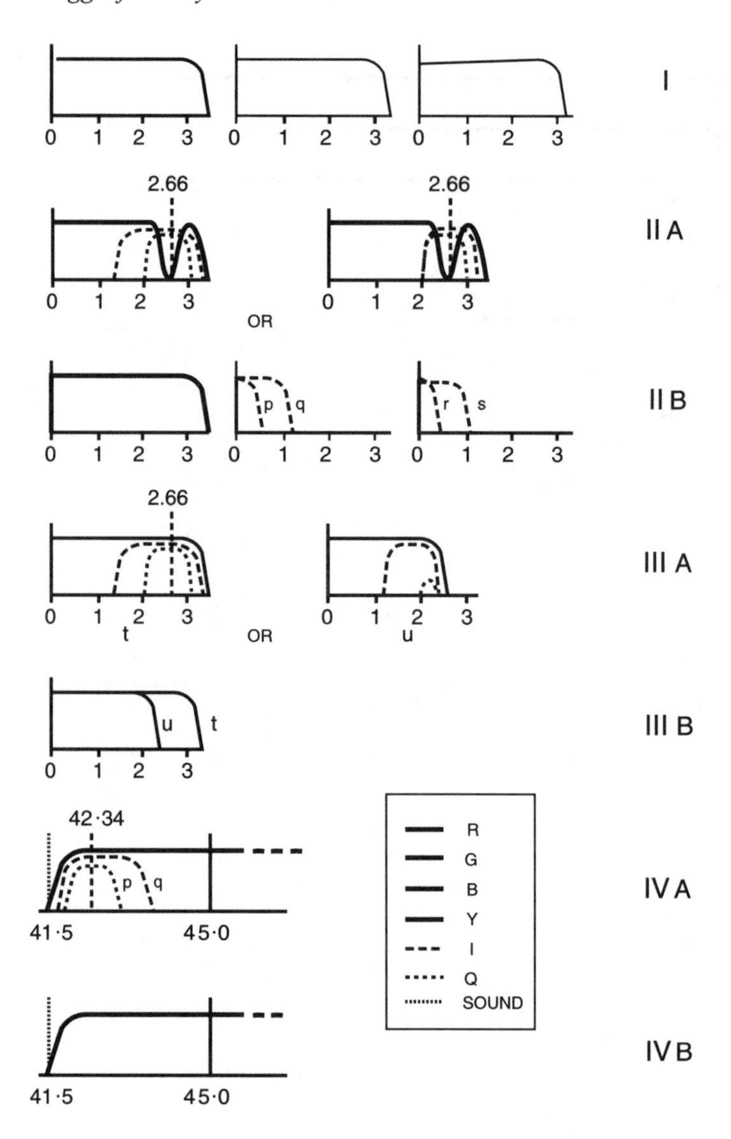

Figure 11.5 Characteristics of the available signals that could be demonstrated by the systems of Figure 11.4 [Source: ibid.*]*

Sub-carrier frequency, and bandwidths allocated to luminance and chrominance information ha[d] been scaled down from American standards in the approximate ratio 3:4, while amplitudes of the quadrature components of the sub-carrier ha[d] the same value relative to the luminance component.

Equipment built to NTSC standards but modified by Marconi's to meet British requirements in one of the possible ways [was] demonstrated along with a 'Wide Band' system which avoid[ed] the cross-talk mentioned above ...

The following variations [were] possible:

(a) The bandwidth of the E_I components [could] be reduced from 1.0 MHz to 0.4 MHz.
(b) The sound carrier amplitude and frequency [could] be varied to investigate sound/sub-carrier cross-talk both on vision and sound.
(c) The synchronizing pulses [could] be operated synchronously with the mains frequency, keeping the sub-carrier frequency unchanged, thus destroying the 'odd multiple of half line frequency' relationship.
(d) The amplitude of various components of the picture signal [could] be changed, e.g., the sub-carrier, the synchronizing pulses, and the 'burst'.

At the 'receiver', the sub-carrier in the luminance channel [was] normally rejected. The sidebands, however, [were] still present and in order to remove these completely, the luminance bandwidth ha[d] to be reduced to about 1.75 MHz. The rejector may be removed to demonstrate the consequent degradation in colour rendition.

Wide Band System (Signal IIB)

Better colour pictures result and compatibility [was] improved if the chrominance information [was] outside the 3 MHz luminance channel. A 3 MHz bandwidth luminance signal E_Y [was] used in conjunction with two colour-difference signals E_I and E_Q which [could] each have bandwidths of either 0.4 MHz or 1 MHz. In the demonstration each of these [was] allocated a separate channel, but in practice, the two chrominance signals could either be modulated on quadrature carriers, or for example each on its own separate carrier.

Monochrome display of the coded signals from IIA and IIB (Signals IIIA and IIIB)

A high-grade laboratory monitor [was] used to display the 'compatible monochrome picture' which would be obtained from either of the coded signals previously described. A low-pass filter of 2.3 MHz bandwidth [could] be inserted, to show the effect of reduced receiver bandwidth on chrominance cross-talk with Signal IIA.)

The BBC must have been delighted by the extensive studies and closed-circuit demonstrations made by Marconi's: at last the prospects for extensive 'over air' testing of colour television became realistic. A request [24] for a quotation for a set of colour television equipment was sent to the company on 2 June, but generously MWT agreed to loan, for two weeks, their colour equipment for test purposes and to second two engineers to assist with its setting-up [25]. Moreover the company was willing to cooperate with the BBC in carrying out transmission tests from Alexandra Palace [26].

Initially the equipment was installed in the BBC's Designs Department at Western House to permit study of the problem of transmitting a colour signal – derived from the scanning of slides and 16 mm films – over the SB (Simultaneous Broadcasting) network [27]. According to Pawley [28], the signals were 'sent round the SB system to Manchester and back with surprisingly good results'. The closed-circuit tests by MWT were repeated with variations by the BBC's Research Department which had also developed but not demonstrated similar experimental equipment (though without a colour camera) [29]. This activity led to colour pictures being televised and radiated for test purposes from the 5 kW transmitter at Alexandra Palace on 6, 7 and 8 October 1954 for one hour from 23.00 hours.

The tests gave stimulus for future progress but highlighted the need for further substantial work to parts of the network. For example: the microwave link between Holme Moss and Kirk O'Shotts showed 'very bad differential phase and gain distortion [which led] to severe picture degradation'.

This promising work led to another working party visiting the United States to observe the functioning of colour television apparatus in established studios. A report was written with the outcome that an Experimental Colour Group – which included a producer, a make-up expert and a lighting specialist – was formed. Experimental apparatus was designed and constructed by MWT, to permit an appraisal to be made of the adapted 405-line NTSC system. This apparatus, consisting of one Marconi colour camera (based upon the RCA TK41 camera with three 3 in. (7.62 cm) image orthicon tubes), one MWT coder and one BBC coder, one Marconi triniscope picture monitor and several shadow mask tube receivers, and a 16 mm film scanner designed by the BBC's Research Department, was set up in the original 1936 M-EMI studio B, at Alexandra Palace, which had been converted to an experimental colour television studio [30]. This had a floor area of *c.* 2,000 ft^2 (186 m^2) and was equipped with a simple lighting installation of about 150 kW input.

On 7 October 1954 the first broadcast of compatible colour television was radiated from Alexandra Palace. Other occasional test transmissions followed until 10 October 1954, when a regular series of test transmissions commenced after the main black-and-white television service had closed down [30]. From this date until early May 1956 there were a total of *c.* 150 hours of experimental transmissions, comprising *c.* 100 hours of test signals (e.g. colour bars), *c.* 35 hours of colour slides, and 12 hours of 16 mm films. The earliest occasion when live studio shots were broadcast was during the 3–7 April 1956 period at a demonstration for the CCIT delegates of Study Group IX, which had been established to consider the merits of the various colour television systems [31].

These colour television signals were radiated from the 5 kW transmitter at Alexandra Palace. To test the quality of the received images, a number of receivers – some fitted with RCA 21 in. (53.3 cm) tri-colour tubes and some utilizing an optical combination of three colour images – were distributed among several observers, including BBC engineers, members of BREMA firms, and some members of the public selected by Audience Research. They had to complete questionnaires and record their opinions on such issues as, for example, colour accuracy, picture sharpness, and compatibility. Fifty thousand observations were received from 424 laymen and 1,237 engineers [32]. Nearly all these observations were concerned with the quality of the compatible monochrome picture when colour was transmitted. (The UK's Television Advisory Committee, like the USA's National Television System Committee, had recommended that colour television signals should be receivable by owners of black-and-white receivers.)

Analysis of the questionnaires confirmed the picture quality was adequate for a public television service. On the one hand, approximately 95 per cent of the viewers saw no deterioration in the compatible black-and-white picture and were satisfied with it: on the other hand, 5 per cent noticed some effect due to 'dot crawl' interference. Of the 5 per cent, 2 per cent were not seriously troubled by it but 3 per cent (mostly engineers) thought it spoiled the picture. Significantly, in the USA a similar dot crawl could sometimes be noticed but it was not such as to stimulate unfavourable comment. On the statistical evidence, the BBC's Engineering Division felt the tests justified an optimistic consideration of a colour television service using the NTSC system because, it was: (1) 'the only acceptable colour system which [could] be applied to Bands I and III: (2) a practicable system which [the BBC could] handle and distribute reliably without excessive additional cost for staff, equipment or services, and (3) a system which [could] give a colour picture with considerable entertainment value'. [32]

Demonstrations of the 405-line colour television system were given in April 1956 to the Postmaster General, the Television Advisory Committee and the BBC Board of Management. These demonstrations and tests gave the programme departments and the engineering department valuable experience in the production of live colour pictures. Moreover, the performances of several experimental colour television receivers, designed by the BBC and by at least six British firms, and all using the RCA tri-colour tube, had been assessed.

The BBC's successes with the adapted 405-line NTSC system posed a policy question for the BBC's Board of Management, namely, whether the Corporation should urge on the TAC, without further delay, that it should recommend the Postmaster General to approve the introduction of the modified NTSC system in Bands I and III in October 1958. Such a move would be likely to meet with the opposition of the Post Office, since it seemed to favour the experimental study of colour television utilizing Bands IV and V. The use of these bands would allow 625-line colour television to be transmitted and lead to a common line standard prevailing throughout the continent. (By 1955, several countries of Europe – Austria, Belgium, Bulgaria, Czechoslovakia, Denmark, the German Democratic Republic, the German Federal Republic, Hungary, Italy, the Netherlands, Sweden, Switzerland and Yugoslavia – were employing the 625-line black-and-white standard – albeit both the CCIR and Soviet versions. Later, this standard was agreed by the European Broadcasting Union (EBU) to be the operative standard throughout Europe.) A common standard would ease the interchange of programmes from one country to another. However, the BBC's director of technical services (H. Bishop) [31] argued that, 'If this attitude prevails and if things are allowed to take their course without any action by the BBC, the introduction of colour is likely to be put off for many years'. Moreover, if the Post Office view succeeded, it seemed to Bishop the consequence would be the complete absence of any colour television for the BBC's transmissions, given the BBC's desire to establish a second programme in Band III. For Bishop there seemed to be just one course to take: the BBC should advocate a decision that would enable it to introduce colour without making the 'tremendous and costly transformation which would be required by a move into Bands IV and V'.

There was some urgency, from the BBC's position, to the resolution of this question of the use of Bands IV and V. The new Television Centre was being planned and important judgements on the allocation of space and resources to black-and-white television programme production, and to that of colour television, had to be made. If Bands I and III could be used for colour television and it could be introduced by the autumn of 1958, then there was little need to equip all the studios with black-and-white television apparatus. Conversely, if colour television were restricted to Bands IV and V, and black-and-white transmissions to Bands I and III the difference between the two facilities would obviously differ.

In 1956 the Television Advisory Committee was asked to give its views on 'whether the existing 405-line standard was likely to remain adequate for all purposes for the next 25 years; and whether there was any reason why the United Kingdom should not adopt 625 lines for broadcasting at ultra high frequencies in Band IV (470–582 MHz) and Band V (614–854 MHz) in this country if it were recommended by the International Radio Consultative Committee (CCIR) as the European standard'. Studies to determine the propagation characteristics of Bands IV and V and field trials to assess the subjective appearance of picture quality were clearly necessary and these were undertaken by the BBC.

Full-scale experimental colour television transmissions began on 5 November 1956 from the 60 kW e.r.p. (effective radiated power) transmitter at Crystal Palace. They were sent out on Mondays, Wednesdays and Fridays after the closedown of the normal television service and were scheduled to run for six months. The tentative timetable was as follows [33]:

	Monday		**Wednesday and Friday**
23.10	Tuning signal	23.10	Tuning signal
23.15	Sound announcement	23.15	Sound announcement followed by slides
23.16	Film and slides	23.16	Studio camera production
		23.40	Film
23.56	Sound announcement and close down	23.45	Announcement and close down

The programme content was changed approximately every fortnight. Some effort was made to produce interesting programmes and a wide range of subjects was covered, including drama, ballet, variety, a mannequin parade and a demonstration of cookery.

Again, questionnaires were distributed to selected persons. These had to be returned to either the Research Department or the Designs Department depending on whether the Monday evening, or the Wednesday/Friday evening, transmission was viewed [34].

The conclusions of the tests were described in an important report [35] of the Colour Unit. Discounting the registration problems of the three-tube camera, with its complicated optical system, and those of the three-gun shadow mask display tube in the receiver, the results were 'extremely satisfactory'. 'The overall accuracy of hue appear[ed] to be extremely good, the un-coded picture being far more faithful than comparable photographic or printing processes. Coded, the pictures [were] nearly as pleasing but [did] not stand a direct comparison with the un-coded ones.' A limitation of the camera system was soon manifested. Whereas the eye is not very sensitive to errors of absolute hue, it is very sensitive to very slight changes in comparisons: therefore, 'cameras had to be very precisely matched and the present equipment [was] not sufficiently accurate or stable in this respect.' 'Certain errors due to the dichroic mirrors were being dealt with by the engineers. The main departures from accuracy occurred on flesh tones and low luminance colours, and many of these errors [were] due to shading and dichroic reflection, not to any fault in the system.' 'Film quality was excellent except in respect of hue. With [the] flying spot scanner there [was] no registration problem in the pick-up device and the definition [was] consequently better. The hue errors [were] not serious and [would] probably become negligible when better films bec[a]me available' [35].

At the end of January 1957, both Houses of Parliament at Westminster had witnessed the new medium on six receivers set up in one of the rooms of the Palace of Westminster [36]. Later a 35 mm film scanner was added to the studio equipment and on 17 April images from this unit were broadcast for the first time.

During the winter of 1957–58 the apparatus was employed with the Band I transmitter to enable experimental transmissions to be radiated into people's homes [37]. The main objectives were: first, to assess critically the quality of the colour pictures that would be received in the average home and to gain further experience of the technical performance and limitations and operation of the camera equipment; second, to explore the artistic possibilities of colour television as a medium for programmes; and third, to make a further appraisal of some aspects of the earlier tests which had been criticised.

These studio broadcasts ended in April 1958 and were replaced in June by outside broadcasts (OBs). An old prewar OB vehicle, which had previously accommodated one of the mobile Band I transmitters, was fitted out with two colour cameras and associated apparatus. It was first used on 25 June 1958 on the occasion of an Institution of Electrical Engineers soirée held at the Festival Hall. The next OB event was the coverage of the Military Tattoo at the White City in August 1958.

Subsequently, the slide scanner and 16 mm and 35 mm tele-cine machines were installed in the Lime Grove studios where, from October 1958, a regular series of weekday-afternoon colour television tests commenced using the Crystal Palace transmitter.

Assessments of the colour broadcasts by both non-technical and technical observers viewing the colour television transmissions at home, substantiated the engineering excellence of the NTSC standard, and provided much valuable experience of the production and operational aspects of running a colour television service.

Contemporaneously with these OBs, the BBC was undertaking experimental studies of: monochrome television broadcasting in Band V from Crystal Palace (they began on 5 May 1958); quadruplex video tape recording (it was first used for broadcast programme purposes on 1 October 1958); and lead oxide vidicon cameras manufactured by Philips in the Netherlands.

The BBC's comprehensive trials led the TAC to recommend in March 1959 that the UK delegation to the IXth Plenary Assembly of the CCIR should be briefed that 'in the interests of frequency planning, the UK would adopt an 8 MHz channel in bands IV and V if Europe generally adopt[ed] this'. The following month at the meeting in Los Angeles all the major European countries concerned said they were prepared to employ this standard and a colour sub-carrier of 4.43 MHz.

In the report [38] of the UK's Television Advisory Committee (TAC), dated 17 May 1960, to the Postmaster General, the committee made several observations on the questions that had been referred to it in 1956. First, the TAC believed the adapted NTSC colour television system was satisfactory and was 'perhaps the only one that could be considered now for use on the present 405-line standards'; though, because of the 'necessary complexities of the present receiving equipment' and the unlikelihood of its being manufactured at a sufficiently low price to command an adequate market price, it was not ready to be brought into service. Second, the Committee felt the definition standards to be adopted in Bands IV and V should be delineated before a decision was reached on the final choice of a colour system. The TAC recommended that colour television should be introduced on a 625-line standard, if this were adopted, and not on the 405-line standard in Bands I and III in the meantime. Third, there was a need for further consideration of the technical details of the colour television standards to be accepted for Bands IV and V, and for further development of colour display tubes before full advantage of 625-line standards for colour television could be realized. From their deliberations the TAC concluded, '...we are of the opinion that present technical and economic limitations make it undesirable to introduce a colour television system in the near future'. Summarizing its views the TAC concluded:

1. The existing 405-line standards would not be adequate for all purposes for the next 25 years;
2. 625-line standards making full use of an 8 MHz channel would give a definite improvement in picture quality over that provided on 405-line standards, particularly with larger pictures (and the tendency over the years had been for screen sizes to increase);
3. the maintenance of 405-line operation would show the United Kingdom to a disadvantage in Eurovision as conversion to a higher standard degraded picture quality; and
4. 625-line operation with the use of an 8 MHz channel would ease the problem of channel sharing with neighbouring countries.

At the European VHF/UHF broadcasting Conference held in Stockholm in 1961, 8 MHz wide channels for use with the 625-line standard were approved in Bands IV and V for utilization throughout Europe.

These colour television activities of the BBC appear to have utilized most of its available television R&D resources and during 1959 and 1960 there was little further development in the field of colour television.

Colour television was also fully discussed by the 1960 Committee on Broadcasting in the United Kingdom. The Committee's Report, known as the Pilkington Report [39] (Figure 11.6) supported the second of the TAC's recommendations; moreover, this

'It's not that I'm against colour television; I just feel that bloodstains in black-and-white are more suitable for child viewers'.

Figure 11.6 Cartoon [Source: Sprod in Punch, *1 March 1961]*

position was endorsed in the government's White Paper (July 1962), which followed. It stated:

1. 'The Committee recommend that colour television should start on a modest scale on 625 lines in UHF, and that meanwhile no public colour service on 405 lines should be authorized.' (para. 56)
2. 'This is in line with Government thinking and it is proposed to authorize the BBC to start transmitting some programmes in colour as part of their second programme (paragraphs 39 and 49), and similarly for any future ITA [Independent Television Authority] second programme. The system of transmission is not yet finally decided.' (para. 57)
3. 'Television programmes are expensive to produce in colour, and the special receiving sets necessary would also cost much more than monochrome sets, particularly to start with. But the Government feels that colour must have a place in

the future pattern of television, and that its introduction should not be postponed indefinitely on the score of cost.' (para. 58)

In giving its decisions the Committee noted:

> We recognise that this means that no part of the country is likely to have a colour service until UHF transmissions are available to it, but our reasons are these. First, cost: it would be wasteful if the broadcasting authorities and the public equipped themselves for 405-line transmission and reception if this were to become obsolete in a few years. Second, the efforts of the broadcasting authorities should be concentrated as far as possible on developing 625-line services. Third, the public should be given every reason for expecting the changeover: a decision to develop a 405-line colour service would raise justifiable doubts.

Curiously, in discord with the Pilkington Committee's recommendation, the BBC in December 1960 requested permission to restart a very limited experimental colour television service on 405 lines to commence in November 1961 [40]. It had carried out publicly announced tests of colour television for six years at a total cost of more than £250,000 and now believed that 'having regard to the high quality of experimental results, a delay until bands IV and V had been introduced would be intolerable . . .'

This brought a verbal blast from C.O. Stanley [41], the irascible and idiosyncratic boss of Pye Ltd. Conditions were 'bad in the television set business', he said, 'and now this morning what is left of profitable television set business will be completely undermined by the BBC rushing into print with a proposal to start colour television by next autumn, on a system which is obsolete and in opposition to the recommendations of the TAC'. He said his company would not manufacture colour sets on 405-lines because it was an obsolete system, and added, 'If the BBC goes ahead with its plans, Pye will apply for permission to send out colour broadcasts on the 625-line system'. Shoenberg [42], of EMI was also opposed to 405-line colour television, as was L.H. Bedford [43] of MWT. His views were based on certain technical factors: (1) 'the degradation of the effective standard of the monochrome service owing to the necessity of restricting the horizontal resolution, or tolerating the dot pattern'; (2) 'the restriction of the luminance bandwidth to 2.3 MHz'; and (3) 'the permanent restriction of the Q channel bandwidth to 0.4 MHz'.

The BBC's request was refused and the government announced in the White Paper of July 1962 that 'colour television should start on a modest scale on 625 lines in UHF, and that meanwhile no public colour service on 405 lines should be authorised'.

By *c.* 1962, the BBC's experimental work on colour television (which had included a major demonstration at the Radio Show, Earls Court in 1961) had led the BBC to the conclusion that, of the NTSC, PAL and SECAM systems, the NTSC specification was the most desirable form for adoption. From the BBC's point of view, a successful colour television service should: give a good quality colour picture under all reasonable conditions of service; be compatible with black-and-white television reception; enable colour television programmes to be distributed over long and complex routes; allow programmes to be produced without too much difficulty, or be much more expensive than existing programmes; and permit colour television receivers – with easily operated controls and some freedom from disturbances caused by propagation effects, interference, etc. – to be manufactured at prices affordable by the general public. To establish the UK position on all these matters the BBC planned to establish a comprehensive programme of experimental work, including field trials on 625-line UHF, in the London area. These were intended to supplement some earlier work, which had been restricted to monochrome transmission, and more especially to obtain information on the problems that might arise from the transmission of colour television signals. It was anticipated the work would be completed *c.* June 1963 [44].

The BBC's view was not shared by the government, which felt the time was not opportune for such a service, and the experimental team was disbanded. The film scanners were moved to the BBC's Lime Grove studios where they provided, for the enlightenment of the television industry, a regular series of day-time transmissions of colour films and slides.

With the development and growth of television broadcasting across Europe, there was an evident need to attempt to select the parameters of any new colour television service so that the networking of, and exchanges of, programmes from one country to another could be accomplished easily, and, if possible, simultaneously with the actual events being televised in a given country. This is the topic of chapter 14.

Note 1: The Paul Raibourn scheme

In this scheme the colour information and brightness information would be transmitted in adjacent channels, so that, to enable a black-and-white receiver to receive colour television, all that would be necessary would be an additional receiver tuned to the adjacent channel, and a tri-colour tube and some additional circuitry. The principal disadvantage of the method stemmed from the possibility of co-channel interference.

At a meeting of Study Group XI held in Stockholm it was agreed internationally that for interference which could just be tolerated for a small percentage of the time in co-channel working (using the half-line offset technique) the interfering signal had to be at least 27 dB below the wanted signal. Using this criterion a television viewer receiving a wanted signal of 100 μV/m would have to be more than 380 miles (611 km) from a co-channel transmitter with an effective radiated power of 100 kW in order to experience interference for just 1 per cent of the viewing time. For a distance of 280 miles (450 km), interference would be experienced for 10 per cent of this time.

Raibourn's suggestion was examined by the BBC's Research Department but it concluded that for the United Kingdom the station separations would be insufficient to prevent serious co-channel interference.

References

1. Farquharson M. Memorandum to Collins, 26 October 1949, BBC T16/47/1
2. Pawley E.L.E. *BBC Engineering 1922–1972*. London: BBC Publications; 1972
3. Barnes G. Correspondence with Box, and Wilson, 19 July to 2 August 1951, BBC T16/47/1
4. Ritchie D. Correspondence with G. Barnes, 26 to 29 October 1951, BBC T16/47/1
5. Bishop H. 'Colour television'. 6 May 1953, BBC R53/40

6. Ashbridge N. Letters to EMI, MWT and Pye Ltd, 22 June 1950, BBC R53/40
7. Collins N. Correspondence with N. Ashbridge and C.O. Stanley, 3 to 17 October 1949, BBC T16/47/1
8. Bishop H. 'Colour television'. 30 December 1953, BBC R53/40
9. Barnes G. Letter to G. Condliffe, 26 June 1953, BBC T16/47/1
10. Postgate. Memorandum to McCall, 29 July 1953, BBC T16/47/1
11. Bishop H. Memorandum to G. Barnes, 10 September 1952, BBC T16/47/1
12. Anon. Memorandum on 'Colour television and its introduction to the UK', 4 November 1953, BBC T16/47/1
13. Head of the Research Department. Memorandum, 1 April 1953, BBC R53/40
14. Bishop H. 'Colour television', 31 August 1953, BBC R53/40
15. Head of Research Department. Memorandum to DTS, 15 December 1953, BBC R53/38/1
16. Anon. Notes of a meeting held on 27 October 1953, BBC T16/47/1
17. Minute, Board of Management, 15 February 1954, BBC T16/47/1
18. Anon. 'Colour television and its introduction to the United Kingdom', 4 November 1953, para. 10, BBC T16/47/1
19. 'Report and Order (Adopted December 17, 1953)'. Federal Communications Commission, p. 669
20. Anon. 'The Marconi demonstration equipment', *Journal of the Royal Television Society*. 1954;**7**(6):241–7
21. Howe A.B. 'Paul Raibourn: suggestion for colour television transmission', BBC R53/38/1
22. Jesty L.C. 'Recent developments in colour television'. *Journal of the Royal Television Society*. 1955;**7**(12):488–508
23. Bishop H. 'Marconi's two-tube colour television camera', 12 May 1954, BBC R53/38/1
24. Wynn R.T.B. Letter to MWT, 2 June 1954, BBC R 53/38/1
25. DCE. Memorandum to HDD, 18 June 1954, BBC R53/38/1
26. DCE. Memorandum to HRD, 16 July 1954, BBC R53/38/1
27. Bishop H. 'Colour television tests', 19 August 1954, BBC R53/40
28. Pawley E.L.E. *BBC Engineering* 1922–1972. London: BBC Publications; 1972. pp. 363–4
29. *Ibid*. p. 463
30. Anon. 'Colour television', September 1955, BBC R53/34/3
31. 'Report of the United Kingdom delegation on the colour television demonstrations given in the United Kingdom, 3rd-7th August 1956 to members of the CCIR Study Group XI', 26 May 1956, BBC file R53/394/2
32. Bishop H. 'Colour television', 10 May 1956, BBC R53/38/3
33. Watson, S.N. 'Programme of experimental colour television transmissions: Autumn 1956', 2 October 1956, BBC T16/47/2
34. Anon. 'BBC Experimental Colour Transmissions: Autumn 1956', BBC R53/38/3
35. Atkins I., Learoyd B. '1956/57 Report – Colour Unit', 12 June 1957, BBC T16/47/3

36. BBC Brochure. 'Colour Television, Demonstration of Transmission and Reception by the BBC, Houses of Parliament, 30 and 31 January 1957'
37. HDD. Memorandum on 'New series of colour tests starting October 1957', 21 June 1957, BBC T16/47/3
38. 'Report of the Television Advisory Committee, 1960', paras. 43–6
39. 'The Report of the Broadcasting Committee, 1960', Cmnd. 1770; 1962
40. Bishop H., Beadle G.C. 'Colour television: experimental transmissions', 8 December 1960, BBC T16/47/5
41. Stanley, C.O. Letter to the BBC, 12 December 1960, BBC R 53/396/1
42. D. Tech. S. memorandum to CE, *et al.*, 19 May 1954, BBC R 53/40
43. Bedford L.H. 'Considerations affecting choice of standards for colour television in the United Kingdom', BBC R 53/40
44. McLean F.C. 'Colour television', 29 April 1963, a Note by the BBC

Chapter 12
Camera tubes for colour television

Mention has been made in an earlier chapter to the genesis of RCA's camera tube – the iconoscope – and the independently developed, but similar, camera tube – the emitron – of EMI. The merits of these tubes led to the growth of high-definition television in the UK, the USA and elsewhere, though they were not ideal and suffered from a number of defects [1]. First, the tube has a much lower sensitivity than would be expected of a charge-storage camera tube, because only a few per cent of the photoelectrons generated by the light flux incident on the photo-mosaic contribute to the picture signal. Second, the use of a high-velocity scanning electron beam, and the consequent liberation of secondary electrons from the mosaic's surface, leads to the production of spurious signals, usually known as 'shading', or 'tilt and bend' signals. Third, because of these unwanted signals, the signal level corresponding to areas of absolute black in the image can vary over a wide range. And, fourth, the incidence of both the light flux and the electron beam on the same side of the target plate produces a tube design that is somewhat awkward for incorporation into a television camera. On the other hand the iconoscope/emitron tube was capable of reproducing very rapidly moving objects with 'quite good definition as a series of distinct images', since the photoelectrons which gave rise to the picture signal were those emitted in about 1/250th of a second before a pixel on the mosaic was scanned. Again, the contrast law – the relationship between the signal output of the tube and the luminous intensity of the light falling on the mosaic – was such that the tube gave useful picture signals over a very wide range of mosaic illumination. This property made it stable in operation and uncritical of the adjustment of the iris diaphragm of the object lens. Furthermore, the use of a high-velocity electron beam led to a good beam focus on the mosaic and hence to a very satisfactory picture definition, though, because of the oblique scanning necessitated by the configuration of the optical and electron lenses, the definition varied over the picture area.

The low sensitivity required a scene illumination which gave a peak-white brightness of *c*. 200 ft lamberts; and the spurious signals called for the employment of operatives to adjust continuously special tilt and bend circuits to annul their deleterious effects on the quality of the displayed image.

Nonetheless, the iconoscope/emitron tube could produce an excellent picture provided that adequate scene illumination was used; rapid changes of scene or scene illumination were avoided; and the scene being televised had a small depth of field.

These limitations imposed appreciable constraints on the artistic freedom of programme producers, especially when the tube was used for outside broadcasts where lighting and object/scene conditions could not be controlled. Ideally, a tube was needed that had an improved sensitivity, a freedom from shading, and generated picture signals that had a fixed black-level. The immediate successor to the iconoscope/emitron tube was the super-emitron/super-iconoscope tube. It stemmed from an idea first proposed by Lubszynski and Rodda [2], of EMI, in 1934 and was described by Lubszynski and McGee [3] in 1938.

Figure 12.1 illustrates the layout of the super-emitron. The optical image was focused by lens 12 onto a transparent, conducting, photo-cathode 5, from which photoelectrons were emitted. These were accelerated by an electric field and focused by a short magnetic field lens 13, to form an electron image on the mosaic 3. The mosaic comprised an array of very many minute secondary electron emitting cells each of which was capacitatively coupled to a signal plate. The electron gun 1 produced a high-velocity beam of electrons which was scanned across the mosaic 3. In operation, the photoelectrons from the photo-cathode bombarded the mosaic cells causing secondary electrons to be liberated. Each photoelectron, of *c.* 500 eV, gave rise to five or more secondary electrons. (These had the same effect as the photoelectrons

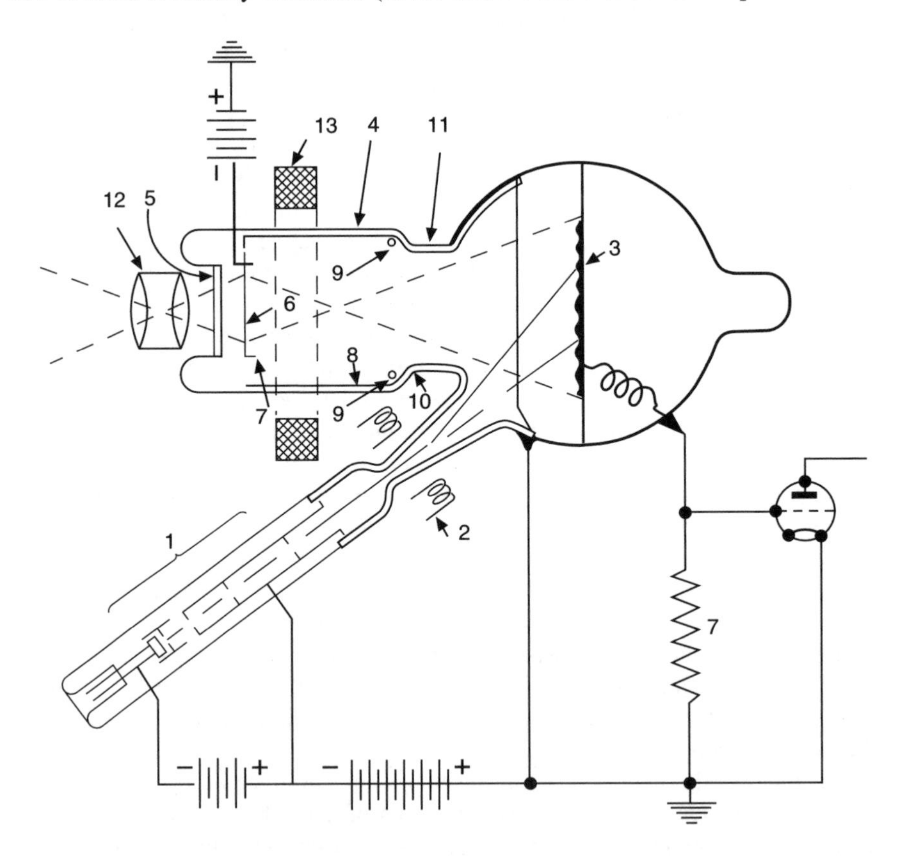

Figure 12.1 *Diagram of the super-emitron [Source:* Electronics, *A.C.B. Lovell, p. 178]*

emitted from the photosensitive mosaic in the emitron tube.) This process caused the mosaic surface to acquire an electrostatic charge distribution but when the scanning electron beam moved over the mosaic the accumulated charges were annulled thus generating small electric currents which flowed from the signal plate through the resistor 7 to earth. In practice elaborate magnetic screening was needed between the image-focusing coils 3 and the beam-scanning coils 2 to prevent interaction between them.

According to McGee, the pictures produced by super-emitron cameras of moving objects were 'not defined so sharply as by emitron cameras; in fact rapidly moving objects showed quite appreciable blurring'. Again, the mechanism of signal generation led to shading in the picture 'since a large bright area of image [contributed] a large number of secondary electrons to its immediate neighbouring mosaic areas, which [were] driven to potentials more negative than would normally correspond to black' [1]. This type of internal shading of the images was, opined McGee, 'inevitable in any pick-up tube using an image section in which the secondary electron emission [was] not completely saturated, such as the super-emitron or the image orthicon [see below]'.

A major advantage of the super-emitron, over the emitron, derived from the separation of the functions of the photo-cathode and the insulating mosaic. The use of a more efficient photo-cathode and a more highly insulating mosaic led to a gain in the conversion efficiency (light into electrons) of 3 or 4. Moreover, because each photoelectron could yield at least five secondary electrons from the mosaic surface, a given light flux could release as many as 20 times the number of electrons from the mosaic as from an emitron tube. In practice the increase in efficiency was used to improve the optical system of the camera. The photo-cathode was reduced in size to ease the problem of accommodating a range of lenses – from telephoto to wide-angle – in a lens turret. This feature was particularly welcomed by producers of outside broadcasts (OBs), especially because the camera gave a satisfactory picture with *c*. 20 per cent of the incident scene illumination required by the emitron.

The super-emitron camera was employed for a public broadcast by the BBC, for the first time, at the Cenotaph Ceremony on Armistice Day, 11 November 1937 [4]. Other OBs followed including the Wimbledon tennis championships, the Lord Mayor's Show, the Derby, the University Boat Race, the Cup Final, Trooping the Colour, the Test Matches from Lord's and the Oval and so on. (Interestingly, the American equivalent of the super-emitron, known as the super-iconoscope, was never used – up to 1950 – for regular broadcasts in the United States. From *c*. 1940 until 1950 CBS [5] utilized the orthicon (see below) and not the super-iconoscope for its direct pick-up colour television work [6]. The orthicon was first publicly demonstrated for this purpose on 9 January 1941.)

During the development of the emitron camera tube, Blumlein and McGee, in 1934, invented the ultimate solution – cathode potential stabilization – to the problem of the spurious signals which were a feature of the emitron camera tube. Since, in this tube, the primary electrons struck the mosaic plate at a high velocity, thereby causing secondary emission, Blumlein and McGee reasoned, independently, that, if the primary beam approached the target in a decelerating electric field and struck the surface with substantially zero energy, no secondary emission would occur, and hence there would be no shading signals. Actually the primary electrons charge the surface more and more negatively until no beam electrons are incident on the surface.

The method of, and a proposed tube for, cathode potential stabilization (c.p.s.) of the target were patented [7] by Blumlein and McGee on 3 July 1934.

Cathode potential stabilization has several important advantages:

1. the utilization of the primary photo-emission is almost 100 per cent efficient, thus increasing sensitivity by an order of magnitude;
2. the signal level is generated during the scan return time, or 'picture black', in the picture;
3. the spurious signals are eliminated; and
4. the signal generated is closely proportional to the image brightness at all points; that is, it has a photographic gamma of unity.

These advantages became of great importance in the operation of signal generating tubes such as the c.p.s. emitron, the image orthicon, the vidicon, the plumbicon, and every one of the Japanese photoconductive tubes up to the advent of the charge-coupled devices.

Electric and Musical Industries did not commence to develop a cathode potential stabilised tube (the c.p.s. emitron) in 1934 because there were very considerable practical difficulties in implementing the principle. Furthermore, at that time the Television Committee, chaired by Lord Selsdon, was taking evidence from Baird Television Ltd, EMI, and others [8] and it appeared that the Committee would recommend the early establishment of a system of medium/high-definition television in the United Kingdom. Because of this, and the improvements in image quality that were being obtained from the experimental emitron tubes, Shoenberg decided to concentrate McGee's group's efforts on these tubes.

Later, *c.* 1935–36, in some early experiments, attempts were made at EMI to stabilise a target at the potential of the cathode of an electron gun by scanning the electron beam over the target in the conventional manner. It was found the mosaic could be stabilised at cathode potential (and the anticipated advantage obtained) only if the angle of scan was kept small. If this condition was not maintained the surface potential of the mosaic 'broke down, always from the outer edges'. Lubszynski, of EMI, suggested a solution in his patent [9] of January 1936: namely, that the scan should be such that the scanning electron beam is always incident perpendicularly onto the mosaic surface. His view was confirmed when a tube was fabricated in which the mosaic was formed on a target that was curved about the point of deflection of the scanning beam as centre. This was clearly an unsatisfactory arrangement since it meant that the optical image on the mosaic had to be curved convex towards the lens [1].

The solution to the problem was presented in a paper by Rose and Iams [10], of RCA, in 1939, and was based on a method of electron focusing that had been described and implemented by Farnsworth [11]. RCA called their camera tube the orthicon. An upgraded version known as the super-orthicon would become, in the 1950s, the standard camera tube for colour television cameras.

Figure 12.2 illustrates the main features of the orthicon. It comprises a cylindrical glass vessel 1 with an optically flat glass window 2 at one end and an electron gun mounted on a pinch 3 at the other end. The electrons emitted from the indirectly heated earthed cathode 4 are accelerated to about 100 V by an anode 6 and are collimated (and the beam restricted in diameter) by an apertured diaphragm 8. A modulator grid 5 controls the beam current. Electrostatic deflection plates 9 are provided for line scanning and a pair of saddle coils 17 for frame deflection. A screen 10 having a long narrow slot constrains the electrons to move only in the direction of the line scan prior to their entering the magnetic field produced by 17. After deflection by the two fields the electrons strike the target 12.

The target consists of a transparent dielectric onto which a transparent photoelectric mosaic is deposited on one side (that facing the window 2 and the lens 14) and a transparent conducting metal layer on the other side which is connected to earth via the resistor 16. In operation the electrodes 6, 8, 10, and 11 are all held at the same potential of *c*. 100 V; this is also the mean potential of 9. This configuration permits the electrons to travel in an equipotential space – aside from the effects of the deflecting potentials. A ring 13 allows the equipotential surfaces in front of the mosaic to be adjusted so that the geometry of the image can be corrected. The glass vessel is surrounded by a solenoid 15, which produces a uniform axial magnetic field which focuses the scanning beam.

Since the electrons strike the mosaic with very low energies, only a small number of secondary electrons are emitted. These are accelerated away from the mosaic and do not generate spurious signals. The charge that any point on the mosaic acquires is strictly proportional to the image brightness at that point, and the totality of the charges over the mosaic is annulled by the scanned electron beam to provide the signal current. Consequently, if a point on the mosaic is non-illuminated, no charge is acquired and the signal current at that point is zero.

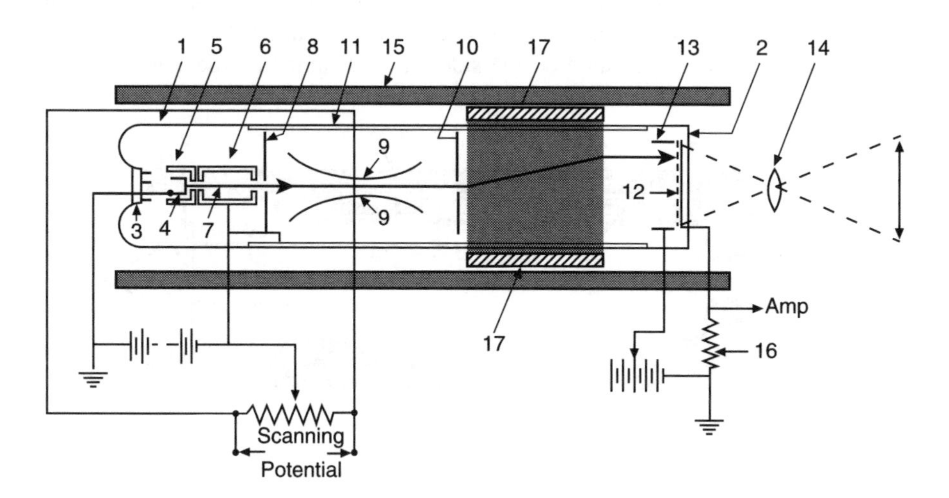

Figure 12.2 Diagram of the orthicon [Source: ibid., *p. 190]*

The orthicon has the following properties: (1) shading signals are not present; (2) a fixed black level is obtained; and (3) almost full charge storage is achieved. The principal defects are: (1) the low sensitivity – due to the attenuation of the light through the conducting metal signal plate, the target dielectric, and the photosensitive mosaic – of just *c.* 2 μA/lumen compared with that of the emitron of *c.* 10 μA/lumen; (2) the decrease of definition from the centre of the picture to the edges and the distortion of the image's geometry, which results from imperfections of the electrostatic line scan electrodes; (3) the production of some spurious signals, 'white ion spot' and 'black bar'; (4) the inoperability of the tube under conditions of intense illumination; (5) the loss of definition resulting from movement of the image, which corresponds to that in a photograph taken with an exposure equal to the frame period, i.e. 0.04 s; and (6) the linear contrast characteristic which must be corrected electrically though this leads to a loss of signal-to-noise ratio. So, 'although the orthicon is 20 times as efficient as the emitron the actual photo-emission from the mosaic is only about one-tenth as much' [1]. It is found that the working sensitivity of the orthicon camera is just a 'little better than that of the emitron'; it is 'certainly less than that of the super-emitron' [1], which was well established in service before the orthicon was developed.

In the United Kingdom work began at EMI in 1945 on the development of the cathode potential stabilised camera tube known as the c.p.s. emitron. Figure 12.3 shows elements of the tube diagrammatically. It comprises a glass vessel 1, an electron gun 2, a metal wall coating 3 (the wall anode), a metal mesh 4 held at the same potential (*c.* + 200 V) as 3, a short cylindrical metal electrode 5 mounted on the glass wall (next to the flat polished glass end window 6), which is at approximately earth potential to decelerate the electron beam and to stabilise the mosaic at or near earth potential, the target 7, which is mounted on the flat inner surface of the window, the focusing solenoid 11, and the scanning coils 12 and 13.

The target 7 is fabricated from a transparent, insulating dielectric such as glass or mica, about 0.004 in. thick, and has dimensions of 45 mm × 55 mm. A conducting transparent metal layer is deposited on the side of the target that faces the glass window, and a mosaic is formed on the side that is scanned by the electron beam.

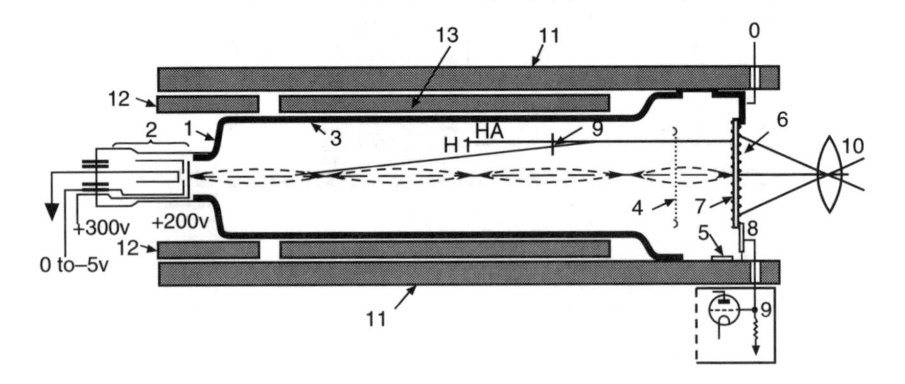

Figure 12.3 Diagram of the c.p.s. camera tube [Source: Proceedings of the IRE, *1950, **38**(6), p. 601]*

The mosaic is coated with an antimony–caesium alloy photosensitive layer formed by evaporating the material through a metal mesh (of 1 million apertures per square inch and a shadow ratio of just 30 per cent) used as a stencil.

In his November 1950 paper on television pick-up tubes, McGee [1] compared the c.p.s. emitron with the emitron, the super-emitron, and the orthicon. His conclusions are given in Table 12.1

The first public outside broadcasts with the new c.p.s. emitron were from the Wembley Stadium and the Empire Pool during the 14th Olympiad held in London in 1948. These cameras were about 50 times more sensitive than the existing cameras and enabled a 'wealth of detail and remarkable depth of field' to be obtained. Much enthusiasm for British television prevailed. Lord Trefgarne, chairman of the UK's Television Advisory Committee, felt moved to express some national fervour when he said, 'In its television transmissions from the Olympic Games, the BBC has just given the world a striking visual demonstration of the technical excellence of the British system, using the latest British cameras, British transmitting equipment and British receivers. The results have been the admiration of our overseas guests, including the Americans . . .'

An interesting application of the c.p.s. emitron stemmed from EMI's work on the investigation of the colour properties of cathode-ray tubes [12]. For this purpose a field-sequential colour television system was used to provide good-quality colour television signals. Later, in the early 1950s, when the importance of compatibility assumed some importance, the Research Department began to consider the possibility of converting the field-sequential signals into simultaneous signals by a technique similar to that employed by the European broadcasting authorities when television images had to be changed from one standard to another. The technique involved the display of an image (of a given standard) on a studio monitor display tube and the televising of the reproduced image by a pick-up camera tube (operating on a different standard). By this means 819-line programmes originating in France could be converted to the British 405-line format. For colour television images it was, of course, appreciated that three converters would be necessary.

During the development of EMI's colour converter, newspapers reported on a similar system being advanced by the Columbia Broadcasting System, which the company called a chromacoder. This was demonstrated in New York on October 1953.

The method, due to Kell [13], as engineered by EMI, utilized three cathode ray storage display tubes and three Emitron camera tubes configured as shown in Figure 12.4 [14]. From the field-sequential camera, which operated on a standard of 405 lines per picture, 150 fields per second, interlaced 2:1, and a bandwidth of 9 MHz, the outputs were applied to gating circuits, which routed the red, green and blue signals separately to the three display tubes. Thus each storage display tube showed an image in one of the three primary colours for 1/150 of a second in each 1/50 of a second. All three camera tubes were scanned simultaneously, using a standard of 625 lines per picture, 50 fields per second, interlaced 2:1, as in a standard three-tube colour camera. Essentially, the converter operated as a frequency divider since the 9 MHz signal of the field-sequential camera was changed into a 4.6 MHz output signal. Other line standards were, of course, possible.

Table 12.1 A comparison between the c.p.s. emitron, the emitron, the super-emitron and the orthicon [1]

	Sensitivity	Definition	Picture geometry	Shading	Stability	Colour response	Black level	Exposure time	Signal-to-noise ratio
Emitron	much greater	better	same	none	poorer	same	fixed	poorer	better
Super-emitron	superior	superior	same	none	poorer	same	fixed	same	better at low light levels
Orthicon	better	better	better	better	better	better	equally good	equally good	equally good

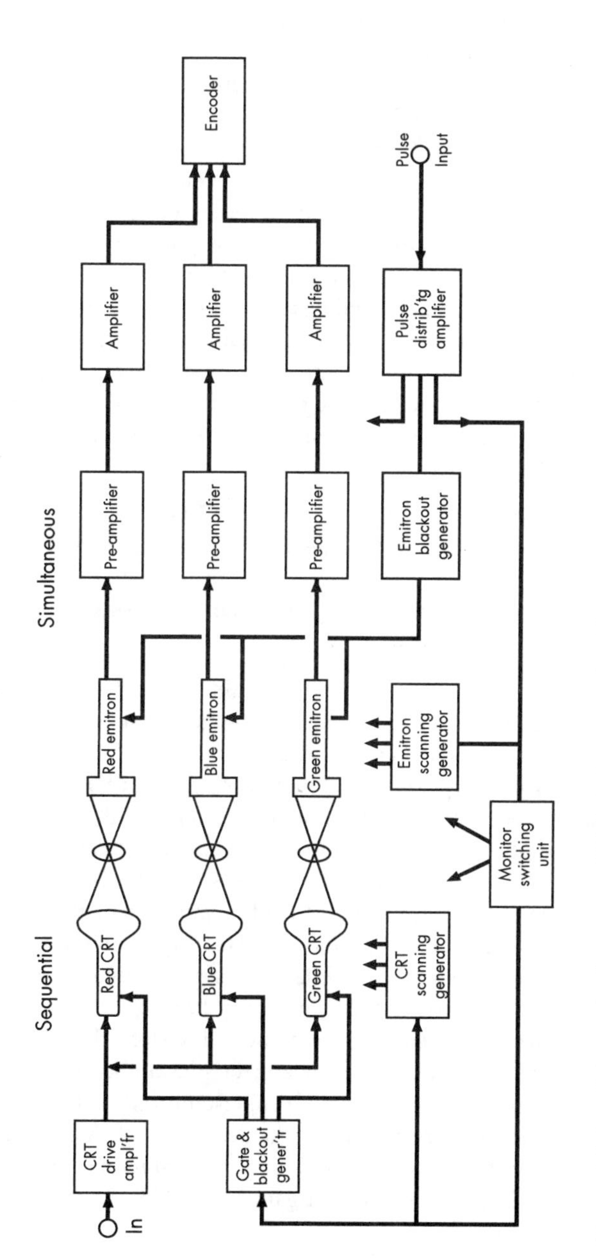

Figure 12.4 *Schematic diagram of EMI's chromacoder [Source:* Journal of the Television Society, *1955, 7(7), p. 307]*

EMI found that its c.p.s. emitron camera tube was very satisfactory for use in the converter since there were no spurious secondary emission effects and the tube was linear in operation. Following a visit by Goldmark (of CBS) to EMI in January 1954, several EMI engineers travelled to the USA and worked with CBS's and GE's engineers on the engineering and manufacture of the Chromacoder for the CBS [15]. This work used the c.p.s. emitron.

Prima facie, at the beginning of 1954, it appeared that the chromacoder would be 'a very strong competitor with the 3-tube camera RCA system, so far as studio equipment [was] concerned' (H. Bishop [16] to the Director General of the BBC). This system had a number of serious disadvantages, which made it difficult to operate. It was large and heavy (>135 kg), and its complex optical system introduced a considerable loss in light efficiency. Lens turret changes were difficult and it was essential that every camera contained three selected pick-up tubes, which were costly to fabricate; and with a multi-camera studio set-up all the colour responses of the filters and tubes had to be carefully matched. 'In a studio with three cameras, ignoring the question of matching the colour filters and registration, there [were] altogether 27 possible variables involved in matching the pictures from the three cameras.' [16]

The BBC's view was probably based on the prices of colour television cameras at that time, namely [17]:

- RCA: cost of supplying one tri-colour camera – £30,000
- GE: cost of modifying a black-and-white camera to give a field-sequential colour television signal – £2,150
- GE: cost of modifying two black-and-white cameras and supplying one Chromacoder – £35,000

From these figures the BBC's deputy chief engineer felt that 'if the results were good the Chromacoder [had] very attractive first cost advantages as well as very considerable savings in running costs' [17]. However, in May 1954, Shoenberg [18], of EMI, informed the BBC's director of technical services that he saw no permanent future for the chromacoder. It was of 'temporary interest only' because the camera problems associated with the three-tube simultaneous camera, on which EMI was working, using c.p.s. emitron tubes, would be 'solved fairly quickly' [18]. Furthermore, the chromacoder system had all the disadvantages of the field-sequential system that had been highlighted at the 1950 FCC Hearing on colour television. Such a system could not compete with the simultaneous system advocated by the NTSC, and EMI abandoned its work on the chromacoder in 1954.

Of the various camera tubes suitable for colour television which had been developed by 1953 none surpassed the desirable characteristics of the image orthicon tube, which comprised an image section (as in the super-emitron) and c.p.s. scanning in one tube. Work on the new tube began during the Second World War, when RCA was given a military contract to develop television systems suitable for reconnaissance and guidance purposes. The principal investigators were H.B. Law, A. Rose and P.K. Weimar [19].

A diagrammatic representation of the image orthicon is illustrated in Figure 12.5. Electrons emitted from the photo-cathode are accelerated to and focused onto the

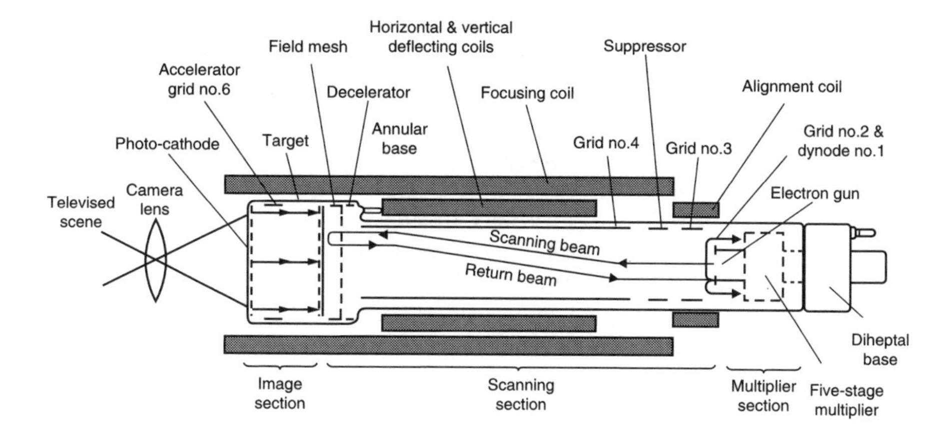

Figure 12.5 Diagram of the image orthicon [Source: Journal of the Television Society, *1963, **10**(8), p. 257]*

two-sided glass target and form an electrostatic positive charge pattern on its image side by the production of secondary electrons, which subsequently are collected by the target mesh, placed several thousandths of an inch away from the target, and held at a slightly more positive potential than the target. (In later fully developed image orthicons the glass plate was less than 0.0002 in. (5.0 μm) thick.) The electrons lost from this side are restored by the scanning electron beam, which is incident orthogonally onto the opposite side of the glass target, the conductivity of the glass being sufficient to permit the image charges to be neutralised in a picture scan time, i.e. 1/30 s in the USA, without degrading the resolution by lateral spreading. No secondary electrons are produced and hence this side of the target is stabilised at the potential of the cathode. The fraction of the beam electrons not needed to discharge the target return to the gun end of the tube, where they enter an electron multiplier. It is apparent that the return beam is modulated by the image signal and that the output from the multiplier must be a maximum when a dark area of the image is scanned and vice versa. Initially the tubes were limited in resolution and sharpness by the 220 mesh target screen which was in sharp focus in the picture. Later, Law devised a new electrolytic process for making high transmission copper or nickel meshes with up to 1,000 wires per inch.

During the Second World War image orthicon cameras were used in radio guided bombs, and after the war they became the mainstay of 'live' television broadcasting for many years. The image orthicon had ample sensitivity for field pick-up of sports events, and it was very tolerant of difficult lighting situations. It was completely stable in the presence of bright lights in the scene, although such overloaded areas would often develop a black halo around them.

The sensitivity of the image orthicon was remarkable. At a public demonstration held in January 1946 the televised subject – a young lady – was illuminated by 'two very large spotlights, then [by a] 60 W lamp, then [by] four paraffin candles, [and] finally [by] one candle.' [20] A magazine report of the event described the reproduced image using the single candle as 'quite distinguishable' [20]. The same magazine said the new tube would be in production in the second quarter of 1946. Subsequently, the image orthicon, with a colour disc, was demonstrated by CBS at the January 1947 Federal Communications Commission Color Television Standards Hearings [21]. And at the 1949 FCC Hearings the image orthicon was also being utilized by CTI (Color Television Incorporated) for its work on line sequential colour television [22]. From *c.* 1955 the tube was used universally, for all television broadcasts, in the USA.

Figure 12.6 illustrates the complexity of the basic design of an image orthicon camera [23]. The light passes through the objective lens, mounted in the lens turret, and is relayed by a unity magnification relay system onto the three image orthicon tubes. Because of the relatively short focal length of the relay system, astigmatism correction plates are included. Beam splitting is achieved by dichroic mirrors mounted in a V formation, and trimming filters are provided to obtain the red, green , and blue characteristics required. Neutral density filters are positioned before the red and green tubes. Optical focusing is accomplished by moving the objective lens.

In operation it is most important that the images incident on the photosensitive targets of the tubes be registered not only geometrically (that is, the scanning linearities should be identical and the optical images superimposed) but also that the images must be registered in the contrast sense (that is, there must be equality of signal tracking of the gray scale over the targets of the three tubes). Poor geometrical registration gives rise to a loss of resolution, and inadequate contrast registration shows up as colour shading – usually in patches – and a varied coloured grey scale. These were the most troublesome defects of camera tubes prior to *c.* 1963 [23].

Apart from these defects, shading errors can arise. They are additional signals in the output of a camera tube that are not a direct function of the light incident on it, and stem from either inherent physical limitations of the operation of the tube, or fabrication difficulties. Both black shading and white shading can arise. A consideration of these various defects is beyond the scope of this book, suffice to say they can be ameliorated by manual controls. In practice, the adjustment of tilt and bend at

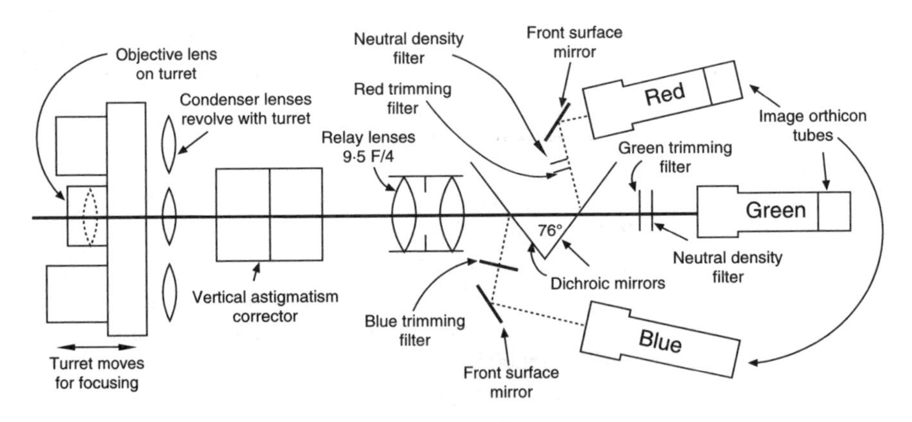

Figure 12.6 Image orthicon colour camera optics [Source: Journal of the Television Society*, 1961,* **9***(10), p. 422]*

both line and field frequencies requires 12 controls for the three tubes of the camera, and if the black-and-white shadings are independently corrected the total number of adjustable parameters is 24. Thus, in a studio using three cameras 72 controls are necessary [23].

Until 1947 all the electronic camera tubes – the iconoscope, emitron, super-emitron, image iconoscope, orthicon, c.p.s. emitron and image orthicon – which had been engineered were complex and expensive. At that time RCA considered that many applications existed for a compact inexpensive camera that would be rugged and simple to operate, and so, from 1947, fundamental studies of photoconductivity and the development of a photoconductive tube commenced.

Actually, some fundamental investigations of such a tube had been carried out by Miller and Strange, of Electric and Musical Industries in 1937. They seem to have been stimulated by the ideas of Campbell Swinton, who attempted, though unsuccessfully, to produce a camera tube by replacing the fluorescent screen of a cathode-ray tube by a layer of selenium, an element with known photoconductive properties. In addition, the patents of Schoultz [24] (1921), the Seguins [25] (1924), and Blake and Spooner [26] (1924) described television pick-up tubes in which selenium is utilized as the photosensitive material.

The experimental tube employed by Miller and Strange [27] comprised an evacuated glass vessel, an electron gun, deflecting coils, a wall anode, and a conducting metallic base (the signal plate) onto which a layer of photoconductive material could be deposited. In operation, as the electron beam scanned the non-uniformly illuminated photoconductive layer, the current flowing in a resistor was modulated. Many materials were tested, including those that were known to be photoconductive and others which might be expected to show an effect.

No attempt was made by Miller and Strange to produce a photoconductive mosaic having storage properties and their work seems to have been restricted to an examination of the various photoconductive materials available (see Table 12.2).

Table 12.2 Photoconductive materials investigated by Miller and Strange [27]

Material	**Observations**
Selenium	Very faint signals for small + or – values of V. Sensitive to yellow-orange region of the spectrum
Zinc sulphide	Strong signals observed in ZnS, ZnS-Cu, ZnS-Mn, ZnS.CdS-Cu. Sensitive in blue-violet range
Cadmium sulphide	Signals produced by red-yellow region of the spectrum
Thallium sulphide	No signals, or faint signals depending on preparation
Antimony sulphide	Fairly strong signals produced by far-red and infrared light
Zinc sulphide	Most sensitive material tested. Sensitivity mainly due to yellow region of spectrum
Cadmium selenide	Faint signals obtained due to red and infrared light
Zinc telluride	Trace of signals present

Iams and Rose [28], in some 1937 experiments, used selenium-sensitized targets of three different types, and targets prepared by spraying either aluminium oxide or zirconium oxide on a metal sheet, treating them with caesium vapour and baking. Their results showed that television video signals could be generated but much research work would be necessary to produce a commercially satisfactory tube.

In RCA's 1947 experimental programme of research on photoconductivity, various materials were deposited by evaporation upon glass substrates, which had been coated with a transparent conductor. The resulting targets were subsequently placed in a demountable vacuum system and tested by scanning with a low-velocity beam. Weimar was a member of RCA's team and has described his reaction to an experiment when he evaporated crystalline selenium:

> We had no expectation then that crystalline selenium could produce a target which could work in the storage mode since its volume resistivity was reported to be $\sim 10^6\ \Omega$ cm. To our surprise the deposited layer of selenium was a beautiful glassy red colour, and gave an excellent picture. Its dark resistivity proved to be very high giving a clean black level, and the resolution and speed of response were excellent. The sensitivity was good, corresponding to approximately one electron per photon for blue and green light, but with very low response for red light. Reference to the then current books on photoelectricity indicated that the red selenium layer was an amorphous form which was stated to be a non-photo-conducting insulator!' [29]

Figure 12.7a shows a cross-sectional drawing of the resulting photoconductive tube which RCA called the vidicon. The first tubes were 1 in (2.54 cm) in diameter, had a target image area approximately 9.5 mm × 12.5 mm and gave a sufficient signal without the need for an electron multiplier. An industrial version of the vidicon was demonstrated at the IRE National Conference in 1950 [30]. (As an aside, for a 625-line television system the size of a picture element is just $9.5 \div 625 = 0.015$ mm, demonstrating the requirement in an ideal colour camera for the registration of the

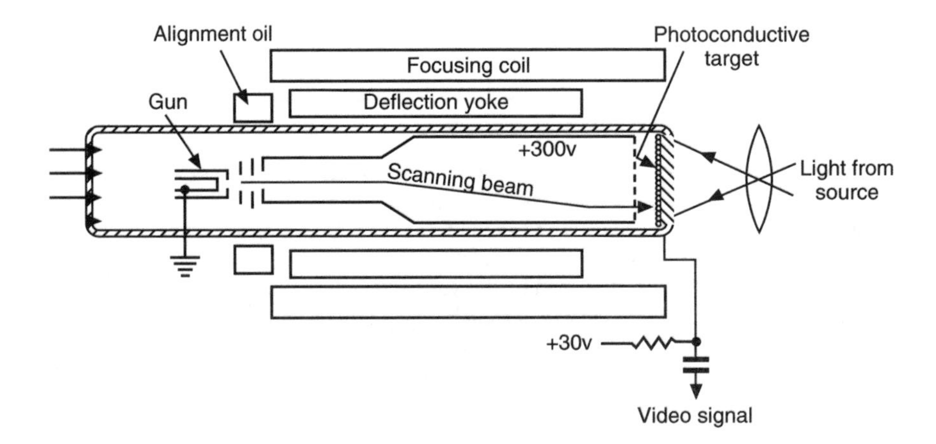

Figure 12.7a Cross-sectional diagram of the vidicon tube [Source: Transactions on Electronic Devices, *1976,* ***ED-23****(7), p. 744]*

three separate image components to be better than *c.* 0.005 mm to prevent degradation of a camera's resolution.)

Experience with the early tubes showed that after several hundred hours of operation the initially defect-free images would show bright spots. It was found that the amorphous form of selenium slowly changed to the more conducting metallic form as a consequence of the temperature rise caused by the electron bombardment. The inclusion of certain additives into the selenium retarded the change but the response time became degraded. Fortunately, Forgue and Goodrich [31] discovered that porous layers of antimony sulphide deposited by evaporation in a poor vacuum produced a stable photoconductor having a good spectral response and a reasonable freedom from lag at normal light levels. Fortuitously, Sb_2S_3 seems to be unique in having a low gamma and a photoconductive property which can be controlled by automatic gain control means. Both of these features would later be of some importance in the design of inexpensive cameras having a fixed iris.

Though Sb_2S_3 was satisfactory for such an application, its sensitivity and lag characteristics precluded its adoption for high quality studio productions. A 1961 report [23] compared the properties of the vidicon and image orthicon as shown in Table 12.3.

The severe problems associated with the registration of the colour images in a three-tube camera (see Figure 12.7b), led RCA to investigate whether the three colour signals could be obtained from a single signal plate [32]. RCA experimentally developed two types (25.4 mm and 50.8 mm) of vidicon colour tube and reported on their work in 1960. Figure 12.8 shows the target structure realized. The signal plate comprises a series of linear multilayer interference filter strips (arranged in the order red, green, blue), which are covered by a semitransparent photoconductive layer. The sets of red, green and blue filters are interconnected at the top and bottom by conductors from which the red, green, and blue image signals are obtained. RCA's experimental tubes had *c.* 290 strips for each colour. In the 25.4 mm tube the signal

Table 12.3 Comparison of image orthicon and vidicon colour cameras

400 ft candles incident	**Image orthicon**	**Vidicon**
Aperture	f/5.6	f/2.2
Registration	A	F
Shading	A	F
Signal/noise	A	VF
Resolution	A	F
Lag	F	A
Stability	A	F
Sensitivity	A	A
Size	A	F
Weight	A	F
Ease of operation	A	F

Ranking: 1 = very favourable (VF); 2 = favourable (F); 3 = acceptable (A)

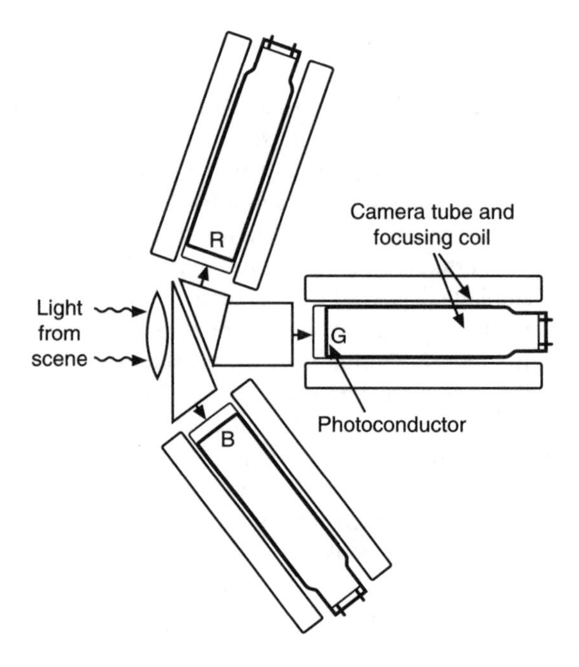

Figure 12.7b Diagram of a three-tube colour camera employing three lead oxide vidicons [Source: ibid.*]*

strips were 0.0002 mm wide with 0.00008 mm spaces between them. The closeness of the strips led to high coupling capacitances and a deterioration of the signal-to-noise ratio (by a factor of about three) compared with the monochrome tube.

In tests it was found the tube gave 'subjectively noise-tolerable pictures' with '700–1,000 ft candle illumination on the scene' and a f/2.8 lens. 'Lag was low, but target defects such as spots and striations were conspicuous because of the high target voltage necessary for maximum sensitivity. In its present state the tube [was] not comparable from the standpoint of sensitivity, colour fidelity, uniformity and definition with the three-tube camera.' [23]

The defects of the vidicon were overcome by the development, in *c*. 1963, of a lead oxide vidicon – known as the plumbicon – by the Philips Laboratories. Apart from having superior sensitivity and lag characteristics, compared with RCA's vidicon, the plumbicon has a linear response which makes it especially suitable for applications in three-tube colour television cameras.

An assessment [33] of the camera tubes available for colour television was undertaken by representatives of the BBC and Marconi's Wireless Telegraph Company in 1964 (see Note 1 at end of chapter for details of Marconi's two-tube colour television camera). They endeavoured to define in very broad terms some of the requirements of an ideal colour television camera. These related to sensitivity, luminance resolution, colour resolution, signal-to-noise ratio, shading, colour fidelity in large areas, and colour fidelity contrast range. Both three-tube and four-tube cameras were considered. On sensitivity, the panel opined that the camera should work with a stop setting

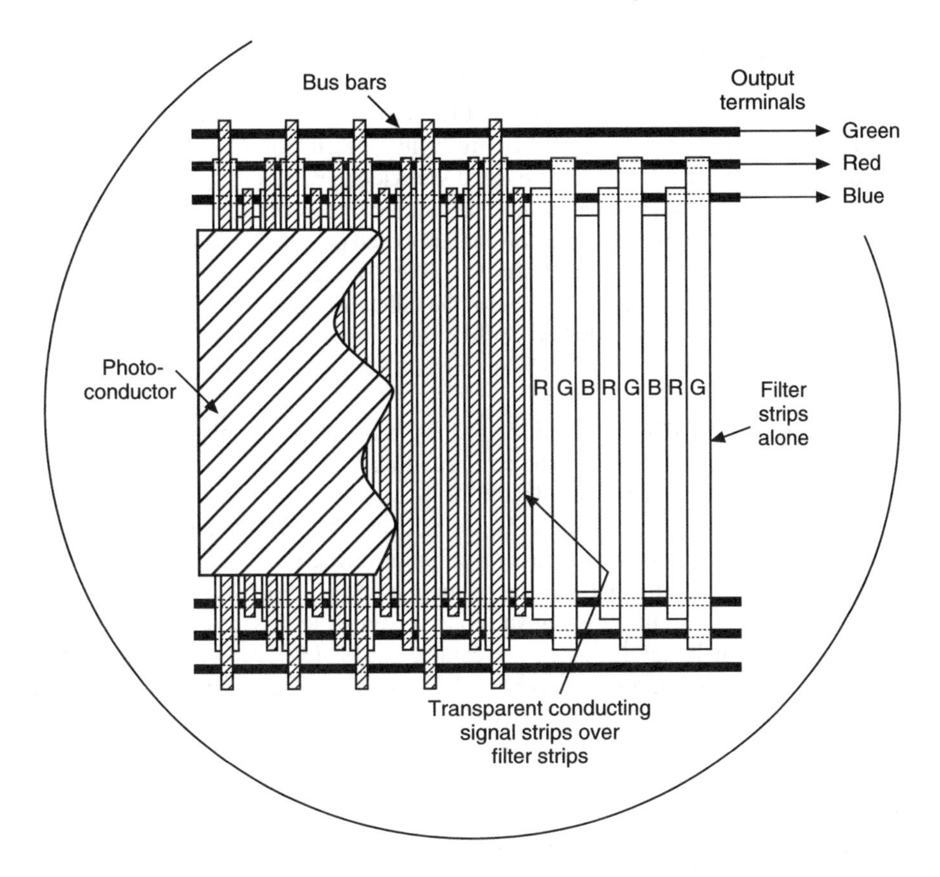

Figure 12.8 Target structure for an early tri-colour vidicon target having 870 internal red-green-blue filter strips and a target output lead for each primary colour [Source: Transactions on Electronic Devices, *1976,* ***ED-23****(7), p. 746]*

of f/8 with respect to an exposure area having a diagonal of 1.6 in. (4.1 cm) with an incident scene illumination of 150 foot candles.

Of the three-tube cameras, those using three 4.5 in. image orthicon tubes had an inadequate sensitivity by a factor of 5–1, and had to be ruled out even though in other respects they gave an excellent performance. Again, those cameras using three 3 in. image orthicon tubes (which was the established practice in the United States) failed to meet the sensitivity requirement by a factor of 3–1 and had a resolution of about half that desired in both vertical and horizontal directions: also they were 'probably somewhat lacking in colour fidelity contrast range. [Hence] this camera [was] obviously a long way from ideal'. Due to lag difficulties the panel considered the three-tube vidicon camera was 'not a competitor for use in a general purpose studio'. The three-tube plumbicon camera was 'probably the best three-tube camera' since it met the 'sensitivity (with something in hand), signal to noise, shading, and colour fidelity contrast range clauses'. However, its lack of red sensitivity was a factor which had to be borne in mind.

Of the four-tube cameras, the camera that comprised one 4.5 in. image orthicon and three vidicons failed to meet the sensitivity specification by a factor of two, and the parameters about which the panel's knowledge was 'so slight as to admit of only the vaguest guesses' included those of lag, colour resolution, registration, and overshoots on compatible pictures. The camera based on one 4.5 in. (11.4 cm) image orthicon and three plumbicons met the sensitivity definition, and 'the problems of lag and colour resolution encountered with the vidicon cameras would disappear, but those of registration and overshoots would remain'. The meeting concluded that the latter camera was clearly much to be preferred if the lack of red sensitivity of the plumbicon was tolerable [33].

Subsequently the BBC decided to 'pin its faith' upon the plumbicon tube. A number of cameras of the three-tube configuration were ordered for installation in the BBC's first outside-broadcast colour units. Soon afterwards plumbicon cameras of the four-tube type became available and these were purchased for use in the production studios. According to the BBC's director of engineering (DE) the choice of which was the better camera remained unresolved until at least 1970. The four-tube camera was more tolerant of slight 'mis-registrations between the images and also to errors in equality of signals due to lag in the camera tubes' [34]. However, the camera was larger, heavier and more complicated than the three-tube camera, and so for the BBC's outside broadcasts the vehicles were equipped with three three-tube Plumbicon cameras.

On the BBC's colour studio operations the DE has recorded (in 1970):

> The decision to use colour cameras fitted with the lead-oxide photoconductive plumbicon had a widespread influence on operations in the colour-production studios. Early experimental colour transmissions had used image-orthicon colour cameras, which required continuous adjustment of the various controls and also a very high level of lighting in the studio. There was a restriction in the style of pictures that could be handled, and the size of the cameras hampered the freedom of movement to which television-programme directors were accustomed in studio monochrome operations.
>
> The plumbicon colour cameras were, however, little, if any, less manoeuvrable than the monochrome cameras, and, although they required a higher lighting level than that to which the studio personnel were accustomed, the change was sufficiently small not to create severe problems. The plumbicon has a linear transfer characteristic, and the dark current is negligible; so that, although it is admirably suited for use in colour cameras in permitting a classical approach to the problems of colorimetry and gamma correction, it is inclined to overload on specular reflections and give rise to what are known as 'comet-tail' effects. New operational techniques had to be devised to overcome this and similar problems.
>
> The stability of the plumbicon camera was soon found to be such that a number of cameras grouped together in one studio could adequately be controlled by a one-man control system. By concentrating all the operating controls in a single desk alongside the lighting controls and by using shared monitoring facilities, a high standard of picture matching could be achieved while still meeting the productivity requirements of an economically operated studio. [34]

For colour film working the selection of suitable tele-cine equipment was less of a problem than that for colour cameras. The BBC had a wealth of experience of flying-spot colour tele-cines and chose these for the start, in 1967, of the BBC2 colour television service. Among the attributes of these tele-cines were the lack of any registration problem, the low running cost, and the good signal-to-noise ratio when reasonably good-quality film was used. With poor-quality film, a number of electrical compensation techniques developed by the BBC were employed; the best known of these was Tarif (television apparatus for the rectification of inferior film). In Chapter 14 the factors that enabled the BBC to achieve its outstanding position in colour television are discussed.

The next chapter considers the various suggestions which were advanced to display the colour television signals generated by the colour cameras.

Note 1: Marconi's two-tube colour television camera

Mention has been made of the Marconi Company's endeavours in the field of colour television. In its two-camera system, one camera tube produced, in a conventional manner, a high-definition monochrome image of 3 MHz bandwidth; the other camera tube generated two low-definition colour signals by means of a static colour filter comprising alternate red and blue strips (of equal widths) placed at right angles to the scanning lines. The outputs from this tube were red and blue image signals having half the definition of the standard monochrome image signal (i.e. 1.5 MHz). [35]

As with the three-tube camera, the signal outputs from the Marconi camera could be easily coded to the NTSC standard, thereby showing an advantage over the single-tube field-sequential camera. Apart from the compactness of the design and the saving in cost compared to a three-tube colour camera, the Marconi camera had an important technical advantage. Because the design was based on the inability of the human eye to see fine detail in colour, the accuracy of registration between the low definition red and blue images and the green image was less than that required for the three-tube camera.

References

1. McGee J.D. 'A review of some television pick-up tubes'. *Proceedings of the Institution of Electrical Engineers*, 1950, **97**, Part III, No. 50, pp. 371–92
2. Lubszynski H.G., Rodda, S. 'Improvements in or relating to television'. British patent application 442 666. 12 April 1934
3. McGee J.D., Lubszynski H.G. 'EMI cathode ray television transmission tube'. *Journal of the Institution of Electrical Engineers*. 1939;**84**:468–82
4. McGee J.D. 'The early development of the television camera'. Unpublished manuscript, Institution of Electrical Engineers library, p. 47
5. Anon. CBS press release, 9 January 1941
6. Anon. 'Description of CBS transmitting and receiving facilities – comparative demonstrations. February 23, 1950', CBS Exhibit No. 301, FCC Docket No. 8736 *et al.*
7. McGee J.D., Blumlein A.D. British patent application 446 661. 3 August 1934

8. Burns R.W. *British Television: the formative years*. London: Institution of Electrical Engineers; 1986
9. Lubszynski H.B. 'Improvements in and relating to television and like systems'. British patent application 468 965. 15 January 1936
10. Rose A., Iams H. 'Television pick-up tubes using low velocity electron beam scanning'. *Proceedings of the IRE*. September 1939;**27**:547–55
11. Burns R.W. *Television: an international history of the formative years*. London: Institution of Electrical Engineers; 1998. Chapter 15, pp. 346–76
12. James I.J.P. 'Recent developments in colour television with special reference to the field-sequential system'. *Journal of the British Kinematograph Society*. 1955;**26**:5–20
13. Kell R.D. US patent application 2 545 957. 1940
14. Anon. 'The EMI experimental colour system', *Journal of the Television Society*. 1955;**7**(7):305–7
15. DCE. Memorandum to DTS, 24 May 1954, BBC file R53/38/1
16. Bishop H. memorandum to Director General, 6 January 1954, BBC file T16/47/1
17. DCE. Memorandum to DTS *et al.*, 8 February 1954, BBC file R53/38/1
18. D Tech S. Memorandum to CE *et al.*, 19 May 1954, BBC file R53/40
19. Law H.B., Rose A., Weimar P. 'The image orthicon – a sensitive television pick-up tube'. *Proceedings of the IRE*. July 1946, pp. 424–32
20. Anon. 'RCA colour television status'. *Electronic Industries*. March 1946, pp. 102, 136, 137, 138
21. Anon. 'Colour television'. *Communications*. 1947;**27**:12–13
22. Anon. 'Colour television'. *Tele-Tech*. October 1949, pp. 18–20
23. James I.J.P. 'Colour television camera problems', *Journal of the Television Society*. 1961;**9**(10):422–30
24. Schoultz, E.G. 'Procédé et appareillage pour la transmission des images mobiles a distance'. French patent application 539 613. 23 August 1921
25. Seguin L., Seguin A. 'Méthode et appareils pour la television'. French patent application 577 530. 28 February 1924
26. Blake G.J., Spooner, H.J. 'Improvements in or relating to television'. British patent application 234 882. 28 February 1924
27. Miller H., Strange, J.W. 'The electrical reproduction of images by the photoconductive effect'. *Proceedings of Royal Physical Society*. 1938;**50**:374–84
28. Iams H., Rose A. 'Television pick-up tubes with cathode-ray beam scanning'. *Proceedings of the IRE*. 1937;**25**(8):1048–70
29. Weimar P.K. 'Historical review of the development of television pick-up devices (1976–1962)'. *IEEE Transactions on Electronic Devices*.1976;**ED-23**(7):739–52
30. Weimar P.K., Forgue S.V., Goodrich R.R. 'The Vidicon, a new photoconductive television pick-up tube'. *Proceedings of the IRE*. 1950;**38**:198. And: *Electronics*, 1950;**23**:70; *RCA Review*. 1951;**12**:306
31. Forgue S.V., Goodrich R.R., Cope AD. 'Properties of some photoconductors, principally antimony tri-sulphide'. *RCA Review*. 1951;**12**:335
32. Weimar P.K. *et al.* 'A developmental tri-color vidicon having a multiple-electrode target'. *IRE Transactions on Electron Devices*. 1960;**ED-7**:147

33. S.N.W. 'Report of meeting between BBC and Marconi Wireless Telegraph representatives', 19 March 1964, BBC file R53/38/4
34. Redmond J. 'Television broadcasting 70–1962: BBC 625-line services and the introduction of colour'. *IEE Reviews*. August 1970;**117**:1469–88
35. H.B. Memorandum on 'Marconi's two-tube colour television camera', 12 May 1954, BBC file R53/38/1. Also: Anon. 'Colour television'. *Electronic Engineering*. July 1954, p. 320

Chapter 13
The development of display tubes

At a September 1949 meeting at RCA Laboratories, Princeton, New Jersey, senior technical staff were informed that colour television was at a critical stage and it appeared the CBS system (which was inexpensive and used a single c.r.t) would be chosen if the Corporation could not evolve a direct view tube. Consequently, RCA management had decided to embark on an all-out effort to develop such a tube. 'Feasibility was to be shown in three months. There was to be no limit to expense, and any manpower that could contribute, anywhere in the company, would be made available. The task of coordinating and organising the activity was assigned to Dr Edward W. Herold.' [1]

Herold joined RCA in 1930, following the award of a BS in physics from the University of Virginia, and engaged in tube research at RCA's Harrison, New Jersey plant. In 1942 he received a MS degree from the Polytechnic Institute of Brooklyn and during that year transferred to the newly established RCA Laboratories where he served as Director of the Radio Tube Laboratory and, later, the Electronics Research Laboratory. Dr Herold had an illustrious career: in 1961 the Polytechnic Institute of Brooklyn honoured him with a DSc for his 'distinguished accomplishments in the fields of electron tubes and solid state devices'. At the time of his retirement in 1972 he was director of technology, research and engineering at RCA.

Herold [2], from a study of the patent and other literature on display tubes, has classified the various methods of displaying a colour television image as follows:

1. the accurate beam-scanning method;
2. the signal control by beam-scanning position method;
3. the adjacent image method;
4. the multiple-colour phosphor screen method;
5. the direction-sensitive colour screen using electron shadowing method;
6. the beam control at phosphor screen for changing colour method.

The first of these is exemplified by the system which Zworykin patented [3] in 1925 (see Figure 3.1), which was based on an application of colour photography to colour television. He suggested placing colour screens, of the Paget type, in front of the camera and display tubes of a black-and-white television system to achieve a simple colour television system. Clearly, extreme accuracy of placement of the screen in two orthogonal directions would be required to prevent colour distortion or colour dilution. Alternatively, the white phosphor screen of the display tube could be replaced by a screen comprising a 'checkerboard' pattern of phosphor 'dots', or

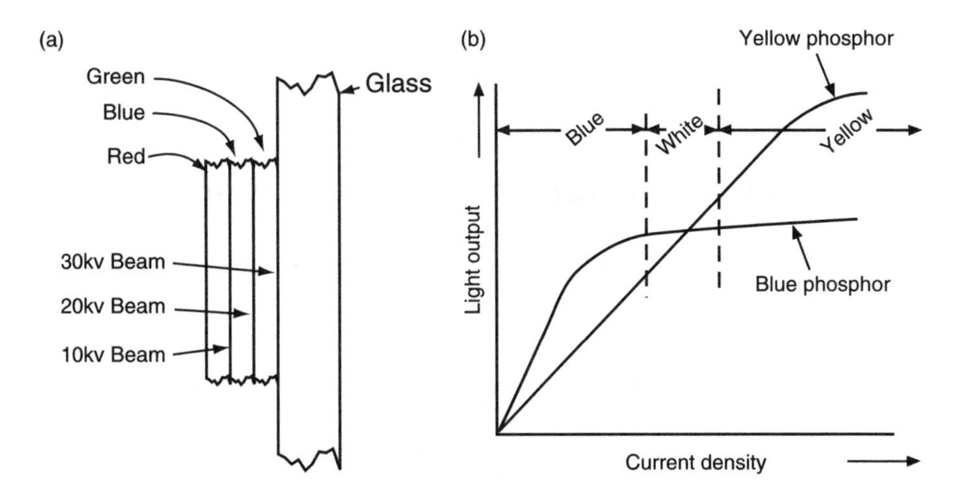

Figure 13.1 *A multicolour phosphor screen: (a) shows a multiple-layer screen with colour depending on beam velocity; and (b) shows how saturation in a two-component screen makes colour dependent on current density [Source:* Proceedings of the IRE*, 1951,* ***39****(10), p. 1181]*

ruled phosphor strips, of the three primary colours, deposited in succession (see Figures 13.2a and b [4,5,6].

The dots or strips may be excited sequentially when only one electron gun and beam are used; or, if the beam is split into three parts each separately controlled, simultaneous display of the three primary colours may be achieved [7,8]. Since the sizes of the colour phosphor dots or strips must be less than one-third of the distance between the scanning lines (when scanned parallel) or less than one-third of a picture element (when scanned transversely) the method poses problems not only of phosphor deposition but also of scanning accuracy.

In practice, high scanning accuracy must be accomplished by means of automatic control and registry by feedback methods and complex circuitry. Many proposals had been published as patents [9,10,11] and others were still, in 1951, being developed in laboratories. For a number of years experiments with line screens, prepared by a three-step phosphor-settling process through a movable mask (consisting of a grid of parallel wires stretched across a frame) were made by D.W. Epstein at RCA Laboratories. Subsequently, at the RCA Victor Division at Harrison, NJ, suitable screens were fabricated in which the three colour phosphor line groups were printed by a silk screen process [12]. A demonstration of the line screen colour kinescope was given to the FCC by RCA on 10 October 1949 [13].

Figure 13.2b illustrates RCA's 16 in. line screen kinescope [13]. The screen size was 7.0 in. × 10.5 in. (18 cm × 27 cm), each phosphor strip was 0.007 in. (0.18 mm) wide and the dark strip between adjacent colour strips was 0.003 in. (0.08 mm) wide, thus giving a triplet width of 0.030 in. (0.8 mm), to permit of a 720-line definition. An advantage of this type of tube is that it can be manufactured with relatively

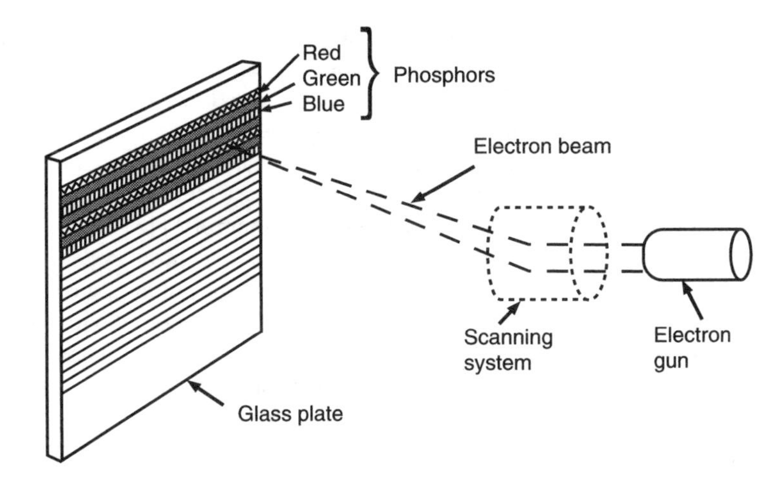

Figure 13.2a A line-screen colour kinescope using one electron beam. To assure correct colours, beam scanning must be highly accurate. No automatic registry means are shown [Source: ibid.*]*

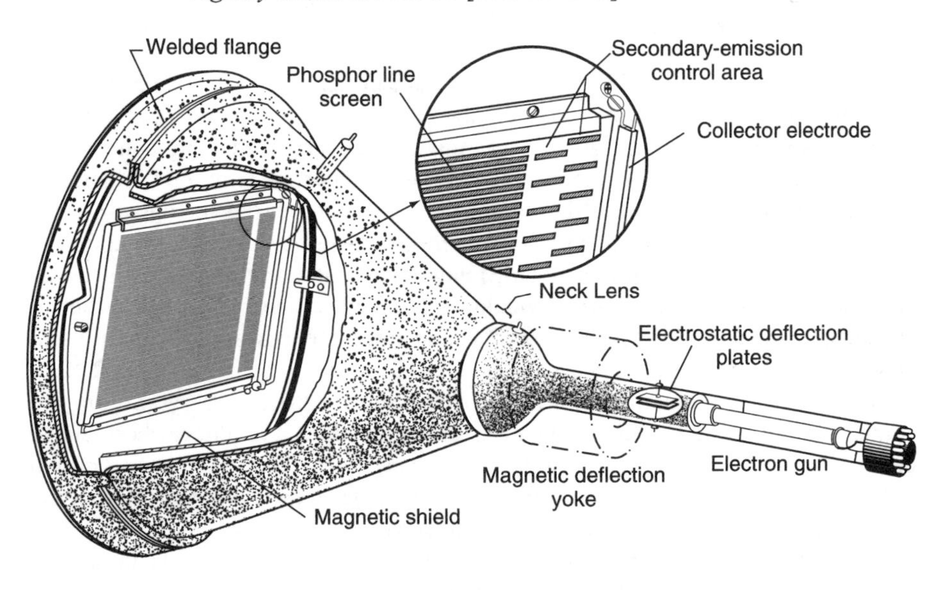

Figure 13.2b RCA's 16 in. (40.6 cm) diameter line-screen kinescope [Source: ibid.*]*

little complexity since it does not need many more elements than a conventional black-and-white kinescope. Among the disadvantages of the line-screen tube are the colour errors when the beam is mis-registered, or incorrectly focused.

In operation, a 'stair-case' generator was used in conjunction with the line-scanning time base to cause the electron beam to scan the phosphor strips. The necessary registration of the path of the electrons with the screen elements was obtained by means of a servo circuit which derived its control information from secondary emission control areas on the kinescope screen.

RCA's work on the line-screen colour kinescope showed that, though the control circuits were 'somewhat elaborate', 'a picture of high horizontal definition and of adequate color quality result[ed]. Improvements in the production of secondary-emission surfaces to give greater and more uniform contrast may be expected to improve color purity and simplify the adjustment of the servo-circuit controls' [13].

In the above method the need for an elaborate servo control system may be negated if the scanning beam can be made dependent on the colour signals applied to the kinescope's control grid: that is, if 'a colour sensitive photo device, or other special signal-generating means built into the screen' enables the kinescope's control grid to switch automatically to the correct primary-colour signal depending on the instantaneous beam position. This 'signal control by scanning position method' has been advanced in several patents [13,14,15], but, as Herold has noted, 'because of the need for an extremely small focused spot, it ['the signal control by beam-scanning position method'] is subject to some of the same disadvantages as the accurately controlled scanning method' [16].

In 'the adjacent image method' two or three complete television images in different colours are combined by means of mirrors or by optical projection [17,18,19]. Figure 5.10 illustrates one version of the method. The kinescope screen has three independent areas coated with red, green and blue phosphors, which can be scanned either field-sequentially or line-sequentially. Of these forms the latter has attracted particular attention since a single line scan traverses all three areas. Subsequent optical registration of the kinescope images permits a colour image to be synthesized. The method is an extension of the Thomas system of colour cinematography [20]. However, because the optical system required is very similar to that needed for the superimposition of the images associated with three kinescopes, each of which displays an image in one of the primary colours, the method does not have a sufficient advantage to make it attractive. Moreover the screen area is not efficiently used. Nevertheless the method has been demonstrated by several investigators [21,22,23], including Baird. As noted previously, he demonstrated, in 1944, a system of two-colour television using a telechrome tube [24,25] in which a thin mica sheet coated on each side by two different phosphors was scanned by electron beams from two independent electron guns. In this tube the two images were combined without the need for an external optical system (see chapter 5).

In 'the multiple colour phosphor screen method' a single phosphor or a multilayer phosphor, when excited by an electron beam, produces a colour that is dependent on either the beam's velocity or its current density. Figure 13.1(a) illustrates the method. The display screen has three superimposed phosphor layers, each responsive to a different primary colour, and is scanned by a single electron beam. In operation, the extent to which the layers are penetrated and excited is a function of the energy of the beam [26]. Either single gun or multi-gun configurations may be utilized.

According to Herold, 'it appears unlikely that such screens can be made to operate with electron-velocity differences of less than around ten kilovolts, so that sequential switching [will be] very difficult at best' [27]. Other difficulties include possible inherent colour dilution and colour fidelity, and problems 'such as scanning amplitude differences' when three electron sources operating at such large velocity differences are employed.

An alternative method is to effect a change of phosphor colour by varying the current density of the electron beam [28,29,30] (see Figure 13.1(b)), but 'the color change in the usual two- or three-component phosphors due to saturation is slight, and high chroma colors are difficult to achieve' [30].

In the direction-sensitive-colour-screen-using-electron-shadowing method, the colour phosphor dot/strip excited is dependent on the direction of arrival of the incident electron beam. The first proposal of this kind to receive considerable attention made use of a non-planar surface (see Figure 5.13), and was suggested by Baird [24,25] for a three-colour television receiver tube. The screen comprised a transparent plate on one side of which the surface was 'ridged', as shown, to enable two primary colour phosphors to be deposited. These phosphors only emit light when bombarded by electrons incident on the phosphors within a restricted range of angles of incidence. The opposite surface of the plate was coated with the third primary-colour phosphor. Thus the method of synthesizing a colour television image was based on the superimposition in good registration of the three independent primary colour images.

A similar tube was patented by C.W. Geer in 1944 and was described by him in a paper [31], which he submitted to the 1949–50 FCC Hearing on colour television. The self-explanatory Figure 13.3a taken from the paper shows the general layout of the tube and the structure of the screen; and Figure 13.3b illustrates Geer's ideas for forming the surface of the screen. His proposed tube differed from that of Baird in having the three electron beams incident onto one side only of the screen.

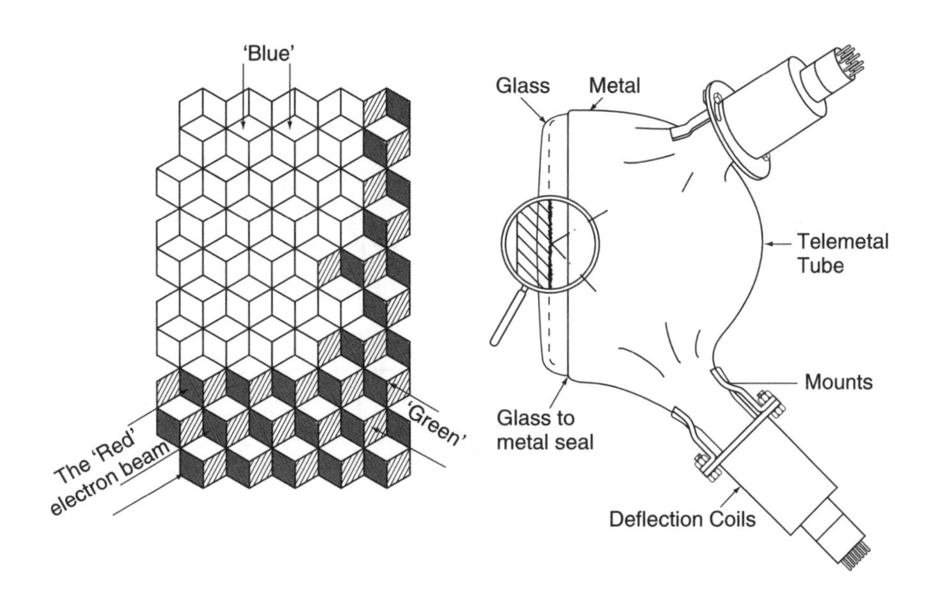

Figure 13.3a Geer's three-electron gun display tube [Source: FCC Docket 8736]

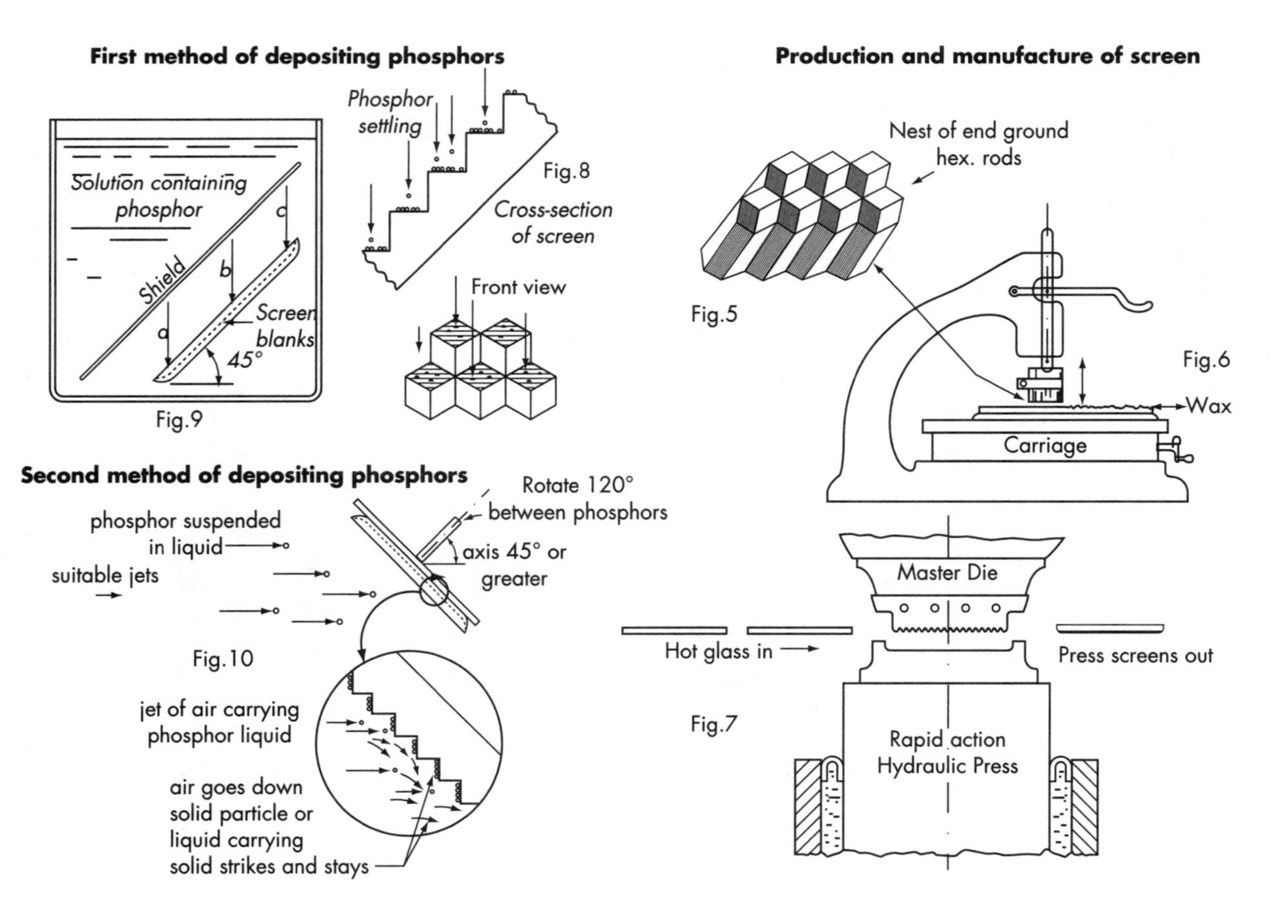

Figure 13.3b Greer's ideas for the production and manufacture of the screen, and for depositing the phosphors upon it [Source: ibid.]

On the methods available for forming the pyramidal surface, Geer noted:

> One method of forming the surface of raised and depressed pyramids is to form that surface on a sheet of semi-hard wax using a needle-like punch with bevelled surfaces on its head so designed that the wax shoved upward on each stroke forms the raised part of the pyramid after the next stroke. The results from this method were good even using crude machinery. [31]

Another method was to make a 'nest of punches and [form] the cast for the die by punching out small areas, say an inch square, at one time, the area formed being the area of the nest of rods used. These rods may be made from hex stock which are individually ground so that their punching ends form a negative of the screen's surface'. [31]

The phosphors, according to Geer, could be deposited either by a settling process, or by spraying (see Figure 10 of Figure 13.3b).

Geer's tube does not seem to have been demonstrated to the FCC; indeed there is no readily accessible evidence that he actually fabricated a tube.

With tubes of the Baird and Geer types the main problems, apart from fabrication, stem from the need to achieve good colour directivity and accurate raster scanning. It is necessary to generate a rectangular raster with three off-axis, keystone-corrected electron guns, in which the three scanning line traversals of the three electron beams must not only be accurately registered, but also the edges of the three individual scans must be coincident. By 1951 these problems had not been solved practically, although RCA, a few years previously, had initiated a study of 'means for reducing the angle of separation of the three beams by using very steep pyramids on the non-planar surface, and also by constructing alternative non-planar surfaces. Unfortunately the deposition of phosphors so nearly parallel to the direction of viewing [led] to so large a light loss that widely spaced guns, with their attendant deflection problems, [might] be essential.' [32] (See Figure 13.4.)

RCA's post-1950 superiority in the field of colour television was due to its development of the shadow mask colour television display tube, which incorporates a version of a direction sensitive colour screen. This tube is one of the truly outstanding inventions of the latter half of the twentieth century and so some mention must now be given of its early history.

Shadow mask tube development at RCA commenced [33] in late September 1949, when it became clear that none of the RCA methods so far advanced was likely to be suitable for the consumer market. They would have led to prohibitively expensive and very bulky receivers. Dr E.W. Herold has written that they produced a direct-view picture of about 12 in. (30.5 cm) diagonal in a cabinet occupying the volume of two upright pianos [1]. Furthermore, the picture quality was not particularly satisfactory: there had to be an alternative method if colour television were to appeal to the public's expectations of its practicality and, obviously, its purse.

A patent search revealed three patents appropriate to the investigations. The earliest, a patent issued on 31 March 1941 and due to W. Flechsig [34], proposed employing the shadow effect of a grill of wires to obtain colour selection at the screen, with three electron beams, one for each colour (see Figure 13.5). His patent indicated how

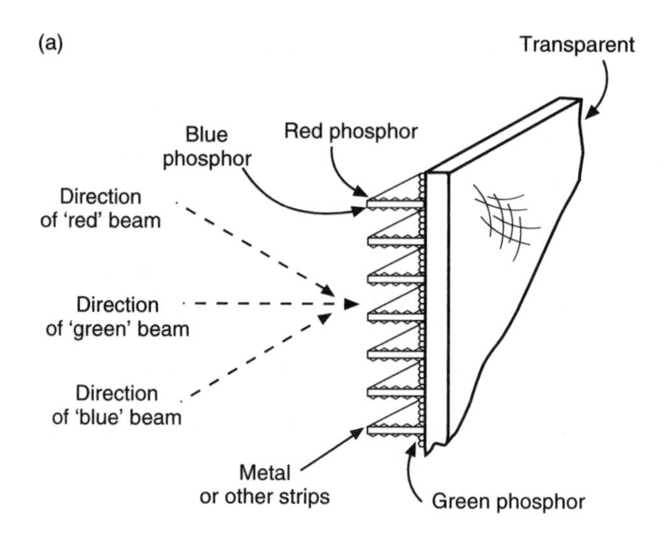

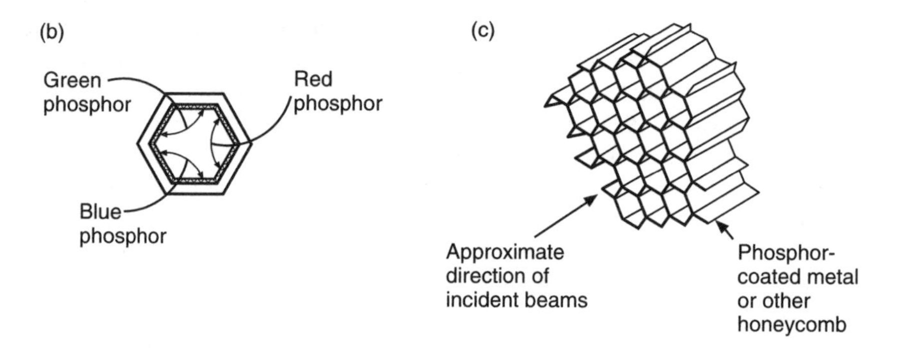

Figure 13.4 Two forms of a non-planar direction-sensitive colour screen [Source: Proceedings of the IRE, *1951, **39**(10), p. 1185]*

the colour phosphors of the screen could be laid down by evaporating the material through the grill from sources placed at the centre of deflection of the electron beams. In Figure 13.5(b) it is shown how one beam may be successively deflected to the positions of the three beams of a multi-gun arrangement, and directed to a common point on the screen. Flechsig proposed using a staircase switching signal to accomplish this operation.

Of the ideas mentioned in the patent, the following were important [35]:

1. the use of a shadowing structure that consisted of a wire grill to permit the excitation of a selected colour at the phosphor screen;
2. the use of three electron beams arranged to converge at the grill and strike separate areas on the phosphor screen beyond the grill as a consequence of the grill's shadowing action;

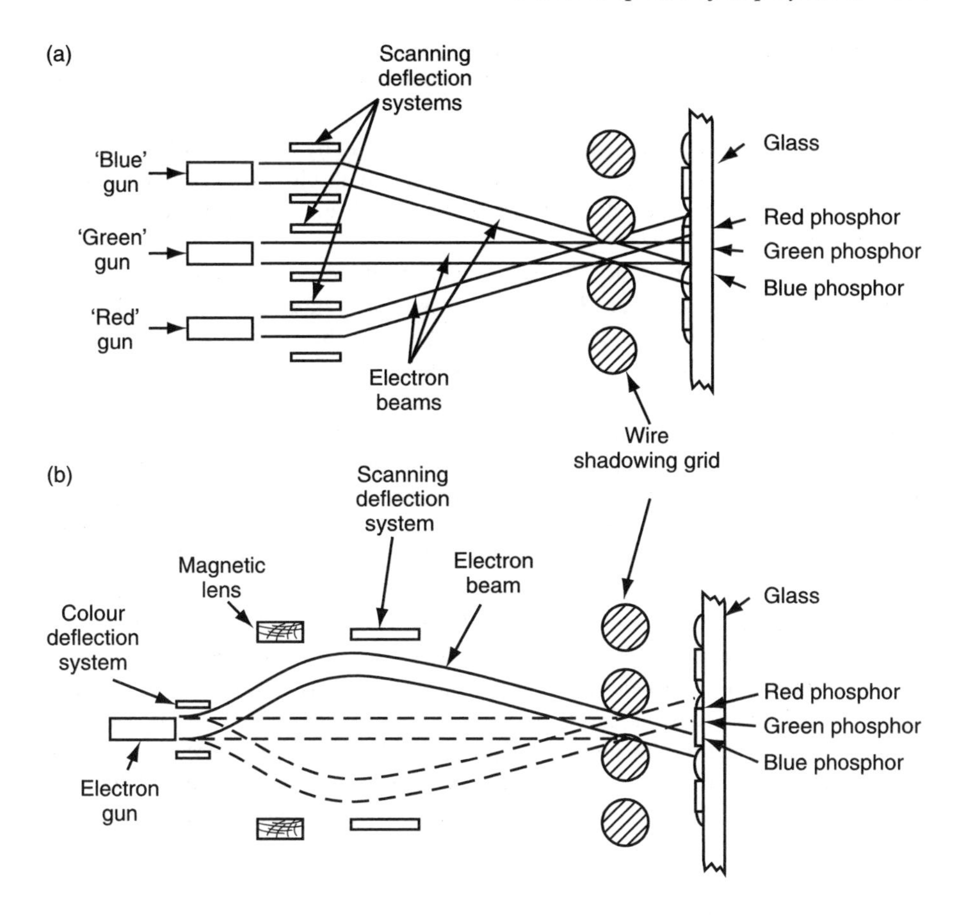

Figure 13.5 *Flechsig's proposal to use colour phosphor strips, shadowed by a wire grid and (a) three electron beams or (b) one beam deflected at the electron gun [Source:* ibid., *p. 1184]*

3. the use of either separate or common deflection means to deflect the three electron beams;
4. the use of a screen coated with colour phosphors arranged in narrow strips and prepared as mentioned above;
5. the concept that the grill openings became cylindrical lenses for concentrating the electrons into narrow planar beams at the screen when the potential of the screen is made positive with respect to the grill;
6. the concept that if the concentration of the electrons is sufficiently large, as a result of point 5, the grill wires may be made very small in diameter and still perform their shadowing action. (In the implementation of this concept, the wires must be coated temporarily to prevent the phosphor strips from being too wide and overlapping: subsequently the coating must be removed.)

But, as Herold has written, it 'seemed clear to many that Flechsig had never tried to make such a tube, or he would have been embarrassed to put his name on the idea' [33].

Another proposal, contained in a patent filed in 1944 by Dr A.N. Goldsmith [36], seemed equally unattractive; namely, to use a mask with thousands of holes instead of wires, together with three electron guns spaced 120° apart and separate deflection systems corrected for keystone.

Of RCA's patents, the only one that related to direct-view tubes was the 1947 application of A.C. Schroeder [37]. He proposed using three closely spaced beams that were deflected by a single deflection yoke with the beams coincident at the screen after passing through the shadow mask. One of the diagrams (see Figure 13.6) of Schroeder's patent shows that when the mask has circular apertures arranged in a hexagonal pattern (illustrated by the solid circles), an array of triplets of colour phosphor dots – red, green and blue – can be produced on the screen. In effect the array consists of three interlaced phosphor dot arrays each capable of emitting red, green or blue light. Consequently, Schroeder noted, when utilized with a properly designed and oriented delta gun cluster, each beam would strike the phosphor dots of only a specific primary colour. The problem seemed formidable: how could several hundred thousand tiny holes in the shadow mask be aligned with an equal number of phosphor dot triplets? No solution was known when RCA began their crash programme in 1949.

Nonetheless, the Schroeder concept was chosen by Dr H.B. Law [38] as the basis for his work. He had joined RCA in 1941, had worked under Dr V.K. Zworykin on the image orthicon camera tube, and had acquired much skill in the fabrication of high-transmission, very fine metal screens and the design and engineering of television cathode-ray tubes. He has described the task that confronted him after the September meeting:

> The problem was: how can the positions, beyond the apertures where the electrons are going to strike [Figure 13.7a] be precisely located: and how can one then place phosphor dots at exactly these locations in a practical and straightforward manner? Then all at once the thought occurred to me that, after deflection, the electrons travel in [a] field-free space so that their paths will be straight and can be simulated by light. Therefore, a light-sensitive material, such as a photographic plate, temporarily positioned in the same location as the faceplate, could record the phosphor-dot positions for a given color if a point light source were placed at the deflection center of the beam for that color. If a photographic plate were used, one could then print a photo-resist pattern on thin metal foil such that the black spots or phosphor-dot locations would not be exposed and would develop out free of resist. Holes could then be etched through the foil where the phosphor dots should be, so the foil could be used as a settling mask. All that would be required in addition was to provide some way to locate the settling mask in the proper position on the faceplate. For this purpose, alignment holes in the mask frame could be used to record alignment marks on the photographic plate at the time of exposure. [35]

All these procedures appeared to Law to be comparatively easy to implement.

In fabricating such a tube he had to have a method of determining the correct orientation of the three electron guns at the stage when they were being sealed in the tube's neck. For this purpose Law devised a structure that he called a 'lighthouse',

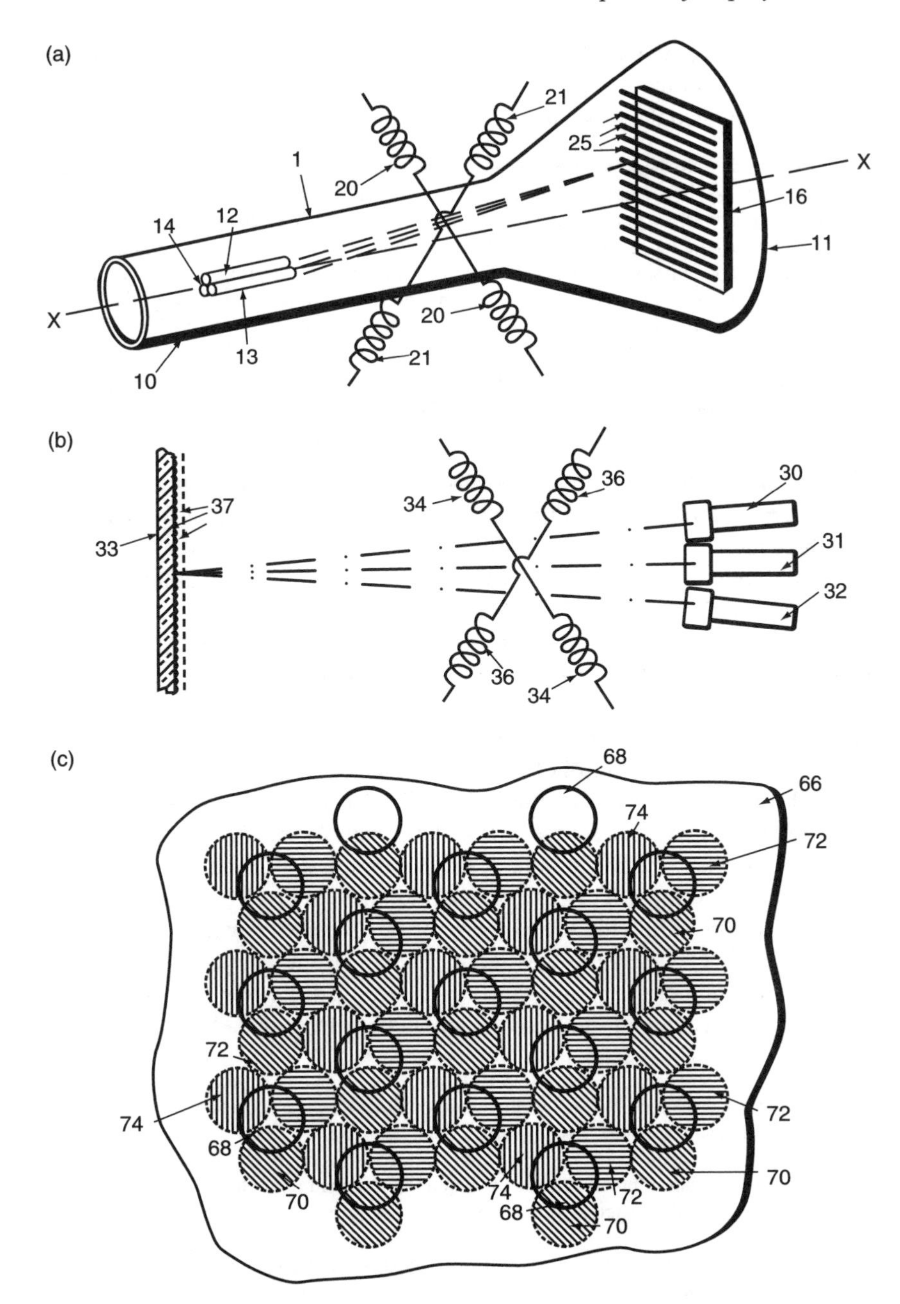

Figure 13.6 Selected figures from the colour tube patent of A.C. Schroeder, showing three beams deflected by a single deflection yoke and the relative arrangement of the mask apertures and the phosphor dots that produces a nested phosphor dot screen [Source: US patent application 2 446 791]

which was attached to the mask frame assembly and carried a small metal plate with three exposure apertures.

Law has recounted the procedure and the first results obtained:

> Three exposures were made on three pieces of photographic film, after which the remaining steps were carried out according to plan; finally the lighthouse was removed after the mask-screen structure had been put in place in the tube. Three individual electron guns were sealed into the tube neck on tungsten rods . . . and the tube put on a vacuum system to pump. Two guns gave enough emission to test but the third was inoperable. Nevertheless, it was a thrill to see the screen change from one color to another by simple adjustment of the grid biases. The experiment was considered a big success and resulted in a number of people at the Laboratories dropping in for a demonstration.
>
> The mask-screen structure was repeated for a sealed-off version and after a couple of tries a tube was produced in which red, green, and blue color fields could be produced and the grids could be modulated with video. [35]

And so, in less than the prescribed three months Law had built a tube with a 7 in. (17.8 cm) diagonal screen that proved the feasibility of a direct colour display tube. In three more months 'a few hundred other people, working seven days a week, had helped produce a dozen or so tubes with a 12 in. (30.5 cm) diagonal picture of remarkable quality'.

A single-gun version of the shadow mask tube was implemented by an ingenious development of Dr R.R. Law [39], of RCA, in which the signals were applied to the gun sequentially and the beam rotated by appropriate deflection circuits synchronized to the sampler to ensure the beam's correct angle of arrival at the holes in the mask. 'Alternatively, by correct control of the beam deflection at the gun, space-sharing of the color phosphors by the beam [allowed] a simultaneous presentation.'

In March 1950, two tubes of each type were taken to Washington and demonstrated on 6 April before the Federal Communications Commission. Each tube had 117,000 triplet groups of phosphors, and had a screen brightness of just 7 foot lamberts, which necessitated an almost fully darkened room for satisfactory viewing.

Prior to the FCC demonstrations, representatives from industry were shown the new direct view colour television receivers. Herold [40] wrote that 'the industry reaction was overwhelming' and quoted a comment from the TV Digest issue for 1 April 1950:

> Tri-color has what it takes: RCA shot the works with its tri-color tube demonstration this week, got full reaction it was looking for . . . not only from . . . FCC . . . and newsmen, but from some 50 patent licensees . . . So impressed was just about everybody by remarkable performance, that it looks . . . as if RCA deliberately restrained its pre-demonstration enthusiasm to gain full impact. [40]

Unfortunately for RCA, in 1950, the FCC had a different view (see Chapter 9).

Figure 13.7b is a cross-sectional diagram of an early shadow-mask tri-colour kinescope [41]. It shows the electron gun assembly, the neck and the glass funnel sealed to the outer metal shell, the magnetic shield of high-permeability metal inside the shell (to screen the electron beams from any external magnetic effects), the shadow mask and the phosphor dot plate (which are located slightly forward of the welding flange and as close to the face of the tube as is practical). To ensure precise alignment

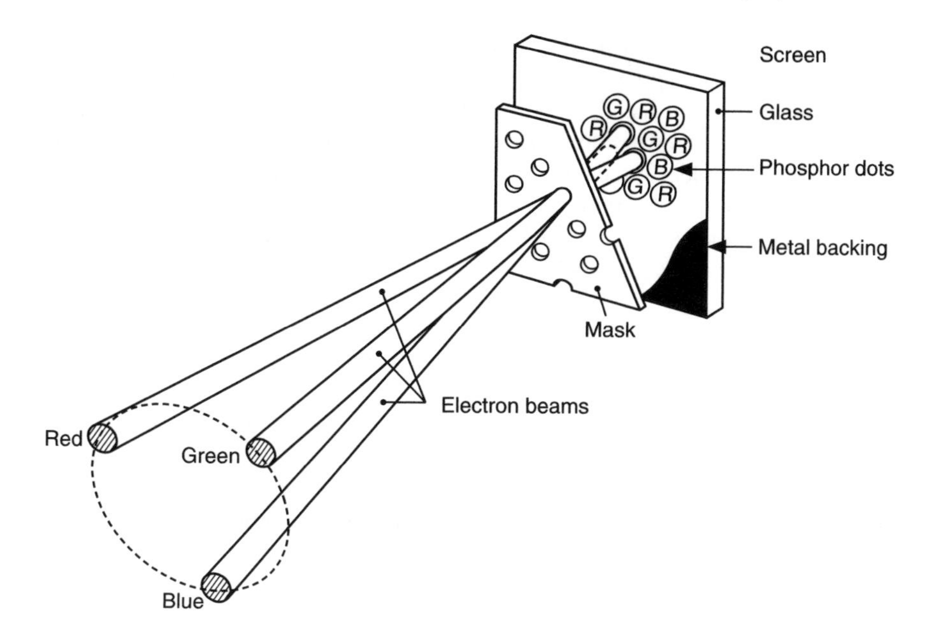

Figure 13.7a *The principle of the shadow mask tube [Source:* Journal of the Royal Television Society, *1967–68,* ***11****(12), p. 279]*

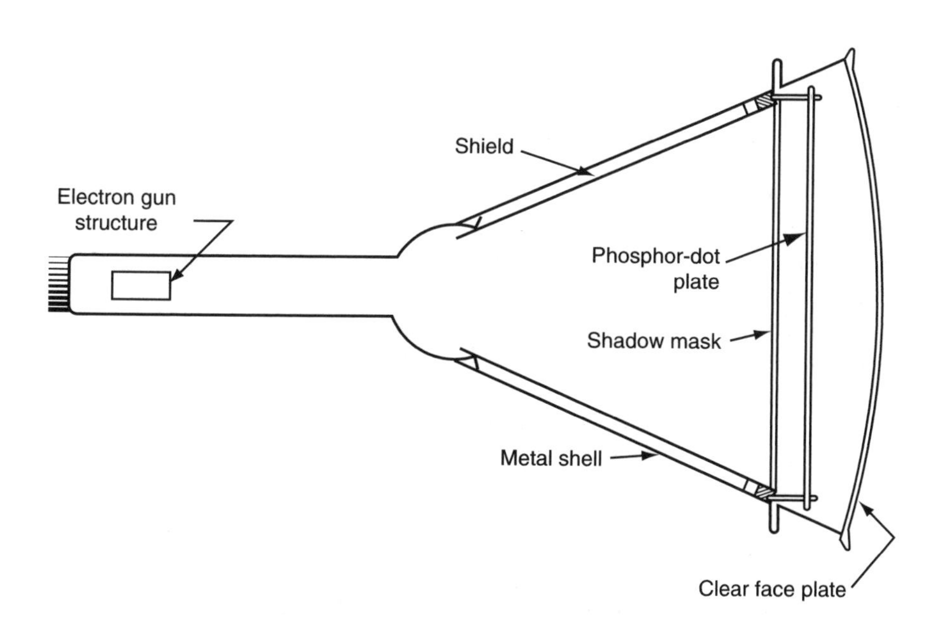

Figure 13.7b *Cross-sectional diagram of RCA's shadow mask tri-colour kinescope [Source:* Proceedings of the IRE, *1954,* ***42****(1), p. 315]*

between the apertures in the mask and the phosphor dots it is necessary to mount the two components together before insertion into the tube. The gun assembly comprises three parallel, closely spaced, electron guns (fabricated as a unit), each having an indirectly heated cathode and four grids – the control grid, the accelerating electrode, the focusing electrode and the converging electrode. The four grids of the three guns open into a common cylindrical electrode, to which they are joined, and, when the tube is in operation, a field exists between the cylinder and the conductive neck coating to establish an electrostatic lens, which serves to converge the three beams at the shadow mask.

The later development of the shadow mask tube centred on improving its performance, designing the tube so that it was more suited to mass production methods, and decreasing its cost.

Enhancement of performance came about as a result of increased phosphor efficiency which arose from improvements in the method of application of the phosphors and by improved screen-processing operations. The maximum energy at which the electron beams can excite the phosphors is limited by the amount of beam energy which the shadow mask can dissipate before it reaches a temperature at which distortion of the mask, and hence mis-registration of its apertures and the associated phosphor dots, occurs. By developing an improved method of screen processing the beam energy that could be dissipated in the mask was increased by 2.5 times and led to a high-light brightness capability of 30 to 40 foot lamberts.

Improvement in the contrast ratio was obtained by: first, using a neutral filter glass instead of the clear glass phosphor dot plate; and, second, by better techniques for aluminizing the phosphor dot plate to provide a more uniform coating of the precise thickness needed for the optimum balance between contrast improvement and light output. These modifications gave a tube that produced a picture having a contrast ratio equal to that of a black-and-white television tube.

Apart from these changes, others included better convergence of the three beams at the viewing screen assembly by the use of improved jigging of the gun parts during the assembly of the gun, and by a redesign of the deflecting yoke to give a better flux pattern within the yoke. Improvements in screen-making techniques and in the construction of the viewing-screen assembly resulted in greatly increased purity of the colour fields, thereby permitting 'excellent colour rendition for colour television as well as reproduction of good black-and-white pictures'.

The shadow mask tube was extensively developed not only in the USA but also in the UK [42] and in Europe. Among the significant areas of interest were the efforts to increase the picture brightness (by the use of new phosphors), and to reduce the length of the tube by increasing the scanning angle to 110°. These matters are beyond the scope of this book.

Instead of using a shadow mask to define those electrons that strike a particular phosphor dot, it is possible to deflect or otherwise control the electron beam in the vicinity of the phosphor screen by means of additional electrodes (see Figure 13.8).

In the earliest proposals the screen comprised repeated triplets of red, green and blue phosphor strips, insulated from each other. The colour strips to be excited were held at a much higher positive potential than the strips of the other colours and so,

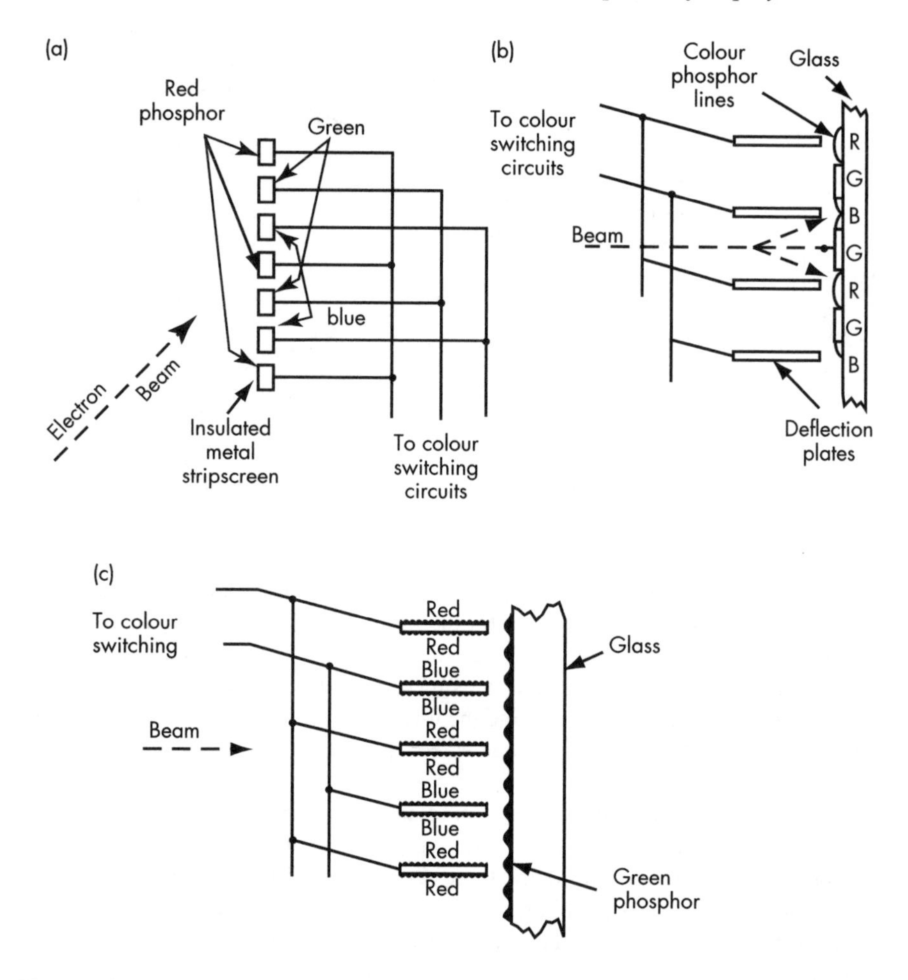

Figure 13.8 Beam control, at the phosphor screen, for changing colour: (a) simple line screen colour switching; (b) deflection switching of colour with line screen; (c) deflection switching without requiring registry [Source: Proceedings of the IRE, *1951,* ***39****(10), p. 1182]*

when bombarded by the electron beam, were able to reproduce the desired colour. Herold has stated that the colour changing circuits must 'operate with voltages of many kilovolts and were difficult to make in practical form'. 'With a sequential presentation, the difficulties increase rapidly as the switching rate is increased, which makes switching least difficult for field or frame color sequencing. There are even greater practical disadvantages when a magnetic field is used for switching color.' [2]

In one variation, due to Bronwell, of the high-voltage switching method (see Figure 13.9a), three insulated, phosphor-coated grids R, B, G, spaced 1–3 mm apart, were suggested. For a 525-line picture definition, each of the red, green and blue phosphor-coated grids would contain 525 parallel coated wires; but according to

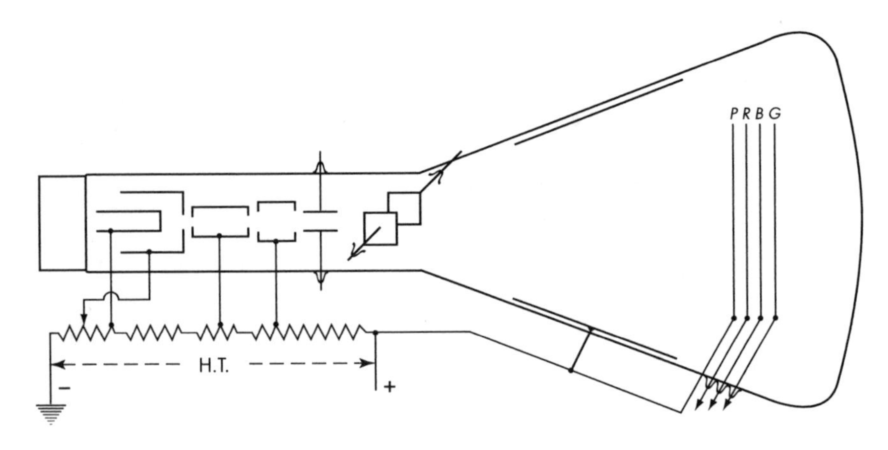

Figure 13.9a *The chromoscope tube arranged for direct viewing [Source:* Electronic Engineering, *June 1948, p. 191]*

Bronwell the screens would be 'relatively transparent to light and electrons, while presenting sufficient phosphor surface for the formation of the picture'. In operation, a high voltage applied to the appropriate grid produced the desired colour. A fourth screen, P, held at a constant potential, was incorporated into the assembly to shield the region between the electron gun and the image screen from potential fluctuations resulting from variations of the colour screen potentials: it prevented defocusing errors.

This method eliminates the fabrication of a tri-colour phosphor screen but problems still exist with the necessity of switching high-voltage control signals to effect the colour changes, and insulating the three sets of grid wires. Furthermore parallax exists because the three colour phosphors are not in the same plane, though this can be negated by the use of a projection system of adequate depth of field, rather than direct viewing.

The chromoscope, as the tube was called, was at one time under development by the DuMont Laboratories but it seems that no tube reached the stage of commercial acceptance.

RCA [44] experimentally investigated two-colour, single-gun; two-colour, two-gun; three-colour, single-gun; and three-colour, three-gun versions, of the grid-controlled colour tube. They included a series of closely spaced screens, each coated with a different primary-colour phosphor and separated from each other by fine mesh control grids. In operation, small changes of potential on the grids control the depth of penetration of the scanning beam(s) into the assembly of screens and permit the individual excitation of any, or a combination of, the phosphors (see Figure 13.9b).

A simpler tube based on the use of two grids and a tri-colour phosphor screen was developed by Chromatic Television Laboratories, Inc. (CTL) [45]. The company was established to further some basic concepts relating to colour television conceived by Dr E.O. Lawrence [46], the inventor of the cyclotron, a Nobel Prize winner, and (in 1953) the director of the University of California Radiation Laboratory.

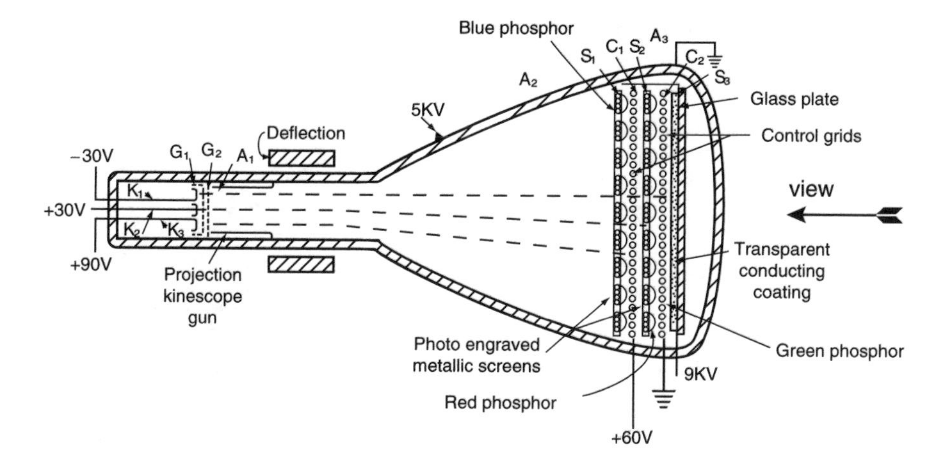

Figure 13.9b *RCA's three-colour, three-gun, grid-controlled colour kinescope [Source:* Proceedings of the IRE, *1951,* ***39****(10), p. 1214]*

Several eminent physicists were associated with Lawrence in his work, including Dr E.N. McMillan, another Nobel Prize winner and co-discoverer of plutonium, Dr L.W. Alvarez, inventor of a ground-control-approach aircraft navigation system, and Dr M.G. White of the University of Princeton.

CTL's research and development work was undertaken in laboratories in New York and Oakland and was financed by the Paramount Pictures Corporation.

Figure 13.10a indicates the principle of the single-electron-gun Lawrence tube (chromatron). The screen comprises a succession of red (R) green (G) and blue (B) phosphor strips, which can be excited in turn by the application of potentials to the 'red' and 'blue' co-planar grids, which act as switching elements and are parallel to the phosphor strips: the width of each strip is equal to one third of the spacing between the wires. Post-acceleration of the electrons is employed between the grids and the screen.

When the electron beam is undeflected the green phosphors are energized, but, when a positive potential is applied to either the 'red' grid or the 'blue' grid, either the red or the blue phosphors emit light respectively. As with all single-beam tubes, no registration problems can arise; hence special correction circuits which may be a feature of multi-beam tubes are not necessary.

In the three-electron-gun version (see Figure 13.10b), post-acceleration is also used. Electron optically the arrangement functions as a system of cylindrical lenses, which allows each of the three electron beams to be focused upon only one kind (red or green or blue) of phosphor strip.

The post-acceleration method has several advantages over the shadow mask method, namely:

1. the transmittance (*c.* 85 per cent) of the colour grid is much higher than that of the shadow mask, leading to increased brightness;

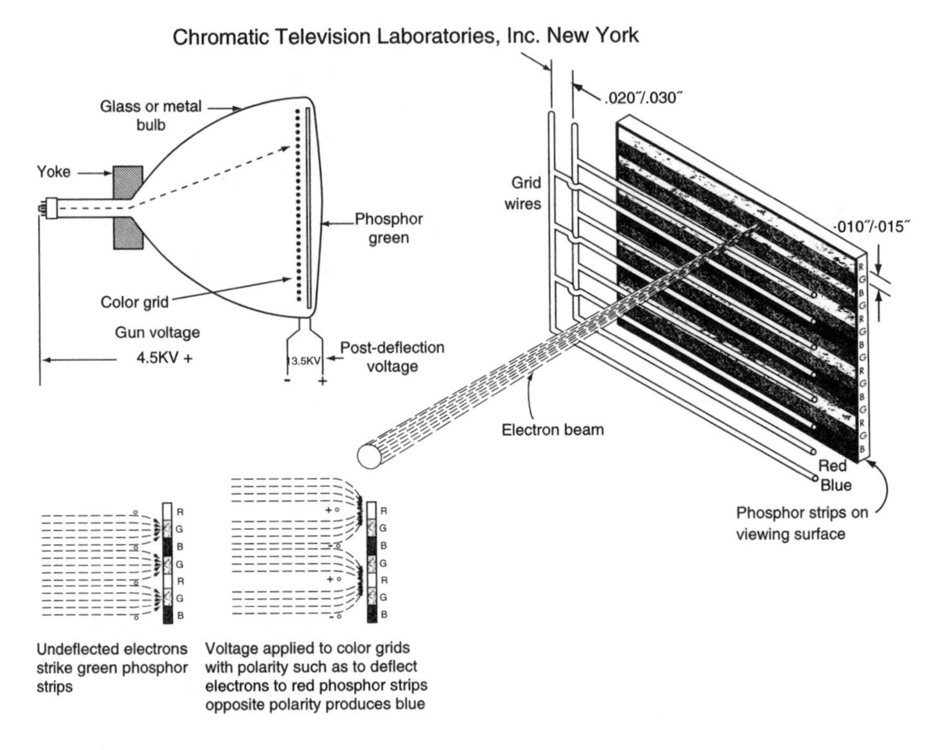

Figure 13.10a Illustration showing the principle of operation of the single-gun Lawrence tube, the chromatron [Source: FCC Docket 8736]

2. the electrode configuration of the tube leads to a smaller deflection power; and
3. the electron guns can be disposed more favourably resulting in less severe registration problems and hence fewer correction means.

CTL's views on colour television were described in March 1953 to the House of Representatives Committee on Interstate and Foreign Commerce, and a 22 in. (56 cm) tube was seen by the Committee on 15 April 1953. The company affirmed that their tri-colour, direct view tube was compatible with monochrome television services and could function on either the approved field-sequential standards or the NTSC standards with 'facility and effectiveness'. In their evidence CTL said, 'the overwhelming reaction of the industry to the tube has been one of admiration for the quality of its color, the brightness of its picture, and the simplicity of its design. Because of [this], the tube offers a unique opportunity for mass production within the means of the public – at once!' [46]

The company's publicity information [45] highlighted the advantages of the Lawrence single-gun tube to the consumer, namely: a larger picture, a more brilliant picture because it made maximum use of the electrons, a stable picture because of the simplicity of the receiver circuits, a better, sharper black-and-white picture, and simpler controls. For the receiver manufacturer the advantages were: a smaller and

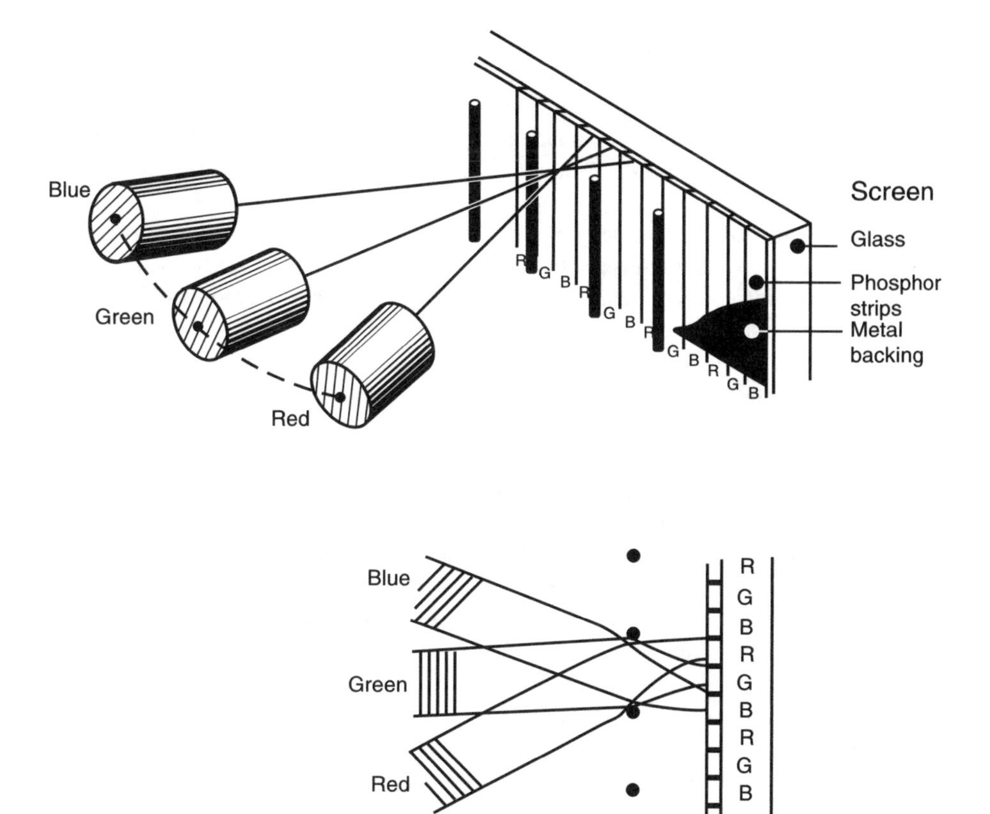

Figure 13.10b Schematic diagram of the three-gun chromatron [Source: Journal of the Royal Television Society, *1967–68,* ***11****(12), p. 279]*

less expensive cabinet because the tube's wide deflection angle made the Lawrence tube a shorter tube, the use of standard deflection and focus components, simple low-cost circuitry and quick circuit alignment, the absence of power supply regulation, colour purifying coils or registration magnets, and mu-metal screens.

All of this seemed to portend a prosperous future for the chromatron's investors. However, by the end of 1955, as a report in Fortune commented, though the tube's pictures were bright 'they probably [could not] yet match those of the shadow-mask tube in color quality and detail. There [were] also manufacturing problems to be solved'.

The chromatron tubes (both three-beam and single-beam) were later subjected to a detailed examination by engineers of the Philips Research Laboratory, Eindhoven. In a paper [47] published in 1967 the authors noted a number of serious disadvantages of the three-beam tube.

1. 'Secondary electrons released at the screen and the grid are conducted to the screen by the accelerating field. Such electrons are substantial both in number

and energy, and they lead to losses of contrast, definition and saturation. This is a basic disadvantage of all forms of post-acceleration.'

2. 'The electrons of the beam being "slow" until they reach the grid, the beam is more strongly influenced by space charges. Sacrifices, either in the quality of the picture (spot width) or in the quality of the beam current must therefore be made.'
3. 'Employment of a colour-selecting grid rules out substitution of electron-optical fine focusing by analogous photo-optical means; that is to say, to bring the phosphor strips into correct position on the screen, the electron beam itself must be used in exposure, which is difficult and expensive in production.'
4. 'Mechanically, colour-selecting grids are highly sensitive: microphony and parabolic distortion, due to the electric field, occur. In addition, the screen of the tube cannot be spherical but must be cylindrical.'

In the authors' opinion the picture quality – sharpness, contrast and colour saturation – of the three-gun chromatron still had to undergo some improvement before it could be successfully employed in domestic colour television receivers. On the single-beam tube they noted that it 'may gain in importance as a tube for the smallest type of receiver provided less than optimum picture quality can be accepted in return for lower power consumption'. Such receivers, in 1967, were being manufactured and marketed in Japan. The subsequent development of beam indexing tubes is beyond the scope of this history of colour television.

By the early 1960s a great deal of R&D effort, and financial resources, had been expended in the United States of America on the evolution of a practical, high-definition system of colour television that could be accommodated within the bandwidth constraints imposed by the regulatory authority. The NTSC system had been developed by the leading engineers and physicists of the television industry and had satisfied – with some reservations – the Federal Communications Commission. There was an expectation that, if the NTSC system could be adopted throughout the world, the interchange of national programmes would be greatly facilitated. However this proved to be a forlorn hope. In Europe, particularly in France and Germany, attempts were made to negate the limitations of the NTSC system, and variants of it, namely, the SECAM and PAL systems, were advanced. One of these, possibly, could have been chosen for service throughout Europe if nationalistic pride had not been a factor in the debate on the choice of system to be adopted. In the next chapter the struggle for unity, of a given system of colour television, is considered.

Note 1: Subtractive colour television

All the colour television display devices mentioned previously have been based on the additive colour-mixing principle. An interesting proposal to use the subtractive colour-mixing principle was proposed by A.H. Rosenthal, of the Scophony Company, in 1938 [48]. Figure 13.11 illustrates his configuration: it utilizes three Skiatron tubes, in which the fluorescent screens of the conventional cathode-ray display tubes are replaced by screen materials that darken when bombarded by electrons. In Rosenthal's arrangement, the materials are chosen so that the darkening produces an absorption

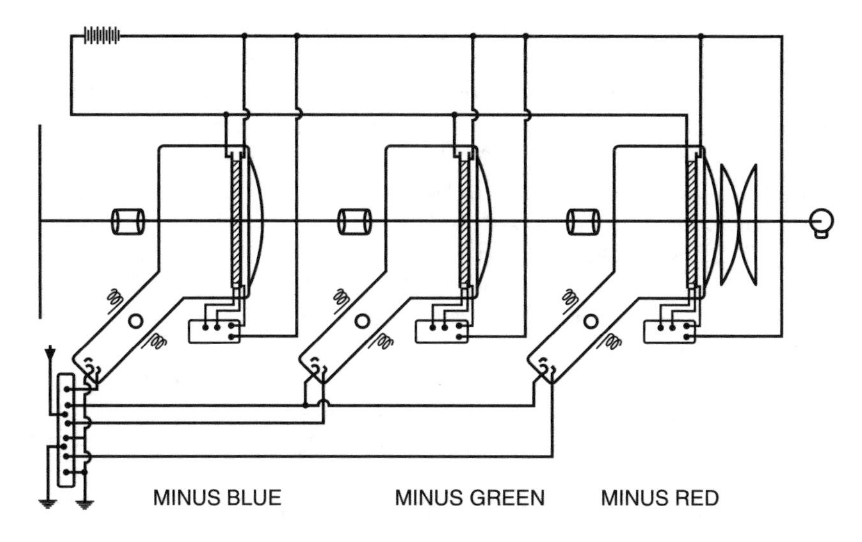

Figure 13.11 *Rosenthal's proposed subtractive colour display scheme [Source:* Journal of the Royal Television Society*, 1954, 7(6), pp. 241–7]*

of red, green, and blue, in the three c.r.t.s respectively, that is the actual colours of the screens when excited are 'minus red', 'minus green', and 'minus blue'. Hence, when the red, green and blue image signals are applied to the grids of the three tubes, and the resulting screen images are superimposed onto a display screen, by a suitable optical system, a coloured image will result. The effective superimposition of the three absorbing c.r.t. screens is analogous to the use of the three subtractive dye layers in modern photographic colour films. No evidence appears to exist to show that Rosenthal reduced his ideas to practice.

References

1. Herold E.W. 'History and development of the color picture tube'. RCA publication, 1974, pp. 1–6
2. Herold E.W. 'Methods suitable for television color kinescopes'. *Proceedings of the IRE*. 1951;**39**:1177–85
3. Zworykin V.K. US patent application 1 691 324. Applied for 13 July 1925
4. Rudenberg R. US patent application 1 934 821. Convention date 5 May 1931
5. Ardenne M. von. British patent application 388 623. Convention date 19 June 1931
6. Kasperowicz H. US patent application 2 508 267. Applied for 26 October 1945
7. Fernseh A.G. British patent application 434 868. Convention date 16 March 1933
8. Hewson B.T., Locan A. British patent application 533 993. Complete specification left 17 June 1940
9. Zworykin V.K. US patent application 2 415 059. Applied for 11 October 1944

10. Stevens W.H. British patent application 803 080. Complete specification left 4 July 1945
11. Huffman C.E. US patent application 2 490 812. Applied for 4 January 1946
12. Freedman N.S., McLaughlin K.M. 'Phosphor screen application in colour kinescopes'. *Proceedings of the IRE*. 1951;**39**:1230–6
13. Bond D.S., Nicoll F.H., Moore D.G. 'Development and operation of a line-screen colour kinescope'. *Proceedings of the IRE*. 1951;**39**:1218–30
14. Weimar P.K. US patent application 2 545 325. Applied for 22 January 1948
15. Goldsmith A.N. US patent application 2 431 115. Applied for 5 August 1944
16. Herold E.W. 'Methods suitable for television color kinescopes'. *Proceedings of the IRE*. 1951;**39**:1180
17. Schlesinger K. US patent application 2 083 203. Convention date October 1932
18. Wilson J.C. US patent application 2 294 820. Applied for 29 April 1941
19. Fernseh A.G. British patent application 432 989. Convention date 16 March 1933
20. Haverline C. Statement to the FCC, 13 December 1946, Exhibit 61, FCC Docket 7896
21. Anon. 'J.L. Baird's improved colour television'. *Electronic Engineering*. January 1943;**15**:327
22. Color Television Inc. Exhibits 237, 259 and 260, 1949–50 FCC Hearings, FCC Dockets 8736, 8975 and 9175
23. Columbia Broadcasting System. Exhibit 210, 1949–50 FCC Hearings, FCC Dockets 8736, 8975 and 9175
24. Baird J.L. British patent application 562 168. Applied for 25 July 1942
25. Anon. 'J.L. Baird's telechrome'. *Journal of the TV Society*. September 1944;**4**: 58–9
26. Szegho C.S. US patent application 2 455 710. Applied for 21 December 1943
27. Herold E.W. 'Methods suitable for television color kinescopes'. *Proceedings of the IRE*. 1951;**39**:1181
28. Bril A., Kroger F.A. 'Saturation of fluorescence in television tubes'. *Philips Technical Review*. October 1950;**12**:120–8
29. Szegho C.S. US patent application 2 431 083. Applied for 3 December 1943
30. Herold E.W. 'Methods suitable for television color kinescopes'. *Proceedings of the IRE*. 1951;**39**:1182
31. Anon. Statement to the FCC on the Geer colour television receiver tube, FCC Docket 8735, **66**, pp. 133–48
32. Herold E.W. 'Methods suitable for television color kinescopes'. *Proceedings of the IRE*. 1951;**39**:1185
33. Herold E.W. 'History and development of the color picture tube'. RCA publication, 1974, p. 1
34. Flechsig W. *Cathode-ray tube for the production of multi-colored pictures on a luminescent screen*. German patent application 736 575. Filed 1938
35. Law H.B. 'The shadow mask colour picture tube: how it began – an eye witness account of its early history'. *IEEE Transactions on Electron Devices*. July 1976;**ED-23**(7):752–9

36. Goldsmith A.N. US patent application 2 431 115. Filed 1944
37. Schroeder A.C. *Picture reproduction apparatus*. US patent application 2 595 548. Filed 1947
38. Law H.B. 'A three-gun shadow-mask colour kinescope'. *Proceedings of the IRE*. 1951;**39**:1186–94
39. Law R.R. 'A one-gun shadow-mask colour kinescope'. *Proceedings of the IRE*. 1951;**39**:1194–201
40. Herold E.W. 'History and development of the color picture tube'. RCA publication, 1974, p. 2
41. Grimes M.J., Grimm A.C., Wilhelm J.F. 'Improvements in the RCA three-beam shadow-mask colour kinescope'. *Proceedings of the IRE*. 1954;**42**:315–26
42. Wright W.W. 'Recent developments in shadow-mask tubes for colour television'. *Journal of the Royal Television Society*. July/August 1971;**3**(10):221–30
43. Bronwell A.B. 'The chromoscope. A new colour television viewing tube'. *Electronic Engineer*. June 1948, pp. 190–1
44. Forgue S.V. 'A grid-controlled colour kinescope'. *Proceedings of the IRE*. 1951;**39**:1212–18
45. Anon. 'The Lawrence color tube', press release, undated, Chromatic Television Laboratories, Inc. New York
46. Hodgson R. Statement before the House Interstate and Foreign Commerce Committee Color Television Hearings, 26 March 1953, 14 pp.
47. Haan E.F. de, Weimer K.R.U. 'The beam-indexing colour television display tube' *Journal of the Royal Television Society*. 1967–68;**11**(12):278–82
48. Scophony and Rosenthal A.H. British patent application 514 776, 1938

Chapter 14
An attempt at unity

By 1962 the time seemed to be ripe for an attempt to be made to coordinate the trials being undertaken in Europe on the several systems of colour television being investigated and to determine whether any one of the systems could be accepted generally throughout Europe, and possibly the world.

The appropriate body which pronounced on matters relating to television for the whole of Europe was the Comité Consultatif International des Radiocommunications (CCIR), an advisory committee of the International Telecommunication Union (ITU), which itself was a specialized agency of the United Nations. Of course, a decision on the colour television system to be adopted by any country rested with the administration of that country, but, as all the administrations concerned were members of the CCIR, they attached considerable weight to its recommendations.

Two other international bodies closely concerned with the question of colour television standards were the European Broadcasting Union (EBU) and the International Broadcasting and Television Organization (IBTO). The EBU, with headquarters in Geneva and a technical centre in Brussels, had, as full members, most of the broadcasting authorities of the European Broadcasting Area (Europe and countries bordering the Mediterranean) except those of Eastern Europe, and, as associate members, broadcasters from widely scattered countries throughout the world. The Technical Committee of the EBU carried out investigations and coordinated the experimental work of its members, as well as participating actively in the international exchange of programmes by means of the Eurovision network. Both the BBC and the ITA were members of the EBU, and from time to time the UK's Post Office was invited to send an observer to meetings of the Technical Committee and some of its subcommittees.

The IBTO, with headquarters in Prague, was an association of the broadcasting authorities of Eastern Europe. Its conclusions appeared to have 'the full approval of the administrations concerned'. Both the IBTO and the EBU sent representatives to CCIR meetings and additionally there was some direct liaison between the two organizations. Recommendations of the CCIR (which had a subcommittee known as Study Group XI, which dealt with television issues) became authoritative (though not mandatory) only when approved by a Plenary Assembly.

The question of European harmony in colour television matters was discussed at the CCIR Study Group XI meeting held at Bad Kreuznach in 1962. It was attended by representatives of several telecommunication administrations, by members of the European Broadcasting Union, and by the radio manufacturers who were engaged on aspects of colour television development, and the delegates (from France, the Federal

Republic of Germany, Italy, the Netherlands, Switzerland and the United Kingdom) agreed the question should be examined.

The EBU's Ad-Hoc Group on Colour Television was set up in November 1962, at the instigation of F.C. McLean, the BBC's deputy director of engineering [1]. Its mandate included not only the coordination of the various trials, but also the preparation of proposed standards relating to colour television for submission to the CCIR. The establishment of such a group was authorized by the EBU Administration Council at its meeting at Istanbul in November 1962. Professor R. Theile, of the Institut fur Rundfunktechnik, Munich, was elected chairman and six subgroups were constituted and chairmen elected, as follows [2]:

Subgroup 1, 'General characteristics' (Dr R.D.A. Maurice, BBC, UK)
Subgroup 2, 'Receivers' (Mr F.C. McLean, BBC, UK)
Subgroup 3, 'Propagation' (Mr K. Bernath, PTT, Switzerland)
Subgroup 4, 'Transmitting equipment' (Mr H. Hopt, ARD, Federal Republic of Germany)
Subgroup 5, 'Programme-distribution networks' (Dr E. Castelli, RAI, Italy)
Subgroup 6, 'Studio and recording equipment' (Mr L. Goussot, RTF, France)

The average membership of each subgroup was ten and the membership of the Ad-Hoc Group was *c.* thirty. From its inception to February 1966 the Group produced 238 documents and met formally on 11 occasions; it also visited various laboratories in Western Europe. Dr Maurice has commented, 'It is only fair to mention that the major contributor to the tests, field trials and mathematical treatments was the BBC. There are more relevant EBU Ad-Hoc documents from the BBC than any other organization'.

So that any potential standardization by the CCIR could apply to as many countries as possible, the delegates agreed to approach the OIRT to sanction an exchange of information among the participants on the systems and trials in progress. Curiously, the UK was almost alone in Europe in being anxious to reach an early European concord on the choice of colour television standards, but other Western European countries were sympathetic to the UK position and were prepared to expedite their consideration of the problem [3].

At this time new systems and variants of the American NTSC system had been and were being suggested and tested. These included the double-carrier system (TSC), the double message system (LEP), the Valensi system, the Henri de France/SECAM system, the FAM system and, more recently, the PAL system; hence there was no immediate consensus that one of these would be generally acceptable throughout Europe. Some of these systems will now be considered [4]. (Non-technical readers may wish to pass over the following section.)

(The following abbreviations are used in the diagrams which illustrate the systems of this section: AM, amplitude modulator; ADM, amplitude demodulator; suppr., suppressed; f_c, carrier frequency; f, line frequency; φ_o, phase shift; O, modulator; τ, delay line; FDM, frequency demodulator; +, adder; –, subtractor.)

In the NTSC system, as noted previously, the colour difference signals are modulated on the colour sub-carrier by a process of quadrature modulation. The resultant output of the modulator is a signal the amplitude and phase of which depend on the saturation and hue, respectively, of the televised colour picture. Thus, as the colour content of a scene diminishes, i.e. the colours become de-saturated, the amplitude of the colour sub-carrier decreases and similarly any distortion products associated with the process, or the transmission, likewise decrease – an advantage when viewing compatible monochrome pictures.

Figure 14.1 shows diagrammatically the encoding and decoding principles. The synchronous detection process at the receiver necessarily requires a signal derived from the encoder and for this purpose a 'burst' of sub-carrier oscillation is transmitted on the back porch of the video signal and is employed to synchronise the decoder's oscillator to the encoder's oscillator.

Figure 14.2 illustrates the relation between hue and phase with reference to the axes of the colour difference signals, and the transformed difference signals (I and Q). If the vector representing the colour signal is rotated as a consequence of differential errors (or phase variations which are a function of amplitude), then, when the received amplitude modulated sub-carrier is demodulated, the two colour components will be incorrectly reproduced. This susceptibility to distortion of the NTSC's quadrature modulation process was the main source of criticism of the NTSC's system and led to various alternative encoding and decoding arrangements. Some of those that have been tested experimentally will now be described.

1. **The two sub-carrier system (TSC) (1956) [5]**
   Philips Laboratories at Eindhoven developed a system that used two different sub-carrier frequencies for the two colour signals (see Figure 14.3(a)). Since amplitude modulation was used, thereby avoiding the need for synchronous detection at the receiver, the modulating signals had to be positive. Consequently colour difference signals such as R-Y, B-Y, or I, Q as in the NTSC system could not be employed.

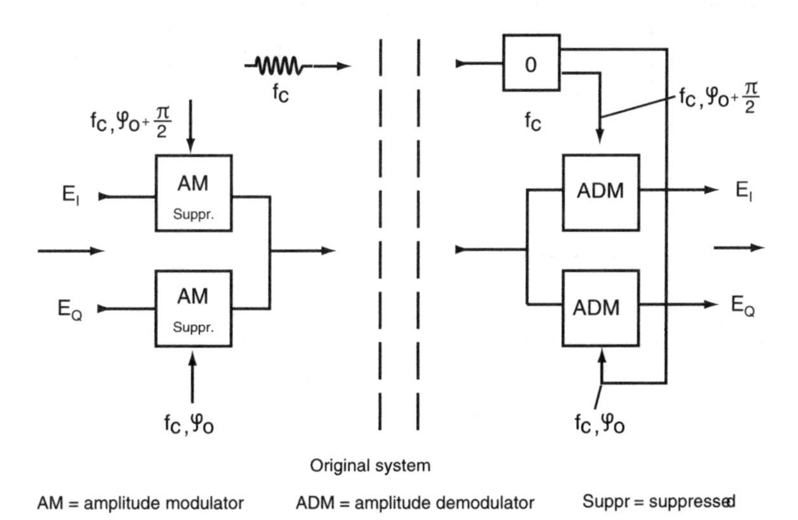

Figure 14.1 Modulation and demodulation processes of the colour signals in the NTSC system [Source: EBU Review, Part A – Technical, *1965, no. 93, pp. 194–204]*

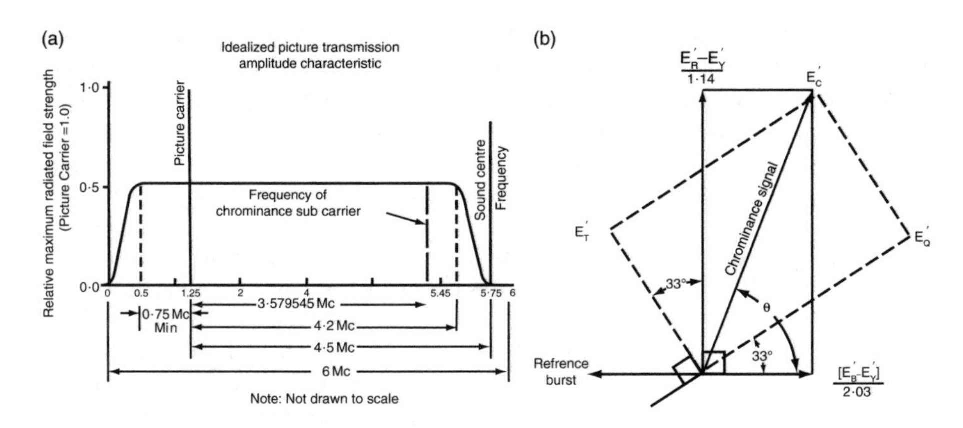

Figure 14.2 The NTSC transmission characteristic [Source: Journal of the Royal Television Society, *1954, 7(6), pp. 236–240]*

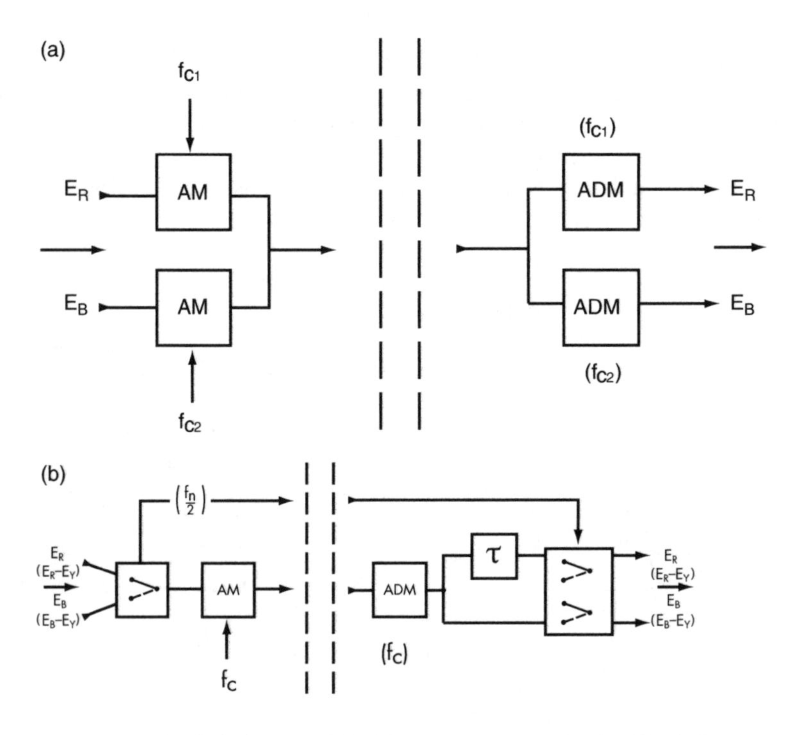

Figure 14.3 (a) Block diagram of the TSC system (1956); (b) block diagram of the H. de France system (1957–58) [Source: ibid.*]*

Instead, R and B colour signals were used directly. Cross-talk effects between each sub-carrier and the luminance signal and between the sub-carriers themselves still existed, but Philips devised an ingenious method of phase shifting (at the transmitter) the sub-carriers on successive scanning fields to minimize the visibility of these effects. The method simplified the receiver sub-carrier detection circuits but new problems were encountered with the compatible monochrome picture and with the stabilization of the ratio of the two colour components which determines hue.

2. **Henri de France system (1957–58) [6]**
   In this system, Figure 14.3(b) only one colour signal at a time modulated the colour sub-carrier. Sequential switching ensured that the two colour signals were applied alternately, following a line-by-line sequence, to the modulator. Amplitude modulation was employed, and so the transmission of a 'burst' of sub-carrier signal was unnecessary since conventional envelope detection could be used at the receiver. Nevertheless, a synchronizing signal was sent during the field-blanking period to control the phase of the electronic switch in the receiver.
   The process necessitated the utilization of a delay line (having a group delay equal to a line period) to store the colour information transmitted during the preceding line period. By its use a continuous flow of colour signal information could be provided to permit the reconstitution of the primary R, G and B signals. Initially, the delay line was operated at video frequencies and was positioned after the detector, but further development, undertaken by the Compagnie Français de Television, Paris, led to the SECAM system.
3. **The SECAM system (1959–60) [7,8]**
   The design and practical implementation of a cheap, ultrasonic delay line able to function at carrier frequencies enabled the delay line to be placed before the detector (see Figure 14.4(a)). The system was given the name SECAM – **SEQ**uential **A M**émoire, or, as some French engineers jocularly remarked, '**S**upreme **E**ffort **C**ontre L'**AM**erique. In a development of SECAM the amplitude modulation process was replaced by one of frequency modulation. To reduce the effects of interference from the sub-carrier upon the compatible monochrome picture various ameliorative measures were taken. These included phase alternation of the sub-carrier signal at each third line, and from field to field; as well as pre- and de-emphasis of the higher video frequency sub-carrier sideband components.
4. **The FAM system (1960) [9]**
   N. Mayer of the IRT at Munich sought to eliminate the deleterious effects of differential phase errors by using both amplitude and frequency modulation of the two colour signals (see Figure 14.4(b)). Synchronizing signals were unnecessary since conventional demodulators were employed in the receiver.
5. **SECAM-NTSC system (1961–62) [10]**
   In this system W. Bruch of Telefunken AG adapted the switching principle to the NTSC method of modulation (see Figure 14.5). Whereas the two colour signals are used simultaneously in the NTSC quadrature modulation process, Bruch applied the signals alternately to the suppressed carrier modulator, i.e. only one colour signal at a time was transmitted. By this means the phase sensitivity of quadrature

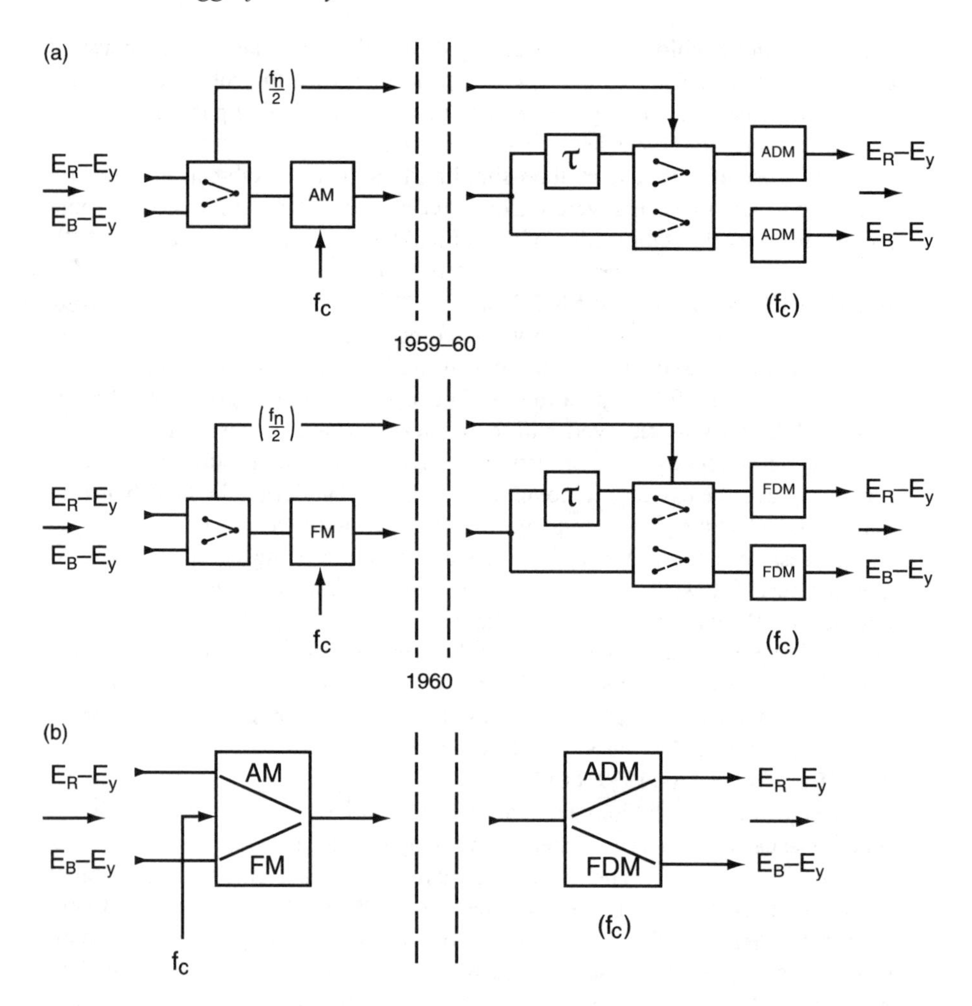

Figure 14.4 Block diagrams (a) of the SECAM system 1959–60 and 1960; and (b) of the FAM system (1960) [Source: ibid.*]*

modulation was negated. Two auxiliary signals now had to be transmitted: (1) a 'burst' of the sub-carrier to lock the receiver oscillator, and (2) a synchronizing signal to control the electronic switch of the receiver.

Initially, the delay line was placed after the demodulator, but later, following the introduction of ultrasonic delay lines, the delay line was inserted before the switching process, as shown in Figure 14.5. Further development of the system led to the PAL system

6. **PAL system (1962) [11,12]**

   In this system the sequential switching principle was not concerned with the *sequential selection* of one of the colour signals; rather it was applied to the phase alternation of one of the two *simultaneously transmitted* colour signals.

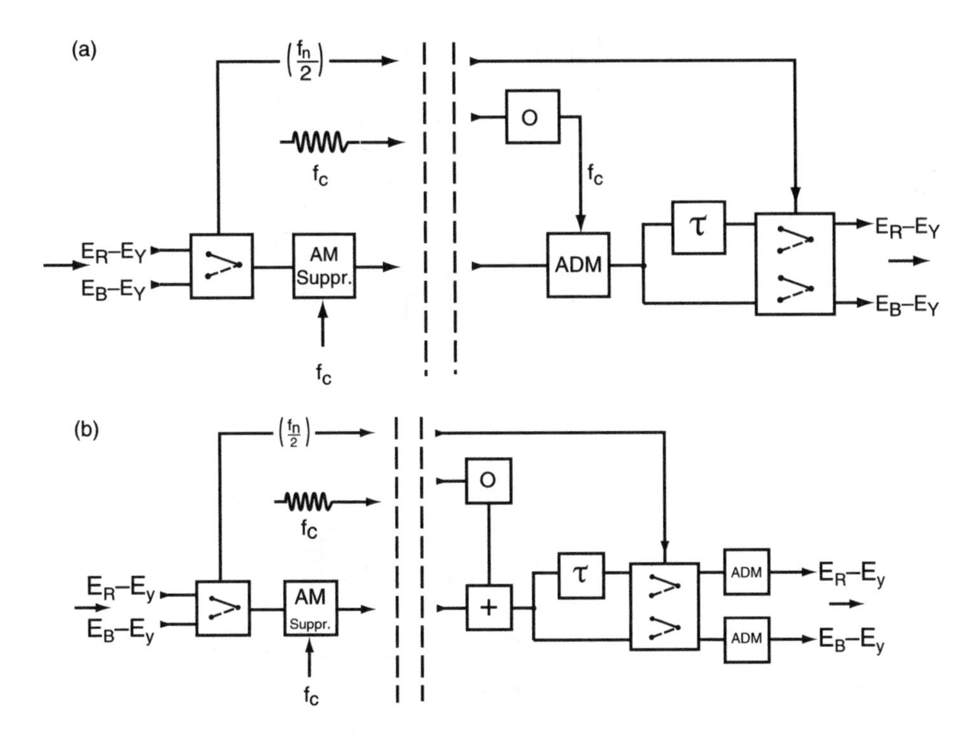

Figure 14.5 Block diagrams of the SECAM-NTSC system (1961–62), first version (a), and the SECAM-NTSC system, second version (b) [Source: ibid.*]*

Referring to Figure 14.6, the E_I colour signal modulates the sub-carrier in a suppressed amplitude modulator as normal, but the phase of the sub-carrier changes between $+\pi/2$ and $-\pi/2$ (with respect to the phase of the sub-carrier applied to the EQ modulator) from line to line. The acronym PAL (**P**hase **A**lternation **L**ine) derives from this modus operandi. This corresponds to alternate changes in sense of the colour vector of Figure 14.2, and so differential phase errors in transmission will alternate in opposite senses from line to line and will average out in an appropriately designed receiver.

Interestingly, the essential idea of this method was well appreciated (particularly by Loughlin of the Hazeltine Corporation) during the development of the NTSC system. The principle of phase alternation was considered as a means of reducing the cross-talk between wideband asymmetric sideband, quadrature modulated colour signals but, at that time, the necessary delay lines were not available. When the alternative (NTSC) method of Figure 14.1 was finally adopted, the principle was not further investigated.

Bruch's contribution to the process was the improvement of some of the techniques required to implement the system, namely, the line-to-line switching, the choice of a suitable new sub-carrier frequency, and the utilization of the colour information

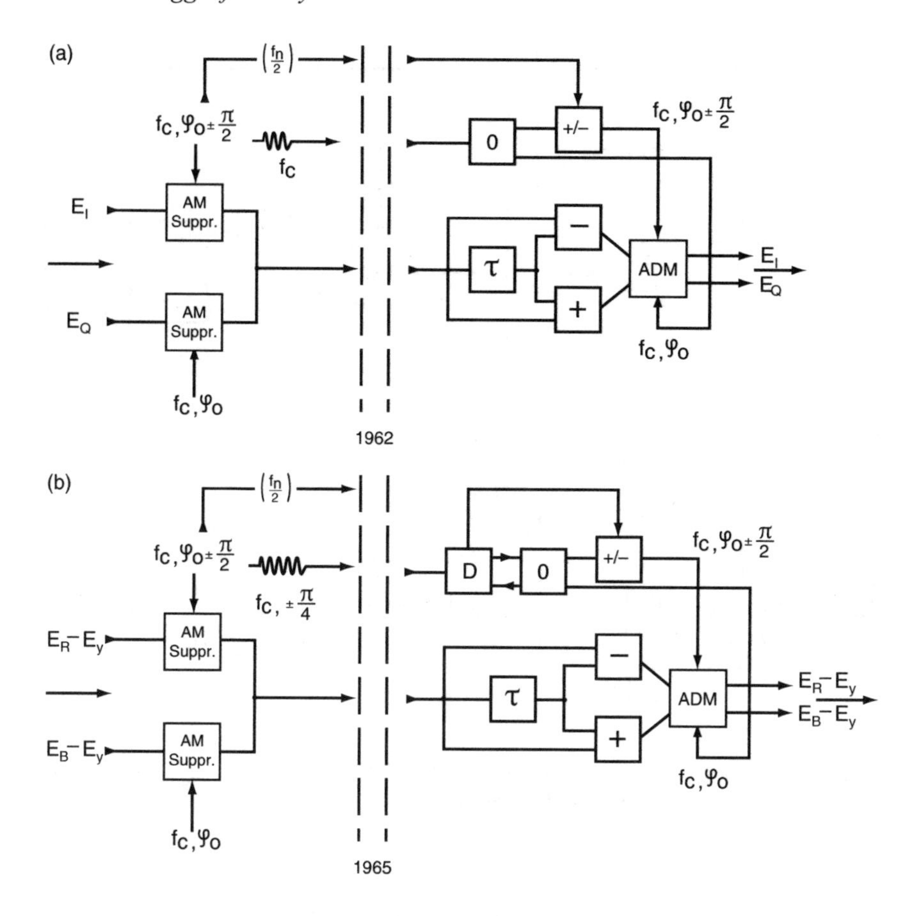

Figure 14.6 Block diagrams of the PAL system (1962), and of the PAL system (1965) [Source: ibid.*]*

contained during preceding line periods for the derivation of hue information free from the effects of transmission errors.

In the PAL system, two auxiliary signals must be transmitted: the 'burst' signal at the sub-carrier frequency, and a signal for phase locking the receiver's electronic switch.

7. **The ART system (1963) [13]**

   This system, Figure 14.7, developed by N. Mayer and G. Holoch of the IRT, utilized a pilot or reference signal, at the sub-carrier frequency, which was transmitted with the QAM colour signal, of NTSC type, to determine the distortion to which the latter signal might be subjected during propagation. The pilot carrier was phase-modulated line by line to enable it to be recovered from the chrominance signal. Following reception and demodulation the reference signal was used to control the oscillator of the synchronous quadrature detector, so ensuring the

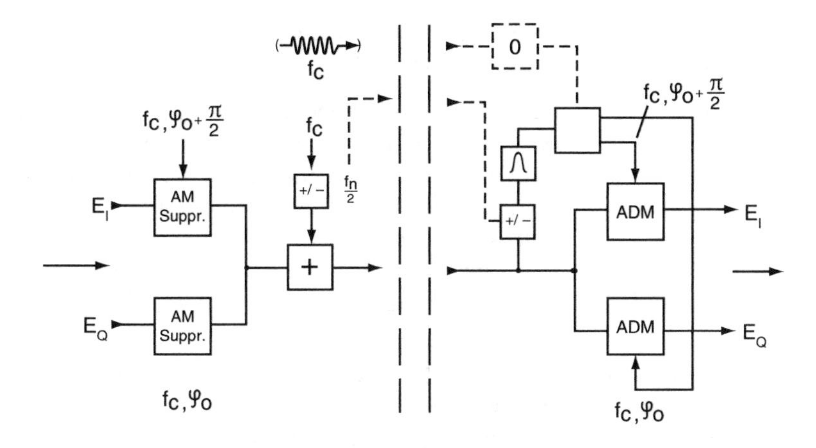

Figure 14.7 Block diagram of the ART system (1963) [Source: ibid.*]*

correct phase lock. The abbreviation ART (**A**dditional **R**eference **T**ransmission) follows from the principle of working.
Clearly, the frequency of the reference signal must be different from any signal component of the modulated colour signal so that the two signals can be separated at the receiver. To aid this process, the reference signal must be modulated by a periodic line-to-line phase reversal.

In 1962, the British Broadcasting Corporation began a series of trials of 625-line monochrome television and NTSC colour television using two channels in the UHF band IV. Six committees, constituted by members from the BBC, ITA, UK Post Office, BREMA, EEA, RECMF, DSIR and Mullard Research, were established to investigate: field strength and propagation; transmitters and transmitting antennas; and receivers and receiving aerials. A transmission signal specification was drafted and a set of questionnaires was prepared to assist in the formulation of subjective assessments. At the same time the first meeting of the Ad-hoc Group was held in London: it lasted from 19 to 21 November 1962. A number of events were organized by the BBC and by the ITA for the visiting delegates, including inspections of the BBC UHF transmitter at Crystal Palace, which was being used for experimental colour television transmissions; and of the mobile laboratories of the BBC and ITA where demonstrations, at various sites, were given of the reception of televised images in colour and monochrome. They also visited the BBC Lime Grove studio centre and watched colour television studio equipment in operation, utilizing the NTSC system, during a programme comprising live and filmed sequences [14].

At the second meeting of the Ad-hoc Group (3 to 5 January 1963, Hanover), the modified version of the NTSC system, known as PAL, was shown at the Telefunken laboratories: it elicited much interest [15].

Further discussions were conducted at the meeting of the EBU Study Group in Paris from 26 to 29 March 1963, and general discussions took place at the EBU Technical Committee Meeting in Edinburgh from 17 to 20 April 1963. There was a desire that the work being coordinated under the auspices of the EBU could be scheduled according to the following timetable [16]:

8 to 16 July 1963	Demonstrations in London to representatives of Western European countries, and possibly to those of some Eastern European countries, of the methods employed in the examination of the problem and of the typical results obtained.
23 to 24 July 1963	Preliminary EBU meeting to discuss progress.
15 to 18 October 1963	EBU meeting to reach agreement on EBU policy on this matter.

Furthermore, the BBC hoped that, if the Special CCIR Study Group XI meeting scheduled for January 1964 determined that the results of the various trials were sufficiently conclusive, the Postmaster General would be willing to authorize the adoption of the colour television standard in the UK by April 1964 'at the latest'. (The Study Group XI meeting had been planned for spring 1965 but, as this date was undesirable from the UK point of view, the UK delegation had issued an invitation to the CCIR to hold a special meeting in the United Kingdom, late in 1963 or early in 1964, at which colour television standards for Europe would be debated.)

The timetable certainly appeared to be optimistic given all the disparate, and nationalistic interests involved, and the experience of the FCC in trying to reach a general standard on colour television. Still, as Medawar once said, 'To deride the hope of progress is the ultimate fatuity, the last word in poverty of spirit and meanness of mind.' Hence, the BBC planned its future colour television service on the basis that a favourable decision would be made.

If the NTSC system were chosen, the BBC would have experimental equipment available and could introduce some 'simple programmes' and films in its services at the end of 1964 and would be ready to launch full-scale colour programmes by early 1966. On the other hand if the decision were in favour of SECAM or PAL, the BBC would need more time to initiate the start of a limited service and the likely commencement date would be some time during the spring of 1965. New equipment would be required for full-scale programme production, which could not begin until the early part of 1966. By this time the BBC anticipated colour receivers could be in quantity production.

During 1963 extensive field trials and laboratory investigations on the NTSC, SECAM and PAL systems of colour television transmission, using the European line standard were undertaken under the auspices of the EBU (see Note 1). Broadcasting authorities, industry, and the administrations in France, the Federal Republic of Germany, the Netherlands, Italy Switzerland and the United Kingdom cooperated and appreciable data were accumulated. This work showed there was little to choose between the three systems when the conditions were favourable for reception. 'In other respects', commented the TAC's Technical Sub-Committee (TSC), 'the three systems differ to an extent which assumed greater or less importance to the weight attached to, for example, the ease of manufacture of domestic receivers, the ease of handling programme material by the broadcasters, or the ease of transmission of the signals from studio to broadcasting stations.' Taking all the available factors and knowledge into consideration, the TSC felt the more significant advantages of the NTSC, SECAM and PAL systems were as follows [17]:

> NTSC [was] slightly the best in respect of compatibility and resistance to noise and interference, and the NTSC receiver contain[ed] fewer components and functional circuits, which suggest[ed] it [would] be the cheapest and the easiest to maintain. Furthermore, it [was] believed that the NTSC system [would] be more readily adaptable to the simpler colour tube (using only one gun) that, it [was] hoped, [would] be produced at some future date.
>
> SECAM require[d] the least complicated equipment for recording on magnetic tape and [was] substantially easier to transmit over the distribution network between programme source and broadcasting transmitters than the NTSC system.
>
> The properties of the PAL system, in general, [were] intermediate between those of NTSC and SECAM.

For the United Kingdom's future colour television broadcasting service, the controlling technical factors that were thought to influence the final choice of system were: compatibility, the cost and reliability of the colour receiver, the receiver's resistance to noise and interference and the system's potential for future development. These factors were of greater importance to the general public than to the broadcasting authorities, and so, if the greatest weight was accorded to them, the conclusion of the TAC was that the order of preference for the systems should be: NTSC, PAL, SECAM.

The meeting of the CCIR Study Group XI, mentioned above, was held in London from 14 to 26 February 1964 [18]. It was attended by representatives of 19 administrations together with representatives from private operating agencies and international organizations. (See Note 2.) Among the points recorded by the UK delegation were: (1) 'only the UK was anxious to reach a decision and . . . most other countries [with the exception of the Netherlands which "supported the UK view that sufficient technical work had already been done to justify a decision in favour of NTSC"] were playing for time'; (2) 'the French delegation took every possible opportunity of emphasizing the advantages of SECAM and playing down the merits of NTSC without committing themselves to use SECAM'; (3) the Eastern Bloc countries required more time to complete their tests and the results would not be considered by the IBTO until June 1964; (4) the opinion of the USSR members 'seemed to be divided'; (5) the response to a questionnaire, from the chairman (Mr E. Esping of Sweden), 'designed to evoke some indication of future policy or preference from, on the whole, reluctant delegations', showed:

1. support for NTSC from the UK, Austria, Denmark, the Netherlands, the USA, and Sweden (with PAL as an alternative);
2. support for PAL from Norway, and Sweden (with NTSC as an alternative);
3. no direct support for SECAM ('even from France'), and
4. support from 11 countries, including France, Germany and Italy, for a deferment of a decision until at least April 1965 (when the next full meeting of Study Group XI was scheduled).

The brief of the UK delegates, as defined by the TAC, had been to stress the importance of compatibility, receiver cost and reliability, and the potential for development of the selected system. They argued strongly for the obvious advantages of the NTSC system and presented 'it as being the only choice that would represent a step towards, and not away from, the creation of a single world standard'.

In the event, no decision was taken at the conference and the NTSC system was criticised primarily on three points. First, 'the need for and difficulty of use of a hue control on the NTSC domestic receiver'; second, 'the difficulty of transmitting NTSC signals over long-distance line and radio links'; and, third, 'the difficulty of recording NTSC signals on tape'.

An appraisal of the London meeting was, of course, presented to the Postmaster General following full consideration by the Television Advisory Committee [18]. The TAC continued to confirm its position with regard to the NTSC system, namely: the system held out the best prospects of any system of promoting the development of a radically simpler and therefore cheaper colour receiver – a matter of some importance to the general public and the growth of a colour television broadcasting service. On the issue of the system's comparative weakness in the field of videotape recording and transmission over intercity links, the TAC had been assured that this point was 'not enough . . . to outweigh its advantages over the other two systems [PAL and SECAM] and [it] expect[ed] technology, in normal course, to improve the application of the NTSC system in both fields'.

So convinced was the TAC of the correctness of its opinion, it informed the PMG that 'there [was] a fair chance that a majority of European Administrations [would] reach a decision on a common system' at the meeting of the CCIR Study Group XI which was scheduled to be held in Vienna in April 1965. 'A meeting of East European broadcasting organizations (IBTO) in Moscow in July may help the prospects of a successful outcome at Vienna.' Nonetheless, the TAC was careful to advise the PMG that the UK, having 'set out to achieve a common colour system with the rest of Europe, should not at the present stage prejudice this objective. In our view a unilateral decision now to proceed with the NTSC system might damage our influence and standing in Europe, and might lead to an indefinite postponement of a European decision, with no guarantee that that decision, when it came, would coincide with ours'.

For the TAC, any unreasonable delay in arriving at a European consensus could have two effects. First, the continuing doubt in the public's mind about the introduction of a colour television service might affect the public's readiness to purchase the new dual standard (405-/625-line) black-and-white sets; and, second, the uncertainty placed the television manufacturing industry in a difficult position regarding when to commence the production of colour television receivers.

All of this led the TAC, with the full support of the BBC, the ITA and the radio industry, in March 1964, to recommend to the Postmaster General that an early announcement should be made that:

1. 'the United Kingdom intend[ed] to introduce a regular colour television service in 1967';
2. 'a final decision on the colour system to be employed [would] be deferred until the CCIR meeting in Vienna in April, 1965'; and

3. 'if the Vienna meeting fail[ed] to agree a common system, the United Kingdom should adopt the system of its choice (in the absence of fresh evidence, this would be NTSC)'.

However, though the Postmaster General of the Conservative government had stated colour television would be introduced in 1967, the change of government in 1965 necessitated confirmation of this policy from the new Postmaster General. The matter was now becoming urgent.

Six weeks after the London meeting, the first Eurovision colour television demonstration was arranged with the support of RTF, RAI, the BBC, the PTT administrations of the countries involved and the EBU International Control Centre in Brussels [19]. Held on 8 April, the private, closed-circuit demonstration had as its objective the exposure of any special problems that might arise in the setting-up, testing, switching and operation of trans-continental colour television links. The demonstration comprised the following items, each of which lasted *c*. 15 minutes:

- Opening announcement from Cologne using PAL
- Contribution from Paris on video tape using SECAM
- Contribution from Rome on film using PAL
- Live studio production from London using NTSC
- Closing announcement from Cologne using PAL
- The measured distortions on the vision circuits were:

Circuit	Differential Gain	Differential Phase
Paris-Hamburg	−31%	−10°
Rome-Hamburg	−5%	+8°
London-Hamburg	−15%	−10°
Cologne-Hamburg	−8%	+4°

A report of the demonstration mentions that it was 'entirely successful and the colour pictures received from all four sources were entirely acceptable. Despite the fact that the differential phase distortion on the London-Hamburg circuit was one of the highest of all, the quality of the NTSC pictures was as good as that of the SECAM and PAL pictures received from the other sources.'

Following the special meeting of the CCIR Study Group XI, the Technical Sub-Committee of the TAC considered its conclusions – especially the points of criticism of the NTSC system, together with reports of developments in the field of colour television – particularly those of the European Broadcasting Union's Ad Hoc Group on Colour Television. Further developments by the Post Office and the BBC substantiated the results of the Eurovision demonstration that long-distance transmission of the NTSC signal was readily possible. Indeed, transmission tests, during which signals were sent from London to Rome and back, and from London to Moscow and beyond, indicated the superiority of the NTSC signal over the SECAM signal [20]. Moreover, BREMA – from tests with the viewing public – confirmed the ease with which a hue control could be provided. On the third point, which related to recording, progress had been such the TSC believed the difficulty was no longer significant.

Because of these new facts, the Technical Sub-Committee reviewed the brief given to the UK delegation to the London meeting. The order of preference for the three systems remained unchanged, and the TSC's view was: 'the arguments in support of a preference for NTSC [were] now stronger than ever while the objections of PAL and particularly to SECAM [had] been shown to be greater than previously thought'. For the Vienna meeting scheduled for 24 March 1965 the UK delegation should [21]:

1. 'press strongly for the adoption of the NTSC type of colour television system as the European standard for broadcasting as the best in respect of both colour and compatible black-and-white quality with also all the advantages that would accrue in having one system of colour transmission throughout the world';
2. 'oppose the adoption of the PAL system for use in Europe as being a system having marked disadvantages in respect of the compatible black-and-white picture. In the event, however, of a strongly supported recommendation to use PAL, the UK should reserve the freedom to use NTSC for broadcasting since the two systems employ[ed] similar techniques and programme exchange remain[ed] relatively simple';
3. 'strongly oppose a proposal to use SECAM as a standard system in Europe. SECAM [was] significantly inferior to NTSC and PAL both in respect of colour and compatible black-and-white quality and moreover its adoption would give rise to difficulties in programme exchange with the rest of the world using the NTSC system. In the unlikely event of its adoption the UK should record its objections and reserve the right to use a system of its own choice when commencing its colour service'; and
4. 'reserve the right, in the event of no agreement being reached for the UK to use the system of its own choice when commencing its public colour television service.'

The TAC's urgency for an immediate decision on the starting date for the UK's colour television service stemmed from its belief that the opinions of those countries about to begin a colour television service would carry most conviction [22]. The Soviets were known to be planning to introduce a colour service to commemorate the 50th anniversary of the October Revolution; and the French were engaged in planning a colour service in time for the opening of the winter Olympic Games at the end of 1967. It was highly likely this service would be based on the SECAM system. There was an additional point: if the PMG did not immediately announce the date of the start of the colour service for 1967 the radio industry would be unable to manufacture the colour receivers in time for its inauguration, and the BBC might not be able to provide the studio and outside-broadcast equipment by the due date. In summary, the chairman of the TAC informed the PMG, 'It is necessary to the best prosecution . . . of the United Kingdom's case for the adoption in Europe of the NTSC system, that [the delegation] should know what date the Government has in mind for the start of colour television.'

In the NTSC system the I and Q chrominance signals were transmitted simultaneously, but in the SECAM system these signals were sent sequentially on alternate lines. The result was the SECAM system had half the colour resolution in the vertical direction of that of the NTSC system, and was thereby the inferior system. Fortunately, because the human eye's acuity of vision in resolving colour information is less than for black-and-white information, the difference in vertical resolution between the two systems was not particularly noticeable except when tests were made using special cards. Nonetheless, as a senior member of the ITA noted, '. . .it seems a thousand pities that anyone with half an eye to the future should wish to adopt a system which has from the start potentially half the vertical resolution of some other competitive system' [23].

There would be another problem – standards conversion – if SECAM were adopted. In converting from one standard to another it is necessary to separate electrically the luminance and chrominance components of the composite signal and re-encode them in the desired form. With an NTSC signal the recovery of the luminance and chrominance components is simple, since they are contained within clearly defined frequency bands: consequently conversion from NTSC to SECAM is easy. The conversion from SECAM to NTSC is more difficult because the colour information is 'all over the place'. Nevertheless it had been demonstrated in the laboratory. For both systems there is some loss of the luminance information – the loss being greater in converting from SECAM to NTSC than in the reverse direction.

In one respect SECAM had an advantage over NTSC, namely, the recording of vision signals on tape. But, as the above-mentioned ITA engineer wrote in March 1965, 'who can seriously doubt that the powerful American technological resources will not come up with the answer before long?'

Mindful of the forthcoming meeting in Vienna of the Colour Sub-Group of the CCIR, scheduled for the 24 March 1965, the French government, the CFT, the CSF and the ORTF mounted a very extensive campaign to advance SECAM's adoption. Most days there were references in one paper or more to the colour question and to the merits of SECAM, but, as the BBC's director of engineering noted (1 February 1965), 'Many of these statements [were] completely incorrect, but it [was] to be expected that they [were] having an appreciable effect on the countries of Eastern Europe.' [24]

Of much concern to the French was the need to persuade the USSR to adopt the SECAM system. If this could be accomplished it was 'certain all the eastern European countries would follow', including the Democratic Republic of Germany. And, if this country chose SECAM, the Federal Republic of Germany would 'certainly follow', 'so as not to further separate the two countries'. Then, 'the rest of Europe (including England) [would] finally follow so as to align themselves with France and Germany. The countries of the Commonwealth, the countries of the community of South America, Asia and Africa [would] surely adopt the system which [had] been accepted in Europe.' For the writer of these comments in the two-page spread in *Candide*, 'It [was] therefore a chain reaction which the French [were] endeavouring to set off, a reaction which could bring a triumphant tomorrow for the French.'

If the SECAM system were adopted it would 'open an enormous market to our [French] patents, to our components, to our machine suppliers, and to our engineers, this without counting the intangible gain in prestige which [would] be enjoyed by the whole of the French industry.' To further their objective the French distributed 110 colour television receivers to some important politicians, including de Gaulle, and engineers, and sent a 'large number of receivers' to Russia. 'Who will win?' the writer ended. 'The Americans or the French? The choice [would] depend upon the courage of the Russians.'

By September 1964 the French had given demonstrations of SECAM in Moscow, Warsaw, Prague, Bucharest and Sofia. A Minister of SECAM Affairs with diplomatic status was appointed [25], and a mission headed by the Minister of Information was sent to Moscow to persuade the Russians to opt for SECAM. The French appeared to feel that their national pride was bound up with the success of SECAM and 'they were evidently determined to press for its adoption by every means in their power, at the political level, regardless of its technical merits or demerits'.

The French government's objective was strengthened on 22 March 1965 with the signing of the Franco-Soviet agreement to concentrate the efforts of the two countries to secure the adoption of SECAM by the countries of Europe. Though the communiqué that was issued was cautiously worded and did not explicitly mention that the USSR was finally committed to the SECAM system, the French press unanimously interpreted the agreement as meaning that the Soviets had decided to adopt the French system. *La Nation* enthused lyrically:

> It was, without question, up to France, country of the arts, to spread colour with this giant brush across Europe from the Atlantic to the Urals. Who would have thought that the fine present offered by the Communist regime to the Soviet people to celebrate the 20th anniversary of the revolution would have been a magic box, a kaleidoscope, of French origin? France, 'grand marchand de couleurs': the role suits her well for it corresponds to her artistic vocation and also to the place which she has taken in the most difficult technical affairs.

Le Monde believed the Soviet Government wished to help 'France in this way to show their interest in the latest French diplomatic initiatives in view of their anxiety about United States policy, and their disappointment over British policy.'

British reaction to the agreement was singled out for sharp criticism by the French press. *Figaro's* Vienna correspondent opined that American disappointment was to be expected, but 'that other countries, in particular England, should show bad temper not to say surliness' was surprising.

The Vienna meeting [26] began with 175 registrants from 25 countries, and ended with 250 delegates from 45 countries – seven countries being represented by proxies. 'There were a score or more countries (for example, Morocco) who were registered by individuals [who gave] as their address the appropriate Embassy in Paris, and these, in turn, held proxy, for example, for Gabon-Madagascar, etc.' The USA, Canada, Brazil, Argentina, the USSR and almost all the countries of Europe sent representatives.

During the first week of the meeting, a point-by-point technical discussion was conducted, based on the detailed studies undertaken by the EBU over two years on the relative merits of NTSC, PAL and SECAM. These merits were described in an important document, 'Report of the EBU Ad-Hoc group on Colour Television', which had been published in February 1965.

At this stage in the narrative it is appropriate that the main findings of the Ad-Hoc Group should be stated so that the discussions at the Vienna meeting may be put in context.

The Report [27] summarized the work of the Group undertaken up to January 1965 to facilitate the selection of a colour television system for countries using the 625-line standards. In its 56 pages the Report made reference to 85 documents received by the EBU from various agencies associated with its work. A simple analysis of these documents confirms Maurice's view that the BBC was a major contributor to the tests, field trials and mathematical treatments of various aspects of colour television.

Organization	*No. of documents listed in Report*
BBC	30
ABC Television/GPO/BREMA	5
Bruck/ZVEI	5
FNIE	12
RTF	1
RAI	8
Haanjtes/Philips	4
ORTF	3
CFT	2
Bernath/Swiss PTT	4
Anon: The Netherlands	2
Mayer/Hopf/Fix/Muller	6
Anon	3

The following points regarding the NTSC, SECAM and PAL systems have been extracted from the Report and are given verbatim to ensure impartiality of treatment.

1. Compatibility

The assessments of observations made on picture monitors using Standards G, I and L have shown that the compatibility of NTSC is 'slightly better' to 'better' than that of the other systems. PAL has a slight advantage over SECAM III.

Further results show the assessments made during observations on domestic receivers. In these results the most favourable assessment is for NTSC and, from observations made on Standard I, the percentage of assessments having a grade greater than 3.5A is noticeably less than that for either of the other two systems for very saturated camera pictures viewed under the conditions laid down by the Ad-Hoc Group. The average absolute assessments for SECAM III and PAL are equal.

Observations made on Standard G and using domestic receivers show that on average PAL is 'very slightly worse' than NTSC.

2. Fundamental quality of the colour picture

NTSC is about the same as PAL with a delay-line receiver and PAL with simplified receiver, but 'slightly better' than SECAM III as regards vertical resolution and horizontal-edge effects; for horizontal resolution and vertical-edge effects PAL with a delay-line receiver was found to be 'slightly better' than NTSC on Standard G. SECAM III is almost 'as good' as NTSC.

3. Receiver cost

The NTSC receiver is generally considered as the cheapest. The difference between costs of receivers for the three systems are contained within −0.35% and +9%.

4. Receiver operation

In an NTSC or simplified PAL receiver a hue control and a saturation control are available and necessary. In the delay-line PAL receivers, no hue control is necessary. In the

SECAM III receiver neither of these controls is essential, but they may be added if desired.

5. Display tube

NTSC is the most adaptable to the single-gun display tube and the SECAM III system has less flexibility in this respect than the other two systems.

6. Differential gain of the chrominance signal

The SECAM III system is about twice as tolerant as the NTSC and PAL systems; some recent tests, however, would indicate that the tolerances for NTSC and PAL could be made less stringent.

7. Phase of the chrominance signal

SECAM III requires no tolerance; PAL with a delay-line receiver is three times as tolerant as NTSC.

8. Differential phase of the chrominance signal

SECAM III and PAL with delay-line receiver are two and a half times more tolerant than NTSC and PAL with simplified receiver.

9. Unwanted attenuation of the upper sideband with respect to that of the luminance signal

PAL with a delay-line receiver is 'much better' and PAL with a simplified receiver is 'better' than NTSC.

10. Co-channel interference in the colour picture

(i) No offset NTSC and SECAM III are both 'slightly better' than PAL with a delay-line receiver.
(ii) Nominal offset (+/− 250 c/s) NTSC is 'better' than SECAM III and 'slightly better' than PAL with a delay-line receiver.

11. Sensitivity to echoes

NTSC is the same as the other systems for weak echoes which do not impair the colour pictures significantly. When the echoes are sufficiently strong and numerous to cause serious impairment, SECAM III and PAL with a delay-line receiver are on average 'slightly better' than NTSC.

12. Sensitivity to random and impulsive noise occurring between coder and decoder of a colour transmission

NTSC is the same as PAL with a delay-line receiver and PAL with simplified receiver and SECAM III is almost as good as NTSC for signal-to-noise ratios giving rise to a picture quality better than 'rather poor'. For signal-to-noise ratios giving rise to a picture quality worse than 'rather poor', SECAM III is 'slightly worse' than NTSC, PAL with a delay-line receiver and PAL with simplified receiver.

13. Studio centres – basic unit

No special difficulties are foreseen with regard to the basic unit; difficulties with regard to the mixer used with the SECAM III system can be overcome by suitable design at the expense of a reduction of luminance bandwidth to about 3Mc/s for the duration of the mix and/or special effects.

14. Video-tape recording

The magnetic recording of NTSC signals requires a complement of auxiliary equipment.

The magnetic recording of PAL signals requires a small complement of auxiliary equipment.

Normal good quality black-and-white video-tape recording machines can be used without modifications for recording SECAM signals.

The quality of the colour picture obtained from a video-tape machine in current use was judged as good with PAL as that with SECAM III and better than that obtained with NTSC. . . .

15. Domestic and international networks

In general, SECAM III and PAL signals are easier to transmit than those of NTSC. However, in certain experiments made on some international circuits, PAL appeared to be somewhat better than NTSC and SECAM III. Several methods of automatic correction of the distortion of NTSC signals . . . have been proposed and used for experimental transmissions of NTSC signals

16. Time during which certain overall tolerances might be exceeded

Following a statistical study, the percentage of the time during which the tolerances appertaining to differential gain and phase could be exceeded is slightly greater for PAL than for SECAM III and could be still greater for NTSC.

17. Direct comparison between NTSC, SECAM III and PAL

Field trials have shown that when the NTSC colour picture quality is 'fairly good' (2.5A) or worse, the SECAM III and PAL systems show an advantage over NTSC. The PAL system showed itself to be the best under these conditions.

(Under the headings 'Ratio of the amplitude of the chrominance signal to that of the luminance signal', 'Group-delay errors of the chrominance signal with respect to that of the luminance signal', 'Transmitters', 'Coverage of colour-television systems' and 'Home receiving tests', the Ad-Hoc Group felt there were no significant differences between the systems.)

Returning now to the Vienna meeting, the replies to a questionnaire issued by the chairman to expose preferences showed a 'fairly even balance' between those who had been persuaded by France and the USSR to support SECAM ('in some cases, if corridor talk can be believed, against their technical judgment', noted one commentator), and those who preferred PAL or NTSC. 'During discussions outside the formal meetings it became clear that many countries were under severe pressure from the French, at government level, to support SECAM . . .' [26]

A second questionnaire again indicated a similar state of opinion, but with a strong swing towards PAL, rather than NTSC, among the anti-SECAM delegates. 'The replies to this second questionnaire showed that the French had succeeded in increasing the support for SECAM, at least on paper, by bringing (literally at the last moment – since the telegraphed proxy from Madagascar came in as the replies were being discussed) in delegations or proxies from a number of African territories, but opposition to it was also strengthened.'

One delegate (probably an official of the General Post Office), in a report on the 'long and hard discussions (France versus the rest)', mentioned, 'France first insisted on the results of the Conference being interpreted as an overwhelming majority in favour of SECAM.' When this failed, France proposed the wording that 'but for the unsatisfactory procedures of the CCIR, it would have become evident that there was an overwhelming majority, etc....' Eventually, the agreed text stated, '...lack of unanimity did not permit the drafting of a Recommendation ... for a single world system'.

The substantive outcome of the meeting could be summarized as follows [27]:

1. A valuable and fully agreed technical report [had] been prepared on the merits and demerits of NTSC, PAL and SECAM III. It [was] obviously a work of compromise in its judgements. Its facts [were] indisputable.
2. France [had] not succeeded (pace USSR) in gaining a single country that [had] experience in operation or research and development into colour television. Her champions [were] her former dependencies.
3. Germany [had] succeeded in attracting Scandinavia, Austria, Switzerland, Italy, Australia, New Zealand to PAL. [Also Norway, Sweden, Denmark, Finland and the Irish Republic were pro-PAL.] In fact there [was] a PAL 'belt' from Sicily to North Cape. But it [was] clear that many aspects – particularly that of receiver design and its effects on choice of system parameters – of the PAL system have not yet been fully studied.
4. USSR (and throughout most cordial relations [had] existed between USSR – UK delegations – whereas the French [had] avoided us) [was] not yet prepared to state a choice and her expressions suggest[ed] that there may be opportunity yet to secure cooperation – at least on some aspects of the problem.
5. Detailed technical discussions suggest[ed] that colour programme interchange between systems of differing standards may be more rather than less difficult than hitherto thought. It could only be made less difficult by deliberate alteration of the characteristics of each system for the benefit of the other (and to the possible detriment of each).
6. It was generally recognized, in the closing moments of the meeting, that the effect of a multiplicity of standards upon domestic receiver design production, sale and usage was most serious.

The pro-SECAM supporters were the Eastern European countries (excluding Yugoslavia, which expressed no preference), France and Spain. Belgium had no preference, and the UK and the Netherlands were pro-NTSC.

With the division of opinion on standards clearly known, the task now of the UK's TAC's Technical Sub-Committee was to assess more fully the operational and technical problems associated with the exchange of colour television programmes between countries having different systems of colour television; and to investigate more closely certain technical aspects of the PAL system. In its work, the TSC was aided by BREMA, which, by the middle of 1965, had concluded that, from the receiver-manufacturing and servicing point of view, the PAL system was completely acceptable.

The TSC reported in May 1965. Its opinion was that:

> ... the balance of technical advantage as between NTSC and PAL [was] not sufficient to affect a choice between them [see Note 3]; considerable operational complications would arise from the co-existence of three different 625-line colour television systems, particularly for a country which exported many programmes; there would be no difficulty in manufacturing PAL receivers in the UK; studio equipment would, in any case, be manufactured and exported for any system, but there [was] a considerable advantage if the number of different types of equipment required [was] kept to a minimum; and one of the major advantages claimed for the SECAM III system – ease of transmission over long distances - [had] now been shown to be false. [28]

A later version of SECAM known as SECAM IIIA was also rejected by the Technical Sub-Committee. The reasons were [29]:

1. 'the SECAM signal contain[ed] less colour information than the other two standards and as a corollary there [was] significant loss of vertical colour definition';
2. 'the mixing of AM [amplitude modulation] and FM [frequency modulation] signals within the vision signal causes complexity in the receiver';
3. 'the signal [was] less suitable for single gun picture display devices';
4. 'its proponents state[d] that the parameters now chosen [gave] the best possible performance and that there [was] no room for further development of the transmitted signal.'

Given these various factors the TSC was now veering towards the PAL system. If it supported PAL there was a prospect the Netherlands would follow and the number of disparate systems would be reduced from three to two. Again, some uncommitted countries, such as Belgium and Yugoslavia, might be influenced to follow the UK's choice and, optimistically, thereby cause a number of East European countries to reconsider their position. If this happened it would lead to 'a split in the SECAM camp'. Finally, the selection of the PAL system would be an insurance against any pro-PAL country changing its opinion [27].

On balance the TAC, acting on the TSC's advice, believed the right course of action was to defer a decision on the choice between NTSC and PAL. This would allow the BBC to undertake additional tests on the PAL system, and enable further soundings of other European countries to be initiated. Accordingly, the Postmaster General was recommended to adopt the TAC's advice. He was again requested to give a decision on when a colour television service in the United Kingdom would start [30].

At this stage in the saga of events (August 1965), the PMG [31] prevaricated. He told Sir Willis Jackson, chairman of the TAC, there would be very great difficulty in announcing such a decision while so many major uncertainties remained, and listed five questions on which he required advice before deciding what the government's policy should be. The questions related to the choices of colour television systems of the other European countries, and whether colour television, in the UK should be transmitted on both 405 and 625 lines per picture. On each of these substantive points the PMG had previously been given advice. However, the TAC again had to give the issues some consideration.

The position in Europe (in November 1965), as revealed by information obtained through diplomatic channels, was still little changed from the attitudes taken up by the delegations at the CCIR meeting in April [32]. And on the issue of line standards, the TAC's 1960 recommendation that colour should be introduced on the 625-line standard only still applied, though the Committee recognized that the Independent Television Authority would be placed at a serious disadvantage if a second independent television channel in the UHF were not allocated to it for a colour television service.

One month later, the TAC recommended to the PMG the adoption of the PAL system in the United Kingdom [33].

When Parliament debated Broadcasting Policy on 3 March 1966 the PMG announced that a colour television service would be authorized to start on BBC2 'towards the end of next year, using the PAL system' [34]. A supplementary licence would be required by viewers of the service, and the BBC aimed to commence with about four hours of colour television a week rising to *c.* 10 hours within a year. He said the cost to the BBC had been estimated to be from £1 million to £2 million a year. Estimates of the price of sets suggested they would start at *c.* £250 each, and on the basis that 150,000 receivers would be in use in two years after the commencement of the service, with a more rapid increase later, consumer expenditure could possibly approach £100 million on sets over the first three- or four-year period of the service.

On the prospect of good exports, the PMG believed a net figure of £10 million in exports over imports was realistic. He noted, 'Sales of sets in America [were] rising so rapidly that the industry there [was] unable to meet the demand. In a few years it will have caught up. Unless we start our industry going now in colour television, when it does come it could simply lead to imports of foreign components.' The TAC, which had been pressing this point for the past two years, must have been heartened by the PMG's statements.

Obviously, the Postmaster General's announcement did not connote the end of European discussions on which system of colour television, if any, would apply throughout Europe. The XIth Plenary Assembly of the CCIR had to debate the issue and this had been arranged to be held in Oslo from 22 June 1966. So, once again, the UK's delegation had to be given a brief, initially drafted by the Technical Sub-Committee of the Television Advisory Committee, approved by the TAC [35], and endorsed by the Postmaster General. Essentially, the accepted brief was not too dissimilar to that adopted previously at the Vienna meeting. The delegation had: to 'press strongly' for the adoption of PAL; to oppose 'strongly' the use of SECAM III; 'not [to] initiate any discussion on the NIR/SECAM IV system but, if such discussion [took] place, [to] sustain the argument that NIR [was], overall, inferior to PAL'; and 'in the unlikely event of there being widespread support for some other system than PAL or SECAM III, reserve the right to use the system of its own choice'.

The Plenary Assembly seems to have been discordant and vexatious at times [36]. A brief report by one of the officials of the CCIR was so biased that the leader of the USA Delegation felt obliged to display his annoyance at its one-sidedness. On another occasion one of the European delegations invaded the rostrum and remonstrated vehemently with the chairman of a Sub-Group. There was an accusation from the head of the USSR delegation that a senior official had exceeded his authority in advancing a certain resolution. There was 'widespread dissatisfaction with the attitude of [a certain delegation] and the way in which the proceedings of Study Group XI were being conducted'. 'There was a most unfortunate scene in which the Chairman of the Study Group was insulted by a representative ... and there was a widespread revulsion of feeling in the Study Group at the way in which the whole question of colour television [had] been handled.' Political machinations and efforts to enhance national prestige and pride are not valid when matters of technical excellence or otherwise are under consideration in a non-political forum.

Discussions during the Plenary Assembly largely centred on a French proposal that SECAM IV should be adopted as a compromise standard. Though the proposal was not put in writing it appeared to be based on an immediate agreement in principle, by the Plenary Assembly, to adopt SECAM IV and to collaborate thereafter in a six- to twelve-month period of development that, the French stated, was necessary. 'This was to be accompanied by (a) a delay of six to twelve months in the introduction of colour television services by those countries which had announced the intention to start such services in 1967, and (b) an "industrial moratorium" on the production and development of colour television receivers and allied equipment'.

This proposal conflicted with the UK government's decision to start a colour television service towards the end of 1967 and the delegation was not empowered to change this policy. Nor could the delegation asseverate that the government would impose constraints on industry that the industrial moratorium seemed to demand. Moreover, the delegation was convinced that PAL was a superior system than SECAM IV. They acknowledged that SECAM IV was a better system than SECAM III and its use would facilitate programme interchanges with the PAL and NTSC systems and promote the television unity of Europe. Both the Federal German Republic and the United Kingdom delegations – which were in accord throughout the discussions – offered to collaborate in joint studies of SECAM IV – but could not agree to adopt it [37].

This was not acceptable to the French, who stated that unless there was general agreement to proceed with their policy regarding SECAM IV they would withdraw their proposals and maintain their support for SECAM III. Since the UK party lacked the necessary authority to depart from its brief it returned to the United Kingdom for consultation.

In London, on 14 July, a meeting of officials of the BREMA, the EEA, the BBC, the ITA, the Treasury, the Ministry of Technology, the Foreign Office and the General Post Office was convened, with Sir Ronald German, the Director General of the GPO as chairman. After a full discussion on whether the UK delegation's brief should be changed, Sir Ronald felt the overall consensus of the meeting was that [37]:

1. 'there was a good deal of doubt about whether in practice the study and development of SECAM IV would be concluded within six months';
2. if the time needed was longer than six months, 'it would not be before the end of 1968 or early 1969, at best, that enough receivers would be ready to permit of a start to the colour service'. In that event, there would be a delay of more than one year beyond the date of commencement of the colour television service announced by the government;
3. 'even if the study and development of SECAM IV did take place it was highly problematical whether in the result there would be a single system for Europe. SECAM IV offered no outstanding technical advantage over PAL, but might be

> marginally worse. It was hard to see why West Germany should sacrifice the PAL system; and France might – having delayed a start here and so reduced the industrial and commercial momentum which had been built up – revert to her own preferred system, SECAM III'.

On this appraisal, the conclusion was clear: the Director General of the GPO could not recommend to the Postmaster General that any change should be made to the UK delegation's brief. The PMG accepted this view.

Subsequently, at the CCIR XI th Plenary Assembly, the French delegation withdrew their proposals. The final outcome of the preferences of the countries represented at the meeting was little different from that of the Vienna meeting and is given below [38].

PAL	**SECAM III**	**Uncommitted**
Norway	France	Israel
Sweden	Greece	Portugal (likely to follow France)
Denmark	USSR and other communist countries	Spain (likely to follow France)
Federal German Republic		
United Kingdom		
Republic of Ireland		
Italy		
Austria (not very firm)		
Netherlands		
Belgium (a slight preference only)		
Switzerland (not very firm; would be influenced by Austrian choice)		
Australia		
New Zealand		

Following the outcome of the Plenary Assembly, broadcasting authorities commenced planning and implementing their new colour television services. In the United Kingdom the first colour television programme was transmitted from the Centre Court at Wimbledon on BBC 2 on 1 July 1967. The *Times* correspondent noted:

> The quality of the colour was good, better in my experience than in Canada and in the United States.... It was certainly more restful than black and white. The close-ups were particularly good and the flesh colours true to life, only when the cameras were used for distant shots, such as for the gaily coloured Wimbledon crowds, did the picture become blurred and the colours take on a gem-like quality. [39]

Initially, the BBC2 service broadcast an average of five hours of regular colour programmes per week but from 2 October 1967 this was increased to 10 hours each week. Then, from 2 December 1967 when the full colour service commenced, more than 80 per cent of the programmes were in colour: they totalled approximately 34 hours per week. The purpose of the Launching Period from 1 July was to enable everyone who was associated with colour television – whether in the studios, the transmitter stations, industry or people's homes – to gain experience in handling the colour signal. According to the BBC's director of television the launching programmes were 'kept deliberately on a minimum scale at the earnest entreaty of the set makers, who were anxious not to see the full impact in December blunted' [40]. He estimated, in August, that the full service would provide *c.* 24 hours of colour television each week, not allowing for additional ad hoc material, and would comprise 3.5 hours of drama, 3.5 hours of light entertainment, 1.25 hours of documentation, 2.25 hours of features and science, 2.25 hours of sport, 1.5 hours of music and arts, 4.25 hours of feature films and tele-films, and 2.75 hours of presentation and of news [41]. Actually, as noted, this output was considerably exceeded and the BBC broadcast more colour television than any other service in Europe.

During the Launching Period the BBC, the radio and television industry, and some trade associations made strenuous efforts to publicise the new colour television service. The BBC and BREMA cooperated in providing a travelling exhibition to present continuous shows of colour television, either direct from the BBC2 transmitter or from a tele-cine generator, from 2 p.m. to 9.30 p.m. every day except certain Saturdays, when the opening hours were from 11 a.m. to 6 p.m. Included in the exhibition, which covered 10,000 ft^2 (930 m^2), were display stands – each measuring 10 ft (3.05 m) in diameter by 11 ft (3.4 m) high – at which some 50 colour television receivers could be seen working. Information displays and colour television demonstrations, together with publicity material, posters, questionnaires, answer sheets, banners, a four-page BBC2 News quarterly, and stickers were all available to give dealers and members of the public details of the new service and the prospects for the future of colour television. The exhibition opened in Croydon on 25 September 1967 and after a few days began its tour of England and Scotland, visiting Bournemouth, London, Birmingham, Newcastle, Manchester, Leeds and Glasgow in turn. [42]

In opening the travelling exhibition in Croydon the BBC's Director General, Sir Hugh Greene, said, 'No TV service in the world has started with so much colour or with such a wide variety of programmes. Apart from plays and feature films in colour there would be, among other things, a late night horror [*sic*].

'I have in the past, in other countries, suffered from blue horses, yellow brunettes and news readers with faces as green as Dr Fu Manchu, to say nothing of the improbably rich flesh tints of United States senators. Colour TV here is not like that.' [43]

He went on to say that he hoped next year's Olympic Games would do for colour what the 1953 Coronation had done for black-and-white television. 'If you haven't got a colour set by October next year, make sure you live next door to someone who has', he said. [44]

At the exhibition, the receivers (of 17 different brand names made by eight different manufacturers) varied in price from £346 10s downwards. The cheapest set was a

19 in. GEC at £257. 5s. (1967 prices). [43] For comparison, in the decade 70–82: coal cost £15 per ton; first-class postage was 2p; 4 lb of bread cost 16p; beer was 6p per pint; sugar was 3p per lb; pipe tobacco was 20p per ounce; school fees in the City of London were £294 maximum; skilled and unskilled males earned on average £20 and £13 per week respectively; and income tax was 40.25p in the pound [45]. Thus the price of a colour television receiver represented a very considerable financial outlay for the average householder.

From the outset BBC2 colour transmissions were available to *c.* 70 per cent of the population; the coverage was expected to increase to *c.* 80 per cent by the end of 1968. In 1968 the number of colour television sets in use was approximately 80,000 and the prospect was that this number would reach 500,000 by the end of 1970 [40].

Details of the changeover from the all-black-and-white services of the BBC and the ITA to colour were given by the Postmaster General in the House of Commons on 15 February 1967. He said, '[The] Government have decided to adopt the recommendations of the Television Advisory Committee and immediately to authorize the BBC and ITA to undertake the duplication on 625 lines and in [the] UHF of their 405-line VHF services. If all goes well the duplicate services should start in London, the Midlands and the North simultaneously within the next three years. In this way, the duplicated programmes, including colour, would be available to nearly half the population from the outset. The expectation is that, by the end of 1971, they would have achieved about 75 per cent coverage.' [40]

Planning for this changeover now progressed under the supervision of a Working Party established by the Postmaster General (PMG). It led to an agreement between the Post Office and the broadcasters on the initial stations to be opened in accordance with a 'rough timetable' that had been announced by the PMG. The first stage of the plan specified that 26 main transmitting stations should be operational before the end of 1971. Of these, the first four were to be Crystal Palace, Sutton Coldfield, Emley Moor and Winter Hill. [40]

For the start of the BBC2 colour television service in December 1967, two large studios, each of 740 m^2 floor area, were completed at BBC Television Centre, London. These were followed soon afterwards by a smaller studio of 350 m^2. Together they comprised a group using similar equipment with a common apparatus area (see Figure 14.8) [46].

The larger studios had a maximum lighting load of 450 kW, corresponding to *c.* 600 W/m^2, which energised the main dual source lanterns. They consisted of an incandescent lamp with two 2.5 kW filaments for spotlight working and four 1.25 kW tungsten-halogen lamps for diffuse lighting. Thyristor dimming was employed in the studios, and in one of them a computer system provided storage and recall for 100 different combinations of each or all of the 290 individual dimmed circuits.

The first Outside Broadcast (OB) colour mobile control room (c.m.c.r.) vehicle (Figure 14.9), equipped with three plumbicon cameras, was completed in May 1967. Two more c.m.c.r.s were completed in November 1967 and March 1968 respectively.

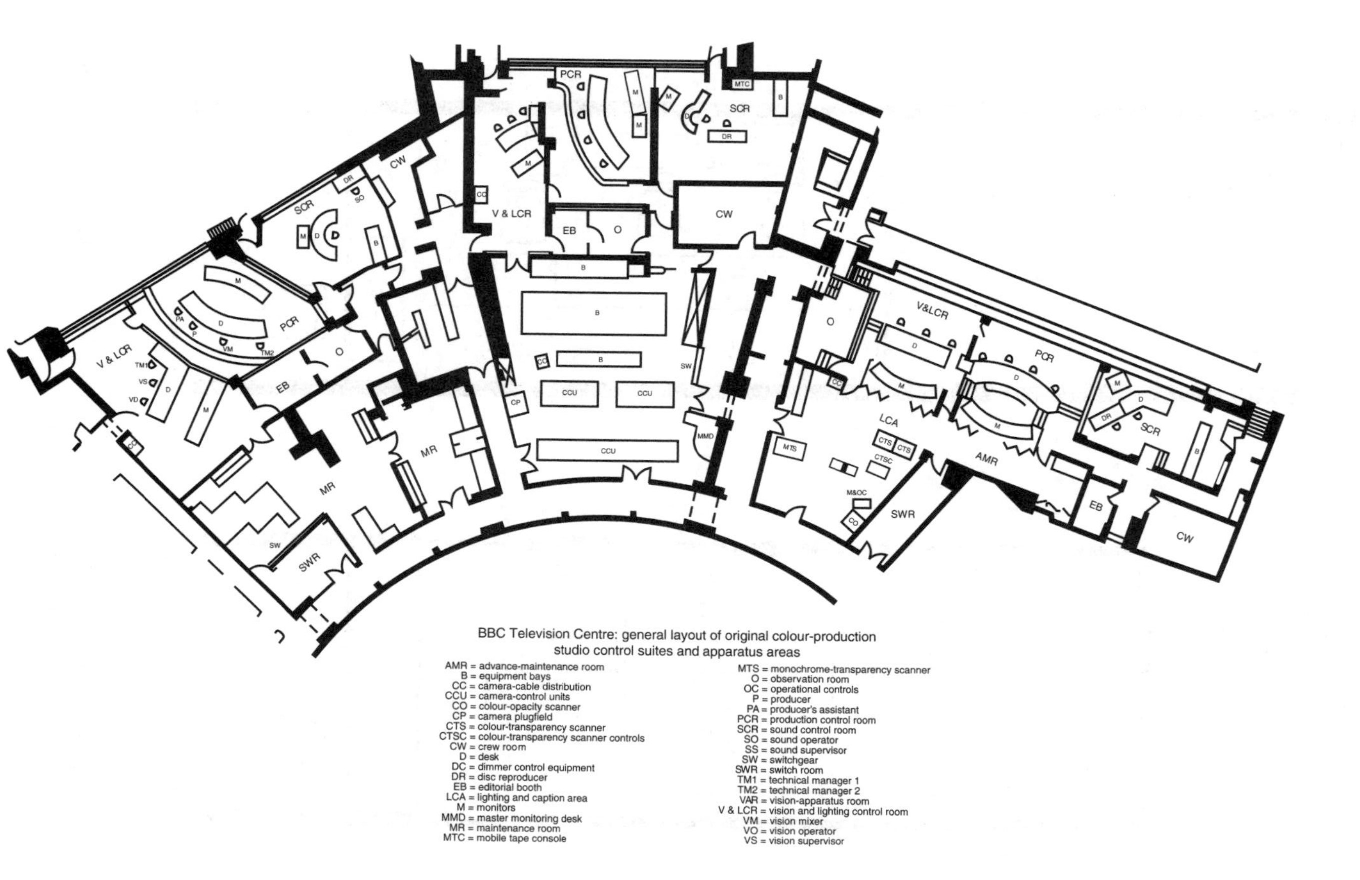

Figure 14.8 BBC Television Centre: general layout of original colour-production studio control suits and apparatus areas [Source: IEE Reviews, *August 1970, vol. 117, p. 1479]*

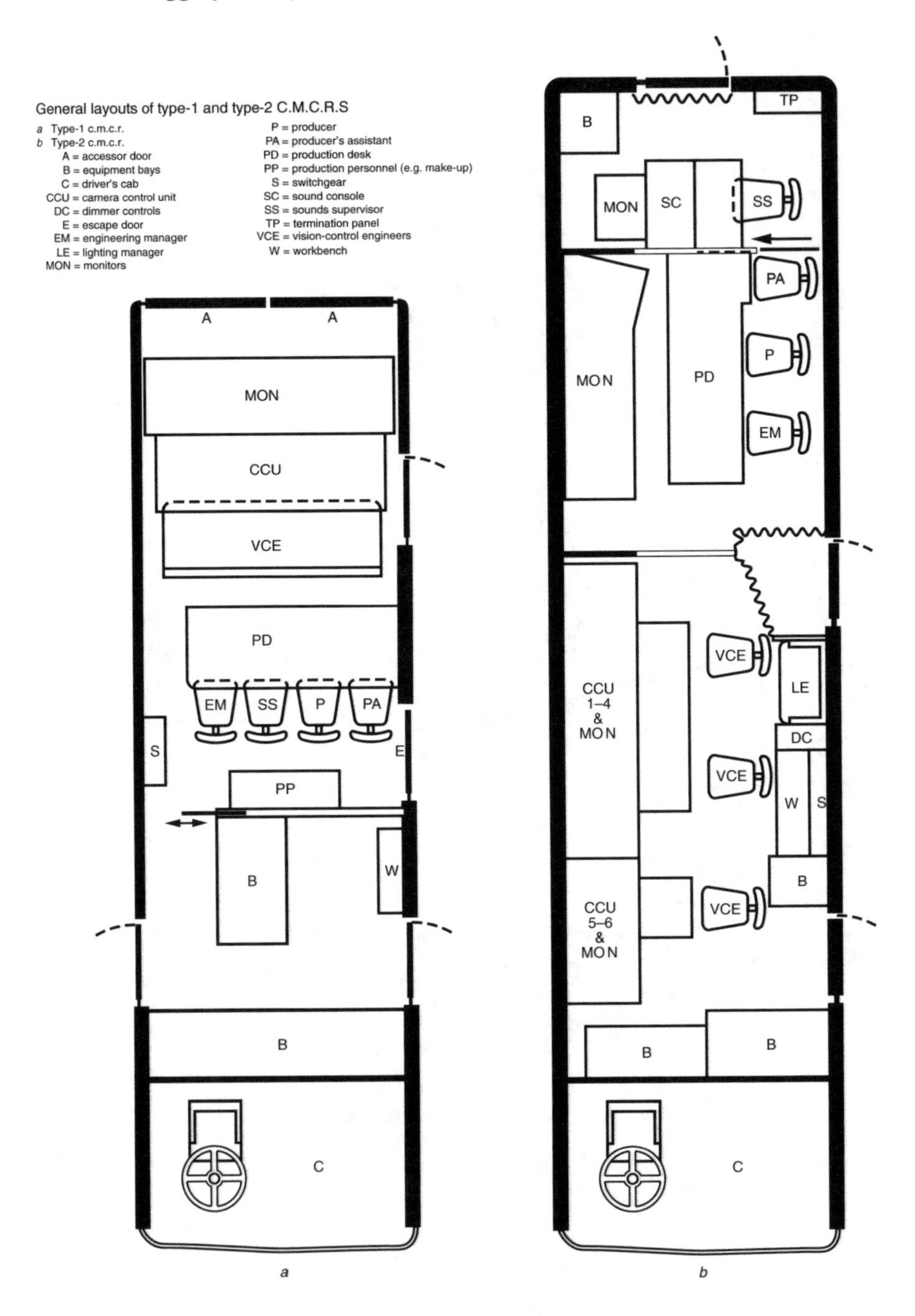

Figure 14.9 General layout of Type-1 and Type-2 colour mobile control rooms (c.m.c.r.s) [Source: ibid., *p. 1480]*

Colour television transmissions on BBC1 commenced on 13 November 1969. In that year it was anticipated the colour output would total *c*. 40 hours per week allocated as follows [47]:

	Hours per week	**Estimated colour differential costs per week, £**
Drama	6.75	6,940
Light entertainment	5	2,690
Documentaries	0.75	1,085
Features and science	1.5	1,595
OB Sports and Events	10.75	1,230
Music and Arts	1.25	925
Current Affairs	3.75	2,690
Feature Films and Telefilms	4.75	1,300
Children's	5	1,365
Totals	39.5	19,820

For the year the additional Programme Allowance cost for colour programmes was £1,030,640, or just over £500 per hour.

Elsewhere, also in 1967, colour television broadcasting began in Germany, France and the USSR. Then, slowly, the NTSC system, and its PAL and SECAM variants, spread around the earth to enable scenes in colour to be transmitted, via satellite links, from one country to any, or all, of the world's nations. The fulfilment of this task was triumph for the engineering excellence of the basic NTSC system; for the collective genius of the many scientists and engineers whose ideas and inventions made the system possible; and, probably above all, for the imagination of Sarnoff of RCA. His belief in the practicability of a viable colour television system and his faith in the ability of his staff to implement such a scheme never faltered. It was Sarnoff who persuaded the RCA board of directors to fund the enormous cost of the necessary research and development programme. In the following chapter the importance of imagination in the progression of engineering is discussed.

Note 1: The EBU Ad Hoc Group on Colour Television

Table 14.1 summarizes the activities of the meetings held by the EBU Ad Hoc Group on colour television.

Table 14.1 The activities of the EBU Ad Hoc Group on Colour Television

Hosts, place and dates of meetings	Activities
BBC and ITA. London. 19 Nov 1962	Inspection of the BBC's UHF transmitter at Crystal Palace, which was being used for experimental colour television transmissions. Visits to mobile labs belonging to BBC and ITA to view pictures received in colour. Visit to BBC Lime Grove studio. Demonstration of colour television studio equipment in operation during a programme comprising live and filmed sequences. 625-line NTSC system used throughout [48].
Telefunken Hanover. 3–5 Jan 1963	Discussion on frequency spacings between the sound and vision carrier frequencies of 5.5, 6.0 and 6.5 MHz. Demonstration of colour television systems proposed by Mr Bruch [49].
RTF, and FNIE. Paris. 26–29 Mar 1963	PAL demonstrated. RTF demonstration comparing NTSC and SECAM systems. FNIE demonstration of long distance transmission – from the Eiffel Tower, Paris to Lyon and back, *c.* 1,000km – of NTSC and SECAM signals [50].
BBC, ITA, ITCA, BREMA, MWT, EMI, and Rank-Bush-Murphy Labs, London. 8–12 and 15–16 Jul 1963	Demonstrations of PAL, NTSC and SECAM transmissions radiated in Channels 34 and 44 by two UHF transmitters of 150 kW e.r.p. at BBC London Television Station, Crystal Palace. Comparison of 10 different makes of colour television receiver. On six pairs of receivers the compatibility of reception of the NTSC, SECAM and PAL systems shown. Effects of interference, noise and phase distortion demonstrated by BBC. Various types of colour television camera displayed [51].
RAI Rome. 15–19 Oct 1963	Demonstrations given of NTSC, PAL and SECAM systems using CCIR Standard G (5.5 MHz vision/sound spacing) over long distances, namely: Rome-Palermo-Milan-Rome (3,400 km); and Rome-Palermo-Limbara (Sardinia)-Rome (3,400 km, including two 240 km paths over the sea). Demonstration of quality of reception when two tandem-connected, rebroadcast transmitters were used [52].
Zurich. 10–13 Dec 1963	40-page report summarizing the work undertaken by the Group from Nov 1962 to Oct 1963 drafted [53].
London. 15 Feb 1964	Report on work effected on PAL, NTSC and SECAM, especially SECAM III since the Zurich meeting. Service trials' results on the three systems reported by the Swiss Telecommunications Administration [54].

Table 14.1 Continued

Hosts, place and dates of meetings	Activities
BBC, London. 27–28 May 1964	Finalization of a programme of trials and meetings with a view to completing the earlier Report of the Group. Collaboration with OIRT discussed. Demonstrations, relating to the transcoding of SECAM colour television signals to NTSC signals, given by ABC Television Ltd at Teddington. A provisional EBU standard colour television test film shown by ABC Television Ltd [55].
BBC and BREMA, London. 26–29 Oct and 5 Nov 1964	Demonstrations given by BBC of effects of multi-path propagation conditions on NTSC colour television transmissions. Further tests on short- and long-distance loop paths using the multi-burst corrector. BREMA displayed several makes of receivers (NTSC type), and demonstrated compatibility of NTSC and SECAM systems. '. . .the delegates appreciated the very good quality of the pictures, which had certainly progressed compared with those shown in London during the demonstration in July 1963.' [56]
ORTF, Paris. 21–28 Jan 1965	Preparation of Final Report based on the tests and documents (195) of the Group. Final Report unanimously approved. Unanimity of choice of system – NTSC or PAL, or SECAM – not possible [57].
Rome. 7–9 Dec 1965	New system (NIR or SECAM IV) noted. Systems of member countries of Group reviewed and examined. Some 'very successful demonstrations [of PAL] were given by Dr Bruch'. The frequency of the chrominance sub-carrier was fixed [58]
ORTF, Paris. 9–10 Feb 1966	Modifications to Report of the EBU Ad Hoc Group unanimously approved. NIR mentioned. Exchange of views on tests of that system undertaken by the BBC, the ITCA and by Dr Bruch [59].
RTI Rome. 3–5 May 1966	Special wide-ranging demonstrations and comparisons given by RTI to members of the EBU who were not represented on the Ad Hoc Group. PAL and SECAM III-b compared [60].

Note 2: CCIR Study Group XI, Sub-Group on Colour Television, London, 1964 [18]

This meeting was held from 14 to 26 February 1964 at Church House, Westminster, London and attracted international interest. The attendance list is included here because it gives some indication of the administrations, broadcasting companies and other organizations that were either contemplating the introduction of colour television in their countries or maintaining a watching brief on developments in this field.

1. *Administrations represented*
Belgium, Denmark, Spain, United States of America, France, Ireland, Italy, Japan, Luxembourg, Norway, Netherlands, Poland, The Federal Republic of Germany, Romanian People's Republic, United Kingdom, Sweden, Czechoslovak Socialist Republic, Union of Soviet Socialist Republics

2. *Delegates from recognized private operating agencies*
Associated Electrical Industries Ltd (AEI), American Telephone and Telegraph Co. (AT&T), British Broadcasting Corporation (BBC), Compagnie Générale de Telegraphie Sans Fil (CSF), Independent Television Authority (ITA), Marconi International Marine Company Ltd (MIMC), Nippon Hoso Kyokai (NHK), Nippon Ninkanhoso Remmei, Oesterreichischer Rundfunk, Radio Corporation of America (RCA), Radiotelevisione Italiana (RAI), Redifon Ltd, Swedish Broadcasting Corporation

3. *International organizations*
International Broadcasting and Television Organization (IBTO), European Broadcasting Union (EBU)

4. *Scientific and manufacturing organizations*
Marconi Italiana, N.V. Philips' Telecommunicatie Industrie (PTI), Siemens and Halske AG, Standard Electrik Lorenz (SEL), Telefunken

Note 3: Limitations of the NTSC and PAL systems [28]

The UK's Technical Sub-Committee considered that the factors affecting the choice between the NTSC and the PAL systems were fairly evenly balanced. From the TSC's analysis the main limitations of these systems (in 1965) were as follows:

Limitations to be accepted if NTSC were chosen

1. The Broadcasting Authorities and the Post Office [UK] would have to operate their equipment and lines within closer but practicable tolerances.
2. In some areas where strong multi-path interference [was] encountered, the picture quality with NTSC could be slightly worse than with PAL.
3. NTSC [was] less likely to be accepted in Europe than [was] PAL; thus the programme interchange between Europe and the UK would possibly become more complicated and the export of complete television receivers would be handicapped.
4. The easier tolerances in the manufacture, installation and maintenance of receivers, and improvements in colour stability, permitted by PAL, would be lost.

Limitations to be accepted if PAL [were] chosen

1. Tests of compatibility using a scale of subjective impairment to picture quality suggest[ed] that the number of unfavourable opinions due to sub-carrier pattern with NTSC may be half the number with PAL. In neither case, however, [was] there likely to be a significant number of complaints, especially bearing in mind that sub-carrier pattern would

be only one of a number of impairments arising in everyday broadcasting. BREMA have agreed that they [would] arrange that the response at the sub-carrier frequency in their new monochrome receivers [would] be acceptable to the public and that on existing receivers, dealers can carry out, if necessary, a simple modification.

2. With receivers designed to take full advantage of the PAL system there would be a small inherent loss of picture colour detail but this is unlikely to be noticeable to the public.
3. Receiver costs to the public would be higher by 3.5 to 4% but some of this would be offset by lower maintenance costs.
4. PAL [was] unlikely to be used in America or in countries in her field of influence and this [would] make programme exchange with these countries slightly more difficult.

The Technical Sub-Committee felt that neither system had an overall advantage but they agreed the two most significant points were:

1. From the point of view of programme origination, recording and distribution, and for receiving, PAL would in most respects present the easier problems to the people concerned, although the problem for NTSC [was] readily soluble;
2. The compatible picture obtained with PAL [was] on average slightly worse than that obtained with NTSC but the number of complaints likely to arise [was] insignificant, and BREMA [were] satisfied that they [could] deal satisfactorily with this problem.

References

1. Maurice D. 'Colour coding systems or the European colour stakes'. *Institution of Electrical Engineers Conference Publication* No. 271 on 'The History of Television'. 1986, pp. 42–6
2. Anon. Report, 'Meeting of the EBU Ad-Hoc Group on colour television'. *EBU Review, Part A – Technical*, No. 76, December 1962, p. 298
3. McClean F.C. 'Colour television. Note by the BBC'. 29 April 1963, ITC A/E/0102
4. Theile R. 'The development of compatible colour television'. *EBU Review*, Part A – Technical, No. 93, October 1965, pp. 194–204
5. Haantjes J., Teer K. 'Compatible colour television – two sub-carrier system'. *Wireless Engineer*. January 1956, pp. 9–82; February 1956, pp. 39–46
6. De France H. 'Le système de television en couleurs sequential-simultane'. *L'Onde Electrique*. 1958;**38**:479–83
7. Cassagne P., Sauvanet M. 'Le système de television en couleurs SECAM – comparison avec le système NTSC'. *Annales de Radioelectricite*. April 1961;**XVI**(64):109–21
8. Townsend G.B. 'New developments in the SECAM colour television system'. *Television Society Journal*. 1963;**10**(6):172–90
9. Mayer N. 'Farbfernsehubertragung mit gleichzeitiger Frequenz und Amplitudenmodulation des Farbtragers (FAM-Verfahren)'. *Rundfunktechnische Mitteilungen*. December 1960;(6):238–52
10. Bruch W. 'Some experiments with modification of the NTSC colour television system'. Presented at the 2nd International Television Symposium, Montreux, 1962

11. Loughlin B.D. 'Recent improvements in band shared simultaneous color television systems'. *Proceedings of the IRE*. **39**;1951:1273–79
12. Bruch W. 'Das PAL-Farbfernsehsystem. Prinzipielle Grundlagen der Modulation und Demodulation'. *Nachrichtentechnische Zeitschrift*. 1964;**17**:109–21
13. Mayer N., Holoch G. 'NTSC-Farbfernsehubertragung mit addition Referenztrager'. *Rundfunktechnische Mitteilungen*, No. 3, June 1965, pp. 157–165
14. Anon. 'Meeting of the EBU Ad-Hoc group on Colour Television'. *EBU Review, Part A – Technical*, No. 76, December 1962, p. 298
15. Anon. 'Second meeting of the EBU Ad-Hoc Group on Colour Television'. *EBU Review, Part A – Technical*, No. 77, February 1963, p. 44
16. McClean F.C. 'Colour television. Note by the BBC'. 29 April 1963, ITC A/E/0102
17. Anon. TAC paper. 21 January 1964, ITC A/E/0102
18. Anon. Report of the United Kingdom Delegation on the CCIR Study Group XI Sub-Group Meeting on Colour Television, London, 1964, ITC A/E/0102
19. Anon. 'Eurovision colour demonstration, Hamburg, 8th April 1964', BBC R53/38/4
20. Mumford A.H. Letter to Sir Willis Jackson, 7 January 1965, ITC A/E/0102
21. Jackson W. Letter to the Postmaster General, 26 February 1965, ITC A/E/0102
22. Jackson W. Letter to the Postmaster general, 30 January 1965, ITC A/E/0102
23. Beresford-Cooke A. Memorandum to the DDG (AS), 23 March 1965, ITC A/E.0102
24. McClean F.C. 'SECAM Press Campaign in France', 1 January 1965, ITC A/E.0102
25. Minute, 12 May 1965, General Advisory Council, BBC R6/29/5
26. Anon. 'Vienna: March 24-April 7, 1965', ITC A/E/0102
27. Anon. Report of the EBU Ad-Hoc Group on Colour Television, Technical Centre of the EBU, Brussels, February 1965
28. Anon. 'Colour Television. Report by TSC', May 1965, ITC A/E/0102
29. Anon. 'Choice of colour system for the 625-line standard', undated but later than May 1965, ITC A/E/0102
30. Jackson W. Letter to Mr A Wedgwood Benn, 7 July 1965, ITC A/E/0102
31. Wedgwood Benn A. Letter to Sir Willis Jackson, 5 August 1965, ITC A/E/0102
32. Birchall J.V.R. 'TAC, Colour television', TAC/256, 17th November 1965, ITC A/E/0102
33. Jackson W. Letter to Mr A Wedgwood Benn, 14 December 1965, ITC A/E/0102
34. Anon. Debate on Broadcasting Policy, 3 March 1966. Postmaster General's announcement about colour television, 3 March 1966
35. Jackson W. Letter to Mr A Wedgwood Benn, 16 June 1966, ITC A/E/0102
36. HS. 'Report to TAC/TSC on CCIR XIth Plenary Assembly, Oslo, 1966, Colour Television', 2 August 1966, ITC A/E/0102
37. GPO. 'Note of a meeting held in Room 109, Post Office H/Q's Building on 14th July 1966', ITC A/E/0102
38. GPO. 'TAC. Summary of the proceedings of the CCIR XIth Plenary Assembly at Oslo 1966 in respect of Colour television', TAC/267, 10 October 1966, ITC A/E/0102

39. Critchley J. 'BBC colours its vision'. Report, *Daily Telegraph*. 3 July 1967
40. Anon. 'The introduction of colour television and the 625-line definition standards for all television services'. Undated paper, TAC(68) 1
41. Anon. '25 hours colour television by autumn'. *Electrical and Radio Trading*. 4 August 1967
42. Anon. 'BBC colour comes to town'. *Electrical and Radio Trader*. 30 June 1967
43. Gander L.M. '100pc more television shows in colour'. *Daily Telegraph*. 26 September 1967
44. Anon. 'More colour programmes from BBC next week'. *Guardian*. 26 September 1967
45. Priestly H. *The What it Cost the Day Before Yesterday Book, from 1850 to the present day*. Hampshire: Kenneth Mason; 1979
46. Redmond J. 'Television broadcasting 70–82: BBC 625-line services and the introduction of colour'. *IEE Reviews*. August 1970;**117**:1469–88
47. Head of Finance Services, Television to Director General. Memorandum, 'Television: colorisation of BBC-1', 30 September 1969, BBC file T39/9/1
48. Anon. Report, *EBU Review, Part A – Technical*, No. 76, December 1962, p. 298
49. *Ibid*. No. 77, February 1963, p. 44
50. *Ibid*. No. 78, April 1963, p. 92
51. *Ibid*. No. 80, August 1963, p. 180
52. *Ibid*. No. 82, December 1963, p. 280
53. *Ibid*. No. 83, February 1964
54. *Ibid*. No. 84, April 1964, p. 91
55. *Ibid*. No. 86, August 1964, p. 192
56. *Ibid*. No. 88, December 1964, p. 296
57. *Ibid*. No. 89, February 1965, p. 52
58. *Ibid*. No. 95, February 1966, p. 40
59. *Ibid*. No. 96, April 1966, p. 90
60. *Ibid*. No. 97, June 1966, p. 134

Chapter 15
Epilogue

The fully compatible NTSC system of colour television is undoubtedly one of the great achievements of telecommunication engineering. As Sir James Redmond, a former director of television, BBC, has observed, 'The end result [of the NTSC's deliberations] was one of the most spectacular conjuring tricks that has been performed on the telecommunication stage; one and a half quarts were successfully squeezed into a pint pot. And it seemed that little or nothing was spilt!' [1]

RCA must be given much credit for the success of the NTSC system. It was RCA who from 1929 embarked on a research-and-development programme to evolve an all-electronic television scheme; it was RCA who engineered the image orthicon camera tube on which colour cameras, for approximately a decade, were dependent; it was RCA who carried out the necessary experimental investigations on display tubes that led to the large-scale manufacture of the shadow mask tube – an outstanding invention; it was RCA who indicated the way forward in the encoding and decoding of the luminance and chrominance content of a televised scene; and it was the directors of RCA who were willing to approve the necessary huge financial outlay required for its colour television policy.

Above all it was the imagination and faith of David Sarnoff, the dynamic and forward-looking president of the Radio Corporation of America, that enabled success to be accomplished. Without his drive, encouragement and vision it is likely that the introduction of a compatible, simultaneous system of colour television would have been considerably delayed.

Albert Einstein once said, 'Imagination is more important than knowledge.' Sarnoff's life's work exemplifies this statement. He was not an engineer or scientist and could not personally have undertaken any practical work in the field of radio or television but, from early in his career, he showed that he did not lack imagination. His early career is an exemplar of what could be achieved by initiative and dogged determination even when an absence of a college/university education might have been considered a hindrance to advancement.

David Sarnoff was born in Minsk, Russia, on 27 February 1891. With his family he emigrated, in 1900, to the United States of America and settled first in Albany, New York, and then in New York City. While at school he helped to support the family by selling newspapers, singing the liturgy in a synagogue and running errands. He left school in 1906 and obtained a position as a messenger boy with a telegraph company. One of his early purchases was a telegraph instrument and with this he soon became proficient in sending and receiving Morse code signals. This proficiency led to Sarnoff's becoming a radio operator with the Marconi Wireless Telegraph Company at shore-based and ship-based radio offices.

After several years he became the operator at the world's most powerful radio station, which was atop John Wanamaker's department store in Manhattan, New York. Here, while on watch on 14 April 1912, Sarnoff picked up the distress signals radiated from the 'unsinkable' SS *Titanic*, and remained at his operator's position for the next 72 hours, receiving and retransmitting the signals from the doomed ship and ships going to the rescue of its passengers.

This disaster marked a turning point in Sarnoff's career. He was rewarded by the company with rapid promotion and by 1916 was the assistant traffic manager of the American Marconi Company (which had E.J. Nally as its vice-president and general manager). At this stage Sarnoff showed his imagination. He considered the desirability of a public broadcasting service being set up by the Marconi company, and, in an internal memorandum to Nally, outlined his suggestion:

> I have in mind a plan of development which would make radio a household utility in the same sense as a piano or phonograph. The idea is to bring music into the home by wireless ... Should the plan materialise, it would seem reasonable to expect sales of 1,000,000 'radio music boxes' within a period of three years. Roughly estimating the selling price at 75 *per set*,75,000,000 can be expected. [2]

The Marconi company did not heed this prophetic forecast at that time. A few years later, following the US government's desire for an American-dominated wireless communications organization, the Radio Corporation of America on 20 November 1919 acquired a controlling interest in the American Marconi company. (The ownership of RCA at this time was, as stated previously, General Electric, 30 per cent; Westinghouse, 20 per cent; AT&T, 10 per cent; United Fruit Company, 4 per cent; others, 36 per cent.) [3]

Sarnoff became the commercial manager of the new corporation and seized the opportunity that neither General Electric nor Westinghouse had grasped and commenced the marketing of a small radio receiving set. 'It was produced and sold amazingly well at a rather high unit price. It [was] considered to be one of the factors that started the broadcasting ball rolling.' [2]

RCA was blessed in having such a far-seeing, knowledgeable, able and aggressive commercial manager in the person of David Sarnoff. In a report to the RCA board of directors, dated 5 April 1923, he wrote,

> I believe that television, which is the technical name for seeing as well as hearing by radio, will come to pass in due course ... It may well be that every broadcast receiver for home use in the future will be equipped with a television adjunct ... which ... will make it possible for those at home to see as well as hear what is going on at the broadcasting station. [4]

His optimistic outlook for the eventual creation and growth of a television industry never wavered.

When, in January 1929, Zworykin met Sarnoff, 'he [Sarnoff] quickly grasped the potentialities of my proposals [for all-electronic television] and gave me every encouragement from then on to realize my ideas'.

Recalling Sarnoff, 'a brilliant man without much education', Zworykin has recorded the following anecdote:

> Sarnoff saw television as a logical extension of radio broadcasting, which was already commercially prosperous. His parting question, after my presentation, was to estimate how much such a development would cost. I said I hoped, with a few additional engineers and facilities, to be able to complete the development in about two years and estimated that this additional help would cost about a hundred thousand dollars. This of course was too optimistic a guess, as Sarnoff has stated that RCA had to spend many millions before television became a commercial success. [5]

Towards the end of 1929 Zworykin lectured on, but did not demonstrate, his research group's kinescope [6]; four years later his team's iconoscope-based camera and kinescope-based receiver were comprehensively field tested in RCA's experimental, all-electronic, 240-line television system. The R&D effort that led to the fabrication of the iconoscope and kinescope tubes established much of the fundamental theory of future colour television camera and display tubes.

Sarnoff's imagination may be contrasted with that of Dr H.E. Ives, the very capable but, perhaps, conservatively minded leader of AT&T's television group.

In 1927 Ives's staff successfully demonstrated small screen and large screen black-and-white television over both radio and line links [7]. The demonstrations were the best that could be achieved with the technology as it existed at that time. Following these demonstrations Dr F.B. Jewett, the president of Bell Telephone Laboratories, believed, as previously stated, that the Laboratories should carry on, as adequately as possible, whatever fundamental work would be necessary to safeguard the company's position. Ives seized the opportunities made accessible to him and during the next three years daylight television, large-screen television, television recording, colour television, two-way television, and multichannel television were all subject to his group's scrutiny and engineering prowess. However, they were all based on the use of mechanical scanners.

Ives was not impressed by the lecture on the kinescope which Zworykin gave in 1929, and felt Zworykin's account was 'chiefly talk'. 'This method of reception is old in the art,' wrote Ives, 'and of very little promise. The images are quite small and faint and all the talk about this development promising the display of television to large audiences is quite wild.' [8]

With the leader of the television group holding such an opinion it was perhaps inevitable that R&D effort on the evolution of an all-electronic television scheme would not be a major part of the group's activities. Instead, in a May 1931 memorandum on 'The future programme for television research and development' [9], Ives's suggestions for future work were all posited on extending his group's basic scheme. He confirmed his support for mechanical scanning by suggesting three special projects all based on this mode of reconstituting an image. These were: (1) the demonstration of reception from an aeroplane; (2) the demonstration of some major outdoor events; (3) the demonstration of reception on film and thence projection in a theatre. All these recommendations if approved would have been founded on existing principles and technology. There was a lack of new ideas in his plans. Ives did not propose an investigation of the high frequency bands of the electromagnetic spectrum for television purposes, even though it was appreciated by 1929 that such bands would have to be utilized if high-definition television were to become a reality. Nor did his group pursue any fundamental work on electronic camera or electronic display tubes. And yet when his group eventually embarked on such a programme an important advance was made.

Essentially, by *c.* 1930 the analysis and synthesis of objects and images by mechanical scanning had shown itself to be unsuitable for the reproduction of high-definition images. This inappropriateness was highlighted from November 1936, when the BBC's London television station was established. Initially, two alternate services were provided by the BBC, utilizing equipment manufactured by M-EMI Television Ltd and by Baird Television Ltd respectively. The M-EMI apparatus was wholly electronic, whereas that of Baird Television employed mechanical scanners for its spotlight camera, intermediate film studio camera, and tele-cine apparatus. Within only two months of the opening of the London station the BBC's studio staff and engineers had decided that the Baird Television company's systems, with the exception of the tele-cine scanner, could not compete on equal terms with the emitron cameras of M-EMI. From February 1937 only that company's equipment formed part of the UK's television service, much to J.L. Baird's dismay [10].

The disappointment suffered by Baird resulted from his ignoring for too long the inevitable move towards high-definition television and the use of cathode-ray tubes in receivers and cameras. In October 1931 he was quoted as saying that he saw 'no hope for television by means of cathode ray bulbs', that 'the neon tube [would] remain as the lamp of the home receiver', and that he was sceptical about the success of the use of short waves in television 'because they covered a very limited area' [11].

By May 1931, when Ives wrote his memorandum, it was evident that new approaches to the problem of seeing by electricity were needed if television broadcasting were to become widespread in popular appeal. This appeal would be encouraged when sporting and athletic events, news reports, plays and so on could be televised to give adequate image detail. But Ives ignored the moves towards electronic scanning which were being advanced by others. He seems to have lacked the imagination of Sarnoff. Of course, Sarnoff, lacking a science/engineering education, could not have had any realistic conception of the formidable engineering difficulties involved in engineering an all-electronic television system. Dr Ives, on the other hand, was a highly competent physicist/engineer who in 1931, following the completion of his group's two-way video telephone link, undertook an important appraisal of the progress that had been made by Bell Telephone Laboratories. He endeavoured to define the course of action that had to be implemented for the future advancement of television. His prognosis was gloomy in outlook and he concluded, 'The existing situation is that if a many-element television image is called for today, it is not available, and one of the chief obstacles is the difficulty of generating, transmitting, and receiving signals over wide frequency bands.' [12]

Ives's engineering conservatism contrasted markedly with the forward-looking approaches of Farnsworth, Sarnoff of RCA, and Shoenberg of EMI. They held a firm belief in the ineluctability of the move away from head-and-shoulders images and towards high-definition television by all-electronic means. They worked strenuously to perfect their companies' electronic camera tubes: the image dissector tube, the iconoscope, and the emitron respectively [13].

The convictions of Zworykin, Sarnoff, Farnsworth, Shoenberg and others was not shared by Jewett. During his testimony before the Federal Communications Commission (the Hearing of 10 December 1936) he expressed himself as being pessimistic on television matters and stated, 'I am very sceptical personally about television.... Now, if it does come – and I am just as clear as a crystal in my own feelings on this thing – if it comes, if these people who are interested in it are right in their belief, and I am wrong ...' [14] Future events would show he was wrong and that he lacked the imagination of, for example, Sarnoff concerning the inevitability of television broadcasting.

Zworykin described, but did not demonstrate, his ideas at a meeting of the Institute of Radio Engineers held on 26 June 1933 in Chicago. The meeting was attended by Dr F. Gray, of Ives's group. He wrote a short memorandum (dated 6 July 1933) on what he had learned from the lecture and on the same day Ives sent a memorandum to Dr H.D. Arnold, the director of research at Bell Telephone Laboratories. 'The device', wrote Ives, 'involves some principles which are new in the television art and represent a considerable advance in the direction of attaining a television transmitting device suitable for general use in such places as a moving picture camera might be used.' [15] He felt Zworykin had taken an important step in the right direction and opined, 'It is not at all improbable that there will emerge from his development a television transmitter [i.e. a camera] which will make high grade transmission of television material for entertainment a practicable possibility'.

Several days later (20 July 1933) Ives, in a memorandum, noted, 'The consensus of opinion at our conference yesterday [on the subject of cathode ray transmitting apparatus] appears to be the feeling that we would do well to start some work along these lines.' [16] On 31 July Gray [17] outlined a possible programme of work and soon a beginning was made on an enquiry into the design of an electronic camera tube. Unfortunately for the prestige of Bell Telephone Laboratories this effort began three years too late. On 20 May 1930 Gray had written a comprehensive memorandum [13] (21 pages long with 11 figures) on a 'proposed television transmitter [camera]' based on charge storage, in which he had put forward 11 methods by which a vision signal might be generated by various photoelectric means. One of his objects in listing the proposals was 'to aid in deciding the direction of certain future researches on television'. Had Gray's inventiveness and resourcefulness been acted upon it is possible the Laboratories, with their excellent experimental facilities, resources and staff, would have demonstrated an electronic camera contemporaneously with Zworykin's 1933 disclosure.

Ives's team seems to have been reinvigorated by its new task. Within a year: Gray [19] had proposed a cathode-ray camera that utilized the stored charge flowing through a photoconductive film during a complete image cycle, and had put forward various ideas for projecting images from the screen of a cathode-ray tube [20], whose optical reflecting power and transparency could be affected by the electron beam (cf. the Eidophor); G.K. Teal [21] had reported on experiments undertaken with new photoelectric emitters suitable for use in an iconoscope; a conference had been held at which seven novel methods for displaying large images in theatres had been discussed [22]; Ives had suggested a very simple form of iconoscope suited for generating picture signals from continuously moving motion picture film; and J.R. Hefele had demonstrated a basic form of electronic camera [23].

By the end of July 1934 photosensitive targets, of the charge storage type, had been prepared by two methods and tested. Twelve months later, on 26 July 1935, Hefele demonstrated to several senior members of the Bell Telephone Laboratories (BTL) a 72-line photoconducting television camera tube. In view of the excellent images displayed on a cathode-ray tube, designed by C.J. Davisson, it was decided to attempt 'immediately' to construct a 240-line camera tube [24]. The demonstration appears to have been the first ever given of a photoconductive camera tube, and showed that BTL had the expertise, the experimental facilities and the essential ideas to engineer an all-electronic television system. It would appear that all of these developments could have been initiated by Bell Telephone Laboratories from *c.* 1930.

A further advance was made when on 9 November 1937 the Laboratories demonstrated the transmission of video signals along a coaxial cable that had been installed between New York and Philadelphia, a distance of approximately 200 miles (*c.* 320 km) [25]. Three years later, on 21 May 1940, 441-line video signals with a bandwidth of 2.7 MHz were sent along the coaxial cable from New York to Philadelphia and back [26]. Demonstrations of such transmissions were subsequently given before the National Television System Committee on 8 November 1940.

Of all the contributions made by the staff of the Laboratories in the field of television, perhaps the most important was the invention by Gray of a method of transmitting two television signals within a frequency band normally sufficient only for one such signal. The method was based on the utilization of the interstices discovered by Gray (and independently by P. Mertz) between the spectrum lines of a television signal, these lines being spaced at intervals equal to the line-scanning frequency. This fact was first mentioned in a memorandum by Hefele and Morrison dated 21 January 1928 and was the basis of a patent [27], in Gray's name, filed on 30 April 1929. Later Gray and Mertz collaborated on the preparation of a paper that is now regarded as a 'classic' in the television literature [28].

Gray's invention did not have an immediate application in 1929. More than two decades later, however, the great value of his disclosure was appreciated when it became necessary to transmit, simultaneously, the luminance and chrominance signals of a colour television system. For this work Gray received, in 1953, some months after his retirement from Bell Telephone Laboratories, the Vladimir K Zworykin Award of the Institute of Radio Engineers [29]. Normally the award is given to the person who has made the most important contribution to television in the year immediately preceding the award. The justification for granting it to Gray was that it was only during the early 1950s that the industry realized the importance of the interleaving principle in designing the NTSC colour television system.

In his biography of A.D. Blumlein the author discusses the nature of genius [30] and the primary factors which can influence discovery/invention, and concludes that they are:

1. commitment and perseverance – the progression of a task even when adversity arises, as exemplified by Baird's work on television;
2. curiosity – the disposition to be inquisitive;
3. imagination – the prime determinant of creativity in science, engineering, and the arts;
4. originality and faith – the investigation of new fields and a belief that success will follow, as illustrated by Zworykin's work on electronic television, and Sarnoff's faith in the ability of the RCA scientists and engineers to find a solution to the colour television problem;
5. insight and intuition – the finding of solutions based on experience, comprehension and perception, as represented by Baird's work on colour television, and Gray's thoughts on the spectral interleaving of two independent signals; and
6. appreciation of the problem or the need – the recognition of which may lead to inspiration and thence to realization, as evidenced by RCA's engineering of its all-electronic, simultaneous colour television system.

Secondary factors include:

1. serendipity/luck/good fortune – the finding of a discovery or useful fact or thing not sought for; and
2. the favourable accident – Goodyear (famed for his investigation of the vulcanization of rubber) once said, 'I was encouraged in my efforts by the reflection that what is hidden and unknown and cannot be discovered by scientific research, will most likely be discovered by accident, if at all, by the man who applies himself most perseveringly to the subject and is most observing of everything related thereto.'

Of all these factors the greatest is imagination. It is the basis of all creative work. The originality that marks the efforts of great engineers and scientists is to be ascribed to a tireless activity of their imaginations followed by a profound intellectual analysis of the possibilities of their ideas. Imagination distinguishes the creative geniuses of civilization from their commonplace fellows. As Chesterton said, 'It isn't that the latter can't see the solutions, it is that they cannot see the problems.'

Problems are part of our existence, their solution leads to truth or enrichment. Mankind advances by the labours of those for whom imagination, faith, and intuition are ever vital components of their human nature.

When Newton was once asked how he made his discoveries he replied, 'By always thinking unto them. I keep the subject constantly before me and wait till the first dawnings open little by little into the full light.' Similarly, single-mindedness of thought and purpose typified the efforts of Baird, AT&T, RCA, EMI and the NTSC.

From these considerations it is apparent that the engineering of the NTSC colour television system was a most outstanding example of collective genius. With its variants PAL and SECAM it has led to worldwide colour television, and has enormously enriched the lives of hundreds of millions of people.

References

1. Redmond J. 'Television broadcasting 1960–70: BBC 625-line services and the introduction of colour'. *IEE Reviews*. August 1970;**117**:1469–88
2. Bitting R.C. 'Creating an industry' *Journal of the SMPTE*. **74**:1015–23
3. Archer G.L. *Big Business and Radio*. New York: American Historical; 1939. p. 8
4. Sarnoff D. Report to the RCA Board of Directors, 5 April 1923, RCA Archives, Princeton
5. Zworykin V.K. 'Electronic television at Westinghouse and RCA'. Unpublished paper, 13 pp., RCA Archives, Princeton
6. Zworykin V.K. Television with cathode ray tube receiver'. *Radio Engineering*. December 1929;**9**:38–41
7. Burns R.W. 'The contributions of the Bell Telephone Laboratories to the early development of television' in Hollister-Short G., James F.A.J.L. (eds.). *History of Technology*, Vol. 13. London: Mansell; 1991. pp. 181–213
8. Ives H.E. Memorandum to H.P. Charlesworth, 16 December 1929, case File 33089, 1. AT&T Archives, Warren, New Jersey
9. Ives, H.E. 'Future progress for television research and development', a memorandum for file, 18 May 1931, Case File 33089, pp. 1–11. AT&T Archives, Warren, New Jersey
10. Burns R.W. *British Television: the formative years*. London: Institution of Electrical Engineers; 1986
11. Burns R.W. *John Logie Baird: television pioneer*. London: Institution of Electrical Engineers; 2000
12. Ives H.E. 'A multi-channel television apparatus'. *Journal of the Optical Society of America*. 1931;**21**:8–19
13. Burns R.W. *Television: an international history of the formative years*'. London: Institution of Electrical Engineers; 1998
14. Anon. 'Observations on Dr F.B. Jewett's testimony before the Federal Communications Commission concerning the division of expenses of the Laboratories as between AT&T and Western and as between local and toll', memorandum for file, 20 July 1938, pp. 1–2. AT&T Archives, Warren, New Jersey
15. Buckley O.E., Ives, H.E. Memorandum to H.D. Arnold, 6 July 1933, Case File 33089, pp. 1–2. AT&T Archives, Warren, New Jersey
16. Ives H.E. Memorandum to O.E. Buckley, 20 July 1933, p.1, Case File 33089, AT&T Archives, Warren, New Jersey
17. Gray F. 'A suggested outline for development work on a cathode ray transmitter'. Memorandum for file, 31 July 1933, Case File 33089, pp.1–3, AT&T Archives, Warren, New Jersey
18. Gray F. 'Proposed television transmitters'. Internal report, 20 May 1930, AT&T Archives, Warren, New jersey

19. Gray F. 'A proposed cathode ray transmitter'. Memorandum for file, 3 January 1934, Case File 33089, AT&T Archives, Warren, New Jersey
20. Gray F. 'Projection of images from a cathode-ray tube'. Memorandum for file, 15 January 1934, Case File 33089, pp. 1–5, AT&T Archives, Warren, New Jersey
21. Teal G.K. 'A new photoelectric emitter for use in the iconoscope'. Memorandum for file, 25 April 1934, pp. 1–2; and 'The potassium-potassium oxide-silver matrix as the photosensitive element of the iconoscope'. Memorandum for file, 14 May 1934, pp. 1–2, Case File 33089, AT&T Archives, Warren, New Jersey
22. 'Notes of a conference held in Dr Buckley's office on large image schemes for theatre showing', 20 June 1934, Case File 33089, pp. 1–8, AT&T Archives, Warren, New Jersey
23. Ives H.E. 'Iconoscope for transmission from motion picture film'. Memorandum for file, 6 July 1934, Case File, AT&T Archives, Warren, New Jersey
24. Nix F.C. 'Photo-conducting television transmitter'. Memorandum to H.E. Ives *et al.*, 6 August 1935, Case File 33089, p. 1, AT&T Archives, Warren, New Jersey
25. Anon. 'Bell labs test coaxial cable'. *Electronics*. December 1937, pp. 18–19
26. Anon. 'Principal Bell System dates in television'. November 1946, AT&T Archives, Warren, New Jersey
27. Gray F. US patent application 1 769 920. Filed 30 April 1929; issued 8 July 1930
28. Mertz P., Gray F. 'A theory of scanning and its relationship to the characteristics of the transmitted signal in telephotography and television'. *Bell System Technical Journal*. July 1934;**13**:464–516
29. A.G.J. 'Early history of Bell System television work'. Memorandum for file, 1954, pp. 1–4, AT&T Archives, Warren, New Jersey
30. Burns, R.W. *The Life and Times of A.D. Blumlein*. London: Institution of Electrical Engineers; 2000. pp. 483–96

Appendix A

The NTSC signal specifications

A.1 General specification

Channel width	6 MHz
Picture signal carrier frequency	1.25 MHz above the lower boundary of the channel
Polarization	horizontal
Vestigial sideband modulation	to be in accordance with Figure 14.2
Aspect ratio	4 (horizontal) to 3 (vertical)
Sound signal carrier frequency	4.5 MHz +/− 1,000 Hz above the picture signal carrier
Sound signal characteristics	frequency modulation, with maximum deviation of +/–25 kHz, and with pre-emphasis in accordance with a 75 μs time constant
Sound transmitter output (e.r.p.)	between 50% and 70% of the peak power of the vision transmitter
Out-of-channel radiation	at least 60 db below the peak picture level at any frequency outside the limits of the assigned channel
Scanning and synchronization	
1. Direction of scanning	from left to right and from top to bottom (uniformly) with 525 lines per frame interlaced 2:1
2. Horizontal scanning frequency	2/455 times the color sub-carrier frequency, i.e. 15,734.264 +/− 0.047 Hz
3. Vertical scanning frequency	2/525 times the horizontal scanning frequency, i.e. 59.94 Hz

4. Color television signal	consists of color picture signals and synchronizing signals transmitted successively and in different amplitude ranges except where the chrominance penetrates the synchronizing region, and the burst penetrates the picture region
5. Synchronizing signals	as specified in Figure 9.1 as modified by vestigial sideband transmission specified in Figure 9.1 and by the delay characteristic specified in III.B

A.2 The complete color picture signal

A. General specifications

The color picture signal shall correspond to a luminance (brightness) component transmitted as amplitude modulation of the picture carrier and as a simultaneous pair of chrominance (coloring) components transmitted as the amplitude modulation sidebands of a pair of suppressed sub-carriers in quadrature having the common frequency relative to the picture carrier of +3.579545 [MHz] +/− 0.0003 per cent with a maximum rate of change not to exceed 1/10 cycle per sec per sec.

B. Delay specification

A sine wave, introduced at those terminals which are normally fed the color picture signal, shall produce a radiated signal having an envelope delay, relative to the average envelope delay between 0.05 and 0.20 [MHz], of zero μsecs up to a frequency of 3.0 [MHz]; and then linearly decreasing to 4.18 [MHz] so as to be equal to −0.17[μs] at 3.58 [MHz]. The tolerance on the envelope delay shall be +/− 0.05[μs] at 3.58 [MHz]. The tolerance shall increase linearly to +/− 0.1[μs], down to 2.1 [MHz], and remain at +/− 0.1[μs] down to 0.2 [MHz]. The tolerance shall also increase linearly to +/− 0.1[μs] at 4.18 [MHz].

C. The luminance component

1. An increase in initial light intensity shall correspond to a decrease in the amplitude of the carrier envelope (negative modulation).
2. The blanking level shall be at (75 +/− 2.5) per cent of the peak amplitude of the carrier envelope. The reference white (luminance) level shall be (12.5 +/− 2.5) per cent of the peak carrier amplitude. The reference black level shall be separated from the blanking level by the set-up interval, which shall be (7.5 +/− 2.5) per cent of the video range from the blanking level to the reference white level.
3. The overall attenuation versus frequency of the luminance signal shall not exceed the value specified by the FCC for black-and-white transmission.

D. Equation of complete color signal

The color picture signal has the following composition:

$$E_M = E'_Y + \{E'_Q \sin(\omega t + 33^\circ) + E'_I \cos(\omega t + 33^\circ)\} \text{ where}$$

$$E'_Q = 0.41(E'_B - E'_Y) + 0.48(E'_R - E'_Y)$$

$$E'_I = -0.27(E'_B - E'_Y) + 0.74(E'_B - E'_Y)$$

$$E'_Y = 0.30(E'_R + 0.59E'_G) + 0.11E'_B$$

The phase reference in the above equation is the phase of the color burst +180°. In these expressions the symbols have the following significance:

- E_M is the total video voltage, corresponding to the scanning of a particular picture element, applied to the modulator of the picture transmitter.
- E'_Y is the gamma-corrected voltage of the monochrome (black-and-white) portion of the color picture signal, corresponding to the given picture element.
- E'_R, E'_G, and E'_B are the gamma corrected voltages corresponding to red, green, and blue signals during the scanning of the given picture element.

The gamma corrected voltages E'_R, E'_G, and E'_B are suitable for a color picture tube having primary colors with the following chromaticities in the CIE system of specification:

	x	y
Red (R)	0.67	0.33
Green (G)	0.21	0.71
Blue (B)	0.14	0.08

and having a transfer gradient (gamma exponent) of 2.2 associated with each primary color. The voltages E'_R, E'_G, and E'_B may be respectively of the form $E_R^{1/\gamma}$, $E_G^{1/\gamma}$, and $E_B^{1/\gamma}$ although other forms may be used with advances in the state of the art.

E'_Q and E'_I are the amplitudes of two orthogonal components of the chrominance signal corresponding respectively to narrow-band and wideband axes, as specified (above).

The angular frequency ω is 2π times the frequency of the chrominance sub-carrier.

The portion of each expression between brackets represents the chrominance sub-carrier signal which carries the chrominance information.

1. The chrominance signal is so proportioned that it vanishes for the chromaticity of CIE illuminant C ($x = 0.310$, $y = 0.316$)
2. E'_Y, E'_Q, E'_I and the components of these signals shall match each other in time to 0.05 μs.
3. A sine wave of 3.58 [MHz] introduced at those terminals of the transmitter which are normally fed the color picture signal shall produce a radiated signal having

an amplitude (as measured with a diode on the RF transmission line supplying power to the antenna) which is down (6 +/– 2) db with respect to a radiated signal produced by a sine wave of 200 [kHz]. In addition, the amplitude of the radiated signal shall not vary by more than +/– 2db between the modulating frequencies of 2.1 and 4.18 [MHz].

4. The equivalent bandwidths assigned prior to modulation to the color-difference signals E'_Q and E'_I are given below:

   Q - channel bandwidth

   at 400 [kHz] less than 2db down

   at 500 [kHz] less than 6db down

   at 600 [kHz] less than 6db down

   I - channel bandwidth

   at 1.3 [MHz] less than 2db down

   at 3.6 [MHz] at least 20db down

5. The angles of the sub-carrier measured with respect to the burst phase, when reproducing saturated primaries and their complements at 75 per cent of full amplitude, shall be within +/– 10° and their amplitudes shall be within +/– 20 per cent of the values specified above. The ratios of the measured amplitudes of the sub-carrier to the luminance signal for the same saturated primaries and their complements shall fall between the limits of 0.8 and 1.2 of the values specified for their ratios. Closer tolerances may prove to be practicable and desirable with advances in the art.

Bibliography

Abramson A. *A History of Television*, 1880 to 1941. Jefferson: McFarland; 1987

Abramson A. *Zworykin, Pioneer of Television*. Urbana: University of Illinois Press; 1995

Abramson A., Sterling C. *A History of Television, 1942 to 2000*. Jefferson: McFarland; 2002

Baird J. L. *Sermons, Soap and Television*. Royal Television Society, London, 1988

Briggs A. *The History of Broadcasting in the United Kingdom*. Vol. V. Oxford University Press; 1995

Burns R.W. *British Television: the formative years*. London: Institution of Electrical Engineers; 1986

Burns R.W. '*Television: an international history of the formative years*. London: Institution of Electrical Engineers; 1998

Burns R.W. *John Logie Baird: television pioneer*. London: Institution of Electrical Engineers; 2000

Burns R.W. *The Life and Times of A.D. Blumlein*. London: Institution of Electrical Engineers; 2000

Burns R.W. *Communications: an international history of the formative years*. London: Institution of Electrical Engineers; 2004

Carnt P.S., Townsend G.B. *Colour Television: the NTSC system, principles and practice*. London: Iliffe; 1961

Eder J.M. *History of Photography*. New York: Columbia UP; 1945

Fink D.G. *Television Standards and Practice*. New York: RMA-NTSC; 1943

Fink D.G. *Color Television Standards – NTSC*. New York: McGraw-Hill; 1955

Goldmark P.C. *Maverick Inventor – My Turbulent Years at CBS*. New York: Saturday Review Press; 1973

Gouriet G.G. *An Introduction to Colour Television*. London: Norman Price; 1955

Hazeltine Laboratories Staff. *Color Television Receiver Practices*. Rider; 1955

Hazeltine Laboratories Staff. *Principles of color television*. New York: Wiley; 1956

Morrell R., Law R.R., Ramberg E.J., Herold E.W. *Color Television Picture Tubes*. New York: Academic Press; 1974

Patchett G.N. *Colour Television, with Particular Reference to the PAL System*. London: Norman Price; 1967

Pawley E.L.E. *BBC Engineering 1922–1972*. London: BBC Publications; 1972

Reed C.R.G. *Principles of Colour Television Systems*. London: Pitman; 1969

RCA Staff. *Television*. Vols. IV (1942–46), V (1947–48), and VI (1949–50). Princeton: RCA

Shiers G. *Technical Development of Television*. New York: Arno; 1977
Wentworth J.W. *Color Television Engineering*. New York: McGraw-Hill; 1955
Wright W.D. *The Measurement of Colour*. London: Hilger; 1964
Zworykin V.K., Morton G.A. *Television. The electronics of image transmission*. New York: Wiley; 1940

Index

Page references may include both text and figures, except for those in italics which refer to figures only.

Adamian, J. 15
Additional Reference Transmission (ART) 274–5
additive three-colour process 3–4, 15, 34–6, 91–4
additive two-colour process 11–15, 54, 84–7, 94–7
Admiralty Research Laboratory 20
advertising and sponsorship 67, 125, 190–1, 194
Air King Products Company 151
Alexandra Palace 63
  first regular high definition service (Nov 1936) 59
  cessation of services during WW2 83
  reopening of services 110
  colour transmission tests 210–11
Alvarez, L.W. 259
American Telephone and Telegraph Company (AT&T)
  exclusive right to manufacture, lease and sell transmitters 47
  investment in low definition television (1925–30) 31
  see also Bell Telephone Laboratories (BTL)
anagylphs 88
Anderson, E.O. *72*, 84
Anschutz, O. 6, 7
'apple' tube 190
Arago, François 2
Archer, F.S. 2
ART (Additional Reference Transmission) 274–5
Ashbridge, N. 60–1, 62
AT&T: *see* American Telephone and Telegraph Company (AT&T)
Atkins, I. 188

Baird, John Logie
  education and early career 20–2
  first experiments in television 22–3
  obtains patent for two-colour television (1925) 15
  first demonstration of television (1926) 23
  obtains financial support 24
  appoints business partner 24–5
  amasses a collection of patents 25, 27
  fails to interest BBC 27
  first demonstration of colour television (1928) 27–9

Baird, John Logie (*cont.*)
obtains patents for colour television 29
continues work on technical problems 29–30
demonstrates stereoscopic television 30
experimental television service transmitted by BBC (1929–32) 30
expresses views on cathode-ray tube receiver 43
weakening relations with Baird Television Ltd 71
reaction to discontinuance of BTL system by BBC 71
reverts to private research and development (1933) 71
concentrates on development of cinema television 71–83
demonstrates two-colour images at Dominion Theatre (1938) 75–7
formation of Cinema Television 78
demonstrates colour television using projection c.r.t. (1939) 82
liquidation of BTL 83–4
continues to work privately 84
demonstrates 600-line two-colour television (1940) 85
appointed technical advisor to Cable and Wireless 87
works on stereoscopic television 87–90
patents additive three-colour television system 92–4, 247
develops telechrome tube 94–6, 246
Baird International Television Ltd 20, 25
Baird Television Development Company 20, 25
Baird Television Ltd
demonstrations to BBC (1933) 59, 61, 62–3
discontinuance of BTL system by BBC (1937) 65, 71, 306
cinema television developments 74–83
formation of Cinema Television 78
goes into liquidation 83–4
Baker, W.R.G. 68, 154, 165
Baldwin, M.W. 171
bandwidth
6 MHz standard 65, 101, 120
8 MHz standard 214
requirement 55, 108
reducing 134, 171–2
Barnes, G. 198
BBC: *see* British Broadcasting Corporation (BBC)
Beatty, R.T. 43
Bedford, A.V. 171
Bedford, L.H. 200, 216
Belin, Edouard 19, 20, 47
Bell Telephone Laboratories (BTL)
first television experiments 20
early demonstrations of colour television 30
work on television in 1920s 31–8, 305
demonstration of colour television (1929) 34–6
work on television after 1930 31, 308
demonstration of cine film transmission 36–8
cathode-ray tube receiver 43
research on colour vision 171
cable networks 186

Bernath, K. 268
Berthon, R. 15
Berzelius 1
Birkinshaw, D.C. 188–90
Bishop, H. 199–200, 211
Blake, G.J. 233
Blumlein, A.D. 61, 223, 309
Bonham Carter, Sir Maurice 74
Bowie, R.M. 166, 182
'bracket standards' 144–5, 148
BREMA (British Radio & Electronic Equipment Manufacturers' Association) 201, 210, 286
British Broadcasting Corporation (BBC)
  black and white television
    lack of enthusiam for Baird's low definition system 27
    first low-definition experimental service (Sep 1929) 30
    demonstrations of rival systems to BBC (1933/34) 61, 62–3
    public television broadcasting policy issues 62–3
    first public television service (Nov 1936) 63
    cinema television demonstration (1937) 74–5
    attitude to cinema television 81
  camera tubes
    first use of super-emitron camera 223
    first use of c.p.s. emitron camera 227
    adoption of plumbicon camera tube 238
  colour tele-cines 213, 239
  colour television
    commences experiments (1949) 197–200, 209–13
    first broadcasts of compatible colour television (Oct 1954) 210–11
    begins full-scale experimental transmissions (Nov 1956) 212–13
    request to begin experimental service refused (1961) 216
    launches colour television service (Jul 1967) 290–1, 292
  colour television standardization
    supports NTSC standard 216–17
    supportive views on chromacoder 230
    contributes to EBU Ad-Hoc Group on Colour Television 268
    trials of alternative systems 275, 276, 306
  Television Centre 197, 200, 292, *293*
British Radio & Electronic Equipment Manufacturers' Association (BREMA) 201, 210, 286
Bronwell, A.B. 257
Bruch, W. 271–2, 273–4
BTL: *see* Bell Telephone Laboratories (BTL)

Cable and Wireless Ltd 87
cable networks 185–6
calotype process 2
camera tubes 221–41
  early experiments
    cathode potential stabilisation 223–4
    charge storage principle 51
    emitron 221–3

camera tubes (*cont.*)
early experiments (*cont.*)
iconoscope 51–2, 221–3
J.L. Baird's system (1939–) 84–6
mosaic elements 4, 51–2, 61
Zworykin's patent application 46
Zworykin's work in RCA (1930–) 51
ideal requirements 236–7
photoconductive materials 233–4, 235
prices 230
types
comparisons 228
emitron 221–3, 224, 226–7
four-tube 238
image orthicon 51, 230–3, 235, 236, 237
orthicon 224–6, 230
plumbicon 236, 237–8, 238
super-emitron 223
super-iconoscope 223
super-orthicon 224
two-tube 239
vidicon 234–6
Castelli, E. 268
cathode potential stabilised emitron 224, 226–7
cathode potential stabilization 223–4
cathode-ray television (early work) 38–9
French work late 1920s 47–8
take up by RCA 49, 51
V.K. Zworykin's work late 1920s 47, 49–51
cathode-ray tube camera: *see* camera tubes
cathode-ray tube display: *see* display tubes
celluloid roll film 6
charge storage property 51
Charles Urban Trading Company 12
Chevallier, P.E.L. 44, 47–8, 50–1
Chrétien, G.L. 1
chromacoder 227, 229
Chromatic Television Laboratories, Inc. (CTL) 258–61
chromatron 259–62
chromoscope 3, 258
chronophotography 6
Church, A.G. 71
cine film transmission
Bell Telephone Laboratories (1930) 36–8
Zworykin (1925–) 49–50
HMV demonstrations (1931/32) 59, 61
British Broadcasting Corporation (1958–) 213, 238–9
Cinema Television 78
J.L. Baird's system 71–83
public demonstrations at London cinemas 74–82
copyright and licensing issues 78, 80–1

projection equipment *77*, 78–9, 81
cinématographe 8
cinematography
early history 4–8
sequence photography 5–6
celluloid roll film introduced 6
first successful cine camera 6
electrotachyscope 6
toothed sprocket wheel 7
electrotachyscope 7
kinetoscope 7–8
cinématographe 8
see also colour cinematography
Cock, Gerald 65, 105
Color Television Incorporated (CTI) 94, 128–30, 138–43, 232
colour cinematography
early history 9–17
first public cinema show 12
Kinemacolor 12–13
colour fringing problem 13–14
subtractive method adopted 15
Technicolor 14–15
colour conversion systems 227, 229
colour photography 2–4, 34
colour standards: *see* standardization
Columbia Broadcasting System (CBS)
broadcasting studios 186
colour television
announces plans 68
frame sequential system 101–4
first broadcast (Aug 1940) 101
field sequential system 104–5, 127–8, *129*, 137–44
use of orthicon camera 223
first public showing of direct pick-up system (Jan 1941) 105–6
begins broadcasting on an experimental basis (Oct 1945) 112–14
commences public broadcasting of colour (1951) 149, 151, 186, 190
acquires manufacturing facilities for receivers 151–2
ordered to stop manufacture of sets for defense reasons (Nov 1951) 152–3
advances chromacoder 227
withdraws from the television scene 191
colour television standardization
presses for utilization of UHF band 111–12
representations to FCC for adoption of CBS system (1946) 116–17, 118–19, 120
demonstrations to FCC 1949–50 Hearings 127–8
system adopted by FCC (Oct 1950) 149
reaction to RCA demonstrations (summer 1951) 154–5, 156
demonstration to FCC (Apr 1953) 158
Columbia Graphophone Company 59
Comité Consultatif International des Radiocommunications (CCIR) 267
Sub-Group on Colour Television 276, 277, 282, 288–9, 290, 297–9
Committee on Broadcasting in the United Kingdom 1960: *see* Pilkington Report

Committee on Interstate and Foreign Commerce 125, 155–8
compatibility of colour with black and white 118, 131, 144, 148–9, 159, 201
Condliffe, G. 198
Condon, E.U. 125
Conrad, Dr F. 46, 49
conversion systems 227, 229, 230
Craven, T.A.M. 65
Crofts, W.C. 5
Crystal Palace 212, 275
CTI (Color Television Incorporated) 94, 128–30, 138–43, 232
CTL (Chromatic Television Laboratories, Inc.) 258–61

Daguerre, L.K.M. 2
daguerreotype 2
Darling, A. 11
Dauvillier, A. 44, 47
Davis, H.P. 46
Davisson, C.J. 308
Davy, Humphry 1
Day, W.E.L. 24
DeCola, R. 182
Dickson, W.K.L. 7
display tubes 243–65
  early experiments
    Bell Telephone Laboratories 43
    methods of focusing the electron beam 43–4
    Zworykin's kinescope (1929) 49–51, 54
  accurate beam-scanning method 243
  adjacent image method 246
  alternative techniques 243–4
  beam control at phosphor screen for changing colour method 256–8
  beam-scanning position method 243–6
  direction-sensitive using shadowing method 247–52
  multiple colour phosphor screen method 246–7
  Philco 'apple' tube 190
  shadow mask tube 190, 249–56
distant vision 19, 44
Dominion Theatre demonstrations 74, 75–7
Donisthorpe, Wordsworth 4–5
dot-interlaced scanning 128, *129*, 134–5
double-sideband operation 112
dry-plate cameras 5
du Hauron, L. and A.D. 3–4
du Mont, Allen B 154
DuMont Laboratories 106, 108, 180, 190

Eastman, George 6
EBU: *see* European Broadcasting Union (EBU)
Edison, Thomas Alvar 7
Electric and Musical Industries Ltd (EMI)
  electronic television development in 1930s 59–65

demonstrations to BBC (1953) 198
colour convertor 227, 229, 230
emitron 221–3, 224, 226–7
see also Marconi-EMI (M-EMI) Television Company Ltd
electronic camera tubes: *see* camera tubes
electronic display tubes: *see* display tubes
electronic scanning: *see* cathode-ray television (early work)
electrostatic focusing 44
electrotachyscope 6, 7
EMI: *see* Electric and Musical Industries Ltd (EMI)
emitron 221–3, 224, 226–7
Empire State Building experimental station 55, 65–6, 154
Engstrom, E.W.
evidence to Committee on Interstate and Foreign Commerce 155
RCA's experimental system 54, 112, 115
work for NTSC 108, 166, 179
Epstein, D.W. 243–4
Essig, S. 52
European Broadcasting Union (EBU) 267
Ad-Hoc Group on Colour Television 268, 275–6, 282–5, 295–7
Eurovision 279
European VHF/UHF Broadcasting Conference (1961) 214
Eurovision 279
Evans, R.M. 2–3

FAM system 271
Farnsworth, P.T. 38–9, 44, 56, 306
Federal Communications Commission (FCC)
early reluctance to set standards 65, 66–8
establishes nineteen 6 MHz television channels 65, 67
1941 Hearings on NSTC report 106, 108–9
1944–5 Hearings 111
new rules on television standards adopted (Apr 1941) 109
approves 480–920 MHz band for experimental television (May 1945) 112
1946–7 Hearings 118–19
denies CBS petition 119
1949–50 Hearings 126–45
First Report on Color Television (Sep 1950) 145
Second Report on Color Television (Oct 1950) 148–9
Order confirming CBS standard (Oct 1950) 149
approves NTSC's signal specifications (Dec 1953) 159, 178
Fernseh A.G. 20
field-sequential system
advantages and disadvantages 105, 117–18
CBS system 103–5, 127–8, *129*, 137–44
conversion to simultaneous signals 227
J.L. Baird's systems 28–9, 73–4
RCA's system 53–4, 55
Figuier 4
Fink, D.G. 114, 126, 181, 204
Flechsig, W. 249–52

Fly, J.L. 67, 68, 101
Forgue, S.V. 235
Fox Talbot 2
France
  support for SECAM 282, 289
  work on cathode-ray television in late 1920s 47–8
frequency allocation
  Europe 214–16
  USA 65, 111–12, 126
frequency interleaving 172–4
Friese-Green, William 7

Gaumont-British Picture Corporation (G-BPC) 71, 74–5, 78, 83
Geer, C.W. 247–9
General Electric Company (GE)
  first television experiments 20
  research carried out on behalf of RCA 47
  right to manufacture receivers 47
  colour television system 107
  post-acceleration tube 190
  two sub-carrier system 201
General Post Office 61, 63
George, R.H. 44, 50–1
German, Sir Ronald 289
Germany 20
Goldmark, P.C. 101, 112, 151, 230
Goldsmith, A.N. 170, 181, 249–50
Goldsmith, T.T. 166, 182
Goodrich, R.R. 235
Goussot, L. 268
Gramophone Company 20, 59
Gray, F. 172, 173, 307, 308
Greene Sir Hugh 291
Greer, Sir Harry 82
Grosser, W. 44

Hackbridge and Hewittic Rectifier Company 94
Hankey Committee 110–11
Harnett, D.E. 181
Haverline, C. 92
Hazeltine Electronic Corporation 171, 175, 179
Hefele, J.R. 308
heliographic copies 1
Henri de France system 271
Herold, Edward W. 55, 243
Hirsch, C.J. 171–2
HMV 59
Hofstetter, M. 34
Holland brothers 8
Holmes, M.C. 86
Holoch, G. 274
Holweck, F. 44, 47–8
Hopt, H. 268
Hutchinson, O.G. 24–5, 27, 30
Hyde, R.H. 178–9
Hytron Radio and Electronics Corporation 151

Iams, H. 52, 224, 234
iconoscope 46, 51–2, *52*, 221–2
  adaptation to colour television 54
image orthicon 51, 230–3, 235, 236, 237
Independent Television Authority (ITA) 215, 275, 292
interleaving principle 172–4, 308
International Broadcasting and Television Organization (IBTO) 267
International Radio Consultative Committee (CCIR) 212, 214
International Telecommunication Union (ITU) 267
ITA (Independent Television Authority) 215, 275, 292
Ives, H.E. 31–4, 36–8, 50, 305–6, 306–8

Japan 194–5
Jenkins, C.F. 19, 20
Jensen, A.G. 166
Jesty, L.C. 200
Jewett, F.B. 34, 305, 307
Johns Hopkins University 34
Johnson, Arthur 94
Johnson, E.C. 125
Joliffe, C.B. 119
Joly, Professor 4
Joy, H.W. 14

Kaar, I.J. 166
Kell, R.D. 227
Keller-Dorian, A. 15
Kelly, W. van D. 14
Kinakrom 14
Kinemacolor 12–13
kinescope 49–50, 54, 244–6
  adaptation as a camera tube 51–4
  grid-controlled 258, *259*
  shadow mask 254–6
kinetoscope 7–8

Kintner, S.M. 45, 46, 49
Kinzie, M.E. 165
Kodacolor process 15–17
Kodak Company 15, 16–17

Laboratoire des Établissements Edouard Belin 44, 47–8
Langer, N. 22
Laroche 2
Law, H.B. 230, 231, 252, 254
Lawrence, E.O. 258
Lawrence tube 190, 259–62
Lee, A.G. 76–7
Lee and Turner process 9–11
Le Prince, Louis-Aimé-Augustin 7
lighting, studio 186, 188, 210, 238, 292
line standards 211–12, 214–17, 287, 292
Loughlin, B.D. 273
Loughren, A.V. 166, 171–2, 181
low definition television 19–41
Lubszynski, H.B. 222, 224
Lumière, Auguste and Louis 8
luminance/chrominance systems 201–4
Lynes, B.J. 72, 73

magazine dry-plate cameras 5
magic lanterns 5–6
manufacture
  receiving equipment
    annual production (1952) 155
    first commercial CBS type 151
  transmitters 47
Marble Arch Pavilion 81–2
Marconi-EMI (M-EMI) Television Company Ltd 59
  demonstrations to BBC (1932–34) 61, 62–3
Marconi Wireless Telegraph Company
  first television experiments 20
  patents relating to charge storage principle 51
  post-war developments 199
  post-war encouragement by BBC 199, 200
  adaptation of NTSC standard 204–9
  two-tube colour television camera 205, 239
Marey, E.J. 6–7
Marx, F. 182
Maurice, R.D.A. 268
Maxwell, James Clerk 2, 9
Mayer, N. 271, 274
McGee, J.D. 61, 222, 223, 227
McIlwain, K. 182
McLean, F.C. 178, 200, 268
McMillan, E.N. 259
mechanically scanned television systems
  BTL (1927) 31–3
  CBS (1940–) 101–5, 112–14
  J.L. Baird (1928) 28–9
  J.L. Baird (1937) 73–4, 75–6
  RCAs two-colour system (1933) 54
Mertz, P. 173, 308
Middleton, W.E.K. 86
Miller, H. 233
mirror drum scanners 73–4
mixed-highs principle 133–4
mosaic photosensitive screens
  colour photography 3–4
  colour television 4, 51–2, 61
Moseley, S.A. 30, 71
motion pictures: *see* cinematography
multiple-layer screens *243*, 246–7
Muybridge, E.J. 5–6

National Broadcasting Company (NBC)
  commences testing of sequentially scanned all-electronic system (1941) 55
  starts broadcasting of colour programmes (1952) 154, 159, 185, 186, 190, 192
  begins the first delayed broadcast in colour (1957)
  studios 186
National Television Standards Committee (NTSC) 68
  colour television standards 106, 108–9, 156–7
    British adaptation 200–4, 206, 209
    consideration by the EBU 276–8, 283–6
    revised version (1953) 165–84
    supported by Television Advisory Committee 278–9, 280
    variants 268–75
  membership in June 1951 180–1
Natural Color Kinematograph Company (NCKC) 12
NBC: *see* National Broadcasting Company (NBC)
NCKC: *see* Natural Color Kinematograph Company (NCKC)
Newnes, Sir George 5

Niepce, Joseph Nicéphone 1–2
Nipkow discs 23, *26*, 28, 32
Noble, D.E. 182
NTSC: *see* National Television Standards Committee (NTSC)

Ogloblinsky, G.N. 47, 49–51
Olpin, A.T. 36
orthicon 224–6
Ostrer, Isidore 71, 75, 77, 78, 82
Ostrer, Mark 75
outside broadcasting 221, 223
  UK 213, 223, 227, 238, 292, *294*
  USA 178, 185
Oxbrow, W. 84

Paget plates 4
Palais-de-Luxe cinema, Bromley 74–5
PAL system 197, 272–3, 280, 283–7, 288
Paul, Robert W. 8
persistence of vision 7, 136
Philco 175, 178, 180, 190
Philips Laboratories
  examination of chromatron tubes 261–2
  plumbicon 236, 238
  two sub-carrier system 201, 269
photoconductive materials 233–4, 235
photoelectric cells 19, 29–30, 32, 36
photography
  colour 2–4, 34
  early history 1–4
  sequence photography 5–6
photosensitive materials 1–3, 51
photosensitive mosaic screens
  colour photography 3–4
  colour television 4, 51–2, 61
Pilkington Report 214–16
plumbicon 236, 237–8
post-acceleration tube 190
Postmaster General (PMG)
  establishes the Television Committee (1934) 63
  position on cinema television 78
  colour television standards 278–9, 287
Powers Photo Engraving Company 34
Prizma process 14
programme schedules
  UK 212, 290–1, 295
  USA 151, 190, 192, 194
projection equipment *77*, 78–9, 81, 82
Pye Radio Ltd 101
  post-war developments 198, 199, 216

Radio (and Television) Manufacturers Association (RMA)
standardization issues 65–6, 66–7, 68, 117, 147
Radio Corporation of America (RCA) 191
black and white television
first television experiments 47
cathode-ray display tube – kinescope 49–50
purchases the rights of George's and Chevallier's patents 51
V.K. Zworykin's group moves to RCA (1930) 51
cathode-ray camera tube – iconoscope 51–2
transmission field tests in the 43–80 MHz band 53
second experimental television system 240 lines 53–4
introduction of public television services (1939) 66
starts marketing of television receiving sets (1939) 66
ceases television operations during war 109
camera tube research
iconoscope 221–2
image orthicon 230–3, 237
orthicon 224–6, 230
vidicon 234–6
colour receiver sales 192
colour television
early work 54–5
two-colour mechanically scanned television system (1933) 54
demonstrates to FCC three-colour, additive, sequentially scanned system (1940) 55, 107–8
starts work on all-electronic system (1940) 55
commences testing of sequentially scanned 343- and 441- line, all-electronic system (1941) 55
demonstrates three-colour television using colour filter drum (1945) 114–15
demonstrates three-colour, simultaneous all-electronic system (1946) 115–16
demonstrates system to 1949–50 FCC hearings 131–6, *137*
FCC views on system 138–43, 148
public demonstrations (1951–52) 153–4
demonstrates compatible colour television (Apr/May 1953) 157–8
promotes colour television transmission 185
colour television standardization
opposition to standards based on mechanical systems 119, 121
opposition to FCC order Sep 1950 based on CBS system 147–8, 149–51
petitions FCC to adopt its colour television standards (Jun 1953) 158
FCC approves signal specification to RCA's system (Dec 1953) 159
contribution to NTSC standard 179–80
credit for the success of the NTSC system 303, 304
controlling interest in the American Marconi company 304
display screens research 243–6, 249–54, 258
expenditure on colour television 191
photoconductivity research 234
Radio Manufacturers Association (RMA)
standardization issues 65–6, 66–7, 68, 117, 147
Radio Technical Planning Board 117
Raibourn, Paul 201, 217
receiving equipment
installation and servicing 191–2

receiving equipment (*cont.*)
manufacture
annual production (1952) 155
first commercial CBS type 151
prices 152, 190, 192, 193, 291–2
sales 67, 190–1, 192, 193–4, 288
see also display tubes
recording equipment 191
Redmond, Sir James 303
Reith, Lord 62–3, 81
Reveley, P.V. 72, 73
RMA: *see* Radio Manufacturers Association (RMA)
Rodda, S. 222
Rogowski, W. 44
Rose, A. 224, 230, 234
Rosenthal, A.H. 262–3
Rosing, B. 44–5
Round, H.J. 51
Rowson, S. 74
Royal Institution 23
Russell, Alexander 28–9

Sala, Angelo 1
Sarnoff, David 47, 49, 66, 149, 303–4, 306
scanning, mechanical: *see* mechanically scanned television systems
scanning sequence 127–8, 131, *132*, 134–5
Schairer, O.S. 45, 46
Schoultz, E.G. 233
Schroeder, A.C. 252, *253*
Schuster, L.F. 90
Scophony Company 74, 78–9, 81, 262–3
SECAM-NTSC system 271–2
SECAM system 197, 271, 280–7, 288–9
Seguin, L. and A. 233
Selsdon, Lord 77
Selsdon Committee 59, 63, 224
Senefelder, Aloys 1
sequence photography 5–6
shadow mask tube 190, 249–56
Shelby, R.E. 182
Shoenberg, Isaac 60, 61, 216, 230, 306
sideband transmission 65, 112, 172–3
Siemens and Halske Company 7
silver screens 82
simultaneous colour system
conversion from field-sequential signals 227
J.L. Baird's system 91–4
RCA's system 115–16, 117–18, 131, 172
Thomascolour 91
Skiatron tube 262–3

Skladnowsky, Max and Emil 8
Sleeper, G.E. 128
Smith, D.B. 166, 182
Smith, G.A. 11
Société Keller-Dorian-Berthon 15
sponsorship and advertising 67, 125, 190–1, 194
Spooner, H.J. 233
Stamford, L. 5–6
standardization
  black and white television
    UK 65, 111
    USA 65–8
  'bracket standards' 144–5
  colour television
    alternative parameters 103
    compatibility with black and white 118, 131, 144, 148–9, 159, 201
    Europe 267–301
    UK 277–80, 286–8
    USA: *see* Federal Communications Commission; National Television Standards Committee (NTSC)
  conversion systems 227, 229, 230, 281
  line standards 211–12, 214–17, 287, 292
Stanley, C.O. 216
Stanton, F. 154–5, 156
stereoscopic television 30, 87–90, 115
Stilwell, G.R. 36
Strange, J.W. 233
studios
  floor space 198
  lighting 186, 188, 210, 238, 292
  UK 210, 213, 292
  USA 186, 188
subtractive colour film processes 15, 92
subtractive colour television 262–3
super-emitron 223
super-iconoscope 223
super-orthicon 224
Swinton, Campbell A. A. 38, 233
Sylvania 174

TAC: *see* Television Advisory Committee (TAC)
Tatler Theatre 78–9, 81–2
Teal, G.K. 307
'teapot' tube 81, 90
Technicolor 14–15
Tedham, W. 61
telechrome tubes 94–5
Telefunken 20
Television Advisory Committee (TAC)
  formation 59, 63
  requested to investigate colour television standards 200

Television Advisory Committee (*cont.*)
considers 405 vs 625 line standard 212, 214
supports NTSC standard 278, 279–80
recommends introduction of regular colour television service 278
considers implications of different standards in Europe 286–7
recommends adoption of PAL system 288
Television Committee (1943) 110–11
Television Ltd 20
Theile, R. 268
Thomas, Richard 91
Thomascolour 90–2
three-colour process 2, 3–4, 15, 34–6, 91–4
transmission standards: *see* standardization
transmission systems
RCA's field tests in the 43–80 MHz band 53
M-EMI's and BTL's equipment at Alexandra Palace (1935/6) 63
RCA's experimental transmissions from the Empire State Building (1941) 55
CBS experimental colour broadcasts (1945) 112
USA networks (1950s) 185–6, *187*
experimental colour transmissions from Alexandra Palace (1954) 209–10
Turner, E.R. 9–11
two-colour process 11–15, 54, 84–7, 94–7
two sub-carrier system (TSC) 201, 268–71

Urban, Charles 11–13
Urban–Joy Process 14
US House of Representatives Committee on Interstate and Foreign Commerce 125, 155–8

Valensi, G. 47
vestigial sideband transmission 65, 112
video recorders 191
vidicon 234–5
vision persistence 7, 136
Vladimir K Zworykin Award 308
Vogel, H.W. 3
von Mihaly, D 19, 20

Walker, P.A. 156–7
Walton, G.W. 74
Warwick Trading Company 11
Wedgwood, Thomas 1
Weimar, P.K. 230, 234
West, A.G.D. 71
Westinghouse Electric and Manufacturing Company (WEM)
first television experiments 20
engagement of V.K. Zworykin 45–6
broadcasting of radio movies 46, 50
KDKA transmitter station 46, 50
research carried out on behalf of RCA 47
right to manufacture receivers 47
work on cathode-ray television in late 1920s 49–50
wet collodion process 2
White, M.G. 259
White City Television Centre 197, 200
Willmer, E.N. 86
Wilson, E.D. 46
Wilson, J.C. 72
Wolverton, C.A. 155, 158
Wood, R.W. 34
Wood, Sir Kingsley 62–3
Wright, W.D. 86

zoetropes 6
zoogyroscope/zoopraxiscope 6
Zworykin, Vladimir Kosma
education and early career 44–6
engagement by Westinghouse 45–6
applies for a patent for electronic camera tube 46
early work on cathode-ray television (1924/5) 46, *48*
advances use of mosiac screen in colour television 4
works on recording and reproduction of sound on cine film 46
visits Europe (1928) 47–8
reports back to RCA on European visit 49
develops cathode-ray display tube – kinescope 49–51
applies for patent for kinescope 50–1
moves to RCA (1930) 51
develops electronic camera tube – iconoscope 51–2
recalling Sarnoff 305
Vladimir K Zworykin Award 308

P9-BTY-606

Created and Directed by Hans Höfer

INSIGHT GUIDES
PROVENCE

Edited by Anne Sanders Roston
Photography by Catherine Karnow
Editorial Director: Brian Bell

HOUGHTON MIFFLIN COMPANY

APA PUBLICATIONS

PROVENCE

Second Edition (Reprint)

Printed in Singapore by Höfer Press Pte. Ltd

Distributed in the United States by:
Houghton Mifflin Company
222 Berkeley Street
Boston, Massachusetts 02116-3764
ISBN: 0-395-66166-8

Distributed in Canada by:
Thomas Allen & Son
390 Steelcase Road East
Markham, Ontario L3R 1G2
ISBN: 0-395-66166-8

Distributed in the UK & Ireland by:
GeoUK GeoCenter International UK Ltd
The Viables Center, Harrow Way
Basingstoke, Hampshire RG22 4BJ
ISBN: 9-62421-087-X

Worldwide distribution enquiries:
Höfer Communications Pte Ltd
38 Joo Koon Road
Singapore 2262
ISBN: 9-62421-087-X

About This Book

It seems strange that so few guidebooks have been written on Provence. Once an independent nation, this southeastern part of France possesses volumes worth of history and folklore. And the Provence of today, with its sensuous appeal, has become, thanks partly to the efforts of Peter (*A Year in Provence*) Mayle, one of Europe's most popular destinations.

Anne Roston, this book's project editor and main contributor, first encountered Provence on typical teenage excursions to Avignon, Aix and the Côte d'Azur, and her knowledge deepened after a visit to a friend's home deep in the Var. A US citizen who has lived on and off in Paris and Helsinki, Roston was the project editor for *Insight Guide: France*. Outside France, her non-fiction work has been as disparate as covering sports in Arizona and the politico-economic scene in Malaysia and her credits include the *New York Times*, the *Washington Post* and *Harper's* magazine, but she now spends the greater part of the year in New York.

The Team

Photographer **Catherine Karnow** was born and raised in Hong Kong and is now based in Washington, DC. France has been a mainstay of her work, and she has held jobs with *Paris Match*, *Le Point* and Magnum Photo's Paris office. Karnow's interests also extend to the cinema – her film short, *Brooklyn Bridge*, opened the Berlin Film Festival in 1984. She has photographed several Insight Guides, including those to France, Los Angeles and Washington DC.

The chapters on the Vaucluse and the Camargue were written by **Peter Robinson**. A British citizen, Robinson has lived for several years near Grasse, while producing educational television documentaries on the Camargue and Avignon. He also broadcasts on Radio Riviera and has been working on a study of Graham Greene.

Subsequent chapters on the Alpes-de-Haute-Provence and the Var were written by another Briton, **Caroline Wheal**. The years Wheal spent, after graduating from Oxford in French, as a translator of scripts for the BBC and French television were punctuated by periods of residence in her favourite part of France, the Var. She is now a London-based freelance journalist, and has edited Virgin Atlantic's in-flight magazine.

Peter Capella, who covered the Alpes-Maritimes, has been a committed Francophile since spending his teenage years in Toulouse. He headed to Provence after university to live in Toulon, where he worked as a radio producer. He is now a reporter in Berne for Swiss Radio International.

Joel Stratte-McClure was "forced" by never-endingly grey days in Paris to move his Franco-American wife and children down to the sunny Côte. From his nest in the hills behind Cannes, Stratte McClure writes for such publications as *Time*, the *International Herald Tribune* and *People*. He is also publisher of *Sophialet*, a newsletter about the Rivieran science park Sophia Antipolis.

For someone like **Caroline Roston**, a folklorist from the University of Pennsylvania who specialises in material culture, Provence is a wonderland. Formerly the assistant to Henry Glassie, Roston, author of the chapter on Folk Art, has written a thesis on agricultural fairs in Massachusetts.

The feature on Cults was written by **Ward**

A. Roston

Karnow

Robinson

Rutherford, a French-educated Jerseyman who divides his time between Brighton and Southern France. Rutherford has written several studies on Druidism, the ancient Celtic religion of which he claims to have found relics in Provence. To his credit, also, are 16 books ranging in topic from crime novels to *The Ally*, a work on World War I.

David Ward-Perkins lives in the hilltop village of St-Jeannet just north of the Côte d'Azur, from where he cooks Provençal dishes for his wife and four children. A founding editor of *New Riviera* magazine, he was also the editor of *Seven Adventures on the French Riviera*.

The sidebars to the Places chapters were some of the liveliest additions. This was partly due to Paris-born, Franco-American **Mary Deschamps**, who wrote the pieces on Christian Lacroix and the bikini. Deschamps is a freelance writer in Paris, focusing mostly on fashion and international lifestyles. She edited *A Touch of Paris* and currently contributes regularly to the *International Herald Tribune* and *Vogue Hommes*.

Claire Touchard's family, which has been in the art world for several generations, bought a home in the Vaucluse during the 1950s, making her a natural for the piece on Artist Colonies. She now works in Paris as a freelance editor and translator.

Although short, **William Fisher's** article on the Cannes film festival is unlikely to go unnoticed. Fisher, the Paris correspondent for *Screen International*, has any number of different projects up his sleeve – including a full-blown book on the festival experience and a "travel" book on Eastern Europe.

On **Rosemary Bailey's** first visit to Provence, she stayed with friends on a rose farm and was immediately fascinated by the flower business. Based in London, she writes for major British publications and has edited Insight Guides to Tuscany, the Loire Valley, Burgundy and the Côte d'Azur.

After securing the text, Roston turned her attention to the illustrative portion of the book. Material from the **Bibliothèque Nationale** in Paris, the **National Gallery of Art**, the **Academy of Motion Pictures**, **Ruth Aebi** of Gstaad, **Gavin Lewis**, and **Jantzen, Inc**. complemented Karnow's photography.

Finishing Touches

A myriad of other people offered invaluable assistance throughout the project. **Ned Walker** and his assistant, **Susan Hirsch**, of Continental Airlines provided Roston and Karnow with seats on their convenient flights to Paris. They also had the pleasure of staying in the ultra-luxurious Sofitels in Marseille, Nice and Cannes, arranged by **Liza Fierro** of Accor in New York. Karnow's rigorous shooting schedule was eased by Hertz rental cars on loan from **Naomi Graham** in Hertz's London office.

This updated edition, checked out on the spot by **Mark Fincham**, was supervised in Insight Guides' London editorial office by managing editor **Dorothy Stannard**.

Wheal

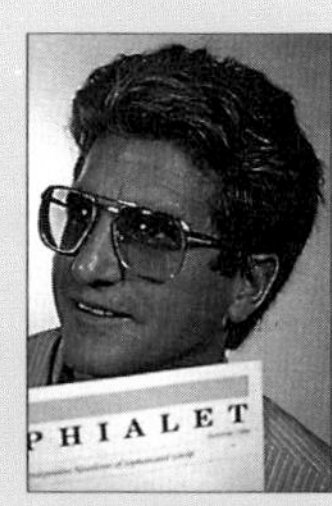

Stratte-McClure

C. Roston

Ward-Perkins

9F
le Kg.

History & People

—by Anne Roston

23 **Bienvenue**

27 **The Land**

32 **Prehistoric Farmers**

35 **The Arrival of Civilisation**

39 **Of Popes and Love Courts**

41 **The Language of the Troubadours**

46 **Unification**

52 **Through an Impressionist's Eye**

56 **The Côte d'Azur Swings into Action**

59 **Coping with Modernisation**

65 **Les Provençals**

Features

71 **Cults**
—by Ward Rutherford

77 **The Cuisine**
—by David Ward-Perkins

85 **Folk Art**
—by Caroline Roston

Places

97 Introduction
—by Anne Roston

103 The Vaucluse
—by Peter Robinson

133 *Artist Colonies*
—by Claire Touchard

139 The Bouches-du-Rhône
—by Anne Roston

144 *The Underworld and Marseille*
—by Peter Capella

166 *Christian Lacroix*
—by Mary Deschamps

181 The Camargue
—by Peter Robinson

193 The Alpes-de-Haute-Provence
—by Caroline Wheal

215 The Var
—by Caroline Wheal

235 The Alpes-Maritimes
—by Peter Capella

254 *The Perfumes of Provence*
—by Rosemary Bailey

259 The Côte d'Azur
—by Joel Stratte-McClure

272 ***Inside the Cannes Film Fesival***
—by William Fisher

278 ***A Short History of the Bikini***
—by Mary Deschamps

Maps

98 All Provence
102 The Vaucluse
106 Avignon
138 The Bouches-du-Rhône
150 Aix-en-Provence
163 Arles
182 The Camargue
192 The Alpes-de-Haute-Provence
234 The Alpes-Maritimes
192 The Côte d'Azur

Travel Tips

282 *Getting There*

283 *Travel Essentials*

285 *Getting Acquainted*

289 *Communications*

290 *Emergencies*

291 *Getting Around*

293 *Where to Stay*

298 *Food Digest*

303 *Things to Do*

304 *Culture Plus*

308 *Nightlife*

309 *Shopping*

313 *Sports*

317 *Special Information*

319 *Photography*

319 *Language*

321 *Useful Addresses*

323 *Further Reading*

For detailed information see page 281

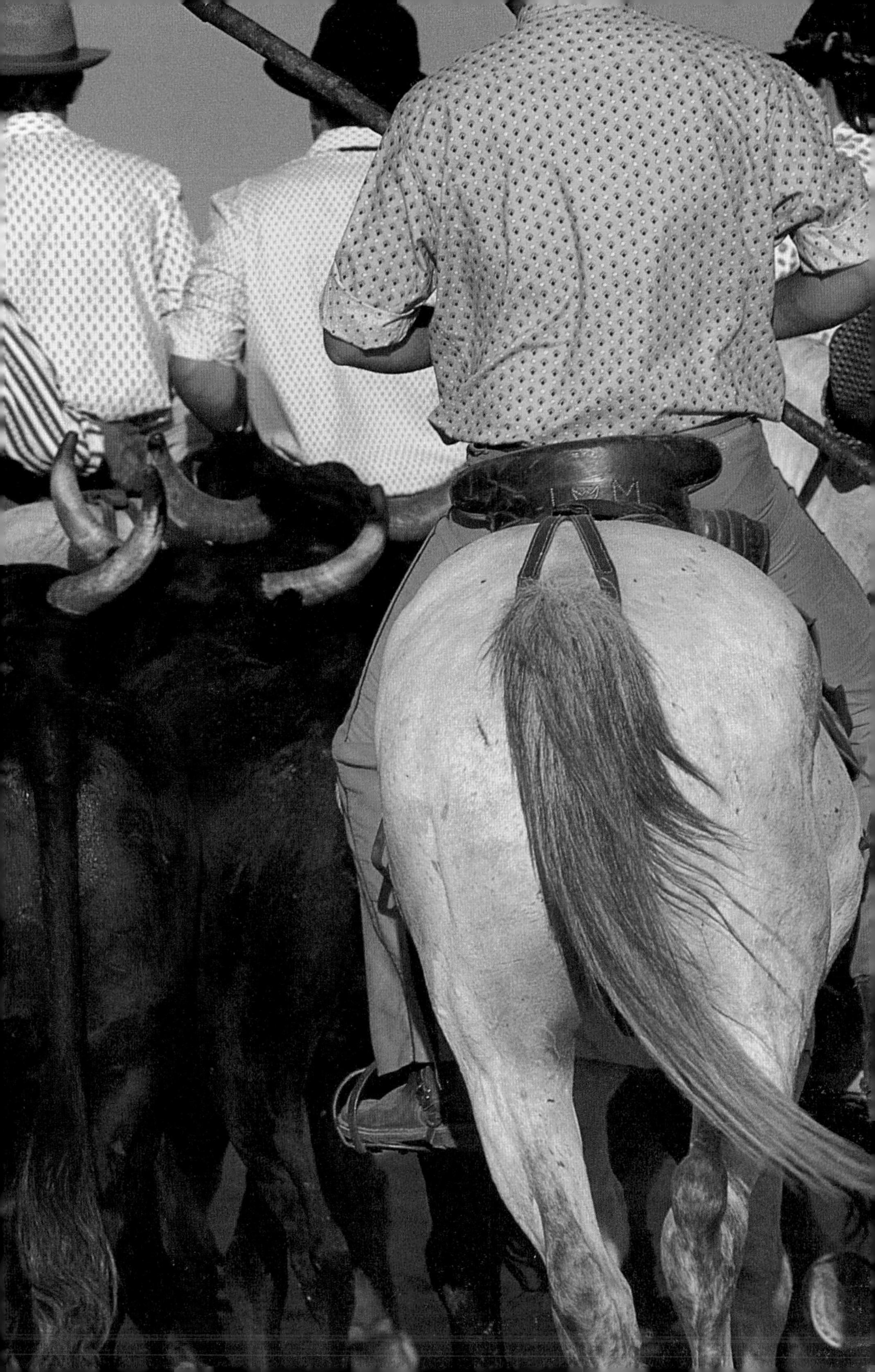

MA 433394
MA706667

DES CALANQUES
MA 596483
A 308 197
MA 481205

Bienvenue

Located in the sunny southeastern corner of France, Provence has a history is as rich as its soil. Prehistoric Ligurians, classical Greeks and Augustan Romans all left their marks here. Later tyrants included the bloody Saracens and the medieval lords of Baux, who founded the infamous "Courts of Love". It wasn't until the 15th century that the Gauls claimed Provence for their own.

Mary Magdalene is said to have spent her last years in the Var, and several popes made Avignon their home during the 14th century. Impressionist painter Paul Cézanne was born and worked in the Bouches-du-Rhône, where Vincent Van Gogh was driven to madness. More artists arrived from the turn of the 20th century onwards, including Dufy, Picasso, Matisse and Chagall. Writers such as F. Scott Fitzgerald and Graham Greene flocked to the Côte d'Azur, which Queen Victoria and international millionaires had already made famous.

What attracted all these different people? It could be the dazzling light that bathes this intoxicating land of black cypress and crooked olive trees. Perhaps it's the sweet-smelling lavender that sprawls endlessly across its fields and hillsides, and the mouth-watering melons and tomatoes its sun-rich soil produces. Or it might be Provence's strategic position between the Mediterranean Sea, the great River Rhône, the Alps and the Italian border.

Thanks to the strong and independent character of the warm-hearted Provençals, the area hasn't become just a graveyard for invaders' memorabilia. Provence's five *départements* and two subregions each have vivid personalities of their own. Even as children increasingly quit the tiny villages where their grandparents herded sheep, mined ochre or grew vegetables to try their luck in Marseille, Nice and Avignon, they bring with them a deep-rooted respect for the land and its traditions.

Preceding pages: Abbey St-Michel-de-Frigolet; garlic; shuttered lives; lavender fields on the Valensole Plain; gathering bulls during a feria du cheval; harbour in Cassis; detail from St-Gilles church. Right, café patrons.

THE LAND

Originally, the entire southeastern corner of France was covered by the sea. Then, 200 million years ago, a rock continent called Tyrrhenia began to emerge in ever-increasing strata of limestone, shale and clay. It was followed – in another 150 million years – by a second land mass. This area was just slightly further "inland" and would become what is today called Provence.

Submarine forces steadily pressed the Provençal land mass up from beneath the sea and into the high folds running east-west above the modern coastline. It was in this way that Mounts Ste-Victoire, Ste-Baume and Ventoux, the Alpille Ridges, and the Luberon and Baronnies ranges were formed.

This new earth continued to develop side by side with the older continent of Tyrrhenia, until the latter dropped back down into the sea about 2 million years ago. Of Tyrrhenia, all that is now left peaking out above the water is Cap Canaille, the Esterel Massif and several Mediterranean islands such as Sicily and Corsica.

Change of direction: Provence has remained geographically more or less the same since then, although, of course, certain natural changes continued to occur and, indeed, are still slowly taking place. The Rhône Valley began to hollow out, and the Rhône itself was rerouted from its original seaward path, via the Durance east of the Alpilles and down through the Crau Plain, to its current position directly south to the sea from Avignon. Over the centuries, under the gradual process of erosion, some of the oldest coastline reverted back to being submarine, and a little more land is reclaimed by the Mediterranean yearly. Today, Provence officially consists of five different *départements*: the Vaucluse, the Bouches-du-Rhône, the Var, the Alpes-de-Haute-Provence and the Alpes-Maritimes. Its more-or-less natural borders are the Rhône to the west, the Mediterranean to the south, Italy to the east and the Baronnies and Alpine mountain ranges to the north.

Popularly, however, Provence is less clearly defined, due to the extensive variety it contains. Inhabitants of the western, more traditional areas like the Vaucluse and the Bouches-du-Rhône frequently exclude the Alpes-Maritimes when discussing eastern Provence, and almost no one would describe the Côte d'Azur as Provençal.

Preceding pages: vineyard village of Gigondas. Left, cherries; and right, the Grand Canyon de Verdon.

The Bouches-du-Rhône is generally considered to be the oldest section of Provence, since the earliest civilised settlements were centred on its shores. Set in the lower half of the western part of the region, this *département* is marked by the Rhône to the west, the Durance to the north and the sea to the south, with a less articulated border by the Massif de Ste-Baume to the east. It is in itself so chock-full of variations that it can be taken as a microcosm for the rest of Provence.

The western part of the Bouches-du-Rhône is characterised by an abundance of plains filled with alluvial deposits. To the north are subcomtadine plains cut off from the rest of the *département* by the ragged Alpilles chain. To the south lies the Crau, a vast field of smooth pebbles polished by the river that once ran through it. South of that, along the coastline and to the extreme west, are the wet plains of the Rhône Delta, commonly known as the Camargue. So unique from the rest of Provence is this latter's sea-level and marshy land that it is often referred to as a region in its own right.

The eastern section of the Bouches-du-Rhône is more mountainous, alternating with synclinal basins. The Ste-Victoire Mountain Range is separated from the Etoile chain by the Arc Valley. Beneath them, the Huveaune Valley and Marseille Basin lean up against the Ste-Baume Massif and the Marseillevey Range.

Most of this area's steep cliffs are composed of limestone, and some peaks reach up to over 1,000 metres. Along the coast, these mountains create narrow valleys penetrated by the sea and called *calanques*.

Overall, the *département* is dry and rugged, with vegetation ranging from cypress trees to Aleppo and Norway pines and low shrubs to heathland; 256,000 acres (103,600 hectares) are under forest.

A combination of abundant sun and mineral-rich soil have helped to make this *département* France's top producer of fruit and vegetables. Although neither olive nor almond trees are indigenous – the first was brought from Greece, the latter from the Orient – they dominate large portions of the 445,000 acres (180,000 hectares) of cultivated land, as do wheat fields. Market goods include garlic, tomatoes, courgettes (zucchini) and asparagus.

The Vaucluse bears a certain resemblance to the Bouches-du-Rhône. It also lies along the Rhône to the west and is bordered on one side (the south) by the Durance. Much of its land is flat plain, called the Comtat Venaissin, and it boasts a wide valley, the Coulon, to the south.

But the Vaucluse is also distinctly different from its southern neighbour. The bulk of its area is consumed by the Vaucluse Plateau, which is hemmed in by the imposing Ventoux Massif to the north. Mount Ventoux is, at 6,263 ft (1,909 metres), the highest peak in the region.

The Vaucluse Plateau merges into the Albion Plateau to the north. The latter is speckled with miscellaneous fissures that lead down to a vast network of underground caves. Rainwater spills down through these holes, collects in the caves, then runs off into another underground network of rivers. The rivers, in turn, flow along beneath the permeable limestone that composes the plateau, emerging here and there in springs like that at the legendary Fontaine de Vaucluse.

The springs are not all that set the Vaucluse apart. Its land is also, overall, more lush and generous than that of the Bouches-du-Rhône. Equally fertile and as highly cultivated, its premiere products correspond to the *département's* more luxuriant nature: melons, wine, lavender and strawberries.

The Var, on the other hand, shares more of the rougher aspects of the Bouches-du-Rhône. Much of its land is acutely dry, and drought is a constant threat to its residents. Wedged in between the Mediterranean to the south and southeast, the Bouches-du-Rhône to the west and Alpine foothills to the north and northeast, the Var is largely occupied by the deep forests of the Massif des Maures. Indeed, the Var is the most heavily wooded *département* in all of France.

Nonetheless, shrubby grasslands climb up and down rocky slopes in some parts, making the Var an excellent spot for raising goats and sheep. By the coast, the massif pushes up against the sea creating a natural *corniche* where many of the famed hilltowns of the French Mediterranean waterfront are perched. Offshore are a number of islands included within the Iles d'Hyères.

Left, Vaucluse vineyards and above, Old Port in Marseille.

Lying to the north of the Var, in between the Vaucluse to the west and the Alpes-Maritimes to the east, are the Alpes-de-Haute-Provence. This is an interesting region that wavers between being Provençal and Alpine. It also possesses one of the greatest natural wonders of the Provence region, the Grand Canyon de Verdon, which is, as its name suggests, an enormous canyon filled with turquoise-blue water from the Verdon River.

As soon as you enter the Alpes-de-Haute-Provence, you know it. Although it grows much the same products as the rest of Provence and is dominated to the south by a plateau (the Valensole) much like the Vaucluse, this region possesses an unmistakably Alpine feeling of physical remoteness. The further north you go, the more Alpine the land becomes, as it gradually assumes the flat-coloured, treeless folds of the Massif des Ecrins in the Alps proper.

The Alpes-Maritimes oscillate between the parched and rocky plateaux of the Var and the full-fledged ridges of the Alpes-de-Haute-Provence. On the eastern side, it is equally influenced by the mellowing Italian Alps. Its coastline is the most individual: first cliffs, then pebbles, then widening out into flat sandy beaches.

Along with the Var, the Alpes-Maritimes lives in constant dread of fire. In August 1987 alone, one fire ravaged more than 20,000 acres of forest, killed two and injured 100, and drove some 2,000 inhabitants from their homes. The dry shrub land makes unfortunate but excellent tinder.

One thing that all the *départements* share is dazzling light from a potent sun that never seems to stop shining. Farmers delight in the long growing season, and painters revel in the magical glow that surrounds the land. Another unifying characteristic is the infamous mistral. This wild and indefatigable northern wind sweeps down over almost the entire region between late autumn and early spring, although the western and central sections are generally the worst hit. Fittingly, its name is derived from the Provençal word for master, *mestre.* Its violent, chilling gales are created whenever a depression develops over the Mediterranean.

Learning about all the different geographies contained within Provence demands diligence, but the information is essential to understanding the Provençal world. Everything – the history, the economy, the characters of the people – centres around the land in this region where the earth is all.

PROVINCIA.
La Provence.
LANGE
DOC.
DAVLPH
CONTE DE
VE
CONTE DE
SAULT
NAN
CIN
LE BERONMON
LA CAM
AN GO
Montauban
Monvoillon
Samllac
Sault
Bedoin
Carpentras
Masan
Requemaure
Ville neufve
Aramon
Avignon
Perne
S. Didier
Cabrieres
Valebergus
Bubon
Laurade
Beaucaire
Nismes
Castelet
Fourques
S. Gillis
Lamote
La Barbon
S. Remy
Orgon
Les Bauls
Aiguines
Durance
Ragues
Allon
S. Canat
Grans
S. Martin
Merveiles
Aix
Rognac
Bove
Magnan
Tour d. Para.
Fos
Les penes
Sauvereael
Trois Mari es
Marseille
Planier
Espoholen
Rieulx
Cassidame
Les Tignes
Gras dorgon
Gras de Paulet
Gras grant
Paulet
Roque de Dour
MARIS MEDITERRANEI
Petrus Kærius Cœlavit
43
42
25
26
Occidens.

28
29
Gap
Embrun
Roßet
Theus
Durance flu.
Vars
Maurin
Tallart
Les Huertns
S. Paol
Vastavon
Pontis
Bacillone
Lausier
Antraygne
Melve
S. Martin
Mialans
Breses
Peivepove
CONTE
Laup
Sambuc
Valernes
Nibles
Gigas
Anset
Barles
Aste
DE
S. Dalmas
Lamoie
Toare
Collobrious
Chanolet
Pors
Villose
BVEIL
Guillumes
Tora
S. Esteve
Chantorci
Archal
Peonie
Peona
Lieusole
Dalluis
Pieri
Digne
Marieault
Aimac
Lagromuse
S. Legier
Bueil
Eora
Lesmees
Mesel
Moines
Annot
Floma
Autvare
S. Salvat.
Bras
Lapel
Val clause
Glandeves
Eetoet
CONTE
Beines
Senes
Nevoilles
Gars
Levillar
Ternafore
Mang
Corbon
43
Roßet
Pimoisón
Tervans
S. Iollian
Lapene
Elevos
Rias
Castelane
Demdes
Le Cor
DE
Greouls
Gillete
S. Crois
Evals
La Roquette
Oriens.
Soleis
Bernon
Les Erres
NICE
Le biose
Leborguet
Lebore
Asprement
Oinguion
Aiguine
Comps
Lebar
Vence
Fallicon
Braudinar
Beudveil
Mons
S. Cesari
Artignosa
Cagne
Barnoul
Enfous
Erapus
Grasse
Benconte
Tor enon
Antibo
LE
Chauvert
Tourtoar
Bagnels
Sillans
Callias
Canes
Flayose
Bras
Montfort
Lanapole
Trans
S. Maxemin
Draguignan
La Grenille
de S. Sosphier
Brignole
Lemuy
TER EL
C. Garaupe
Torenet
Camp
Freiuls
S. Marguerie
S. Honorat
Cabaße
de Teule
Roquebruns
C. Rouls
Pignans
Grimault
P. Dagat
PARS
C. S. Vincent
Cogolin
Goulfe de Grimault
Laverne
Turris
La Molle
S. Trompes
C. de Helbe
Romatuelle
Gaßin
Hieres
C. de Portes
42
Milliaria Gallica Communia
Les Isles Dor Stecades
C. Taliat
Milliaria Germanica commu
C. Lardier
2
4
6
Titan
28
29

Provence has never lacked new admirers eager to conquer its land. Ligurians, Celts, Greeks, Romans, Teutons, Cimbrians, Visigoths, Saracens, Franks – they all had their day under the warm Provençal sun.

The region's strategic position along the Mediterranean, providing the northern Europeans with access to the sea and the southern Europeans, Africans and Middle Easterners with access to the northern Europeans, brought century after century of new colonisations and settlements before France became the stable entity it is today.

The great Rhône, which rushes down through Provence and tumbles out into the Mediterranean, offered added incentive to settlers. The Greeks supposedly used the river as a trade route to tin-producing Cornwall, and the Romans were quick to recognise its value as a means of communication between their many conquests. Not until the invention of the railway did the Rhône's attraction to traders and travellers diminish.

Even in prehistoric times, the area was popular. Archaeological sites at Châteauneuf-les-Martigues and the Baume Longue within the Bouches-du-Rhône area link Provence with Mesolithic development. The nearby coastal shell-midden and fishing sites of Ile Maire and Ile Riou have offered up pottery belonging to the middle Neolithic period of the fifth millennium BC.

Unfortunately, further study into the earliest settlements faces an unsurmountable obstacle, for most of the Stone-Age sites were submerged during the post-glacial rise in sea level. Scientists, nonetheless, are able to agree on two interesting facts.

The first is that – like their descendants – these early Provençals were as involved in growing cereals as they were in hunting. All the elements of an agricultural economy were in place by 5000 BC, although it was probably not until 4000 BC that agriculture became the firm subsistence base.

Secondly, also like their descendants, communications between different settlements around the area were fluid. These early inhabitants were remarkably mobile. Most Stone-Age settlements were found in the arable lands of the Provençal basins and on the coastal plain of Languedoc, but caves discovered further inland that seemed to have been used as transit sites suggest that there was a well-spread economic system.

More evidence of early human habitation, in the form of rock carvings and skeletons, has turned up in the region, dating back to over a quarter of a million years BC. Curious tourists can visit the lagoon-surrounded spot of St-Blaise, near Martigues, where excavations have revealed a parade of inhabitants from 700 to 50 BC. For those less inclined to wandering around half-ruined spots in the hot sun, the Sault Museum has a fine collection of Stone-Age artifacts.

Historically, however, the area didn't become active until the advent of the Ligurians. It's difficult to pinpoint the origin of these people. Some believe all Ligurians descended from the Iberians in Spain, others insist that only some were linked to the Iberians. The term usually refers to anyone living around the Mediterranean as the Neolithic Age spilled over into the Iron Age.

The Ligurians certainly didn't act like a structured group. They lived in small and disparate village settlements and scratched whatever living they could from the rocky land around them. Early cultural bonds between the different Ligurian outposts, which stretched from Catalonia through Provence to the base of the Swiss Alps, can be whittled down to a propensity for flat burial graves and a special skill with razor design. Eventually, however, the Ligurians were "Celticised," although genetically they were related to neither the Gauls nor the Celts.

Even if the shape of their heads may elude historians, we do know that these later Ligurians were a tough people whose taste for piracy annoyed the Romans. Posidonius, the 2nd-century BC historian, said of them: "...daily hardships are such that life is truly difficult for these people whose bodies, as a result, are skinny and shrivelled. It sometimes happens that a woman gives birth to her child in the fields, covers the little one with leaves, and then returns to her work so that a day will not be lost."

One thing the Ligurians did enjoy was a fair amount of anonymity from outside influences. The Phocaean Greeks would change that, however, in the 6th century BC.

Preceding pages: Provence under the Romans. Left, prehistoric graffiti in the Mercantour. Above, Ligurian and Roman ruins at Glanum.

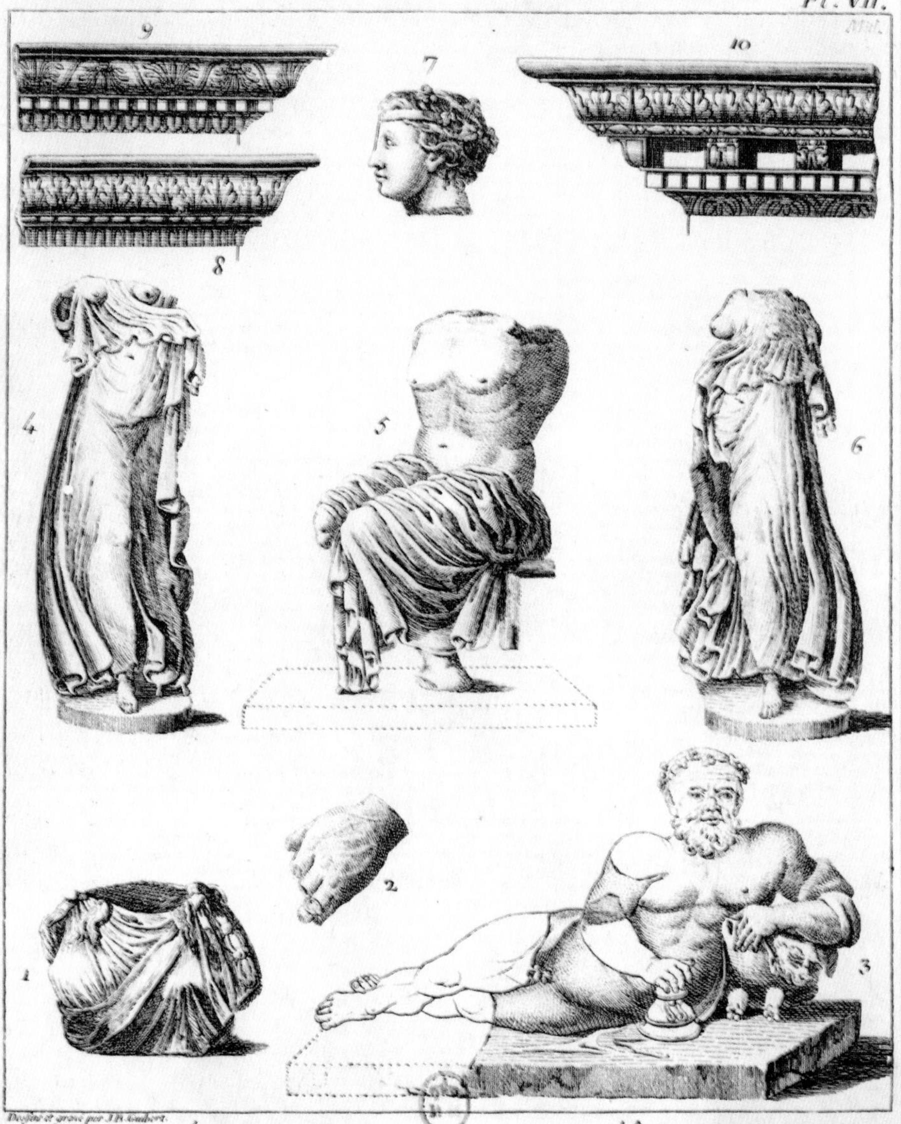

FRAGMENS DÉCOUVERTS, en 1788 et 1789, DANS LE THÉÂTRE ANTIQUE D'ARLES.

THE ARRIVAL OF CIVILISATION

Ici
Vers L'An 600 Avant J.C.
Des Marins Grecs Ont Abordés
Venant De Phocée
Ils Fonderent Marseille
D'Où Rayonna En Occident
La Civilisation
– Plaque on Quai des Belges, Marseille

Marseille, it might be argued, was the birthplace of Western civilisation on the European continent. It was to this marshy bay that Ionian Greeks from Phocaea set sail in the early 6th century BC.

According to legend, the Phocaeans cemented their interest in what was to become Marseille with a marriage. Gyptis, the daughter of the local Ligurian king, was in the process of choosing a husband when they arrived. A gathering of hopeful suitors waited as she decided to whom she would give the cup of wine that indicated her choice. When Protis, the captain of the Greek sailing party, stepped forward, she handed him the chalice.

The Phocaeans formed a happy alliance with the locals. They had come for economic purposes, not battle, and quickly set up a successful trading post. Their new colony, Massalia, soon became the most important commercial centre along the coast and a flourishing republican city-state in its own right. During its heyday in the mid-4th century BC, the Phocaean presence spread throughout the area. Some of the cities founded were Nikaia (Nice), Antipolis (Antibes), Citharista (La Ciotat), Olbia (Hyères) and Athenopolis (St-Tropez).

Pytheas, the Greek responsible for our modern method of measuring latitude and for explaining the tides, was actually a 4th-century Massalian. He expanded the Phocaean world view by visiting the British Isles and the Baltic. Another Massalian navigator, Euthymenius, explored the west coast of Africa around 350 BC.

During this period, Massalia was still very much a Greek city, although independent in nature.The late 3rd century BC, however, witnessed a budding relationship with Rome. When Hannibal, making his way up from Spain towards the Alps, appeared on the plains of Provence with his elephants, the Phocaeans were quick to join forces with the already outraged Romans. This alliance grew steadily to the point where, in 212 BC, Massalia switched over to the Roman form of municipal government.

Left, Roman fragments from Arles.

For the moment, the friendship with Rome proved beneficial. The northern and warlike Celts, Ligurians and Celto-Ligurians had been watching the prosperous Phocaeans with envy for some time and banded together in 125 BC to seize the Massalian riches. Overwhelmed, the Phocaeans were able to turn to the Romans for help.

In response, Caius Sextius Calvinius marched over a whole Roman army. A particularly bloody three-year war ensued but, in the end, the Romans prevailed. By that time Marius had discovered that he liked the area. He stayed on to found a city called Aquae Sextiae (Aix-en-Provence) on what was formerly a Celtic stronghold.

The Romans, as a sign of generosity, decided not to disturb the Massalian control of the coastal strip that reached from the Rhône to what is now Monaco. They increased their own interest in the region, however, and went on to defeat in 102 BC some more northern barbarians, the wild Teutons and savage Cimbrians.

Before too long, the Romans had replaced the Phocaeans as lords of the region, which they called Provincia Romana and from which the modern name "Provence" is derived. Massalia, however, continued to thrive and to operate fairly autonomously. Now no longer just a trading post, the city became famous as an intellectual centre, whose universities rivalled those of Athens.

The good times were not to last. Massalia made the fatal error of siding with Pompey against Caesar during Rome's civil war. After his victory, in 49 BC, Caesar punished the city by making it a Roman vassal. He further crushed the colony by strengthening the port of Arelate (Arles) and using its inhabitants to besiege Massalia.

Arles was not a new town, having been

founded back in the 6th century BC by the Greeks. During the 2nd century BC, Marius had built a canal that connected Arles's position on the Rhône to the Fos Gulf, making the city a handy river and maritime port.

The waterway had helped exploit the Arlesians' particular talent for building rafts supported by goat bladders. To add to its fortunes, this city was situated beside the highway that linked Italy with Spain, the Roman Aurelian Way.

With the fall of Massalia, Arles became the most important city in the area. Rome converted the whole south of France into a Roman province and made Arles its capital, with a secondary city at Cimiez above Nice.

The Romans would remain in Provence for another 600 years. Mostly these were years of peace and prosperity. Temples and bridges and baths and theatres were built, to the delight of modern-day tourist associations. During the reign of Augustus in the latter half of the 1st century, culture and the arts as well as trade expanded considerably. Slowly, due to their firm control and imposing political machine, the Romans managed to unite what had once been a scramble of disparate trading posts. One sign of this was the growth of a regional language, Provençal, derived from Latin.

As legend tells it, the 1st century also saw the arrival of The Word in the northern Mediterranean. Many believe that it was to these shores that Mary Magdalene and Martha – along with her resurrected brother Lazarus; Mary Jacoby, sister to the Virgin Mary; Mary Salome, mother to the apostles James and John; Saints Maximinius and Sidonius; and the servant Sarah – fled after the crucifixion of Christ.

Buffeted by the waves, without sail or oar, this saintly company arrived on the sandy shores of the town now called, appropriately, Les-Stes-Maries-de-la-Mer. Supposedly, they erected a small chapel to the Virgin, then went their separate ways. Mary Magdalene, for one, found a cave up in the hills of Ste-Baume where she lived as a hermit until her death 30 years later.

Whether or not Saint Trophimus was among this group differs according to the source, but he is widely credited with the introduction of Christianity to the northern countries. Presumably sent by his friend, Saint Peter, he headed up to Arles where he built the first Christian church in Gaul – named St-Trophime and still standing today.

By the time the Roman Empire fell in AD 476, Roman civilisation had left a permanent

Above, Charles Martel squashes the Saracen army.

mark on the land. For one thing, the entire region had been Christianised. The frugal and industrious character of the Provençal people today is also widely accredited to their Roman ancestry. The Phocaeans, and more specifically the Massalians, were enthusiasts for the joys of the flesh. The Romans, however, were much more attentive to cultivating the hard dry soil and building up sufficient fortifications, which probably had something to do with their ability to dominate the area for so long.

But the fall did come – and with it the Visigoths. Their interest in Arles came as no surprise, since they had been routinely attacking the city since the earlier part of the fifth century. They quickly assumed all of the area south of the Durance River, while the Burgundians took over to the north.

The Visigoths didn't do much for the region, but they weren't to last long either. In 507, they were firmly trodden upon by the Franks. To escape these northern strongmen, they ceded their Provençal conquests to the Ostrogoths from Italy, who also appropriated all the Burgundian lands for themselves.

The Franks did not agree that the Ostrogoths deserved Burgundy and proceeded to make it their own in 534. The Ostrogoths incurred further problems by alienating the Byzantine Empire. In return for Frankish neutrality, these early Italians had to hand over their Provençal territory in 536.

The Franks now had an awful lot of land to supervise – too much for one authority to manage. They decided to split up the territory, uniting northern Provence with Arles, Toulon and Nice to Burgundy but giving the strip including Marseille, Avignon and much of the coast to the kings of Austrasia.

Representatives, known as *patrices*, were appointed within each region by the central Frankish ministry. The *patrices*, however, proved more interested in promoting their own independence and didn't hesitate to bond with foreign elements if it would protect their own autonomy.

One such group were the Saracens. For the first couple of centuries under Frankish rule, Provence had fared pretty well commercially. Nonetheless, their integration with the rest of France was minimal at best, and when the Arabs arrived in Provence in 732, not all of the natives complained.

In 732, two Saracen armies beseiged Arles. Encouraged by their victory, they continued northward until they encountered the famed French hero Charles Martel in Poitiers. He expelled these heathen intruders from all of Provence in 729.

Martel was less than happy with the Provençal ambivalence towards the Saracens, and throughout the battle against the Moors, he treated the region as hostilely as an enemy. Between his armies and those of the Saracens, countless Provençals were slaughtered and whole cities were destroyed. Avignon was virtually wiped out.

Pepin the Short succeeded his father, and after him came Charlemagne. Each was careful to keep Provence under strict Frankish rule, but this did not prevent the return of the Moors in 813. Their new tactic was to approach by sea rather than via Spain, and it proved to be an effective strategy for about two more centuries.

Meanwhile, in the north, Charlemagne had created the Holy Roman Empire. With the Treaty of Verdun in 843, Provence was turned over to his grandson, Lothair I, as part of the "Middle Realm." When Lothair died, the region was labelled a separate kingdom and passed through the hands of several unmemorable rulers: his son Charles; the emperor Louis II, ruler of Italy; Charles the Bald; Boso, ruler of the Viennois; Boso's son Louis, who was to become king of Italy in 900; then Hugh of Arles, who handed it over to Rudolf II of Jurane Burgundy in the mid-10th century.

All these kingships and dukedoms did little to secure the area. As if they hadn't suffered enough under the Saracens, the Provençals soon had to deal with the Normans, who set out, in 859, to stake a claim in the south land. They levelled both Arles and Nîmes as well as much of the countryside before the fierce opposition of Gérard de Roussillon made them decide to head back to Scandinavia.

The departure of the Normans didn't give the region much respite, however, since the Saracens were soon to return. During the last part of the 9th century, the Saracen hold reached its apogee. The entire coastal area fell to their powers, and they marauded right on up into the Alps. Not until 1032 did William the Liberator – abetted by an attack of plague – finally eliminate the Saracen menace from Provence once and for all.

OF POPES AND LOVE COURTS

Although contained within the Holy German Empire, Provence was a fairly independent entity by the year 1000. Threats from outsiders subsided, giving the major cities of Arles and Avignon a chance to smooth their feathers and the entire region the opportunity to cultivate an identity.

The departure of the Saracens and Normans also left the Provençals with time and energy for internal squabbles. When Boso, Count of Arles in the mid10th century, died, he left control of Provence to his two sons, Rotbald and William. The former was Count of Arles and the latter Count of Avignon, and they ruled jointly over the region from their respective seats. By 1040, however, Rotbald's male line was extinct, and his inheritance moved into the female side of the family. This was the beginning of a long and unfortunate pattern. If Emma, the first lady in question, had known just what future troubles she was bringing by marrying into the house of Toulouse perhaps she would have opted for virginal sainthood instead. Unfortunately, she didn't.

When Bertrand of Arles died in 1093, with him passed the last direct male heirs from the line of Boso. Gerberge of Arles, sister of Bertrand, stepped forward to claim control, followed by Alix, heiress of Avignon. They were soon joined by Raymond IV, grandson to Emma of Toulouse.

The previously intact countship of Provence fell into turmoil. First, the countship of Forcalquier was carved out of the region and given to Alix. Then, in 1112, Gerberge turned her rights over to her daughter Douce. Not one to forsake family tradition, Douce complicated matters by marrying Catalan Ramon Berenguer III, count of Barcelona.

Within a year, Ramon had changed his name to the more-Gallic Raymond Berengar I and assumed the title of count-marquis to Provence. The Toulousians, outraged at his usurpation, contested his claim. They were soon joined by the house of Les Baux, whose own Raymond had married Douce's younger sister, Stephanie.

Left, ruins at Les Baux. Above, fresco from Pope's Palace in Avignon.

The Baux family was no insignificant enemy. The Provençal poet Mistral would later describe them as "eagles all – vassals never." Not known for their modesty, they traced their genealogy back to Balthazar, the Magi king, and implanted the star of Bethlehem on their arms. Before they got around to claiming control of Provence, *grâce à* Stephanie, the Bauxs had already subjugated, during the 11th century, some 80 towns and villages. As their power increased, they gathered up such titles as Prince of Orange,

Viscount of Marseille, Count of Avellinoad and Duke of Andria.

At the height of its splendour, the airy perch of the Baux fortress, which rises from a bare rock spar at 650 feet, was famed for its "Court of Love." Here, troubadours milled among the inhabitants – as many as 6,000 – composing passionate poems in praise of well-bred ladies. In return, the poets would receive a peacock's feather and a kiss.

But being hosts to the Court of Love didn't stop the Bauxs from being the bloodiest, cruellest group around. Their history is an unending catalogue of savagery. One prince suavely massacred the entire town of Courthézon; another was carved up in prison by his own wife; a third besieged the castle of his pregnant niece for the purpose of violating her bedchambers.

Perhaps most famed was the husband of Berangère des Baux who slew the poet Guillem de Cabestanh and gave his heart to his wife for dinner. Unknowingly, she ate the sweet meat and drank a goblet of his blood. In a rare show of conscience, after discovering the ingredients of the meal, she declared that so lovely were that meat and that drink that none other would defile her lips again – and threw herself off the top of Les Baux.

No one wanted to parley with the Bauxs, but the Barcelonians and the Toulousians did manage to sign an amicable treaty among themselves in 1125. This agreement gave everything south of the Durance River to the former and everything north of the river to the latter, with Avignon, Sorgues, Caumont and Le Thor to be shared. They seemed content with this solution, but the Bauxs continued to challenge the Barcelonians.

Throughout this period, the German kings – who, after all, were the real rulers of Provence – were in a constant state of confusion about whom they should recognise as vassal for the region. The greatest still visible evidence of the era's rampant instability are the numerous defensive *villages perchés* dating from that time. The lack of stern central authority promoted the rise of the feudal system. Grandiose castles were built, and the arts found ready patronage within courts other than Les Baux's.

The troubadours emerged as a direct result of feudalism. Wealthy heiresses, daughters to territorial lords, were married off for political and economic reasons rather than those of mutual lust. Once wed, they were left mostly to their own designs and, not unnaturally, welcomed the attentions of their husband's courtiers. The songs of love they inspired were considered a mere convention and harmless – although the courts were far from being models of chastity.

The arts were not all that flourished within this period of lean control. Towns such as Arles, Avignon, Nice, Tarascon and Marseille reorganised as communal regimes under consular government and became independent forces in their own right. The opening up of the Levant by the Crusades and the disappearance of the Arab threat provoked a revival in commerce that added to their wellbeing.

A succession of Raymonds from Barcelona and Raymonds from Toulouse and

Above, St-Michel's 12th-century tympanum in Salon. Right, the Provençal language.

The Language Of The Troubadours

Never content to go along with the crowd, the people of Provence early on developed their own dialect and stuck with it even after the rise of the French-speaking Gauls. Provençal is a Romance language derived closely from the Latin spoken by the Roman conquerors of the 1st century BC.

In fact, the staying power of the language is good evidence of the profound cultural effect of the Romans. Many centuries after they had faded from power and had been replaced first by varied Germanic invasions then by the Franks, the Provençal language clung to its Latin attributes. Hardly any of the Germanic traits evident in French can be found in Provençal.

The feudal organisation of the Midi during the Middle Ages resulted in the fragmentation of Provençal into a variety of related but individual dialects. Nonetheless, a unified literary language, "Classical Provençal," was recognised from the 11th to 15th centuries. It was in this language that the poems of the troubadours were written.

The oldest piece of Provençal verse still extant is a 10th-century refrain attached to a Latin poem in a Vatican manuscript. But the earliest complete works to have survived the centuries are a set of poems by William IX, duke of Aquitaine. These poems consist of 11 different pieces, each written in stanzas and meant to be sung.

Almost certainly, William did not pioneer such work. More likely, his noble rank helped preserve his poetry when that of his predecessors and contemporaries fell into oblivion. The true creators of the troubadour-style literature were probably of much more modest social standing.

It is widely believed that the troubadours developed from the lowly class of jugglers. First they combined poetry with their acts and then, evolving into a more refined group of court residents, composed and recited their poetry unaccompanied by tricks. These poets were encouraged by the lonely wives of their noble patrons to develop the love songs that became the mark of the troubadour, and they eventually created the medieval convention of "courtly" love. Naturally, it's to be expected that some of the devotion lavished on the ladies of the court by the poets did not go unreturned. In general, however, the poets' lesser rank forced them to couch their words of praise in the most respectful manner, and their attentions were considered innocent – even convenient – by the men of the court, who more often than not had married for economic and territorial reasons rather than for those of love.

By the end of the 11th century, the troubadours had become a vital institution of a much higher status than their ignoble predecessors. Among the most famous were Bernart de Ventadour, patronised by Eleanor of Aquitaine; Arnaut Daniel, inventor of the sestina; Jaufre Rudel de Blaye, poet of the *amor de lonh*, or "love from afar"; Giraut de Borneil, creator of the *trobar clus*, or "closed" style; Folquet de Marseille, a troubadour who became a monk, then an abbot, and finally bishop of Toulouse; Bertran de Born; and Peire d'Alvernhae. All in all, the works of over 460 troubadours have been preserved.

With the rooting out of the Cathares during the earlier part of the 13th century came the rise of the French Crown's control over Provence and the fall of the feudal lords. The troubadours were no longer able to make a happy living, and the use of Provençal as a literary language declined drastically.

Over the next few centuries, a handful of valiant artists continued to write in Provençal, and the literature saw a mild revival during the 16th century in what is generally referred to as the Provençal Renaissance. Nothing, however, could compare with the glory of the medieval period.

Nothing, that is, until the advent of the chauvinist society of the Félibrige, established by Frédéric Mistral and a handful of concerned Provençal writers in the mid-19th century. Mistral was influenced by the schoolmaster and poet Joseph Roumanille – a native of St-Rémy, who turned to composing in his dialect so that his mother would be able to read his work – and he dedicated his enormous talent towards "stirring this noble race to a renewed awareness of its glory." His devotion was rewarded in 1905 with a Nobel Prize for literature.

Raymonds from Provence, plus a Spanish Alfonso, traded the countship throughout the 12th century. Eventually, the three claimant parties decided to split the region up definitively. Toulouse became Languedoc, Barcelona was named Catalonia, and Forcalquier evolved into a new, more limited Provence.

In 1209, just to add a little more spice, a huge French army stormed down south to rout out the Albigensians. Also known as Cathares, these ill-fated people supported a radical departure from traditional Christian dogma and were considered dangerous heretics. They also possessed a fair amount of widely coveted treasures. Although the Albigensians were located within Languedoc, the Toulousian connection to Provence and, more particularly, Avignon caused that city to be besieged and belittled by Louis IX of France in 1226.

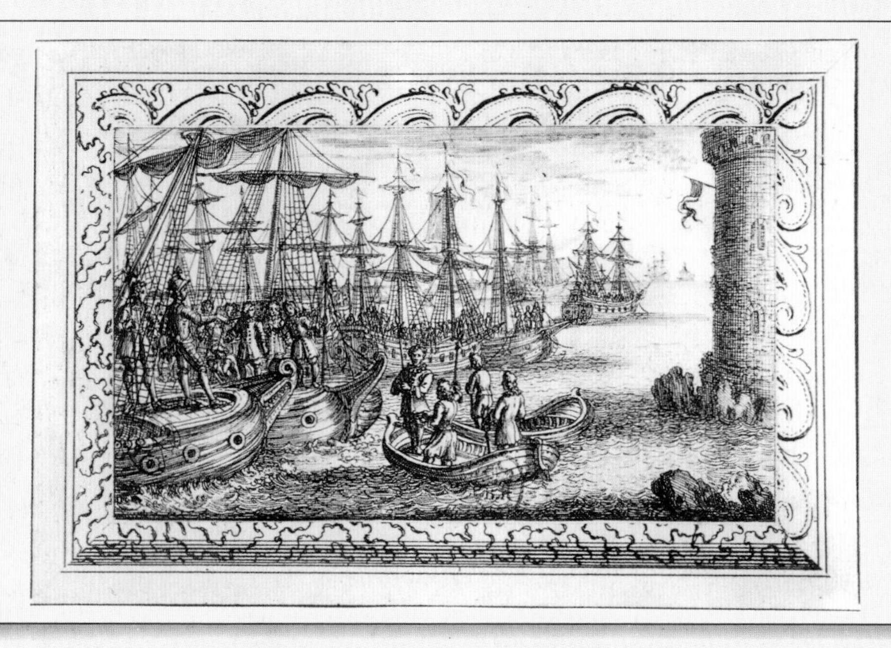

After the Albigensian episode, the Toulousians were eager to get back on the good side of the French kingdom. In 1229 Raymond VII of Toulouse wed his daughter to Louis IX's brother, Alphonse of Poitier. By way of additional placation, the Toulousian territories, including the Comtat-Venaisson region north of the Durance in Provence, also were passed along to the French.

Unlike the area under Toulousian control, that under the Provençal countship experienced a greater taste of authority with the advent of Raymond Berengar VII. In between proving his might, this Raymond begat a bevy of daughters, all of whom made brilliant and fateful matches. The eldest, Marguerite, became queen of France after marrying the prince who would later be known as Saint Louis. The second, Eleanor, landed King Henry III of England. The third, Sancie, married Richard, Duke of Cornwall, who would eventually become the Germanic emperor. This left Beatrice, the youngest.

Although her three older sisters must have had ample charm to marry as they did (some credit must be given to the stature of their homeland, Provence, at the time), Beatrice was considered to be the sweetest and most beautiful of all. Crowned suitors flocked to her side. Yet, in the end, she chose Charles of Anjou, brother to the French king.

Beatrice's sisters ridiculed her unadorned pate until 1265, when Charles was crowned king of Naples and Sicily. He also became king of Jerusalem and, when Raymond VII left Provence to Beatrice, he was named Count Charles I of Provence. The sisters were no longer laughing. Even less pleased were Charles's new subjects, who considered him a foreigner – ever a dirty word in Provençal. In between embarking on Crusades, Charles had to concentrate on squashing rebellions, with Marseille the first to go.

Marseille, that sturdy determined city, had prospered during the Crusades. Its citizens had established new trading centres in Asia Minor and North Africa that resulted in increased revenues, and Saint Louis's departure in 1248 from nearby Aigues-Mortes on his seventh Crusade had bolstered the economic activity back home. They held out against Charles the longest but eventually were subdued along with fellow rebels Arles and Avignon. Charles's final triumph in 1257 marked the end of the era of quasi-autonomy.

The new French dynasty provided long-missing stability. Charles's involvement with Naples and Sicily also heralded a revived link with Italy. Most of the succeeding noble families set up housekeeping in Italy and appointed administrators to lord over the state in their stead. In 1271, Philip III of France turned over the Comtat-Venaisson – inherited after the Albigensian Crusades – to the Rome-based Holy See. And Charles II, who took over in 1285, instigated many legal reforms based on Italian models. His successor, Robert, king of Naples and count of Provence from 1309 to 1343, installed *syndicats* or municipal councils.

Another example of the new link with Italy is the visit of Petrarch. The famous Italian poet made his residence in Fontaine de Vaucluse from 1337 to 1353. During that time he composed the 366 poems and sonnets of the *Canzonière*, inspired by the beloved Laura. He had already caught his first glimpse of her in a church in Noves while visiting Provence 10 years earlier.

When Robert died in 1343, his granddaughter Jeanne took over. Jeanne is alternately known as "the Good Queen" and as "the Wicked Queen." She appears to have possessed some of that old Massallian attitude towards free love and to have exercised it liberally during her lifetime. Married first to Andrew of Hungary, it is believed she had him assassinated. She soon moved on from her second husband, Louis of Tarento, to wed the Adonis-like Jaime II of Mallorca. Eleven years her junior, Jaime was unfortunately mad, due to having been imprisoned in an iron cage for 13 years. Her fourth spouse, however, Otto of Brunswick, seems to have lived and died normally.

Left, Saint Louis leaves for the Crusades from Aigues-Mortes. Above, Petrarch found love in Provence.

Despite her many opportunities, Jeanne never did come up with a male heir. Her cousin Charles of Durazzo became the logical successor but, based on a deep-rooted dislike for him, she adopted Louis of Anjou, brother to King Charles V of France, and named him heir instead. This created new possibilities for internal strife. It also incited Charles of Durazzo to live up to her estimation of him. In 1382, he had her kidnapped to a lonely castle in the Apennines and, tied to her bedposts, strangled. Another war for control of the dukedom ensued.

Meanwhile, the region was in the process of recovering from an invasion by Languedoc armies, who took Tarascon and beseiged Arles and Aix before Neapolitan forces finally overcame them. To pay for these wars – and some say to be pardoned for killing her first husband – Jeanne had had to sell Avignon to the Holy See in 1348. Avignon had already been the seat of the papacy for some 40 years.

In 1309, Pope Clement V, a native Frenchman, had decided that he had had enough of the constant bickering of Rome. When King Philip the Fair of France invited him to return home, he quickly gathered his robes and headed for the Comtat Venaisson on the lower Rhône. His removal and that of his court to Avignon began what would be called the "second Babylonian captivity of the church." Five more popes would follow in his lead. The third, Benedict XII, was widely criticised for being avaricious, egotistical, uncharitable and prejudiced. He also instigated the erection of the famed Palais de Papes with its large number of secret escape routes. The enormous Palais was well over an acre in size, supposedly to exemplify the absolute power of the church. But Benedict's successor, Clement VI, decided that it still didn't properly represent the grandeur of the church. He added a more ornate section,

bringing the total size up to 2.6 acres (1 hectare). Clement VI was also the one to purchase Avignon from Jeanne for 80,000 gold florins.

In 1376, the Florentines sent a Dominican sister from Siena to intercede with the Pope in their favour, having perpetrated a failed attempt at annexing the papal territories. Saint Catherine succeeded not only in gaining their pardon but also in convincing Gregory XI, last in this string of French popes, to return to Rome. Gregory agreed to the homecoming in the belief that reinhabiting Rome would improve negotiations with the Byzantine church, thus increasing the possibility of holding onto the papal territories. He may well have been right, but this didn't mean that the Palais des Papes would remain empty for long.

The next elected pope, Urban VI, an Italian, so displeased a large number of French cardinals that they decided to call for a new vote. They selected a new Pope, a native of Geneva to be called Clement VII, in 1378. The next year, Clement VII retreated to Avignon, and the Great Schism, lasting from 1378 to 1417, was born.

Now there were two popes – one in Avignon and one in Rome. This confusing situation continued for several bitter decades, although not all of these years were spent in Avignon. Benedict XIII was driven out in 1403, and the city withdrew from the controversy and fell back into a peaceful life. Its heightened stature as the seat of the papal court, however, never diminished.

Meanwhile, the rest of Provence had degenerated into a state of constant civil war following the death of Jeanne. The area was already reeling from the arrival of the first great plague in 1348. A semblance of peace and order was restored only after the fortuitous murder of Charles of Durazzo and some skilful manoeuvring on the part of Louis of Anjou's widow, Marie of Blois, in 1348, whose son, Louis II, proved to be a capable ruler. Louis also fathered the most beloved count of Provence, René of Anjou, popularly known as "Good King René."

To this day, René is remembered with love and admiration by the Provençals. A self-styled philosopher and humanist, he devoted himself to promoting the economic revival of Provence and its ports and to reinstating interest in the arts. He showed himself to be a tolerant, amicable leader, whose steady control permitted cultural life to flourish.

René's talents, however, did not extend to the martial arts. He failed so miserably in battles elsewhere that he was impelled to relinquish his right to designate his own heir. Instead, he was forced by the crown to name the unlikely Charles III, comte de Maine and chronic invalid.

Charles assumed the throne in 1480 but died a year later. He left his countship to Louis XI of France.

Above, Pope Benedict XII's palace in Avignon. Right, the beloved "Good King René".

RENATVS DEI GRATIA. SICILIÆ. etc REX

The first steps towards unification with France were taken when the three Estates of Provence (the Clergy, Nobility and Commoners) accepted Louis XI, King of France, as their ruler in 1482. Five years later, their union was cemented by a treaty supposedly designed to protect the autonomy of Provence. The constitutional equality of the countship and the kingdom was ratified, and the formerly independent state was allowed to retain a substantial number of individual liberties and traditional customs.

Marseille had had its day under the Greeks, as had Arles under the Romans. With the advent of the counts of Provence and the papal residency, Avignon had sprung forth in glory. Under Good King René, however, Aix-en-Provence emerged as the star city of Provence. After losing Anjou to Louis XI, René and his beloved second wife, Jeanne, retired to this lovely town, remaining there until his death in 1480.

A metropolitan see since the beginning of Christianity and home to a venerable university, Aix did not diminish in importance with the passing of the good king. The French chose to recognise its eminence and treated it as a capital. Not surprisingly, it was in Aix that the treaty of 1487 was signed.

It was also in Aix that Louis XII decided to establish a parliament, of French origin, 14 years later. The Parlement de Provence was introduced as a supreme court of justice with limited political authority and as a place where all the estates would meet. Most of the native Provençals, however, felt that its real purpose was to impose stiff tax increases.

The affiliation with France may have finally brought a firm authoritarian command to the region, but it did not usher in any newfound peace. The martial escapades of Francis I engendered the invasion of Provence by German Emperor Charles V in 1524 and then again in 1536. Meanwhile, annoyed by constant friction with Provence, Francis I increased the curtailment of the region's independence. In 1535, he issued the Edict of Joinville, which severely limited the freedom of the Estates, squashed the Provençal-originated *conseil éminent* and strengthened the control of the hated Parlement over the local judiciaries.

In the end, Provence was left with more or less the same status as any other French province. But even this did not satisfy Francis: four years later he unfurled his Edict of Villers-Coterets, which introduced a new tactic in the suppression of Provence. This decree installed French as the language to be used in all administrative laws in Provence, supplanting the native Provençal dialect. Although the ancient language was hardly wiped out by this one action, the new law did initiate its eventual decline.

Despite the repression by France, individual families continued to thrive. Large numbers of châteaux, still standing as silent witnesses today, were built, and small fortunes amassed. Concerned that these wealthy inhabitants might try to rival the French nobility, the Parlement turned its Gallic eye towards their belittlement. Deciding that the châteaux, with their rounded corner towers, were potentially aggressive as well as presumptious, they commanded that all towers be truncated to the level of the main roof. The stunted leftovers are still visible, dubbed *poivrière* (meaning literally "pepper pots") by the locals.

The arts and intellectual life also continued to prosper. The troubadours had faded from eminence, although the 16th century saw a certain revival in native literature, mostly in the form of religious mystery plays, and the period is sometimes called the Provençal Renaissance. New champions sprung up in their place, often pioneering more scientific fields.

Nostradamus (Michel de Notredame) was born in St-Rémy-de-Provence in 1503. He first studied medicine and, after extended travels, developed a successful remedy against the plague. He kept the cure a secret, however, which proved unpopular with his colleagues. They voted to have him expelled from the medical profession.

Nostradamus next turned to astrology. In 1555, he produced the famed *Centuries*, a book filled with predictions on the future of the world, written in cryptic quatrains . Interpreters claim that he foretold such 20th-century events as World War II and the holocaust, as well as the 1963 assassination of US President John F. Kennedy.

Left, Adam de Craponne brought water to the Crau Plain. Above, Provençal astrologer and prophet Nostradamus.

A second, less sensational but equally productive Provençal was Adam de Craponne (1527–76). Born in Salon-de-Provence amid the dry Crau plains, De Craponne was the gifted civil engineer largely responsible for bringing fertility to the region. He designed and had constructed the irrigation canal that diverts water from the Durance River through the Lamanon Gap.

Up north, the French continued to devise ways to fetter Provençal activity. Not content to concentrate on local legislation or the lives of the up-and-coming, Baron Megnier d'Oppède led one of the bloodiest campaigns in the history of France against a religious sect called the Vaudois.

The unfortunate Vaudois were descendants of the Waldensians, who had originally formed in protest against the growth of papal wealth and property under Pope Sylvester. Waldensian dogma related vaguely to the Manichaean system of duality with its emphasis on severe ascetism. An earlier branch had already been literally burnt out of existence – in the form of the Cathares.

Most offensive among Waldensian tenets were their beliefs that any layman could consecrate the sacrament and that the Roman church was not the true church of Christ. Nonetheless, up until this point, the Vaudois had managed to escape the iron arm of the Church by living quietly and fairly inaccessibly amidst the Luberon mountains. The French did not let this deter their zeal, and the Vaudois fell victim to endless massacre. They had been virtually decimated by the year 1545.

The slaughter of the Vaudois was only a trial run for what was to come. The brutal

Wars of Religion, brought on by the Reformation and occurring between the Catholics and the Protestants, consumed the greater part of the second half of the 16th century. The region witnessed countless atrocities until finally the arrival of a second, equally horrible enemy – the second plague, in 1582 – caused the Provençal "heretics" to weaken. Under that devastating epidemic, even argumentative Marseille finally submitted to Henry IV of France.

Marseille had steadily added to its wealth and significance since the profitable years of the Crusades. The port city reached new heights during the early 1500s. Being the bullwark of France's Mediterranean coast, it had been virulently threatened by Charles V of Germany but had managed to summon up a successful group campaign against him. Even the fair damsels of the city are said to have participated in the defence, as commemorated by the Rue des Dames.

Triumphant in its resistance to Charles, the city's pride, infused with an inherent spirit of determination and independence, found new fodder. Again and again during the next couple of centuries, Marseille would emerge as the most vociferous of political entities, generally leading the other main cities of Provence in revolt.

With the Protestants under control, with many dead and with Marseille temporarily enfeebled, the 17th century rolled in on a more closely aligned Provence. Local authorities, however, displayed no sudden affinity for the Crown. Minor rebellions continued to flare up, often led by the bullheaded Marseillaises. Then, when in 1630 the French government tried to abolish the Estates' control over taxation, the parliament at Aix simply refused.

Under the threat of overwhelming Gallic violence, the parliament voted in an extraordinary subsidy the following year, but the Estates had already sounded their own death knell. After a particularly obstreperous showing in 1639, the Estates were simply no longer invited to attend the assemblies of parliament. Not until a century and a half later was their voice reinstated.

New dissension against the French emerged in the form of the Provençal Fronde. Unlike its northern brother, which pitted itself against royal absolutism from 1648 to 1653, the Provençal Fronde concentrated its venom on hostilities between the parliament and the governor. Only later did the struggle become one between the Sabreurs (adherents of the Rebel princes) and the Canivets (royalists).

The Southern Fronde was supressed in 1653, but discontent persisted in Marseille. The defiant city revolted in 1659, only to be stripped of all its remaining rights after its defeat the following year. To ensure his control over the uppity Marseillaises, Louis XIV personally ordered the construction of Fort St-Nicholas, from which both the town and port could be closely monitored. A year later, undaunted by their neighbour's lack of success, the citizens of Nice also took up arms. The town was subsequently occupied by the royal army as well.

In 1673, the beleaguered Louis XIV established the *généralité* of Aix, under a superintendent whose job it was to manage the province. From 1691 on, this office was united with that of the premier presidency of the parliament, giving the Crown effective legal control over the region.

The French king also continued his interest in Marseille, ironically to that city's ultimate benefit. He expanded the urban area threefold and encouraged the growth of seaborne trade by making it a free port. Marseillaise commerce continued to blossom during the 18th century, reaching its peak just before the Revolution of 1789.

It is probably due to the greatly broadened status of the Marseillaise port that the third and deadliest of the plagues hit Provence in the early 1720s. It was brought by trading ships from the east, and more than 100,000 people died – half of them from Marseille alone. Whole towns disappeared. Traces of an enormous wall built in the desperate effort to contain the disease can still be seen near Venasque in the Vaucluse.

The newly organised authority under the *généralité* led to improved relations between Provence and the French crown during the early 18th century. Invasions into the region by Austria and the Savoyards in 1707, then again in 1746, also helped force internal squabbles to a back burner. Toulon and Nice suffered particularly badly during the foreign incursions. Both ports were hardily ravaged by the Austro-Sardinian maritime forces before their final defeat in 1747.

By this time, the Crown was facing more serious upheaval than that provided by the arrogant Provençals. The discontent that would lead to the permanent removal of monarchical rule in France was brewing throughout the country.

The people of Provence embraced the ensuing Revolution with open arms and an extraordinary excess of blood-letting, even for that violent era. Their interest in the new constitution stemmed not so much from a desire for social reform as the hope that it might offer a chance to regain lost powers.

Left, protective clothing against the plague. Above, bust of "La Marseillaise."

Marseille was an especially enthusiastic adherent. Before long, a guillotine graced the city's main street of La Canebière, and the cobblestones ran red with spilt blood. As many supposed Royalists were decapitated in Marseille as in Paris itself.

On 11 April 1789, the Société Patriotique des Amis de la Constitution was formed on the Rue Thubaneau in Marseille, and it was there that Rouget de Lisle's *Le Chant des Marseilles* – now the bloody national anthem for all of France – was first sung. Meanwhile, it was from the small nearby town of Martigues that the *tricolore* – now France's national flag – originated. Martigues consists of three boroughs, each with

its own standard. Ferrieres's is blue, Ile St-Genest's is white, and Jonquières's is red. United, they created the red, white and blue flag adopted by the Revolutionaries.

The French Revolution in 1789 did not, however, return to Provence its autonomy of earlier centuries. Instead, the region lost what little independence it had been able to retain. The local government was completely dissolved in 1790, and the region was divided into three *départements*: the Bouches-du-Rhône, the Var and the Basse-Alpes. The Vaucluse was added three years later, after the French annexation of the papal territory in the Comtat-Venaisson.

Among Provence's greatest contributions to the Revolution was the young Corsican captain, Napoleon Bonaparte. The famed Napoleon first came to notice in Toulon, when he managed to wrest that important port city from an occupying Royalist-Spanish-British force in 1793. Three years later, as a commanding general at the age of 27, it was from Nice that Napoleon set off on his first glorious Italian campaign.

Although the Provençals had embraced Napoleon the military genius, they were much less enthused about Napoleon the Emperor. This same people, such eager participants in the early years of the Revolution, soon became staunch supporters of the returning Bourbon monarchy. Once again, blood flowed in the streets, as those suspected of being anti-Royalist were slaughtered with the same intensity as the anti-Revolutionaries had been under Jacobinism.

The Provençal antipathy towards Napoleon proved to be justified, for his ever-escalating wars finally resulted in the bequeathal of much of the eastern territory – Nice and the land up as far as the Var River – to Sardinia during the Congress of Vienna in 1814. It was not until 1860 that Napoleon II managed to return these lands to France.

Nonetheless, the 19th century brought Provence the most peaceful period it had ever experienced. The subsequent revolutions of 1830 and 1848 and the new regimes that each heralded roused little interest from the southern regions. Instead, the Provençals seemed content to concentrate on expanding their economic structure.

The martial lull also permitted the arts to flourish in a way that they hadn't since the days of Good King René. Poetry was rediscovered, and painters from colder climates became aware of the advantages of the Mediterranean environment. It was amidst this atmosphere that the Félibrige arose.

Founded in 1854, the Félibrige was created by the literary masters of the region out of concern for the preservation of their native tongue. Stripped of any political autonomy, the people of Provence had been further slighted by the removal of their dialect as Provence's official language. Unsurprisingly, the language had since then begun to slide into oblivion.

The idea behind the Félibrige was to encourage a revival of the language through the promotion of literary works written within the dialect rather than French. They also compiled a vast dictionary of Provençal called *Lou Tresor dóu Félibrige.*

At the head of the Félibrige was the renowned Provençal poet Frédéric Mistral. In 1859 he published the verse romance, *Mireio*, later set to music by Gounod, whose depiction of Provençal life gained recognition for the Félibrige. A half-century later, in 1905, Mistral was awarded the Nobel Prize for literature. Ever true to the cause of artistic revival in Provence, he used his prize money to resuscitate the ethnographic Museon Arlaten in Arles.

Preceding pages: ***Farmhouse in Provence*** **by Van Gogh. Left,** ***Houses in Provence*** **by Cézanne. Above, the poet Frédéric Mistral and his wife.**

Aside from Mistral, the original Félibrige included Joseph Roumanille who had been Mistral's own teacher and inspiration, the well-respected Avignonais printer Théodore Aubanel, and four contemporary Provençal poets of slightly lesser talent. Native painters also brought glory to the region. One most famous son of the south was Paul Cézanne. The child of a wealthy hatter-turned-banker and one of his work girls, Cézanne was born in Aix-en-Provence in 1839. He was educated at the local Collège Bourbon, where he developed a close friendship with another student, novelist Emile Zola. Although Zola was not born in the area, he also spent much of his youth in Aix.

While Zola committed himself to artistic pursuits, Cézanne struggled to become his father's successor. Eventually, his family came to the realisation it would be better to allow him to pursue his painting. In 1863, he ventured north to join Zola in Paris. Once there, Cézanne's distaste for the "old school" soon led to his association with the revolutionary group of painters now known as the Impressionists. However, he eventually split from that group as well, believing their work to be too casual and lacking in an understanding of the "depth of reality."

Cézanne returned to Aix in 1870, where he would spend the rest of his years until his death in 1906. Some of his most celebrated works were completed during that time, based on Provençal models. Among these paintings are the series after Mont Ste-Victoire, which stands some 20 miles (32 km) east of Aix, and the portrait of Geffroy. His studio in Aix has been preserved exactly as it was and is open to the public.

A second painter to immortalise the beauty and simple lifestyle of the south was not a Frenchman, much less a Provençal. Tired of the dark and dreary lands of the north, Dutch-born Vincent Van Gogh became possessed by the dazzling light and colours of the Mediterranean.

Van Gogh first arrived in Arles in 1888. It was in this ancient city that he would paint such masterpieces as *The Arlesienne*, *The Starry Night* and the series of *Sunflowers*, each of which offers a vivid account of some aspect of Provence. It was also here that he drove himself to madness through too much sun and absinthe and too little temperance.

After convincing Paul Gauguin to join him in what he hoped would become a working community of "Impressionists of the South," Van Gogh's health deteriorated drastically. The two exceedingly strong-minded men clashed constantly and worked with equal fervour. Finally, after a particularly violent argument, Van Gogh suffered the first of his "fits" and sliced off his earlobe.

In April of 1889, Van Gogh voluntarily repaired to a mental asylum at nearby St-Rémy. He was soon allowed to paint once more and continued with his studies of the area. At first, his work was confined to the asylum garden, but even these are among some of the greatest works of the modern age. In May of 1890, however, he chose to return north to be closer to his brother, Theo, and farther from the blinding sun, which he was convinced had contributed to his mental fragility. Two months later, he committed suicide by shooting himself in the breast.

The turn of the century witnessed many more painterly visitors. Martigues hosted Corot and Ziem, while Cassis boasted a long list of luminaries, including Dérain, Vlaminck, Matisse and Dufy.

Less artistic types were also busy. The industrial revolution, which came fairly late to France, did not leave Provence untouched. By the 1870s, industrialisation had carved a distinct foothold in the region. Between 1876 and 1880 alone, the primarily rural *département* of the Vaucluse lost some 20,000 inhabitants to the cities. Although still overwhelmingly agricultural in nature, the economy was changing.

The onslaught of phylloxera, which destroyed half the vineyards of the Ardèche and massive numbers of orchards in the Rhône Valley in 1880, contributed to the imminent overhaul of local economies. New economic avenues were explored and exploited. By 1901, extensive mining of the red and yellow cliffs of upper Provence for ochre was underway and, by 1914, the Vaucluse was exporting 56,000 tons of ochre a year. This, in turn, changed the population by encouraging the immigration of potential labourers, many of them Arabs.

Another new industry that encouraged, indeed thrived on, foreign interests was that of tourism. A scattered number of English aristocrats had begun wintering on the Riviera during the late 1700s, but not until Lord Brougham took a fancy to the town of Cannes did tourism become a profitable staple in the gross regional product.

Nonetheless, agriculture remained the backbone of society. The size and distribution of fields changed little since the beginning of the 19th century. Even today they closely resemble the patterns begun in the 1500s. Most farms are fairly small. Generally, one central tract circles a farmhouse while other fields are interspersed among the holdings of other farmers.

World War I drained the countryside of its manpower, adding to the decay of the agricultural society. It also damaged the blossoming ochre-mining industry by immediately cutting off the Russian market. By 1917, all foreign markets had disintegrated, leaving the industry in dire trouble.

After the war, bauxite and ochre mining both climbed back to their former levels of production. Then a new catastrophe hit: the Depression. Much of the ochre had been used for house paint, and the collapse of the construction industry meant that demand plummeted. The ochre industry never managed to recover. World War II closed off foreign markets again and, shortly after its conclusion, the United States developed a synthetic ochre. Although still active, the mines of the Vaucluse would never again be the source of bounty they had once been.

World War II destroyed more than the ochre industry. Although a greater proportion of youths were killed during the first war, the region had remained physically untouched. But during World War II the Provençals found themselves living on a battlefield for the first time in over 150 years.

Left, Generals Diethelm, De Lattre and Cochet, Marseille-Toulon, August 1944.

On 21 June 1940, Mussolini struck out against the thin Alpine guard. Encouraged by the collapse of the French army under the Germans, the Italian dictator felt confident of his own victory, but the general armistice of 25 June cut short his onslaught. The region fell into an unoccupied zone during the ceasefire, but this semi-independence did not last. When the allied North African landings were launched in the winter of 1942, the Germans marched on Toulon and Marseille. Eager to get in on the deal, the Italians took over the Côte d'Azur. After the collapse of the Italian regime 10 months later, the Germans took possession of the entire area.

The Provençals were swept into the thick of things. Those who had not already been conscripted into the French army banded into fierce local Resistance groups. Particularly active in the mountainous regions, their guerrilla tactics brought relentless reprisals from the occupying forces. Meanwhile, the American Air Force began to mount counter-attacks against the Germans, causing additional damage to the once peaceful towns and destroying numerous historical monuments. While this went on, many innocent civilians lost their lives.

Finally, General de Lattre de Tassigny's 1st French and General Patch's 7th United States armies landed on the Dramont beaches just east of St-Raphael on 15 August 1944. By 15 September, most of Provence had been freed from the occupational armies and by April of the following year, the enemy had been eliminated completely.

Economic recovery came quite rapidly after the end of the war. The coastal areas, in particular, enjoyed a new rush in tourism, brought on by the sun-starved years of the war. But the bitter emotional scars caused by the years of occupation and by the widespread loss of life and limbs were not so quick to evaporate.

The Provençals had long become accustomed to their little niche in the southeastern corner of France and, even after losing all political autonomy, had felt fairly safe and separate from the crazy doings of the rest of the world. World War II shattered their sense of security and anonymity, resulting in an attitude of "who cares because who knows if tomorrow will even come?" – similar to the response that the threat of nuclear war produced in many of the next generation.

THE CÔTE D'AZUR SWINGS INTO ACTION

Today, the Côte d'Azur conjures up images of movie stars and suntans, roadsters and Monaco's royal family, casinos and pleasure boats. But the Riviera, as the Italians affectionately dubbed this strip along the Mediterranean, was not always considered so glamorous.

When English novelist Tobias Smollett came to Nice in 1764 to cure his bronchitis, he wrote home that it was a land of little but rude peasants and persistent mosquitoes. Nonetheless, he admitted, the gentle climate was remarkable. His praise for the weather encouraged a few brave Englishmen to visit, but the French stayed clear away, preferring the fashionable spas of Normandy.

The region remained more or less ignored until the advent of Napoleon I several decades later. His unsuccessful martial escapades resulted in the annexation by the King of Sardinia of everything east of Nice and south of the Var River. However, his construction of the *grande corniche,* which linked Nice to Italy, would bring the area its greatest break. In 1834, Lord Brougham, the English Lord Chancellor, and his daughter were forced to wait in Cannes while their favourite Italian winter playground struggled with a bout of cholera. Brougham was less than thrilled by the detention, but his tune changed drastically after spending a little time in the tiny fishing village. Eight days after his arrival, he bought land and ordered the construction of a villa.

The rage had begun. Within a year, a tiny but extant winter colony had sprung up in Cannes. Within three years, 30 new villas had been built and, the following year, the harbourside avenue now called La Croisette was created. Five years later, the appropriately named Promenade des Anglais was constructed in Nice to facilitate the newcomers' morning constitutionals.

The second half of the 19th century saw a Côte d'Azur in full swing. Any halfway fashionable Briton knew to winter in Nice, Cannes, Menton or Beaulieu. The elaborate Belle Epoque villas for which the Côte is still famous sprung up all over.

Russian society joined the throng. In 1856, King Victor-Emmanuel of Sardinia, still ruler of Nice, invited Empress Alexandra for a visit. Her approval encouraged the Russian nobility to add their own mansions to the coastline. So hectic did the waterfront become that a municipal order had to be decreed in May of 1856 forbidding anyone to bathe without underwear.

On 24 March 1860, the signing of the Treaty of Turin finally returned Savoy and the county of Nice to France. Although this made little difference culturally, it did have a huge economic impact, as tourism in the region was fast becoming one of Provence's most profitable businesses.

Until this time, the Côte had been appreciated mostly for its therapeutic virtues. This was long before Coco Chanel had made sunbathing chic and instead, "taking the waters" was the rage. But, in 1863, when Charles III of Monaco agreed to inaugurate the now-famous Monte Carlo casino, the serenity disappeared. The rigid morality of the Victorian Age proved to be no match against the seductive powers of the Riviera. Even the presence of Queen Victoria herself, who often vacationed along the Côte, didn't stop the mounting excesses of the wealthy Europeans who had made it their favourite home away from home.

Stories of the era are legion. One night, for example, the Russian Princess Souvranoff won 150,000 gold francs at the Niçoise casino. To celebrate, she rented a villa for 7,000 francs, but when her invitees contin-

American actress Grace Kelly.

ued to party past the witching hour of 7am, she simply bought the house for an additional 120,000 francs.

Along with the rich, the turn-of-the-century brought painters to the area. Signac, Matisse and Renoir all found inspiration on her shores. Earlier, the composer Berlioz, the violinist Paganini and the writer Guy de Maupassant had already discovered the delights of the sea air.

The Roaring Twenties brought newly made millionaires from the United States to the Côte, as well as the infamous Gertrude Stein and the wild F. Scott Fitzgeralds. The arrival of the Americans marked another change in atmosphere. Jazz clubs popped up all over, and its spirit encouraged Fauvists like Dufy and Bonnard to move in.

Perhaps the biggest revolution came in 1925 when Coco Chanel hit the Côte and created three of what were to become mondial institutions: the tank-style bathing suit, her now-classic Chanel No. 5 perfume, and the suntan. Until then, most people had come to winter on the Riviera. With Coco's help, summer tourism began.

When the stock market crashed in 1929, American visitors declined, but the French increased and the British remained faithful. The area thrived until World War II. It took something as momentous as a world war to dissipate the power of the Riviera, and dampen things it did. Afterwards tourism picked up again, but things never would be exactly the same.

Chic artists and glamourous movie stars replaced aristocrats as the new lords of the Côte. The first official International Film Festival of Cannes was held in 1947, the same year as Picasso moved to Antibes. Between 1948 and 1951, Matisse decorated the Chapel of the Rosary in Vence, and from 1957 to 1958 Jean Cocteau worked on Menton's town hall. French starlet Brigitte Bardot made history (and St-Tropez's name) when she appeared in the then-racy *And God Created Woman* in 1957. Prince Ranier of Monaco married the American movie star Grace Kelly.

This trend has continued up to the present. Nowadays, only movie moguls and oil tycoons are able to afford the sky-high prices of the good life on the Riviera. But, a new element has appeared, along with dozens of unimpressive condominiums and modern hotels. The Côte has become a place for tourists from all walks of life. And come they do, in droves.

The Belle Epoque Grand Hotel of Nice.

Coping With Modernisation

World War II heralded a new attitude throughout all of France, not just in Provence. What was once a boastfully agricultural country hardily embraced industrial expansion, pursuing such advances as the Concorde supersonic aircraft and nuclear power. Even agriculture became industrial in practice, and farmers were encouraged to modernise their operations with new equipment.

The change in attitude towards farming had no small effect on the already poor region of Provence, which had historically centred its economy around agriculture. The very essence of the Provençal character was – and is – to be found in the earth. Suddenly, they discovered their pride as well as their security threatened.

Some farmers tried to adapt, going deeply into debt to buy the new equipment. But the new tools proved effective only for extensive farms and not the mom-and-pop type that characterise those of Provence. These same farmers found themselves abandoned by the government, which was more interested in pumping funds into industrialisation then propping up Provençal farmers. Some tried to band together in cooperatives, but their independent natures proved unable to bear joint planning on a large scale.

In the 1960s, a few groups of French-style hippies made the move to the country. Most, however, became disillusioned and returned to their urban lives after a number of years. With them went many of the dissatisfied offspring of farmers, hoping to find "better lives" in the cities. Provence remained agricultural at its core, but for the next couple decades that core struggled.

By the early 1980s only 10 percent of the natives worked the land, compared with 35 percent 40 years earlier. Nonetheless, the decade rejuvenated hope for the Provençal farmer. The mondial back-to-nature trend has filtered into the French mentality, and people have returned to using natural substances. Now, in the 1990s, importers around the world are looking for pure high-quality goods that may cost more but taste better, are healthier or carry more prestige.

Left, corporate offices in Sophia Antipolis. Above, fishing is still an occupation in Marseille.

Vintners have always fared the best, since no substitute has ever been found for the intensive human care required in wine-making. Now, the small Provençal farmer seems optimistic that demand for other "specialised" products – such as fresh-pressed olive oil and all-organic herbs – will rise.

Despite continuing problems in several areas, Provence continues to be the top producer and exporter of fruits and vegetables in France, with 445,000 acres (180,000 hectares) of land under cultivation. Animals are also raised locally – particularly sheep, whose high-quality wool is greatly prized.

Close to the farmer is the fisherman. Methods may have changed in some areas, but along the coast at dawn you will still see plenty of locals readying their small boats for sail. The famed fish market of Marseille continues to convene each morning, and popular Mediterranean specialties like sardines, tuna, eel and red gurnet are caught by large trawlers and small-time anglers alike.

The coastline has also become extremely profitable in the areas of shipbuilding and oil refining. The latter, which is centred around the wide Berre Lagoon, has in turn led to a large quantity of dependent petrochemical facilities in Fos-sur-Mer, Lavera and Mède.

Industrial expansion in the Berre Lagoon area may have gained for the locals a prosperity unknown to their inland brethren, but it's been done at a great price. The countryside resembles no other part of the earthy Provence, as industrial estates consume the horizon. Also, local critics claim, no new jobs were actually created by the expansion. Instead, they say, the jobs simply changed from the more traditional and appealing ones of the past, such as boat-building and fishing, to the more impersonal ones connected with big industrial plants.

The industrialisation of Fos has brought the sea in about 1½ miles (2 km) closer. Meanwhile, even more ominous changes hang over this city. When Dutch engineers came down to help build the Port-de-Bouc they had to undertake massive dredging of the sea floor. While doing so, they discovered giant holes lining it, and it is believed that the whole city will eventually and inevitably drop down into the sea.

The enormous Berre Lagoon itself has remained quite beautiful, but it is surrounded by ugly row houses and low-hanging smog. It has been France's principal petroleum port for the past 60 years, but the construction of a new port in Martigues and the South European Oil Pipeline during the 1960s even furthered its eminence. It now pumps more than 65 million tons of oil a year.

The effect on the environment is unarguable. In 1965, you could swim or fish happily in two of the lagoons that lie between the Berre Lagoon and Fos-sur-Mer: the Etang de Lavalduc and the Etang d'Engrenier. Then the waters of the Lavalduc Lagoon, which come from Manosque, were discovered to be incredibly salty, 60 percent in fact. Profit-minded engineers drained its waters and placed them within the Engrenier Lagoon, from which they could more easily produce liquid industrial salt. Both were blocked off from the canal that linked them to the grand *étang*. Now, they are so polluted that you can see their filth from miles away.

Of course, not all industrialisation in Provence has produced such dire results as have those surrounding the Berre Lagoon. In 1952, the Donzère-Mondragon Power Station was inaugurated and, four years later, France's first nuclear reactor began operating in Marcoule. More power was created by the hydroelectric installations at Avignon, Caderousse and Vallabrègues along the Rhône, and St-Estève, Jouques, Mallemort, Slaôn and St-Chamas along the Durance.

As seems appropriate, most of the industrialisation of Provence has focused on using already existent local products. Cork is gathered from the Maures Forest. Ochre and bauxite continue to be mined, although their contribution to the economy is less than it once was. In the Camargue and by Hyères, salt is pumped from seawater between March and September and, once crystalisation has been completed, gathered between September and October. Olives are used not only in making martinis and salad oil, but lamp and oven fuel, soapmaking and, at its most extreme, fertiliser. The production of soap in France continues to be dominated by the *savonniers* of Marseille, although generally within large factories rather than the old-fashioned soap *ateliers*.

Lighter industries also depend on local raw materials. Almonds are used by the confectioners of Aix-en-Provence to make *calissons*, and fruits, especially berries, and vegetables are canned. The chic sparkling mineral water of Perrier is tapped from an underground spring near Nîmes.

Tourism continues to dominate the economy of the Côte d'Azur region and some of the more famous cities. In recent years, however, some smaller towns have begun to put more effort into wooing tourists as well.

Left, flamingos mingle with factories in the Camargue. Above, the Cadarache power station.

Meanwhile, as unskilled, factory-type jobs have grown and discontent in North African countries has come to the fore, the number of illegal or, at least, immigrant workers has also swelled. In a parallel movement, protectionism against the new foreigners has reared its head in the form of National Front leader Jean-Marie Le Pen.

Le Pen was not born in Provence, nor does he embody any particularly Provençal attributes. Nonetheless, this controversial political leader has found a remarkably large following in the region, especially in the area around Marseille. His policies, viewed by many as outrageously bigoted, are centred around using all "foreign" elements in France as scapegoats; they struck a chord with some disgruntled Provençals, and in recent elections, Le Pen has received an especially strong show of support from the southern regions.

Despite the many problems of modern Provence, however, the area does appear to be on the upswing. It has taken a bit longer for this region, deeply rooted for so long in the agricultural life, to come to terms with modernisation. But recent years have seen a more optimistic Provence learning to adapt to the changes – on its own terms, of course, and without losing its identity.

Haricot

LES PROVENÇALS

What is a *vrai* Provençal? Everyone has their own opinion. One person says it is someone whose grandparents were native to the region. Another says it is someone who lives and works in the same village where he was born and where his parents and his parents' parents were born, lived and died.

The explanations become more complicated. An Avignonais is different from the rest of the Provençals – "they are more stuck in their own world, prosperous, cosmopolitan." The people from the Alpilles are "both more conservative and more open than those from the Luberon," due to their greater poverty and dependence on outside markets. And those from the Alpes-Maritimes, even the Var, are they really Provençals at all?

Certain characteristics are true for the entire region. The Provençals are overwhelmingly a rural people with enormous respect and love for the land on which they live. Even the city dwellers know how to cultivate an excellent garden and can sense the oncoming change of weather.

Also, many Provençals are distant and, one might even say, suspicious at first encounter. But, if the visitor persists in trying to be friendly, he or she will find that the Provençals will quickly turn into the most generous and hospitable people imaginable.

Other similarities can be found in their appearance. Provençals come in many guises, especially since today many are sunseekers from the north or returned colonists from Algeria. For the most part, however, you can expect them to be on the smaller side, perhaps stocky, and darker than a typical northern Frenchman. They are, after all, Mediterraneans.

However, if there is one thing that marks the Provençals, it is the richness of their traditions – and their pride in them. Centuries of complicated history, foreign invasions and determined independence from the rest of France have left their mark.

Provence is one of only four areas in France to retain its own regional dialect. Not everyone speaks Provençal, but it is taught as an option in most schools, and in 1987, for the first time, a *faculté* in the Provençal language and literature was created at the University of Aix-en-Provence. The push to preserve the language is on, begun by the determination of the Noble Prize-winning native poet Mistral at the turn of the century.

Religion plays a big role in the life of the Provençal, although at times it may seem almost paganistic in interpretation. Harvest spirits and the beneficence of God, storytale monsters and Christian saints walk hand-in-hand, often enough combined within the very same legend or *fête*. Superstition also rages, even in the homes of the sophisticated. Never leave an iron out on the table of a Provençal hostess – it will undoubtedly bring her bad luck.

Preceding pages: festival participants in Arles. Left, farmer sells his produce in Vaison. Above, young girl of North African descent.

Traditions are taken seriously in this region. Consider the example of Christmas. Whether it be that of a poor rural farmer or a wealthy urban banker, nearly all Provençal homes still follow certain Christmas rituals that are unique to the area.

Christmastime begins on 4 December with *le jour de St-Barbe*. On this day, the children of each household plant a grain of wheat (or a lentil or chickpea) in a saucer. If it sprouts, this predicts good fortune for the coming year; if it doesn't, the household should prepare itself for a rocky time.

Originally, of course, the *blé de St-Barbe* was a purely agricultural ceremony. But, like many other Provençal customs, it has evolved to combine a curious blend of ancient agricultural rites with more recent Christian concerns. The superstition of St-Barbe itself is derived from a pre-Christian Middle Eastern ritual that referred specifically to the harvest year ahead. Today, however, when other lifestyles exist, it has been adapted to offer a more general omen.

Alegre!
Dieu nous alegre!
Cacho-fio ven
Tout ven ben!
De veire l'an que ven
Se sian pas mai
Sieguen pas men.

which translates roughly as:

Joy!
God gives us joy!
Cacho-fio comes
Everything goes well!
Let us see the year which is coming
If we are no more
Let us not be less.

The way the log lights is considered a portent for the year to come, and it should continue burning until the *jour des Rois*.

Around the same time as the planting of the grain of wheat, the crèches and *santons* are brought out of storage. These beloved clay Christmas cribs are set up in a specially designated corner, where they will remain for the rest of the season.

All this is just a preliminary for the rites of Christmas Eve. On that night, 24 December, everyone in the family gathers around for the *cacho-fio* (more or less, "hidden flame"). The very oldest and very youngest members of the household bless, with an olive branch soaked in wine, a log cut from a fruit tree. The elder then lights it, speaking the following words in Provençal, when they can:

After the *cacho-fio*, the family regroups around the *gros souper* in the same room as the crèche has been set up. The table is set with three cloths, commemorating the Birth, the Circumcision and the Epiphany, and is lit by three candlesticks, each containing one sole candle, which represents the Holy Trinity. In the middle are the saucers of germinated wheat from the day of St-Barbe, surrounded by a ribbon.

Although *gros souper* translates as "large supper," it is, in fact, a lean meal that traditionally included no meat, although only the most religious families still respect that stricture. It is followed by the *treize desserts* (13 desserts) after the 13 who attended the Last Supper. These desserts generally include raisins, almonds, dried figs, plums, apples, pears, candied lemon, quince jam, oranges, *nougats noirs* and *nougats blancs*, small round cakes, *pompes à huile* and wine.

Afterwards, Christmas carols may be sung – in Provençal if possible – and then those who are still practising Christians head off for Midnight Mass. Before going to sleep, the adults place presents by the crèche to be opened on Christmas morning.

The most dedicated Provençal families will don traditional clothing for the festivities of Christmas Eve. This "costume" recalls the style of the 18th-century bourgeois and is also worn during some of the many annual folkloric festivals. The colourful and lively festivals are a joy for the tourist trade, but Provençals are quick to remind that the *fêtes* are not "shows" or "charades" put on for the entertainment of visitors but something that they take seriously.

A large number of the festivals for which the Provençals are so famous are, unsurprisingly, connected to the earth around them: flower, olive and wine festivals, rites that beckon rain (such as the pilgrimage of St-Gens) or which give thanks for the harvest (such as the "sacrifice of the last blade"). Whole books have been written on the mythology of the Provençal calendar.

Of course, not every Provençal cares for gardens, crèches and traditional costumes. After World War II, many young Provençals left their home towns, rejecting the lifestyle of their parents for the ways of the big cities. In turn, they either lost interest in, purposefully shed or simply forgot the old ways.

But the traditions and costumes and festivals are far from being a specialised matter. The continuation of an independent and individual regional life is a very live issue among many Provençals. Groups are forming yearly with the specific purpose of protecting the Provençal traditions. In Arles, a "queen" is chosen every three years on the basis of her ability to uphold these traditions (i.e., speak the language) rather than her beauty or long legs. The phrase "a true Provençal" will be spoken with pride.

So what is a *vrai* Provençal? Perhaps the best description is this: "A true Provençal is someone who holds dear the traditions of Provence – its language, legends, costume – and actively works to preserve it."

Left, rice grower in the Camargue. Above left, St-Tropez denizen and right, a former "Queen of Arles" in costume.

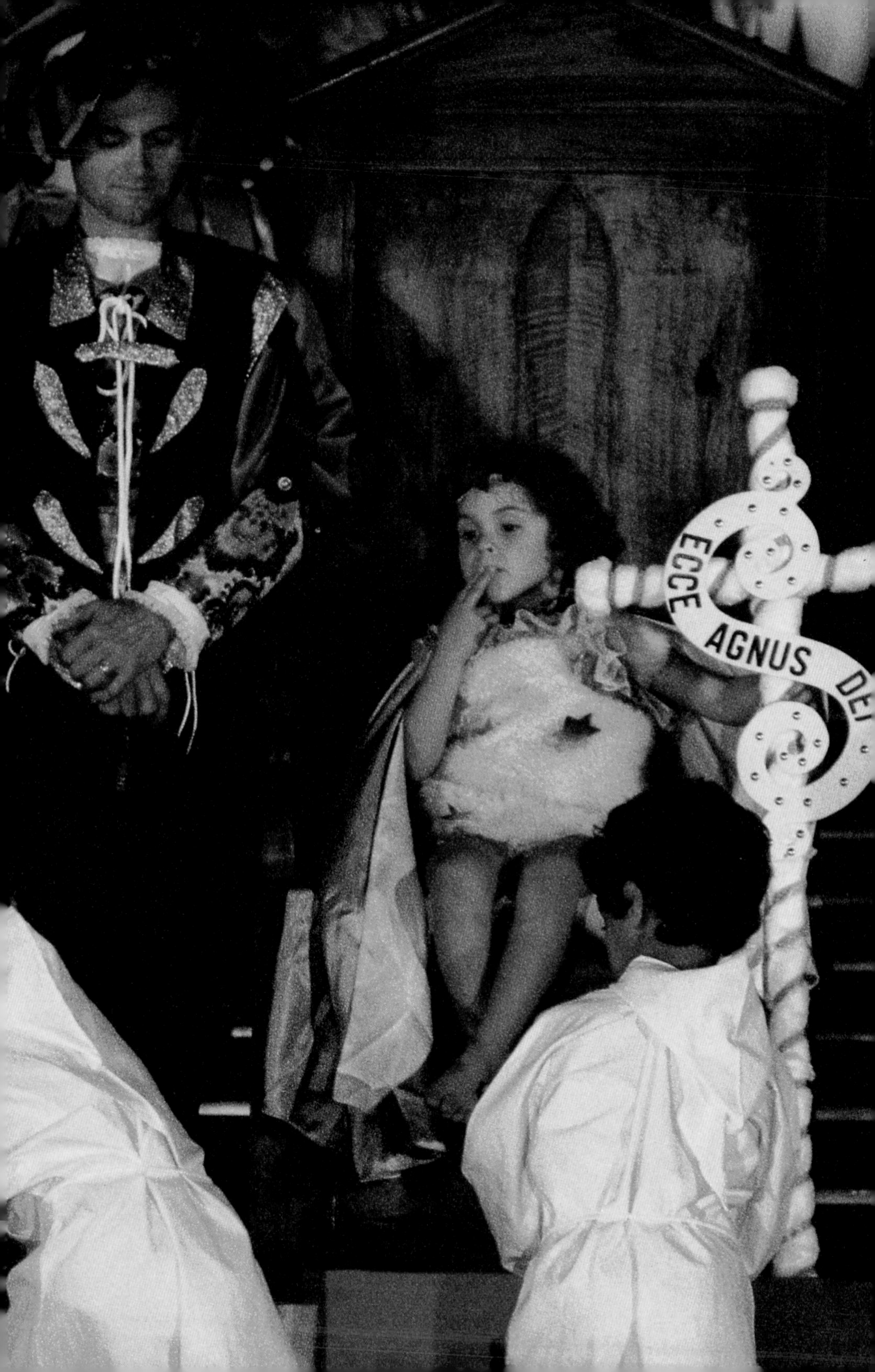
ECCE AGNUS DEI

CULTS AND PAGANISM

So strong are religious feelings in Provence that belief in divine will and miraculous intervention can be found behind even the most mundane activities.

In most places, for example, the decision to build a bridge would be made by local elders. At Avignon, however, the 12th-century bridge spanning the Rhône – the *Pont d'Avignon* of the song – came to be built after a shepherd-boy, Bénézet, claimed he had been commanded to do so by a voice he recognised as Christ's. Overcoming initial ecclesiastical hostility by singlehandedly lifting a stone big enough to form one of the bridge's piers, he set to the task. He was soon joined by a band of volunteers who called themselves the Brotherhood of the Bridge; together they completed the structure in a record eight years.

Other stories emphasise the religiosity of the Provençals by associating their region with the *dramatis personae* of the New Testament. Thus, Mary Magdalene, fleeing persecution in Judea with her (presumed) sister Martha and brother Lazarus – as well as Mary Jacoby, mother of Saint James the Less, Mary Salome, mother of John and James the Great, and their Egyptian servant, Sarah – were said to have been washed ashore in an open boat at the place now known as Les-Stes-Maries-de-la-Mer.

They celebrated deliverance by erecting a rough stone oratory dedicated to the Virgin. It was replaced in the 12th century by a fortified church that today houses a carving of the boat with the two Marys standing in it. In the carving, Mary Magdalene holds a chalice-like object often taken to be the Holy Grail, which she is credited with having brought to France, but which may represent the "precious ointment" with which she anointed the feet of Christ. This would link it with St-Baume on whose mountainside she passed the last 33 years of her life in a cave. Though *baoumo* is the Provençal word for a cave, in the high French used by ecclesiastics *baume* means "balm."

Meanwhile, Mary became a cave-dweller, and Martha is said to have journeyed inland along the course of the Rhône until she reached the town of Tarascon. She found its inhabitants terrorised by the dragon-like Tarasque, who made a habit of dining off anything that crossed its path. Undaunted, she singlehandedly sought it out and, confronting it with the Cross, reduced it to docile submission.

In 1474, Count René of Provence decreed that the miraculous conquest of the monster should be commemorated with an annual fête. Still held on the last Sunday in June, its climax is a procession in which a representation of the Tarasque, jaws snapping and tail whipping like a Chinese New Year dragon, parades through the town.

Appropriately enough, Saint Martha's remains lie in the town's church. They are, as one might expect, by no means the only holy relics to which the Provençals lay claim. The body of Saint Anne, apocryphal mother of the blessed Virgin, lies in the Apt Cathedral and that of Mary Magdalene in the Basilica of St-Maximin. The remains of Saint Sarah, adopted as the patron saint of gypsies, are to be found in Les-Ste-Maries, whose church has become the focus of a mass pilgrimage of gypsies each 24 and 25 May.

Most relics are credited with miracle-working properties, but the most powerful of all is the Holy Bit, made at the order of Helena, mother of the Emperor Constantine, out of two nails believed to have been extracted from the Cross. It is reputed to drive out evil spirits and cure eye disease, neuralgia and haemorrhages. Now kept at the Church of St-Siffrein (or Siegfried) in Carpentras, it is brought out on Good Friday and on 27 November.

Pagan whispers: Despite all this piety, the visitor, chancing upon a local festival, may well notice some startlingly discordant elements. Why are masked and costumed mummers permitted to intrude on so solemn an occasion as a Corpus Christi procession? Isn't the noisy explosion of blunderbusses during the *bravades* (the typical saint's-day processions on the Côte d'Azur) reminiscent of the means used as far away as Africa and China to frighten away evil spirits?

Preceding pages: Fête de St-Jean in Valréas. Left, traditional offering to a newborn baby.

The suspicion that one is witnessing something that owes little to orthodox Christianity becomes even stronger at feasts such as Saint Marcellus at Barjols. Each 16 January the reliquary of this stern 4th-century pope is taken from the church to oversee celebrations that include the ritual blessing and slaughter of a garlanded ox. Its carcass is then paraded about the town before being roasted whole as the townsfolk dance round the fire to the sound of flutes, tambourines and, as at the *bravades*, volleys of shots.

These are, of course, survivals from a distant pre-Christian past, also to be found in innumerable Provençal folk customs. Take for example the practice of preserving half-burned logs from the bonfires lit on Midsummer Eve – itself a pagan tradition – as protection from lightning.

Or consider the custom of giving the newborn an egg, some salt, a piece of bread and a matchstick. The egg as a symbol of the continuity of life can be traced back to the ancient Middle East and survives both in the Jewish Passover meal and the Easter egg. The same archaic tradition regarded salt as the symbol of prosperity and bread as the staff of life. The matchstick, undoubtedly a later addition, is said to ensure that the infant grows straight.

Hints of a pagan heritage are often to be found mixed in with pious legend. Saint Martha, on reaching Tarascon, was taken for the goddess Diana, which provides good evidence that a cult to the pagan deity existed there. Another old story tells how Saint Trophime, who came to convert Arles, consigned the goddess of love, Venus, to the nether regions.

Even edifying stories like that of Avignon's Bénézet contain their primitive themes. One is that of the shepherd as a link to the divine. Naturally, Joan of Arc hearing heavenly voices as she tended her sheep springs to mind; but Apollo, the most mysterious of the Greco-Roman gods and the inspirer of oracles, expiated his slaughter of the Python by serving as a shepherd in Thessaly and thereafter acquired the title of "protector of flocks." The popular Mesopotamian Shamash, with whom Apollo has many similarities, played a similar role.

Another prominent pagan theme is that of the hero proving himself through supernatural strength and, in particular, by lifting a heavy stone. It was by this means that Theseus asserted his right to the Athenian throne and that King Arthur did the same for the British one. The antiquity of this theme is proven by the fact that the Norse Odin underwent the same test and that something very like it is shown on a Hittite stone carving.

Celtic links: How did such patently alien ideas originate? The principal answer is that, perhaps more than any other region of France, Provence has always been a crossroads. From the 6th century BC, its first inhabitants, the Ligurians, played hosts to Greeks who crossed the Mediterranean to establish trading posts along the coast from Marseille to La Ciotat. Not long afterwards invading Celts swamped the Ligurians, and then in the 2nd century it was the turn of the Celts themselves to be invaded, this time by the legions of Rome.

Each visitor left a legacy. The bullfighting found here, as in Spain, may well have come from the Greeks. The ox sacrifice at the Feast of St Marcellus suggests Mithraism, the Persian cult popular in the Roman army up to its conversion to Christianity.

Concrete evidence for this mixture of religious influences comes from excavation sites, such as that on the Glanum plateau, a short distance south of St-Rémy-de-Provence. Artifacts found there include the remains of temples, baths, a forum and a triumphal arch erected in the time of Caesar Augustus (31 BC to AD 14) as well as a temple dedicated to the Phrygian Great Mother, Cybele, whose practices, forbidden to Roman citizens, included self-castration by her male devotees.

But, of all the very ancient pre-Christian peoples, those who left the most indelible paganistic footprints wherever they trod were the Celto-Ligurians. Their remains, too, have been found at Glanum. Among them is a lintel supported by pillars, which, as is typical of Celtic sacred places, lies close to a natural spring. Skull niches cut into the lintel itself are testimony to the custom of head-hunting.

Similar discoveries have been made at other sites, such as Mouriès, Roquepertuse and St-Blaise. At Entremont, 15 male skulls have been uncovered, some retaining the nails by which they had been affixed.

Celtic traces to be found throughout Provence also include linguistic ones. For example, the word *aven*, used to describe the holes bored into the calcareous rock by rainwater, means "a well." The *esprit fantastique* bears a striking resemblance to the British Puck and to those mischievous fairies, the *sidhfolk*, descendants of ancient gods, to be found in rural Ireland.

But most Celtic of all is the very ability to absorb outside influences, Christian as well as pagan, and make them their own.

Left, votive statuary to the Virgin Mary beside Mt. St-Victoire. Above, reliquary of Mary Magdalene in Maximin-Ste-Baume.

D'OLIVES
VIERGE
DES BARONNIES
47 F
FERME DE CLOSONNE
"Elevage au Grain"
26170 VERCOIRAN
75 28 13 42
HUILE D'OLIVE
VIERGE
DES BARONNIES
-47 F-
1ère pression à froid
100 cl -
FERME DE CLOSONNE
"Elevage au Grain"
26170 VERCOIRAN
75 28 13 42

VIERGE
DES BARONNIES
47F
FERME DE CLOSONNE
"Elevage au Grain"
26170 VERCOIRAN
75 28 13 42

THE CUISINE

The traveller to Provence is faced with a delightful choice when looking for a restaurant. After eliminating the obvious tourist-traps, only two kinds of restaurant remain: cheap and excellent – and out of this world. If you are wondering how to recognise the former, don't worry. Just look for the paper tablecloths, hard wooden chairs and hand-written sign that says something like:

Chez Claude
Cuisine Provençale
Menu 70F et 90F
Tomates Farcies ou Soupe de Poisson
Dorade Grillée ou Daube Maison
Fromage Dessert

Such signs, with appropriate variations, hang over 100 restaurants from Marseille to the Italian frontier. Inland, towards Avignon or Sisteron, the soup will be vegetable, the fish will be trout, and the prices will be slightly cheaper. In all cases, however, the service will be warm and friendly, the portions generous and often the meal will be served on a sunny terrace, under the vines or in view of the sea.

In selecting a meal, the first-time visitor to Provence should choose a dish containing tomatoes. If any vegetable or fruit symbolises Provençal cuisine, it is the tomato, the term for which, in Provençal, can be literally translated as "love apple." Anyone who has had more than three meals in Provence can attest to its local popularity.

In addition, if any method of cooking symbolises Provence, it is the stuffed vegetable – the *farci*. Like everyone else, the Provençaux stuff meat, fish and fowl, but the *farcis* they do with vegetables are special. They stuff eggplants with onion and tomato, onions with garlic, and cabbages with sausage and parsley. A traditional recipe, adopted by the fashionable chefs of the *nouvelle cuisine*, is courgette flower *farci* – stuffed with the flesh of the courgette itself. As for the tomato, it is the queen of the *farci*.

The *farci* may be a fairly simple dish to prepare, but the *soupe de poisson* (fish soup) is not. The latter is a dirty unappetising brown liquid, served in a big white tureen from which escapes the delicious, inimitable smell of myriad tasty sea creatures that go into its making. These are all the *poissons de roche* (literally "rock fish") that hide in the shadow of the indented Mediterranean coastline – including the *rascasse*, the ugliest fish on God's earth, bony *girelle* and *rouquier*, crabs and eels – and are caught at night from the little wooden boats called *pointus*. Until dawn when the nets are pulled in, not even the fishermen know which of the 20 or so varieties they will bring to shore. In towns like Beaulieu-sur-Mer, where the fishing fleet cater to at least a dozen restaurants, the competition is fierce. As the boats chug home around the headland, the restaurant chefs position themselves on the quays, to get the pick of the night's catch.

A first-rate *soupe de poisson* can take hours to prepare. In the traditional manner, the fish is crushed, then hung in linen sheets from the kitchen's rafters to squeeze out the rich juice. Nowadays, some cooks have replaced the sheet with a sieve that presses the juice out more efficiently. But many fishermen's wives, the acknowledged experts, refuse to believe that the taste is the same. They prefer to stand on chairs in their kitchens twisting linen sheets rather than make changes to a recipe handed down through several generations.

Another controversy is that of the *rouille*, a thick red spicy sauce whipped up into a kind of mayonnaise and served in most restaurants with *soupe de poisson*. You spread the *rouille* on the croutons that you float in the soup before sprinkling grated cheese on top. Most diners look forward to the *rouille* and consider it to be an integral part of the soup. However, the supporters of linen sheets claim that the *rouille* is a recent addition and clashes with the rich taste of the fish. In older recipes, they point out, you just rubbed a little garlic on the croutons to bring out the taste.

This may not seem of great importance to the traveller, but it is a subject of heated discussion in all the seaports of the French Mediterranean. Indeed, it is almost as heated as the rivalry between the supporters of

Preceding pages: olive oil. Left, M. Fonfon, a celebrated bouillabaisse chef.

soupe de poisson, as served in Nice and the eastern ports, and those of *bouillabaisse*, the speciality of the Marseille coast. The great difficulty in this comparison, however, is that *bouillabaisse* is not strictly speaking a soup. The fish are served whole on a side plate, and the liquid they were cooked in is placed in a dish in the centre of the table for all the diners to dip their hunks of bread into. (Or, in some kitchens, the broth will be ladled into each diner's bowl, then large pieces of fish will be carved up and placed within the bowls.)

Bouillabaisse is a dish to be tasted in the restaurants just off the old port of Marseille, where the cook struts past the tables in his undershirt and the clients wear striped jerseys or heavy black suits with a bulge under the armpit. Unfortunately, most of the restaurants of Pagnol's old port are slowly being turned into dry-cleaners or croissant-bars. The modern visitor has to go a little more out of his or her way, perhaps to one of the creeks or beaches where you can sit at a terrace, watch the boats and smell the sea – such as the Vallon des Auffes – to enjoy a good old-fashioned *bouillabaisse*.

Wherever, it doesn't come cheap. A good *bouillabaisse* or *soupe de poisson* is rarely billed at under 80F a head – even in the cheapest restaurants. And most restaurants with bargain-priced all-inclusive menus buy it frozen from wholesalers in Toulon or Nice who prepare it in the traditional way but on a large scale.

It takes a connoisseur to tell the difference between the fresh and the mass-produced *bouillabaisse*. However, that little difference is the one that separates the cheap and excellent from the out of this world. For the Provençal, the restaurant that serves frozen soup *manque de coeur* (literally "lacks heart"), an accusation that lies heavily against him.

Simple and fresh: So, what makes a real Provençal dish? On one level, it's the choice of the ingredients: the best olive oil, from a first pressing of the olives; good firm cloves of garlic – preferably from the little heads that are purple ("like wine stains," as they say in the markets), not the white lumpy ones that look like potatoes; little tasty courgettes (zucchini) and tomatoes, not the big watery vegetables imported from the north. Then come the traditional herbs – thyme and rosemary in sprigs and fresh basil in season.

All this is prepared *avec coeur* ("with heart"), which means the willingness to take extra pains and to work hard for little material reward. The cuisine explains something basic about Provençal life – about the precarious climate where the summer months may go by without a drop of rain and where an exceptionally cold winter may bring to nothing five generations of back-breaking labour. The Provençaux are small cultivators that fish and farm mostly in family units, and the produce of the land is valued tree by tree and row by row. When the Provençal family gathers around the table, the exact origin of each dish is known.

The purpose of Provençal cooking is not to disguise but to accentuate the flavours and to make the diners aware of each ingredient. Consequently, the cuisine includes none of the rich creamy sauces of central and northern France. The ultimate example of this emphasis upon freshness and simplicity is the Christmas meal that culminates in the *treize desserts*. The 13 desserts are actually 13 simple little side dishes, most composed of a single ingredient such as apples, figs, raisins, pears, almonds, walnuts, candied quinces or prunes. To mix all these ingredients, in the spirit of the English Christmas pudding, would be considered a kind of sacrilege. Even today, the spirit remains the same. The great chefs of southeastern France, such as Roger Vergé of Le Moulin de Mougins, return again and again to the simple side dishes of Provence.

Left, the ubiquitous Provençal tomato. Above left, pompes à l'huile from Marseille; right, herbes de Provence.

Vergé is also aware of the influence of time and place on how a meal is received. "It is particularly important to serve the wine cool and abundantly," he says, "and to place the table in the shade of a large tree. The dish is already suffused with sunshine, leading to thoughts of a siesta."

Just as the rosés of the south are dependent on hot summer days to bring out their flavour, so are the local dishes. It is no doubt for this reason that Provençal cuisine is less well-travelled than that of Lyon and Paris. A *ratatouille* in London or New York is little more than a limp vegetable stew. In Provence, however, *ratatouille* is an explosion of rich vegetable tastes.

Storytelling meals: The Provençal *ratatouille* is also the memory of a bustling market under the parasols, where you elbowed your way towards the piled heaps of aubergine and tomatoes. It is the sprig of thyme that your neighbour passed you over the garden wall, in exchange for a couple of freshly laid eggs. In other words, each dish in Provence is a story, something you can recount over dinner to your guests.

"Tante" Jeanne is the greatest storyteller of all. Jeanne runs a kind of restaurant near Apt, 30 miles (50 km) east of Avignon and

just north of the Montagne du Lubéron. She lives not far from the hamlet of Bouaux, at the end of a small dirt road which has no name, and she has no telephone. Nevertheless, on Sundays, the field beside her house becomes a parking lot for Mercedes and Porsches, as well as many more-modest vehicles. These cars will have come from as far away as Nîmes or Aix, and their owners will be there by personal introduction only. Jeanne is – by choice – in no directory or guide of any kind.

Lunch will begin precisely at 11.30, and latecomers will not be served. No choices will be offered: what is served to one is served to all. There will be only one wine: a Côtes du Lubéron bottled at the vineyard, and with great delight Jeanne will announce that only she and the owner's family have access to this *cuvée*. She will also narrate how she negotiated a certain vegetable out of season or a cut of beef that no butcher this side of the Rhône serves over the counter. The diners, in their Sunday finery (casual dress is not approved of), sit on hard wooden chairs at a single long table and respond in kind, praising each ingredient or comparing dishes with their own family recipes.

Jeanne's restaurant is a Provençal secret, but one that nobody bothers to keep because only the initiated will take the trouble to find it out. The Provençaux learned long ago that the best way to ensure a discerning, appreciative clientele is to open your shop or restaurant on the top of a mountain or in a shady backstreet. Those willing to follow the clues in this article and read between the lines will be able to find Jeanne's. On the other hand, if you are willing to go to that much trouble, you will make your own discoveries in Provence, which will be even more wonderful in that they are your own.

Choosing a restaurant: The casual tourist will do fine eating "Chez Claude" in excellent, moderately priced restaurants close to the town centres: in Nice, on the sunny Cours Saleya; in the old town of Antibes; in the Suquet of Cannes, behind the port; in the back streets of St-Tropez; in Arles, in the quarter running west of the Roman Arena; in Avignon, south of the Palais des Papes. In fact, in any town or village, in the plains or the mountains, where the tablecloths are clean and the waiters look cheerful, you can be guaranteed an honest meal and often one far beyond your expectations.

Those who are more demanding should ask advice of the locals they meet. They should talk to butchers and grocers, who know where the fresh produce is served. They should watch for the restaurants that need no advertising – little doorways in shady streets with small signs, packed with local diners and accompanied by chalk boards announcing the day's special – and consult departing diners, who are always glad to give an opinion.

Finally, once they've found a spot, they should put themselves in the hands of the waiter and defy him to produce a meal that will delight and surprise. Few waiters thus addressed will be prepared to disappoint. They may even serve you dishes that are not on the menu. As elsewhere in France, restaurant management is not so much a means of livelihood as a profession of faith.

Departmental variations: In Nice, a best-selling publication is *La Cuisine du Compté de Nice* by the former mayor Jacques Médecin, and one of the best known personalities of the city is probably "La Mère Barale". This *restaurantrice extraordinaire* once threatened to stop serving the author of the famous Gault et Millau gastronomic guide if he continued to include her in the guide. "I am getting old," she said, "and I have enough customers." The honest critic took the risk and printed her name once again on his list of top recommendations.

Mme. Barale can even make a gourmet dish of stockfish (*estocaficada*, in dialect). This terrifying Niçois speciality is based on dried cod, soaked in water for several days, then cooked with onions, leeks and tomatoes. From an incompetent kitchen, it can smell of boiled manure and taste of glue. *Chez* Barale, it has a delicate flavour that sets off the taste of the vegetables.

Left, goat cheeses. **Above**, market at Malaucène.

Barale's other speciality is ravioli. Some readers may assume that ravioli is a typically Italian dish. Anyone in Provence will put you right: *au contraire*, it is a Provençal dish exported to Italy, originally called *ralhola*.

The ravioli of Provence is stuffed with all manner of vegetables, meats and cheeses. The flavour will vary from town to town, subtly different according to the vegetables grown or the animals raised in each region. In the Camargue, the land of cowboys, you can get an excellent beef ravioli, served in a thick meaty sauce. In the rich plains between Nîmes and Aix, the stuffing is an assortment of vegetables. In St-Tropez, you can taste a *ravioli aux sardines*. As in Italy, the dishes of Provence are essentially local, reflecting the everyday reality of generations of farmers, traders and fishermen.

In the Alpes-de-Haute-Provence, on the high plateaux between Sisteron and Gap where the grass is frozen in winter and burnt away in summer, the shepherds drive their flocks across rocks and scree. The sheep, with their long legs and hard bitter eyes, have little in common with their fat complacent cousins in the lush valleys below.

In taste, also, there is no comparison. An *agneau de Sisteron* has a spicy herby flavour and is never oily or heavy. A southern chef will choose no other kind of lamb for his *gigot de mouton à la Provençale*, a leg of lamb that is boned then stuffed with herbs, garlic and sausage, cooked in white wine, and served in a sauce made of tomatoes, aubergines and green peppers.

Variants of the *gigot* can be tasted at most of the better restaurants in the foothills of the Alps, such as the excellent Auberge de Reillanne. In the mountains, lamb tends to be served stuffed or in a garlic sauce. In the plains near Nîmes or Arles, it is more often cooked on a spit or in a stew with potatoes and parsley.

The truth is that the roots and emotional connotations are what make the Provençal cuisine. The landscape, the love, the struggle and the victory are part of the recipe.

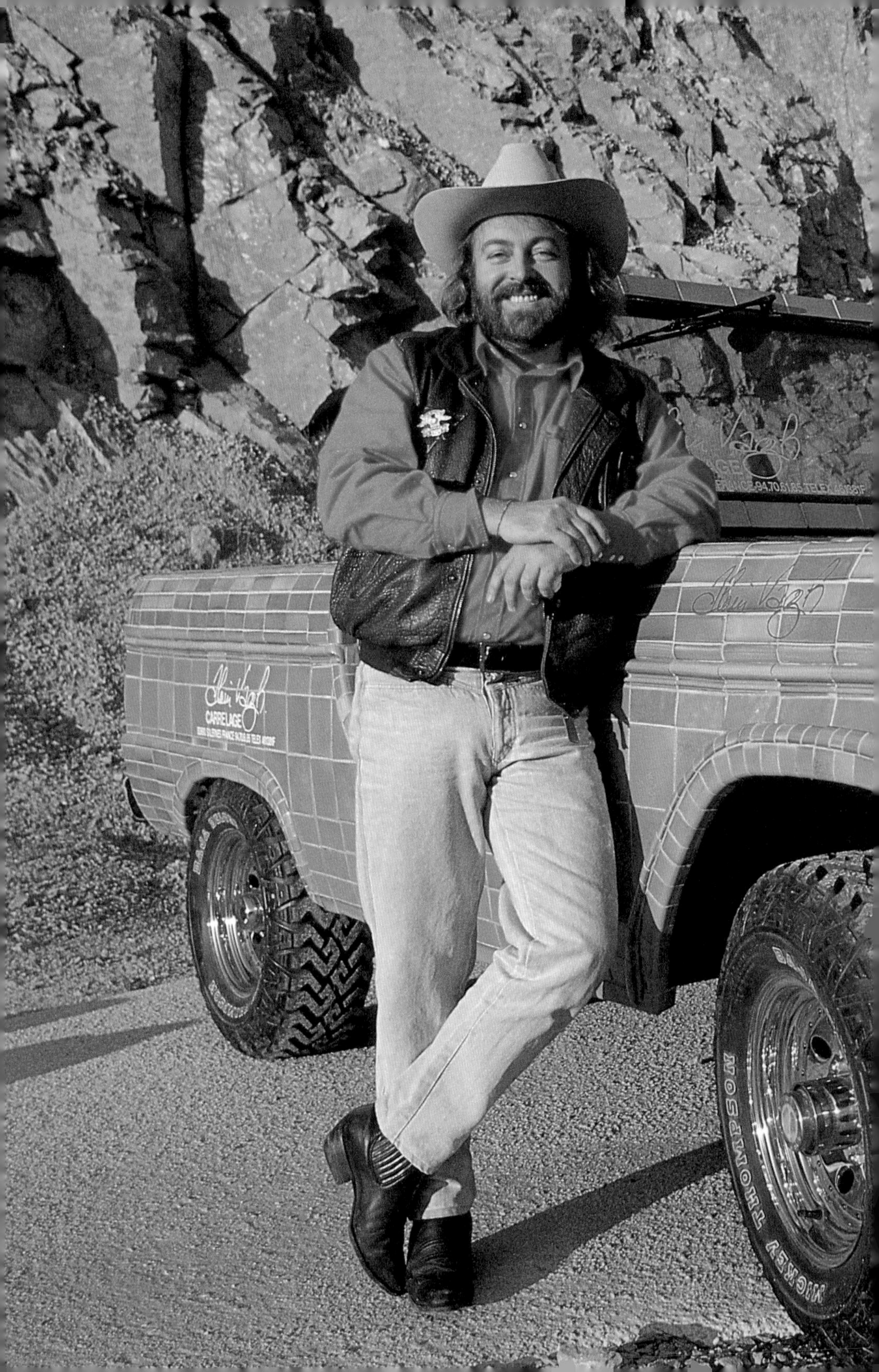
CARRELAGE

CARRELAGE
5842 TD 83

大和

When you bring home a piece of craftwork from Provence, you bring home more than just a pretty pot or well-made belt. Most Provençal artisans rely on local materials and regard their work as pieces of art. And more importantly, just as with so many other aspects of Provençal life, folk art has emerged from long-time regional traditions and is pursued and respected as such.

No folk art displays the importance of tradition more strongly than *santons* (little saints). These symbolic clay figures are used in a crèche, the scene that is laid out in Catholic homes at Christmas to represent the Nativity. Formed out of plaster-cast moulds and baked, their clothing and other details are then painted on with acrylics. They generally range in size from less than an inch (two cm) to five inches (13 cm), and differently sized figures are commonly placed in the same crèche to create perspective.

The most important figures, as in any crèche, are those of the Nativity itself: the infant Jesus, Mary, Joseph, the angel and the three wise men, all clothed in their Biblical costumes, plus the ox and ass. However, instead of Jerusalem, the Provençal crèche places the birth of Jesus in a typical and bustling 19th-century Provençal village. Clay models of stone houses are set up around the central stable. Moss and pebbles are used for the hillside; rosemary and thyme twigs act as trees.

A supporting cast of figurines represents all the tradesmen and society that one would find in a 19th-century Provençal town. They wear the traditional clothing of the period and carry a tool or product that symbolises their trade or social position: the garlic farmer; the fish merchant; women carrying newly drawn water; the gypsy; the coquette; the parish priest. In some *ateliers*, more than 100 subjects are created.

The original *santonniers* (or santon makers) came from Marseille in the beginning of the 19th century. They modelled their figurines after the village characters that were portrayed in *pastorales* (the popular Nativity plays) and in *crèches parlantes* (talking, mechanical marionette crèches), which were commonly staged in the streets of Marseille during the Christmas season. The famous Pastorale of Maurel, written in 1844, is still played at Le Cercle de la rue Nau in Marseille, and you can see a wonderful collection of crèche marionettes at the Musée du Vieil Aix.

Unsurprisingly, Marseille and the nearby cities of Aix-en-Provence and Aubagne remain the *santon*-making centres of Provence. The great *Foire aux Santons* takes place in Marseille for two weeks each December. This fair has become an integral part of the Christmas festivities, and most towns in the Bouches-du-Rhône have their own smaller version each year.

Although the figurines can be bought inexpensively from craft shops and at summer craft fairs, the best place to buy them is at the workshops of the *santonniers* themselves. Almost all have an attached boutique, and you can often take a look at the workshop too. There are a handful of recognised *maître santonniers* working today, whose work can be distinguished by all connoisseurs based upon their individual styles and repertory. The two best known, both winners of the prestigious "*Meilleur Ouvrier en France*" medal, are M. Paul Fouque, who works with his gifted daughter, Mireille, in Aix-en-Provence, and has the largest number of models in his repertory; and M. Marcel Carbonel, who heads the largest workshop in France and has a fun museum of crèche figures from around the world attached to his Marseille workshop.

Brilliant fabrics: Another traditional decoration that has kept a strong contemporary following are the printed cottons of Provence. Known as *indiennes* when they were first produced throughout France during the second half of the 17th century, they were made following the methods and designs of the imported *toiles peintes* of India. In 1686, the Royalty declared a halt to the importation and production of the Indian fabric. However, as part of the Comtat-Venaissin, Provence was not on royal land. The industry, therefore, grew in this region throughout the

Preceding pages: tilemaker beside his jeep of tiles in Salernes. Left, Collet pottery in Vallauris.

17th and 18th centuries, with factories in Orange, Avignon, Tarascon and Aix.

These fabrics, which formerly served as women's shawls and are now mostly used for interior decoration, are brilliantly multicoloured prints in kaleidoscopic floral and geometric patterns. Originally, the dyes were obtained from natural materials and each colour in a design was applied with separate woodblock impressions.

Only one factory has survived the mechanisations of the textile industry: a company based in Tarascon and owned and operated by a member of its founding family, Charles Demery. It trades under the name of Souleiado – a Provençal term taken from the works of poet Mistral and meaning "the rays of sun that pierce through clouds."

Decorative pottery: The region has produced earthenware since the days of the Romans. Pottery workshops complete with kilns dating from the 1st century AD have been excavated in La Butte des Carnes, near Marseille. A model of one of these kilns can be seen at the Musée de l'Histoire de Marseille, along with examples of amphoras and urns from the same period.

Over the centuries, pottery continued to be made in individual workshops, both for domestic purposes and for international export through the Mediterranean seaports. Then, in the 19th century, factories began to open in answer to the ever-growing demand for thrown or moulded cookware and for building materials such as tiles and bricks. The 20th century's economic and martial turmoil and the advent of aluminium and plastic, however, led to the eventual closure of these factories.

In Aubagne, formerly a centre for earthenware manufacturing, one such factory (called Poterie Provençale) still runs, turning out the kind of handmade, mass-produced casseroles and coffee pots made in the 19th century. But generally the days of utilitarian cookware production are gone. Now, most pottery is primarily decorative.

Three basic types of pottery are now produced. *Terre rouge*, the coarsest and humblest, is made primarily from the red clay of the region. This type of pottery can be recognised by its reddish-brown colour (if it is covered with an opaque glaze, look at the underside of the base to see the material from which it is made), its rough texture and its weight – it is never thin and fine, for this kind of clay cracks easily.

Objects such as unglazed moulded flower pots and tiles are still made in *terre rouge* for household use, while a wide variety of thrown forms such as jugs, bowls and mustard pots are sold as partly decorative, partly useful objects. Be careful, however, to check that they are really hand-done.

Roadside pottery stores, where they exist, are usually outlets for small family-run workshops. One such store is La Ceriseraie, in Fontvieille, a modest establishment where M. Monleau makes *la cuisinerie* (cookware) as well as *santons* out of *terre rouge*, and his prize-winning daughter Chantal creates *la décoration*, or *faïence*.

Traditionally, if *terre rouge* was the poor man's pottery, faïence was the rich man's. *Faïence* is made from the finest indigenous clays combined with various mineral elements to produce a greyish-white body that is formed in moulds, covered with a (generally milky-white) enamel shell, decorated with engraved or painted designs and glazed to a high sheen. The drawings will depict anything from entire hunting tableaux with garlanded rims to a simple initial.

Faïence, introduced to France by Italian ceramicists during the 16th century, was originally the unique province of aristocrats. By the 18th century it had become available to all as a popular form of decorative pottery. Elaborate tableware forms such as scalloped plates and platters, pierced bowls and tea infusers, which had been put into use by the aristocracy, joined figurines as decorative pieces displayed on walls or mantlepieces.

Faïence can also be found in simple utilitarian forms, but you can always distinguish it from plain pottery by its enamel porcelain-like finish. Good collections of Provençal *faïence* are owned by the Musée Arbaud in Aix-en-Provence, the Musée Cantini in Marseille and the Musée de la Faïence in Moustiers-Ste-Marie.

The town of Moustiers-Ste-Marie, in the Alpes-de-Haute-Provence, has been a centre of *faïence* since the 16th century. The "Moustiers-style," which is now made throughout Provence as well as in its namesake town, follows the patterns created by the grand 17th and 18th masters who lived and worked in the small town.

Perhaps the most distinctive motif in the "Moustiers-style" is the 18th-century *décor à grotesques* that represents fantastic monsters, the donkey musicians, monkeys and plumed birds. To find it, you might want to visit the town's most celebrated *atelier*, Segriès, which is regularly frequented by the fashion house of Hermès and other members of the modern-day aristocracy.

The third type of pottery is *grès*, made out of grey clay from central France. Contemporary artists introduced it to the region be-

Left, Camarguais cowboys wear shirts of traditional fabric. Above, Biot glass.

cause its harder consistency lends itself better than *terre rouge* to art pottery.

Grès comes in a wide range of original forms and is decorated with a rainbow of polychrome glazes that range from gentle variegated green-pink to earthy brown-speckled blue. The artists who use this medium draw inspiration from the pottery of the Orient as well as Provence and strive to reinterpret the classic beauty of traditional craftsmanship with originality.

Two leading exemplars are P. Voelkel, who works in St-Zacharie and displays his work both there and in Cassis, and Roger Collet of Vallauris. Although long a pottery centre, Vallauris became internationally famous after Picasso joined the local workshop "Madoura" in the 1940s, and it is now crammed with cheap tourist ware. Collet, a master of both *grès* and porcelain, was a colleague and friend of Picasso and is one of only a handful of reputable potters left in the town. However, you can always go to the Musée Nationale de Ceramique, situated next to the church, to see the products of Vallauris's illustrious past.

The bubbles of Biot: The nearby town of Biot offers a refreshing contrast to the shoddy tourism that has overwhelmed Vallauris in recent years. In Biot, a unique system is used to create swirls of bubbles in clear or iridescently coloured blown glass.

Founded in 1956 by Eloi Monod, a potter and engineer who wanted to revive the craft of glass-blowing in Provence, the original Biot glass factory has grown from one blower, Raymond Winnowski, and one glassmaker, Fidel Lopez, to 70 workers. The workshop where all the glass is blown is open to visitors, with the lusty bare-chested Winnowski presiding amidst the glowing coals. The adjoining store offers an array of beautiful yet surprisingly sturdy glassware in more than 100 different forms: glasses and goblets, jugs, vases and bowls, as well as many miscellaneous items such as decanters, oil lamps and knife stands.

Creating new traditions: Today, most handicraft artisans working in Provence are not natives of the region. In the 1950s, artists as well as movie stars were drawn to the Alpes-Maritimes. The 1960s, a period of social upheaval in France when many sought a "return to nature," brought an even greater wave of artists – known as the "Romantics of '68" – to the entire region.

Trained in art schools, the new artisans freely experiment, creating new decorative forms and introducing innovations to the traditional utilitarian ones. Visiting art galleries in towns such as Aix-en-Provence, Avignon, Marseille and along the Côte d'Azur, walking through the many craft fairs such as the *Foire aux Croutes* on the Cours Mirabeau in Aix, or stopping by communal workshops scattered around the countryside, such as Visse and Rosen's San Francisco-style *atelier*, Li Mestierau, at the foot of the Alpilles, one will find an eclectic variety of forms and styles. Many have no specific link to traditional Provençal folk arts: painted silk, woven murals, mobiles, olivewood sculptures, jewellery, watercolours and pastels, leather goods and wooden toys, as well as more traditional Provençal handicrafts such as pottery and puppetry.

Despite the great changes brought on by the industrial age and the advent of the new artists, the traditional handicrafts of Provence have not been lost to the present, and the richness of the craft tradition continues to be both a foundation and source of inspiration for the Provençal artisan.

Above, Moustierware motif. Right, M. Fouque, master santonnier, at work.

LA MARINE

PLACES

Provence consists of five departments with two sub-regions. Of them, the Vaucluse is known for lush landscapes, fruits and the sophisticated papal city of Avignon. The ancient Bouches-du-Rhône combines dry wide plains of olive trees and cypress with a Mediterranean coastline and includes the distinctly Roman city of Arles, the fountain-bedecked university town of Aix-en-Provence and the seaside metropolis of Marseille. Within it, the Camargue is an ecologically fascinating mixture of marshland, salt plains and beaches, populated by flamingoes and French-style cowboys.

Crowned by the naval capital of Toulon, the southern reaches of the Var also lie along the Mediterranean, while massive inland forests make it the most heavily wooded *département* in France. To the north, the Alpes-de-Haute-Provence rises to Alpine heights, with fortified villages clinging precariously to mountainsides. Its neighbour, the Alpes-Maritimes, is equally renowned for its hilltowns, becoming increasingly Italian in character towards its eastern border. As one nears the sea, one enters a different world: beaches and harbours, yachts and villas – the Riviera.

Along with disparities, certain very Provençal characteristics can be found all over. Summer festivals abound, as do fresh produce and a hearty but delicious cuisine. All year round, the legendary light of Provence, which has long lured artists here, dazzles mountains, plains and seaside alike. The people are warm and down-to-earth, and a love for folkloric tradition is reflected in much of their daily lives.

There is no question that where you spend your time will colour your impression of the region. But, wherever you go, you can find a little bit of the magic of Provence.

Preceding pages: the Gorges de Cians; evening on the Cassis waterfront; hilltown in the Var. Left, Niçois flower seller.

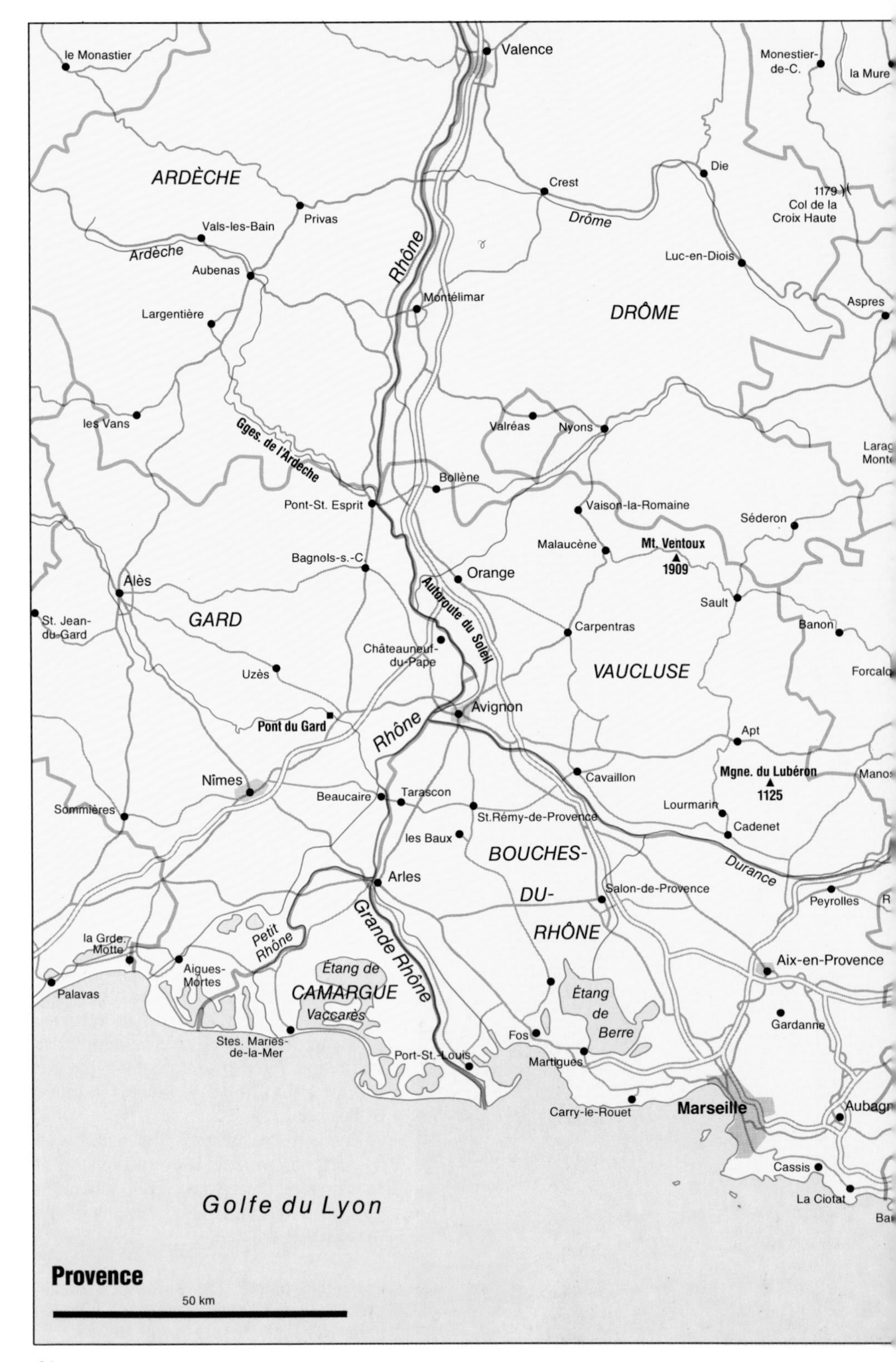

le Monastier
Valence
Monestier-de-C.
la Mure
ARDÈCHE
Die
Crest
1179
Col de la Croix Haute
Privas
Drôme
Vals-les-Bain
Ardèche
Rhône
Luc-en-Diois
Aubenas
Montélimar
Aspres
Largentière
DRÔME
les Vans
Valréas
Nyons
Gges. de l'Ardeche
Bollène
Pont-St. Esprit
Vaison-la-Romaine
Séderon
Malaucène
Mt. Ventoux
1909
Bagnols-s.-C.
Orange
Alès
Autoroute du Soleil
Sault
St. Jean-du-Gard
GARD
Carpentras
Banon
Châteauneuf-du-Pape
Uzès
VAUCLUSE
Avignon
Pont du Gard
Rhône
Apt
Nîmes
Cavaillon
Mgne. du Lubéron
1125
Beaucaire
Tarascon
Sommières
Lourmarin
St.Rémy-de-Provence
Cadenet
les Baux
BOUCHES-
Durance
Arles
DU-
Salon-de-Provence
Peyrolles
Grande Rhône
RHÔNE
la Grde. Motte
Petit Rhône
Aix-en-Provence
Aigues-Mortes
Étang de
CAMARGUE
Vaccarès
Étang de Berre
Palavas
Gardanne
Stes. Maries-de-la-Mer
Fos
Port-St.-Louis
Martigues
Carry-le-Rouet
Marseille
Cassis
La Ciotat
Golfe du Lyon
Provence
50 km

Massif des Ecrins
Mt. Pelvoux
3914
P. d'Olan
3564
PARC NATIONAL DES ECRINS
V. Chaillol
3163
Briançon
Abriès
St. Véran
Orcières
Embrun
Gap
Col de Vars
2111
Lac de Serre Poncon
Savines-le-Lac
Durance
le Lauzet-Ubaye
Barcelonnette
1991
Col de Larche
(Colle della Maddalena)
ALPES-DE-HAUTE-PROVENCE
Sisteron
Digne
Colmars
Verdon
Auron
Isola 2000
Valberg
Mézel
St. André-les-Alpes
Barrage de Castillon
1124
Col de Toutes Aures
Entrevaux
Moustiers-Ste. Marie
Castellane
Lac de Ste. Croix
Verdon
ALPES-MARITIMES
Var
Vence
Grasse
Aups
Barjols
VAR
Draguignan
Brignoles
Massif des Maures
la Garde-Freinet
Grimaud
Ste. Maxime
St. Tropez
la Croix-Valmer
Cap Camarat
Hyères
le Lavandou
Toulon
la Tour-Fondue
Port Cros
ILE DU LEVANT
Porquerolles
ILES D'HYÈRES
Fréjus
St. Raphaël
Massif de l'Esterel
Miramar
Cannes
Juan-les-Pins
Antibes
Nice
Villefranche-s-Mer
Beaulieu
Monte-Carlo
MONACO
Menton
Ventimiglia
San Remo
CÔTE D'AZUR
Mediterranean Sea
ITALIA
(ITALY)
Vinadio
Cuneo
Mondovi
Borgo S. Dalmazzo
Argentera
3297
PARC NATIONAL DU MERCANTOUR
St. Martin-Vésubie
Lantosque
Utelle
Sospel
Tende
Saorge
Breil-sur-Roya
UK.
BELGIUM
LUX.
F.R.G.
FRANCE
SWITZERLAND
ITALY
PROVENCE
MON.
SPAIN
AND.

EST
statues mystéri
ature et effort
NOUVELLE VIE
FLASHES
« Marcl ou rouille ! »

Découverte du cadavre d'une femme à Bollène
Les colis étaient bourrés de bijoux
Vaucluse Matin
le dauphiné
5e FESTIVAL INTERNATIONAL DE FOLKLORE
Une page quartet Cagnes
La rupture

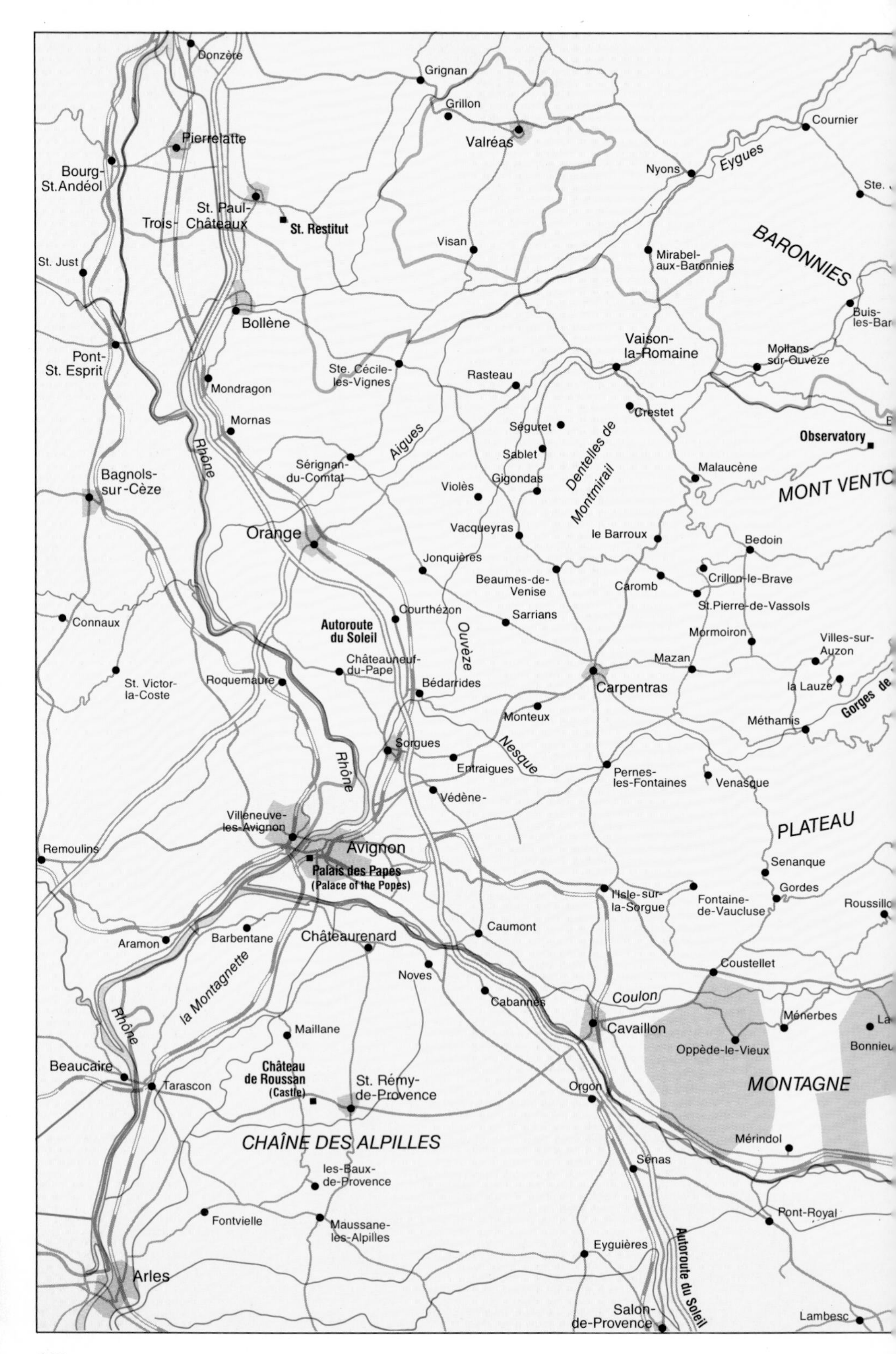
Donzère
Grignan
Grillon
Valréas
Cournier
Pierrelatte
Bourg-St.Andéol
Eygues
Nyons
Ste.
St. Paul-Trois-Châteaux
St. Restitut
BARONNIES
Visan
Mirabel-aux-Baronnies
St. Just
Buis-les-Bar
Bollène
Vaison-la-Romaine
Mollans-sur-Ouvèze
Pont-St. Esprit
Ste. Cécile-les-Vignes
Rasteau
Mondragon
Crestet
Séguret
Mornas
Observatory
Sablet
Dentelles de Montmirail
Aigues
Rhône
Sérignan-du-Comtat
Malaucène
Bagnols-sur-Cèze
Gigondas
Violès
MONT VENTO
Vacqueyras
Orange
le Barroux
Bedoin
Jonquières
Beaumes-de-Venise
Crillon-le-Brave
Caromb
St.Pierre-de-Vassols
Courthézon
Sarrians
Connaux
Autoroute du Soleil
Mormoiron
Ouvèze
Villes-sur-Auzon
Châteauneuf-du-Pape
Mazan
St. Victor-la-Coste
Roquemaure
Bédarrides
Carpentras
la Lauze
Gorges de
Monteux
Méthamis
Sorgues
Nesque
Entraigues
Pernes-les-Fontaines
Venasque
Rhône
Védène-
PLATEAU
Villeneuve-les-Avignon
Avignon
Remoulins
Palais des Papes
(Palace of the Popes)
Senanque
Gordes
l'Isle-sur-la-Sorgue
Fontaine-de-Vaucluse
Roussillo
Caumont
Aramon
Barbentane
Châteaurenard
la Montagnette
Noves
Coustellet
Cabannes
Coulon
Ménerbes
La
Rhône
Maillane
Cavaillon
Oppède-le-Vieux
Bonnieu
Beaucaire
Château de Roussan
(Castle)
St. Rémy-de-Provence
Tarascon
Orgon
MONTAGNE
CHAÎNE DES ALPILLES
Mérindol
Sénas
les-Baux-de-Provence
Pont-Royal
Fontvielle
Maussane-les-Alpilles
Eyguières
Autoroute du Soleil
Arles
Salon-de-Provence
Lambesc

THE VAUCLUSE

In a 1988 survey conducted by *Le Point* magazine, the **Vaucluse** topped the list of 95 *départements* in France for cultural activities, second only to Paris. However, any traveller journeying through this region's beautiful countryside will quickly become aware of a richness far beyond staged events.

Each town and little village within the Vaucluse guards its own special character, overtly expressed through unique and traditional *fêtes* and customs. More intriguing differences exist between mountain and valley people, town and rural life, and even the Provençal and French mannerisms, all contributing to the joy of discovering the region.

A similar diversity in terrain provides the visitor to the Vaucluse with a real choice of things to do. You can go mountain climbing or bird watch, pony trek or canoe, taste some of France's finest wines in Châteauneuf-du-Pape or indulge in a cuisine that draws on an abundance of locally produced herbs, fruits and vegetables. Or you can simply sit back and, as you drive, let the fragrance of lavender, the freshness of rosemary and the softness of pine waft through your open car windows.

It is important to choose your season carefully in keeping with your plan of action, for – unless you intend to visit one of the many summer festivals – there are good reasons to go in early spring or late autumn. During those seasons, the region is less hot, less crowded and more naturally "Provençal." Both the blossoming of springtime and the autumnal greens and browns have a magic that is a feast for the eyes and does wonders for the soul.

Whenever you go, the colour and texture of a land that embraces the snow-capped peak of Mt-Ventoux, the cherry blossoms of Malaucène, the ochre cliffs of Roussillon and, of course, the light that has inspired so many painters should be sufficient in-

Preceding pages: catching up on the news.

centive to wander away from the main cities and motorways to explore. And, as a general rule in the Vaucluse, the adventurous spirit is rightly rewarded. Some of the best finds are still tucked away along mountain tracks or forest-lined backroads.

From the north in Valréas to the west along the banks of the Rhône, southwards to the Durance River and east through Mt-Luberon, the Vaucluse has virtually natural borders. Historically, the region has always been a sort of crossroads. For centuries a trampling ground of marching armies, it has seen conflict between Romans and Gauls, Saracens and Christians, Catholics and Protestants and, more recently, between French Resistance groups and Germans. These many different conflicts have come together to shape the rich architectural heritage of the region.

Nowhere is this more apparent than in the papal city of **Avignon**, *préfecture* of the Vaucluse and gateway to Provence. Strategically situated near the junction of the Rhône and Durance rivers, Avignon was first established by the early Gauls as a tribal capital, and it was known to the Greek traders of Marseille. During the Roman period, however, it was overshadowed by Orange and (although some Roman artifacts can be found in the **Musée Lapidaire**, a former chapel of a 17th-century Jesuit college), most buildings that you can see today belong to a medieval past.

Avignon is, in every sense, a walled city, and 3-mile fortifications enclose its inner core. A walk along these ramparts, broken up by 39 towers and seven gates, reveals a cornucopia of historic buildings, churches and palaces.

Rabelais called Avignon "*la villa sonnate*," because of the number of steeples that adorned its skyline and, today, buildings of a religious character still outnumber the secular. On a less lyrical level, the battlements now also bear witness to the modern growth of the city. Factories and modern suburbs extend outwards from under the shadow of the wall to accommodate the city's population of over 100,000 inhabitants.

The Pont St-Bénezet.

The **tourist office** provides information on two good methods of getting to know the **old town**. The first is via a miniature train, which covers all the key points and the shopping zone. Later, having saved your energy, you can return on foot to explore in greater depth the buildings that most interest you.

Alternatively, the tourist office also offers an excellent guide service that takes you on either a half-day or full-day walking tour of the town, with commentary available in most of the major European languages. For the tour, it is best to book in advance as space is restricted. If you are travelling as a family or small group you can arrange to have your own guide.

Whether you go by bus or by foot, you must not miss the famous bridge, **Pont St-Benezet**, immortalised in the popular children's song ("*Sur le pont d'Avignon...*"), although today only four of its original 22 arches remain. Inspired by heaven, St-Benezet built the bridge at the end of the 12th century, but it was destroyed during the Albigensian War and reconstructed in 1234. The little **St-Nicholas Chapel**, which still stands on one of its piers, was also altered during this period and a Gothic apse was added to it in 1513.

There are four other places in Avignon that every visitor should see: the Papal Palace, the Cathedral of Notre-Dame-des-Doms, the Church of St-Didier and the Calvet Museum.

The **Palais des Papes**, symbol of the papal residency in Avignon (1309–77), is easy to find. As they say in Avignon, all roads lead to the Papal Palace. The Route Rapide, Rue Vieille Porte and Rue de la Monnaie all merge into its frontal square. Seen from this spot, the off-yellow stone, set against a clear blue sky, makes an imposing picture.

When you face the facade, with its two flanking towers, it becomes immediately clear that the palace was built for defensive purposes as well as residency. The former aspect of the structure was necessary, because the original city walls, which were built by Innocent VI and Urban V, were not enough to prevent the bands of wandering knights that plagued the countryside from attacking the palace.

You enter the palace through the **Porte des Champeaux**. Once you step into the **Grande Cour** (or "the great courtyard"), this fortress palace will quickly take on the sense of being a city within a city. Already you will be able to feel all the splendour that once belonged to Avignon.

The immense inner courtyard also acts as a link to the splendour of contemporary Avignon. It is here that many of the open-air theatre performances of the famed summer festival are staged.

It takes about an hour to walk through the palace, which is divided into two sections: the **"old" palace**, built by Pope Benedict XII between 1334 and 1342, and the **"new" palace**, begun under his successor, Pope Clement VI, and completed in 1348. You may want to buy a simple map at the little shop that stands by the entrance, but you won't need any help to recognise the extreme difference in style between these two popes. The old palace has a monastic

Wedding guest in Avignon.

simplicity and austerity that reflect the sombre character of Benedict XII. The new palace, on the other hand, is brightly decorated with fantastic fresco work and flamboyant ceilings, which have much to say about Clement VI, who was a patron of the arts and lover of the high life.

It is easy, almost fun, to get lost in all the rooms, but there are some slightly obscure things that shouldn't be missed. Be sure to look in on the row of portraits of the popes who resided in Avignon that hang in the **Aile du Consistoire** (Hall of the Consistory). Try to decide which one you would like to be seated next to at a sumptuous feast, such as were held upstairs in the **Grand Tinel** (Banqueting Hall).

Photographers might also want to note the great shot of the old town that can be taken from the window on the stairs that mount to the kitchen tower.

The most fascinating room in the new palace is the **Chambre du Cerf** (Deer Room). There your eyes will feast on the frescoes of Giovanetti, painted in 1343. These superbly colourful scenes of hunting, fishing and falconry, and of fruit-picking youths bathing in what looks like a modern swimming pool, give a valuable insight into life at the papal court in the 14th century.

If you were in any doubt about Clement VI's sumptuous tastes, the star-studded ceiling of the **Aile de Grande Audience** (Audience Hall), done in shades of blue and old gold, should dispel any last reservations. And, if it doesn't, the magnificent walls and windows of the **St-Martial Chapel** will.

Eventually, you will find the guardroom that leads back to the entrance. Before leaving, it is worth taking one last glimpse across the great courtyard. Now that you have experienced the interior, the definitive contrast in style between the old and new palaces should really take hold. It does not take an architectural expert to recognise which is which.

However, both parts of the palace are far from modest, although each in their own way. Their grandeur was received

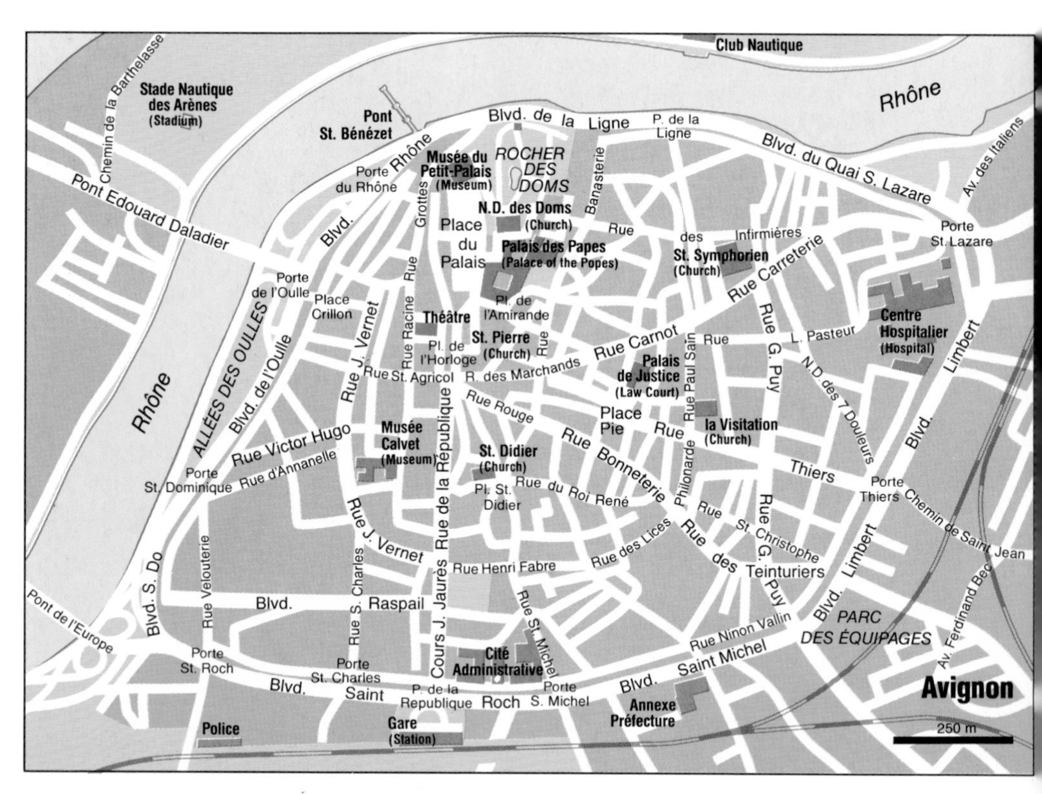

not without controversy. Petrarch, no lover of the court of Clement, called the palace "the habitation of demons," while Froissart, the well-known medieval chronicler, praised it as being "the finest and strongest in the world."

By way of contrast, the **Cathedral of Notre-Dame-des-Doms**, where John XXII is buried, is strictly a serious affair. This Romanesque, 12th-century building has undergone many architectural changes, but it still retains its original spiritual simplicity. Unfortunately, the addition in 1859 of a gilt cast-iron statue of the Virgin gives it something of the look of a wedding cake.

A church that hasn't changed much, since its consecration in 1359, is the **Eglise St-Didier**. It was the largest church to be built in the town during the Avignon papacy and owes much of its clean-lined attractiveness to the influence of ("old" palace) Pope Benedict XII. The elegant restriction of internal decoration is a reflection of the "purity" that Benedict supported.

Upon entry into St-Didier's first chapel, you will come upon one of the church's vantage points: an early Renaissance work by Francesco Laurana called *The Way of the Cross*. In 1953, more 14th-century wall paintings of the Crucifixion were uncovered, which are now slowly being restored. Another interesting feature is the hexagonal Gothic pulpit that stands in the centre of the church, a certain indication of the important role sermons played during the 14th century.

Shopping facilities in this area are good, and there are several spacious pedestrian zones that are lined with boutiques as well as open-air bars and cafés. As a general rule, the prices become more reasonable the further you go from the Palais des Papes. However, in July and August, all the streets of the old town (as in most of Provence) are hot and crowded, and even the smarter restaurants hereabouts do not offer the quality of menu or service that their exteriors promise.

During any month, a visit to the **Musée Calvet** will be rewarding. The

The Palais des Papes.

museum, which can be found on the Rue Joseph Vernet, was named after a doctor who was also an archaeologist and bibliophile. When he died, he left his large library, art collection and funds to start the museum.

Inside, you will discover a comprehensive study of the French and Avignon schools of painting and sculpture from the 14th to the 20th century. David, Gericault, Delacroix, Corot and Manet are just a few of the important painters whose works hang in this nicely lit museum. (Canvases from other painters of the Avignon school can be found in the **Petit Palais**.)

Musée Calvet's archaeological collection is a treasure trove of objects from the neo-Gothic period. Its holdings include finds discovered in the 1960s that have changed theories about the origins of Avignon and have suggested that its culture has much earlier roots than was previously believed.

A cultured city: Avignon's past is in many ways celebrated by the city's renowned summer festival. Started in 1947 by Jean Vilar, it continues Avignon's long-time tradition of cultural excellence. With the exception of Petrarch, who quite openly hated the city, many writers have sung its praises. For Stendhal, it was "*la ville de jolies femmes*" (the city of pretty women). The British philosopher John Stuart Mill chose to be buried within its walls. Frédéric Mistral, the great Provençal poet, was educated at the Old Lycée in Avignon and began his most famous poem, *Mireille*, while living here. So, it seems natural to find dramatists, actors, poets and musicians flocking to the city during the 20th century.

The festival runs from the first week of July to the end of August. Performances include classical theatre such as Shakespeare and Racine, modern theatre such as Becket, "off-Broadway" or fringe-type productions, street theatre, puppets and marionettes, dance, mime and café cabarets. Some orchestral concerts are held and, recently, a French-American film workshop was added to the agenda.

Cherries.

Given their own dramatic flare, the Avignon popes would undoubtedly have approved of the festival as well. At any rate, many of the shows are given in the court of their one-time residence. Other performances make imaginative use of local churches and cardinals' houses to stage their productions.

Don't be surprised, however, while sitting in a boulevard café during July or early August, to find yourself suddenly surrounded by a troupe of strolling players in colourful medieval costumes, carrying drums, pipes and lyres and acting out a full-blown drama. This is part of the charm of the summer festival. Gaiety exudes from every quarter of the city, and modern-day troubadours (who often sound more like Bob Dylan than Raimband d'Orange) will unannouncedly serenade your table and then, with the jingle of silver, move on to the next restaurant.

The town has a busy, but breezy and relaxed, atmosphere during this period. The serious productions arranged by the festival mix comfortably with more informal and often quite satirical street theatre. Just wandering around the old town will probably give you a couple of opportunities for entertainment, and a few minutes on the central **Place d'Horloge** is sure to gain you some sort of impromptu performance.

A good way for visitors to find out what, where and when is officially going on is to visit the **English Bookshop**, on the second floor of 23, Rue de la République. Here you can pick up a free copy of the *Langue Provençal Journal*. Despite its name, this small paper is written in English, and it gives a brief breakdown of events. In the same building is the **French-American Centre**, so Anglophones who opt to stay for the season and want a crash course in French will be all set. Francophones should head over to the tourist office, where there are hand-outs that list the festival productions and programmes of events in French.

During other times of the year, there are still plenty of things happening in Avignon. January welcomes an eques-

Bringing up baby.

trian fair, February an antique fair and mid-November heralds the Baptème des Côtes du Rhône, which is the festival celebrating the yearly first tasting of the regional wine. And regularly, throughout the year, there are musical soirées at the **Minit Conservatory**, open markets, up-tempo disco at the **Embassy Club** and quiet entertainment at the **Piano Bar** of the Hotel Mercurie next door to the palace.

Sporting types will enjoy a swim in the olympic-size pool at **Ile de Barthelasse**. Two other spots that welcome day visitors for a refreshing dip on a hot day are the **Squash-Racket Centre** and the **Golf Club** at Châteaublanc. Boat cruises along the Provençal waterways can be arranged at Allées de l'Ouile and 32 Boulevard St-Roch. Both places offer regular trips.

Across the Rhône: If you want to escape the hustle and bustle of Avignon altogether, you may prefer to stay across the river at **Villeneuve-les-Avignon**. This much smaller city has a rich and celebrated history all its own but lacks the hectic frenzy of its neighbour.

Towering over the attractive little town is **Fort St-André**, which guarded the frontier of France when Avignon was allied to the Holy Roman Empire. The structure is everything a fortress should be. Built in the second half of the 14th century by John the Good, it still retains a military prowess so palpable that you feel it the moment you pass through its magnificent defensive entrance with its strong powerful twin towers. The massive crenellated walls represent one of the finest examples of medieval fortification extant.

The other militaristic building in Villeneuve is the isolated **Tower of Philip the Fair**, which was once the starting point of the St-Benezet bridge. Both the St-André fort and this tower give excellent views over the papal city. On a clear evening you can see some magical twilight colours, as the sun sets on golden stone and the hefty silhouette of Mt-Ventoux in the distance.

Villeneuve rose to splendour during the Avignon papacy. As the papal court

Herbes de Provence at market.

grew in glamour, the number of adjunct cardinals also rose. Finding that Avignon had no more suitable space available, many of these cardinals chose to build their magnificent estates in Villeneuve. At one point, there were as many as 15.

Ironically enough, the one structure to survive from that period is the simple and austere **Chartreuse de Val de Bénédiction**. This Carthusian monastery was founded in the mid-14th century by Pope Innocent VI (whose tomb lies in the adjacent church) to commemorate the general of the Carthusian Order, who had himself been elected pope in 1352 but had refused the position out of a sense of humility.

Although you can't visit any of the cardinals' homes, you can treat yourself to a feast for one at the **Hostellerie La Magnaneraie**. This hotel's quiet restaurant can be found on the terrace, set against the fragrance of pines. Equipped with a swimming pool and tennis court, the hotel also provides a comfortable and convenient base from which to tour the region. Otherwise, Villeneuve offers a reasonable choice of smaller hotels, and there is an excellent campsite on the outskirts of town.

Itineraries: There are three basic journeys that can be easily made from Villeneuve. The first is towards **Châteauneuf-du-Pape**. This trip involves crossing the busy bridge to Avignon, but you can avoid the main-town traffic by joining N7 immediately and heading off in the direction of Sorgues. At Sorgues, join D7 going west.

On either side of the road, rows of green vines rise out of what looks like a beach of pebbles. In fact, this land was once washed by the waters of the Rhône. Today, the curious landscape marks the beginning of the villages whose vineyards are permitted to call themselves under the Châteauneuf-du-Pape appellation.

The legal history of the area is interesting, for it marked a major change in the history of French wine. In 1923, the local wine-growers applied to the courts for exclusive use of the Châteauneuf-

Medieval St-André Fort at Villeneuve-les-Avignon.

du-Pape designation. Until that time, there were no such restrictions on the labelling of wines. After a slow legal process, six years later, judgement was passed in their favour by an Orange court. The action gave rise in 1935 to the *appellation contrôlées* that appears on French bottles of wine today.

The vineyards of Châteauneuf-du-Pape are planted with 13 different grape varieties. The best wines achieve their complexity and character from blends of these grapes. Most red wines are the result of a blend of at least four grape types. Grenache is the dominant red-wine grape, followed by Mourvèdre, Syrah and Cinsault.

The end-product is what Alphonse Daudet nicknamed "the king of wines and the wine of kings." Châteauneuf-du-Pape is a supple, warm and full-bodied wine (made into reds and whites but no rosés) that goes well with strongly flavoured dishes like game, red meat and pungent cheese. This noble wine, perfumed with the scent of the Garrigues, can be sampled at any of the local vineyards. Particularly impressive are the cellars and vineyard of **Domaine de Mt-Redon**.

Above the village of Châteauneuf-du-Pape itself are the ruins and tower of the 14th-century **château** built by Pope John XXII as a summer residence. From here, you can look right down the valley of the Rhône over the red-tiled roofs of the village all the way to Avignon. Unfortunately, the château's strategic importance led to its almost total destruction in 1944, when German troops fought a scathing battle against the Resistance forces.

More opportunities to visit *caveaux de dégustation* (wine cellars where tasting is available) lie along D92 towards the market town of **Courthézon**. One valuable tip to remember about visiting the wine cellars is that it's always a good idea to offer to pay. This indicates to the proprietors that you are a serious taster, so you will be offered the better wines. And chances are the owners will refuse payment anyway.

Real wine lovers should head a little

Left, schedule of corridas at Nîmes. **Right**, young matador.

further along the route to **Beaumes de Venise**, whose wines are excellent. This is the spot made famous by troubadour Raimband d'Orange, who wrote and sung about medieval life in its castle. You'll enjoy the drive, for it is an especially pretty one, on a road lined with vineyards and neat rows of cypress.

Jews, truffles and wagoners: Back in Courthézon, it's about 5 miles (8 km) to **Sarrians**, where there is a fine example of late 18th-century architecture in the eccentric **Château de Toureau**. From here it's a straight run along D950 to **Carpentras**. Along the way, any of the side roads that lead in the direction of the **Auzon River** will take you to beautiful shady groves perfect for escaping the heat of the midday sun and enjoying a picnic.

Carpentras's strange collection of different architectural styles – from Roman to rococo – reflect its history. The town's name is said to have been derived from the Latin word "*karpenton*," meaning a two-wheeled cart drawn by horses. And, indeed, for centuries it was famous for its wagoners. Even before it was conquered by the Romans, however, it was called Carpentoracte and was the tribal capital of the Celto-Ligurian Memini. The conquest itself is recorded in the carvings on the **monumental gate** in the courtyard of the **Palais de Justice**.

By the 4th century, Carpentras had become a bishopric. Then, from 1274 until 1797, it was elevated to the position of capital of the Comtat Venaissin, and as such it was a part of the Holy See. During the first couple of centuries of this period, a thriving Jewish population enjoyed a liberal freedom of worship and were known as the "*juifs du pape*" (or, papal Jews). In testament to this singular period of religious tolerance, France's oldest **synagogue** stands behind the town's Hôtel de Ville. Built in the 15th century, one floor of the building now acts as a museum, but you may have to request permission from the curator to gain entry.

Also dating from the 15th century, the Romanesque **St-Siffrein Cathedral** is

Bullfight in the Nîmes arena.

further evidence of the church's good relationship with the Jewish community of that time, for the south portal is known as the **Jew's Gate**. Note the marble sphere that is depicted as being gnawed by rats and stands above the gate. Historically, the sculpture is interpreted as connoting God's anger, for the rat was considered the spreader of plagues. And, over the centuries, the Vaucluse was no stranger to the dread arm of the plague.

One man who was a good deal less than kind to the Jewish population was the Bishop d'Inguimbert. In the 18th century, he decreed that the Jews had to remain within their own ghetto and couldn't traffic with the better houses. Nonetheless, he played an important role in the cultural and architectural development of Carpentras.

D'Inguimbert ordered the building of the elaborate rococo **Chapelle Notre-Dame-de-Sainte** and founded the **Hotel-Dieu Hospital** in 1750. He also left the city a famous collection of books, on view at the **Musée des Beaux-Arts**. This is by far the most important library collection in all of Provence. It contains over 150,000 books and 2,300 manuscripts, including rare editions of Petrarch and an autographed score by J.S. Bach.

Music is further remembered in Carpentras during the Offenbach Festival. This exciting event not only stages operas but also includes visits from international ballet companies. The delightfully cool, open-air theatre where the concerts are held is a pleasant extra in the hot summer.

During intermission, you can enjoy some of Carpentras's sweet specialities. The town makes its own sort of caramel candy, called *berlingots*, and is also known for its strawberries. But most of all, Carpentras is rightly famous for its truffles, which some call "*les perles noires du comtat*" ("the black pearls of the county").

On the return journey along D942 towards Villeneuve and Avignon, make a short detour at **Monteux**, southwest along D13, to the **Saule Pleureur**. Set

Picnickers.

in the Quartier Beauregard, this restaurant has established a well-deserved reputation for gastronomic delight, using mostly local produce.

Back on the main road of D942, you will pass through **Entraignes-sur-Sorgue**. This village was a stronghold of the Templars during the 12th century, and one tower still stands of the fortress that this powerful group once commanded there. From here, the road leads directly back to Avignon, from where you simply cross back over the Rhône to reach Villeneuve.

A visit to Roman Gaul: The second journey you might take from the Villeneuve-Avignon area is to **Nîmes**. Although this will take you technically outside the borders of the Vaucluse and, actually, even Provence, it's a great introduction to Roman Gaul and is generally included in tours of the area.

To reach Nîmes you pass the **Pont du Gard**. This bridge and aqueduct are a testament to superlative Roman technology. The aqueduct spans the Gardon Valley; 2,000 years after its construction it remains in excellent condition

Vineyards slope away on either side of the route, and it is here that the Tavel wines are grown. However, modernity is the prevalent force in the area. Even as you approach the outskirts of the city, you will become aware of the emergence of modern architecture.

Urbanisation had already taken a firm hold of Nîmes 100 years ago. When Henry James sojourned here in 1884, he described the square where he stayed as having "the air of Brooklyn and Cleveland." When you gaze at the readily visible skyscrapers to the southwest, it is easy to agree with him that Nîmes's "only treasures are its Roman remains, which are of the first order."

The Roman ruins in Nîmes are wonderful. Nîmes was the first French city to be colonised by the Romans and for many years was their link with Spain. As a result, a number of impressive structures were built in the city by the Romans, several of which still stand.

The most outstanding monument is the **Amphitheatre**, built in AD 50. Just

The Pont du Gard.

slightly smaller than its counterpart in Arles, it was originally capable of seating 21,000 spectators at a time. It also possessed a gallery where slaves could sit and was designed so that it could be flooded for aquatic events.

Over the centuries, the amphitheatre experienced a variety of indignities. Visigoths substantially altered its form for use as a fortress. Then, it suffered further changes to accommodate a village for 2,000 poor people, including the addition of many houses and a chapel. Finally, during the 19th century, attempts were begun to unearth the original structure, which was by then hidden under 25 ft (8 metres) of rubble.

Today, the amphitheatre has been returned to something of its early glory. It now welcomes Spanish, Mexican and Provençal matadors in both traditional bull-fighting (where they do kill the bull) and Provençal-style (where they don't). During the summer, its stands are filled with eager fans.

Like the amphitheatre, the **Maison Carrée** had to go through a variety of different incarnations before arriving at its present state. This 1st-century BC temple was even used as a stable for a while, as well as a town hall and a monastery church. Although its original dedication is a matter of discussion – some say to Juno, others say to Jupiter or to Minerva – it is generally considered to be the best preserved of all Roman temples still standing.

The building, which has for centuries been known as the "Square House" despite the fact that it is twice as long as it is wide, now houses a **Museum of Antiquities**. Among its many Roman holdings is the lovely Venus of Nîmes.

On a hot summer day in Nîmes, you may well have found yourself quenching your thirst with a bottle of Perrier. This natural mineral water comes from a spring just a little distance southwest in the direction of Vauvert. To find out more about the strange and wonderful history of what is now one of the most popular drinks in the world, you should visit the plant at **Vergèze**.

It was Hannibal and his 30 elephants

Vineyard owner.

who supposedly first discovered the spring here, in 218 BC. Some 300 years later, the Romans also were enchanted by the cool, naturally sparkling water. Then, in 1863, Napoleon III took up its case. He decided the water should be bottled "for the good of France."

However, it was an Englishman, St-John Harmsworth, who, in 1903, first put the water in a bottle and marketed it, spurred on by a meeting with a Dr Perrier in Vergèze. What they did was to collect the abundant natural gas, found in the underground lake formed by the spring, and reinsert it into the water under pressure.

For a striking contrast to the modernity of the Perrier plant, take a look at the adjacent château with its Louis XIV architecture and its attractive gardens.

More of Rome: The last excursion from Avignon is a full-day's outing and brings together two of the most important Roman towns in France: Orange and Vaison-la-Romaine. This trip is sure to make you realise just how extensive was the culture and civilisation brought by the Romans and how it was adapted to fit the region.

Tiled roofs of Beaumes-de-Venise.

Take the N7 direct for 15 miles (24 km) to **Orange**. In the 2nd century, this southern city was of much greater importance even than Avignon. Today, it is still a busy place, partly because of the autoroute that passes by and brings a lot of through-travellers stopping the night, and partly because of the large daily market that is held between mid-April and mid-October.

The city's name actually has nothing to do with the fruit but, rather, is connected with the Royal Dutch House of Orange, who inherited the city in 1559 from the Chalon family. Its three most outstanding monuments – the **Arc de Triomphe**, the **Ancient Theatre** and the **Forum** – however, date back to the earlier days of the Romans, and each, in its own way, represents the force of Roman colonisation.

The Arc de Triomphe comprises three archways. The decorative friezes and carvings tell a proud tale of Roman success on land and on the sea, but their message is subtle. The weaponry, shields and helmets depicted on the lower part of the side arches are Celtic, and their positioning suggests Rome's triumph over Gaul.

The Ancient Theatre, which at one time could hold up to 10,000 spectators, is still in use. Orange's summer theatre festival and other cultural events all year long are staged here, while the statue of Emperor Augustus sits in the central alcove overseeing things as though nothing had changed. The acoustics are excellent in this well-preserved auditorium, but don't forget to bring a cushion, because the stone seats are hard and cold.

The Forum stands next door to the theatre. It is one of only three Roman gymnasiums still extant. Unfortunately, it suffered some drastic changes during the period of William of Orange, but you can still get a fairly good idea of its original size and shape.

For a clear explanation of how Roman colonial policy worked in Gaul, visit the **Musée Municipal** just opposite the Forum on the Rue de Pouillac.

Inside is a marble tablet, discovered during an archaeological dig in 1963, on which were recorded land holdings. This AD 77 example of land registry has markings that entend some 510 sq. miles (850 sq. km) from Bollène to Auzon. It not only reveals what areas were settled but also to whom the better plots belonged. This finding has given archaeologists a superb avenue for investigating the nature of early property ownership, on both a geographical and sociological scale.

The museum also has two other sections. One specialises in the history of Orange and the other in paintings. Of special note are the gallery's portraits of members of the Royal House of Orange and the paintings of British artist Sir Frank Brangwyn.

At the treaty of Utrecht in 1713, Orange was yielded to France. When Louis XIV said that the city's theatre possessed "the most beautiful wall of the kingdom," it is quite likely he was expressing his joy in having finally won over this valiant town as much as his wonder at the survival abilities of Roman architecture.

There is much more, however, to Orange than its Roman heritage. If you take the time to stroll through its streets, you will stumble upon some delightful old houses on attractive tree-lined squares and avenues. A particularly pleasant walk leads you up the hill behind the theatre, where the ruins of the **château** built by the Orange family can be found. It is easy to understand why they chose to locate their castle here, for it offers a magnificent view of the town.

Peaks of lace: The drive on D975 to Vaison-la-Romaine passes one large vineyard after another and encompasses some of the most picturesque countryside in the Vaucluse. On the way are several short detours well worth taking, for they will rapidly transport you to another century.

D8 brings you down to the hamlet of **Vacqueyras**, near to the already mentioned Beaumes-de-Venise. Before reaching it, however, you might prefer to switch onto D7 towards the beautiful

Wine château in Gigondas.

medieval village of **Gigondas**. The excellent wines produced here are well-enough established to be pricey, but there are still several lesser *caves de dégustation* were you can pick up a bottle or two without feeling you've spent a fortune.

Continuing along D7 will head you towards the prototypically Provençal village of **Sablet**, which in turn leads into **Séguret**. This charming village, with its steep streets and old gate, fountain, belltower, church and castle, looks like a film set for *Ivanhoe*.

In the village is a nice terraced restaurant called **Le Mesclun**. From it you can look down into the valley and point to the vineyard whose wine you would like to try. The lavish portions and mouth-watering desserts are as good an excuse as any for lingering in this calm and tranquil spot. As an added surprise, the owner has his own collection of Belle Epoque paintings.

The sharp limestone peaks that frame the area are **Les Dentelles de Montmirail**. *Dentelle* means lace in French, which should give you a good hint as to this rock formation's appearance. The jagged edges that jut into the clear blue sky present an attractive challenge to climbers, many of whom make special trips here just to tackle the slopes. The less energetic will also find the area great for rambles through the surrounding pine and oak woods and over the wild but beautiful countryside.

From Séguret take D88 back to D977 and you will soon reach **Vaison-la-Romaine**. By a direct route Vaison is only 17 miles (28 km) from Orange but, unlike Orange, Vaison's Roman past is very much a 20th-century discovery. Excavations only began in 1907. Vaison is divided into two distinct parts. The Roman town was built on a flat area of land on the east bank of the River Ouvèze. This is where the bulk of Vaison is located today. Walk down through the town and cross the Roman bridge and you will see the medieval village perched on a rocky outcrop. Clearly with the fall of the Roman empire the villagers no long found it safe to live on the exposed river bank and retreated to the more easily protected hill across the river.

The digs in the ancient part of the city have been done in such a way that you can begin to visualise something about life in Roman Gaul. There are two open sites. The first lies within the Puymin Quarter and is composed of a park dotted with attractive cypress trees and a handful of buildings. Uncovered have been the **House of the Messii**, the **Portico of Pompey** and the **Nymphaeum**. The portico is a type of pillared hall, and the Nymphaeum is the source of the town's water supply.

Reproductions of the statues that are housed in the adjacent **museum** decorate the park's **promenade**, bringing it some life. Actually within the museum are the originals, plus a helpful historical map of the province of Gallia Narbonensis and a variety of antiquarian exhibits such as jewellery, weapons, coins and ceramics. The imposing statue of Tiberius and two larger-than-life marbles of the Emperor Hadrian and Empress Sabina have a lot to say

Tea room in Séguret.

about the one-time arrogance of Rome.

The second excavated spot lies to the southwest of the Puymin site. Here, the ruins of a **Roman villa** and a well-restored and paved **Roman street**, whose mosaic floor leads to the arch of a former **basilica**, give the visitor an impression of the size and layout of the commercial part of the town during the Roman times.

The last major ruin in Vaison is the **Roman Theatre**. Like the one in Orange, it provides the stage for much of the town's summer festival, although it is much smaller in size. Cut out of rock from the north side of the Puymin Hill, it has been appreciated by theatre lovers for over 25 years for its dramatic, operatic and balletic productions. The festival begins during the first week of July, usually with a colourful and expensive folkloric and international gala, and runs through August.

Vaison's **Cathédrale de Notre-Dame-de-Nazareth** is a well-preserved, 12th-century church whose arcade pillars possess beautifully decorated capitals. Other attractions include a walk through the maze of the old streets to the **haute ville** (upper town), where you will find an interesting if dilapitated church and a ruined château that was once the country seat of the counts of Toulouse. Looking east from the top of the upper town will give you one of the best views out over the valleys and foothills that lead to Mt-Ventoux. This spot is slowly becoming an artists' colony and might be a good place to purchase local pottery and paintings.

A day-trip from Vaison will take you across the Aigues River north to **Valréas**. Valréas marks the northernmost frontier of the Vaucluse, in which department it is included for historical reasons despite the fact that it is surrounded by the Drôme River. The witty and literary Madame de Sevigné spent considerable time in this old Roman town and eventually built a château in nearby **Grignan**. There is more than a slight ring of truth to her criticism of the climate, since the mistral wind is par-

Girls' club of Vaison in local costume.

ticularly virulent here. However, she did have much praise for the local food and the immediate countryside, which is brightened with fields of bright yellow sunflowers.

Valréas and **Grignan** are still small, medieval-seeming towns, but the network of roads and number of trucks that roll down them are grim reminders that you are returning to the very different lifestyle of the 1990s.

As you drive back south towards Vaison, you will pass a number of new communities like those of **Pierrelatte** and **Bagnols-sur-Cèze**. These settlements have sprung up as a result of the construction of nuclear and military centres (which, of course, you won't be able to see from the road). All the original locals that know a thing or two have moved out.

After a hot and exhausting day amidst the ruins of Vaison or the secret industrialisation of the north, you may want to stop the night at **Les Terres Marines**, the local health spa. To reach it, take D98 south in the direction of Malaucène, go west onto D76 then look out for a right-hand turn about 2½ miles (4 km) along the road for Crestet.

Les Terres Marines may be difficult to find, but once there you will discover that it offers just about everything a weary traveller could desire. It is set in a truly peaceful spot, with a refreshing swimming pool shaded by oaks and a selection of magnificent walks. It also has a sauna, solarium, hot tub with jet streams and fully equipped gymnasium. If your back is sore from too much driving, a soothing massage by the resident physiotherapist will get you feeling back to normal. And if you feel that you've been appreciating the wines of Châteauneuf-du-Pape too liberally, you will be rejuvenated by the restaurant's tasty health-food menu.

Hilltop villages: After a visit to the spa, you should be ready for a few days in the nearby mountains. This part of the Vaucluse is worth devoting a good piece of time to, for there is a lot to see concentrated in a relatively small amount of land. Among other things,

Evening in Serignan.

the **Mt-Ventoux** area gives you the chance to do some horseback riding, climbing or hiking, any of which can easily run into a full-day activity. Moreover, driving around here takes a little longer than elsewhere, because many of the attractions involve fortified hilltop villages that often can only be reached by winding roads, followed by a walk on foot to a lofty castle or church.

Crestet, south of Vaison, is just such a hilltop village. It is set above olive groves and has both charm and a ghostly quality. Upon first entering, you may think that this 12th-century village has no inhabitants. Then, at twilight, you will see the odd light or two in a window – but, still, you won't hear a sound.

To visit the town, you have to leave your car at the outlying park and set out on foot. A climb up the cobbled alleyways brings you to the **church** and, after a further ascent, to the town's **castle**. From the lookout point on top are some stunning views across the valleys and the peak of Mt-Ventoux. Walks around other parts of the village will also offer some green and ferny spots to picnic.

Get back on D938 and head south to **Malaucène**. The fun main street bustles with cafés and restaurants where, during the season, you can sample some of the cherries for which this town is a capital. Malaucène also produces some delicious local honey.

The town's **church** offers a good example of how closely connected war and religion have been in this part of France. Built by Clement V in the 14th century, it is both fortified and castle-like in appearance from the exterior. As the iron-plated front door suggests, the church was the place of refuge for the townspeople during the times of war. Today, it functions on a more traditional level, and you will have to attend a service to hear the tone of the impressive 18th-century organ.

If you head out west from here in the direction of Mt-Ventoux, you will be following the route that Petrarch took in 1336 when he undertook to climb the legendary mount. First, however, you might want to make a quick dip south to

Horseback riding in the northern Vaucluse.

Caromb. This route, D938 to D13, will take you over a scenic backroad. Even if you don't stop along the way, you will be able to see, on both sides of you, a series of little fortified villages. Almost unchanged from the time they were first built, they crown the surrounding hills.

Caromb is another traditional example of the medieval hilltop village of this region. Long a vineyard haven, the town's well-preserved exterior walls enclose another typical feature of the area: a bell-towered **church** with an 18th-century wrought-iron cage for the bell, designed to protect it from the mistral. Inside, the church is furnished with frescoes and woodcarvings. Don't forget, however, if you elect to stay in the town's hostel, that the church's bell chimes all night long!

Another fortified town in the Mt-Ventoux area is **Le Barroux**, whose **château** and fine Romanesque **church** illustrate the locale's transition from royal to country seat.

The base of Mt-Ventoux is surrounded by many more pretty villages. **Mazan**, **Crillon-le-Brave**, **St-Pierre-de-Vassols** and **Bédoin**, to name a random few, are all charming in their own individual ways, and many are wine-tasting centres for the produce of the Mt-Ventoux vineyards. A good number of these *villages perchés* (literally, "perched towns") are best investigated on foot. Because of their defensive physical characteristics and dramatic locations, they fire the imagination with colourful pictures of knights, sieges and other scenes from the Vaucluse's turbulent past. A visit here will bring to life the songs of the troubadours.

Mt-Ventoux: After you've had your fill of picturesque hill villages, rejoin D974 for the best route to Mt-Ventoux. Placed along the immediate ascent, the small, Romanesque, 11th-century **Chapelle de Notre-Dame-du-Grosseau** is worth a look. On a hot day, drink from the nearby Source du Grosseau. The taste of the fresh water in this mountain pool far exceeds the promises of the average Coca-Cola television commercial.

Mt-Ventoux divides into two very

Lunch along the Ouvèze.

distinctive sections: the lower area, which lies below the ski resort and is ferny and forested and full of beautiful spots where one can stop to enjoy a packed lunch; and the upper, very harsh, limestone area that approaches the summit. When the sun reflects off the white of this higher part, the glare is so great it can leave you momentarily blinded. The peak is snow-capped for about three-quarters of the year.

There are two interpretations of the origin of the mount's name. One suggests that it is a derivation of the Celtic phrase "*ven top*," which means "white mountain." The other associates it with the French word "*venteux*," or "windy." And, indeed, there are always winds blowing around the Mt-Ventoux area. In fact, it is wise to be wary of this, for the gusts can become quite powerful as you climb, and lighter cars can easily be pushed all over the place.

The peak is ever-white, whether it be from snow or limestone, and has a certain science-fiction quality about it. This is emphasised by the placement of a radar station from which protude numerous television aerials and next to which is an observatory.

However, a stop in the **Chapelle de la Ste-Croix**, which stands nearby, should give you the chance to spend a moment meditating on the glories of nature rather than the intrusions of man. From up here, at 2,000 feet, it is also easy to appreciate why the Vaucluse has always been such an important crossroads. On a clear day – although mist and haze are more the norm – you can see the Rhône, the Alps and the Mediterranean. Some even claim to have spotted the Pyrenées.

The drive downhill towards **Sault** takes you through a nature reserve of particular interest to botanists. The 19th-century Provençal entomologist Jean-Henri Fabré discovered a rare species of yellow poppy growing here as well as a collection of other plants to which he attributed unusual medicinal qualities. One hundred years after his death, practitioners of various forms of alternative medicine – generally of the

The misty heights of Mt-Ventoux.

homeopathic variety – are now proving his theories to be correct.

Fabré also was influential in re-foresting this area, and the region is a paradise for nature lovers. As Fabré observed at the turn of the century, "half a day's journey, in a downward direction, brings before our eyes a succession of the chief vegetable types, as we should find in the course of a long voyage from north to south along the same meridian." And it is true. Pine trees give way to oak, then you spot fields of wild thyme and, eventually, in the lower valleys, the resilient mauve of lavender.

A benign statue of Fabré overlooks the main thoroughfare of **Serignan**, well to the west by Orange. The unassuming home and rocky outdoor "laboratory" where this celebrated scientist, affectionately dubbed by the public *L'homme des insectes* ("the insect man") conducted his research are also here and are open to visitors.

Not surprisingly, the Ventoux area is equally popular with ornithologists. They come in early summer with an eye to spotting sub-alpine warblers, sea crossbills and mountain thrushes. However, the decline in species of all sorts of wildlife, due to the rash of forest fires in Provence, is unmistakable. Campers, climbers and picnickers should take extra care not to leave matches or cigarettes smouldering.

The climate of **Buis-les-Baronnies** (actually in the *département* of the Drôme), nestled between Mt-Ventoux and the Baronnies Ridge to the north, is especially suitable for growing apricots, olives and almonds. And, in early July, the fragrance of lime blossoms fills the town. Not surprisingly, in the old quarter, many of the shops sell excellent samples of Provence's most genuine souvenirs: lavender soap, lime-blossom *eau de toilette* and olive oil.

The Gorges de Nesque: If you don't want to go north from Mt-Ventoux, you should probably head southeast towards **Sault**. During the 15th century, Sault was a baronial seat, but now the only remnants of glory are the towers of the 16th-century castle. The Roman-

Mountain life in Buis-les-Baronnies.

esque **Eglise de St-Saveur** stands guard over the town, which acts as a marketplace for the surrounding lavender mills.

The drive along D942 to **Monieux** is lush and green and lined with fields of lavender. If you stop in the town to lunch at the delightful **Les Lavandes**, you will be able both to enjoy a spectacular view from the terrace and mingle with the riders coming in fresh from the mountains. Watch as they tie up their horses in the square adjacent to the restaurant. There is a very pleasant feeling about the buzz of this village as it continues a time-honoured existence.

The Monieux road takes you through the **Gorges de Nesque**, which at all times of the year offer breathtaking views and a sensational cascade of colours. In some spots, the gorge plunges to depths of over 300 metres, and scrub and rocks camouflage profound caves. Despite the fascination of the mountain views, the driver should be sure to pay attention for tunnels and hairpin bends.

The Nesque Gorges have their own share of attractive Provençal villages. **Villes-sur-Auzon** has, like Mazan, had its fair share of disasters in the form of invaders, sieges and outbreaks of the plague. On top of that it bore being the battleground of Baron des Adrets. Nonetheless, it managed to survive and, today, has a steady if small population of a little less than 1,000 inhabitants.

The close-knit community lives off the cultivation of the vineyards that surround the town. You may want to stop at one of the town's *caves de dégustation* to try the local Côtes du Ventoux, a light-coloured "café" wine, which is best drunk when young and which, since 1974, has qualifed to bear an *appellation contrôlée*.

Like many mountainous areas in Provence, the Nesque Gorges make for great climbing and hiking country. An excellent starting point for these activities is **Le Hameau de la Lauze**, but you may well have to ask directions to find this rustic hostel. Located approximately 4 miles (7 km) from Villes-sur-Auzon, its turn-off is most obscurely

Festival in Valréas.

marked and at first looks like no more than a path.

Once you find the way, the winding drive down rough pebbled road, through woods, scrub and gorse brush, gives a clue to the adventure in store. And, indeed the owners of this *ferme-auberge* (literally, "farm-inn") mean what they say when they describe themselves as having a "*cadre rustique*" ("rustic setting").

La Lauze supplies guides, maps and excellent advice for hiking and pony trekking. Local walks through the wild fauna, flowers, herbs and oaks, olives, beeches, cedars and pines that dot the mountainside are sensational. You will find that often, on a hot summer's day, the temperature is just a little bit cooler than that of the valleys and villages of lower altitudes and that the air is fresh and invigorating.

The menu at La Lauze is strictly mountain men's food, and sleeping conditions would probably better appeal to an overnight commando unit. Nevertheless, the spirit is great, and the experience and expertise of the hosts have saved many an amateur explorer from spending an unexpected night on an exposed cliff face.

To the South: As the descent from Mt-Ventoux will have revealed, the countryside of the Vaucluse can change rapidly. Further proof of this will come along the journey south and slightly west to **Venasque** through the **Plateau du Vaucluse**.

Venasque stands quietly tucked away between two small but steep hills, cupped to the east by a deep and dense forest, in which it is easy to become lost. The village has a pleasant and unaltered charm that offers a good example of the benefits of careful restoration. Once the capital of the Comtat Venaissin, to which it gave its name, it still retains an identity of its own, although rather drastically subdued. In the early summer, it comes alive as a market centre for cherries and during the rest of the year has a couple of buildings worth viewing.

The **baptistry** within the **Eglise de Notre-Dame** is one of France's oldest religious buildings. Built during the 6th century and remodelled during the 11th, it still causes dispute among historians as to its specific function.

Also worth noting is the **Chapelle de Notre-Dame-de-Vie**. Constructed in the 17th century on top of a 6th-century site, it houses the tomb of Boetius, bishop of Carpentras and Venasque. The tombstone is an excellent example of Merovingian sculpture.

From Venasque, follow D4 to D177 south through the plateau. Looking down over the **Valley of the Senancole** will reveal the path to the **Abbé de Senanque**. The curved road at its side leads to an area for parking your car, about 1¼ miles (2 km) away from the abbey itself.

The 12th-century Cistercian abbey stands at the edge of a grey mountainside and is surrounded by oak trees and lavender. Its one curiosity is the manner in which the limestone from which it is constructed changes colour as the day heats up. The buildings seem to change from grey to a deep yellow.

Senanque possesses a natural serenity. Although there have been no monks living in it since 1969, it still has the feel of an abbey, and there is a remarkable quietness reflected in the pastoral setting, the simplicity of the building and the austerity of its interior. If you find the mood persuasive, you can retire to a room that has been set aside for meditation. This room also happens to be the coolest spot in the monastery, so you can emerge from it with both the spirit elevated and the body refreshed.

Now a cultural centre, the abbey has established a permanent exhibition on comparative religions. Among its holdings are a fascinating series of detailed photographs of the nomadic tribes of the Sahara and clear documentation of the origins of monasticism. During the summer months, musical concerts are held here. The acoustics are excellent, and Gregorian chants echo through the building as though nothing has changed since the abbey was founded in 1148.

The nearby town of **Gordes** is linked to Senanque by a narrow, winding and bleak road. But don't expect to find it deserted. During the summer, this town is one of the most popular tourist spots in the Vaucluse.

At first glance, the village of Gordes looks as though it is about to slip off the small mountain on top of which it is spread. Over the centuries, however, this peculiar location has contributed to the village's natural defence. Gordes has had a turbulent history, especially during the 16th-century Wars of Religion and then again during World War II when the town was a stronghold for the Resistance movement.

Most of the tourists come to see the "ancient" village of 20 restored *bories* (drystone huts) that lies just beneath the central town. This group of curious, wind-resistant dwellings in round or rectangular shapes are believed to date from the 17th century, but don't be surprised to find 19th-century tools and furniture in some. Others are privately owned and have been modernised for use as holiday homes.

As a bridge to the contemporary world, stop in at the **château**. The castle itself is Renaissance and old but, in more recent times, five rooms were set aside for the contemporary artist Vasarely. Within them, he established a study centre for the "interaction of the arts and sciences."

Overall, the château has a sobering effect. First built during the 12th century, it was radically altered during the 16th century to emphasise the strength of its fortifications. Moreover, within the **Vasarely Museum**, some of the kinetic objects on view have a disturbingly mesmerizing effect. It is with relief that one returns to the sunshine.

Summer homes, artists and pastis: From here, the drive south along D104 will take you into the foothills of the **Lubéron**. This third, and last, mountain range in the Vaucluse is also a general favourite. It is an increasingly popular spot for well-heeled Britons and Parisians to have summer homes. Appreciation for its natural beauty spurred the creation of a 247,000-acre (100,000-ha.) **Regional Park** in 1977.

Outside the park, the foothills make for great walking country. But remem-

Playing *boules* beneath the plane trees.

ber, come autumn, that the hunters here take their sport most seriously. Once a "boar drive" is under way, it's like driving in Paris – pedestrians are considered fair game. After the hunting season has opened, be sure to walk with your eyes and ears open.

On a lighter note, *pastis*, the legendary drink of the Provençal, had its origin in this part of the Vaucluse. First blended within the small villages of the Lubéron, this herbal brew is supposed to be the secret of the Provençal people's reputation for good health. It is nowadays usually associated with Ricard, a licorice-flavoured drink commonly referred to as the "good-natured thirst quencher."

The force of tradition is strong in the Lubéron. The mountain village of **Lacoste**, where stone quarries are still worked, is a good example of how slowly change comes to some parts of the Vaucluse. Rural life here very much resembles that depicted in the Pagnol films, and Lacoste's major acknowledgement of the 20th-century comes in the form of the floodlit *boules* court behind the local church.

Elsewhere in the village, traditional life continues much as it has for the last few decades: hanging out in the cafés, sipping *pastis*, playing the popular card game of *belote*. Perhaps all that is missing is the Marquis de Sade, who lived here from 1774 until his arrest in 1778. The ruins of his huge **château** look down on the village and, as the guide is wont to say, "If walls could talk…"

From the heights of **Bonnieux**, southeast along D109 to D3, you can see the junction of the Vaucluse Plateau and Mt-Ventoux. On one side of you, through an attractive mist, you'll look out over a forest of cedars and, to the other, south towards the Crau Plain, you can catch the glimmer of the Beise Lake and, beyond it, the Mediterranean.

The easternmost section of the Vaucluse, from Bonnieux to **Oppède-le-Vieux**, has the distinct touristic advantage of being less crowded during July and August than many other parts of the *département*. Yet, the area is as

Church in Lacoste.

rich in both natural beauty and historical interest as many of the more accessible parts of the Vaucluse.

Many of the stories written by the renowned Provençal novelist Jean Giono (1885–1970) dwelt upon the dying out of local villages in the 1920s and 1930s, and the "Petite Lubéron" could easily have gone that way. In recent years, however, a new brand of year-round resident has brought a fresh spirit to these small villages.

Writers and artists from all over the world, attracted by the tranquillity and the climate, have undertaken the careful restoration of many of the houses. Their tasteful work shows a great understanding and respect for the Lubéron's past, and their appearance has done much to improve the region's economic future.

All over, the soft yellow and red shades of the houses are both startling and attractive. Magic descends with the setting of the sun, as the colours become illuminated with unearthly light.

Particularly beautiful is the southerly **Roussillon**. This hilltop village, just east of Gordes, is constructed of an incredibly brilliant red, set off by the grey-green landscape that surrounds it. First made known as the subject of sociologist Laurence Wylie's *Village in the Vaucluse*, under the pseudonym of "Peyrane," Roussillon became the site of a real-estate boom in the 1960s. Although it is still charming, visitors of today will have to read the book to find its impoverished heritage.

The capital of the Luberon is **Lourmarin**. The countryside that surrounds this small "city" is colourful all-year-round, because the large variety of trees, wild flowers and vegetables that thrive in the region benefit from having the longest growing season in France.

Lourmarin could also claim to be a sort of cultural capital for the Vaucluse. On the west side of town, and open to the public, is the house where that most famous French author Albert Camus (1913–60) once lived. He also chose to be buried here, and his grave lies amidst a tangle of wild rosemary. The Provençal writer Henri Bosco (1888–1976)

Herbal vinegar.

is buried here as well. Reading his novels is a good way of understanding the area, for they often draw nostalgically on his memories of the childhood he spent here.

More art is still produced at the 16th-century **château** just outside the village. It was restored in 1920 by Laurent Vibert, and today acts as a study centre for artists, musicians and writers. Musical concerts and exhibitions are held here regularly, and an organised tour of the château is available. If you take the tour, be sure to ask for an explanation of the graffiti, scrawled on one of the walls, that depicts a small sailboat surrounded by mysterious birds with human faces. Before the château was taken over by Vibert, it had acted for many years as a stopping place for gypsies on route to Les-Stes-Maries-de-la-Mer. Many people say that the graffiti is a curse put on the place by them after their expulsion.

A little detour 6½ miles (10 km) east will bring you to the delightful 10th-century village of **Ansouis**. Like many of the tiny Provençal towns, its narrow, twisting and climbing streets were certainly not designed for motorised traffic. Unless you are driving an ultra-compact car, you may want to park in the central square and do your exploring by foot.

Ansouis's **château** is an interesting combination of styles, for it has been in the same family for centuries, and each member to live there has left his or her own mark. It commands the top of the hill where the town stands. Also a novelty is the **Musée Extraordinaire**, dedicated to marine life and, in particular, to the underwater creatures that once occupied the Vaucluse.

A market town: From Ansouis, follow the bank of the Durance west along D973 to **Cavaillon**. On the way, you may be tempted to buy fresh melons from a roadside vendor. Cavaillon is the melon capital of the Vaucluse – indeed, its justly famous melons are appreciated all over Europe.

The fertile fields that surround this town contrast with the rocky terrain of

Picking tilleul, which is used for tea.

much of the land that you have been passing through and make for an excellent crop of fruits and vegetables. A typical morning sight is that of the farmer on his Velosolex motorcycle with a sidecar carrying huge baskets of some type of produce, as he heads for Cavaillon's marketplace.

Outdoor markets are a year-round feature of life in this *département* where so many of the fruits and vegetables are grown that feed the rest of France. Far from a side attraction, the markets remain the centre of business as well as social activities for many of the area's towns and villages.

Whether you speak French or not, you can gain real insight into the Provençals and their way of life by visiting their markets. These people take their food seriously, and the beginning of any good Provençal dish invariably involves fresh tomatoes, peppers, herbs and garlic. Colourful fruits decorate the stalls, and you can generally find specialised stands selling local spiced sausage and goat's cheese.

Cavaillon's market is far from an exception, and people travel from all over to shop here. But melons are not all that is legendary about the town. Popular mythology suggests the area around Cavaillon was once flattened by a monster called "the Coulobie." Basically an oversised lizard, he later either flew away to the Alps or was chased away by St-Véran, the patron saint of shepherds.

The Coulobie is the subject of a painting by Mignard to be found in the **Notre-Dame and St-Véran Cathedral** at Place Joseph d'Arrard. The building itself is an unhappy mixture of styles with a 16th-century facade and some unfortunate 19th-century additions. However, its little 14th-century side chapel was nicely restored after the Wars of the Religion in the 17th century and has a particularly attractive interior.

Like Carpentras, Cavaillon "tolerated" a significant Jewish community during the time of the Papal residency in Avignon, although the number of its members never exceeded 300. A first **synagogue** was built during the 16th

Below, the red dirt of Roussillon. **Right**, sketching in Lourmarin.

ARTIST COLONIES

The first artist colony in the Vaucluse dates back to the 14th century, when Petrarch informed his readers of an enclosed valley where "water sprang from the mountain." This magic spot has since been named the Fontaine de Vaucluse.

The valley of the Calavon lies north of the Luberon Mountains, along with a score of famous hilltowns such as Roussillon, Gordes and Ménerbes. Although seemingly calm, this same valley has been the site of frequent wars and invasions since before the birth of Christ. The local population was, therefore, not too surprised when, in the 1940s, the region underwent a full-blown invasion by foreign artists and intellectuals.

The locals watched these painters and writers at first with amazement and then with curiosity and commiseration. Indeed, the humble native residents found themselves able to relate to the barely marginal existence of the artists. Little by little, they allowed their mountains to become the heart and inspiration of an entire new subculture in the area.

The transformation really began with World War II. Many artists and intellectuals took refuge in Provence, where they could find beauty, peace and affordable housing. One such artist, Consuelo de St-Exupéry, wife of the novelist Antoine and an exile from Paris, wound up with a group of artists and architects in Oppède.

This small village had fallen into deep decay, and its original inhabitants had all left to make new homes further down the valley. Undaunted, the colony set about turning the ruins into studios. Although, after the war, the group dispersed and the buildings fell back into ruin, it was not long before other artists took their place in Oppède and neighbouring villages.

Painters were not the only artists to take refuge in the Vaucluse during the war. The playwright Samuel Beckett spent from 1942 to 1945 in the red village of Roussillon, where ochre and iron had been mined for more than 1,000 years. Unfortunately, his visit wasn't as successful as that of his peers. He was so bored by village life that he had a nervous breakdown. He returned to Paris the minute peace was declared.

Meanwhile, Gordes was also being discovered by artists. The hilltop village had been abandoned in 1904 after an earthquake changed the water table but, to the wartime artists, the love of the site superceded the need for water or electricity. In 1943, the famous artist and theoretician André Lhote moved to the town and tried to start an art school. Then, a few years later, painter Jean Deyrolles saw Gordes on his way to Collioure. He decided to stay and was soon joined by his students from the Berlin Academy of Art.

Deyrolles's connection with the Denise René art gallery attracted even more artists, including Hungarian-born Vasarely, one of the most famous of the "op-artists." Vasarely leased the château that stood in Gordes from the town and took responsibility for its restoration. Once it had been returned to its original glory, he transformed it into a museum.

Even though the Lubéron region never really produced an artistic movement, it continued to be a hotbed of inspiration and to attract many artists long after the war had ended. Among these was the surrealist painter Bernard Pfriem who founded an art school in Lacoste in the 1950s that is now affiliated with the Cleveland Institute of Art. The school's dedicated outlook has drawn some of the best instructors to be found in the US. Visiting lecturers have included such personalities as Man Ray, Max Ernst, Henri Cartier-Bresson, Ernst Haas, Peter de Francia, John Rewald and Stephen Spender.

In more recent years, photographers have also discovered the advantages of the Lubéron. Denis Brihat started a school of photography in Bonnieux, Jean-Pierre Sudre founded another in Lacoste, and the French National School of Photography is now in Arles. The region's exceptional light causes the number of shutterbugs steadily to swell.

In addition, many traditional crafts can be studied from masters living in the Lubéron. Among these are pottery, weaving, lacemaking, basketmaking and iron work. Music is also represented by the Deller Academy of Early Music, located in Lacoste and open in summer.

century, but it later fell to pieces, and the elegant building that now stands was constructed in its place in 1772.

A small **Musée Judeo-Comtadin** occupies what once was the bakery where the Jews made their unleavened bread. It stands on the ground level of the synagogue and contains, as well as the original oven, Jewish prayer books and Torahic relics.

Cavaillon's newly constructed **cultural centre** has its feet firmly planted in the 20th century. Apart from staging jazz concerts during the town's annual summer festival, it boasts a lively year-long programme of performing and visual arts.

Also striving to keep up with modern times is the bustling little city of **Apt**. Its particular contributions to the economy of the region are crystallised fruits and preserves, the making of truffles and the bottling of lavender essence. In addition, it is one of the few places in the region still to mine and refine ochre.

On a more spiritual level, Apt is the seat of the **Ancienne Cathédrale Ste-Anne**. The main structure dates from the 12th century, but the Royal Chapel, which is also the major point of interest, was erected in 1660, the year that Anne of Austria arrived in pilgrimage. It is now the destination of an annual pilgrimage by Catholic devotees that takes place on the last Sunday in July.

All over Provence, you will see the word "*mas*," a regional dialect word meaning "farmhouse." Many of the most convivial places to stay are renovated *mas*, for they are often conveniently and pastorally located, personably run and give a nice feeling for the traditional lifestyle of the area.

A good example of the *mas* hotel is in **Gargas**, just outside Apt. The **Mas de la Tour** waits at the end of a winding, dusty road off route D101. This 12th-century farmhouse has been converted into a more-than-comfortable hostel with a swimming pool, a restaurant and 12 clean rooms.

Before leaving the area, you will probably want to visit **Fontaine de Vaucluse**. It's a physically delightful

Markets are the centre of activity.

but busy spot – over a million tourists visit each year. It also is a good place to begin or end a trip to the Vaucluse, for it brings together many of the different aspects of the region's history, geography and mythology.

The town's main architectural sights – a war museum, ruined castle and 11th-century church – don't add up to anything very startling, although the Romanesque **Eglise de St-Véran** possesses the coffin of the 6th-century bishop of Cavaillon, said to have freed the area of the Coloubie monster.

The real attraction is the "**fountain.**" Fed by rainwater that drains through the Vaucluse Plateau, it is thought to be the source of the Sorgue River. It emerges from within a cave and, in winter and spring, the flow can rise to 32,985 gallons (150 cubic metres) a second. Needless to say, it is one of the most magnificent natural spectacles in Provence, indeed, all of France.

Emerald green water gushes forth from the underground river believed to be its source. Such noble investigators as Jacques Cousteau have come to try to substantiate this claim, but by far its most celebrated observer was Petrarch. The poet lived here from 1337 to 1353, between stays in Avignon, and it was here that he composed his celebrated sonnets to Laura. As with so many places in the Vaucluse, poetry and legend play large parts in the village's popularity.

Two-and-a-half miles (4 km) from the fountain is the quintessentially Provençal *village perché* of **Saumane-de-Vaucluse**. Once again, some of the key names from Provence's past crop up in this town's history. The village was a papal gift by Pope Clement V to the de Sade family. Restoration has begun on their impressive, pine-enclustered **château**.

Nature, however, seems to have had the last word here, for all the trees have been permanently marked by the fierce gusts of the mistral. But this is true for so many places in the Vaucluse, where the real wonders are natural ones that owe nothing to human beings.

Left, fruit vendor; right, peaches.

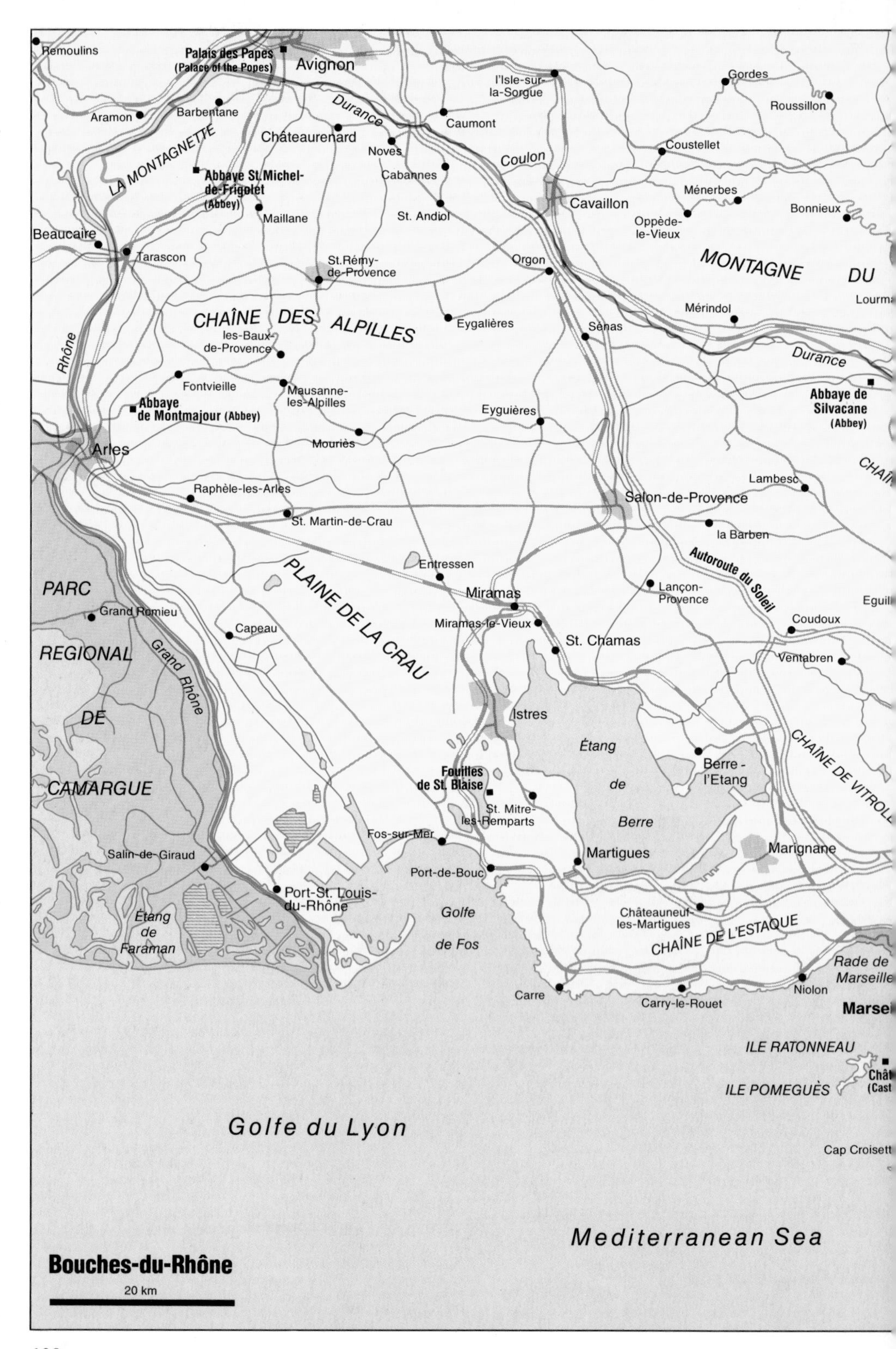
Remoulins
Palais des Papes
(Palace of the Popes)
Avignon
Aramon
Barbentane
Durance
Châteaurenard
Noves
l'Isle-sur-la-Sorgue
Caumont
Gordes
Roussillon
Coustellet
LA MONTAGNETTE
Abbaye St.Michel-de-Frigolet (Abbey)
Maillane
Cabannes
St. Andiol
Coulon
Cavaillon
Ménerbes
Oppède-le-Vieux
Bonnieux
Beaucaire
Tarascon
St.Rémy-de-Provence
Orgon
MONTAGNE DU
CHAÎNE DES ALPILLES
les-Baux-de-Provence
Eygalières
Sénas
Mérindol
Rhône
Durance
Fontvieille
Mausanne-les-Alpilles
Abbaye de Montmajour (Abbey)
Eyguières
Abbaye de Silvacane (Abbey)
Arles
Mouriès
Lambesc
Raphèle-les-Arles
Salon-de-Provence
St. Martin-de-Crau
la Barben
Autoroute du Soleil
PLAINE DE LA CRAU
Entressen
PARC
Miramas
Lançon-Provence
Grand Romieu
Miramas-le-Vieux
Capeau
Coudoux
REGIONAL
Grand Rhône
St. Chamas
Ventabren
DE
Istres
Étang
de
Berre
CHAÎNE DE VITROLL
CAMARGUE
Fouilles de St. Blaise
Berre - l'Etang
St. Mitre-les-Remparts
Fos-sur-Mer
Martigues
Marignane
Salin-de-Giraud
Port-de-Bouc
Port-St. Louis-du-Rhône
Golfe de Fos
Étang de Faraman
Châteauneuf-les-Martigues
CHAÎNE DE L'ESTAQUE
Rade de Marseille
Carre
Carry-le-Rouet
Niolon
ILE RATONNEAU
ILE POMEGUÈS
Golfe du Lyon
Cap Croisett
Mediterranean Sea
Bouches-du-Rhône
20 km

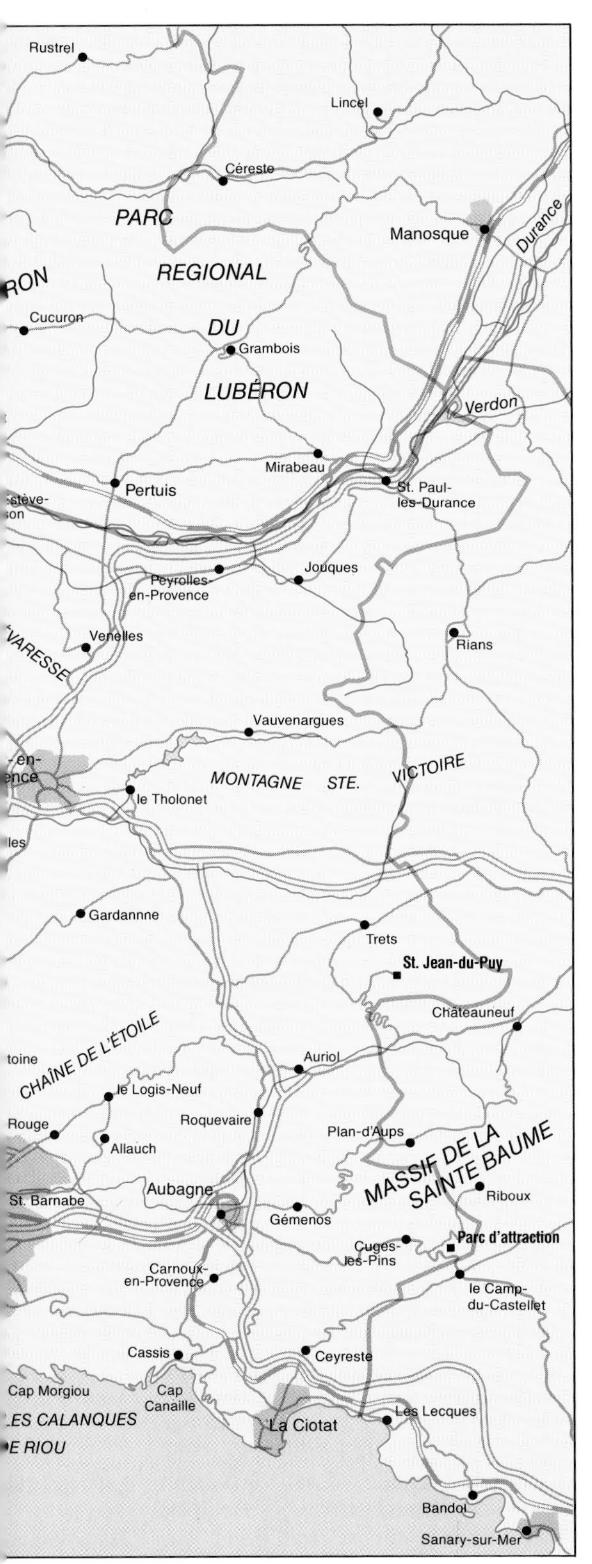

THE BOUCHES-DU-RHÔNE

The privileged position of the Bouches-du-Rhône (literally, "mouths of the Rhône") between the Mediterranean and one of France's greatest rivers has made it a virtual museum of European history. But visitors shouldn't expect to see the Parthenon or the Villa di Borghese gleaming on a distant hilltop. Instead, they will encounter acres of arid land dotted by crooked olive trees and dark vineyards, alternately scorched by a merciless sun and threatened by the violent gales of the mistral.

It takes an informed eye to appreciate the historical bounty of the Bouches-du-Rhône. Many of the monuments are either in ruins or hidden amidst the desolate countryside. Other sites become interesting only after learning the myths and traditions that surround them. Nonetheless, a visit to the right place at the right time with the right information can transform a mass of rock into a touchstone for the magical, mystical world of this *département*.

Although a fairly small *département* in size, the Bouches-du-Rhône can be divided into at least three distinct areas. The divisions centre around the major cities: Marseille, the second largest city in France, and its industrialised coastal territory; Aix-en-Provence, an elegant and bourgeois university town, and its slightly mountainous and fertile surroundings; Arles, ancient and distinctly Roman, set in the midst of dry, flat and poor plains with menacing Alpilles Ridges to the north. A fourth area, the marshy Camargue, is so individual that, although administratively a part of the Bouches-du-Rhône, it deserves (and has been given) a chapter of its own.

Provence's oldest city: It was 2,600 years ago that Greek sailors from the Ionian city of Phocaea first sailed into the natural harbour of **Marseille**. According to legend, they arrived just as Gyptis, the local Ligurian princess, was selecting a husband. Suitors stood in a

Preceding pages: splash of sunflowers.

circle awaiting her decision when, eschewing local blood, she handed the chalice that marked her choice to Protis, captain of the Greek entourage.

From their union flowered a prosperous Greek trading post, called Massilia. Over the centuries, the Massilians knew numerous high and low points of fortune, all the while developing a strong and independent character still apparent in their modern descendants.

Today, the city continues to be a major sea port and harbour – with a somewhat salty reputation. The latter is due to the fact that, as a centre for trade and commerce, a certain amount of drug traffic and black marketeering inevitably passes through its ancient waters. Also, the city is one of the major French entry points for North African immigrants. Their emergence as an integral part of the French work force has engendered a strong local following of the reactionary National Front party headed by Jean-Marie Le Pen.

To make matters worse, the proud Marseillaises have wasted little effort in trying to refute their well-publicised problems. The overall result is that most visitors to Provence consider Marseille a spot either to avoid or to pass through as quickly as possible, looking over their shoulders and clasping their wallets all the while. In truth, however, Marseille is a quite fascinating city, with its pale colours and strong personality, and no more dangerous than any other major urban centre.

Left, the Old Port in Marseille.

Any visit to Marseille should begin at the **Vieux-Port**. Here, pleasure boats share the waters with fishing rigs, and a lively fish market heralds each morning along the wharfs. Although the city's other harbours now handle incoming commerce, the Vieux-Port is Marseille's original and oldest harbour – and still its most picturesque.

At the very base of the Vieux-Port docks a ferry that taxis visitors out to a tiny rock island just off the coast. This is the site of the **Château d'If**, from where the legendary Count of Monte Cristo was fabled to have made his escape. The castle is currently empty and, in truth, not terribly interesting, but Marseille's city council has decided to give it some spice by recreating its interior, complete with furnishings and wax figures. Until the project has been completed, visitors can enjoy a great view of the city, a little swim off its rocky shores or one of the nighttime performances held here during the summer.

On the northern side of the port is the **Quartier Panier**. Many of the city's immigrants live in this, the oldest, quarter and, as there is no public nightlife, it is not a wise place to wander around after dark. During the day, however, it is an interesting (and safe) area to explore, with its narrow climbing streets and colourful inhabitants – as long as you don't mind getting lost at least once.

Within the Quartier Panier are several of the city's museums. Not to be missed is the **Musée de Vieux-Marseille**, which will give the newcomer a fine introduction to the folk life of Provence. Housed within the remarkable Maison Diamantée (1570–76), this museum is dedicated to Provençal culture, with entire rooms of traditional costumes, tarot cards, crèches etc. Also on display are a variety of engravings and models of Marseille over the centuries.

As long as you are in the area, you might as well look in on the **Musée des Docks Romains**, which shelters a series of uncovered Roman docks on the very spot where they were excavated. The open amphores once stored grain, olives and wine for the Romans.

On a more modern scale, and also in the neighbourhood, is the **Galeries de la Charité**. Built in the 17th century as a hospital by the grand architect of Marseille, Pierre Puget, the recently restored buildings are now the site of several galleries of contemporary art. In the very centre stands a chapel with an interesting oval-shaped domed ceiling, and behind the chapel is a small outdoor theatre active in the summer.

Ancient walls: Much of Marseille's ancient ramparts were destroyed during the continual battles of the centuries – and most of whatever was left was decimated by the Germans during World War II. But, among the modern buildings and bobbing shipmasts, a few ancient walls still stand.

Recent excavations at the **Centre Bourse** have unearthed fortifications, wharfs and a road, dating from somewhere between the 3rd century BC and the 4th century AD. Called the **Jardins des Vestiges**, the remains now form a public garden beside the Centre Bourse shopping mall and the **Musée d'Histoire de Marseille**.

The museum is itself a fascinating place, filled with an excellent selection of Ligurian, Greek and Roman artifacts found over the ages in Marseille. Everything is extremely well-documented, such that Francophones can read all about the restoration of the ancient Bateau de Lacydon while examining the boat or the Greek method of firing ceramics while viewing a Hellenic kiln.

Be sure to note, directly on the left when you first enter the museum, the terrifying "Gros Repasse." Once the doorway to a Ligurian sanctuary, frontal ledges are filled by the skulls of unlucky enemies. Also worth an extra

Jazz musician Kid Creole being interviewed on the quay.

minute, for French speakers, is a stop in the *videothèque*, where videotapes concerning Marseille, on anything from soapmaking to World War II newsreels, can be requested and viewed free.

Running straight up from the centre of the harbour is the **Rue de Canebière**. Despite its fame, this is an unattractive street, lined with cheap department stores and unsavoury characters. The cool crowd should head right past it to the **Cours Julien**, where the young hipsters hang out. Around this square are the prerequisite cafés, numerous second-hand bookstores and a number of alternative clothing shops.

Equally *branché*, but a little less youthful, is the pedestrian area around the **Place Thiars**. Located on the south side of the harbour off the Quai de Rive Neuve, its streets are lined with galleries, shops and outdoor cafés. Newly installed parks and fountains make it a sympathetic spot to lunch or take an afternoon stroll.

Also on this side of the harbour, at 47 Rue Neuve St-Catherine, is a small shop of special interest to the *santon* enthusiast. At the **Atelier Marcel Carbonel**, visitors are welcome to watch some of the best-regarded crèche figures being made and view M. Carbonel's personal collection of Christmas cribs from around the world.

Just down the street, at 136 Rue Sainte, is another enclave of Marseillaise tradition, **Le Four des Navettes**. First created in 1781 in this very shop, *navettes* are biscuits cooked in the shape of the boat that is said to have brought Mary Magdalene and a host of other saintly Marys to the shores of Provence 2000 years ago. Made according to a secret method, they are the special mark of the Fête de la Chandeleur, which takes place on 2 February annually. But biscuit fanciers can buy them for two francs each all year round.

The Rue Sainte ends at **St-Victor's Basilica**. In the 3rd century BC, the site was a Greek burial ground and, in the 1st century, it became a Roman monument. Then, during the 3rd century AD, a Christian cemetery began to develop

Mansion on the Corniche J.F.K.

around the tombs of two martyrs who were slain during the persecution of Dece in AD 250.

The church itself was built in the 5th century by St-John Cassian, an Egyptian monk who named it after a third local martyr, Saint Victor, the patron saint of sailors, millers and Marseille. Saint Victor had been ground to death between two millstones. The primary structure now standing is an 11th-century reconstruction, whose high Gothic arches resound with distant taped music. Deep below, in the crypt, lies the aptly solemn original church. Ancient sarcophagi, some hidden in candle-lit coves, line the walls; in the centre stands the sombre tomb of the two martyrs.

Even more eerie, in a rocky alcove to one side, is the 3rd-century tomb of Saint Victor himself. His grave image, carved into the wall on the right as you enter, watches over the shadowy corner. This is not a place for those with a fear of ghosts, for they command the dark and hollowed crypt.

Once back on the port, if you still have an appetite, you might want to try the *bouillabaisse* dining experience. Anyone seeking to understand Marseille should attempt this traditional fish stew – but be fair warned. Only a few restaurants still offer an authentic *bouillabaisse*, and it won't be inexpensive. At the same time, don't expect a refined meal. *Bouillabaisse* was created by fishmonger's wives trying to use up the catch their husbands hadn't sold, and in basic ingredients and recipe it hasn't changed much.

If you can still stand after three bowls of fish soup and are in the mood for some night-time panorama, climb to the site of **Notre-Dame-de-Garde**. No matter where you go in Marseille, you can't miss noticing the airy 19th-century basilica. Erected on a bluff of 530 feet (162 metres), it offers, without question, the best view over the city.

But the most beautiful view in Marseille is to be found along the **Corniche President-John-F.-Kennedy**. No one should miss taking a drive along here. Following the coastline, this 3-mile (5-

Left, city dwellers. **Right**, baker at the Four des Navettes.

THE UNDERWORLD AND MARSEILLE

As a thriving port, Marseille has long had to bear the associations that come with trade and the passing sailor. By the 1930s, such was its tough reputation that every employee in the town hall was said to have a criminal record. The statistics were far from conclusive, but the city nevertheless cemented its reputation as France's capital of organised crime and corruption.

Today, the town hall and the *milieu* (underworld) coexist in an unconcealed loathing that, admittedly, sometimes gives way to pragmatism. This is because, although Marseille has traditionally produced strong mayors, they have little control over a feature that permeates society in southern Europe: clientelism. A favour rendered is a favour returned. A *milieu* "businessman," for instance, will during an election campaign be approached for favours by candidates from all sides. Yes, perhaps he can help, he tells his supplicants – in return for guarantees of protection and a job for a friend.

If the circumstances sound like New York in the 1920s, then there will be no surprise to find that connections between the two cities abound. Marseille was a transit port for Sicilian emigrants on their way to New York. Many never reached Ellis Island and, instead, found hope in France's second largest city. Some brought the code of the mafia; it was to mix with the traditions of the native Marseillaises and the Corsican clans which had come to the mainland in search of jobs. Over generations, this heady brew took to an extreme the Phocaean city founders' vision of Marseille as a trading post.

In 1971, the "French Connection" was broken up, exposing a huge web of drug smuggling with Marseille as the linchpin. The process had been for the illicit raw materials to be hidden in cargo ships from Pakistan and the Middle East, refined in the south of France and shipped out to New York. By then, much of the money had been laundered, finding its way to bars, nightclubs and casinos all the way up the coast.

Of course, busting the French Connection didn't put an end to all criminal activity in Marseille. The *milieu* continues to function with panache. Several years ago, the police seized 40 fruit machines in a clampdown on illegal gambling. Their faith in the security of local warehouses was minimal, so the evidence was stored overnight in the main police station. When they returned the next morning, 36 of the machines had been meticulously stripped down, the mechanism removed for further use elsewhere.

And the dashing brigands of one day frequently turn into blood-thirsty killers the next. *Règlement de comptes* (literally, the settling of accounts) spill into the street, from one-off assassinations to massacres like that of 10 people in the Bar du Téléphone in 1977.

Attempts are often made to strip the core from the *milieu*. The imprisonment in 1983 of one of the gangland bosses, Gaetan Zampa, was thought to have ended the *guerre des clans*, a war of underworld families. Instead, the *milieu* restructured. Its vocabulary now has taken on the jargon of business efficiency, and the *parrains* (godfathers) have virtually disappeared. Five groups are said to share Marseille, occasionally cooperating on joint ventures. Operations aim to maximise revenue with the lowest possible profile. Labour-intensive prostitution and racketeering are out; fruit machines and bank robberies – the *milieu* pioneered the technique of tunnelling in through the sewers – are in.

When the "new" underground does come out of its shell, it is ruthless. An examining magistrate, Juge Michel, was shot dead in 1983 as he rode home on his motorcycle. It was the first time anyone had dared to touch a senior judicial figure in the city. Two people were sentenced for the murder five years later, but the identity of whoever ordered the killing remains unknown. However, Michel's investigations into drug smuggling, and the massive surge in local drug addiction, leave the police in little doubt that the drug trade has come back to roost in Marseille.

Yet, it would be unfortunate for tourists to fear the darker side of the city. Visiting the Louvre is potentially a more dangerous pastime; for despite its reputation, Marseille actually has a far lower crime rate than Paris.

km) road looks out over the sea, the offshore islands, the surrounding *massif* and the winding shore ahead. It also is adorned by some of the most beautiful war memorials to be found anywhere. Of special note is the one that commemorates the Oriental troups of World War II. Even the staunchest of hearts will heave a little beneath its outstretched arms.

The road eventually leads to the **Plage de Marseille**. The beach here has been split into two parts, the first half being pebble and the second sand, and the water is quite clean. Bearing a striking resemblance to Southern California, this area is a lively and chic place to dine at night.

Not just a fishing town with fjords: Hardcore beachgoers may want to continue down the coast another 15 miles (23 km) to **Cassis**. First of the Rivieran resorts, Cassis possesses none of the glamour or urbanity of St-Tropez or Cannes – and therein lies its special charm. It also is blessed with the coolest water along the French Mediterranean, thanks to a series of mainland streams.

Modern painters once spent many summer days in this delicate port, and it is easy to see why. For sheer natural beauty, it is hard to beat – especially at dusk, when irridescent shades of light dust over the harbour.

Nonetheless, it would be stretching it to call Cassis a simple fishing town. The fact is, during the high season, it is as crowded and overbooked as any of the Mediterranean resorts. The trick is to come before or after the rush, in May, June or September, when the sun's still very hot and the masses are gone.

The end-all of any visit to Cassis has to be a bathing holiday on one of its beaches. The town possesses three, the best being the **Plage de la Grande Mer**. But most visitors make at least one trip to the town's spectacular *calanques*.

The first inlet, reachable by car, is **Port Miou**, and it is dedicated to harbouring yachts. The second is **Port Pin**, so named for the shrubby pines that decorate its rocky walls. It must be approached either by boat or foot, as

Left, heroin. Below, the Vallon des Auffres boasts bouillabaisse and crime.

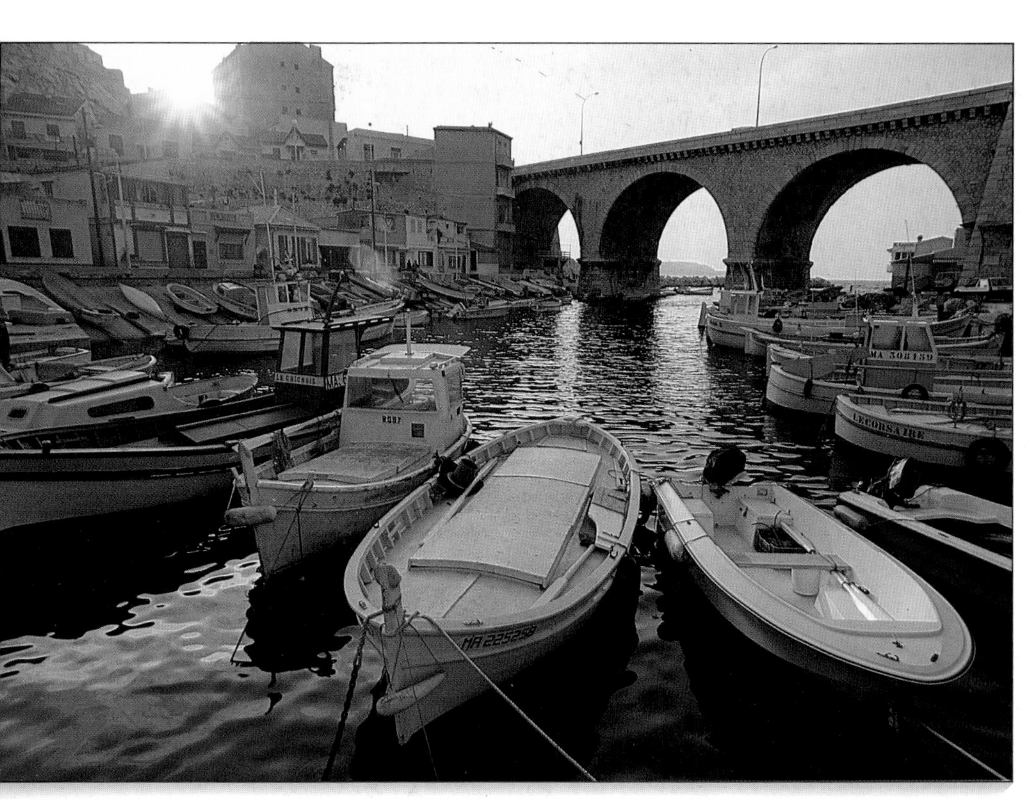

must be the last and most incredible of the *calanques*, **Port En-Vau**.

Port En-Vau is breathtaking. Similar in appearance to a Norwegian fjord, sky-high white cliffs cut directly down into water of the deepest blue-green imaginable. A sandy white beach adorns one end. The view from the top is something never to be forgotten, but be warned – getting there isn't easy and neither is getting down. The paths that lead to both of the last two *calanques* are more or less unmarked, and *ingénues* may find themselves lost on arid clifftops.

In the long run, all but the most adventurous would be better off taking a native guide or a boat to Port En-Vau. And remember, if you're there during the high season, none of these places are going to look like paradise – unless you like naked Scandinavians and countless families with little dogs and picnic baskets. Even the waters start to float with empty wrappers.

Cassis has a casino and a couple of little nightclubs, but don't expect to find a wild nightlife here. This is a resort mostly for families or young marrieds, and after-dark activities centre around the harbourfront and its row of excellent restaurants. While enjoying a meal along here, be sure to sample the delicious local white wine.

Friendly and unpretentious: Dominating the little village is **Cap Canaille**, which at 1,400 feet (416 metres) is the highest cliff in all of continental Europe. For a panoramic thrill, take a drive along the **Route des Crètes**, which climbs up, over and down to the neighbouring city of **La Ciotat**. From summer homes to clinging vineyards to rubble, the road seems to disappear up into the sky. The view is fabulous, as is the lightheadedness it gives the viewer.

Down on the other side of the cliff, La Ciotat has a very different character. First called Citharista and an outpost to Marseille, its harbour now rather resembles a resort in Florida. The people carry a friendly unpretentious air about them, and the long flat beachfront is lined with souvenir shops.

On vacation in Cassis.

Some of the town is quite pretty – in particular, around the old port and 17th-century **Notre Dame de l'Assomption**. Unfortunately, the huge shipyards that build oil and methane tankers at one end of the waterfront keep La Ciotat from achieving ultimate picturesqueness.

Like Cassis, La Ciotat possesses a set of *calanques*, but **Muguel** is not for swimming and **Figuerolles** is polluted and ugly. A more enjoyable tourist attraction is the beachfront monument to the Lumière brothers. These two illustrious sons of the city were the inventors of the moving picture.

Travelling inland along A50, north of La Ciotat and east of Marseille, will bring you to the hometown of another inestimable Provençal native: Marcel Pagnol. The renowned writer and film director was born in **Aubagne** in 1895. But, unless you are a real Pagnol aficionado or are seeking out one of the crèchemakers who have settled here, avoid this town. It is dirty and unattractive, and the inhabitants seem to carry a chip on their shoulders about it.

Movie fans would do better to drive through Cuges-les-Pins and up to **Riboux**, where the pastoral scenes of the films *Jean de Florette* and *Manon des Sources* were shot. One can easily imagine Manon tripping along the wild and rugged mountainside with her goats – and, even if not, you'll enjoy a great view and hair-raising drive.

Afterwards, you might make a stop at the **O.K. Corral Amusement Park**, set at the base of the road that leads up to Riboux. Here you will find France at its tacky best. The entire park has been done around a Western motif, including, of course, a "genuine" Wild Wild West show.

An elegant university town: Although physically but a stone's throw north of the coast, Aix-en-Provence is a million miles away in character. Bastion of culture and bourgeois niceties, seat of one of France's finest universities since 1409, Aix embodies grace and gentility.

The Romans first founded Aquae Sextiae – so named for its hot springs – in 122 BC, after conquering the nearby

The Calanque de Port-Miou.

Ligurian settlement of **Entremont**. It was overshadowed, however, by Massilia and Arles until the 12th century when the beloved "Good King René," count of Provence, made it his city of preference. René's death brought Provence under the rule of the French crown, and Aix was made the seat of a parliament designed to keep the region under Gallic control. By the 17th and 18th centuries, it had become the leading city in the Bouches-du-Rhône.

Much of Aix's renowned architecture dates from this period of prosperity. Buildings erected during the 17th century can be distinguished by the single iron balcony placed over their central front door, frequently embellished by a corniche underscored with little teeth. The 18th-century structures, on the other hand, have iron balconies on all the windows and the door on the right.

The architecture also bears witness to the eventual decline of the city's power. All over you find homes whose windows were closed in, because taxes were assessed according to the number of windows a house possessed and owners gradually were unable to pay them. Eventually, most of the grand homes were repossessed by the banks.

During the Industrial Age, Aix fell back under Marseille's shadow. Over the past century, some new industry has managed to find a niche here – it remains the European capital for prepared almonds (some of which can be sampled in the local specialty: white iced diamonds of almond paste, called *calissons*) – but mostly, now, Aix is a university town. Of its 150,000 inhabitants about 40,000 are students.

City of fountains: The best time to visit Aix is May, and the worst time is July. But whenever you choose to come, the best place to begin your investigation of the city will be at the **Cours Mirabeau** with its famed fountains.

The Cours Mirabeau is a long and wide avenue, lined with cafés on one side and venerable old addresses on the other. Four rows of benevolent and stately plane trees run down the centre of it, punctuated by three 19th-century

The Fontaine de la Rotonde.

fountains. The most celebrated of these is the **Fontaine de la Rotonde**, which dominates the western entrance to the avenue. It is a source of joy to all viewers – and of terror to any motorist battling the roundabout where it stands.

The avenue's original purpose was to separate the southerly **Quartier Mazarin** from the northerly Quartier Ancien. Only parliamentarians and nobles were allowed to grace the Cours Mirabeau's elegant thoroughway, and the Mazarin Quarter was designed specifically to accommodate them. Unsurprisingly, many lovely parliamentarian homes can be found within this quarter.

Of these homes, the **Hotel Arbaud** is generally considered the most beautiful. It was built in 1730 directly on the Cours by the first president of the parliament. A good runner-up, however, is located within the quarter, at the corner of Rue Joseph Cabassol and Rue Mazarin. The only house in Aix aside from the Hotel Arbaud to have been designed by a non-local, the **Hotel de Caumont** was constructed in 1720 by the illustrious architect of Versailles. It is now the Ecole Superior d'Art et de Danse de Milhaud. Go inside to see the foyer fountain and the Atlanteans that hold up the ceiling. Many of the older houses in Aix possess similar interior fountains, remembrances of the city's aquatic origins, and the Atlanteans are a popular local motif, borrowed from the Genoans.

East down the Rue Cardinal lies the sweet **Fontaine des Quatres Dauphins** (Fountain of Four Dolphins). Built in 1667, it was the first fountain in Aix to be placed in the middle of a street rather than against a wall, giving it an unprecedented decorative purpose. Alain Delon fans will be interested to know that the beloved French celebrity owned the home on the northeast corner until the mid-1980s.

The **Musée Paul Arbaud** stands close by the Fontaine des Quatres Dauphins, at 2a, Rue du 4-Septembre. Immediately as you enter its large foyer, you will be struck by a plethora of reliefs and capitals from different ages.

The ladies of Aix.

Be sure to check out the "Seven Deadly Sins," hanging directly on your right.

The first room in the museum has one of the most important collections of Moustiers-Ste-Marie-style *faïence* to be found in Provence. Among the numerous examples of traditional work are such whimsical pieces as that marking the hot-air balloon ascent of 1783.

Going up the stairs, note the 18th-century, hand-painted, ornamental trimming. This was clearly not the home of a miser. Indeed, the house itself might be considered the most interesting exhibit. The detail upstairs is incredible, with acorns and leaves carved directly out of the wood portals and exquisitely handcrafted fireplaces.

Another museum within the Mazarin Quarter is the **Musée Granet**. Its main and first floor are devoted to a great deal of fairly unremarkable paintings. Even the eight Cézanne canvasses, of which the museum is so proud, look rather like leftovers. Its true interest lies in its basement where an excellent collection of lengthily documented Celto-Ligurian artifacts have been assembled, with maps and text describing early geological and prehistoric activity.

Next door is the 13th-century **St-Jean-de-Malte**, former priory of the Knights of Malta and the oldest Gothic building in Aix.

Old but lively: When you cross over the Cours Mirabeau to the **Quartier Ancien**, you will instantly notice a change in atmosphere. Whereas the Mazarin Quarter oozes a sense of serenity, the streets of the Old Quarter are filled with the lifebeat of the city.

This part of town also possesses a lion's share of interesting museums and splendid architecture. The **Hotel Boyer d'Eguilles**, about two blocks in from the Cours, at 20 Rue Espariat, combines both. Dating from 1675, the architectural plan is Parisian while the exterior decor is Italian. The **Museum of Natural History** lies inside, with a number of large dinosaur eggs as its star attraction.

Unfortunately, internal political turmoil during the 1980s and the lack of a sturdy mayor to champion the upkeep

Aix's Hôtel de Ville and clocktower.

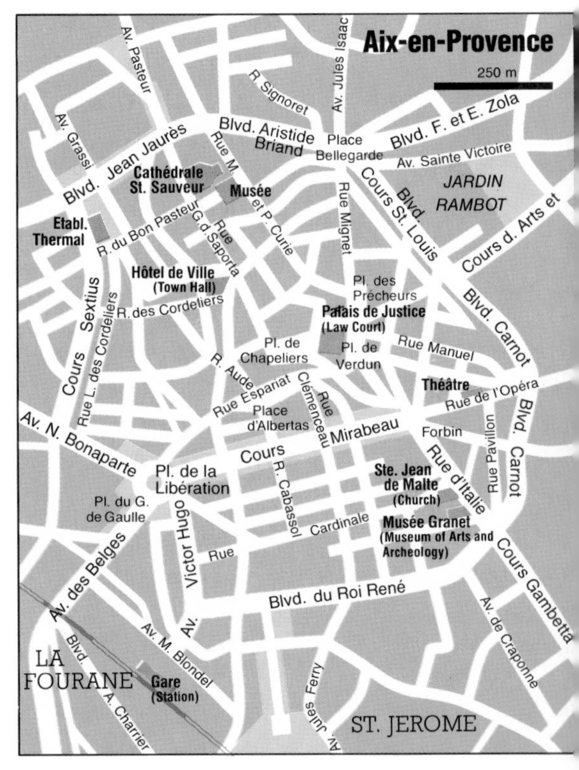

of Aix's landmarks have begun to tell on the Hotel Boyer. But, down the street at the charming **Place d'Albertas**, there are no such signs of decay.

When the parliamentarian Albertas built his home in the 1720s, he also bought all the land in front of it – to "protect his view." This land, over time, came to be called the Albertas Square. The four contiguous buildings that enclose it were not constructed until 1740. The fountain was added in 1912.

The Place d'Albertas lies at the base of the **Rue Aude**, home to many of Aix's most fashionable stores and best-loved tourist sites. However, the first square that you come upon when heading up it, **Place Richele**, may seem slightly incongruous. Young rockers and their impromptu concerts have claimed the Place Richele for their own, and the square is more or less the centre for illegal substances in the city. But, remember, Aix is primarily a town for university students.

Gentility is restored at the **Place d'Hotel de Ville**, just up the Rue Aude. The liquid fare at the many little tables that fill the square is strictly a nice cup of tea or harmless *ballon de rouge*, and it makes for a relaxing place to pass some of the afternoon.

On the western side of this square is the **Town Hall**. When going in, be sure to glance up at the beautiful wrought-iron work above the entrance, which fans out in a representation of the sun. Crossing the pretty paved courtyard will lead you into the foyer, whose elegant staircase was the first to be built in the double style in France.

To the south side of the square is the **Post Office**. It was first begun in 1718 as a granary, but work was shut down two years later because of the plague. Another 40 years passed before work was resumed, and by then it had been designated as a public structure. For this reason, it possesses windows and decorations such as one wouldn't ordinarily find on a lowly granary. Particularly engrossing are the cavorting figures, intended to represent the Rhône and Durance rivers, that drape themselves

M. Fouque's *santons*.

into the stone beneath the central eave.

On the north side of the square is the whimsical **Tour d'Horloge**, dating from 1510. This tall clock tower houses four different statues, each of which marks a season and appears alone for an appropriate three-month stretch. In the summer, you will see a woman holding wheat. Autumn shows the wine harvest. Winter has a wood bearer, and spring appears as a woman carrying fruit and a young salmon.

Passing under the arch of the Tour d'Horloge and up the Rue Gaston-de-Saporta will bring you to the **Musée de Vieil Aix**. This is a funny little museum, badly lit with what at first seems like an odd collection of riff-raff. Don't let this put you off. To the left is a fun hall of mechanised crèche marionettes, and to the right is a fascinating room documenting the *jeux de la Fête-Dieu*.

The Fête-Dieu, half-religious and half-profane, takes place during June and is a long procession in honour of the battle of Christianity against paganism. Its origins lie in the 13th century, but it wasn't until 1836 that the marionettes of Bontoux, displayed within the museum, were created. Also shown are some of the participants' masks and a large painted screen. Don't miss the emphatic devil masks hanging up on high or the devil puppets with red faces, white teeth and pitchforks.

The **Cathédrale St-Sauveur** heads the Rue Gaston-de-Saporta. At first on this site there was just a small Roman church. Then, in 1170, after the Crusades had caused the population of Aix to swell, the little church was deemed no longer big enough. This led to the planning of the cathedral.

Soon after work had begun, however, the plague struck, and it was closely followed by the 100 Years' War. All in all, 140 years went by before construction could be continued and, as a result, the cathedral is an eclectic structure. Its facade combines the 12th century with the 16th, the belfry belongs to the 15th and the Gothic nave to the 16th.

Slightly more coherent is the cathedral's 5th-century **baptistry**. It encompasses all that is left of the initial Roman settlement in Aix and is one of only three original baptistrys (the others are in Fréjus and Riez) left in France.

The baptistry is a fascinating example of symbolic architecture. The exterior plan and cupola are cubic to represent the four apostles. The interior pool is octagonal with eight marble columns for the "eight" days of the week. Water flowed in from the west and exited to the east, in keeping with the direction of the sun and its light. Similarly, the two westernmost columns were formed of black granite (for the darkness of unenlightenment), and it was between these that the to-be-baptised would enter the pool. The other six columns, through which the newly baptised exited the pool, were of green marble (for the light of redemption).

Country cottages: One last – but definitely not least – local architectural feat is the **Pavillon Vendôme**. To reach it, you must quit the central city, if just barely, and head down the Rue Célony. Its original owner, Louis de Mercoeur, duke of Vendome and governor of

Atlantean holds up the Pavillon Vendôme.

Provence, had it built in 1665 outside the central confines of Aix – for very specific reasons.

Louis de Mercoeur had the good fortune to be the grandson of King Henry IV and the nephew-in-law of Cardinal Mazarin. After his wife died, however, he had the misfortune to fall madly in love with Lucrèce de Forbin-Soliers, "La Belle du Canet." Although Lucrèce was available, being a widow, she was not of a high-enough social stature to become the official consort of the duke. Desperate, Louis had this second home built outside the town, so that they might continue to rendezvous in secret. Eventually this wasn't enough, and he decided to marry her anyway, at which point an outraged Mazarin made him a cardinal and thereby elevated his status to a place where any legitimate union was forever impossible.

Outside, the Pavillon looks much like an English country home, enhanced with *quatrepartite* formal gardens. The only disturbance to the symmetry of the grounds and building is the unfortunate third floor, which was added during the 18th century.

The house itself is replete with memories of the starry-eyed couple. Groaning Atlanteans hold up front portals that are decorated by the figure of Spring – said to have been modelled after Lucrèce. Inside, portraits of the lovers hang side by side.

A small room in the back of the first floor is equipped with a video that tells the entire story of the Pavillon in French, German and English. The ceiling of this same room bears a gay fresco of giggling and naked cherubs. It was only revealed recently, having been covered over during the days when the home was a convent school. Apparently, the nuns that ran it considered such frolicking inappropriate for their students' dining room.

Artists in Aix: The **Atelier Paul Cézanne** lies behind a little wooden gate, just north of the Old Quarter. The renowned modern artist was born in Aix in 1839 but left during the 1860s to join the Impressionists in Paris. He soon

The Vasarely foundation.

became disenchanted and returned to his hometown in 1870, where he remained until his death in 1906.

Cézanne had his studio built in 1900, and it was here that he painted his last works. The large windows are testimonials to the fact that he himself designed the building. Inside are a couple of reproductions, his easel, a rucksack, a cape, his books and many of the objects that he used in his still lifes. Cézanne fans will enjoy matching the models with the paintings, but anyone with a knowledge of Cézanne should find a visit here fascinating.

To see more of his inspiration, drive out of the city along the Route de Tholonet, D 17. This will bring you to the foot of Cézanne's perhaps most-famous model, the huge and silvery **Mont St-Victoire**.

Another artist who left his mark on Aix was Paul Vasarely. A museum to his work (which he himself built, named the **Fondation Vasarely** and donated to the city) stands just to the southwest of Aix on the Jas de Bouffan. The building is unmistakeable – for it was designed in geometric shapes with black dots on white squares and vice-versa.

Mt. St-Victoire.

Even the rooms inside are hexagonal. All of the 42 works, which the artist described as "mural integrations," were painted directly onto the high walls in 1975. Also on view are some 800 "experiments" and a gallery of contemporary exhibits. Visiting this incredibly dated museum is like wandering into a wonderland of 1970s op-art.

Painting is not the only art on display in Aix. During the summer festival, original productions of opera and chamber music can be seen almost every night. The Met it isn't, but the offerings are earnest and entertaining and sometimes even quite good.

The Aix festival is just one source of nocturnal distraction. Jazz clubs, encouraged by the large student population, swing all year round, as do a number of rock clubs and discos. Restaurants abound, and the cafés that line the Cours Mirabeau are always full of people-watching pastis drinkers.

It's easy to fall into being a night owl in this young city. Be sure, however, not to sleep through the market that overtakes the Old Quarter every Tuesday, Thursday and Saturday morning.

The *marché d'Aix* is undoubtedly one of the largest and most colourful markets to be found in all of Provence. In front of the Hotel de Ville are the flower stands. Over by the Palais de Justice are produce on one side and second-hand goods on the other. In between, the narrow stone streets spill over with additional merchants selling everything from kitchen utensils to underwear.

North of Aix: On days when there is no market and you feel like getting out of the city, you may want to head up north towards the **Durance River** and the solemn **Abbaye de Silvacane**. Built by the Cistercians in 1144, Silvacane's name is derived from the Latin for wood *(silva)* and reed *(cane)* after its marshy surroundings.

True to the Cistercian spirit, this abbey is anything but frivolous in appearance. No stained glass, no statues and no decoration whatsoever were allowed.

Add to its original austerity an empty interior and a poor job of exterior reconstruction and Silvacane makes for a rather bleak destination.

The area up along the Durance, however, does provide a nice break from the dry plains of Aix. As you head north, the land begins to rise, coniferous trees appear and green surrounds you. Occasional meadows, deep reservoirs and distant mountains are a welcome sight. You are now heading into wine country.

Those with a yen to visit a working wine château from the *Côteaux d'Aix* should follow the river east to the cute little town of **Jouques**, with its gay central fountain. The young owners of the **Château Revelette**, who speak French, German and English, are happy to offer the curious a tour of their *cave* and a taste of their grape.

Afterwards, head south on A11 through the **Forest of Peyrolles**, past numerous fields, vineyards and sheep-crossing signs, and you will find a number of shaded groves suitable for picnicking. You'll need some fortification before embarking on the climb up over the mountains and back down towards Aix. The curvy, shoulderless road resembles nothing more than a goat path that has been mildly picked over to accommodate modern civilisation. Even its surface undulates.

Once you've hit the crest of the **Col de St-Buc**, the road begins to wind down through cliffs of sheer rock, cut into natural tiers. At the bottom, head just west and you will come upon the **Château de Vauvenargues**. Perched on a little hill, but within a deep valley and surrounded on all sides by green mountains, the château is heartbreakingly picturesque. Apparently, Picasso thought so too, for it was here that he chose to spend his final years. His tomb lies within the park, but the grounds are not as yet open to the public.

Continue west, in the direction of Aix, and you will pass the **Barrage de Bimont**. This dam makes for a nice afternoon trip of its own, with its tremendous James Bond waterworks and large nature reserve. Hikers will find

Inside the Château Revelette.

numerous trails to choose from, several of which lead to the **Barrage Zola**, designed by Emile Zola's father in 1854.

Heading West: Another day trip from Aix lies to the west of the city on the road to Arles. However, a visit to the impressive **Château de Barben** and its **zoo** best serves to show the public the perils of tourism in Provence.

Physically the château is a lovely place. Originally built in the Middle Ages, it was enlarged during the 14th and then the 17th centuries. The formal gardens were designed by Versailles's own Le Nôtre, and in their centre is a sweet fountain, which Napoleon's little sister, Pauline, liked to bathe in *au naturel*. To protect her modesty as she frolicked, servants would hold sheets of black cloth up around it.

Inside are a number of rooms worth seeing, such as Pauline's delicate boudoir, with its hand-painted wallpaper. Unfortunately, however, management of the château is atrocious, and unpleasant guides rush eager visitors through with all the warmth of prison guards.

On the other hand, the guides seem delightful in comparison to the adjacent zoo. Go only if you feel like climbing dozens of perilous stone steps to find some ill-looking animals, plopped down in bare cages with no attempt to create a "natural habitat."

Olive country: Situated on the edge of the arid Crau Plain, **Salon-de-Provence** likes to call itself the "crossroads of Provence." And, indeed, it lies within 30 miles (50 km) of Arles to the west, Avignon to the north, Aix to the east and Marseille to the south.

The town is also at the heart of the olive-growing country, making it France's number-one marketplace for that silky oil. Add to this the 50-year-old presence of the officer's training school for the French Air Force, and Salon comes up as a fairly *puissant* spot.

In fact, however, Salon is a very laid-back city with a warm, small-town feel. Gay standards hang from the 17th-century **Town Hall**, and more flags decorate the heavy arches that give passage-

Life at the Barben Zoo.

way into the cobbled streets of the central **old town**. Above the arches, imposing towers mark what remains of ancient ramparts.

The old town is dominated by the grim **Château de l'Emperi**, built between the 10th and 16th centuries and around which Salon originally sprang up. Inside, the **Musée de l'Emperi** showcases France's largest collection of military art and history. Under the vaulted Gothic ceilings await copious exhibits of saddles and sabres, guns and medals, toy soldiers and life-size figures. The tour culminates with several rooms dedicated to the Napoleonic era, including among its treasures the brilliant general's short blue bed.

The museum has been marvellously arranged, with great attention to presentation and documentation. And, while only a quarter of the collection is on view at any given time (all objects date between 1700 and 1918 and pertain to French – not just Provençal – history), the tour still seems gargantuan. Serious military history buffs would make the trip to Salon just to visit this museum.

Spiritualists, on the other hand, should come to see the **home of Nostradamus**. The renowned physician-turned-astrologer settled in the town, birthplace of his wife, in 1546, and it was here that he both wrote his famous book of predictions, *Centuries*, and spent the last 20 years of his life.

Salon has made much of this famous inhabitant but, if truth be told, he wasn't so warmly welcomed during his lifetime. The townspeople held him in suspicion, partly because of his star-watching and partly because of his conversion to Judaism. Indeed, he was only saved from condemnation as a sorcerer by his ties to King Charles IX.

His house is a modest affair, recently turned into a museum but without any objects to exhibit. All his original manuscripts are under lock and key in the mayor's office, and most of his household objects were lost in a fire. Restoration, however, is at long last under way, and the authorities intend eventually to move the priceless manu-

Left, mural of Nostradamus in Salon. Right, statue of Adam de Craponne.

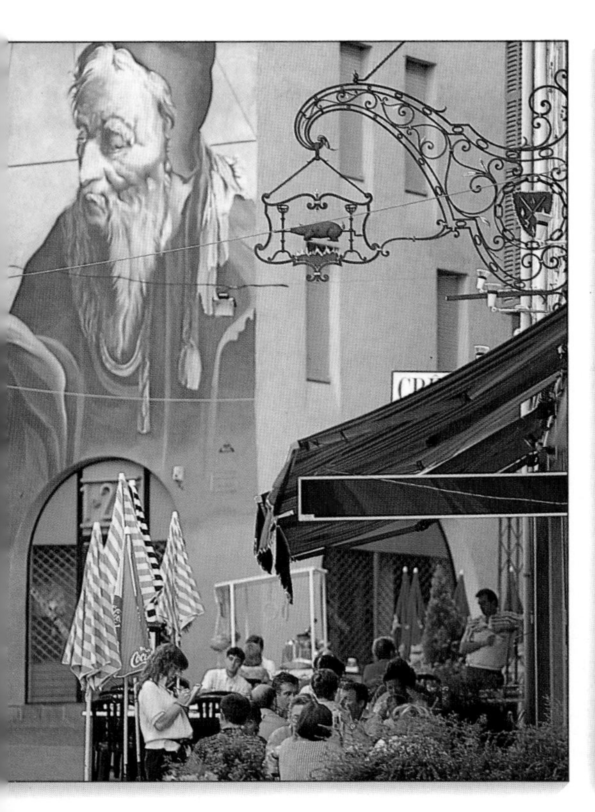

scripts into their rightful home.

Also within the ancient ramparts of the old town is the beautiful **Eglise St-Michel**, dating from 1220 to 1239 and built in the Roman Gothic style. The church possesses two belltowers, one a delicate 15th-century addition, the other contemporary to the construction of the building. Inside, cobwebs hang from the skylights above a tiny cool chapel. Be sure to note the stone tympanum, which shows Archangel Michael killing a snake above the paschal lamb, that decorates the wooden front door. It pre-dates the church, having been carved in the 12th century.

A second church lies just outside the ancient walls of the original citadel. **St-Laurent** was built between 1344 and 1480 and presents a much grander spectacle than its forerunner. Done in true Provençal Gothic, its exterior presents a simplicity that its sombre and majestic interior reiterates.

Only the largest of chapels within is lit. Placed directly across from the entrance, it is dedicated to the Virgin. A smiling, 16th-century alabaster statue commemorates her maternal piety.

The **tomb of Nostradamus** stands beside the statue of the Virgin. On the wall behind it is an inscription worth translating: "Here lie the bones of Michel de Nostradamus, alone at the judgement of humans worthy of knowing the stars of the future. He lived 62 years, 6 months and 17 days. He died in Salon in 1566. May future times not trouble his repose. Anne Ponsard, his wife, wishes him true happiness."

A second famed inhabitant, Adam de Craponne, has been remembered with a statue on a small square across from the Town Hall. De Craponne was the gifted 16th-century engineer who brought fertility to the region by building an irrigation canal from the Durance River down through the dry plains of the Crau and out into the sea.

Although Salon has served as a centre for the olive trade since the 15th century, insiders flock to three neighbouring towns to buy their oil. You may want to also but, before going off in

Olive country.

search of that perfect bottle of olive oil, it is important to know what to look for. First, check whether the bottle is marked *premier pression à froid vièrge extra*. If it isn't, don't buy it. If it is, then hold the bottle up to the light. The best oil will be a delicate green in colour.

South of Salon is the first town, **La Fare**, which many say offers the best oil in all of France. The local oil mill and wine cooperative are open to the public, and products can be bought on site. If you can, of course, it is best to go during the harvest season.

Other locals swear by the olives of **Aureille**, a half-forgotten town on the road westward to Avignon in the midst of grassy plains and vast expanses of olive trees. Aureille is a poor town, without any of the gussied-up look of most of the touristic circuit. At the same time, it is a pleasant enough place with a friendly central bar and a ruined 11th-century castle rising up on a hill above it that is only accessible by foot. For those in search of the "real" Provence, it can be both a relief and point of interest.

Its olive mill, just off D17, can also be visited by the public.

A third town with a claim on the olive market lies more within the jurisdiction of Arles than Salon. Like Aureille, **Mouriès** gives the outsider an opportunity to generalise about the Provençal character. Although hardly more than a village, with just one main street, Mouriès boasts several cafés, a handful of markets and two hotels. It also has two olive oil mills, one of which is a local cooperative. The other, on the Cours Paul Revoi, is open to the public with on-premise sales.

Since Mouriès is just outside Arles, it's a good place to remember if you have a car and can't find a room in the city during one of its festivals.

Bastion of tradition: For many, Arles *is* Provence. No other city in the region offers a more colourful atmosphere or possesses a greater awareness of its Provençal heritage.

The proud inhabitants of this old city are the first to tout themselves as not only "*vrai* (true) Provençals" but "*Ar-*

.andowner
›nd
'aughter
›utside
›rles.

lesiens," said with a special emphasis to drive home the significance of the phrase. Societies dedicated to upholding local customs flourish, and a "queen of Arles" is selected every three years on the basis not of her beauty but of her knowledge of Provençal traditions and her ability to speak the regional dialect and dress in its costume.

There is something almost otherworldly about Arles. It bears none of the shrewd urbanity of Marseille or the cultured sophistication of Aix. Instead, it seems to live in its past, haunted by the spectre of the Romans, populated by fierce upholders of tradition and deeply affected by its proximity to the wild marshes of the Camargue.

Founded by the Massilians as a trading post in the 6th century BC, Arles's position at the crossroads of the Rhône River and the Roman Aurelian Way made it a natural choice for development by the Romans.

It grew slowly for several hundred years, with some special help from Marius in the 2nd century BC. Then, during the struggle for power between Caesar and Pompey, Arles got its lucky break. Marseille made the fatal error of showing friendship towards the latter, and Caesar turned to Arles, which was already known for its skilled boatmakers, with the request that 12 war vessels be built for him within 30 days. The city complied, and Arles's good fortune was sealed. After squelching the claims of Pompey in 49 BC, Caesar squelched Marseille and designated Arles as the first city of Provence.

The city continued to prosper and under Honorius in 418 was made the administrative capital of the territory for Gaul. Although its predominance would waver over the following centuries, Arles remained a central maritime and river port up until the advent of the steam engine, when transportation by train overtook the tugboat method.

Train travel spelt disaster for the Arlesian economy, which had from the very beginning depended on its port activity for importance. As gateway to the Camargue (the top producer of rice

The old tow[n] lies along the Rhône.

in France), Arles has managed to recoup some of its value as a trading centre, but its days of eminence are long gone.

Lacking any major industry or university, the Arles of today relies heavily on the glorious Arles of yesteryear for its livelihood. Tourism abounds, and nary an ancient monument exists without a cardtable and guardian beside it demanding 10 francs. But don't let this dissuade you. The ghosts of the past are far more powerful than the souvenir sellers, and a feeling of mystery and romance still surrounds the city.

At the heart of Arles is the **old city**; a maze of narrow stone streets that mingle crumbling Roman edifices with sturdy medieval stonework. It is bordered to the south by the wide Boulevard des Lices, to the east by the remnants of ancient ramparts and to the north and west by the grand pathway of the Rhône. Surrounding all of this is the fairly innocuous spread of the more modern city, including the large Trinquetaille section.

Most visitors will first arrive in Arles via the **Boulevard des Lices**. If you have a car, it might be a good idea to leave it right away in the municipal parking garage to be found on this street's south side. Almost all of the city's touristic sites are within easy walking distance of each other, and the narrow lanes of the old town are not hospitable to automobiles.

The Boulevard des Lices is a busy spot with many cafés and the local tourist office, but it lacks both the charm of the Cours Mirabeau and the commerciality of the Rue Canebière. Its highest points come on Saturday morning during the extensive produce market and every first Wednesday of the month when a wonderful flea market spills over its sidewalks. At other times, it serves mainly as a thoroughfare past the ancient part of the city.

Before heading into the old city, take a quick detour south along the **Avenue des Alyscamps** to the site of the same name. Not much is left of this ancient necropolis, but no first-time visit to Arles is complete without having taken a stroll down the leafy **Allée des Sarcophages**, painted so vibrantly by Vincent Van Gogh. As you walk down the tomb-lined lane, notice the plaque that reads: "Van Gogh. Here, struck by the beauty of the site, he came to set up his easel."

The cemetery was begun by the Romans but by the 4th century had been taken over by Gallic Christians. Its fame spread during the Middle Ages, but the necropolis gradually fell into decline as its stone began to disappear; some of it went towards the building of other religious structures, some to antique dealers, some as presents to illustrious guests of the city.

All the tombs are now empty, and the stone shells that are left bear an uncanny resemblance to a long line of molars. Nonetheless, there is a peacefulness under the poplars that guards the solemnity of the spot and makes it a favourite for meditative afternoon talks.

At the very end of the lane stands what remains of **St-Honorat des Alyscamps**. Once upon a time, a large Romanesque church, St-Honorat now functions mainly as a backdrop for

The reigning 'Queen of Arles."

some of the city's theatrical activities.

To enter the old city, return to the Boulevard des Lices and walk a block inwards, up the cobbled Rue Hôtel de Ville. This will bring you to the spacious **Place de la République**. The Town Hall stands at one end, but more impressive is the **Eglise St-Trophime**. Its front portal dominates the eastern side of the square with a beautiful severity befitting a church dedicated to the saint who is credited with having brought Christianity to France.

The doorway of St-Trophime's is itself a glorious example of Provençal Romanesque style. Constructed between 1152 and 1180, its intricate tympanum shows the Last Judgement being overseen by a barefoot and crowned Jesus. He is flanked by the four apostles (Matthew with wings, Mark as a lion, Luke as an ox and John as an eagle). To the left of this group, the elect are presented to Abraham, Isaac and Jacob by an angel. To the right, the damned are refused admission to Heaven by an angel brandishing a sword of fire.

Inside, the lofty tone continues. The church's broken-barrel vaulted nave is, at over 60 feet (20 metres), the highest in Provence. By the left of the entrance, a fourth-century sarcophagus serves as a baptismal font. Further in, the life of the Virgin is depicted on a huge 17th-century d'Aubusson tapestry.

The **St-Trophime Cloisters**, which stand to the right of the church, are more cheerful. The sound of birds fills the air as you enter, and in warm weather pink flowers dot its bright courtyard, which is enclosed by solemn 14th-century figures posted as columns.

Work began on the cloisters during the second half of the 12th century, but was interrupted for about 100 years and they were not completed until the 14th. The northern and eastern galleries belong to the earlier period, the western and southern to the later.

Be sure to climb up to the very top before leaving. As you step out onto the "roof," you will be struck by the brilliant sun, reflected off the white stone. Look down into the courtyard, then to

Festivities in the Roman amphitheatre

the lofty belltower circled by birds. Delicate columns and the stone tiles that were used for eaves complete the exquisiteness of this lofty perch.

Across from St-Trophime is the **Musée d'Art Païen** (pagan art), housed in an unconsecrated 17th-century church that, despite its date, appears curiously Gothic in style. Its logical counterpart lies around the corner; the **Musée F.-Benoit d'Art Chrétien**. This museum of Christian art stands within a 17th-century Jesuit chapel and is particularly replete in sarcophagi.

A third museum, the **Museon Arlaten**, waits a couple of blocks away down the Rue de la Republique. Native-born poet Frédéric Mistral founded this ethnographic museum with the money he received for winning the Nobel Prize in literature, with the stipulation that it be dedicated to all that is Provençal. Accordingly, the name is Provençal (for "Musée d'Arles"), all the documentation within is in dialect (as well as in French) and even the *gardiennes* are dressed in traditional costume.

Room after room lead you through the world of the Provençal. One hall is filled with costumes and explanations of how they should be worn, others have amulets and magic charms, *tambours* and other instruments, nautical and agricultural equipment, wallhangings and paintings. There even are several life-size re-enactments of Provençal domestic scenes. It is the sort of place where you can easily spend a lot of time.

The *museon* is not the only mark left on the city by the poet. A formidable **bust of Mistral** looks down over the **Place du Forum**, just a stone's throw away. Most of the nightlife that doesn't involve a festival or cinema takes place in this lively square that overflows with a variety of cafés. Grab a table, if you can find a free one, but don't expect the food you are served to be gourmet. Bullfighting aficionados may prefer to head into the Tambourin, where the local champions patronise; the walls are lined with their autographed photos.

If you continue towards the river, you will hit the **Thermes de la Trouille**.

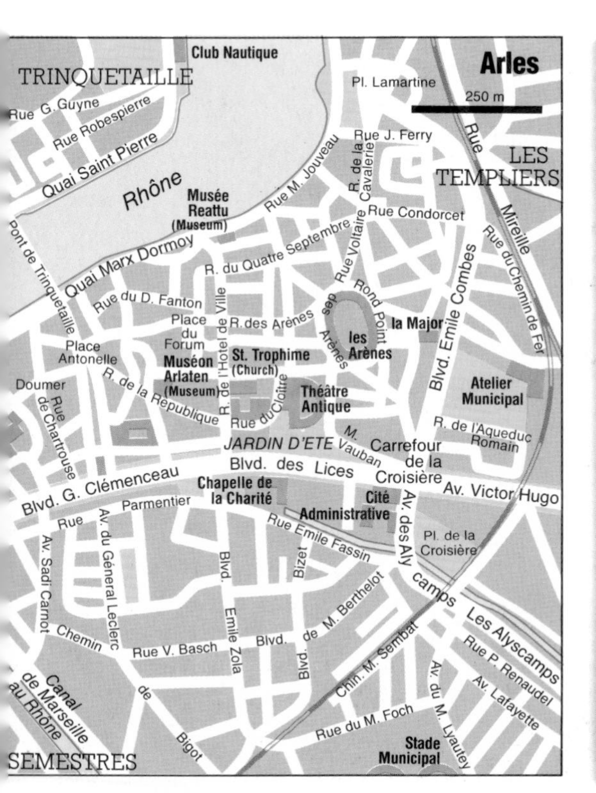

Barmaid.

These 4th-century baths were once the largest in Provence, but little of them still exists. The **Musée Réattu**, across the street, is far more interesting.

The Réattu Museum bespeaks eclecticism. It isn't worth paying the entrance fee of 13 francs to see most of the paintings, many of which are badly in need of restoration and some of which have not even been attributed. However, the collection does include some intriguing modern sculpture from the region, including one fascinating metallic piece with "living beads," and a collection of 57 Picasso sketches. The latter, dating from 1971 and donated by the artist himself, shows a somewhat more personal Picasso than one normally gets to see. The sketches may not be great art, but as free-form drawings they are a good deal of fun. Also of interest are the Salle Henri Rousseau and an extraordinary chapel-like room, vaulted by stone megaliths and containing a 1975 hanging-rope sculpture by Joseph Gran Garriga.

If the inevitable press of tourists has begun to get to you, steal a quiet moment walking down the ramparts that line the Rhône. Here, with the wide river on one side of you and crumbling stone facades on the other, you may recapture a sense of the Roman spirit that once dominated this city. (Try not to look at the new five-storey tenements on the facing bank.)

The **Amphithéâtre** is another place to chase Roman ghosts. Enormous in size – 440 by 345 ft (136 by 107 metres), 11,500 in total, and capable of accommodating more than 20,000 spectators – this Roman arena is both larger and older than its cousin in Nîmes, having been built in the 1st century AD. A visit to its tower offers a view out over the expanse of the red-roofed old city, its modern environs, the Rhône and the Alpilles Ridges to the north.

Bullfights of various types are still held here. Saturday afternoons find locals munching peanuts and drinking sodas as they watch their favourite *cocarde* champions cavort in front of snorting bulls. The first of May wel-

Local painter.

comes the Fête des Gardians, when cowboys from the Camargue gather for a yearly spectacle. More annual bullfights take place during the Feria Pascale of Easter and the Prémices du Riz in mid-September.

Chances are that, if you enter the old city on a warm summer night, you will be greeted by the sound of flamenco, opera or jazz wafting out from within the walls of the neighbouring **Théâtre Antique**. During the summer days, this 1st-century-BC theatre is little more than some piles of rubble, a siding of worn stone seating and two half-standing columns. At night, however, the Théâtre is transformed into a magical backdrop for the Festival d'Arles.

Like all the other cities of Provence, Arles puts on a special cultural festival during the summer. The Festival d'Arles is particularly successful, wisely taking advantage of Arles's many antiquities as ultra-atmospheric stages for everything from folkloric dance to classical music to high-fashion shows.

The festival is accompanied in July by another widely applauded annual event: the **Rencontres Internationales de la Photographie**. This "festival" brings together photographers from around the world to compare techniques, peruse exhibits and enjoy nightly spectacles. Workshops with respected masters are also available. Like the festival, the Rencontres use many of the Arlesian monuments and museums to enrich their productions.

Artists who come to Arles with the intention of making a pilgrimage to the haunts of Van Gogh may be disappointed, however. The house where he and Gauguin once lived, worked and argued no longer stands, and not much has been done elsewhere to commemorate his days down south. Devotees will have to content themselves by seeing some of the sites he once painted, the fabulous light he once saw and the copious fields of sunflowers from where he once gathered bouquets for still lifes.

Into the Alpilles: The dry flat land that surrounds Arles, circled by jagged, white stone peaks called **the Alpilles**, is

Fields of sunflowers surround Arles.

CHRISTIAN LACROIX

When a creation by *haute-couture* designer Christian Lacroix is modelled down the catwalk, a great gust of sunshine goes through the room. For it is no secret that Lacroix, one of the fashion market's most innovative and eclectic modern talents, is strongly inspired by his native region of Provence. Indeed, Lacroix's madcap designs are a pure reflection of his love for the sunniest area in France. His magic paintbrush is reminiscent of the colours one finds in Van Gogh's paintings: bright sunflower yellows, poppy reds, hot pinks, purples and blues, as well as swirling prints.

Since the mid-1980s, the outlandish decor of Christian Lacroix's *couture* house – designed by Elizabeth Garouste and Mattia Bonetti – at number 73 of the very trendy Faubourg St-Honoré in Paris, has welcomed the *crème de la crème*. International names such as Ivana Trump and Lucy Ferry (wife of rock star Brian) are just two among the host of cosmopolitan beauties to cross its threshold. And, in 1987, the designer, atill in his mid-thrities, made the cover of *Time* magazine. But, even with his newfound fame, Lacroix admits he often feels the need to go back to his childhood Provençal roots and draw breath.

"I've always been crazy about terracotta floors, primitive people, sun and rough times," says the 37-year-old designer. "This is my real side – goat cheese and bread, elementary things. I am fascinated with Paris, its elegance, its women, even its artificiality. But with my heart and skin I love the south – bullfighting, pleasure, music, nature, the sea."

As an adolescent, Christian Lacroix grew up in Arles, which he still speaks of with a feverish nostalgia for his golden youth: the now-deserted Rue de la Roquette, then alive with gipsies; the Rue des Porcelets, where he bought the *fougasse aux gratillons*, a delicious bread that Marcel Pagnol often talks about in his novels; the Rue de la République, where he used to be fascinated by the shop of *santons*.

Some of Lacroix's favourite landmarks around Arles include the entrance hall of the Hôtel de Ville, the St-Trophime Cathedral and the Notre-Dame-de-la-Majour Church near the Arena. Following the sinuous Rue Parade leads to the Musée Réattu, in the former palace of the Maltese knights. When the young Lacroix played truant from school, it was often to go and look at *L'Arlésienne* and *L'Atelier de couture en Arles*, painted by Antoine Raspal in the 18th century.

"Our local hangout was Le Mallarte," remembers Lacroix. "It was the bourgeois café in town, where we liked to spend long hours practising the art of conversation, seated at smoky tables in our navy pea coats and long hair." Nowadays, he is more likely to dine at Vaccarès, a popular local restaurant whose terrace overlooks the Place du Forum and whose other famous patrons include writers Michel Tournier and Yvan Andouard.

When he can escape from his numerous obligations as the new darling of *haute couture*, Lacroix can generally be found somewhere in the golden triangle of Provence – and not just in Arles. The little towns he so cherishes are scattered along the bright road of the Apilles, from Maussane to Mouriés, from Les Baux to St-Rémy-de-Provence, and eventually down into the Camargue region.

"In fact, Arles is really a passage between my two worlds: Provence and the Camargue," the designer says. "On the one side, the laughing universe of the Alpilles commemorated by Alphonse Daudet; on the other, as soon as one crosses the Rhône and enters the Trinquetaille, where my grandparents lived, it is already the door to the Camargue and its flat countryside. Since I am both Cévenol and Provençal, I am very sensitive to this double aspect."

It's just a short drive south to the region of the Camargue and the town of Stes-Maries-de-la-Mer, where Lacroix remembers with great fondness spending long lazy summers. "I have always loved the swampy landscapes of the Camargue," he says. Proof of this is the lovely *mas* (the local name for a traditional Provençal farmhouse) that he bought once he had established himself in Paris. "I come here as often as possible with my wife Françoise and renew my relationship with my friends, to whom I am just Lacroix, the old pal from 20 years ago."

filled with sunflowers – not to mention cypress groves and olive orchards. Arles holds reign over this region, generally referred to as the Pays d'Arles.

Although most are extremely poor, the natives of this dusty land bear the mixture of pride and generosity intrinsic to all that is Provençal. It is a rewarding area to travel in, filled as it is with anecdotes and reminiscences.

If you don't have a car and are energetic, you can try cycling, but be wary of the hot midday sun in summer and the steep inclines of the Alpilles. Local buses will get you where you want to go, but they will take both time and money. As a final choice, try hitchhiking. It's never the best idea but probably safer and easier here than in most places.

Just 4 miles (7 km) northeast of Arles, past a long field of peppers, lies the **Abbaye de Montmajour**. Founded in the 10th century by Benedictine monks, on an island amidst a swamp that was passable only by raft, the monastery fell into decline by the 1600s. Total collapse came during the next century when the head abbot, Cardinal de Rohan, became entangled in a scandal concerning Marie Antoinette.

In the 1790s, the property was bought and then stripped by antique dealers. Restoration was begun in 1872, but there is still something like a movie facade about the abbey: grand from the front, empty at the back. To add to its problems, it was occupied by foreign forces in 1943 and burnt a year later.

Nonetheless, the plan of the abbey is fabulous. In fact, it is touted as being, along with the Abbé St-Gilles, the most elaborate of any Romanesque church in Provence. The central crypt, which is above ground and inscribed with cryptic inscriptions, forms a perfect circle with five evenly radiating chapels. And the enormous 12th-century **Eglise de Notre-Dame** possesses a never-completed nave that is considered one of the masterpieces of Romanesque art. Its thick echoing walls and floor would have made a perfect stage for the dances of Isadora Duncan.

Left, Lacroix fashion show in the Theâtre Ancien. Below, the Rhône near Tarascon.

Stepping out of the cloister and onto the ruins offers a wonderful view of the outlying countryside, with the city of Arles in the distance. For an even grander panorama, climb to the top of the imposing tower erected by the Abbot Pons. From there, you can see the nearby village of **Fontvieille**.

Following the road just northeast will take you right into this pleasant little town. As you enter, note the modest oratory, erected in 1721 in thanks for the end of the plague. Similar oratories mark each corner of the town.

Fontvieille is a charming spot, but one that has clearly been marked for tourism, with prices to match. Its primary attraction is the **Moulin de Lettres**, as they call the mill where Daudet presumably wrote *Letters from My Windmill*. The fact that the windmill actually belonged to friends (not the author) and probably witnessed little penmanship (although Daudet did visit it) has done little to deter the hordes of tourists that crowd its tiny interior.

Set in a picturesque spot, the mill does contain some interesting features, such as a round ceiling marked in compass fashion with the names of the winds that sweep the hill. Unfortunately, you can't hear these winds over the din of tourists. The narrow subterranean museum is equally disappointing.

Today, a number of literary immigrants continue to make Fontvieille a writer's enclave. Most notable among them is the (once-Parisian) author Yvan Audouard. Other artists have also found the town an amenable haunt, such as the Swiss painter Carl Liner. The photographic community is admirably represented by the Rockport Workshops held in town every September.

Fontvieille is also a popular spot for summer homes, and quite a few Germans, with some Swiss and English, have found a niche here. But weekend visitors won't find the town lacking for traditional charm, and if you arrive on the first Sunday of August or on Christmas Eve, you can indulge in some trueblooded folkloric revelries.

Another place to indulge is the

Plane trees line many roads in the Pays d'Arles.

Château d'Estoublon Mogador on the outskirts of the village. This traditional 14th-century wine château, complete with geese and kittens playing in the front yard, welcomes visitors. The Lombrage family sell their own red and rosé wines, plus their prize-winning olive oil, right on the property.

Alphonse Daudet also left his mark on the neighbouring town of **Tarascon**. This small city was founded by the Massillians in the 3rd century BC and became a Provençal legend in AD 48 when its horrible resident monster, the amphibious Tarasque, was tamed by the sweet Saint Martha. Widespread fame came in the personage of Tartarin, fictional "hero" of the book of the same name by Daudet.

Tartarin was Daudet's revenge against what he saw as the smallness of the provincial spirit. Despite a superficially sympathetic aspect, the character was a liar, a braggart and an all-in-all ridiculous figure. The people of Tarasque, from where Tartarin was supposed to have come, have never been able to live this reputation down. Without too much prodding, they'll tell you just how little they appreciate being known as the compatriots of Tartarin.

This hasn't prevented them from creating a tiny museum to the character, the **Maison de Tartarin**. Opened in 1985, the Tartarin House features wax figures in re-enactments of scenes from the book. It's kind of a silly place, really only for those who know the book well.

All visitors to Tarascon will want to stop by the **Collegial Ste-Marthe**. Like many other churches in the region, this imposing edifice presents an interesting eclecticism of architecture, for one half was built during the Romanesque 12th century and one half during the Gothic 14th century. Unfortunately, most of the bas-reliefs that once decorated its doorway were destroyed during the Revolution.

Nothing, however, has touched its remarkable crypt. Descending into its tiny alcove, you are immediately overwhelmed by the cool hush. A button on one side illuminates the oratory just

Tomato picker.

long enough to expose the relics of the saint, hostess to Christ at Bethany, passenger on the *navette* from Jerusalem and tamer of the menacing Tarasque. Her crypt is one of those unique places where the unseen overtakes the everyday. From the moment you enter its dark and damp confines, you feel a shroud of sanctity fall over you. It's not grandiose – but it is convincing.

Grandiose would more accurately describe the feudal **Château du Roy René**. Built directly on the banks of the Rhône during the first 50 years of the 15th century, it proudly rivals the Château of Beaucaire, which stands across the river on the opposite shore.

The interior of the castle is mostly empty. In the first gaping hall, a changing exhibition of modern abstracts creates a striking effect against the severe medieval walls. Aside from that, the castle has no furnishings but six magnificent 17th-century tapestries that recall the glory of Scipion.

The city puts on several *fêtes* each year but none more celebrated than that of the *Tarasque*. Among the activities that accompany this festival, which takes place during the last weekend in June, is a colourful parade led by a *papier-mâché* fascimile of the monster.

From Tarascon, travel north along D35 to reach the 10th-century **Abbaye St-Michel de Frigolet**, where a handful of monks still dwell. The land seems slightly less forbidding here, as the Alpilles meld into the gentler **Montagnettes**. These hills cover the area between the curve of the Rhône and the east-west stretch of the Alpilles and harbour numerous alcoves of cypress and poplars, olive and almond trees, straggly pines and fragrant herbs.

Barbentane is a friendly little town just north of the abbey, with a warm and comfortable atmosphere. Despite its modesty, a wonderful **château**, built in 1674 and still occupied by the marquis of Barbentane, stands right off its main street. Step through the gates and you'll wonder whether you have suddenly been transported to the Ile-de-France.

Luckily for the public, the marquis and marchioness have fallen on hard times, forcing them to open their doors to the curious. Guided tours are now available at scheduled times.

Outside, the subtle evidence of decay is everywhere. The lawns are slightly overgrown, and moss sprouts from the heads of classical statues. The backyard is lost in weeds. And, somehow, this only adds to the château's charm.

The interior is one-in-a-million. The grandson of the first owner spent 20 years as ambassador in Florence and brought home from his Italian voyages masses of marble and 18th-century furniture. Everything within is original, and many items, such as the wrought-iron bannister, are handmade.

Among the most amazing articles are: the marble floor and circular mosaic ceiling of the "statue" salon; the hand-painted 18th-century wallpaper of the "Chinese" sitting room; the Italian-style wall fountain in the parlour (it was once the dining room, but lack of servants and distance from the kitchen forced the marquis to switch their uses); the Provençal-style library and upstairs

Column from the Abbaye St-Michel-de-Frigolet.

bedroom, also with original hand-painted wallpaper. The beds were short because people never lay on their backs (considered the position of death). Above the stairwell is a stone-carving of the blustry mistral.

From the front door you can see the ominous **dungeon**, but it is not open to the public. Also within sight is the crumbling **Notre-Dame-des-Graces Church**. A more contemporary attraction on the outskirts of town is the **Provence Orchidées Greenhouse**. Here, you will find an extensive offering of prize-winning orchids for sight and for sale. It is best to come between October and May.

Before heading back south, you may want to make a quick sweep east through **Noves**. The town today boasts little to attract tourists (all that remains of its château are segments of the 14th-century ramparts), but it does have an almost mythic past. Founded in the 5th century BC, it was here that Petrarch first laid eyes on his beloved Laura, about whom he would later write some of the world's most famous love poems. The wine grown on the neighbouring hillsides is called, most appropriately, the *Cuvée des Amours* (Love Vintage).

Another town of literary significance lies south along D5. **Maillane** was both the birth and burial place for Frédéric Mistral. The house where the poet lived from 1876 until his death in 1914 has been turned into a museum, containing his desk and gloves as he left them, many of his books, and portraits of him, his wife and friends. In the cemetery down the street rests his tomb.

D5 runs directly into **St-Rémy-de-Provence**, the mini-capital of the area. St-Rémy makes a fine base for any vacation in the Alpilles, especially for those wishing to avoid the greater bustle of Arles. It is right in the middle of the region, possesses many attractions of its own and is, all-in-all, an extremely pleasant place to be.

St-Rémy is also the place to come for those interested in herbalism, for which it is the recognised centre. During the 1960s and 1970s, the number of *herbo-*

Young monk.

ristes in Provence declined, due to the growing popularity of chemical substitutes, but recent years have seen a resurgence in the use of natural products. Locals are optimistic that this will, in turn, help boost regional economy.

Wholesalers head for the **Herboristerie Provençale**, just outside the old town on Avenue Frédéric Mistral. Amateurs are welcome there too but might prefer one of the town's many "herb boutiques." An excellent one exists on Boulevard Mirabeau. Called **Les Herbes de Provence**, this pretty little store offers a variety of luxuriant herbal and flowery bath products as well as sacks of freshly dried basil, thyme, rosemary, etc.

Herbs are only one aspect of St-Rémy's popularity. Like most of the larger towns in the Bouches-du-Rhône, at the heart of the (relatively) new city lies a rather older city built in a round and criss-crossed with narow stone streets. In St-Rémy, these streets are labelled with names in Provençal as well as French, and several of the houses have been turned into museums.

One old house that hasn't been converted but is still worth taking a peek at stands along the Carriero di Barrio de l'Espitau. The building has long been boarded up, but an exterior plaque proclaims its significance: "Here was born, on the 14th of December in 1503, Michel de Nostradame, alias Nostradamus, astrologer."

The cavernous **Collegiale St-Martin** is less modest. The ceiling of this church is decorated with faded blue paint and a sprinkling of gold stars and lit by numerous stained-glass windows. What really catches the eye, however, is the enormous organ that dominates the church's anterior. Known to be outstanding among its peers, free concerts are given on it every Saturday.

Crossing behind the church leads you to the **Musée des Alpilles**, arranged within the 16th-century mansion of Mistral de Mondragon. This ethnological museum contains scores of photographs, portraits, costumes and household objects linked to the surrounding

Herbal bath oils from St-Rémy.

region. One room features the minerals mined within the Alpilles: bauxite, limestone, sandstone, etc. The building, which extends over an old alley as a courtyard, is also remarkable.

Next door, the venerable 17th-century **Hôtel de Lubières** offers informal exhibitions. One side is filled by modern paintings, and the other shows local works and sculptures in wood and other natural materials.

Across the street stands the **Hôtel de Sade**. Built during the 15th and 16th centuries, it is now an archaeological museum, housing all the objects discovered at the nearby site of Glanum. Its contents include everything from votive altars to pottery to bones.

Glanum and **Les Antiques** lie just outside the city. Once the site of a major Phocaean trading post, constructed beside a sacred spring, they are the town's top tourist attractions.

Les Antiques stands on the right-hand side of the road and comprises a commemorative arch and a mausoleum. The arch, situated across what once was the road connecting Spain to Italy, was built during the reign of Augustus to indicate the entrance to the city of Glanum. The mausoleum is a funereal monument of three levels that was dedicated to Caius and Lucius Caesar, the two grandsons of Augustus who died prematurely. The mausoleum has been particularly well-preserved, and both it and the arch are amazing pieces of Roman stonework such as you would more expect to find in an Italian city than upon an Alpilles plain.

Glanum, across the road, is slightly less approachable. Originally Celto-Ligurian, a sanctuary city and then a flourishing point of commerce during the Gallo-Greek years and later Roman times, it was virtually destroyed by invaders in the 3rd century. Excavations continue to uncover archaeological treasures, but the layman will need a good text and plan of the site.

Posted in between Glanum and the centre of town is the **Clinique de St-Paul**, where Van Gogh commited himself after slicing off his earlobe in Arles.

Roman frieze on Les Antiques.

The tiny cloisters of the medieval **St-Paul-de-Mausole Monastery**, which houses the clinic, can be visited, but the still-active rest home can not.

One pleasant way to spend a meditative afternoon in St-Rémy is fishing. Follow the Avenue Antoine de La Salle, just down the road from the clinic, until you see the sign for Le Lac on the Chemin de Barrage. At its end lies the **Peiroou Lake**. The Lac de Peiroou, which means chauldron in Provençal, is reserved for fishing – you're not supposed to swim there – and behind it is a nice park for picnicking.

Or, if you are feeling really lazy on a hot afternoon, stroll up to the sandy square in front of the *syndicat d'initiative* and you will find the spot where the locals gather daily to play *boules*.

A feudal city like no other: When you're ready for a little change of scenery, take D99 west and switch on to D27 south. This perilous route climbs up into the peak of the Alpilles, overlooking the aptly named **Val d'Enfer** (Valley of Hell). Narrow stone cliffs cut into the winding road on one side, and on the other you will find a sheer drop. If possible, keep both hands on the steering wheel, and pray you don't meet another car.

From the **Cave de Sarragon**, the aerie peak dominated by Les Baux comes into view. But first, before making the prerequisite tour of this Provence-style ghost town, stop in on the **Cathédrale d'Images**. Movie buffs will recognise the spot as the site of both Cocteau's *Orpheus* and *Antigone*.

Even as you near the huge covered quarry-turned-theatre, the cool air of its interior reaches out to envelop you. When you step inside its darkened caverns, cold blackness surrounds you. Then, as your eyes begin to adjust, you realize that continually changing slides are being projected in monumental sizes over the rocky crevasses of the walls. Some photos even drape across the floor, so that you feel as though you are walking upon whatever they portray. Meanwhile, music echoes through the hollowed chambers. The whole ex-

The heights of Les Baux.

hibit will belong to a single theme, say Saharan Africa one year, Van Gogh another. No one should miss a visit to the "Cathedral."

Originally, the quarry belonged to the fabled citadel of **Les Baux**. Set on an imposing site way above rocky outcrops that tumble into deep valleys, the site of Les Baux has been occupied for the past 5,000 years. Its fame, however, comes from the period, between the years 1000 and 1400, when the bold and arrogant Lords of Baux made their presence felt throughout the region.

The feudal life of Les Baux offers a fascinating paradox. Here, "amidst this chaos of monumental stones and impregnable fortresses, inhabited by men whose roughness was only matched by that of their suits of armour, the 'Respect of the Lady' and ritual adoration of her beauty was born." The patronage of the troubadours by these violent lords engendered the first "Court of Love," and much of France's literary tradition emerged from their bloody citadel.

Today, more than a million and a half tourists a year come to see where once the poets roamed and the lords raged, with July and August being close to unmanageable. Craft shops and ice-cream boutiques spill out of many of the old buildings, and even parking costs ten francs. The impenetrable Les Baux has become the ultimate in tourist traps.

The spot divides into two parts: the ancient city, containing the ruins of the feudal court, and the village, where some 60 people still live and the shops and galleries now stand. In winter, the number of inhabitants drops to about 40, and ice and snow ravage the exposed peaks. Over the years, Les Baux has become somewhat of a retreat, but the number of year-round artisans is currently on a decline.

The village boasts three museums. The **Hotel de Manville** is a good place to start, since it contains the historical museum of Les Baux. The three rooms offer thorough documentation (in French) on the annals of Les Baux from neolithic times to more modern days. The last room shows photographs of

he good
ife at the
)ustau de la
aumanière.

Princess Grace and Prince Albert Grimaldi of Monaco to whose family Les Baux belonged from 1642 to 1791.

Down the street, the **Hôtel des Porcelets**, built in the 16th century, houses a museum of fairly unremarkable contemporary art. But, next door is the best-preserved of all the monuments in the old city: the **Eglise de St-Vincent**. Its central nave dates back to the 12th century, and its stained glass is a striking *mélange* of the very ancient and the ultra-modern. Across the way is the charming 17th-century **Chapelle de Penitents Blancs**, decorated by Brayer. The apt theme of its mural, painted in 1974, is the shepherds of Provence, with an Alpilles Nativity. Finally in this little corner, cut directly into the rock face, is the **Galerie St-Vincent** with more modern paintings and sculptures.

A last museum, of archaeological and lapidary objects, occupies the **Tour de Brau**. Built at the end of the 14th century, the tower also marks the entrance into the **Cité Morte** (the ancient city).

A tour of the Cité Morte begins with the 12th-century **Chapel de St-Blaise**. Beside it lies a small modern cemetery filled with lavender and butterflies. Continue further up to the top of the ruins where stands a round tower from which the Baux family must have surveyed the valley with watchful eyes. The view is incredible. In the distance, the Alpilles sweep across the countryside. Pepper fields, olive groves and tiny hillside swells, all framed by white stone outcrops, nestle below.

More of the ancient city can be found along the northern ramparts. The energetic can climb them to attain even wider views or to investigate what is left of the 15th-century castle, the 10th-century keep and the Paravelle Tower.

Hugging the cliffs below the old village, the world-renowned **Oustau de la Baumanière** restaurant can compete with any dining establishment in Provence – or Paris – for quality. The proud chef uses mostly local produce and much of the menu is regional, but that doesn't mean the fare is peasant style. On the contrary, if you want to dine here, make a reservation, pack your pearls and *stuff your wallet.*

If the Alpilles have left you hot and thirsty, you may want to head back towards the Mediterranean. One choice would be the area around the **Etang de Berre**. Most of this section is highly industrialised, but the old port of **Martigues** still possesses some of the charm that enticed such painters as Corot, Renoir and Dufy to sojourn there. It makes a nice place to lunch.

Martigues is also near one of the most important archaeological discoveries in the Bouches-du-Rhône. Excavations at **St-Blaise** have uncovered several superimposed Celto-Ligurian cities as well as Greek ramparts. The site is open to the public, but it isn't well-marked and the uninitiated may easily lose their way (and understanding).

Another easy excursion from the Pays d'Arles is the **Camargue**. Indeed, many young Arlesians drive down nightly to enjoy its discos. Daytime visitors will find plenty of distraction as well when they are transported into what seems like a whole new world.

Left, bust of Mistral in Maillane; right, cooling down at a café.

RESTAURANT LE COMMERCE
FISCHER GOLD
CAFE DU COMMERCE

RES
MAUSSANE
PAUL RICARD
C
ST. MARTIN DE CRAU
S. GRAPPELLI
TRIO
19
FOYER RURAL

MA VIE
POUR
MA
FAMILLE

THE CAMARGUE

The **Camargue** lies directly to the south of the Pays d'Arles and, although officially a part of the Bouches-du-Rhône, is very much in a world of its own. Almost triangular in shape, bordered to the west by the Petit-Rhône, which flows to the sea past Les-Stes-Maries-de-la-Mer, and to the east by the Grand-Rhône, which runs down from Arles, its unusual natural environment has helped create for it a unique history.

Most of the region is marsh and lagoon, creating a perfect setting for one of France's greatest wildlife reserves. The area also provides a spiritual home for the country's gypsies, stages a national windsurfing championship, offers the joy of riding across beaches and salt plains, and provides the tourist with a choice of historic castles and churches to visit. It can be an equestrian's dream, a birdwatcher's paradise, a historian's treasure chest, a gourmet's delight or a windsurfer's playground.

At the same time, without advance reservations, the Camargue can turn into a tourist's nightmare. The famous horizons that inspired the Provençal poet Frédéric Mistral and the stunning light that fascinated Vincent Van Gogh also have encouraged an onslaught of touristic commercialism, resulting in the construction of simulated-rustic hotels and brash campsites and leaving behind a pile of left-over debris generated by summer holidaymakers.

To get the best from the Camargue, which covers over 480 sq. miles (800 sq. km), takes careful planning. Bear two things in mind. First of all, it is not always possible to wander off the main track without special permits. Secondly, the multiplicity of attractions makes it exceptionally popular with tourists, particularly families with children, in both July and August. During the high season, demand frequently exceeds both space and services available.

Preceding pages: vacationing couple. Left, white horses graze beside a *mas*.

Despite the growth of tourism, there is still much worth seeing in the Camargue and, with a little forethought, you can get around most of its downfalls. Any exploration of the region entails a certain amount of unavoidable zig-zagging, due to the layout of the land, so some people may prefer to stay on its edge near Arles and visit during a series of day trips.

Once within the region, travellers will find themselves with a wide selection of ways to get around. The curious blend of marsh, swamp and salt plain, which have in turn led to the establishment of copious wildlife parks, means that one can either journey by boat, by jeep, by horseback or on a bicycle. Real nature lovers will probably find that some areas can best be discovered on foot.

Even if you decide only to spend a few days in the region, visits to two or three well-chosen spots can reveal an amazingly rich variety of wildlife. And the real appeal of the Camargue lies in its wildlife: white horses, black bulls, wild birds, salt-water vegetation and swamps and lakes.

Getting to know the area: For a good, over-all introduction, take N570 west from Arles to the **Mas du Pont du Rousty**. This combination sheep farm and regional museum contains a permanent exhibition illustrating the history and traditional way of life of the Camargue. Best of all is its explanation of the different terrains in the region, which describes how the wildlife of fresh-water marshes is dissimilar from that of the salt lagoons and that of the coastal dykes. (At the same time, as you actually start to move around, don't be surprised to see a flock of pink flamingoes – inhabitants of the marsh – in full flight over an industrial salt plant in the southeast sector. It may seem incongruous, but it will just be another one of those startling contrasts that make the Camargue such an exciting region.)

For a hands-on look, take the signposted footpath that lies to the rear of the museum and leads to the beginning of the marshes. The walk will take you

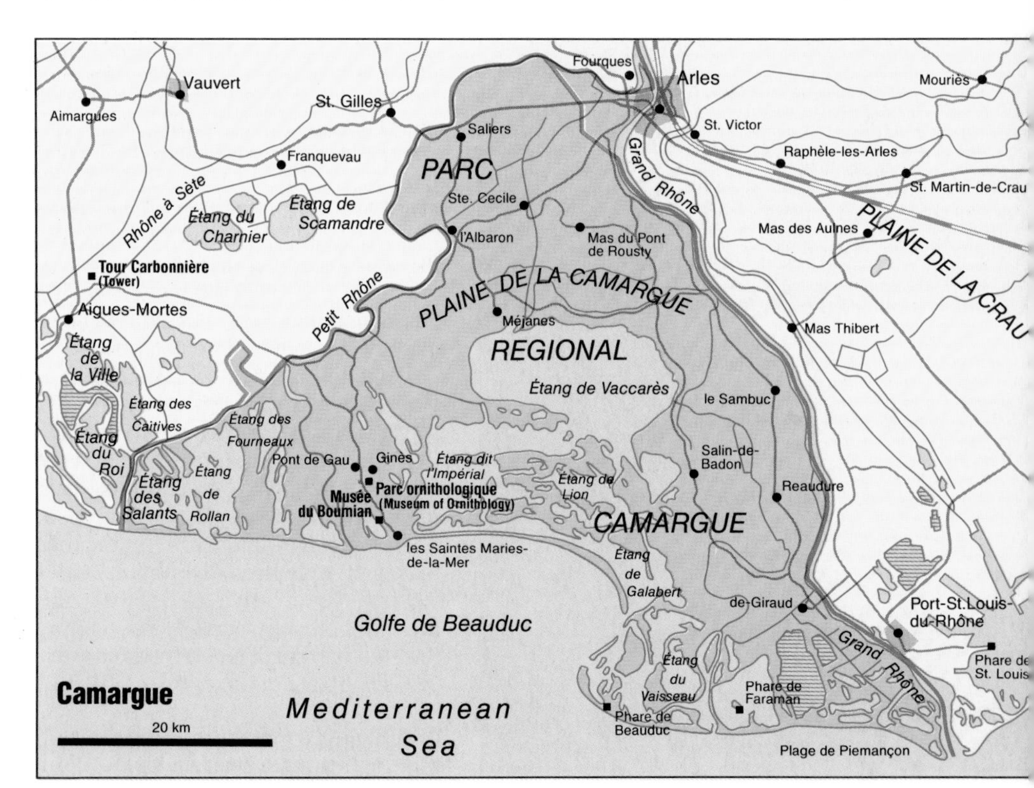

about an hour and give you a personal feel for these strange flatlands.

At first, you will be struck by the loud croaking of frogs, then the colourfulness of the flowers, which comprise everything from alofile plants to tamarisk and, during the warmer six months of the year, daisies, irises and asphodels. In addition, you will see a wide selection of birds: the curious black-winged stilts with their long pink legs, herons and all sorts of wading birds such as the ringed plovers.

The Camargue is a great place for birdwatchers. If you happen to be one, remember that you will find a lot more feathered activity in the early morning – which also is much the most pleasant time to be out rambling, especially during the hot summers. Springtime, of course, is the ideal season, for you will both get to hear their singing and see the large variety of migrants that visit the Camargue during their trip back north.

Once you've left the Mas, continue west along N570 towards **Albaron**. This route will take you past both paddy and wheat fields, land that has been recovered from the swamps over the past 40 years. Along the way, you will notice constant advertisements for a roadside ranch, complete with a wagon wheel and mock "western" frontage. This is just one of the many places in the Camargue that run pony trekking and safari trips into the marshes. The trips last from an afternoon to a full day.

At Albaron, a right turn onto D37 will lead you into **St-Gilles**. The carvings on the western front of this busy little agricultural town's 12th-century church are mild masterpieces. Upon entering the structure, you will find a remarkable stone spiral staircase, which has managed to survive despite extensive destruction during the Wars of Religion.

French cowboys and white horses: If you had taken a left-hand turn in Albaron onto D37, you would have entered the **Rhône Delta**, where lies the **Etang de Vaccarès**. Going in this direction, even from a car, will provide you with excellent views of pink flamingoes and other water fowl. On foot, a

Wedding at St-Gilles.

perceptive eye should be able to spot a beaver or a turtle or a scampering little lime-green tree frog. Here, too, the famous black bulls of the Camargue roam, and sometimes (but not as often as postcards may lead you to believe) you will catch a *gardien* (the French-style cowboy) astride his white horse.

Some say that the famouse white horses of the Camargue are the descendants of the wild horses that populated prehistoric Gaul and are depicted in the ancient cave paintings at Lascaux. Others claim that they are of North African origin, and even others suggest that they were first introduced from Tibet.

To get a closer look at them, you may want to check in at one of the Camargue's "ranches," where the *gardiens* take you out on horseback to visit the marshes. Extensively publicised is the *mas* belonging to Paul Ricard at **Méjanes**. Here, they've set up a well-organised trekking centre, and the *gardiens* who accompany you are particularly knowledgeable and authentic, for they work during the winter months as herdsmen. However, Ricard's is extremely commercial, and the bullfights and equestrian displays staged there are strictly for the tourists.

If you choose to travel without a guide, remember that the lagoon is a protected area. Any kind of hunting or fishing and even the picking of plants and flowers is prohibited.

Motorists will find few alternative routes once in the area beside the lagoon. This shouldn't matter for, as one follows the curve of the *étang* east, the view over seascapes, sandbanks and lagoon is magnificent. Butterfly lovers will be delighted to discover a rich variety of brightly coloured little creatures, including the Southern White Admiral, Swallowtails and Spanish Festoons. Don't be surprised, however, to learn that this (and all of the Camargue) is also the land of the mosquito.

Eventually, you will start to see powdery white hills and salt pans decorating the landscape. This southeastern corner of the Camargue, leading down to **Salin-de-Giraud**, is the French capital

Flamingos on the *étang*.

for salt production, and over half of the country's supply comes from right here. As might be expected, salt is a major industry for the region.

This area is also the gateway to the Camargue's beaches. After a long hot drive, few things are more refreshing than racing into the Mediterranean, cooling off in its inviting waters and then collapsing behind a sand dune for an afternoon's siesta. One word of caution, however, if you've parked your car out of sight. Sadly, the last few summers have witnessed a deplorable increase in thefts by seaside resorts. Best is to leave your car visibly empty and to carry any valuables with you.

Continue along D36 to what becomes a much narrower track, and you will reach the **Plage de Piemenson**. This beach represents a different sort of "natural" paradise, for its frequenters are all *"naturistes"* or nudists. Anyone familiar with beaches along the Côte d'Azur may well wonder why the local authorities have bothered to make any distinction between "nudist" beaches and "normal" beaches, but they do. And, it's true, there is something startling about watching middle-aged couples barbecuing in the nude. But you will soon get used to it. Besides, playing *boules* in the nude is more than fashionable, it is positively trendy.

If you don't feel up to baring all, you will find plenty of other sandy stretches without nudists along the same coastline. The **Plage d'Arles** will give you more chances to glimpse sea birds, such as the yellow-legged gulls, slender-billed gulls or the tern as it dives into the water for a fish. A fine breeze blows across these waters, making them particularly good for windsurfing.

If you are lucky, you may be able to find, along one of these beaches, a private boat operator who can take you directly over the sea to **Les-Stes-Maries-de-la-Mer**.

However, if you prefer to drive here, return by D36 to D37 up to Albaron, then head back south on the other side of the Etang de Vaccarès along D570. At twilight this journey is a bewitching

'oung
ypsy.

Camargue experience: the setting sun casting its colours over the lagoon.

A saintly city: Les-Stes-Maries-de-la-Mer is one of those places that has been very much influenced by its natural environment. During the Middle Ages, it stood several miles inland, and even as recently as 100 years ago its outlying borders were still a good 600 ft (200 metres) from the sea. Today, however, it is a flourishing resort, right on the border of the water.

The encircling seas have also brought scores of eastern visitors over the centuries. The town owes its name to an early group of these, who were by far the most famous of its settlers: Mary Jacoby (sister of the Virgin Mary), Mary Magdalene, Mary Salome (mother of Apostles John and James), Martha and risen-from-the-dead Lazarus, along with their Egyptian servant Sarah.

According to legend, in about AD 40, these early Christians had set out to sea from Palestine in a small boat without sails. They were miraculously washed ashore at this spot, safe and sound. In thanks, they built a chapel to the Virgin, and Mary Jacoby, Mary Salome and Sarah remained in the Camargue as evangelists.

After their deaths, the two Marys that had stayed in the Camargue and Sarah became the subject of pilgrimage. During the 9th century, a **church** was built in place of the simple chapel and fortified against invading Saracens. Excavations beneath this structure, begun in 1448 under the behest of Good King René, led to the discovery of a still-extant well and a spring filled with "the fragrances of sweet-scented bodies." In this same spot were uncovered the remains of the saints.

Their tomb was placed in the the church's **High Chapel of St-Michel**. The stones on which the bodies had rested are kept below in the crypt and have become "miracle" stones. They are said to heal painful eyes and to have the power to cure sterility in women.

The servant, Sarah, has attracted a cult all her own. Considered the patron saint of gypsies, she is remembered on 24 and 25 May each year with two special nights of celebration. Gypsies from all over the surrounding regions descend upon the town for horse racing, running of the bulls and a whole variety of other festivities.

In the evenings, the fiesta spirit runs wild, and the frenzied excitement of gypsy music and dancing fills the air. At the same time, during this period, the town has become a rendezvous spot for the bullherders of the Camargue. Needless to say, in recent years, the festival has become a major media event, attracting television teams, film crews and even some politicians.

Another day of celebration comes on the last Sunday of October. This is the special feast day for Mary Salome, and it is commemorated by colourful processions around the **Church of Notre-Dame-de-la-Mare**. During the summer months, when there isn't a festival but lots of tourists around, this church is still very crowded. If you decide to visit it anyway, climb to the roof for a stunning panoramic view across the sand hills and down to the waves as they splash against the sunlit shore.

During the holiday season, Les-Stes-Maries-de-la-Mer is unquestionably a commercial resort. Its busy pedestrian zones are lined with souvenir shops, and the many reasonably priced hotels and camping sites do well. There are many places where you can rent horses and boats, swim or play tennis during the day and, at night, some of the area's hottest discothèques swing into action. Bullfights are held in the local arena.

A less strenuous attraction is the **Baroncelli Museum**, housed in the one-time domicile of the Marquess of the same name. Inside is her collection of varying types of literature on the Camargue. Also on display are examples of local furniture, fauna, artwork and *gardien* paraphernalia.

If the bustle is starting to get to you, escape by booking a boat trip along the Petite Rhône. Go down to the landing stage near **Baroncelli-Jauron's Tomb** to join one of these excursions, which leave at regular intervals. Photographers will be thrilled by the action shots they can take of Camargue wildlife from the water, especially of the marsh birds. Everyone will enjoy a new opportunity to see the herds of bulls and horses that run along its banks.

If you would prefer to find some quiet beaches in the area and are willing to challenge weathered tracks off the main road, head west out of town. Along here you will be able to find a place to park. Leave your car and climb the sand dunes, and you should be rewarded with some clear and empty spaces far away from the crowds.

Van Gogh, who was fascinated by the light in this region, painted the famous *Boats of the Stes-Maries* on these beaches. Of course, that was before campsites, caravans and water scooters had intruded on the seascape. If he arrived today, Frédéric Mistral, who set the tragic end of his famous Provençal work *Mireio* on these sandy shores, would no longer be able to write of it "no trees, no shade and not a soul."

As you leave Les-Stes-Maries-de-la-Mer area by the main route of D570, you will pass by the **Musée Boumian**. Its much-touted waxwork tableau, which attempts to recreate a romantic version of the Camargue, is actually well-worth missing. You may, however, want to stop in on the **information centre** at **Ginès**. An informative sideshow here explains about the indigenous flowers, plants and fauna that grow in the outlaying land.

Western Camargue: Continue north on D38, then turn left onto D58, and you will enter the plain of **Aigues-Mortes**. The drive along here will take you across a part of the Camargue that has not radically changed for years and years. From time to time, you will spot an authentic *mas*, single-storey cabins with thatched roofs that over the years still defy the mistral.

The plain borders on a rich agricultural area that produces grapes and cereals. The **Tower of Carbonnière** at the end of the drive along D58 will give you the chance to look out over these fields, to the foothills of the Cévennes in the northwest, the Petite Camargue in the east and across the salt flats of Aigues-

Enjoying a game of *boules* at the Plage de Piemançon.

Mortes to the south. Built as an outpost to the garrison of Aigues-Mortes during the time of the Crusades, the tower guards what once was the only route in and out of town.

The area around Aigues-Mortes is made up of salty lagoons and water channels where the Rhône meets the sea. The town itself was built amidst the marshes, salt flats and lagoons in 1241. It was named for its location (Aigues-Mortes means "dead waters") as it contrasted with a town, **Aigues-Vives** ("living waters"), that lay in the hills about 12 miles (20 km) north of it.

Good fortune from an architectural point of view has limited the town's growth, and there has been little expansion beyond its walls. What stands today is a well-preserved medieval town that has changed little on a superficial level. The ramparts remain intact and still readily dominate the surrounding countryside. With a little concentration, it is easy to imagine this town as the springboard from which Saint-Louis (King Louis IX) launched the Seventh Crusade in order to liberate Jerusalem.

When photographed at twilight from the southwest, the **battlements** at Aigues-Mortes make a fabulous picture, as sharp silhouetted shapes emphasise the strategic position of this curious medieval garrison town. If by now you have become adept at bird-spotting, climb up and take a look out over them. It is not uncommon to see kestrels, marsh harriers and the colourful bee-eaters with their blue, yellow and green plumage as they somersault after tasty insects.

Once you've passed through the gateway, you will find a number of attractive narrow streets criss-crossing the town. They lead to the **central square**, which is overlooked by a statue of Saint-Louis, the patron of the town who granted it its special rights and privileges. Although you will find the square lined by cafés and restaurants, you can get a better meal by wandering down the adjacent and more secluded side streets. Look for a restaurant that specialises in local dishes like *Gardiane de Taureaux*

Preparing for the *feria du cheval*.

(a delicious type of beef stew cooked with olives) or local seafood dishes.

After dining, you may want to climb the **Constance Tower** and **ramparts**. They provide an excellent overview of Aigues-Mortes. Look down to the northeast sector of the town, and you will see what originally was all one farm owned by the Knights of St-John until the Revolution.

The town has a population of 5,000 and, although its importance as a seaport has diminished, it is still a centre for salt production and vine growing. Visits to "Les Salins du Midi" (The Salt Marshes of the South) are organised in July and August on Wednesday and Friday afternoons by the **tourist office**, which stands in the central square.

The local wine, Listel, is grown commercially and is remarkable for being a *vin de sable*. This means that it is made from vines that grow directly out of the sand. The white is called *gris de gris* and the red *rubis*. Nearby *caves* offer free wine-tasting.

he
ceanfront.

The horses that play such an important part in the life of the Camargue have also been woven into one of legends that surround Aigues-Mortes. Young children are told that if they misbehave they will be dragged off by "Lou Drape," a horse that is supposed to hover over the ramparts at night. It is said that its body can lengthen to carry away as many as 100 naughty children on its back into the marsh, where it will then devour them. Even today myths, legends and popular superstitions are an integral part of life in the Camargue.

A little bit of Languedoc: Le Grau-du-Roi and Port-Camargue, although only 6 miles (10 km) west of Aigues-Mortes, actually belong to the region of Languedoc-Roussillon, but they may well be of interest to travellers who wish to discover the Camargue from the sea. The lagoons and mooring facilities of these two ports make them attractive harbours, with excellent opportunities for watersports and fishing.

Le Grau-du-Roi is a narrow coastal strip that separates the Etang du Repausset from the sea. It has two very different identities that contrast old and new architectural styles. The eastern quay is an active fishing port with canal connections to Aigues-Mortes some 3½ miles (6 km) away, and it can be a useful mooring for people entering the Camargue by boat. The quayside cafés serve excellent fish, and you can actually see the clams being shovelled out of the water before they arrive on your table. Across the bridge to the west lies a new development of holiday apartments that manages to blend into the old town without making too much of a sharp contrast. Despite modernisation, Le Grau-du-Roi is still a pleasant town to stroll through on summer evenings.

The **Port-Camargue**, which lies to the south, has excellent boating facilities and broad sandy beaches. It is also extremely practical for sailing parties as boats can be tied up outside the front doors of the recently constructed holiday apartments. A first-class sailing school is based at the marina and offers instruction for complete beginners or for advanced students interested in competitive sailing.

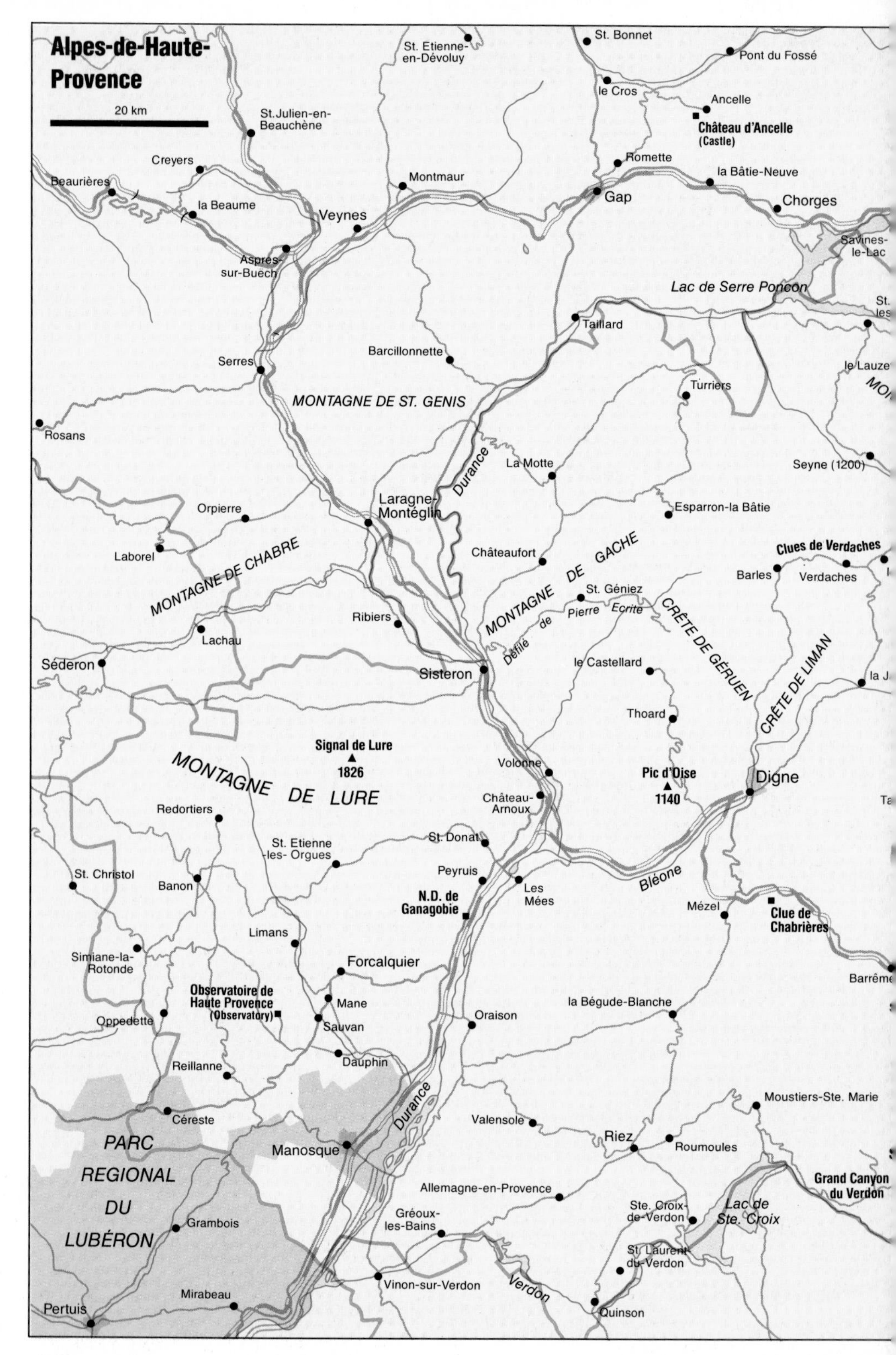

Alpes-de-Haute-Provence
20 km
St. Etienne-en-Dévoluy
St. Bonnet
Pont du Fossé
le Cros
Ancelle
Château d'Ancelle (Castle)
St.Julien-en-Beauchêne
Creyers
Romette
la Bâtie-Neuve
Montmaur
Beaurières
la Beaume
Gap
Chorges
Veynes
Aspres-sur-Buëch
Savines-le-Lac
Lac de Serre Poncon
Taillard
Barcillonnette
Serres
le Lauze
Turriers
MONTAGNE DE ST. GENIS
Rosans
Durance
La Motte
Seyne (1200)
Laragne-Montéglin
Orpierre
Esparron-la Bâtie
Laborel
Châteaufort
Clues de Verdaches
MONTAGNE DE CHABRE
MONTAGNE DE GACHE
Barles
Verdaches
St. Géniez
Ribiers
Défilé de Pierre Ecrite
CRÊTE DE GÉRUEN
Lachau
Séderon
Sisteron
le Castellard
CRÊTE DE LIMAN
Thoard
Signal de Lure 1826
MONTAGNE DE LURE
Volonne
Pic d'Dise 1140
Digne
Château-Arnoux
Redortiers
St. Donat
St. Etienne-les- Orgues
St. Christol
Peyruis
Banon
Les Mées
Bléone
N.D. de Ganagobie
Mézel
Clue de Chabrières
Limans
Simiane-la-Rotonde
Forcalquier
Mane
Observatoire de Haute Provence (Observatory)
la Bégude-Blanche
Oppedette
Sauvan
Oraison
Dauphin
Reillanne
Durance
Moustiers-Ste. Marie
Céreste
Valensole
PARC REGIONAL DU LUBÉRON
Manosque
Riez
Roumoules
Grand Canyon du Verdon
Allemagne-en-Provence
Ste. Croix-de-Verdon
Lac de Ste. Croix
Gréoux-les-Bains
Grambois
St. Laurent-du-Verdon
Vinon-sur-Verdon
Verdon
Mirabeau
Pertuis
Quinson

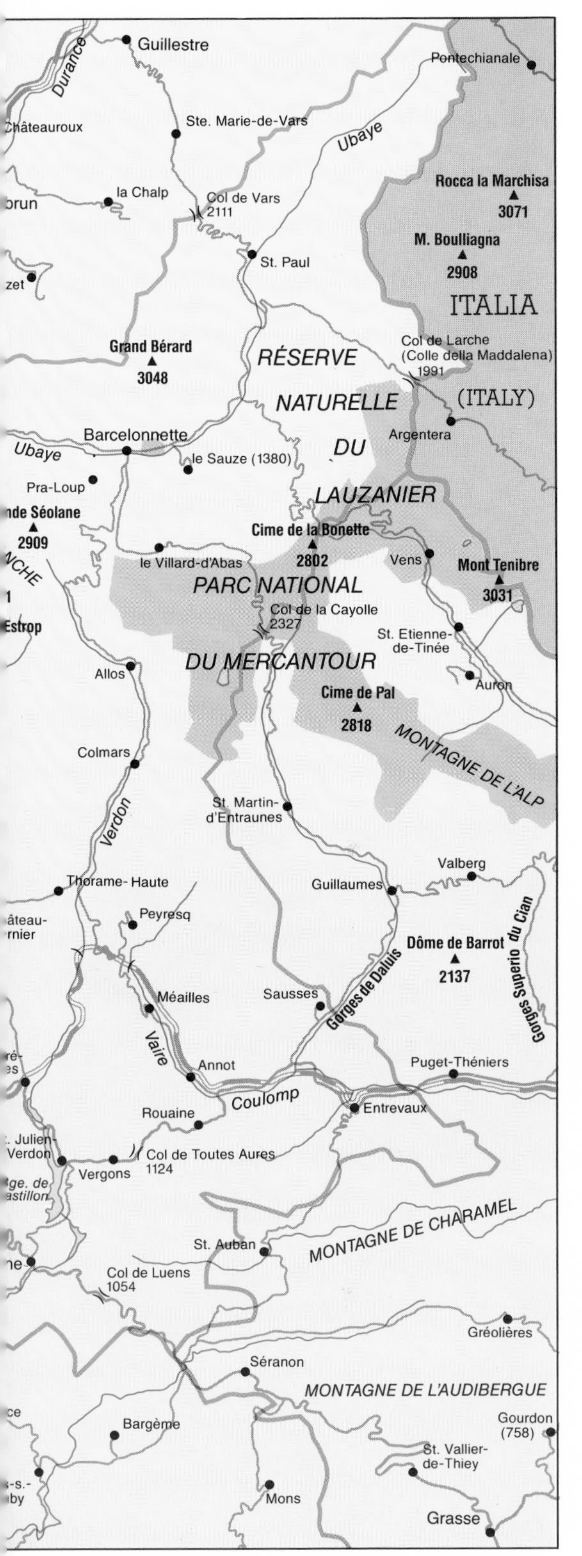

ALPES-DE-HAUTE-PROVENCE

Perhaps Provence's least-known *département* and certainly the least densely populated, the **Alpes-de-Haute-Provence** seem at first glance to miss out on much that makes the rest of Provence so attractive. Bypassed by the main artery of communications down the Rhône Valley, this vast region lacks the beaches and glitz of the Côte, the gems of art and architecture of the Vaucluse and the fertile vineyards and perfect *villages perchés* of the Var.

Instead, the Alpes-de-Haute-Provence show a wild, rugged face. The turbulent sweep of the Durance and the desolate Valensole Plain, brought to life in summer by countless rows of lavender bushes, dominate the western region. Vertiginous gorges to the south rise to heady Alpine heights as you move north and east. In between are azure lakes, bizarre rock formations and inaccessible mountainous terrain.

Simplicity and practicality characterise the region's architecture. Stark defensive citadels crown many towns, and villages are constructed to give maximum protection against the elements. The much-vaunted climate – balancing on a happy axis of Provençal warmth and cool Alpine clarity – can turn to harshly low temperatures in winter.

Ask any local to sum up life in the Alpes-de-Hautes-Provence, and he or she will give the same answer: "*pauvre*" (poor). Like its architecture, the region's cultural heritage is marked by simplicity and austerity. Until recently, the inaccessibility of much of the country meant that progress arrived slowly and ancient traditions endured. Life followed the rural rhythms of the movement of sheep herds (known as *transhumance*) and was characterised by constant battles against the elements.

Nonetheless, the Alpes-de-Haute-Provence have witnessed a surprising amount of migration in the past 100 years. In common with the rest of inland

Preceding pages: the famed Alpine sheep.

Provence, the 20th century has left many villages abandoned, as locals move south to work in the more prosperous areas around the coast.

In the past 30 years, the Alpes-de-Haute-Provence have changed beyond recognition. The Durance River has been tamed by an elaborate system of dams, providing fertile ground for fruit cultivation along the valley, and is soon to be served by a new autoroute along its length. New jobs in hydroelectricity and at the Cadarache nuclear research centre have been created.

Tourism, though a good deal slower to develop here than in the more obviously enticing parts of Provence, has gained impetus. The spa towns of Digne and Gréoux-les-Bains attract health-cure seekers, the northern Alpine resorts lure skiers throughout the winter, and wind-surfers and hikers swarm over the vast lakes and austere peaks.

The French press have devoted considerable space to the Alpes-de-Haute-Provence as the next region waiting to be "discovered." Wealthy Parisians now spend the summer in Manosque and increasing numbers of city-dwellers from Marseille buy holiday homes in the mountains. The *pauvreté* that has marked the lives of many of its inhabitants is still in evidence, but the impressive, singular beauty of the Alpes-de-Haute-Provence is increasingly attracting profitable new visitors.

Border town: Sisteron, the natural gateway to Provence, seems hardly Provençal at all. This pretty, tidy town, dominated by a stark towering citadel, has a definite air of Dauphiné about it. Just to the north, the olive trees stop and, to all intents and purposes, you say goodbye to Provence.

Sisteron's setting is impressive – on the left bank of the Durance and looking across at the craggy **Rocher de la Baume**. For centuries, people have passed through this strategically placed "gateway," lending it a feeling of lively, if not exactly cosmopolitan, activity. A bustling market, some light industry and a steady wave of tourists along the N85 maintain that atmosphere today.

Tending the flock.

In the past, such comings and goings brought Sisteron both good times and bad. Plague and typhus, carried by passing travellers and armies, decimated the population in the 14th, 17th and 18th centuries. Napoleon stopped here for lunch on a misty March day in 1815, when returning from exile on Elba. Still standing in the Rue Saunerie is the now-private **Bras d'Or Inn**, owned at the time by the grandfather of the Sisteron novelist Paul Arène. A plaque commemorates the site where the imperial lunch was quaffed.

The town's **citadel**, originally built in the 13th century then redesigned by Henri IV's military engineer Jean Erard in the 16th, is Sisteron's most obvious crowd-puller. An impressive collection of keeps and dungeons, watchtowers and crenellated battlements stands on a massive rock topped by a small chapel.

Its complex, fascinating history as a fortress and prison pales next to the blow dealt by our own age. On 15 August 1944, the Allies bombed it in an attempt to speed up the retreat of the German occupying army that had taken refuge there. More than 300 people died, and a quarter of Sisteron's fine medieval town was destroyed.

Visitors to the citadel today can listen to a dramatised version of the fateful day, relayed through small speakers dotted round the ramparts. The questionable quality of the sound effects does nothing to dim the poignancy of the tragedy.

Fortunately, a good deal still remains of the **old town** that sits huddled at the foot of the citadel rock. A mixture of modern boutiques and pleasant old-fashioned shops line its narrow streets, known here as *andrônes*, from the Latin word for "alleyway" or the Greek term for "between two houses," depending on which linguist you ask.

The town's bakers sell the local speciality, *fougasse à l'anchois*, a delectable bread dough smeared with anchovies and sold by the kilo. Butchers display the justly famous Sisteron lamb, given a fragrant herbal flavour by the wild thyme and rosemary culled from

Sisteron.

pastures of the upper Durance Valley.

North of Sisteron, on the D3 to **St-Geniez**, look out for a rock slab covered with Roman inscriptions. The **Pierre-Ecrite**, carved in the 5th century, records the conversion to Christianity of Claudius Dardanus, prefect of the Gauls. Tantalisingly, it mentions a nearby "city of God," which apparently was founded by Claudius, though archaeologists have been unable to fix its exact setting.

For centuries, the Durance was known as one of the "three scourges of Provence" (the *mistral* and the Parlement at Aix completed the triumvirate). In recent years, its unpredictable surges have been harnessed by a series of major dams beginning at **Serre-Ponçon** on the northern border of the Alpes-de-Haute-Provence. Apple and pear orchards, created over the past 30 years on the newly irrigated alluvial plains, have brought a refreshing vigour to the once-ailing economy hereabouts. And hydroelectric power, nicknamed by locals *la houille blanche* (white coal), has brought added prosperity to the region.

Nonetheless, the Durance remains tortuous, complex and peppered with islets. It is one of the few rivers in Provence that contains more than a thin trickle of water in summer.

Limestone penitents: As the Durance snakes its way south and meets the Bléone arriving from the west, a sprawling collection of light industrial plants appears. Across the apex of the confluence, south of Château-Arnoux, the rocks of **Les Mées** stand eerily over miles of flat maize fields.

The village of Les Mées itself does not tempt much, but the staggering **"Pénitents,"** the name given to the curious row of limestone pinnacles outside it, do. The smooth, dolmen-shaped formations, some as high as 330 ft (100 metres), rise sheer and bare out of what seems like a dwarfed, stunted forest – some alone, some clustered in groups – until they close ranks to form a single mini-*massif*.

The legends that surround them have a sadness that echoes their lonely as-

Serre-Ponçon.

pect. During the Saracen invasions, a group of monks from the Montagne de Lure are said to have been attracted by the beauty of some Moorish girls. As the cowled, disgraced figures were banished from the village, Saint Donatus turned them to stone, in punishment for their impropriety.

The religious theme continues on the opposite bank of the Durance, in the foothills of the **Montagne de Lure**. The evocative ruins of the **Church of St-Donat**, reached by the narrow D101, sit isolated in thick oak woods. Formerly the retreat of Saint Donatus, a sixth-century monk from Orléans, the church is today in a sorry state, its floor covered with rubble and graffiti etched into its venerable stone. The eight mighty pillars that support its ancient vaulting and a faded-white apse, decorated with pale terracotta-coloured stars, hint at its former glory. It is one of Provence's few remaining examples of early medieval Romanesque architecture.

More obviously impressive is the nearby medieval **Priory of Ganagobie**. Like St-Donat, it was in a sad dilapidated state, until restoration work began in the 1960s. Now its principal attractions, other than the wonderful view of the Durance Valley, are its magnificent zig-zag west portal and the stunning 12th-century mosaics, in red, black and white, that decorate the interior.

Alpine summits: The vast, forbidding Montagne de Lure is a continuation of the vine-rich Ventoux Range to the west and is bordered by the Lubéron Mountain to the south. Its heights are reached via the little town of **St-Etienne-les-Orgues** that lies at its feet. The prosperity of this pretty village was traditionally based on medicinal remedies concocted from mountain herbs and sold by travelling pedlers.

A road lined with fir trees and, in summer, purple fields of lavender leads out of the village. Soon, the lavender stops, and the dense oak and fir forest of the mountain proper appears. Though seemingly deserted, this route becomes an animated pilgrims' way in August and September each year, as locals con-

A core crop.

tinue the centuries-old tradition of pilgrimage to the isolated **Chapel of Notre-Dame-de-Lure**, located halfway to the summit.

Also founded by the reclusive Saint Donatus, Notre-Dame-de-Lure has none of the architectural distinction of the Eglise de St-Donat, but its setting is ample compensation. In summer, local people and tourists picnic in the shade of the splendid lime trees that shelter the church's entrance.

The familiar pattern of decay of so many of Provence's ancient religious sites has in this instance been broken by the restoration work currently being carried out by a religious group called the Community of Jerusalem. Groups of young Christians, camped in the only building adjoining the church, chisel enthusiastically at the stonework or sit in circles reading the Bible. The purity of the air, which seems positively cool here even in the height of summer, adds to the healthy atmosphere.

Somewhat less wholesome is the fact that one of the church's few pieces of ornamentation, a statue of the Virgin, was stolen in 1985. A rather sad photograph, propped up on the statue's empty plinth, is all that remains.

After Notre-Dame-de-Lure, the road to the summit rises in steep curves, edged in summer by seas of purple larkspur. The **Signal de Lure**, the mountain's summit (5,990 ft/1,826 metres) boasts an undistinguished *station de ski* and some of the most impressive panoramas to be found anywhere in Provence. Views of the Cévennes and Mt-Ventoux are interrupted only by the sight of soaring black buzzards.

The wild isolation of the Lure inspired many of the novels of Jean Giono. A native of Manosque, Giono made his name in the first half of this century with a series of novels on village life – and death – in the Alpes-de-Haute-Provence. Giono's Provence has little in common with the smiling, sunny nostalgia depicted by Pagnol or even Daudet. His central themes were the progressive abandonment of the small mountain villages of the region,

Young goatherd girl.

the inexorable destruction of the traditional patterns of rural life and the rough sensuality of man's communion with nature. In *Regain* (1930), the fictional village of Aubignane, already reduced to just a handful of inhabitants, eventually houses just a solitary man, determined to stay in his native village.

Many have tried to locate Giono's settings, mostly without success. Nonetheless, it is widely believed that Aubignane is based on the ruined village of **Redortiers**, north of **Banon**. Whether or not intimations of Giono appeal, it is worth making a visit to Banon to sample its renowned goat's cheese wrapped in chestnut leaves and to Redortiers to see an example of Giono's reality.

A 1980s angle on Giono's literary theme is provided by the changes that have taken place in the beautiful, remote village of **Simiane-la-Rotonde**. Simiane, dominated by the strange "rotunda" that gives the village its name, is today almost entirely composed of *maisons secondaires* (second homes). Outside the summer months, its closed shutters and deserted streets have more than a little in common with the desolate ruins of Redortiers.

oat cheese
f Banon.

An unshaken heritage: No such cultural instability dogs the everyday life of one of the most appealing towns of the *département*, **Forcalquier**. The locals here happily welcome visitors during the summer months but don't allow the brief invasion to disrupt the rhythm of their prosperous lives.

Forcalquier is at its best on Monday – the day of its wonderful weekly market. Countless stalls, groaning with local produce or stacked high with better-than-average *objets artisanaux,* crowd the spacious **Place du Bourguet** and the labyrinth of streets around it. The market acts as a magnet to people from the region who come to buy, sell or simply gossip in the pleasant market-place cafés.

Forcalquier's air of independence has its roots in the 12th century when the town was an independent state. During this period, the town served as the capital of Haute-Provence, making it a centre for culture and trade and a favourite residence of the counts of Provence.

The origin of the town's name, however, goes back much further. Situated on the Via Domitia, one of the three great Roman roads in Provence, Forcalquier is said to be derived from *furni calcarii*, Latin for the limestone kilns that the Romans hewed into the hillside.

Forcalquier boasts a clutch of reasonably interesting "sights": the **Couvent des Cordeliers**, a Franciscan convent, restored in the 1960s and now part privileged private housing, part visitable monument; the austere **Cathedral Church of Notre Dame** on the main square; and the 19th-century **chapel** on the hill above the old town with a *table d'orientation* to help you identify the surrounding mountains.

But the real flavour of Forcalquier can be tasted more sharply simply by taking an aimless wander around the splendid alleyways of the **old town**. Almost every narrow street here is lined with fine stone doorways and arches that are decorated with chiselled plaques, scrolls and intricate relief.

Forcalquier seems a place where the exigencies of the 20th century have had to work hard to dislodge a firm historic anchor. However, preservation of the cultural heritage that emanates so naturally from Forcalquier's people and buildings has its official home a few miles south of the town.

The **Priory of Salagon**, to be found just outside the village of **Mane**, has been converted into a major centre for local studies and research. As well as fascinating permanent exhibitions covering local history and a botanical garden planted with the medicinal plants that once played a significant role in the region's economy, serious lectures are given throughout the summer on aspects of local rural life, architecture and customs.

Salagon's principal aim is to inform and enrich the lives of the people of the region. As such, it is an unmissable detour for visitors looking for a real insight into the *Pays de Forcalquier*.

Those with a taste for imposing country seats should combine Salagon with a visit to the **Château de Sauvan**, arguably the *département*'s finest 18th-century building. The Alpes-de-Haute-Provence is not an area renowned for its châteaux, and many are little more than small stately homes. Nonetheless, Sauvan possesses a classical elegance that merits its local title of "*Le Petit Trianon Provençal*."

Other delights in the region include the small hilltop village of **Dauphin**, the pretty series of *pigeonniers* in and around **Limans**, and the **Observatory of Haute-Provence**, which attracts astronomers from all over France.

A boomtown: Manosque has an elegance and sense of refinement that sets it apart from the rest of the Alpes-de-Haute-Provence. Largely as a result of its position – a plum site on the Durance within easy reach of Forcalquier, the Montagne de Lure, the Valensole Plain and the Grand Canyon du Verdon – Manosque has become the principal economic axis of the *département*.

Twenty years ago, it was a sleepy

Dusk at Montfuron.

town with a population of less than 5,000. Today, that number has increased more than fourfold. Blossoming agriculture in the Durance Valley (Manosque is known for its yellow peaches), the proximity of the Cadarache nuclear research centre that has drawn scientists and their families to settle here, and the relentless flow of tourists have combined to give Manosque an energetic air of modernity.

Fortunately, progress has not seeped fully into the bricks and mortar of the town, which remains essentially medieval in layout and character. Manosque covers a small hill above the Durance; "a tortoise shell in the grass," in the words of Jean Giono.

Two imposing stone archways, the tall crenellated **Porte Saunerie** and the **Porte Soubeyran**, stand guard at the entrance to the **old town**. These are linked by the town's main artery, the Rue Grande. Encircling old Manosque is a busy ring road that has replaced the town's ancient ramparts.

Stepping through either of the *portes* (gates) brings you into a pedestrian area of narrow streets, honey-coloured churches and fountain-filled squares. Outwardly, Manosque seems to have survived the onslaught of its astonishing growth, though the rows of chic modern shops and small pretty squares converted to parking lots jolt you back to the future.

Manosque *la Pudique* (the Modest), the nickname given to the town in the 16th century, seems rather inappropriate today. The name derives from a tale, almost certainly apocryphal, concerning a visit by François I. The young king showed more than a passing interest in the daughter of a local dignitary. To repel the king's advances and preserve her honour, the young girl promptly disfigured her face with sulphur.

Giono's magnificent friend: Giono, of course, is what most Frenchmen think of when they think of Manosque. It was here that he lived, wrote and died between 1895 and 1970. The sweep south from the Montagne de Lure, through Manosque, and west to the vast plains of

Inside Château de Sauvan.

Valensole and Puimoisson mark the boundaries of Giono's Provence.

Giono called the wide, flat **Valensole Plain** his "magnificent friend." Today, it is France's principal centre of lavender cultivation, but it wasn't always so. Lavender production is a 19th-century phenomenon, and the almond trees that blossom so spectacularly on the plain in early spring are a much more ancient element of the economy.

Summer is the time to see the Valensole in all its purple magnificence. On the cusp of July and into August, around the time of the harvest, the dusky violet rows stretch away to the horizon. The unmistakable odour, so redolent of a grandmother's linen cupboard, fills the air, while wooden trucks piled high with vast purple bunches lumber their way along the roads of the high plateau.

The few villages that huddle on the plain, surviving the winter lashings of wind and hailstorms, are for the most part unremarkable. The exception is **Riez**, notable for its four **Corinthian columns** – remains of a Roman temple of Apollo – standing alone in a field just outside the town. The ancient **Merovingian baptistry** is Riez's other great "sight." But what is most appealing about the town, particularly in its **old quarter**, is a pleasing scruffiness and an attitude of *laissez-faire* that pervades the old streets and makes the people seem relaxed and welcoming.

By contrast, **Digne-les-Bains**, capital of Alpes-de-Haute-Provence and last stop on the Train des Pignes, is not scruffy at all. A genteel spa town, spacious and airy but without much architectural distinction, it lies northeast of Riez in the imposing Pré-Alpes de Digne. It was here that Hugo set the first chapters of *Les Misérables*.

But Digne is not simply a town of fictional *malheureux*, rheumatics and departmental officials. The shady **Boulevard Gassendi** tempts with the pleasant cafés along its length and, in early August, Digne asserts its position with a boisterous festival and procession called the *Corso de la Lavande*.

Four days of revelry climax in a grand

Alpine peaks north of Digne.

procession. An array of huge flower-bedecked floats, some modern in theme, some traditionally Provençal, glide down the main boulevard to uproarious applause, preceded by the town's sanitation department trucks, which douse the streets with lavender water. Cunning Dignois reserve their tables at the boulevard's cafés for lunch and a good view of the parade.

Lavender, or its more common hybrid *lavandin*, is sold in Digne in every conceivable form. Soap, essence and sachets of the dried plant are all worth buying.

Those with no wish to take the waters, either of the thermal or lavender varieties, should head south towards the outskirts of town to Digne's eccentric **Alexandra David-Néel Foundation**. The woman who gives her name to this cultural-centre-*cum*-museum was a Parisian adventurer who spent much of her life travelling in remote parts of Asia, including perilous trips to the forbidden Tibetan capital of Lhasa. Seduced by the beauty of the Alpes-de-Haute-Provence, which she called a "Himalayas for Liliputians," she bought a house in Digne in 1927 and named it *Samten Dzong*, meaning "the fortress of meditation."

-residence ibetan onk at the avid-Néel oundation.

When she died in 1969, aged 101, this remarkable woman left the house and its fascinating contents to the city of Digne. As in her lifetime, the Foundation continues to attract visitors and Buddhist pilgrims to view the collection of objects and documents that she acquired during her peripatetic existence.

Architecturally, Digne's greatest attraction is the Lombard-style **Cathedral Church of Notre-Dame-du-Bourg**, north of the Boulevard Gassendi on the outskirts of town. Look for the 13th-century portals in stunning blue and white limestone. Digne also boasts an averagely interesting municipal museum with local history exhibits and a collection of butterflies (for which the area was once well known).

More specialised is the well-ordered **Centre de Géologie**, signposted on the opposite bank of the Bléone, with an astonishing range of fossils and animal skeletons from prehistoric times.

Natural wonders: Near the southern borders of the Alpes-de-Haute-Provence lies the *département*'s greatest natural site, the **Grand Canyon du Verdon**. Even if you see little else in the area, on no account miss this. Natives of Arizona may find the name unlikely, but this massive crevice scored deep into the limestone is one of the most breathtaking of natural phenomena, at least in the Old World.

First stop on the Canyon trail is **Moustiers-Ste-Marie**, where an astonishing backdrop of craggy cliffs provides a taste of the glories to come. Moustiers is an attractive small town perched on the edge of a ravine. High up behind the town, the two sides of the ravine seem held together by a massive chain that is around 720 ft (220 metres) long. Suspended from the chain, like a pendant on a giant's necklace, is a man-sized metal star.

This curious piece of adornment has inspired poems by Mistral and Giono and all manner of speculation as to its origin. But local historians have settled on the legend that it was placed there by a knight returning from the Crusades, in fulfilment of a religious vow. In any event, the present star dates only from 1957, when the mayor of Moustiers decided that the town's most famous piece of jewellery was due for a facelift.

More obviously visible is Moustiers' major claim to fame – its *faïence.* The narrow village streets are crammed to bursting with shops and studios producing the white-glazed decorative pottery.

Established in the late 17th century by Antoine Clérissy, the recipe for the white glaze was said to have come via an Italian monk from Faenza. The industry prospered for the following 200 years, counting Madame de Pompadour among its customers. Changing fashions, however, brought about its decline in the 19th century, and the art was not revived until the mid-1920s.

Sadly, many of the items produced today have fallen prey to the depressing combination of high prices and low quality. Though it is still possible to seek out good examples from among the vast range available, perhaps more rewarding is a visit to the **Musée de la Faïence** in the Placette du Prieuré.

Opposite the entrance to the museum, housed in a vaulted crypt, is the stunning three-tiered belfry of the Romanesque village church.

A fork in the road less than 2 miles (3 km) south of Moustiers offers a choice of routes to the Grand Canyon du Verdon. Heading south, the road skirts the spectacular man-made **Lac de Ste-Croix**, a 7-mile (11-km) stretch of perfect azure water created in the 1970s by the installation of a hydroelectric dam. The route then doglegs east to follow the southern ridge of the canyon, known as the **Corniche Sublime**. Alternatively, the old road to Castellane (D952) snakes along the northern ridge, veering away from the canyon edge around halfway along its length.

Both routes have their virtues. Choosing the Corniche Sublime means consistently spectacular views, most notably those at the **Balcons de la Mes-**

On the road to Annot.

cla, where the Verdon converges with the smaller canyon formed by the River Arturby. The left bank has its own impressive *belvédère* (viewing point), the **Point Sublime**, a dizzy 600 ft (180 metres) above the river, and offers two chances to climb down to the Canyon floor, at Chalet de la Maline (branch off from the village of La Palud) and just east of the Point Sublime (follow signs for Couloir de Samson). Both routes require at least a half-day trip.

Statistics on the Grand Canyon only hint at its magnificence. Twelve and a half miles (21 km) in length and 5,000 ft (1,500 metres) deep, the vertiginous limestone cliffs are gouged by the waters of the Verdon River, in turn glassy and crystalline or tumbling in white-water chaos. Colours are unreal: jewelled emerald and turquoise contrast in autumn with the ochres and russets of surrounding deciduous trees.

Less than a century ago, the canyon was largely deserted, with only a few local peasants still eking out meagre livings as woodcutters. By the late 1940s, the touristic potential of the canyon was accelerated by the construction of the Corniche Sublime. Tourism has now moved in wholesale, and those in the know make strenuous efforts to avoid the area completely during the month of August.

Nowhere are the crowds of summer visitors more oppressive than in **Castellane**. Like Moustiers-Ste-Marie, proximity to the canyon has brought welcome prosperity to the locals, at the expense of over-commercialisation. Queues of cars choke Castellane's narrow streets, beachballs and hiking paraphernalia fill its shops, and the native hospitality becomes, at times, understandably strained.

Seasonal demands aside, Castellane has a pleasing, if small, **old quarter**, lying to the north of the town's bustling hub, Place Marcel-Sauvaire. Market days (Wednesday and Saturday) enliven the town throughout the year, and the *Fête des Fétardiers*, held annually on 31 January, commemorates the lifting of a Huguenot siege in 1586. The

eft, loustierware om the telier de egriès. **ight**, utumn in loustiers.

chapel of **Notre-Dame-du-Roc**, perched daringly on a high rocky outcrop, makes Castellane's skyline its most appealing feature.

On to higher ground: The Verdon River, so resplendently enclosed by the Grand Canyon in the south, keeps its luminous colour in its northern reaches. The **Lac de Castillon**, in the upper Verdon Valley, has much in common with its larger cousin, the **Lac de Ste-Croix**. Also man-made, it was created by a dam that was constructed and managed by Electricité de France. The dam was completed in 1947, making it the first of five built to harness the waters of the Verdon. Ste-Croix, completed in the 1960s, is the most recent.

The lakes that have resulted (Chaudanne, Ste-Croix, Castillon) perform the dual function of conserving precious water for this region of Provence and attracting tourism. Edging the Lac de Castillon, bizarre vertically grooved rock formations alternate with dazzling white sand spits that provide pleasant spots for swimming in summer.

In a broad valley of orchards and lavender fields at the head of the lake sits the small summer resort of **St-André-les-Alpes**, a haven for windsurfers, hang-gliders, anglers and *sportifs* of all sorts. This small village has not a great deal to recommend it architecturally or historically but is redeemed by the genuine hospitality of its small population. In summer, the village somehow manages to find room for four times its normal number of inhabitants. Nonetheless, tourism has not left as deep a scar here as in Moustiers or Castellane.

St-André's tiny station forms a stop on the route of the Train des Pignes. In addition, a *Belle Epoque* steam train, dating from 1909, rattles through on its way from Puget-Théniers to **Annot** every Sunday in summer.

East of St-André, the route to the charming small town of Annot leads through one of the many *cols* (high passes) that pepper this mountainous eastern side of the Alpes-de-Haute-Provence. Moving north and east, towards

Summer bubbles.

the "Grands Alpes" and the Alpes-Maritimes respectively, the *cols* increase in frequency and, at times, almost alarmingly in altitude. The **Col de Toutes Aures** (Pass of All Winds), however, is a relative small-fry, at a mere 3,688 ft (1,124 metres) high. Towering over it is the massive **Pic de Chamatte** which, at 6,160 ft (1,878 metres), dwarfs even the Montagne de Lure in the west.

This is a richly forested landscape, in marked contrast to the bleak beauty of the canyon region. Shaley precipitous slopes are cloaked with vast sweeps of dark evergreens parading down the steep cliff faces. Minuscule villages such as **Vergons** and **Rousine**, consisting of little more than a handful of dry-stone houses protected by weathered shutters, prefigure the Alpine experience to come.

Annot in the Vaire Valley is typical of this area's dual aspect – warmly Provençal yet markedly Alpine. Its wrought-iron balconies and stone *lavoirs* are as classically Provençal as anything found further south. But the majesty of the Alps leaves its mark in the pure clarity of the air and the steely grey waters of the numerous streams that tumble through the town.

In winter, Annot sleeps under thick coverings of snow. Although the oldest populated centre in the area, Annot was unreachable by carriage as recently as 1830. In common with much of the rest of the Alpes-de-Haute-Provence, access was provided only by the network of mule tracks that crisscrossed the mountainous landscape.

For years the town has attracted painters. East of the sizeable main square, planted with a fine esplanade of ancient plane trees, the narrow streets of the picturesque **old town** climb steeply in medieval formation. Vaulted archways and a predominance of carved stone lintels bearing their 17th and 18th-century dates of construction line the tall houses of the Grand Rue. At the top of the old town, the narrow streets converge on a pretty square with the parish church and surrounding houses

Ste-Croix de Verdon on the Lac de Ste-Croix.

painted in many different pastel hues.

The quaintly picturesque lure of Annot is not simply confined to its evocative old quarter. Just outside the town to the south is a vast cluster of massive rocks scattered far across the hillside. Known as the **Grès d'Annot,** these house-sized sandstone boulders seem to have been flung to earth from the cliff behind by an angry giant. Locals have built houses directly beside them, often using their sheer faces as an outside wall. Local legends, featuring troglodytes and primitive religions, surround them to this day.

Annot's local industry is based on two thriving factories: producing biscuits and, more traditionally, meat products. The town is also a popular centre for summer excursions and, like St-André, a stop on the route of the Train des Pignes.

Those with stout hearts and boots can follow the spectacular hiking trails into the nearby high-altitude **Coulomp Valley**. Alternatively, the Vaire Valley, running due north from Annot, is the setting of the typical mountain village of **Méailles**, perched high above the river. Still further north lies the ancient village of **Peyresq**, in ruins until 1955 when a group of Belgian students decided to transform it into a tiny but thriving cultural centre.

Into the past: Cross the drawbridge of the **Porte Royale** into the fairytale town of **Entrevaux**, and the 20th century recedes as if by magic. Situated on the eastern fringes of the Alpes-de-Haute-Provence and for centuries a frontier post defending Provence from Savoy, Entrevaux exudes a history more redolent than any town in the *département*. Seventeenth-century ramparts (the work of Louis XIV's masterful engineer, Vauban), turrets, drawbridges and a deep moat formed by the Var River cocoon the town. Cars are firmly relegated to the busy Nice road on the opposite bank of the river.

Like the ramparts, the majority of Entrevaux's houses, together with its cathedral church, date from the 17th century. A **citadel**, testimony to the

The fairytale town of Entrevaux.

town's key strategic position, is reached by an ascending path of nine zig-zag ramps, a remarkable feat of engineering that took 50 years to complete.

As well as with its history, Entrevaux entices with cultural and folkloric events. In August, a two-week music festival of 16th and 17th-century music is organised by an English resident of the town. The festival of John the Baptist, held annually on the weekend closest to 24 June, sees locals in traditional costume celebrating with a mass, dancing and a procession to the isolated **Chapel of St-Jean-du-Désert**, 4 miles (7 km) to the southwest. The popularity of the saint is such that similar festivals on a smaller scale are held concurrently in many of the small villages nearby, each with its procession to a chosen chapel "in the desert."

One delightful curiosity brings the visitor to Entrevaux back into the modern age. The minuscule **Musée de la Moto** is devoted entirely to the history of the automobile. Run by a former Grand Prix mechanic, it houses more then 70 machines dating from 1905, all still in working order.

Entrevaux also boasts a thriving Centre of Mycology and Applied Botany, which attracts mushroom experts from all over France in the autumn. But perhaps of greater interest to the casual visitor is the tiny main square, **Place St-Martin**. It boasts one pleasant café, a clutch of pretty chestnut trees and a butcher who dispenses the local speciality, *secca de boeuf*, a type of dried salt beef, delicious when eaten with olive oil and lemon juice.

If Entrevaux seems uniquely untouched by the proximity of the Alps, **Colmars**, 18 miles (30 km) to the north, could hardly be more Alpine. Colmars is a fortified town crowned by two massive medieval castles, and, like Entrevaux, a former frontier post between Provence and Savoy. But there the similarity ends.

Approaching Colmars along the attractive D908 from St-André, the sloping roofs of wooden Alpine chalets seem a universe away from the dry-

ocal
utcher.

stone *bories* of the Lure or the long, low *mas* of the Var.

Inside the well-preserved ramparts of the town, once again the work of the architect Vauban, the Alpine feel only increases. Houses are constructed with tidy wooden balconies, known as *solerets* (sun traps), locals sport jaunty Alpine caps that would not seem out of place in Bavaria and shop windows display amber bottles of *génépi* liqueur, made from Alpine flowers. Only the small fountains and the geraniums that line the balconies in summer confirm that this is Provence.

Colmar's name stems from Roman times when a temple to the god Mars was erected on the hill (*collis Martis*) that today forms the backdrop to the town. Of the two castles that dominate Colmar's skyline, the Fort de Savoie is the more imposing. Reached by a covered alleyway, the fort is a fine piece of 17th-century military engineering.

Sleepy in winter, Colmar's unspoilt charm and beautiful setting make it a popular centre for family holidays, particularly among residents of the major coastal cities. A *bravade*, more rooted in tradition than the celebrated St-Tropez version, takes place in June.

The French Alps proper lie hidden behind the **Col d'Allos**, 15 miles (25 km) north of Colmars, The high-altitude **Lac d'Allos**, ringed by snow-capped mountains and the breathtaking **Col de la Cayolle** (7,635 ft/2,327 metres) form stunning diversions along the route to the Alps, though they are frequently inaccessible in winter.

Barcelonette, squeezed into a narrow glacial valley surrounded by towering peaks and completely Alpine in character, is the northernmost town in Provence. It owes its unlikely name to its 12th-century rulers, the counts of Barcelona, who originally christened it Barcelona.

The Hispanic connection does not end there. Early in the 19th century, a period of migration to Mexico began, prompted by the success achieved by three local brothers who opened a textile shop there. Many followed, some of whom later returned to Barcelonette to build the incongruous Mexican-style villas for which the town is famous.

Less than a century ago, the villages of the Ubaye Valley remained spectacularly remote, ancient traditions persisted and locals spoke *le gavot*, an Alpine version of Oc. The **Musée de la Vallée**, on the Avenue de la Libération, provides fascinating accounts of the region's chequered history.

Today, though the "Mexican connection" remains strong, with frequent cultural exchanges and shops selling Mexican artifacts, a new flavour has emerged to dominate the atmosphere and preoccupations of the town: skiing. From December to April the modern ski resorts of **Super-Sauze** and **Pra-Loup,** just south of Barcelonette, pack the Ubaye Valley with international tourists and bring a welcome upturn in local fortunes.

In many ways, the history of Barcelonette is the history of the Alpes-de-Haute-Provence in microcosm: poverty, migration, gradual prosperity. It is an evolution that continues today.

Left, mountain wildflowers. Right, minstrel in Entrevaux.

Var
20 km
Reillanne
Puimoisson
Moustiers-Ste. Marie
Durance
Valensole
PARC
REGIONAL
LUBÉRON
Manosque
Riez
Allemagne-en-Provence
la Bastide--des-Jourdans
Gréoux-les-Bains
St. Martin-de-Brômes
Lac de Ste. Croix
Ste. Croix-de-Verdon
Grand Canyon du Verdon
Point Sublime
Falaise des Cavaliers
Vinon-sur-Verdon
Verdon
Mirabeau
GRAND PLAN DE CANJUERS
MONTAGNE DE BARJAUDE
Montmeyan
la Verdière
Aups
Tavernes
Grottes (Cave)
Tortour
Rians
Fox-Amphoux
Château du Picasso (Castle)
Salernes
Barjols
Vauvenargues
MONTAGNE STE. VICTOIRE
Draguignan
Cotignac
Entrecasteaux
Lorgues
Châteauvert
Ollières
Carces
Argens
Abbaye du Thoronet (Romanesque Church)
Trets
St. Maximin-la-Ste. Baume
Vidauban
le Vieux Cannet
Brignoles
Châteauneuf
le Luc
le Cannet-des-Maures
Montagne de la Loube
830
MASSIF DE LA SAINTE BAUME
Besse-sur-Issole
Aille
Roquevaire
St. Pilon
la Roquebrussanne
Gonfaron
Carnoules
Gémenos
Signes
Méounes-lès-Montrieux
Aubagne
Chartreuse-Montrieux-le-Vieux
Chartreuse-Montrieux-le-Jeune
Gapeau
Cuers
Pierefeu-de-Var
Collobrières
Chartreuse de la Verne
le Castelle
Solliès-Pont
La Ciotat
les Lecques
Mt. Faron
542
Bormes-les-Mimosas
Cavalière
Ollioules
Bandol
Hyères
le Lavandou
ILE DE BENDOR
La Seyne
Toulon
Sanary-sur-Mer
Port-de-Miramare
Six-Fours-les-Plages
Carqueiranne
la Garonne
Rade d'Hyères
Cap Blanc
les Sablettes
ILE DES EMBIEZ
N.D. du Mai (Church)
Golfe de Giens
ILES D'HYÈRES
Cap Sicié
Giens
ILE DE PORQUEROLLES
Porquerolles
PARC NATIONAL DE PORT-GROS
Cap d'Arme

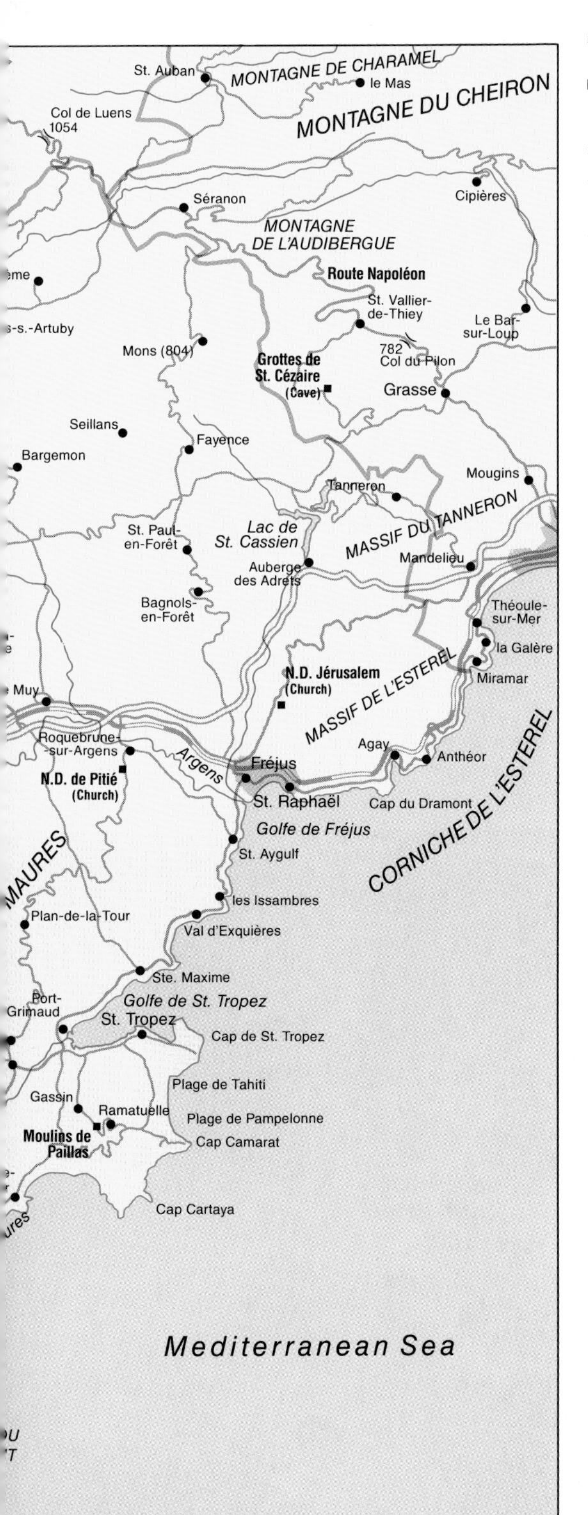

THE VAR

The Var is a palette of varied landscapes that connect the sea with the mountains and plains of the north. The richly forested *massifs* of Ste-Baume, Maures and Esterel that dominate the bottom section by the coast give way to acres of vineyards in the centre of the region. Further north, substantial hills are studded with small hilltop villages, waterfalls, caves and more vines. Here, in the least spoilt section of the *département*, the atmosphere of relative calm is sharply different from the noise and frenzy of the coast.

The coastal area is a broad spectrum of sights and sounds: the cosmopolitan urban sprawl of Toulon, the wild natural beauty of the Iles d'Hyères, the pleasing gentility of summer resorts such as Hyères, Sanary-sur-Mer, and Bandol, the Rivieran splash of St-Tropez.

Tourism, of course, stands on an equal footing with wine production in the list of the Var's major industries. Many *visages pâles* (or "pale faces" – a local appellation for summertime visitors) never venture north of the busy A8 autoroute that bisects the *département* from east to west. However, a multitude of restored farmhouses and elegant modern villas serve as summer homes in the hills from St-Tropez to Aups. In central and upper Var alone, the number of *résidences secondaires* is estimated at more than 25,000.

Nonetheless, inland Var still offers a multitude of quiet secrets to those prepared to explore beyond the autoroute.

The Var lacks none of the Provençal clichés: pretty squares dappled by the shade of plane trees, clusters of medieval architecture, limpid pastel light. As they say in Lorgues, "*Ici on vit vieux et heureux*" (Here you live a long and happy life). Few would disagree.

The big city: The *préfecture* of the Var, **Toulon** is a major port and urban centre and, as such, not subject to the seasonal ebb and flow of population experienced

Preceding pages: the Massif des Maures.

by much of the rest of the *département*.

Clustered round a deep natural harbour and enclosed by a crescent of high hills, Toulon is also France's leading naval base. Many tourists bypass it, however, for its fetid narrow alleyways and chaotic suburban sprawl offer little enticement to the casual visitor. What they don't realise is that, along with all the low-life trappings of a major seaport, Toulon has its own share of grand buildings, chic boutiques, lively fish-markets and a sense of cosmopolitan energy matched only by Marseille.

The main route through the city, the **Boulevard de Strasbourg**, bisects the grid of tall apartment blocks, department stores and imposing administrative buildings that make up 19th-century Toulon. South of the **Place Victor Hugo**, with its fine theatre and opera house, lies the seamier side of the city – a warren of narrow streets leading down to the **old port** (*darse vieille*). Unsalubrious bars rub shoulders with dingy restaurants that dispense the Toulon version of *bouillabaisse* (made with potatoes) and North African specialities. Scruffy kids play football in the shadow of peeling tenements, and Arab *chômeurs* (unemployed workers) watch the world go by.

But the *basse ville*, as the Toulonnais call it, is not all sleaze. Small attractive squares (such as the Place Puget, Place Trois Dauphine and Place Camille Ledeau) draw elegant locals and some tourists to their designer shops and shaded cafés. Superb fish, vegetable and flower markets characterise mornings in the **Cours Lafayette**. And the patchwork streets lead inexorably towards Toulon's *raison d'être*: the port.

In fact, much of the *basse ville*'s bad press undoubtedly comes from the Avenue de la République, a filthy swarming roadway lined with hideous concrete post-war architecture. Damage caused by Allied bombings and the German razing of the harbourfront "for reasons of hygiene" hasn't been well-disguised. Historically, Toulon's era of major expansion occurred during the 17th century, though it was a base for

Shell girls welcome tourists to Toulon.

the royal navy as early as 1487. Under the orders of Louis XIV, the arsenal was expanded and the city's fortifications enlarged. A century later, the city unwisely took the side of the English against the Revolutionary government and was promptly brought to heel by a young Napoleon Bonaparte. The English fleet was roundly defeated in 1793, ensuring that Napoleon's name would never be forgotten.

High on the list of Toulon's many visitable sights are the comprehensive **Musée Naval**, the excellent contemporary art collections of the **Musée de Toulon** and the evocative, writhing figures of Puget's sculpted *atlantes*, now affixed unceremoniously to the modern façade of the city hall annex on the Quai Stalingrad.

For an excellent overall look, take the cable car ride at **Mont Faron**, just outside the city. From the top you will have a terrific bird's-eye view of Toulon, its satellite towns and the sea.

Calm resorts: To either side of the steamy cityscape of Toulon lies a trio of attractive coastal resorts. Bandol and Sanary to the west and Hyères to the east can't claim to be as lively or as chic as St-Tropez and Cannes, but many prefer them for that very reason.

Screened from the ravages of the mistral by an arc of wooded hills, **Bandol** has attracted numerous visitors to its sandy coves and pleasant promenades since the beginning of the 20th century. Among its more famous guests have been New Zealand author Katherine Mansfield, who wrote *Prelude* in the quayside Villa Pauline in 1916, and the celebrated Provençal actor Raimu. His cliffside villa, the Ker-Mocotte, is now a luxury hotel.

Modern-day visitors, a large proportion of them French, come for the town's three sandy beaches, lively harbour and air of calm sophistication. Another plus is Bandol's vineyards, which produce wines (particularly reds) rated among the best in Provence.

About a mile (2 km) off the coast of Bandol lies the tiny island of **Bendor**, enterprisingly transformed into a holi-

Toulon boy scouts.

day village by the pastis magnate Paul Ricard in the 1950s. On the island is a hotel, a clutch of rather expensive cafés, a recreation of a Provençal fishing village and Ricard's pride and joy – a museum devoted to the pleasures of alcohol. The grandly titled Exposition Universelle des Vins et Spiritueux contains more than 8,000 bottles of wine and spirits from all over the world, plus a selection of glassware. Though the island has an air of artificiality, its shady paths, lined with mimosa and eucalyptus, and its tiny sandy beach are reason enough for a summer excursion.

Just 3 miles (5 km) from Bandol is the pretty pink-and-white resort of **Sanary-sur-Mer**. Like its neighbour, Sanary benefits from a sheltered position supplied by its backdrop of hills, known as the Gros Cerveau. A number of *pointus* (old fishing vessels) add spice to its attractive harbour.

Artists and writers began to flock to Sanary in the early 1930s, inspired by Aldous Huxley's presence there. They were soon joined by a group of German intellectuals, headed by Thomas and Heinrich Mann, who fled to the town after Hitler's rise to power in 1933.

The island of Porquerolles.

The nearby **Sicié Peninsula** makes for a worthwhile trip from Sanary. At its southern point, the **Chapel of Notre-Dame-du-Mai** sits on a high clifftop that drops sharply towards the sea. The view from here takes in *calanques* and the Iles d'Hyères.

To the east is **Hyères**, the most substantial of the three major resorts that surround Toulon and, in many ways, the most interesting. Hyères was the first resort to be established on the Côte, setting a trend that spread rapidly east from the late 18th century onwards. The list of famous consumptives, or merely pleasure-seekers, drawn to its balmy winter climate reads like an international Who's Who: Queen Victoria, Tolstoy, Pauline Bonaparte, Aubrey Beardsley, Edith Wharton, etc. Robert Louis Stevenson, though desperately ill during his stay, wrote: "I was only happy once – that was at Hyères."

By the 1920s, however, medical opinion had switched its allegiance to the curative properties of mountain, rather than sea, air. This and the increasing popularity of the "real" Riviera to the east swiftly relegated Hyères to a distinctly unfashionable position.

In many ways, the town's lack of chic has become one of its most attractive qualities. Busy all year round, Hyères plays host to several sporting and cultural events (such as the sailing regatta and the festival of cartoon animation) and has a thriving agricultural economy (peaches and strawberries) independent of tourism. The **vieille ville** is an appealing small medieval quarter, topped by a park and a ruined 14th-century castle with the spacious, flag-stoned **Place Massillon** at its heart.

Modern Hyères has an elegant *Belle Epoque* feel and some interesting examples of neo-Moorish architecture. A 19th-century taste for things "oriental" (inspired by Napoleon's Egyptian campaign) led to the construction of **La Mauresque** (Avenue Jean Natte) and **La Tunisienne** (Avenue de Beauregard) in the 1870s. Minarets, Arab arches and tropical date palms add a taste of the exotic.

Outside the town lie the vestiges of ancient Hyères (called Olbia), once an important Phocaean and Roman port. Antiquities excavated from the site can be seen in the **municipal museum**.

Isles of gold: Jutting out into the sea, the **Presqu'île de Giens** is a joyless, flat peninsula composed of salt marshes and some good beaches surrounded by ugly campsites. Salt collection, a local industry since pre-Roman times, continues today at the Côte's only remaining productive marsh, the Marais Salins des Pesquiers.

The main reason for taking this uninspiring route is to catch a boat from **La Tour-Fondue** at the southeastern tip to the **Iles d'Hyères**. (Boats also sail from Le Lavandou and Cavalaire).

This group of three subtropical islands, also known as the **Iles d'Or** (Isles of Gold)**,** is a haven of unspoilt natural beauty. **Porquerolles**, the largest and most accessible, has a small town with cafés and some fabulous beaches amidst dense vegetation. **Port-Cros**, more rugged and mountainous, is one of only six designated national parks in France. The third, **Levant**, is 90 percent inhabited by the French military and, therefore, mostly out of bounds. All that exists on it is a small nudist colony known as **Héliopolis**, founded in the 1930s and clinging to the bare and dramatic island's western tip.

Unless you're a naturist or a botanist, Porquerolles is probably the best choice for a visit. Its village, established as a small military base in the 19th century, is more colonial than Provençal in character. From here, you can hire bicycles to tour this eucalyptus and pine-clad island where cars are not allowed. Major spots of beauty include the lighthouse at the **Cap d'Arme** and **Plage Notre-Dame**, which is also excellent for swimming.

On Port-Cros, the terrain is much more challenging, and strict rules against smoking and the lighting of fires must be observed. A small **tourist centre**, open in summer, provides maps and

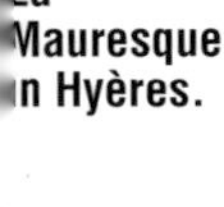

La Mauresque in Hyères.

guidelines. Perhaps the most rewarding walk (around two hours for a roundtrip) is along the **Vallon de la Solitude** that cuts across the southern end of the island. At its start, the island's only hotel, the 18th-century **Manoir d'Hélène**, serves a fine lunch.

Flower Village: Back on the mainland, a number of towns line the coastal road that leads inexorably to St-Tropez. This is not, strictly speaking, the Riviera, but you'd be hard pressed to tell the difference. A case in point is fashionable **Bormes-les-Mimosas**, a hilltop village enjoying stunning views of the Iles d'Or and the sea.

Recent critics have accused it of being over-prettified, but Bormes remains, despite the rather precious tag of "Les Mimosas" that was added to its name in 1968, one of the jewels of Provence. Bougainvillea, mimosa and eucalyptus (some of which sadly was destroyed by the harsh winters of 1984 and 1985) make this the archetypal *village fleurie*.

The names of the picturesque narrow streets – for example, Roumpi-Cuou (neck-breaker) and Plaine-des-Anes (donkey's sorrow) – colourfully suggest their steepness. Look out for the painted sundial on the **Eglise St-Trophime**, the fine medieval château at the top of the old town and the small museum on the Rue Carnot that contains some terracottas by Rodin.

South of Bormes, connected to the exclusive promontory of Cap Benat by a dyke, **Fort de Bregançon**, the "Camp David" of the French president, sits proudly on its own islet. Understandably, it can't be visited.

Dark forests: The **Massif des Maures**, a splendid tract of mysterious deep forest, much of it inaccessible, stretches roughly from Hyères to Fréjus and from the A8 autoroute down to the Mediterranean. Its name comes from the Provençal word *maouro*, meaning dark.

The Maures, together with the neighbouring Esterel *massif*, are components of the oldest geographical area in Provence. Schist, shot through with sparkling mica, makes up the high hills

Wine country above the coast.

and deep ravines of the massif. Cork oaks, Aleppo pines and chestnut trees form its dense vegetation.

The area was ruled by invading Saracens for more than 100 years, beginning in the early 9th century. Pillaging the countryside as far afield as Lake Constance, their reign of terror lasted until 972 when Count Guillaume destroyed their bastion at La Garde-Freinet.

In more recent times, the area has been ravaged no less destructively by the constant plague of forest fires, sparked spontaneously or by human carelessness. The desiccated forests quickly transform into monstrous walls of flame, leaving nothing but charred scrub in their wake. More than a million acres of the Var – France's most forested *département* – have been destroyed in the last 50 years. Preventative measures, such as the creation of fire-break paths and radical pruning of vegetation have had an impact since their introduction, but the light aircraft of Canadair scooping water from the sea to douse flames remains a common sight. It is an annual battle, still being fought.

The Massif des Maures has few towns of any size, though its wooded hills conceal the private estates and villas of the rich and sometimes famous. **La Garde-Freinet**, an ancient Saracen stronghold and, in a sense, the capital of the Maures, is little more than a large village. Its traditional industry of cork production was a skill acquired from the otherwise barbaric Saracens in the ninth century. By 1846, the town produced more than three-quarters of France's bottle corks but, like many traditional industries, decline set in after the 1950s. Since the early 1980s, however, the industry has diversified into other uses of cork, and neatly stripped trunks of cork oaks line the *massif*'s serpentine roads and tracks.

In the town, the only obvious evidence of cork production are the cork bowls and ornaments on sale to tourists in the pricey boutiques of the Rue St-Jacques. St-Tropez chic has left its mark here, accelerating particularly in the last decade. Arty café loungers at the **Claire**

Firefighters in the Maures Forest.

Fontaine, epicentre of the village, are now more likely to read the trendy national newspaper *Libération* than the local *Var-Matin*. And existence of a "cocktail bar" – once an impossibly foreign concept – testifies to the proximity of the Côte.

Mercifully, La Garde-Freinet still has a stunning setting, charming stone *lavoirs* and unpretentious back streets. "Real" Garde-Freinet congregates in the **Cercle des Travailleurs Bar** on Place Neuve, where local men drink pastis under the watchful eye of a splendid Marianne bust.

South of the village, a walk to the ruins of the **Saracen fortress** is rewarded with great views of the massif and the sea. Recent archaeological digs have unearthed a collection of pottery and stoneware that are now on display in the tiny **Musée du Freinet** at the lower end of the village.

The rollercoaster road heading south from La Garde-Freinet cuts through the centre of the Maures towards the peninsula of St-Tropez. *En route* is the picture-postcard hilltop village of **Grimaud**, a maze of pretty streets awash with bougainvillea and oleander. Wander along the Rue des Templiers, for its medieval arcades, then up to the evocative ruined castle, destroyed on the orders of Cardinal Richelieu.

Cogolin, 1¾ miles (3 km) to the south, is a lively small town known for the manufacture of briar pipes and carpets. It is also a considerable producer of good quality wines. Visits can be made to the pipe and carpet factories.

Once on the peninsula, resist the magnetic pull of St-Tropez and head inland to investigate two perfect *villages perchés*, **Gassin** and **Ramatuelle**. Both have stunning views over the Baie de St-Tropez and numerous tiny streets, some no more than an arm's length across. Between the two villages, on the highest point of the peninsula, are a group of ancient windmills, the **Moulins de Paillau**, and a superb panorama.

Chestnuts and Carthusians: While the eastern section of the Maures is tinged with the atmosphere of the Côte, the

The Cercle de Travailleurs bar in La Garde-Freinet.

villages to the west remain largely unscathed. **Collobrières**, bisected by a small river, is shady and peaceful even in August. This town is famous for its *marrons glacés*, along with all manner of confections made from the local chestnuts. Unlike the nougat of Montélimar or the *faïence* of Moustiers, Collobrières keeps its most famous export well hidden, but a small shop, the **Confiserie Azuréenne**, attached to the main factory, is a good source for buying chestnut delicacies.

Deep in the forest east of Collobrières is one of the real treasures of the Maures: the **Chartreuse de la Verne**. Founded in the 18th century, the Carthusian Charterhouse suffered centuries of fires and Protestant attack. The majority of the still existing buildings date from the 17th and 18th centuries, though the medieval cloisters and a sparsely restored monk's cell are evidence of more ancient origins. The various signs enjoining visitors to silence are a reminder that, after years of neglect, the Chartreuse is now occupied by nuns from the Order of Bethlehem.

Priest at the Chartreuse de la Verne.

Squeezed between the Maures and Esterel massifs are the major resorts of **Fréjus** and **St-Raphael**. For many years these towns were playgrounds for the rich, but today the choking traffic that pours through them during summer has dissolved much of their appeal. Unfortunate high-rise developments have aggregated the two towns together into one urban mass that is overcrowded in the summer and grim in the winter.

Of the two, Fréjus, 1½ miles (3 km) inland, is certainly the more interesting, largely because it contains more vestiges of the Roman past that both towns share. The **amphitheatre** on Rue Henri Vardon, though substantially damaged, is still in use as a setting for rock concerts and bullfights. Here, those with a taste for the drama and cruelty of the *corrida* can watch a spectacle with all the trappings of its Spanish equivalent. And, in the centre of town stands a medieval **cathedral**, begun in the 10th century and bordered by fine 12th-century cloisters that feature a fantastical ceiling decorated with animals and chimera. Also within the cathedral complex is an unmissable 5th-century **baptistry**, one of the three most ancient in France. All are reached from an entrance on Place Formigé.

Having devoted time to Fréjus's Roman and medieval attractions, it would be a shame to miss a couple of very unGallic curiosities nearby. A **Buddhist pagoda**, just off the N7 to Cannes, commemorates the death of 5,000 Vietnamese soldiers who perished in World War I. And a prettily faded and dilapidated Sudanese **mosque** lies unprepossessingly in the middle of an army camp off the D4 to Bagnols.

Swimming beaches can be found south of the town at Fréjus-Plage, though they are as crowded and unappealing as anywhere else you might go on the Côte in high season.

St-Raphael's rail terminus accounts for much of its liveliness, though it fell out of fashion as a tourist base some years ago. Around the middle of the last century, its attractions were much vaunted by the journalist Alphonse

Karr, whose enthusiasm lured the likes of Dumas and Maupassant to winter here. Now, though less trendy, St-Raphael has no lack of visitors, even if they're just here to lose a few francs at the rather hideous casino on the seafront. A **Museum of Underwater Archaeology**, close by a Romanesque parish church, is probably the town's only worthwhile "sight." It contains a good collection of Roman amphorae and a display of underwater equipment.

Heading west along the Gulf of St-Tropez, **Ste-Maxime** offers a taste of what St-Raphael was like in the not-too-distant past. The archetypal Côte resort, its palm trees, promenades and parasols offer the classic *corniche* experience. Though suffering as much as its neighbours from accelerating suburban sprawl, Ste-Maxime's neat marina and relatively golden beaches are still undeniably attractive. When quayside strutting and café lounging loses its appeal, head just north to visit the **Museum of Sound and Mechanical Instruments** in the Parc de St-Donat. It houses a unique collection of old phonographs, bizarre music boxes and a turn-of-the-century dictaphone.

More forests: Highwaymen and sundry criminal types ruled the impenetrable reaches of the **Massif de l'Esterel** for many centuries. Separated from the neighbouring Massif des Maures by the Argens Valley, the Esterel is more sparse in vegetation – a result of the devastating forest fires that swept the area in 1964. In addition, a disease has decimated the indigenous sea pines. But you will still find it an appealingly wild region only steps from the glitter of the Côte.

The red porphyry rock of the Esterel tumbles down to the sea in a dramatic sweep of high hills and ravines. The coastal road, the **Corniche d'Or**, is one of the least crowded sections of the Côte, though the familiar pattern of private villas and large hotels blocking views and access to the sea can be terribly frustrating even here. Head to the section from Cap Roux to Anthéor for some welcome relief and to the inviting

Along the Côte d'Or.

coves around Agay for swimming.

Inland, panorama lovers should head for **Mont Vinaigre** (at 2,060 feet/628 metres, the highest point of the massif) or, for wonderful sea views, the **Pic de l'Ours**. Skirting the northernmost section of the Esterel along the N7 brings you to the marvellous **Auberge des Adrets**, an authentic 17th-century coaching inn and one-time haunt of Gaspard de Besse, the Robin Hood of the Esterel. Gaspard spent many profitable years ambushing the mail and passenger convoys that plied the routes southeast of Mont Vinaigre. He was arrested with his accomplices in 1780 and executed the following year.

Central Var: The central area of the Var is dominated by the fertile **Argens Valley**, which runs horizontally across the *département*. To the south and north are the wooded, vine-covered hills that produce the majority of the A.O.C. Côtes-de-Provence rosés and reds. To most tourists, however, the area is little more than a transport corridor *en route* to the Côte. **Le Luc**, where the autoroute and N7 intertwine, is a small market town which has been sadly overrun, like many of its neighbours, by the pressure of passing traffic. Its rich history as a Roman spa town and Protestant refuge can be traced in the local museum, the **Musée Historique du Centre Var**, housed in the 17th-century **Chapel of Ste-Anne**. Look out, too, for Le Luc's best-known landmark, a 16th-century hexagonal **tower**.

As well as its wine, olives and chestnuts, Le Luc gained fame during the 1800s for the health-giving purity of its mineral water, *eau de Pioule*, which is still bottled at source on a small scale today. Perhaps less medically sound was Le Luc's reputation two centuries earlier as the centre for what was considered to be the most effective cure for whooping cough. Provençal superstition maintained that children could be cured by being passed seven times under the belly of a donkey. Le Luc's donkey had such prestige that children from Draguignan and even Cannes were brought to suffer the ordeal.

larbourfront n St-Tropez.

Outside Le Luc is **Le Vieux Cannet**, one of the prettiest of the region's hill villages, clustered around an 11th-century church.

Further west along the autoroute lies **St-Maximin-la-Ste-Baume**. Pilgrims have poured in to this town since the fifth century to view one of the greatest of all Christian relics – the presumed bones of Mary Magdalene. After the so-called Boat of Bethany supposedly landed in the Camargue, its saintly crew dispersed to preach the word of God throughout Provence. Mary Magdalene is said to have made her way to the Massif de la Ste-Baume where she lived in a dank cave for more than 30 years. She died in the town of St-Maximin, where her remains were jealously guarded by the Cassianites.

Work began on the magnificent **basilica** that now contains the relics during the 1200s. It is considered to be one of the most impressive examples of Gothic architecture in Provence. Inside, a tiny blackened crypt, etched with centuries of grafitti, houses the relics. Visitors peer through iron bars to catch sight of the holy remains, wedged into a somewhat macabre gold setting.

The 19th-century writer Prosper Mérimée, in his role as inspector of monuments, dismissed St-Maximin as a dreary place and, excluding the basilica, it's tempting to share his view. But the *vieille ville* does have some worthwhile medieval arcades (Rue Colbert) and the interesting remains of a small Jewish community.

To the south of St-Maximin, in the ancient limestone mountain range of Ste-Baume, is the even more evocative cave where Mary Magdalene is said to have spent those last years. Reaching its entrance involves a strenuous climb through dense forest. This forest, in particular the reaches lying below the holy cave, was a magic and sacred place to Ligurians, Gauls and Romans, and the towering beech trees and lush undergrowth still seem bewitchingly sylvan today.

One hundred and fifty stone steps lead up from the shade of the forest to

Tilemaker and son.

the cliffside cave. Inside the vast dark recess, now filled with altars and saintly effigies, the old stones drip water. A final effort will bring you to St-Pilon, which is nearly the highest point of the *massif*. Mary Magdalene was said to have been lifted up to this peak by angels seven times a day during her years of cave-dwelling.

North of the autoroutes: Located in the northwest corner of the Var and dubbed the "Tivoli of Provence," **Barjols** is a small industrious town filled with streams, fountains and masses of peeling plane trees. In fact, what is reputedly the largest plane tree in France, 40 feet (12 metres) in circumference, casts its shade over the most famous of Barjols's 25 fountains, the vast moss-covered **"Champignon" fountain** in the tiny Place Capitaine Vincens.

Though the Tivoli tag attracts a good number of summer visitors, Barjols is in reality more of a unpretentious, workaday town than a successful tourist trap. Its prosperity was originally based on its tanneries, the last of which closed in 1986. Barjols is still, however, known for the manufacture of the traditional Provençal instruments, the *galoubet* (a three-holed flute) and *tambourin* (a narrow drum), which are played simultaneously by a single musician.

Barjols's **old quarter** is being extensively renovated and, as such, is in a state of flux. Ancient cobwebbed hovels alternate with low medieval archways in the dusty alleyways around the former college **Chapel of Notre-Dame-de-l'Assomption**. The church is all that remains of what was during the Middle Ages the favoured school for the children of the counts of Provence.

An undistinguished square close to the chapel hides one of Barjols's best treasures, the magnificent entrance to the **House of the Postevès**. The impressive entrance was sculpted during the Renaissance.

Barjols's claim to fame rests largely on its Fête de St-Marcel, which is arguably the most ancient and picturesque festival in Provence. Held on the weekend nearest to 16 January, the Fête des

he flutes of
arjols.

Tripettes celebrates the town's victory over the rival village of Aups in securing the relics of Saint Marcel for its own chapel. The day that the relics arrived, in 1350, just happened to coincide with the long-standing pagan practice of sacrificing an ox for a village feast. Eventually, secular and Christian festivities were combined into a single ecstatic festival.

Today, the festival consists of noisy processions to the sound of *galoubets* and *tambourins*. Every four years (1994, 1998, etc.) an ox is roasted on the Place de la Roquière and shared out among the revellers.

The real Var: Between the empty northern expanses of the **Canjuers Plain**, occupied by the French military, and the autoroute to the south lies a string of towns that come closest to representing the "authentic" Var.

Aups, on the fringes of the plain, is an access point for the Gorges du Verdon (on the Var's northern border) and a busy market town in its own right. The town is crowned by a lovely 16th-century clocktower decorated with a sundial. Aups, and the northern part of the Var in general, has a strong tradition of republican resistance and was the scene of many popular uprisings in the mid-19th century. The portal of the town's **Church of St-Pancrace** proudly bears the republican inscription: *Liberté, Egalité, Fraternité.*

Salernes, 6 miles (10 km) to the south, is a larger and more sprawling town with one of the best markets in the area. Like Aups, the town has been a centre of political resistance, especially during World War II. Above all, though, Salernes is known for its tile-making, the most prolific in the *département*. Around 15 factories still function today, many using traditional wood-fired kilns.

Salernes itself lacks the classic Provençal prettiness of many of the towns further east, but nearby are some unmissable gems of hilltop village architecture which are worth seeking out. Minuscule **Fox-Amphoux** crouches on a high hill to the west, its pretty streets

Tiled kitche[n] in Salernes.

clustered round a Romanesque church.

More commercial, but even more spectacular, is **Tourtour**, "the village in the sky." An irresistible *mélange* of breathtaking views and medieval vaulted passageways keeps its appeal even in the crowded summer months. Sadly, Tourtour's finest "monument" – two venerable elms planted in the main square to commemorate the birth of Louis XIV – recently fell prey to disease. They have been replaced by olive trees.

From **Cotignac**, a superb collection of 16th-century houses dominated by two ruined towers, the D50 snakes east to **Entrecasteaux**. The elegant **château**, centrepiece of this small village, was built for a local nobleman in the 1600s. Scandal ruined the family during the next century, when the residing *seigneur* shot his wife, and the château fell into disrepair.

Then, in 1974, it was bought by painter, soldier and adventurer Ian McGarvie-Munn, a larger-than-life Scotsman who married into the Ecuadorean political hierarchy. The family set about the castle's restoration, despite the death of McGarvie-Munn in 1981 and considerable local hostility.

Today, it is a delightfully eccentric combination of architectural monument, art gallery, hotel and private residence. Wander through the family's kitchen to airy rooms hung with modern art and filled with McGarvie-Munn's collections of artifacts from around the globe. Only the gardens disappoint.

South along the **Bresque Valley** from Entrecasteaux stands what is indisputedly the most beautiful building in the Var. The austere simplicity of the **Abbaye de Thoronet** is enough to turn non-believers into pilgrims. Together with Senanque and Silvacane, the abbey is one of the Provençal trio of so-called "Cisterian sisters," built in the 12th century to the ascetic precepts of the Cistercian Order. Only the play of light and shadow decorates the perfectly proportioned chapel, cloisters and chapterhouse.

Neglected since before the Revolu-

athroom at he Château d' ntrecasteaux.

tion, the abbey was saved from ruin by Prosper Mérimée who urged its restoration in the 1840s. Work continues today, but a new threat to the abbey has surfaced in recent decades. Bauxite mining, the lynchpin of the local economy, has destabilised the abbey's foundations. A pressure group based in St-Raphael is currently campaigning for its protection.

Once a capital: Draguignan, the Var's only major inland town, was capital of the *département* until 1975, when it lost the title to Toulon. Its name recalls a dragon that was said to have terrorised the town in the 5th century. For this reason, the mythical creature can be seen in stone crests on many of the town's medieval gateways and houses.

The small city's greatest attractions lie in the tiny **medieval quarter** beyond the grid of 19th-century boulevards, designed by Baron Haussmann. The Rue de la Juiverie still has the remains of a synagogue facade, a relic of the important Jewish community that thrived here in the Middle Ages. Nearby is an attractive clocktower built in 1663, with a classic wrought-iron campanile.

After the sure delights of Draguignan's bustling market days, head out of town for an eclectic collection of visitable sights. Just to the northwest on D955 is the mysterious stone monolith of the **Pierre de la Fée** (the "fairy stone"), a vast slab of neolithic rock on three mighty stone legs. Following this road further north brings you to the verdant **Gorges de Châteaudouble**, a smaller version of the Verdon Gorges, cut by the River Naturby.

East of Draguignan, on the D59, is an **American military cemetery**, a legacy of the bitter fighting that raged around Le Muy in August 1944.

Nearby **Trans-en-Provence** boasts a cascading waterfall, a delicately painted Hôtel de Ville and a truly original curiosity on its outskirts. The "airborne well" (follow signs out of Trans for the *puits aérienne*) was conceived in 1930 by an eccentric Belgian inventor as a moisture-capturing solution to the region's thirst. The bizarre, beehive structure never fulfilled its laudable aim and exists now as a unique and entertaining folly.

The northeastern Var: The villages of eastern Var catch much of the appealing flavour of the Alpes-Maritimes. **Fayence**, the largest town in the area and a major centre for hang-gliding, has little overt charm, but its satellite villages are among the prettiest in Provence.

The German artist Max Ernst made **Seillans** his home, spending the last years of his life in a villa at the top of the village. An atmospheric, painterly place, Seillans tumbles down a sheer hillside in a mass of pink and ochre stone and steep cobbled streets. Its boundaries are marked by the hilltop château, and the cool, shady Place du Thouron at the lower end of the village. Cars are banned within the ramparts.

The town, whose name is derived from the "pot of boiling oil" they poured over the heads of unwelcome Saracens, is known for its perfume-making and flower cultivation, a thriving industry since 1884. In recent years, Seillans has instituted an annual flower festival, a suitably chic affair for this attractive and prosperous village. Visit the **Eglise de Notre-Dame-de-l'Ormeau**, moments away on the low road, for its stunning colourful altarpiece, carved from wood by an unknown 16th-century Italian artist.

Less exclusive, but equally tempting, are the villages of **Bargemon** to the west of Seillans and **Mons** and **Bargème**, north of Fayence. Mons has all the ingredients of a perfect Provençal village, plus a superb view from the Place St-Sebastien of the Italian Alps, the Iles de Lérins and Corsica. Bargème, much more spartan, is splendidly isolated. It is the highest village in the Var, perched on a peak of the **Montagne du Brouis**.

Here, the surrounding landscape is sparse and dreary, and the resulting poverty gave locals a sorry reputation for crime and dishonesty in the last century. A decade ago towns like Bargème were virtually deserted, but the energetic restoration now being carried out in so much of the Var signals a future more shining than its past.

Right, the original Statue of Liberty stands in St-Cyr.

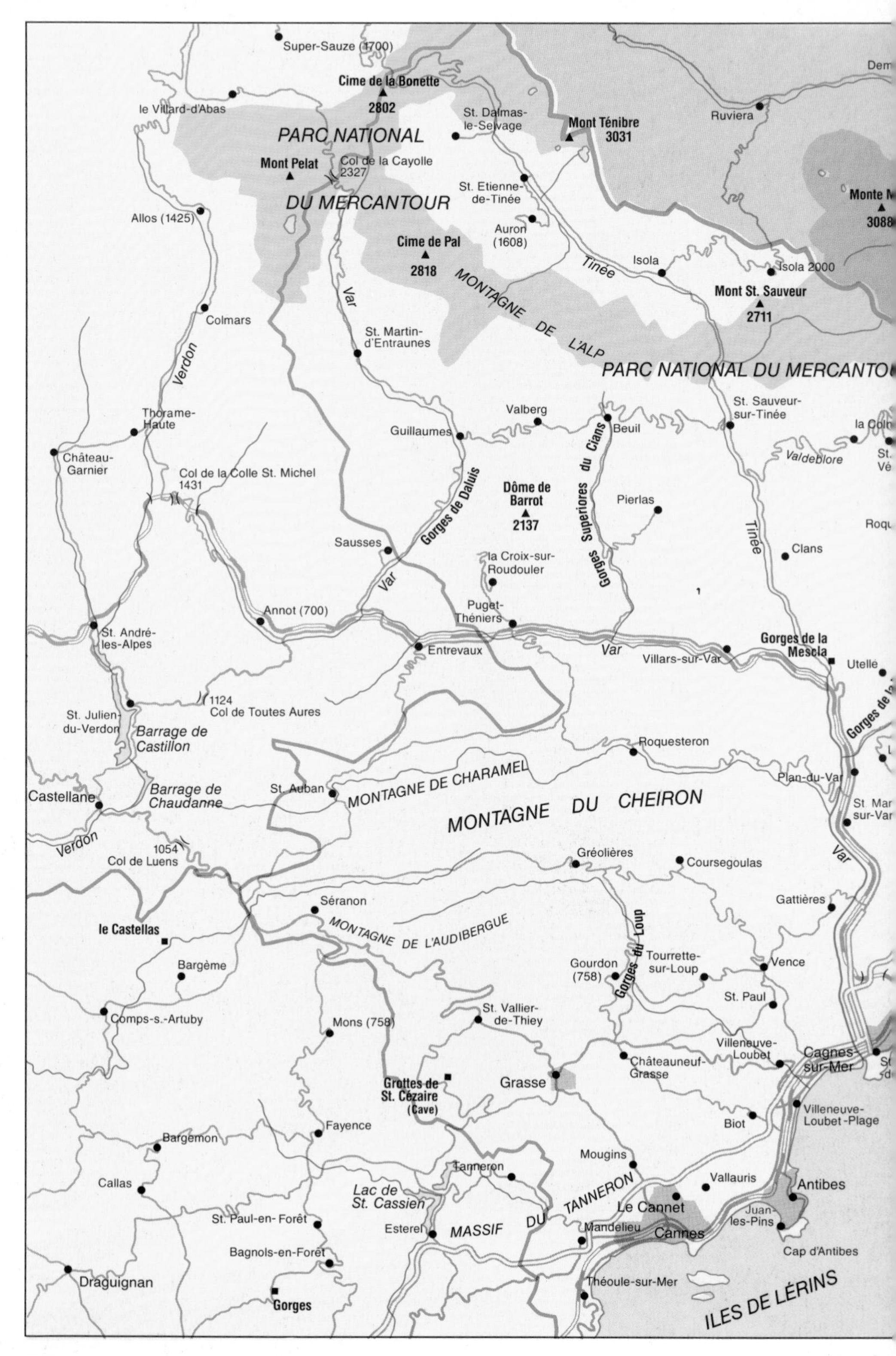

Super-Sauze (1700)
Cime de la Bonette
2802
le Villard-d'Abas
PARC NATIONAL
DU MERCANTOUR
Mont Pelat
Col de la Cayolle
2327
St. Dalmas-le-Selvage
Mont Ténibre
3031
Ruviera
St. Etienne-de-Tinée
Allos (1425)
Auron
(1608)
Cime de Pal
2818
Tinée
Isola
Isola 2000
Mont St. Sauveur
2711
MONTAGNE DE L'ALP
Var
Colmars
St. Martin-d'Entraunes
PARC NATIONAL DU MERCANTO
Verdon
Thorame-Haute
Valberg
Guillaumes
Beuil
St. Sauveur-sur-Tinée
Château-Garnier
Valdeblore
Col de la Colle St. Michel
1431
Dôme de Barrot
2137
Gorges Superiores du Cians
Pierlas
Gorges de Daluis
Sausses
Tinée
Clans
la Croix-sur-Roudouler
Annot (700)
Puget-Théniers
St. André-les-Alpes
Entrevaux
Var
Villars-sur-Var
Gorges de la Mescla
Utelle
1124
Col de Toutes Aures
St. Julien-du-Verdon
Barrage de Castillon
Roquesteron
Castellane
Barrage de Chaudanne
St. Auban
MONTAGNE DE CHARAMEL
Plan-du-Var
MONTAGNE DU CHEIRON
Verdon
1054
Col de Luens
Gréolières
Coursegoulas
Var
Séranon
Gattières
le Castellas
MONTAGNE DE L'AUDIBERGUE
Gorges du Loup
Tourrette-sur-Loup
Gourdon
(758)
Vence
Bargème
St. Paul
Comps-s.-Artuby
St. Vallier-de-Thiey
Mons (758)
Villeneuve-Loubet
Châteauneuf-Grasse
Cagnes-sur-Mer
Grottes de St. Cézaire
(Cave)
Grasse
Villeneuve-Loubet-Plage
Biot
Fayence
Bargemon
Mougins
Tanneron
Vallauris
Antibes
Callas
Lac de St. Cassien
TANNERON
Le Cannet
St. Paul-en-Forêt
Esterel
MASSIF
DU
Mandelieu
Cannes
Juan-les-Pins
Bagnols-en-Forêt
Cap d'Antibes
Draguignan
Théoule-sur-Mer
Gorges
ILES DE LÉRINS

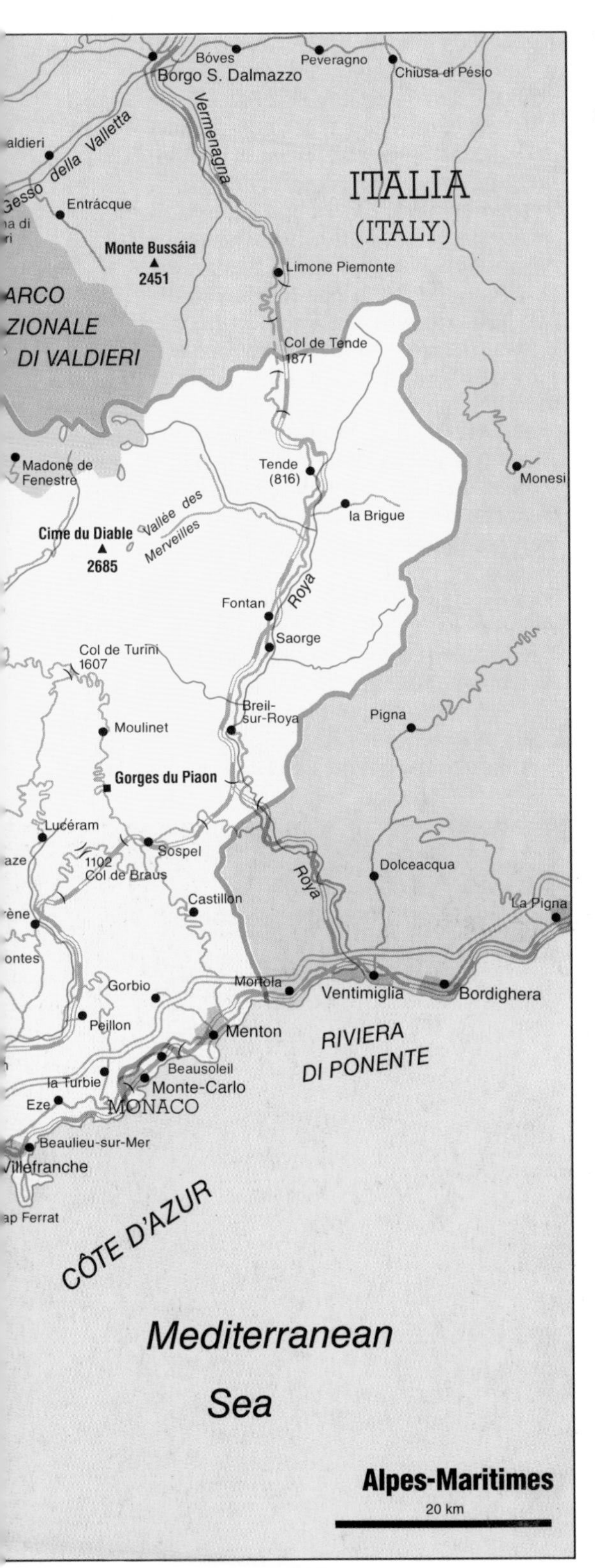

THE ALPES-MARITIMES

It would come as a surprise to many of the tourists who regularly fight for a deck chair on the Côte d'Azur to find that the Riviera has a peaceful back garden. Not a typically French garden, of the geometrically trimmed and designed variety that the nobility used to keep by their châteaux in the Loire Valley. Nor, despite the Riviera's British patronage, one of those rambling but manicured gardens to be found at a country manor. Instead, a neglected and untidy garden of a character all its own.

Ninety percent of the *département* of the **Alpes-Maritimes** is forgotten. During the summer months, a combination of heat haze and smog along the overburdened Côte hides the *arrière-pays* (or back country). The rugged hills and snow-capped Alpine peaks are only revealed during other seasons, affording a tempting peek at what the rest of the region has to offer.

And yet, getting to this peaceful hinterland from the Côte takes but a matter of minutes. The **Vallée des Merveilles** (Valley of Wonders) lies barely 40 miles (70 km) north of Nice, close by the Italian border. Its solitary population of ibex, goats and the occasional lynx are hidden from prying eyes by rugged mountains that shelter a remarkable display of the prehistoric version of graffiti: rock drawings. In fact, 35,000 such drawings were etched into these rocks from the Bronze Age through to medieval times.

The valley is accessible only by foot, preferably for those with limbs and lungs strong enough to endure a day or two of solid walking and with a stock of hiking gear to match. There are no four-star hotels, luxury flats or villas with swimming pools here. If an overnight stay is necessary, the solid virtues of a mountain refuge will have to do. You will find no buses or trains to hop on to if you get tired and no cafés to stop at if you get thirsty. And, in many cases,

Preceding pages: Saorge perched village.

there aren't even any paths, let alone signposts to look at if you get lost.

For the inexperienced, a mountain guide is essential, and the tourist offices in **St-Dalmas** and **Tende** will help with this. Hardened hikers, on the other hand, can probably make do with a small scale map. They should head between the ominous Mt-Bégo and the 9,626-ft (2,934-metre) **Grand Capelet** and if they stumble across several lakes on the way, they'll know they can't be far wrong. There are mountain refuges on either end of this 40-mile (60-km) valley. If a thunderstorm strikes, don't panic. For the shepherds working in the valleys below over 3,000 years ago, Mt-Bégo was like a temple, a place where sheep were sacrificed in an attempt to appease terrifying storms. In our contemporary and less mysterious terms, it simply acts as a lightning conductor.

The Vallée des Merveilles is the centrepiece of the **Parc National du Mercantour**, a huge national park that cloaks the northern edge of the Alpes-Maritimes. As a whole, it is rather less foreboding than the Vallée, but it is still a deserted mountain area and comes with the same warnings. Snow and ice are present from October through June, and the weather can change without warning or be different from one valley to the next. Even in summer the local newspaper, *Nice-Matin*, is full of stories of those lost for days and of the injured and even the dead. The locals and the gendarmerie will not laugh at people wearing heavy gear in summer, nor will they shrug at those asking about a mountain guide.

The Mercantour is the lifeblood of the Alpes-Maritimes. The sources of all the rivers that flow through the *département* can be found here – the Tinée, the Vésubie and the Var, which after successive re-drawings of departmental boundaries is now located outside the *département* to which it gave its name. Yet, for all its rigour, the Mercantour is now much more civilised than when it was the protected hunting ground of the kings of Italy.

Although the Comté de Nice was re-

Hotel sign near St-Etienne-de-Tinée.

turned to France in 1860, the Mercantour remained an Italian enclave until 1946. After it was seceded, however, hunting was banned and the whole of the park became a nature reserve. Botanists now believe that over half of the 4,000 or so species of wildflowers to be found in France grow in the Mercantour. Equally, of course, another reason for the survival of such natural treasures is the minimal impact that human civilisation has had on the region – so far.

To get around the Alpes-Maritimes, some form of personal transport is essential. Public transport is sparse, centred on coach routes to some of the larger towns and the skiing resorts. Only two railways cut their way inland. One runs from Nice to Sospel and Breil and on to Cuneo in Italy, with about four trains a day. The other, dubbed the Petit Train des Pignes by the locals, is a more occasional route running from Nice up the valley of the Var River towards Levens. Attempts to revive it in recent years have stumbled.

Train travel inland is a commodity that has been neglected by the regional council – thankfully. Soon after the first trains crossed the Massif de l'Estérel in 1863, the Côte d'Azur began to take the crowded appearance that it revels in today. But in the *arrière-pays* there are plenty of ruined railway viaducts and stretches of disused track to testify to a different attitude. The most striking is the Pont du Loup, which straddles the southern edge of the Gorges du Loup. It was bombed by the retreating German army at the end of World War II and never revived.

Travel up the main road – the N202, nicknamed the Route d'Hiver des Alpes – from the coastal town of St-Laurent-du-Var, and the reasons for the absence of public transport become clear. Nearer the coast, the highway on the eastern bank of the Var River is wide and straight, but as you gradually move inland it becomes narrower and starts to wind. Within 15 miles (25 km) you find yourself at the base of a gorge, one of the chief geological features of the *département*. These sheer valleys carved by

Street talk in Lucéram.

torrents are the only way for transport to reach the north.

From Plan-du-Var northwards, N202 gives off into several smaller roads, following tributaries of the Var, which head down the narrow gorges and into the mountains to the skiing resorts. All are dead ends or circular routes that eventually return to the N202 as it heads east to the Lac de Castillon, in the neighbouring *département* of the Alpes-de-Haute-Provence. The way out of the north of the Alpes-Maritimes is blocked by mountains.

This upper limit of the Alpes-Maritimes is marked by the medieval town of **Entrevaux**. Its fortress, with walls running up the hillside, was designed by Vauban, who was also responsible for the fort overlooking the port at Antibes. Entrevaux's violent history gives a rough guide to the many different nationalities that have had an influence on the region. The town's roots are to be found among the Ligurian tribes, with the area subsequently invaded by the Greeks, the Celts and the Romans. Christianity took hold but only survived until the 10th century, when the Saracens invaded the region. Entrevaue became Spanish for a few years in the 16th century, only to end up in French hands a few years later.

In summer, the N202 is virtually deserted. But, in winter, the 85 percent of the population of the Alpes-Maritimes that lives near the coast suddenly acquires a taste for the wilderness of the *arrière-pays*. The right bank of the Var becomes a fashionable artery on weekends as the Niçois head to the skiing resorts: **Auron** and **Isola 2000** for the *sportif* in search of a conversation piece; **Valberg** and **Beuil** for the less pretentious or those more inclined to cross-country skiing.

Auron lines up a full range of slopes and trails for the inveterate skier, as does Isola 2000. However, modern chalets and flats are not what the back country is about. It may be fun for a while, but mountains that have had trees and rocks gouged out, with ski lifts dotting the pastures, are the inland

Ski chalet in Auron.

equivalent of the concrete structures on the coast.

Isola 2000 has no history beyond skiing. It borrowed its name from the village below and added a futuristic sounding number – the station's altitude – to match the space-age architecture of the buildings. It seems lurid next to the charms of other more historic villages in the area that double up as ski resorts. Beuil, for example, was once the seat of the Grimaldi family. During the Middle Ages, it was also the capital of the region.

Head down the mountain from Auron into the old town of **St-Etienne-de-Tinée**, and you rediscover rural France, with stone buildings and churches that enclose the most anonymous forms of artwork the region has to offer. The **Chapel of St-Sebastian** hides some of the frescoes that are the unsung treasures of the Alpes-Maritimes. They may be a far cry from the Picassos on the coast, but the artist was undoubtedly aware of their presence and value.

Most of the frescoes are hidden inside the simple stone churches found in villages all over the Alpes-Maritimes, such as the **Chapelle St-Antoine** in **Cians**. Many were painted on ceilings and walls by artists whose names have not survived the five centuries that their masterpieces have endured. Others are by figures whose fame has not gone beyond the art world, such as Louis Bréa, possibly because their work lacks the transportability of framed canvas. This makes it all the more unique. Few museums in capitals of the world can boast of having a Bréa. And the work of other Provençal artists such as Giovanni Canavesio or Baleison is not to be found in a Sistine Chapel.

Après-ski in Auron.

The isolated **Chapel of Notre-Dame des Fontaines** represents another common feature of the *arrière -pays*. There is every chance that a visitor will have to ask for the entrance key to the chapel at one of the two inns in nearby **La Brigue**. In this understated region, if a visitor finds a chapel locked, the normal procedure is to ask about the key at a nearby restaurant or *mairie*.

The greatest pieces of art in the region, however, have to be attributed to Mother Nature. One of her more unique works can be found on the way to the ski slopes at Beuil and Valberg. The **Gorge du Cians** makes a change from the usual – though inevitably spectacular – narrow limestone rift valley, with deep red slate as its main material. The mix of red from the slate and the green from the vegetation makes for a colourful contrast that does not abate even when the weather is dull; elsewhere, grey limestone bears an uncanny resemblance to grey clouds.

From here the **Vallée de la Tinée** becomes disappointing, though the **Gorges de la Mescla** at its southern end have spectacular slabs of rock overhanging the road and the river. But it does enjoy the saving grace of having been visited by the Provençal artists, as nearly every village church will testify.

To revive one's enthusiasm for the area, the **Vésubie Valley** is on hand. Within easy reach of the coast, it takes you from a Mediterranean landscape (pines rooted on parched rock and dusty soil, olive trees) into a fresher Alpine scenery (tall dark-green pines, waterfalls near La Boréon and Roquebillière, green pastures).

A few miles to the east of St-Martin-Vésubie is **Valdebore** and the small skiing resort at **La Colmiane**, complete with a few old chalets. Visitors like to call the area "Little Switzerland," which produces a quizzical look on the face of many locals – if only because the translation, "*Petit Suisse*," is a brand name for a type of cottage cheese. But it is, in fact, an apt reminder that the region north of Nice was part of the kingdom of Savoy for several centuries. The same kingdom covered what is today the western half of Switzerland.

The valley is rich in revolutionary history as well. Republican soldiers on their way south to Nice in 1793 were attacked here by a small local army, which used the area's geography to its advantage. They literally sneaked up behind the enemy and pushed them off the cliff face. The location of this minor setback to the French Republic is known as the Saut des Français, or "the Frenchmen's Leap."

The Vésubie Valley is also lined by several crumbling and precarious mule tracks once used by smugglers. One of the substances transported along the tracks was salt, brought in by traders from Italy anxious to avoid the *gabelle*, a tax on salt imposed shortly after the Revolution.

St-Martin-Vésubie offers cool respite as a mountaineering centre in summer. It is a sleepy and unpretentious small town disturbed only by the rush of water down a one-metre-wide canal that bears a distinct resemblance to an overgrown gutter. All its native inhabitants seem to have the rough and thick skin that goes along with decades spent in the mountains, yet are unmistakably Mediterranean by dint of their olive complexion.

In summer, they sit by and watch the hikers leave the Bureau des Guides de la Haute-Vésubie, generally on a trip into the nearby Mercantour. Their winter routine changes little, though much of

The Madone d'Utelle.

the watching is done from behind closed windows (temperatures can be over 27 degrees centigrade colder than in the warm season). Summer for a Vésubian starts on 2 July, when a procession wends its way out of St-Martin carrying an 800-year-old wooden statue of Notre-Dame-de-Fenêstre 7 miles (12 km) to the sanctuary at **Madone de Fenêstre**. Winter begins when the procession makes the return journey in the third week of September.

Utelle, at the foot of the Vésubie Valley, has a similar religious vocation. Its sanctuary, the **Madone d'Utelle**, is on the hill above the village, at a height of 3,786 ft (1,154 metres). The view is impressive, taking in the whole of the south and west down to the coast. Looking to the north and up the valley, you can see the gradually increasing height of the mountains.

The areas nearer the coast are already becoming more developed. The greater width of the lower valleys has allowed for the sprawling and ramshackle development of large towns like **Grasse**. On a day without wind, Grasse lives up to its reputation as the perfume capital of France; a sweet aroma lingers in the air from the three large perfume factories dotted around the town.

Fragonard, **Galimard** and **Molinard** offer guided tours for visitors, but much of the finely tuned creation of scents for the Paris couturiers is carried out in small unnoticed laboratories on the Cours Honoré-Cresp. And don't expect much antiquity beyond the huge copper vats used to distill perfume. Their modern factories have to keep up with a cost-efficient 20th century. History is provided at the **Museum of Perfume** – another of those sights where, if the door is locked, a call 50 metres down the road at the **Musée de l'Histoire de Provence** gains admission.

Beyond perfume, Grasse is little more than a provincial town. Social climbing is literally that in Grasse. The rich live further up the hill, cloistered in red villas surrounded by cypress trees, peering out occasionally at a panorama of Cannes. Down below, in the darkened narrow alleyways of the **Old Town**, urchins scurry from one set of steps to the next, occasionally tripping over the odd shopper. Washing hangs from windows, and families gather outside doorways to chat.

Grasse suffers from faded glory; the days when it welcomed regular holiday visits from the British monarch, Queen Victoria, or Napoleon's sister, Princess Pauline, are distant. The elegant cypress trees are confined to the gardens of the villas above and the remaining palms are scraggy. The casino is being rebuilt in a bid to recapture some of that swinging clientèle that made the Côte d'Azur a cliché.

Grasse may sound unprepossessing, but it acts as a useful start for the art-lover's trip through the Alpes-Maritimes. Famous artists came and went through the region, leaving their mark in the form of museums and collections. Several are to be found in the lower part of the *département*; the civilised flower beds in the garden, there only to be shown off to visitors.

he altar
ithin.

To go with the lavender, mimosa and jasmine that grows on the surrounding terraced hills is the town's **Fragonard Museum**. Jean-Honoré Fragonard was as torn between Paris and Grasse (where he was born and where he died) as he was torn between the charms of several wealthy consorts. Paris offered life, while Grasse offered light. But Grasse has managed to hold on to many of the 18th-century artist's floral and voluptuous works, thanks to a series of paintings commissioned by Madame du Barry. They also possess the *oeuvres* of his less-renowned son and grandson.

All in all, this is not a town that shouts its interest at you. You have to find the hidden gems. In a dark corner of the **Cathédrale Nôtre-Dame-de-Puy** hang two Rubens paintings that have belonged to the town for over 150 years.

If the tourist had not been invented, towns like nearby **Vallauris** would not exist. And, if Pablo Picasso had not lived, the tourist would never have come to Vallauris. Today, Vallauris is a largely residential town, hidden in a valley of its own barely two miles from the sea. The mimosa on the surrounding hillsides struggles for a revival after having suffered the winter of 1984, the sharpest known to a living Azuréen. At least the orange groves survived, leaving some idea of the summer light that lured Picasso away from the more arid inland sprawl of Mougins.

Ceramics have been the livelihood of Vallauris for four centuries. The main shopping street is thick with tacky shops selling plates, bowls and other artifacts. Uphill, the Place de la Libération contains Vallauris's two museums. The **Musée National Picasso**, with the *War and Peace* mural that the artist finished in 1952, is one. (Not to be confused with the larger Picasso Museum in Antibes. Picasso left some 60,000 works of art altogether and a will worth over 1,200 million francs when he died in 1973 – more than enough to fill several museums.)

Next door, the other, the **Musée Municipal**, devotes itself to *"l'art ceramique,"* and exhibits some of the work

Painting perfumed soaps in Eze

that the effusive Catalan artist accomplished during his six years on the Rue de Fournas. The Madoura Pottery – a few yards away from the summer bustle on the main Avenue Clemenceau – retains the copyright to Picasso's ceramics, dating from the friendship between the artist and the workshop's owners, the Ramie family.

Fernand Léger adopted **Biot**, a *village perché* (perched village) in a valley about 2 miles from the sea, as his outpost. Shortly before his death, he bought the land that was later used by his wife, Nadia, to build a museum for his works. The **Musée National Fernand Léger** is a cubist temple, with a vast ceramic mosaic on the outside, and the permanent home to nearly 400 of the artist's paintings, carpets, stained glass and ceramics.

The stark features of the Léger museum, the creation of a Niçois architect, stand out against the impractical charm of the perched village of **Biot** – 500 years old, maybe, but also lived in, which explains why all the houses are restored. The steep cobbled alleyways are lethal once the rain falls, allowing the visitor to slip and slide with his precious cargo of Biot glass and pottery. Biot is now built on *l'artisanat*, and sells to tourists.

Yet, a brief look through one of the gaps between the ramparts shows still another landscape for the Alpes-Maritimes. Its own private valley with a view reminiscent of Tuscany: cypress trees, the occasional goat wandering in between the villas and swimming pools.

St-Paul-de-Vence, though further inland, shares a similar view. Countless celebrities chose the valleys below Vence and St-Paul to live in – a cheaper alternative to the hills of pricey Cannes. The Mediterranean vegetation is still there, emphasised at times by the cultivated sprawl of cyclamen and primula in the gardens. The curious bright blue dots add to the impression of colour brought by the flowers in the spring and summer light, though the feeling dwindles somewhat when a more attentive glance reveals that they are swimming pools.

St-Paul is a larger version of Biot, albeit with more cachet. Rather than pottery, the emphasis here is on artists' studios and medieval history. There should be no fear of the cannon that menacingly points towards you at the entrance to the village; the favourite ammunition in one minor skirmish with the inhabitants of Vence was cherry stones. Nevertheless, the village had a more serious function through the centuries as a fortress guarding the entrance to the Var region.

Art has lived with St-Paul for hundreds of years. The church houses a Murillo, a Tintoretto and works by other Italian masters. Nevertheless, it wasn't until the 1920s that the village became known as a centre for art. The **Auberge de la Colombe d'Or** was a meeting place for a bevy of artists – including Picasso, Utrillo, Bonnard, Chagall, Modigliani, Soutine – who left behind canvasses, as the tale goes, sometimes to settle long-running accounts. Whatever the truth, the Colombe d'Or has now assembled its own priceless – and private – collection of paintings and its own ceramic mural by Léger. It would have pains, however, to rival the collection to be found further up the hill.

The **Fondation Maeght**, arguably the most interesting of all the museums that the Côte d'Azur has to offer, rivals the best art collection in the world. It is unique in that the whole place was created as a temple of modern art and that many of its works were specially commissioned.

The building's brick and white concrete exterior was designed by the Spanish architect Jose-Luis Sert, who was heavily influenced by Le Corbusier, and is partnered by Miró sculptures. Indeed, all the structural features were devised to promote an understanding of 20th-century art, and the architect worked directly with some of the artists, principally Miró and Chagall.

Aimé Maeght's position as an art dealer undoubtedly helped him in his efforts to assemble the collection by 1964. Few people can boast of having a Chagall with the personal dedication "*à Aimé et Guigette (Marguerite)*

Maeght." Fewer still could have had the friendships, knowledge and money to integrate mosaics by Miró and Braque or a garden full of Giacomettis into the whole and then to continue collecting works of the following younger generations. Some of the holdings move out on special exhibitions, but the collection is so vast there is always something to replace it. Equally, there are special exhibitions from February to October.

Back in St-Paul, lesser mortals can at least sit opposite La Colombe at the **Café de la Paix**, which looks as if it provided the model for the French café in countless Hollywood productions. Amid all the contemporary richness and revelry, it is worth remembering that St-Paul's history is rather more killjoy. When King François I built the ramparts of the village in the 16th century, he uprooted all the inhabitants and packed them off to live in nearby La Colle-sur-Loup.

Over 100 years later, new residents arrived in the form of the monks of the Ordre des Pénitents Blancs. The order, created by the Bishop of Grasse, lent its name to countless chapels and churches throughout the region. Mercifully, the monks, clad in hooded white gowns, no longer parade eerily through the streets to celebrate the Lord's Supper.

Although **Vence** is larger and less artistic than St-Paul, the modern-day art pilgrim will end his or her journey by making straight for its **Chapelle du Rosaire**. Designed by Henri Matisse in 1951, the chapel is probably the most recent example of the kind of religious patronage that brought artists like Michelangelo to decorate churches throughout Italy.

In design it is Provençal and simple. Rows of tall and narrow arches form the windows, and the tiled roof is dominated elegantly by a tall and thin wrought-iron cross. Overall it is strikingly pure and bright, with the only colour in its white interior brought by the stained-glass windows. Matisse called it his most satisfying work.

Vence is a large town (21,000 inhabitants) that is lively all year round. Its

Vallauris master potter, Roger Collet.

origins date back to the 5th century, and its turbulent history matches that of Entrevaux: Ligurians, Lombards, Romans, countless Christians and Saracens plus the Germans and Italians of World War II all passed through the town, and some destroyed (the Lombards flattened the town on the two occasions that they occupied it). The 15th-century walls of the *vieille ville*, within the modern town, betray its former feudal vocation.

Yet, the historical instability of the town has lent it curiosity value. The **Cathedral** in Place Clemenceau was built on the site of the Roman temple and mixes a patchwork of styles and eras. It possesses a simple baroque facade, some Byzantine stonework, Gothic windows, Roman tombs and a mosaic by Chagall. Indeed, Vence's religious offerings are as refreshing as the fountain in the Place du Peyra and the bright atmosphere in the summer. Despite its popularity, the town retains a purer feeling of southern France, with its vegetable and fruit markets and the restaurants spilling into the squares. For once, the tourist becomes incidental.

Tourette-sur-Loup apparently has found a simple way of keeping the hordes at bay. Traffic can only pass through and is prevented from venturing into its steep and narrow alleyways, allowing geraniums to flower on the stone walls in the spring and summer. This *village perché* overlooks the southeastern end of the Loup Valley, though the Loup River itself is a short distance further down through olive trees and, in spring, fields of violets.

Actually, the reason behind Tourette's relative emptiness, even in the height of summer, is probably just that it lives in the shadow of its more illustrious neighbours, Vence and St-Paul. Arts and crafts are carried out without the glare of renown, although with equal skill. The potters, painters and wood-carvers have to be sought out. Most welcome the public, but their only sign is normally the open door of a village house. Few attempt the external displays to be found in Vallauris or Biot.

ubuffet
ountain in
ne
oundation
laeght.

The town's history is more tranquil as well, having attracted few of the attacks by Saracens or Lombards experienced by its other perched neighbours. Today, the only time the village ventures to cry its fame comes with the Festival des Violettes in March, a processional celebration of the flowers in the surrounding countryside. But March is still the low season as far as tourism is concerned, with only a trickle of outsiders making their way to Tourette for the festivities.

There are no ramparts, walls or battlements surrounding Tourette. The only war-like evocation comes with the 15th-century château, once the entrance to the village and now in the middle and occupied by the Mairie. A belfry marks the archway that leads through to the cobbled hills of the **Old Village**. The church stands in the main square and houses a Bréa triptych – an apt reminder that Tourette is on the edge of the *arrière-pays*.

Descending into the river valley at Pont-sur-Loup, you find yourself at the mouth of the imposing **Gorges du Loup**, the nearest of the inland rift valleys to the coast. It is also one of the most spectacular, with the torrential river scything its way through a deep gorge of grey rock and lush vegetation.

The **Cascade de Courmes** is halfway along the gorge, a waterfall that drops about 160 ft (50 metres) into a pool by a roadside tunnel. A pathway runs underneath the cascade of water, but years of erosion have made it mossy, smooth, slippery and treacherous. The start of the path is about 65 ft (20 metres) to the right of the waterfall, the steps sealed off by a rusting metal barrier. Another path, for the agile, begins on the opposite side of the road by the tunnel entrance and affords a different view of the waterfall after a short scrambled climb. Both are dangerous ventures.

Continuing north, the road crosses to the other side of the gorges. A glimpse of the river from the bridge that links the cliffs (be careful that you don't lean against a barrier that is rickety and rusting) bears witness to the link between the spectacular and the dangerous. Forty metres below, the Loup plunges through the rocks while the rusting hulk of a car's wreckage lies beside it.

More pleasant thoughts are revived by the smaller **Cascade des Desmoiselles**. And further up are nice spots for trout fishing. Trips in the river valley are punctuated by warning signs about sudden rises in the water level, depending on the mood of a small hydroelectric station upstream.

If the term *village perché* needed a perfect example, it would find it in **Gourdon**. One whole side of the town teeters on the edge of a rocky cliff, a natural rampart obviating the need for any fortress walls to repel invaders. The feudal **château** was restored most recently in the 16th century. Inside you are treated to a display of all the charms of feudality: weaponry, a dungeon – complete with a bed, apparently to let the torturers' victims rest in between sessions – and a small collection of medieval art and memorabilia.

The château is reached after a brief walk (no cars allowed) up into the village. A few steps more allow you to run the gauntlet of the main street, lined with small shops full of crafts and postcards. The main square lies just down this way, with an anonymous and simple 11th-century church. But the view from the end of the square is Gourdon's real delight: the Loup Valley to the sea, the Massif de l'Estérel to the mouth of the Var, hills and valleys. Apart from defensive reasons, the *villages perchés* were placed as they were so as to have a clear line of sight to allied villages, to signal impending invasions. But, nowadays, the small terrace next to the square has been taken over by a nice little French restaurant.

The road from Gourdon winds down to Grasse. It's narrow at the top as might be expected and at the bottom too. Curiously, in the middle, it widens into a large expanse of smooth pitch-black tarmac, though the road never takes on the kind of traffic that could warrant such an extension. A freak of planning, and a warning for the future.

In 1988, several villages around the Loup Valley began to voice concern at projects for a new autoroute running inland, ostensibly to relieve the load on the coastal motorway and to ease access to the Côte d'Azur. Flysheets on the matter stuck to local telegraph poles will undoubtedly be a feature of the region for years to come.

Finding a straight stretch of road on the eastern side of the River Var, where the Alps spill down into the sea, is virtually impossible. Immediately behind Nice, a cluster of rocky hills surround the **Mt-Chauve** (or "bald mountain"). The name was aptly chosen. Vegetation stops dead partway up, leaving a bald patch at the top.

The "monk's head" is occupied by a disused fort with a 360-degree view. From here, the distance between villages appears short, and you may optimistically think of a quickly completed excursion through the dusty hills. But you will find that the miles are often tripled as you slowly swing through a pitted series of hairpin turns from one commune to the next.

A Renaissance fountain in the Place Republicaine of **Contes** offers some refreshment. And therein hangs a tale of how the God-fearing inhabitants of this small village tried to chase the devil away. The devil, being a thirsty soul, was known to quench his dry palate at the village fountain. The Contois conspired to capture him by smearing glue over the square, and he was banished from the town.

Contes is rare in that it is not at the top of a hillside but on a spur that juts out into the Valley of the Paillon. The château is at the bottom of the village and retains its Provençal name, Lou Castel. But Contes and its neighbours have grown. The area is more densely populated than the rest of the inner reaches of the *département*, as the villages within 15 miles (25 km) of Nice begin to assume the role of dormitories.

Few hamlets have stranger names than Coaraze, a derivation of the Provençal for *queue rasée*, or "shaved-off tail." Here, the devil appears anew. In search of a new home after his banishment from Contes, he entered this medi-

'uppy love
n Utelle.

eval village. The inhabitants once again took exception to his presence and this time captured him by his tail. He managed to struggle free, but his tail was ripped off in the process.

Despite its medieval setting, Coaraze has taken on an artistic zest that is more akin to the towns on the other side of the Var River. On the wall alongside the cobbled stairway leading up to the church are three ceramic sun-dials, one by Jean Cocteau, the writer, filmmaker and artist.

On the edge of the village, the **Chapelle Bleue** furthers the strange mix. From the outside, it appears to be no more than one of the interminable number of ageing chapels in the area. But the inside has been well-restored, with post-war frescoes decorating the walls. Everything, including the light through the stained-glass window, is blue, which is a reminder of the Côte d'Azur rare in these parts – here, the light rarely takes on the clarity for which the region is famous.

Peillon offers far more classical charms. It is medieval once again, a period of history that seems rather sombre and full of strife. Peillon is painfully thrust on to the jagged edge of a small mountain side, the houses clustered together in such a way that from a distance the village appears to be a castle, with the church standing proud in the middle like a turret. Barely 9 miles (15 km) from the avenues of Nice, the little town has no streets, just alleyways and steps.

West of Peillon, within reach of the international border, the Italian influence starts. **Tende** actually was Italian for a long time – the town was handed back to France in 1947 after a plebiscite – but its inhabitants always professed to feel French. It looks Italian, however. The atmosphere is medieval, but not with the dismal stone walls of so many of the neighbouring villages.

Many houses have plastered walls, painted with matt colours that fade under the onslaught of the sun. Others use a green-hued schist, as befits an Alpine area using local materials. Indeed, the style is more Romanesque, and each has

Farmers' wives in Berthmont-les-Bains.

a balcony and an overhanging roof.

Tende, like any self-respecting town in a predominantly Catholic region has a church. But it is a different type than its western neighbours, with a belfry shaped like two stacked barrels and a small roof. Built in the 15th century, green columns protect its doorway.

La Brique, a brief excursion away, offers a similar history. In fact, so does most of the **Roya Valley**. This valley, which heads south from Tende towards Breil, was retained by the Italians until October 1947, because it was the only link with the Mercantour and the hunting grounds of the king of Italy. The current border has been designated according to the water table. No tributaries cross the frontier on either side.

Saorge is a spectacular sight. Squeezed onto a cliff face at the entrance of a gorge, it is the only town in the region to have lent its name to a gorge. It takes a trek by foot to reach the centre, through some of the steepest cobbled alleyways one could imagine in an urban, albeit small, area. Some people do still live here, but the young are quick to move out. The contemporary exodus contrasts with ancient times, when it seemed like everyone was trying to invade the fortified village – unsuccessfully. Such was its reputation for impregnability that they eventually all stopped trying.

Saorge-style ravioli.

The steepness does not rule out the existence of a Renaissance church, **St-Sauveur**, with an altar of red and gold. Bear in mind that the church organ dates from the 19th century, and some poor mules had to carry it up the hill. Saorge has not forgotten the contribution the animals made (and still make) to life in the village. One of the village festivals is dedicated to Saint Eloi, patron saint of the mules.

Visitors may moan about the steep alleyways but, for the Franciscan monks that dwell in the monastery, it helps create a tranquil and serene atmosphere. The monastery rests in a small square at the top of the village, among a cluster of olive trees. Below it, the valley apparently has a good echo, a feature that allowed 20th-century soldiers to communicate with a nearby château from the ruined castle above. The Franciscans only returned to their home in 1969.

By some quirk of administration and diplomacy in 1947, the frontier south of Breil was not arranged according to the border rules applied to the section further north. The Roya flows happily through the Italian border and into the sea at Ventimiglia. So does the main road, and **Breil-sur-Roya** is the last chance to turn off to head towards Sospel and Nice.

Breil produces olive oil and is the centre for a type of olive that is only found in the Alpes-Maritimes, *the cailletier*. The *département* is also unusual in that it cultivates olives up to a height of 2,600 ft (800 metres), twice as high as anywhere else. Production of olive oil, however, is quite limited, about 285 tonnes per year. Most of it is used to make high-quality cooking oil, and little of the production falls to industrial use. Breil itself lays claim to more than 40,000 olive trees and celebrates a lively grand Fête de l'Olive at harvest time, some time in October.

The town of Breil makes for a restful stop. The drama of Saorge is past, and here the urban centre lies on a flat expanse of valley, with the Roya running wide and calmly through it.

Sospel shares the tranquillity and the olive groves, if not the river. The Bevera, a tributary of the Roya, runs through Sospel instead, leaving a series of islets around an old bridge. The bridge is an oddity, with a toll gate in the middle that was reconstructed after World War II, when it was destroyed in a bombing raid. North of the bridge is the oldest part of the town, a close cluster of buildings with wooden balconies. By the river there is still an old sheltered washing trough filled with water even at the height of summer.

The setting for Sospel is Italian: the valley with its olive groves, the rivers meeting at the town with their islets, the toll gate on the old bridge. The impression is reinforced on the southern side by the square around the church. Of no particular shape and bare, it boasts a baroque church on one end. All the buildings are plastered and painted in weathered shades of orange, yellow, ochre or red, with contrasting green shutters. This is a style predominant in the east and all the way down through Italy, but which disappears just a few miles to the west.

The trick is to head south towards the coast at Menton, a town with similar architectural appeal. The road climbs to the **Castillon Pass** then winds interminably through a small dry valley. Along the way, **Piène** offers a reminder of the relatively underdeveloped world we are leaving. An Italianate square similar to Sospel but smaller, its unkempt buildings only escape tattiness by virtue of being baked by the sun. The church is odd on the outside, possessing a bell tower with strange pointed horns on all four corners. And tacked beneath the arch of the belfry is a modern clock. The rest of this dusty, deserted and humble village, however, seems to have missed modernity.

Winter's dusk.

Twelve miles (20 km) beyond the Castillon Pass, you drop below the motorway into the populous Menton, a shock after the quiet villages of the interior. You might prefer to rise away from the coastal sprawl towards the sanctuary of villages like **Castellar**. Moving west, with the ridge overlooking the Mediterranean on your left, you pass simple hillside villages like **Ste-Agnès** and **Gorbio**.

Here, within a few kilometres of one another stand two different worlds. To the south is the developed coast, peppered with villas and shops in between the pine trees. Just north perches the simplicity, quietness and relative poverty of the back country.

La Turbie is on the edge of the over-developed world. Sometimes it is caught in the swirl of clouds, lending a rather more sinister vein to the ruin of the **Trophée des Alpes** and its demonstrations of trained eagles. Set on the Grand Corniche (known as the Via Julia by the Romans and Napoleon's favourite way to Italy), the Trophée is the bait, and the Grande Corniche is the fishing line. The tourists reel themselves in.

The Trophée des Alpes originally was presented as a reward by the Roman Senate to the Emperor Augustus for his successful campaign against the remaining rebellious tribes of Gaul. Only one side with four columns remains to bear witness to the imposing 50 metres the monument used to measure. This section was restored by archaeologists, using original stones, at a height of 105 ft (32 metres). Inscribed inside are tributes to Augustus and a list of the 44 conquered tribes. Adorning the Trophy's walls are quotes from Virgil and one from Dante's *Purgatory*.

Below the monument lies the **Eglise St-Michel**. This church is a recent 18th-century offering, with a domed top to the bell tower and a clock. Baroque styles use more colour, and this one is no exception. Inside, the red marble is extensive, bordering on the gaudy, insolent style that might befit a film star's villa in St-Tropez.

The Grande Corniche is the highest of

Villa in La Turbie.

the three corniches that run parallel to the coast. Of them, it's also the longest route from Monaco to Nice, though not the slowest since heavy traffic clogs the coast road.

The Moyenne Corniche is the quickest and probably the most stylish of the three routes. Alfred Hitchcock once caught a bus here, playing an extra in one of his films, *To Catch a Thief*. In that same movie, Cary Grant lived alongside the road, a retired cat burglar, residing in this part of the Côte d'Azur because he wanted to avoid the bustle of the coast and the big towns. Who could forget Grace Kelly – the Hollywood star who was having a real-life romance at the time of filming in 1956 with Prince Rainier of Monaco – as she swooped along the curves of the Moyenne Corniche in her roadster? She died tragically on that same corniche in 1982.

Whether Hitchcock stopped and wondered if he should fit the village of **Eze** into his film is not recorded. Eze village is impaled on a rocky spike by the road, a dramatic location suited to cinema's master of suspense. It is about 1,300 ft (400 metres) above the sea, which in summer shimmers silently and distantly below. Corsica is said to be visible from the highest point.

The German philosopher Friedrich Nietzsche is thought to have found the inspiration for his final book, *Thus spake Zarathustra*, while walking down a pathway to the sea here. Nietzsche retired from active life in 1869 due to syphilis and spent the rest of his time in Italy, with winters in Nice. By the time he started writing *Zarathustra*, he was already edging into insanity, as illness ate away at his brain.

Nietzsche's presumed path took him through an area of bushes and pine that is now bare, with a yellowing and, in parts, charred appearance. In 1986, the area's worst forest fire swept along this part of the coast. Its flames crossed the Moyenne Corniche, and the blackened landscape of sparse pine it left behind serves as a tragic reminder. The Alpes-Maritimes suffers less from forest fires than the neighbouring Var, but fires are still a summer hazard in this tinder-dry landscape – often fuelled by no more than a castaway dog-end.

The entrance to Eze through a small and easily lost archway comes after a short winding climb from the main road. Few other *villages perchés* command the popularity of Eze, and pushing through the narrow streets at the height of summer can be trying. The architecture (that medieval stone is familiar) hasn't changed, it has just been restored and cleaned up to house the souvenir shops. Where Eze gains is with the smell that lingers in its confined space. There are several restaurants in the town and a *crêperie* and bakery. They don't need to advertise – by the time you have been through the church and the chapel and the famed **exotic garden** at the summit of the hill, you'll be at their doorsteps, no matter how expensive the visit may be.

Eze also houses at least one expensive hotel. Honeymooners need look no further. The rooms match the price, as does the splendid view over the sea

***Boules* player.**

from the windows. Breakfast? Why, "*croissants sur la terrasse, bien sûr*," as the air warms up in the sun, and the turquoise water glimmers silently below. There are only two real reasons to stay in this hotel: frivolous indulgence and romance.

The strategic skills of the medieval architects who built the *villages perchés* are to be admired. But their efforts seem to have been somewhat wasted. Few of the villages were impregnable and, like Eze, most were at varying times occupied by the region's invaders. In the case of Eze, the multitude of changeovers has left historians confused about its actual origins. Some say the site was found by the Ligurians in 500 BC and the village built later during the Roman occupation. Others believe the village was founded by the Saracens. At the very least, all agree that the latter occupied it at some point.

A more recent creation for this part of the back garden of the Côte d'Azur is Eze's cactus patch, or "*jardin exotique*." This garden was planted in 1950 around the ruins of the château, which had been dismantled under the orders of Louis XIV in 1706. Round cacti, hairy cacti, bristling cacti and flowering cacti share the garden with one-inch-long ants and the snails that have managed to snuggle in between the plants prickly spines.

Five items characterise the Côte d'Azur: the sea, the light, the weather, the culture and the crowds. Inland, you simply take away the sea and the crowds and add the feudal history that created the *villages perchés* owned by the fiefdoms: the Grimaldis, the Lascaris around Tende and the Villeneuves near Vence. But if war and religion moulded the development of the area, so did its natural wilderness. By simply being mountainous and rugged, it has preserved its beauty and kept crowds away.

In 1988, the regional council decided to promote tourism in the *arrière-pays*, in an attempt to revive the impoverished economy of the area. Fortunately for those who do visit the area, their effort has so far failed.

Corniche above Monte Carlo. Following pages: the "nose" at Fragonard; bottles of Orange Blossom perfume.

The Perfumes Of Provence

As you drive down the Autoroute du Sud, the first anticipatory whiff of Provence wafts in through the car window – air laden with the scent of yellow broom, thyme, mimosa, lavender. More than anything, it is this heady scent for which Provence is known by poets and tourists. So, it is entirely appropriate that the Provençal town of Grasse has been the centre of the world's perfume industry for the past four centuries.

The gentle climate, rich soil and cradle of mountains that protect it from the north wind make Grasse an ideal place for flower production almost all year round. Golden mimosa blooms in March. By early summer, there are acres and acres of fragrant roses, ready to be picked. Jasmine appears in the autumn. And, high above the town, the mountains are terraced with row upon row of purple lavender.

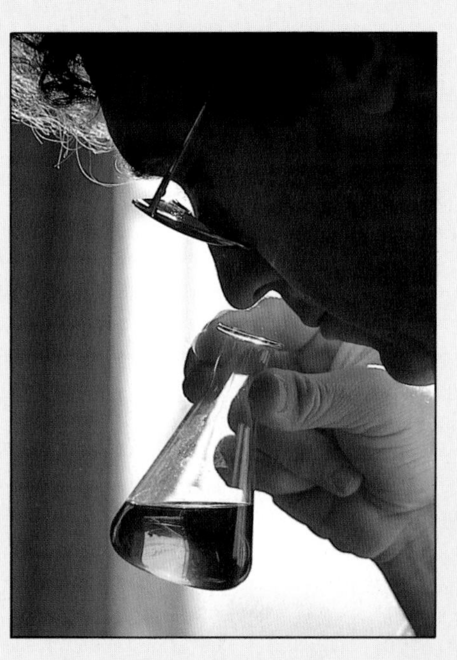

The perfume industry in Grasse originated with an immigrant group of Italian glovemakers in the 16th century. They discovered the wonderful scents of the flowers in the area and began perfuming their soft leather gloves, a favourite way to use perfume (along with pomanders and scented handkerchiefs) at a time when the odour of the general populace definitely required masking.

Demand for the floral perfumes steadily grew in the 18th and 19th centuries, and Grasse as a perfume mecca prospered. But local production of raw material declined after World War II. Competition from countries such as Turkey, Egypt and Bulgaria, where labour costs are much cheaper, proved decisive, and the gentle climate – so ideal for growing flowers – attracted many wealthy people to the area, pushing the price of land sky-high and causing many acres to be sold off as building plots.

Today, Grasse is better known for improving raw materials imported from other countries. Nonetheless, you can still see vast mountains of rose petals, vats of mimosa or jonquils and spadefuls of violets and orange blossom just picked and waiting to be processed each morning.

The flowers must be picked early, when the oil is most concentrated, and delivered immediately. It takes enormous quantities of the blooms to produce even the tiniest amounts of "absolute" perfume: about 750 kilos of roses for just one kilo of rose "absolute," about 4,000 kilos to produce one kilo of "essential oil."

There are a number of different methods used to create the "absolutes" or "essential oils," which the perfumer combines to create a fragrance. The oldest method is steam distillation, which is now used mainly for orange blossom. Water and flowers are boiled in a still, and the essential oils are extracted by steam – the heat breaks open the cells of the petals and leaves, and the essential oil floats on top while the water sinks to the bottom.

Another ancient method still used today, though it is very expensive, is "*enfleurage.*" The flowers are layered with a semi-solid mixture of lard, spread over glass sheets and stacked in wooden tiers. When the fat is thoroughly impregnated with the perfume, the scent is separated out by washing the *axonge* with alcohol. *Enfleurage* is used particularly for jasmine and tuberose.

A more modern method is extraction by volatile solvents. Perfume can be extracted from flowers and plants by immersion in alcohol or a volatile hydrocarbon solvent such as petroleum ether. Only the perfume, colour and natural wax dissolve, leaving behind the cellulose and fibre components of the plant. After recovery of the solvent by distillation, the product is perfume. The wax is separated, leaving a final concentrate called the "absolute."

The highly trained perfumers or "noses" of Grasse can identify and classify hundreds of fragrances. In creating a fragrance, a perfumer is rather like a musician, using different "chords" of scent to blend together in harmony. The desired result is a complex perfume that will radiate around the body in a slow process of diffusion – what the French call *"sillage."*

A good perfume may include hundreds of different ingredients to achieve the correct balance, using powerful animal fixatives like ambergris, civet and musk to capture the delicate, ephemeral fragrances of the Provençal hillsides.

EAU DE FLEURS D'ORANGER
RÉCOLTE 1988
A. MONTEUX
PARFUMEUR DISTILLATEUR
VALLAURIS (Alpes Mmes)
EAU DE FLEURS D'ORANGER
RÉCOLTE 1987
A. MONTEUX
PARFUMEUR DISTILLATEUR
VALLAURIS (Alpes Mmes)

THE CÔTE D'AZUR

The French Riviera – that fabled stretch of Mediterranean seaside that runs from St-Tropez to the Italian border – is more steeped in myths, sensuality and surprise than just about anywhere else on earth. The name alone seems to excite the senses, and almost everyone, whether they are long-time residents on the Côte or they merely once spent a week in Cannes in the 1960s, likes to boast familiarity with the territory.

Whatever their experience and exposure, most people are confident they can recount the myth and recognise the reality of the Riviera. And, at any rate, the components of the myth are easy enough to enumerate.

Great Gatsby parties at private villas, with vast lawns sloping down to the deep blue sea at Cap Ferrat and the Cap d'Antibes. Breaking the bank at the casino in Monte Carlo, before returning to a luxurious suite (with a view of the château, *bien sûr*) at the Hôtel de Paris. Getting a *laissez-passer* (general pass) to the Cannes Film Festival each May and basking in the late afternoon sun in Nice's Cimiez Arena every July while listening to laid-back jazz.

The stories go on and on. Living next door to Brigitte Bardot in St-Tropez and descending to the Place des Lices for an afternoon game of *boules*. (Officially the French Riviera is in the Alpes-Maritimes *département* of France, but both friend and foe tend to include St-Tropez, which is in the Var, and Monaco, which is an independent principality, when they throw the term around.) Or, for the artistic, enjoying the light (Renoir said it was "the light, the light" that made the Riviera so special) and striving to become an artist to rival Matisse, Picasso, Signac and other past painters of the region.

The myth is, in fact, still reality. All of the above occur, or exist, today as much as they did in the past. Jet-setting, party-

Preceding pages: Menton. Left, Rivieran sisters on the go.

ing, gaiety, artists, casinos and Brigitte herself are not yet dead.

Not long ago, everyone who's anyone showed up dressed for the 1920s at a Great Gatsby party given at the Villa Araucaria in Cannes, a humble Riviera home that went on the market for US$8 million. American television superstar Bill Cosby regularly vacations at the luxurious Hôtel du Cap on the Cap d'Antibes and feasts on fresh, albeit expensive, fish at the nearby Bacon Restaurant. Princess Caroline spends a healthy part of the summer yachting on the Mediterranean, and Palm Beach in Cannes continues to feature cabarets under the full moon. Most people who live here don't even pinch themselves to make sure they're not dreaming.

Indeed, even its detractors agree that – from the point of view of weather, transport, services and, yes, light – the Riviera is certainly one of the best places to live or visit in Europe. The views, whether of distant corniches or near nude bathers, are splendid. The rocky cliffs, the hilltop towns, the markets, the national parks and a vast number of other amenities transform even the least romantic tourist into an unquenchable poet. Dufy's seascapes actually come to life.

When you have arrived for an afternoon of Alpine skiing in March, at the small resort of Gréolières, which is a mere 45-minute drive from Cannes, and you look down from the top of the mountain and see spring blossoming on the hills and summer at the beach, you have to marvel at your good fortune. The Riviera, myth and all, is quite a piece of work.

Relaxed but state of the art: It's no problem getting to the Riviera, because the Nice-Côte d'Azur International Airport, just a five-minute drive from the centre of Nice, is the second largest in France. Or you can arrive by the highways that link southern French cities with every neighbouring European country. Or take the high-speed train (TGV) from Paris or the milk train in from Italy or Spain.

Once you're here, for business or fun, you will be surprised by the modernity to be found around every corner. Telephone density in the region is among the highest in France, the number of hospital beds per inhabitant is well above the norm and there are excellent educational facilities, including American and international schools.

But the myth is not damaged by these state-of-the-art banalities. Even obstinate Parisians admit the pace is slower, the food is healthier and the grass is greener (everyone has automatic sprinkler systems, although these do nothing

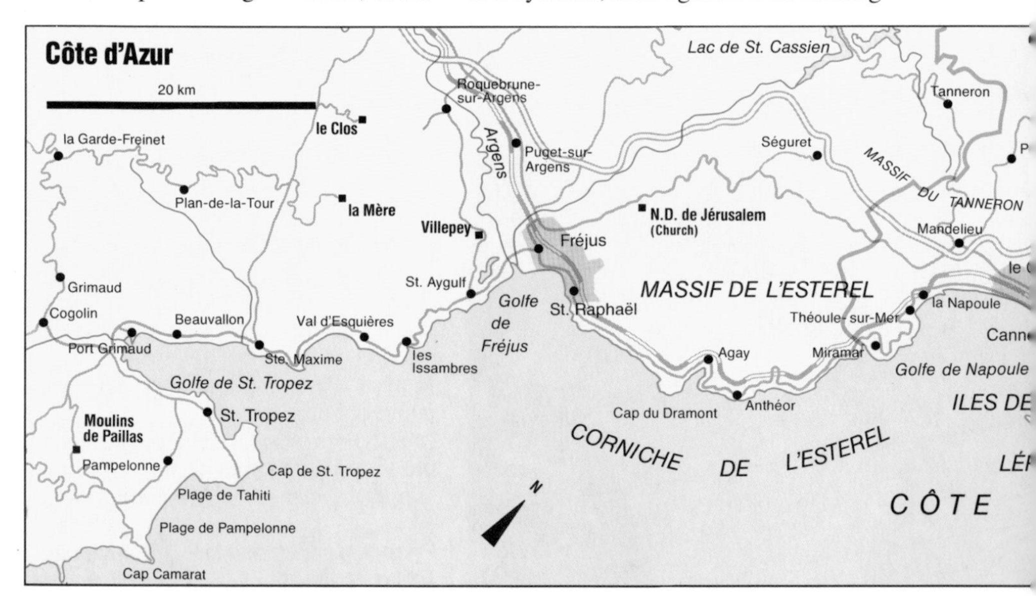

to combat the serious fires that occur every few years).

The relaxed pace is the key, of course, and both business and pleasure usually involve a round of golf, a sail on the Mediterranean or a run on the ski slopes rather than simply a sedate dinner at a restaurant (though there is no shortage of noteworthy business lunch and dinner spots) or a chat in an office (yes, there *are* offices).

Things are so easy going that the general mood is often referred to as "Mediterranean mellow." No one dares to look unduly hurried. Most local businessmen take their ties off between May and September and work in their shirt-sleeves. Frenchmen on the coast, largely because of the exposure they have had to an international population since the days of Queen Victoria, are much more likely to invite you to their homes for a pastis or Perrier than their Parisian counterparts.

And everyone still takes the traditional Mediterranean late-afternoon stroll along the palm-lined Promenade des Anglais in Nice or the Croisette in Cannes – those stretches of street paralleling the usually azure sea where fashionable swimwear (or the lack of it) is constantly on parade.

And everyone can get by here – although there's no question that this is one place where the rich really are different from the rest of the world.

The Riviera, its beauty so democratic in nature, is a great equaliser between the rich and the poor. You can spend a fortune on a meal at the Louis XV in Monte Carlo (indisputably the top restaurant on the Riviera and much better than the over-touted Moulin des Mougins) or get a perfectly enjoyable pizza and *salade Niçoise* or a catch-of-the-day almost anywhere – though there are frequent exaggerations about the freshness of the fish.

You can get a suite at the Byblos in St-Tropez, the Negresco in Nice, La Reserve in Beaulieu or the Hermitage in Monaco for a small fortune, or you can camp in the delightfully rustic Estérel (the comparatively virginal coast between Cannes and St-Raphael) for a pittance. You can shop on chic streets – the Avenue des Beaux Arts in Monaco, the Rue d'Antibes in Cannes, the Rue de France in Nice – or find lively, colourful vegetable markets and hidden antique stores in the smaller villages.

But is the Riviera really this seductive? Would Picasso have been more productive had he not spent much of his creative life in Vallauris and Mougins? Would Anthony Burgess and Karl Lagerfeld perhaps be less flamboyant if they didn't live in Monaco? Would Ein-

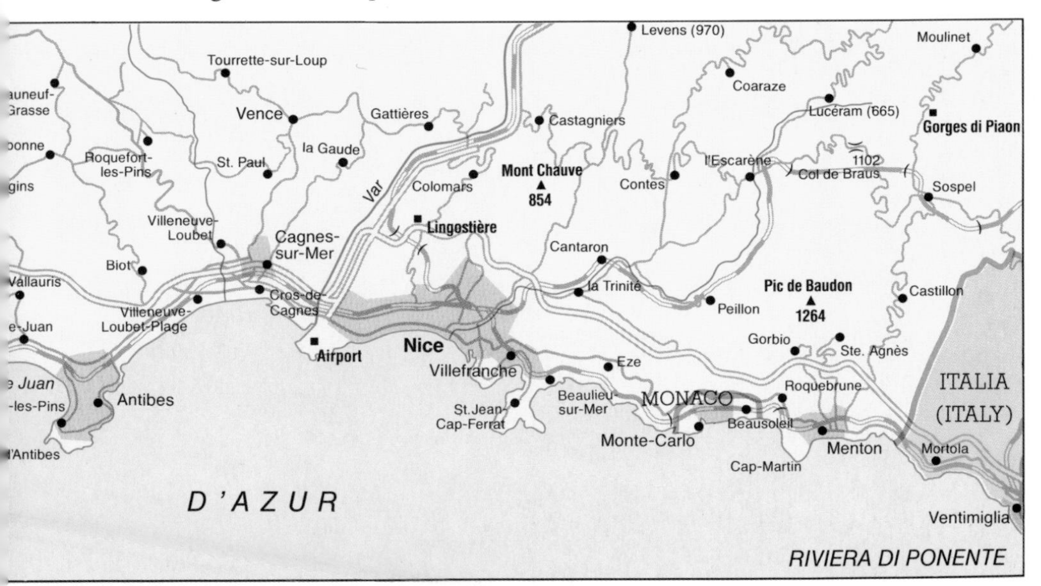

stein have been more creative if he were working with the many scientists and boffins based at Sophia Antipolis, the booming and fascinating 10,000-acre (4,000-hectare) technology park located just off the autoroute between Nice and Cannes?

No place for paleskins: Most inhabitants of the Riviera don't hide their pleasure in being able to boast the distinction of residing here throughout the year. While they each may have their own pet peeves (for example, it is virtually impossible to find a resident who doesn't complain about the influx of summer visitors), they generally tend to enjoy reinforcing Riviera myths and often make pale-skinned visitors from the north feel a little unwelcome.

The first thing locals or foreigners living on the Riviera let visitors know is just how *nice* it is to live down here. They enthusiastically recite the professional and personal conveniences of the area and immediately make you feel like an idiot for commuting in New York or London, contending with the traffic problems in Los Angeles or living with lousy weather in Paris, Stockholm or Moscow. But, face it, they're right. The Riviera *is* a good place to live and work, because most professionals (artists, bankers, brokers and lawyers) have chosen, and often claim they have made serious professional sacrifices, to make it home (which the French refer to as a "principal residence," in contrast to the many "secondary residences," or vacation houses, also to be found on the Riviera). Very few people are here against their will, which makes it both a productive and fun area.

The Riviera may not yet have turned into California, but the differences are fading fast, thanks to the preponderance of fast-food havens and autoroutes. The people aren't quite as beauty and health conscious as their sun-loving brethren in southern California, but the skin cancer scare certainly has not hit and being white and fat is definitely *out* between May and October. It is only a bit less *out* during the winter, when you can get by, just, with a facial tan from skiing.

A business lunch.

Indeed, there is even some slight discrimination on the Riviera if you show up at the beach in the middle of summer without a suntan and noticeably overweight. How much? Best not to try to find out unless you're one of the many northern businessmen with a good expense account to shield you.

At the same time, fortunately, the people living in the south are not sports fiends. They don't bore you with how far they have jogged or with what's new in weightlifting equipment. But it's rare to find someone who is not at all interested in one sport or another: The opportunity for exercise is so vast that it's hard to pass up.

In fact, it's difficult not to be tanned and in relatively good health when you pass a lot of time in the south of France, because so much of Rivieran life is spent outdoors. Even the most highly stressed businessman (there may be two or three around) will spend enough time in the sun to protect his tan.

The Mediterranean is itself the scene of numerous sports antics. Although polluted in parts (the new water-treatment plant near the Nice airport is intended eventually to make everyone forget pollution), it is still ideal for diving, swimming (a favourite, non-polluted spot is on the Estérel) and windsurfing (the winds in the summer are not overly strong, and most locals prefer the Ile de Giens near Hyères). There are numerous marinas at which to keep, rent or buy a boat, and one popular and somewhat relaxing sport is, in fact, simply just walking through the docks to look at other people's crafts.

Of course, you don't need to go in for exerting activities at all, if you don't want to. The most relaxed sport is *boules,* and every year the local daily, *Nice-Matin*, features a large colour photograph of celebrities playing in front of the Café des Arts in St-Tropez. There are also fine marked hiking trails and numerous camping sites throughout the back country. You can either follow the national *grands randonées* or take a stroll along the Balcon de la Côte d'Azur.

Left, limos. Right, Carlton Hotel.

If you decide that all activity is too strenuous, you can just sit on a beach. There's no doubt that boy and girl watching is also a big sport, and a lot of people get exercise just hanging out at places with weird names like "Waikiki Plage." Naturally, not even hanging out comes free, and it can cost over 50 francs to get onto a private beach and lie on a mattress under an umbrella for a full day.

The downside of Paradise: The French Riviera may be more than the Cannes Film Festival, the celebrity-filled Principality of Monaco, the exquisite Chantecler Restaurant on the Promenade des Anglais in Nice and bronzed beauties sunning on the beaches near Antibes. But do keep in mind that there are always pimples in paradise and that even the Riviera has its sour spots. The people attracted by the myth is the major problem.

Northern Europeans traditionally scurry to the Mediterranean sun and beaches during their August vacation. The resulting congestion on roads, on beaches and in restaurants exasperates most local residents even if it titillates crowd-seeking tourists. Scandinavians, presumably alone and contemplative during their harsh winter, seem to get a particular thrill out of body-to-body sunbathing.

And, while the Riviera does have its share of splendid villas, the cramped little housing developments (read secondary residences) make parts of it resemble an upmarket Calcutta. The closest thing to a building code is the presence of the Mediterranean itself, which, fortunately, limits expansion to the south (though Nice, Monaco and other communities have worked around this inconvenience by building on reclaimed land).

The much-touted slow pace also means local services have their ups and downs. Anyone with a New York mindset had better quickly adopt a more Mediterranean-mellow mentality. Shops still close between noon and three, and no one would ever pretend to be in a hurry or unduly respect punctu-

Monte Carlo casino.

ality. The exception, of course, are the drivers. They tailgate and run red lights more consistently than any other nationality this side of Taiwan.

Naturally, the area, which has a number of inordinately wealthy inhabitants, has attracted more than its share of human sharks. Some are selling a load of questionable investment opportunities. Watch out. The sun has got to a lot of them, and the casinos (in Monaco, Nice, La Napoule, Beaulieu and other famous hot spots) may be a better bet for doubling your money than some of their risky ploys.

Build a house, for example, and you might find that many of the contractors go bankrupt during the construction phase and leave you holding the plans. Watch out for scams in the antique markets (the best and most reliable is on the Cours Saleya in Nice) and on the roads, where stolen goods, especially rugs, are sold for a pittance. Legal or illegal, smart or stupid, cash is preferred for most services, to avoid the value-added tax (TVA) and income declarations. In fact, it is rare to find the tradesman who prefers a cheque.

Even the beaches and sea aren't always what a lot of people expect. The beaches in Nice are covered with large flat rocks rather than sand, and some doctors advise against swimming in the sea despite the renowned new sewage disposal system. Nor is the weather all it is cracked up to be. It is extremely frivolous and can often be overcast, or pour, for days at a time. The mistral is a vicious wind, and snow has made seaside skiing possible for two out of three winters during recent years.

Indeed, some days the Riviera looks a bit worn and, like the rest of us, it is aging. In fact, you might even say it is over 100 years old. The name "Côte d'Azur" (somehow translated into "French Riviera") was coined by French poet Stephen Liegeard a century ago, and during the summer of 1988 the local authorities (headed by Jacques Médecin, mayor of Nice for almost 25 years) tried to create something special for visitors to the south. They acknowl-

Hanglider above Monaco.

edged the "100th Year of the French Riviera" with a number of special events – and omnipresent flags, shirts, bumper stickers and billboards.

Take the best with the worst and, once you wind up, there is something for everybody on the Riviera.

The big olive: The geographical focal point of the summer frenzy is **Nice**. The city attracts a healthy share of the region's eight million annual visitors, is the area's administrative capital and is occasionally alluded to as "The Big Olive." Most visitors do not come to the Riviera just to see Nice, but few of them leave without having included it in their itineraries.

Nice is not nearly as chic as neighbouring Monaco, nor as quaint as perched villages like **Eze** (a delightful spot with no cars and intimate touches like the fact that your luggage is taken up to the hotel on the back of a mule). Nor is Nice as celebrity-filled as St-Tropez. But it is the Riviera's largest city (almost half of the area's one million people live in and around it) and in the midst of a noticeable transformation to increase its amenability.

Founded by the Greeks, Nice was first embraced by English lords and Russian aristocrats as a winter resort. The boom in bathing and summer tourism didn't begin until the 1920s. Today, it blends a traditional Mediterranean-mellow lifestyle with a big push towards conventions (it is France's second largest convention city after Paris) and high-tech business (income from which now rivals the Riviera's annual revenue from tourism).

The state-of-the-art Acropolis convention centre and the budding $200-million crane-filled Arenas business complex, going up opposite the airport that was itself built on reclaimed land, are forcing inhabitants to readjust their priorities. The result: improved services, more hotels and restaurants, and an increasing number of off-beat and inviting activities.

Like most French towns, Nice has also created downtown pedestrian zones. Many of the mod shops and fast-food outlets are on the chic cobblestoned Rue de France, but the best way to discover the city is simply to get lost in the maze of narrow cool streets in the colourful **old city**, "le Vieux Nice," with its *trompe l'oeil* paintings, hole-in-the-wall boutiques selling perfume and carnival masks, and wealth of moderately priced restaurants featuring local specialities.

The **Cours Saleya**, filled with antique stalls and flower markets, is the centre of action in Old Nice. Locally grown flowers, olives, vegetables and herbs are sold, and early risers can hit the fish market on Place St-François.

Feeling hungry? Ilene Médecin, the American-born wife of the former mayor, says the *trucha omelette*, made from pine nuts and a spinach-like vegetable called *blettes*, is out of this world at Le Safari on the Cours Saleya. But *trucha* is just a starter. *Socca* (a traditional Niçoise chick-pea pancake cooked in an enormous round dish), pizza, *salade Niçoise* and other garlic-scented fare can also help fill up a dull afternoon and empty stomach.

Quiet beach in Menton.

Good, moderately priced restaurants featuring local fare include Acchirado (39 Rue Droite), Nissa-Socca (5 Rue Ste-Reparate), Le Demode (18 Rue Benoit-Bunico) and La Taverne (3 Rue St-François). For a more Michelin-starred cuisine go to Chantecler in the Hotel Negresco (37 Promenade des Anglais), where chef Jacques Maximin makes tantalising dishes using local ingredients, herbs and spices.

The only way to stay cool after dinner is at "in" discothèques like the usually private La Camargue (Cours Saleya), Chez les Ecossais (6 Rue Halévy) or L'Aventure (12 Rue Chauvain), all of which open around 11 p.m. Disco-goers usually leave at 4 a.m. and head to the beach for a pre-dawn dip, an act that reminds visitors to France that they are in Nice – not Paris.

The sun sets late during Niçois summer nights, and the promenade provides romantic views, despite occasional interruptions from landing aircraft.

Princes and tennis stars: Cannes and Monaco are Nice's biggest rivals and are somewhat similar to one another. If Nice is generally regarded as tough and feisty (which is an exaggeration), Monaco and Cannes are considered chic and chi-chi (in fact, another exaggeration).

Monaco, Europe's second smallest independent country after the Vatican – its 460 acres (186 hectares) make it smaller than New York's Central Park – has an idyllic perched-on-the-Mediterranean setting that makes it an unofficial part of the Riviera. But Monegasques go out of their way to stress their independence.

The Principality is often considered a hotbed of gossip, intrigue and royal shenanigans. But don't go to Monaco expecting theatrics – even though columnists certainly spare no ink regaling readers with tales of high-spending gamblers breaking the bank at the casino (almost never), beautiful princesses dancing into the night (rare), the rich and not-always-famous frolicking on and in the Mediterranean (occasionally) and small apartments selling for big millions (almost always).

ypical ousing in Ionaco.

Contemporary Monaco is both a more sedate-than-scandalous playground in the sun and a serious business community. The streets can be safely walked at midnight (wearing real jewellery makes a part of the myth that is Monaco reality), the omnipresent police force is more helpful than intimidating, the public services are exceptional and even the discothèques (like The Living Room or Jimmy'z) are rarely raucous. There is a definite lack of violence, screeching police cars and dens of iniquity.

Monaco is still ruled by Prince Rainier, whose family, the Grimaldis, have been running Monaco since 1297, when François Grimaldi snuck in disguised as a Franciscan monk. Rainier gets through the year on a $7 million budget and shows little sign of wishing to relinquish power to his son, Prince Albert, who is the Riviera's most sought-after bachelor.

Meanwhile, since her mother's tragic death in an automobile accident in 1982, a maturing Princess Caroline, now mother of three, has assumed the role of being Monaco's most important cultural and social force. As head of a variety of organisations, like the Monte Carlo Ballet and the Princess Grace Foundation, she is often seen performing official functions about town.

Princess Grace, the former American actress Grace Kelly who married Rainier in 1956, is entombed in the cathedral located on "the Rock," as the hill dominating Monaco is called.

The complexion of Monte Carlo is changing. There are a dwindling number of ageing dowagers and wealthy Greek yacht owners (if you want to feel like Niarchos, who keeps his yacht here, the 200-ft/60-metre *Sea Goddess* can be rented for $50,000 a day), but there is steady growth among members of the burgeoning financial community and still some star-studded foreign residents like fashion designer Karl Lagerfeld, writer Anthony Burgess and tennis player Boris Becker.

Monaco is worth a visit just to see the cool, cobble-stone streets on the Rock,

Waiting for the bus.

the luxurious suites in the Hermitage Hotel and Hotel de Paris, the basement aquarium at the majestic **Oceanographic Museum**, Charles Garnier's historic late 19th-century **casino** and Prince Rainier's pink-tinted **castle**, where there is a daily changing of the guard at 11.55am.

Sit on the terrace at the renovated Café de Paris in the Place de la Casino, rent a private *cabana* on the *plage* at the Monte Carlo Beach Hotel or drop in on the fascinating **Museum of Dolls and Automata**, with its miniature 18th- and 19th-century houses and collection of 400 dolls, and you might even want to make Monaco your own third or fourth residence.

Really want to live here? You'd be buying a veritable jewel on the Mediterranean. Twenty percent of Monaco is already built on reclaimed land, but even this unreal terrain constantly rises in value. A big penthouse (500 sq. metres) in Le Florestan sells for over $7.5 million. The many cranes in evidence are putting up new and luxurious apartment and business complexes like the "Monte Carlo Palace."

acades of illefranche.

There are only 30,000 people in Monaco, of which an even smaller 5,000 are Monegasque citizens. Most of the working population arrive from France and Italy by car or train each day, but you can also get there by helicopter from Nice Airport. Among the new attractions is the **Princess Grace Rose Garden** in Fontvieille, the latest addition to the Principality built mostly on reclaimed land.

If you really want to partake of the myth that is Monaco, shop in the **Place du Casino** and on the **Avenue des Beaux Arts**, only 500 ft (150 metres) long, at Cartier, Bulgari, Buccelati, Ribolzi and other extravagant boutiques. Nabil Boustany, the Lebanese entrepreneur who has completely rebuilt the Metropole Hotel, also has a number of glittering boutiques in the adjoining marble-floored, chandelier-laden commercial gallery.

Want to blow a fortune before you go to the casino? The Louis XV Restaurant on the ground floor at the Hotel de Paris, which chef Alain Ducasse has wondrously transformed into the Riviera's culinary hotspot since it opened in May 1987, is actually worth its weight in gold (or French francs, which is what the Monegasque use for currency).

Other good places to be seen dining are the Roger Vergé Café in the Winter Sporting Club's shopping arcade (Vergé owns the three-star Moulin de Mougins) or at Rampoldi's (very "in"), Le Saint Benoit (good fish and a view of the palace), La Coupole (where Yves Garnier is probably the second-best chef in town), and the Grill Room on top of the Hotel de Paris.

More than 250,000 visitors flock to Monaco each year, and hundreds of day-trippers arrive in tour buses to look at the sights. The heavily touristed Monaco of today is more likely to include attendees at an insurance conference than gigolos and high rollers heading for the casino tables.

The chi-chi: Cannes, which got its name from the canes and reeds in surrounding marshes that no longer exist, is "twinned" with Beverly Hills and often puts on some of the same airs. But it is, in fact, a small manageable town with a nice shopping street (**Rue d'Antibes**) lined with both internationally known and locally based boutiques, a pleasant beachfront (**La Croisette**), a pretty **carousel** behind the municipal casino, and mythically named hotels like the Carlton and the Majestic.

Every visitor to Cannes should take a walk along the **old port**, mingle on the shopping streets in the **old town** (there is a calming view from the dominating citadel and an interesting selection of archaeological artifacts in the **Castre Museum**) and get in some beach time. During the high season you may want to head for the water early because by mid-morning the sea will already be spilling over with bathers. And, of course, finding a parking spot on those days becomes an all-day event.

Summer is not the only time to be wary of in Cannes. The town also tends to become obnoxiously overcrowded during the Film Festival in May. The latest attempt to control traffic by changing the one-way streets has done little to eliminate congestion.

The two best restaurants in Cannes are the Royal Gray, featuring *nouvelle cuisine* from a chef trained by Michel Guerard, and La Palma d'Or at the Martinez Hotel, where you can eat on the terrace on the first floor – and get above the traffic.

Chic getaways: Go outside these three "metropolitan" areas and you, like everyone else, will soon discover your own favourite town on the Riviera. Each little village, usually with cool squares and fountains, has distinctive characteristics and idiosyncracies that either attract or repel visitors.

Most of the towns on the Riviera have pleasant pedestrian zones, medieval architecture and exceptional views. Again, all this seems irrelevant when you can't find a parking place in the middle of summer. But during the winter they are, must one say it, quaint.

St-Tropez (generally referred to simply as St-Trop) still features the most

Beach vendor in St Tropez.

celebrities *per capita* of any spot on the Riviera. Its 6,000 permanent residents (many of whom rely on tourism for a considerable portion of their income) take the summer fashion show in their stride and understandingly tolerate the gatherings at the Senequier Restaurant along the port and dancing until dawn at the Caves du Roy. Yes, Brigitte Bardot might have closed her fashion boutique, but she still protects her animals in her villa, La Madraque.

Ask the residents for the best restaurants, and they'll mention Le Chabichou (expensive) or Auberge des Maures, L'Escale and La Fregate. Culture in St-Tropez? There is the **Maritime Museum**, with everything you might want to know about Mediterranean shipping, and the "Annonciade" in the 16th-century **chapel**, along with a collection of modern French masters.

But St-Trop in the summer, when the population is about 100,000, is the Riviera at its worst. Beaches (with names like Tahiti Beach, Tropezin and Tabou) are polluted with bronzing bodies, and even the sea, the majestic Med, reeks of suntan oil.

Matisse would probably no longer title his painting of St-Tropez *Luxe, Calme et Volupté*, though he would presumably be pleased that there is still not a major highway or rail service into the town. If in St-Trop, the best time of day is dawn near the port or by the colourful markets in the **Place aux Herbes**. To get away, take a stroll in the nearby pine forests or the vineyards on the adjacent peninsula.

Another chic "getaway" (meaning place to be seen) spot is **Port Grimaud**, a modern city built to resemble a contemporary Venice that serves as a second home for the likes of actress Joan Collins. It was designed in 1966 and depicts the type of Venice that Walt Disney might have created. At least, however, there is no entrance charge.

Antibes (the name means "the city opposite," referring to its proximity to Nice), is more authentic and somewhat distant from the hustle-bustle. It features seafront **ramparts**, an old **fort**,

Sophia Antipolis.

Inside The Cannes Film Festival

The Cannes Film Festival ranks with the truffle as France's most misunderstood export item. Just as that fetid tuber, rooted out by swine, is erroneously viewed as an aphrodisiac, so too the Cannes Film Festival is falsely perceived as a leading event in the cultural life of the seventh art.

In fact, it is a gruelling two-week gathering each May of celluloid salesmen and a marathon homage to bad taste and conspicuous consumption that resembles nothing so much as the Loyal Fraternity of Water Buffalo assemblies once attended by Fred Flintstone and Barney Rubble.

If you're lucky, you've come to Cannes during one of the other 11 months. Consult your calendar. If you have erred, take the next train out to Nice. Should you choose to stay, there are a few things you should know. May weather in Cannes is cool and capricious. But the world's film industry doesn't throng to the city to catch rays and surf. Perhaps *you* do not tire of topless beaches and *bouillabaisse*, but those Riviera specialities retain little charm for the motion picture executives who have been coming to the festival year-in, year-out since it began in 1947.

One look at the row of tacky billboards that will disfigure the Croisette, Cannes's palm-lined main street, for the next fortnight should make things clear: for visiting movie moguls, Cannes is above all else a convention centre, complete with its very own casino, seasonal commerce and attendant sleaze.

Nonetheless, Cannes is a convention centre with a difference. Hoteliers will remind you that you are in the official "sister city" of Beverly Hills – usually when you mention the exorbitant price of rooms. However, it also goes a long way toward explaining life in Cannes. As in that pricey L.A. suburb, snootiness and exclusivity come a-tumblin' down when a paying customer comes to town. Where else could the likes of Pia Zadora hold a group of adults in thrall?

Thus Cannes, the jewel of the Riviera, plays fickle host year round to trade events in the advertising, television and recorded music industries. To the executives who attend them, these acronym happenings (with names like MIPTV, MIDEM and SPONCOM) spell an extended opportunity to cut deals and take advantage of the swollen expense accounts that have helped make show business legendary for its vulgarity.

But none of these events manage to excel the *Festival international de film* – as your haughty hotelier will undoubtedly call it – in the sway it holds over the popular imagination. Nor can they touch its sales volume, conservatively estimated at $3 billion each year.

And still it remains misunderstood, this most famous festival. When it comes down to it, the uninitiated have only an inkling as to the real reason why over 12,500 film producers, directors, distributors and actors, over 3,000 journalists and no less than 2,500 hangers-on are attracted to the French Riviera every year for two weeks.

Common fallacies attribute these attendance figures to the allure of celebrities or the opportunity for participants to appear important for those few brief days. But these explanations leave many questions unanswered. All those ruthless women and cynical men with name tags, the mountains of printed hokum, the watts and volts squandered in an effort to keep Pia Zadora and her ilk illuminated during their press conferences – what does it all mean? Why doesn't the whole place just turn into a pillar of salt?

The answer is held in the palm of the invisible hand of capitalism.

Sure, there are movies here, hundreds of them – without question, the widest selection available anywhere in the world at any one time. (Yet it is possible – indeed, not uncommon – for a film executive to serve the industry's obligatory two-week May sentence here and not see a single one in its entirety.) Sure, there's the prestigious Golden Palm competition, the countless retrospectives and the priceless photo opportunities. Even the world's press is out in force – more often than not at the Press Bar, wrestling with their daily deadline for copy.

But all the cloying puffs about "art" and "culture" that billow around Cannes for two weeks, like so much smoke from a clove cigarette, don't emanate from the real fire at the festival. That inferno is located in the belly of the beast known by regulars here as "The Bunker" – that is, at the film market or *marché du film*, located in the basement of that unsightly concrete convention centre that dominates the coastal landscape of this little harbour town.

For the film market is, ultimately, the force that drives the motion picture industry and the real reason that everyone who's anyone in show business is present in May. Indeed, sellers and would-be sellers of motion pictures from countries the world over come here to hawk their wares like so many carpet merchants at a Turkish bazaar.

But how can you tell the players from the fans in the basement, what with everybody smiling incessantly? Lucky for you, the name tags will be colour-coded. Don't worry about recognising any faces – none of the real celebrities would be caught dead down here in the basement. The stars only give press conferences and an occasional photo session on the beach. And the big-fish producers do business over cocktails at the Carlton, Martinez or Majestic. (Go, *have* a drink or two for $10 a pop at one of these elegant *Belle Epoque* hotels, and appreciate too just how much an expense account separates the men from the boys in filmdom.)

No, down in the basement what you'll find are the official representatives of national film industries taking savage bites out of their countries' GNPs, faceless salesmen from assembly-line celluloid factories the world over and mustachioed peddlers who keep their collection of film reels underneath their trench coats.

Pity the first-time Cannes festival-goer who makes the mistake of breaking his piggy bank to bring his "personal statement" film here in search of an appreciative audience.

We're talking about the small-potatoes producer. He's the one who went way over his small-potatoes budget, renting equipment and making prints of his film. He never got around to paying his technicians or actors. But he doesn't really care. He's finished his film, and now it's for sale and that's what matters. Maybe he takes out more advertising than he can afford in the daily trade papers here like *Screen International*, the festival's journal of record. Next there's the question of whom to invite to the screenings. One problem is all those damn journalists trying to see movies for free. Of course, they, too, have their place in the film industry. But down in the basement, journalists are viewed less as paying customers than simple freeloaders. To enhance their discredit,they're likely to be staying at flea-bag flophouses far from the Croisette. Producers never shake the journalist's hand at the market if they can avoid it.

The hand they seek is that of the distributor, typically the least interesting person from the point of view of conversation but the most important in terms of box-office receipts. The distributor's the one who puts the film into theatres, thereby enabling the public to pay to see it. (And if they pay enough, maybe, someday, the aforementioned technicians and actors will see some remuneration for all their efforts.)

The distributor plays for keeps: He or she is too busy brooding over *Screen International*'s list of yesterday's box-office figures even to crack a smile at the latest Pia Zadora story going around. This one silently emerges from the screening room and shakes his or her head – or sits down to sign a contract with the producer. If the latter, the producer may join the hard-working journalists at the Press Bar and apologise profusely for not shaking their hands earlier.

Now, you, like our neophyte producer friend, know where it's at in Cannes at this, the mad, lecherous, granddaddy of all film festivals. And, also like our novice, you shouldn't go away disappointed, now that you've frittered away the last of your piggy-bank savings in this overpriced seaside resort.

If you came to rub shoulders with the stars, don't let a line of truncheon-wielding *gendarmes* stand between you and Miss Zadora. If the flicks are your game, don't be put off by the fact that all accreditation for the festival is arranged out of Paris six months in advance. After all, you're in Cannes, the closest thing in Europe to Sunset and Vine. And anything is possible in the cinema.

Above all, remember: it's never too late. In the month of May, there are at least a dozen daily trains that can get you out of here and safely to Nice in an hour.

tree-lined avenues and a splendid cobble-stoned **old town** with a seductive array of markets.

Picasso lived here, in the **Château Grimaldi**, for six months in the 1940s, and the château now houses his famous *Joie de Vivre* (the term that many people use to define the allure of the Riviera). Another famous visitor to Antibes was Napoleon, who was imprisoned in Fort Carré.

The nearby **Cap d'Antibes** features the jet-setters' favourite, Hotel de Cap, and Bacon, a great fish restaurant. **La Garoupe**, the beach on the Cap d'Antibes that Gerald and Sarah Murphy used to rake clean every morning, is still *the* place to enjoy the sun. **Juan-les-Pins**, like Nice, has an annual jazz festival and dozens of discothèques, but it is too congested nowadays to be considered similar to the fashionable resort it was in the 1920s when the "Lost Generation" of Americans flocked here.

Other towns are much less congested and worth dropping into during a drive along the Med. **Villefranche** has a "Tex-Mex" restaurant called Texas City, Henry Crews's château is still a site to see in **La Napoule**, **Cap Ferrat** has its charming port of St-Jean and **Beaulieu** is still the closest thing to the *Belle Epoque*. Here there are, at dusk, intimations of Somerset Maugham. Inhabitants often refer to the part of the coast between Nice and Monaco as "little Africa" in reference to its bougainvillea, palm trees and other vegetation.

Menton holds an annual lemon festival when floats parade through the town, but its waterside Promenade du Soleil and 17th-century fort containing the **Jean Cocteau Museum** are away from the glamour.

To avoid the crowds, head for an island, like the **Iles de Lerins** off Cannes. Frequent boats go to **Ste-Marguerite**, the larger island where stands the fortress of "Man in the Iron Mask" fame, and to the smaller **St-Honorat**, with a guest-accommodating Cistercian monastery. Many visitors, though, prefer the islands off **Hyères**, which boast a nudist colony.

Rivieran business executive.

Tomorrow's Riviera: Though the myth lives on, the Riviera is in the midst of a major transformation which will, one day, perhaps, put the fabled days of yore to rest once and for all.

The Riviera has become the frontier of new business. As already noted, income from science, service industries and light industry now rivals the $2 billion annual revenue from tourism that has been the area's long-time economic mainstay.

During the 1970s and 1980s the then mayor Jacques Médecin decided to make a concerted effort towards diversification away from tourism and towards attracting educational institutions, scientific ventures and light high-tech industry. They encouraged the construction of new industrial zones, created the financial structure to attract large and small companies and gave a special boost to science and education.

Today, there is a pleasant blend of Mediterranean climate and lifestyle with state-of-the-art technology, enlightened business attitudes and a well-established communication, educational and scientific infrastructure. Even Cannes is setting aside some land for high-tech industries.

Want to go to school here? It's almost free if you're French and much less than an American university if you're not. The University of Nice, with a student body of 20,000, sets the pace for higher education (about 10 percent of the students are foreign), and numerous graduate schools – such as the renowned Ecole des Mines – round out the educational opportunities on the Riviera.

There is also competition between the different cities, towns, technology parks and industrial sites to attract tourism and new industry. Municipal and local agencies will sell you on the investment potential of the Riviera for your company – though they admit that there are not the government subsidies you will find in Ireland or other parts of France. The "quality of life," rather than financial incentives, are the main selling points.

Besides business in general, business

Promenade des Anglais in Nice.

tourism is also budding, and companies flock to the Riviera for conferences and conventions. The **Acropolis** – the Nice arts and convention centre – actively competes with the **Palais de Festival** in Cannes, which hosts the Cannes Film Festival, for international conferences.

But even smaller communities – like Menton, Grasse, Beaulieu, Sophia Antipolis and Antibes – have their own conference facilities, and visitors are accommodated in numerous hotels (there are 1,680 hotels on the Riviera) of varying price. For these and other reasons, American and British firms, scattered in different towns along the Mediterranean, contend that there is little strain adapting to the environment.

Merrill Lynch and Citicorp are located in Monaco, and the Coopers and Lybrand accountancy firm is represented by an associated firm in Nice. NCR, Rank Xerox and other high-tech leaders have sales offices in the region, while the UNISYS International Management Centre is in St-Paul-de-Vence.

Monaco, due to its relaxed tax laws, is the centre of most financial activity, while **Sophia Antipolis** is already home to numerous international advanced technology companies, service and engineering firms, educational and training centres, research laboratories and light production facilities.

There are some 7,500 people living and working within Sophia Antipolis (combining the Greek word for wisdom and the name of the nearby city of Antibes), in more than 400 companies, and this figure is expected to quadruple. Sophia Antipolis is now generating service industries, including travel agencies, public relations companies, banks, conference centres and a 35-court tennis club.

Mediterranean salad: The new Riviera could, in fact, turn out to be better than the old. One positive aspect is the mix of professionals who are making the place a permanent home. There is an engaging blend of French and international business people (Merrill Lynch in Monaco has 27 employees representing 12 different nationalities) in a wide variety of industries ranging from perfume (which still accounts for over 25 percent of all exports from the Riviera) to fashion (among which are Chacok and Jitrois) to multinationals like Air France, IBM, Digital Equipment Corp., Texas Instruments, Wellcome and numerous others.

In addition, the Riviera's service sector, employing two-thirds of the population, has kept up with the economic growth. This transformation is beginning to wipe out the "resort" feel of the region – except during the summer. For example, the area has a banking network employing almost 6,000 people at over 440 branches.

Want to settle here? If you've got venture capital, there are a number of worthwhile young start-up companies on the look-out for funding. Entrepreneurs flock to southern France because of the large affluent market, its talented labour pool and available financing.

There's no question that the Riviera will keep moving in a business-oriented direction. And what this does to the myth remains to be seen.

Left, chic visitor. Right, Givenchy shop window in Monaco.

GIVENCHY

A Short History of the Bikini

"For the first time in history, the entire staff of the European edition and the foreign service of the *New York Herald Tribune* now in Paris insisted yesterday on covering the same assignment. Each was so determined to do that job that, for the sake of organisational morale, they were all assigned to the story. It turned out to be an exhibition of the world's smallest bathing suit, modeled at the Piscine Molitor."

– *Paris Herald Tribune*, 1946.

When the first two-piece bathing suit was unveiled in Paris on 5 July 1946, it had such an impact that its designer, Louis Réard, named it "bikini" after a recent atomic explosion in the Pacific Ocean. Its arrival launched the beginning of a new era in bathing attire and a new attitude towards women's bodies. Today, it seems amazing that such a small amount of fabric could be interpreted in so many different ways and at the centre of so much controversy.

In the late 19th century, bathing suits came in two pieces, but they were very different. Composed of a long tunic and knickers – generally of wool or serge – their integral purpose was to cover up the body. In fact, for a long time, bathing suits were designed for beachwear rather than for swimming.

It was only at the turn of the century that one-piece garments appeared. The first rib-knit, elasticised, one-piece bathing suit was made in the United States in 1920 by the Jantzen Company. This change of attitude from prudishness to practicality can be attributed largely to swimmer Annette Kellerman, who joined many competitions against men in swimming events held in the River Thames in London, the Seine in Paris and in the English Channel.

In 1924, designer Jean Patou added a fashionable touch by introducing bathing suits bearing Cubist-inspired designs. And, of course, the ubiquitous Coco Chanel was also instrumental in the bathing suit revolution. During her much publicised affair with the Duke of Westminster, she was one of the first women to sport a tan, acquired during long vacations in Spain and on the Riviera, where she wore the most up-to-date swimsuits and beachwear.

But, of the many designers who contributed to promoting the bikini over the years, Jacques Heim was the most innovative. Having opened a chain of boutiques selling sportswear as early as 1946, he popularised the use of cotton for beachwear when he used the fabric in his couture collections.

With the end of World War II, beaches on both sides of the Atlantic reopened and seaside vacation became once again possible. But, although dressier swimsuits in brighter colours were emerging, as well as backless models in the new lightweight fabrics, the attitude towards baring one's skin was still prudish. Peggy Guggenheim told of being fined on a Spanish beach where she was vacationing with painter Max Ernst in the early 1940s. The local *carabinieri*, having gone to the trouble of measuring her back *decolleté*, deemed it indecent.

American *Vogue* began giving advice on how to choose a good suit: "Don't spoil the looks of your perfect dive with a fluggy-ruffles, cutie-pie suit...don't go in for violent water sports in a suit with trick fastening...take off your girdle before you try it on for it's likely to give you delusions of grandeur about your figure...don't overdo the little girl angle when you are chronologically or anatomically unsuited ...don't stop at one bathing suit..."

But it was really only in the 1950s that the attitude towards the bathing suit started to

Early ad for Jantzen one-pieces.

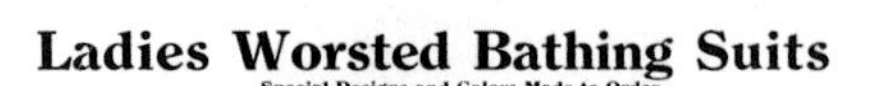

ease. Even though Esther Williams and her nautical extravaganzas did a great deal for the glamour of sophisticated swimming attire, it was not until Brigitte Bardot came on the scene that bikinis found their role model.

Right from the start, the French Riviera was the focal point of the bikini's development, particularly in St-Tropez. In 1956, Bardot was photographed there wearing a gingham bikini decorated with frills. Instantly, gingham became the rage.

During the 1960s, swimwear was designed to emphasise the body with underwiring for the bosom. "The new way for a bikini," said British *Vogue* in 1963, "is little-boy shorts and a built-up bra." But, by the end of the decade, when smaller busts became the look, the costumes were more severely cut.

And bikinis were designed at their scantiest: "the minimum two-piece for a perfect tan, leaving the least possible marks from sunbathing" (French *Vogue*, 1969). The arrival of nylon also was important: "The quick-change *maillot* – when it's dried in the sun it's a sinuous velvety black, and when it's soaked with water it glistens like a seal on the rocks. In the miracle fibre Vyrene." (British *Vogue*, 1965).

In keeping with the hit song "She wore an itsy-bitsy, teeny-weeny, yellow, polka-dotted bikini," two-piece suits inched their way to nothingness. By the mid-1970s the more daring sunbathers had even begun to remove their tops. The arrival of the monokini (just a bottom) was inevitable.

While going topless is still taboo in many countries, it is accepted easily in most of Western Europe, especially on the French Riviera. However, the one-piece is finding new interest. Take it from Felix Palmari, the knowledgeable owner of the Tahiti beach club, one of the oldest "private" beaches on the "Route des Plages." When Felix opened his concession on the 6-mile (10-km) stretch of pure white sand, he was alone. Today, it is staked out with sprawled bodies in every stage of undress and dozens of exotically named concessions renting parasols and mattresses.

"We've seen it all," says Felix, who remembers how beautifully deserted the beach was when he arrived here 40 years ago. "Fashions move, but I think that women today are less into showing off their nudity. They take off their tops in order to tan, but there is a definite comeback of the one-piece bathing suit."

Brigitte Bardot.

INSIGHT GUIDES

Travel Tips

TRAVEL TIPS

Getting There

282 By Air
282 By Sea
282 By Rail
283 By Road

Travel Essentials

283 Visas & Passports
283 Money Matters
284 Health
284 What to Bring & Wear
284 Animal Quarantine
284 Customs
285 Porter Service
285 Reservations
285 On Departure

Getting Acquainted

285 Government & Economy
286 Geography & Population
286 Time Zones
286 Climate
286 Culture & Customs
286 Weights & Measures
286 Electricity
286 Business Hours
287 Holidays, Festivals
289 Religious Services

Communications

289 Media
290 Postal Services
290 Telephone & Telex

Emergencies

290 Security & Crime
290 Medical Services
291 Lost Luggage

Getting Around

291 Orientation, Maps
291 From the Airport
291 Domestic Travel
291 Water Travel
292 Public Transport
292 Private Transport
292 On Foot, Hitchhiking

Where to Stay

293 Hotels
297 Campgrounds
297 Gîtes Ruraux
297 Youth Hostels
297 University Residence

Food Digest

298 What to Eat
298 Where to Eat
302 Drinking Notes

Things to Do

303 City
303 Country
303 Tour Operators
304 Tour Guides

Culture Plus

304 Museums
306 Galleries
306 Concerts
307 Dance, Opera
307 Theatre, Cinema

Nightlife

308 Cafés & Bars
308 Discos
308 Nightclubs & Cabarets
309 Gambling

Shopping

309 What to Buy
310 Shopping Areas
312 Markets

Sports

313 Participant
316 Spectator

Special Information

317 Children
317 Disabled
317 Students
318 Courses

Photography

319 Pilgrimages

319 *Language*

Useful Addresses

321 Tourist Offices
321 Tourist Information
322 Consulates

Further Reading

323 Other Insight Guides

324 *Art/Photo Credits*

325 *Index*

GETTING THERE

BY AIR

It used to be that travelling to Provence meant time, and visitors either had to board a train, a yacht or a motor car (and sometimes all three) before arriving at their destination. Those days are long gone. Nice airport is now the busiest French airport after Paris, welcoming more than 4 million passengers a year.

Thirty regular airlines link Nice to 79 towns in 34 countries, with direct flights to and from New York, Los Angeles, Frankfurt, London (up to 24 a week) and Paris (up to 113 a week), not to mention Beirut, Abidjan and 73 other regional and international destinations. The national airline, Air France, is just one of many good places to check when looking for one of these direct flights.

If you plan to visit the eastern part of Provence rather than the Côte d'Azur or Alpes-Maritimes, you may want to fly into the Marseille-Provence airport. Although also international, this airport is slightly smaller than its Riviera cousin, and some passengers may need to arrange connecting flights through Paris. Truly small airports also exist in Nîmes, Toulon and Fréjus. Air Inter and UTA are the obvious carriers to contact for any internal flights.

Travellers planning to connect with one of these smaller airports, fans of the picturesque and speedy train, those on a tight budget or those determined to visit the city of lights before heading to the region of light will want to fly into Paris first. Numerous airlines service both the Orly and Charles-de-Gaulle airports, and all are very close in price and service.

One of the best airlines offering flights to Paris, at least for travellers from the United States, is Continental Airlines. Scheduled flights between New York and Paris leave daily. They are no more or no less comfortable than any of the other American airlines, but Continental's prices are very competitive with those of its peers.

BY SEA

True romantics may want to sail into Provence, and there is no better way than on a yacht. There are no formalities for arrivals at a sea port, but be sure to have the registration certificate of your craft on board. Once there, you will find no shortage of mooring space along the coast, with 126 yachting ports and 20,416 berths in all. Nonetheless, in high season, if you have your heart set on a spot along the Croisette, you should reserve ahead.

For further information and a complete list of places from where to rent a yacht, contact:

Syndicat National des Loueurs de Bâteaux de Plaisance, Port de la Bourdonnais, 75007 Paris. Tel: (1) 45.55.10.49. Telex: 203963 F.

(Also, *see below "Getting Around: Water Transport," "Things to Do: Tour Operators"* and *"Sports: Sailing".*)

Those without a little sloop of their own and no wish to lease one should check with one of the larger cruise lines. Ports of call for cruise ships and shipping companies include Marseille, Toulon, Nice, Cannes, Villefranche and St-Tropez. (*Also see "Getting Around: Water Transport" and "Things to Do: Tour Operators".*)

BY RAIL

The Société Nationale des Chemins de Fer (SNCF), with its ultra-modern trains, offers without question one of the nicest ways to go anywhere in France. The trip down into Provence is particularly pleasurable. From your comfortable window seat, you will get a chance to experience the incredible change in light and landscape as your train plunges into the south.

Your trip will also be incredibly fast – possibly almost as fast as flying if you count getting to and from the airport, waiting for your bags, etc. SNCF is the proud parent of the fastest train in the western world, the Train Grand Vitesse (TGV), and its incredible 162 mph (270 kph) service includes Paris to Nice. The cost is a bit higher, and you do have to reserve your seats ahead of time.

SNCF's Motorail service is particularly convenient for Europeans – including residents of the British Isles. Chances are that you will want a car to get around in Provence, and the Motorail service allows the visitor to enjoy a berth on a train-auto-couchette while his/her own vehicle occupies a nook on the same train. Motorail is even offered for the train-ferry-train route from Great Britain.

SNCF has also designed a service for cyclists. Under the Train & Velo system, you register your bicycle half an hour before leaving, for a cost of 30 ff, and it's transported with you. It will take about half an hour to retrieve it once you've arrived. For an additional 25 ff, SNCF will collect your bike from your home and/or deliver it at your destination.

One last advantage in taking the train is the number of discount fares available. The France-Vacances pass (sold outside France only) allows adults to travel for either 4, 9 or 16 days of unlimited travel at a discount. Americans can also purchase France Rail 'n' Drive passes, which offer couples a flexible schedule of combined second-class train travel and car rental. And SNCF honours the Eurailpass and Eurail Youthpass.

REMY XO BECAUSE LIFE IS WHAT YOU MAKE IT

Swatch. The others just watch.

seahorse/fall winter 94-95

shockproof
splashproof
priceproof
boreproof
swiss made

swatch
SCUBA 200

Some tips for travelling by train: don't forget that you need to punch your ticket into the little boxes stationed at the platform entrances before boarding the train. When reserving for the TGV, be sure to indicate whether you want *fumeurs* (smoking) or *non-fumeurs* (non-smoking), if it makes a difference to you. Also, if you wish to have breakfast on the TGV, you would do well to reserve it as well (the cost will be additional to the cost of your ticket). Finally, be sure to keep an eye on your suitcases when the train is waiting in a station, if you've left them in one of the between-car baggage nooks. Although the trains are clean and civilised, it is not unknown for thieves to hop on and off during stops.

For further information, passes or tickets, you can contact:
Paris: 10, Place de Budapest, 75436 Paris. Tel: (1) 42.85.60.00.
London: French Railways Ltd, 179 Piccadilly, London W1 0BA. Tel: 071-499-2153.
New York: French National Railroads, 610 Fifth Ave, New York, NY 10020. Tel: (212) 582-4813.
Los Angeles: 9465 Wilshire Boulevard, Beverly Hills, CA 90212. Tel: (213) 274-6934.
San Francisco: 360 Post Street on Union Square, San Francisco, CA 94102. Tel: (415) 982-1993.
Chicago: 11 East Adams St, Chicago, IL 60603. Tel: (312) 427-8691.

In Miami: 2121 Ponce de Leon Boulevard, Coral Gables, Miami, FL 33134. Tel: (305) 445-8648.
In Eastern Canada: 1.500 Stanley Street, Suite 436, Montreal, Quebec H3A IR3. Tel: (514) 288-8255.
In Western Canada: 409, Granville Street, Suite 452, Vancouver, BC V6C 1T2. Tel: (604) 688-6707.

BY ROAD

BUS

Several bus companies operate coach services between Provence and international destinations. Here are two Nice-based lines that have stops in Barcelona, Brussels, Geneva, London, Madrid, San Remo, Valencia, Ventimiglia and, by extension, Morocco, Portugal and southern Spain:
Intercars Euroline: Gare Routière de Nice, Promenade du Paillon, 06000 Nice. Tel: 93.80.08.70.
Phocééns Cars: 2, Place Massena, 06000 Nice. Tel: 93.85.66.61.

From London, here are two:
Euroways: 52 Grosvenor Gardens, Victoria, London SW1 0AU. Tel: 01.730-8235. London to Paris, Aix-en-Provence, Antibes, Cannes, Juan-les-Pins, Marseille and Nice.
Riviera Express: 8 Park Lane, Croydon, London CR0 1JA. Tel: 01.680-8787. London to Aix-en-Provence, Cannes, Ste-Maxime and Nice.

CAR

It is not a bad idea for Europeans to drive down into Provence, since they will probably want to have a car once they arrive. One extremely comfortable way to do this is by the aforementioned Motorail service offered by SNCF, where your automobile is stored upon the same train where you are given a couchette. This service is available on ferry routes as well.

If you would rather drive yourself, you will find that the autoroutes of Provence are swift and well-maintained and provide easy connections with the rest of Europe. Tolls are, however, both frequent and very expensive. (You don't just pay once when you get off, but rather in stages *en route*.)

There are a couple of popular scenic routes that tourists might want to take, although they aren't as fast as the Autoroute du Soleil. The most famous of these is the Route Napoleon, tracing the path that the emperor himself once took on his march into the Alps. It follows from Digne to Grasse to Cannes and is indicated on most maps. A second route originating in Digne goes through the Vallée du Var and takes you by Puget-Theniers into Nice. Travelling laterally, the Route Nationale 7 (RN 7) will give you a look at many pretty towns.

TRAVEL ESSENTIALS

VISAS & PASSPORTS

All persons entering France must possess a current, valid passport. France now also requires that all visitors (including those from the US and Canada) possess a visa, except for citizens of EC countries, Andorra, Monaco and Switzerland. Visas can be obtained from any French consulate for a nominal fee (varies according to country). Unless there is something suspect in your background, the process of awarding visas is immediate. However, you must be sure to get it *before* your departure.

MONEY MATTERS

Exchange Rates: The French unit of currency is the (French) franc, and it is equal to 100 centimes. The exchange rate *(le change)*, however, can vary greatly according to its current strength and that of foreign currencies. The US dollar has ranged in value from a meteoric 10 ff: $1 to a dismal 4.50 ff: $1 since 1980. At the time of going to press, in late 1992,

$1 would buy around 4.50 ff and £1 sterling around 9.30 ff.

You can exchange money at any Bureau du Change, Office du Tourisme or Syndicat d'Initiative and many banks and larger hotels. Tourist boards and hotels are good bets for weekends, but their rates will not be as good.

CREDIT CARDS

If you pay by credit card remember that the company figures the exchange rate according to the date they *process* your charge, rather than the date you *made* the charge.

Although most hotels and "better" restaurants in Provence do accept one or more credit cards, it is always wise to check ahead of time to avoid any unforeseen problems. The cards most widely accepted are: American Express, Carte Bleue, Eurocard, Diners Club, Mastercard and Visa. If you plan to rely on plastic, don't go with only one of the latter two cards. Also, keep in mind that many of the nicest places – the little-known kind that are small, family-run, traditional – and, of course, markets don't accept any at all.

TRAVELLER'S CHEQUES

Traveller's cheques are an extremely safe way to carry around funds and, unlike credit cards, can easily be turned into hard cash. They are accepted just about everywhere as long as you have proper identification with you and they aren't in an obscure currency from an obscure agency.

HEALTH

Don't be surprised if you return home from Provence *healthier* than when you left. Plenty of exercise, fresh food, stress-reducing wine and general calm are a way of life in Provence, and most visitors find themselves quickly adapting. Only those with high cholesterol should (as ever when in a French region) be wary, although the emphasis on fish and away from cream sauces helps that too.

On a technical level, the water is safe to drink, and fewer pesticides and preservatives are used on the vegetables and fruit than in many other countries. If you buy directly from the greengrocer, you will find indicated those foods that have been treated, and at a market you can ask. Nonetheless, always wash or peel them before eating.

The greatest peril that the newcomer to Provence faces – and it *is* a serious one – is over-exposure to the immensely powerful sun. A bad sunburn or sunstroke can ruin an entire trip, not to mention your skin and health permanently. Bring a sunblock of 15, even if you're going specifically to get a St-Tropez tan (the lotion will only prevent you from burning, not from tanning) and buy a big hat as soon as you arrive. Try to avoid major sightseeing outings on foot or bike at midday.

WHAT TO BRING & WEAR

Aside from the above-mentioned sun protection, what you will want to bring depends on your plans. No matter what, you will probably need good walking shoes or sneakers and, if you're coming during the warmer season, pack a bathing suit. Many visitors like toting a camera to capture the famed Provençal beauty.

Even in the larger cities, modes of dress are more casual than you would find further north. For fine dining, however, you will want something respectable. Keep in mind that the hinterland of Provence is fairly conservative, though Aix-en-Provence, Avignon and Marseille are cosmopolitan. On the Côte d'Azur anything goes – as long as it is *à la mode*.

If you are going in the winter or early spring be sure to bring some warm clothes and windbreakers to protect you against the fierce mistral wind. Any Alpine or elevated destination will be chill at night.

ANIMAL QUARANTINE

The French are fanatical dog lovers, so don't be startled if you find pooches in even some of the better restaurants. Likewise, many hotels accept pets for a slight additional cost.

Nonetheless, there are certain restrictions regarding bringing pets into the country. Animals under three months old are not allowed in at all. Owners are allowed to bring in a maximum of three pets (dogs and/or cats) of which only one can be a puppy. Each animal should be accompanied by a rabies vaccination certificate or an official statement that the animals are in good health and that they are coming from a country free from rabies for more than three years.

If for some reason you feel the need to bring in more than three pets, your request must be addressed to:

Ministère de l'Agriculture, Direction de la Qualité, Bureau des Echanges Internationaux, 175, Rue du Chevaleret, 75646 Paris Cedex 13.
Tel: (1) 45.84.13.13.

CUSTOMS

There are few limitations on what can be brought into France. Any items destined for personal use (including bicycles) are admitted without formality provided the quantity or type of good does not indicate an intention to carry out a commercial transaction. Also, there are no particular customs formalities for small quantities of food products brought into France by tourists.

If you cross the border between France and Belgium, Germany, Luxembourg or the Netherlands you may proceed without stopping as long as you carry in your car only EC nationals and live animals and goods that can be allowed in without special formality. In order to do so, you must affix on your windscreen a green disc of at least 8 centimetres in diameter and drive slowly through the border controls.

The following items are subject to special formalities: gold currency may be brought in only with special permission from the Banque de France. Other articles of gold (i.e. jewellery) should not exceed 500 grams in weight. Arms and ammunition can be brought in only with an import authorisation issued by the customs headquarters, after approval by the appropriate ministry. For pets, see notes in the previous section.

The following items are strictly prohibited: narcotics, copyright infringements, fakes and counterfeits, and weapons and ammunition (unless an import authorisation is shown).

The following table covers tax-free allowances:

Commodity	**EC**	**non-EC**
Cigarettes	300	200
Cigarillos	150	100
Cigars	75	50
Smoking tobacco (gms)	400	250
Wine	5 litres	2 litres
either/or		
drinks over 38.8 proof	1.5 litres	1 litres
drinks 38.8 proof or under	3 litres	2 litres
Perfume	75 g	50 g
Toilet water	3/8 litre	½ litre
Coffee	1000 g	500 g
Tea	80 g	40 g

Passengers under 17 are not allowed to carry any tobacco or alcohol.

PORTER SERVICES

In the airports, you can generally hire your own baggage trolley for free or a few francs. Just look for the trolley corrals. Trains are more difficult. In general, porters have their own official scale for tips, but it is usual to give them about 5 ff extra.

RESERVATIONS

Most trains (except for the TGV or sleeping cars) do not require reservations, nor do most buses except for the tour lines. Boats vary according to size and distance. Of course, all planes do, and it is wise to confirm them ahead of time.

ON DEPARTURE

Value Added Tax (VAT) can be deducted from your purchases and/or refunded to you upon departure depending on the quantity you have spent in any one store (2400 ff for EC citizens, 1200 ff for other countries). Look for the sign *Détaxe à l'Exportation*. For a refund, you must be sure to keep all your receipts and have a VAT form filled out at the store before leaving the country. Then you must remember to show your receipt and the goods purchased before checking your bags at the appropriate office in the airport.

The purchase of precious metals, jewellery, works of art, collectors' items and antiques are subject to a capital gains tax of 6–7 percent, which you have to pay when leaving the country unless you can prove with a bill that the purchase was made from a dealer or that tax has already been paid.

For detailed information, contact a customs information centre:

In Marseille: tel: 91.08.60.50
In Paris: tel: (1) 42.60.35.90

GETTING ACQUAINTED

GOVERNMENT & ECONOMY

Provence is one of the 22 separate regions that make up France and comprises five of the country's 96 *départements*. Its capital is Marseille – France's largest seaport and second largest city, with a population of over 1 million.

France itself is a republic, led by a president elected popularly for a term of seven years, a prime minister who heads the government and its ministries, and a parliament (or government) composed of the National Assembly and the Senate. Provence, of course, sends representatives to the national parliament as well as having local bodies of its own. It has traditionally been a seat for communism although facets of the radical right also have a strong following.

For centuries, Provence's economy has been agriculturally based, producing a lion's share of the country's fruits, vegetables, herbs, olive oil, rice, sheep, etc. Many of these products are, in turn, used in such local industries as soapmaking. Provence also has vital interests in fishing and oil refineries. Over the last century, the economy has been overwhelmingly assisted by a growth in tourism. The Côte alone acquires about 10 billion ff annually from the millions of tourists who visit it each year.

GEOGRAPHY & POPULATION

Provence is located in the southeastern corner of France. It is bordered to the west by the Rhône River, to the north by the Baronnies Range and the Hautes-Alpes, to the east by Italy and to the south by the Mediterranean Sea. It comprises five different *départements:* the Bouches-du-Rhône, the Vaucluse, the Alpes-de-Haute-Provence, the Var and the Alpes-Maritimes. The Bouches-du-Rhône contains an area frequently referred to as a region in its own, the Camargue.

The landscape is widely varied – from seaside cliffs and sandy beaches to dry wide plains and salt marshes to opulent Alpine reaches and jagged limestone mountains. Much of this land is sparsely populated with heavy concentrations of people in the major cities. For example, the Bouches-du-Rhône (the largest of the Provençal *départements* and third largest in all of France) possesses some 1,700,000 inhabitants – however, over a million of them live in Marseille alone.

TIME ZONES

All of France is located within the same time zone. It is six to nine hours ahead of the United States, five to eight hours ahead of Canada and one hour ahead of the United Kingdom. In winter, it is within the same time zone as the rest of continental Europe; however, from April until September, it is one hour ahead of Central European Time.

CLIMATE

Provence is renowned for its sunshine. There are 300-plus sunny days a year and less than 800 mm/yr of rain, but, when it rains, it pours. Average temperatures range from 45°F (7°C) in winter to 76°F (24°C) in summer. Winters are mild and sunny, and summers hot and dry with long days.

Climate will vary, of course, between the high country and low country. Nonetheless, all the region experiences the sporadic virulence of the legendary northwest wind, the mistral, during the late autumn to early spring.

CULTURE & CUSTOMS

The Provençals may at first appear standoffish, but persevere and they will become the friendliest and most generous people you could hope to meet.

Never forget that the Provençals spent very many centuries as an independent nation and are fiercely proud of their rich heritage. They behave as any modern Western Europeans, but don't be surprised to find even the more cosmopolitan folk adhering to old superstitions and rites. You can't be expected to know them all, so just be polite, respectful, willing and friendly, and – unlike the Parisians – they'll forgive you any gaffes.

Tipping follows much the same rule as in the rest of France. Taxi drivers get about 10–15 percent of the amount marked on the meter. Hotel staff, depending on the elegance of the establishment, expect 5–10 ff for every item of baggage. You can leave 5–10 ff for the chambermaid daily, but you don't necessarily have to tip the concierge. Restaurants generally include tips in the bill *(service compris)* at 15 percent, so you don't have to worry about this. The same goes for café waiters. Hairdressers get about 10 percent with another 10 ff to the shampooist and the manicurist.

WEIGHTS & MEASURES

The metric system is used in Provence, as it is in all of France.

Units of Length:
1 km = 0.62 miles (1 mile = 1.6 km)
1 m = 3.28 feet (1 ft = 0.36 m)
1 m = 1.09 yards (1 yd = 0.91 m)
1 cm = 0.39 inches (1 in = 2.5 cm)

Units of Capacity:
10 litres = 2.2 imp. gallons
10 litres = 2.6 US gallons

Units of Weight:
1 kg = 2.2 lbs (1 lb = 0.45 kg)
1 g = 0.035 oz (1 oz = 28 g)

ELECTRICITY

Provence runs on 220–230 volts AC. Current alternates at 50 cycles, not the 60 in use in the US, so US citizens should bring a voltage transformer. Outlet prongs are shaped differently, so an adapter may be needed as well.

BUSINESS HOURS

Shops: are generally open from 9am–noon and 2–7pm with the prerequisite two-hour midday break for lunch and siesta. Everything except a handful of ambitious *boulangeries* and greengrocers will be closed on Sunday, and many stores will be closed on Monday as well.
Banks: are open Monday through Friday from 8.30am–12 noon and 2.30–4pm, but watch for long weekends.
Post offices: are open from 8am–5pm; all but the largest branches are closed from noon–2pm. If you get desperate, Telecom Service, 12, Boulevard Charles-Nedelec, in Marseille is open all night for telephone, telegram and postage stamp services.

HOLIDAYS

Official national holidays are 1 January, Easter Monday, Ascension Thursday, 1 May (Labour day), 8 May (V Day), Whit Monday (in the beginning of June), 14 July (Bastille Day), 15 August (the Assumption), 1 November (Amnesty Day), 11 November (All Saints' Day), Christmas and 26 December (Boxing Day). Everyone takes these days off.

Provence is the land of *fêtes*. Check local schedules (or the one below) and stock up on petrol and provisions.

FESTIVALS

The rich Provençal heritage has left in its wake an abundance of colourful festivals. With a little planning, you should have no trouble including at least one in your visit. Below is an extensive, though not exhaustive, list:

1–31 January to February: Maurel Pastoral, Marseille; Month of the Sea Urchin, Sausset-les-Pins; Equestrian Fair, Avignon; Music Convention, Cannes; Car Rally, Monte-Carlo; Mimosa Festival, Mandelieu.
6 January: Fête of St-Clair, Allauch.
17 January: Fête of St-Marcel, Barjols.
31 January: Fête des Petardiers Castellane.

Early February: Fête de la Chandeleur, Marseille; Antiques Fair, Avignon; Votive of Ste-Agatha, Maillane; Corso Carnavalesque, Pelissane; Mardi Gras, Corso Coudoux; Mimosa Festival, Cannes; Mardi Gras Lemon Festival; Menton.
4–14 February to March: International TV Festival, Monaco; Carnival, Chât.-l-Martigues.

4–14 March: Nice Carnival, Nice; Ski Yachting Cup, Cannes; Cycling Race, Paris-Nice; Mandelieu Festival of the Gourds, Nice.
25 March: Commemoration of Mistral Maillane.
26–29 March: Grand Prix of Magic, Monte-Carlo.
27 March: International Marathon, Nice
End month: Corso Fleuri, Pelissane.

April to October: Cocardes and Corridas Arles Art Festival at the Castle Mandelieu.
11–26 April: Antiques Fair, Antibes.
Second week: Spring Fête, Martigues.
13–17 April: Springtime of Arts, Monaco.
15 April: International Tourn. Water Polo, Antibes.
Last two weeks: Sailing Regatta, Hyères.
18–26 April: International Open Tennis Cham., Monaco.
18 April: Floral Symphony, Antibes.
25 April: Fête of St-Mark, Meyreuil.
Last Sunday: Fête of the Cowboys, Arles.
End month: Fête of St-Mark, Eyguières.
End month: Fête of St-Mark, Villen.-l-Avignon.

Beginning May to June: Fête of the Good Angel, Rognonas; International Film Festival, Cannes; Rose Festival, Grasse; Concours de Farandoles, Salin-de-Giraud.
1 May: Carreto dis Ases, St-Rémy-de-Prov.
First Sunday: Fête of Ste-Croix, Le Tholonet.
First two weeks: Fête de Mai, Nice.
9–17 May: Old Cars Rally, Antibes.
Second Sunday: Fête of the Horse, Barbentane.
Sunday & Monday after 15: Fête of St-Gens, Monteux-le-Baucet.
16 May: Vine Growers Festival, Antibes.
16, 17, 18 May: Bravade, St-Tropez.
Mid-May: Fête of Bresson Blazets, Salon-de-Provence.
24 & 25 May: Gypsy Pilgrimage, Stes-Maries-de-la-Mer.
25 May: Votive Fête, Peypin.
26 May: Baroncellienne Day, Stes-Maries-de-la-Mer.
Last Sunday: Fête of the Horse, Châteaurenard.
Second & third Sunday: Corso Fleuri, Lambesc.
28 May: Vow of the Echevins Mass, Marseille.
30 & 31 May: Racing Cars Grand Prix, Monaco.

1 June: Fête of the Grape Harvest, Boulbon; Fête des Barjaquets, Rognac.
Beginning of month: Lei Farandoulaire, Sestian Aix-en-Provence; Fête at the Windmill, Fontvieille; Fête des Chirons, Miramas; Fête des Canourgues, Salon-de-Provence.
First Sunday: Wine Festival, Courthèzon; Fête of St-Elgius, Mollèges; Courses de Vachettes, Cabannes; Fête of the City, La Ciotat; Festivals de Quartier, Marseille; International Theater Festival, Antibes; Provençal Festival, Hyères; Spanish Bravade, St-Tropez; Fête of the Sea, Toulon.
June to July: Bullfighting, Barbentane.
June to September: Bullfighting, Nîmes.
11–14 June: Fête du Cordage, Tarascon.
13 June: Pilgrimage, Cuges-les-Pins.
14 June: Fête of St-Anthony, Meyreuil.
Mid-June: Grande Fête Votive, Stes-Maries-de-la-Mer.
22–24 June: Fête des Aires de la Dine, Salon-de-Provence.
23 June: Fête of St-John, Valrèas; Fête of St-John, Cabannes; Fête of St-John, Istres.
23, 24, 25 June: Fêtes, Martigues.
24 June: Fête of St-John, Entrevaux; Fête of St-John, Aubagne; Fête of St-John, Les Baux-de-Prov.; Fête of St-John, Eygalières; Fête of St-John, Fontvieille; Fête of St-John, Mallemort.
Sunday after 24: Fête of St-John, Allauch.
Third week: Fête des Carabins, Fos-sur-Mer.
29 June: Fête of St-Elgius, Eyrargues.
Last Sunday: Parrot Festival, Bollène.
Last Sunday: Fête of St-Peter, Menton.
Last week: Fête of the Sea, Cassis.
Last week: Fêtes Traditionelles, Arles.
End month: Music Festival, St-Rémy-de-Provence.

End month: Fête of St-John, La Ciotat.
End month: Fête Votive, La Gavotte.
Last Sunday: Country Festival, Salin-de-Giraud.
Last Sunday: Fête of the Tarasque, Tarascon.

July: Fête of St-Elgius, Maillane; Fête of St-Elgius, Rognonas; Music Festival , Uzès; Music Festival, Toulon.
July to August: Offenbach Festival, Carpentras; The Season at Aix, Aix-en-Provence.
16 & 17 July: C Music Festival, Entrevaux.
July and August: Cultural Festival, Vaison-la-Romaine; Jazz Festival, Cavaillon; Courses de Vachettes, Salin-de-Giraud; Music & Theatre Festival, Sisteron.
July to September: Bullfighting, Arles.
1–15 July: Local Festivals, Vauvenargues.
First week: Provençal Week, Aix-en-Provence; Cocarde d'Or, Arles.
First weekend: Fête de Bel Air, Salon-de-Provence.
First Saturday: Venetian Festival, Martigues.
First Sunday: Fête of St-Elgius, Châteaurenard; Local Festival, Cadolive.
Early July: Provençal Festival, La Ciotat.
Second weekend: Local Festival, Salon-de-Provence.
8–14 July: Garlic and Local Products Fair, Cabriés.
9 July: Annual Fête, La Ciotat.
10–24 July: Artisan's Fair, Riez.
11–14 July: Feria du Cheval, Méjanes.
14 July: Wine Festival, Vacqueyras; Arrival of the Bulls, Tarascon; Local Fête, Noves; Local Fête, Maillane; Grand Fête of Summer, St-Andiol; Grand Fête, St-Rémy-de-Prov.; Fête of Summer, Lamanon.
Weekend before 15: Linden, Lavender and Olivetree Festival Buis-les-Baronnies.
Second Sunday: Fête Votive, Carnoux.
Mid-month: Fête of the Sea, Fos-sur-Mer; Fête de Notre Dame, Carpentras; Flower Festival, Seillans.
Mid-July to August: Dance & Drama Festival Avignon; Music Festival, Aix-en-Provence; Festival of the Arts, Orange.
16, 17, 18 July: Fête of Ste-Severe, Villars-Colmars.
17–25 July: International Jazz Festival, Antibes.
Third week: Picodon Fête with Goat, Cheese Competition Saôu.
21, 22 July: Fête of Mary Magdalene, Ste-Baume.
21–25 July: International Fireworks Festival, Monaco.
22 July: Fête of Ste-Madeleine, St-Maximin; Provençal Dance, Fontvieille.
Third Sunday: Fête Virginienco, Stes-Maries-de-la-Mer.
Fourth Sunday: Wine Festival, Cairanne.
Last Sunday: Fête of St-Elgius, Graveson.
28 July: Local Fête, Sanary.
29 July: Fête of St-Peter, Auriol.
End month: Festival of Folklore, Aubagne; Local Festival, Seillans.
Last week: Market of Provence, St-Rémy-de-Prov.

August: Jasmine Festival, Grasse; Music Festival, St-Maximin.
First two weeks: Fête of Hte-Provence, Forcalquier.
Early August: Fête of St-Elgius, Cuges-les-Pins; Fête of St-Stephen, Istres; Fête St-Peter, Marseille; Local Fête, La Garde Freinet.
First Sunday: Jasmin Festival, Grasse; Fête Votive, La Barben; Fête Voti ve, Fontvieille; Lavender Corso, Digne.
First weekend: Fête du Q. Fontsainte, La Ciotat.
First Sunday after 6: Grape Festival , Fréjus.
7–21 August: Artisan's Fair, Riez.
10–15 August: Local Festival, Fos-sur-Mer.
14 August: Fête des Facteurs, La Ciotat.
14–15 August: Pilgrimage to Notre Dame, Garde, Marseille.
14–17 August: Feria Saintoise, Stes-Maries-d-l-M.
15 August: Grande Corrida, Arles; Pilgrimage to N.D. de Lure St-Etien.-l-Orgues; Fête Votive, Carry-le-Rouet.
Mid-month: Fête of St-Elgius, St-Etien.-du-Gré; Fête du Logis-Neuf, Allauch.
22–23 August: Rally Cross Int'l, Antibes.
24 August: Garlic Festival, Hyères.
Third week: Festival Provençal, Séguret.
Third Sunday: Fête Votive, Meyrargues.
28, 29 August: Fête of Notre-Dame, Maillane.
30 August–9 November: Holiday on Ice, Nice.

Early September: Fête Votive, Marignane; Fête of Ste-Rosalie, La Fare-l-Oliviers; Fête of Cassis Wines, Cassis; Festival of Animation, Hyères; Jazz Dixieland Festival, Mandelieu.
First weekend: Autumn Festival, Courthèzon.
First Sunday: Fête Votive, Graveson.
6, 7, 8 September: Fête for Mistral, Maillane.
8 September: Pilgrimage to N.D. Galline, Marseille; Fête of La Diane, Moustiers-Ste-M; Pilgrimage to N.D. de Lure St-Etien-l-Orgues.
Mid-month: Fêtes of the Rice, Arles; Fête for Mistral, Aix-en-Provence; Fête du Jumelage, Salon-de-Provence.
Second Sunday: Fête of St-Maurice, Pélissane.
Fourth Sunday: Fête Votive, St-Rémy-de-Prov.
End month: Fête of St-Michel, Mallemort; Fête for Grape Harvest, Auriol.

October: Niss'Artisanat Crafts, Nice.
First Sunday: Votive Fête of St-Batie, Jouques.
Beginning month: Provençal Day, Mallemort; Grand Fête of St-Hubert, Vauvenargues.
8 October: Car Rally of Antibes, Antibes.
14–18 October: Mountain Film Festival, Cannes.
Sunday closest 22: Blessing of the Sea, Stes-Maries-d-l-M.

November: Chestnut Festival, La Garde Freinet; International Crafts Show, Cannes.
1–7 November: Birds Exhibition, Antibes.
First or Second Sunday: Chestnut Festival, Isola.
Mid-month: Baptême des Côtes du Rhône, Avignon;

Commemorative Ceremony, Stes-Maries-d-l-M.
18, 19 November: Monaco Feast Day, Monaco.
End November–January: Santons Fair, Marseille.

December: Sacred Music Festival, Cannes; Italian Films Festival, Nice; International Circus Festival, Monaco.
2–6 December: Underwater Film Festival, Antibes.
5 December–10 January: Luna Park Fair, Nice.
Mid-month: Veillée Calendale, Martigues.
24 December: Provençal Midnight Mass Every Town & City; Shepherds Festival, Les Baux; Mystery Play, Séguret.
20 December–January: Crèche i.e. Arles, Auriol, Marseille, Allauch.
End month–January: Pastorales i.e. Barbentane, Marseille, Cassis.

The summer cultural festivals in Provence are terrific. Following is a list of names and addresses, from where you can get information or purchase tickets. (*See also Culture Plus.*)

Aix-en-Provence: Festival International d'Art Lyrique et de Musique, Palais de l'Ancien Archevêche, 13100 Aix-en-Provence. Tel: 42.23.34.82; telex: FESTAIX 410.065 F.
Arles: Dance, Music and Photo Festival, Bureau du Festival, 13200 Arles. Tel: 90.96.47.00.
Avignon: Festival d'Avignon, Bureau du Festival, BP 92, 84006 Avignon. Tel: 90.86.24.43.
Orange: Chorégies d'Orange, Maison du Théâtre, BP 180 - Place Sylvain, 84105 Orange Cedex. Tel: 90.34.24.24.
Vaison-la-Romaine: Hôtel de Ville, Bureau du Festival de Vaison-la-Romaine, 84110 Vaison-la-Romaine. Tel: 90.36.24.79.

RELIGIOUS SERVICES

Roman Catholicism is the predominant religion in Provence, as in all of France. However, all religions are represented and in most cities possess places of worship.

COMMUNICATIONS

MEDIA

Radio & TV: You will find a wide selection of different radio stations, some of them very good. In English, along the Côte d'Azur, is the Radio Riviera. French television is improving, but don't expect to find channels in your native language or to find TVs in any small to mid-size hotels.
Newspapers & Other Publications: Train stations are good places to find large selections of international publications. Otherwise, *tabacs* and kiosks will have all the local and national dailies. Widely available (in English) is the *International Herald Tribune*. London papers are sold on the Côte.

LIBRARIES

Keep in mind that the word *librairie* in French means bookshop and *bibliothèque* means library. Aix-en-Provence, with its university, has no shortage of libraries, unlike Marseille or Avignon. The Inguimbertine Library in Carpentras is world famous, comprising some 230,000 volumes plus musical scores, 14 to 16th-century Book of Hours and 19th-century prints and drawings. Check with your local tourist office for a *bibiliothèque* that might contain books in your language.

BOOKSHOPS

Unsurprisingly, the university town of Aix-en-Provence is an absolute mecca for bookshops. Indeed, due to its proud cultural heritage, most of Provence's larger towns have a nice selection of bookstores, many of which carry ancient, priceless editions. As ever, check with the local tourist office.

Anglophones will like: the English Bookshop, 23, Rue de la Republique, 2nd floor in Avignon; the Paradox Bookstore, 2, Rue Reine Jeanne in Aix-en-Provence; and the Cannes English Bookshop, 10, Rue Jean-de-Riouffe in Cannes.

French speakers should check out Les Arsenaulx, 25, cours d'Estienne d'Orves in Marseille. After selecting a book from their well-chosen collection of upscale books, you can settle down at the adjacent teashop and read.

POSTAL SERVICES

Unlike the phones, postal service in all of France is excellent and reliable. All but the very smallest post offices (PTT) are open from 8am–7pm Monday through Friday, and 8am–12 noon on Saturday. Stamps can be purchased there, at a tobacconist or at many hotels. Post boxes are yellow. Air mail ranges in price (1991) as follows:

	up to 20 g	to 50 g	to 100g
United Kingdom	2.50 ff	4.40 ff	6.20 ff
United States	3.40 ff	5.50 ff	8.00 ff
Canada	3.40 ff	5.50 ff	8.00 ff
Australia	3.40 ff	6.00 ff	10.00 ff

You can also send and receive money from a PTT.

TELEPHONE & TELEX

The telephone service has much improved over the past few years, but it still leaves much to be desired – particularly in the provinces. To top this off is the locals' strange attitude towards telephones. Ask to use one, and you may feel like you've asked to borrow the Rolls-Royce.

Don't be surprised to find direct-line telephones in bedrooms of only the hotel chains or most expensive establishments. The rest of the time, you'll have to go through a central switchboard and often on a house phone downstairs. This can cause inconvenience in late-night calling and confusion when it comes to paying your bill. It is best to pay for each call as you make it or keep very strict records.

You will have to dial certain prefixes (area codes) before most numbers. Ask your concierge or host for assistance.

A good place to find fax and telex services are at major business hotels, though fax bureaux are becoming increasingly common. Otherwise, enquire at the post office.

EMERGENCIES

SECURITY & CRIME

Aside from the port cities (notably Marseille and Toulon) and along the Côte d'Azur, Provence tends to be filled with poor but honest folk. Nevertheless, it is always wise to exercise caution. In particular, *never* leave anything behind in a car or unattended in a station.

If you do run into trouble, you should inform the local police – but don't expect too much. You might also want to contact the local tourist board. For grave problems, you should not delay in calling your nation's consulate.

LOSS

If property is lost in a subway or on a bus, go to the terminal point and, if you're lucky, it will be waiting for you there. Otherwise, ask a conductor for the number of the lost and found. For property lost on a train, ask the conductor or general information officer for assistance.

MEDICAL SERVICES

Not all the smaller towns in Provence have pharmacies or immediately accessible doctors, so do bring any prescriptions or medicines you generally need. Cautious anglophones may want to become members of the International Association for Medical Assistance to Travellers (IAMAT) before leaving home. They print a directory of English-speaking physicians, available from 736 Center St, Lewiston, NY 14092. Your hotel concierge or local tourist board can help you find a local doctor, dentist or pharmacy.

IN AN EMERGENCY

In Marseille
SOS Doctor, tel: 91.52.91.52
SOS Cardio, tel: 91.52.84.85
SOS Pediatrics, tel: 91.26.19.19
SOS Centre Antipoison, tel: 91.75.25.25
SOS Dentists, tel: 91.25.77.77
SAMU (for serious medical emergencies), tel: 91.49.91.91

In Aix-en-Provence
SOS Doctor, tel: 42.26.24.00

In Antibes: SOS Doctor, tel: 93.33.40.20
In Nice: tel: 93.83.01.01
In Cannes: tel: 93.38.39.38

LOST LUGGAGE

Contact your airline or local train station as appropriate for any luggage forgotten while in transport. In the case of the train, don't be surprised to discover that you must pay for a ticket and travel personally to the spot where your bag is being held. For luggage forgotten in a hotel, contact the concierge.

GETTING AROUND

ORIENTATION

Provence is in the southeastern corner of France. While there are several major cities along the coast and the Rhône River, much of the region is not reachable by train or plane. Tiny hillside hamlets, forgotten plains' towns and mountain-perched villages make for some of the most rewarding visits but can only be reached by car or bicycle.

MAPS

The regional Michelin maps are without question the handiest and most thorough maps available. You will find them all over, in bookshops, tabacs, train stations, etc.

FROM THE AIRPORT

If you fly into the Nice Airport, you can rent a car, hail a cab or take the STANCA bus, tel: 93.21.30.83, to: Cros de Cagnes, Villeneuve-Loubet Plage, Antibes, Juan-les-Pins, Golfe-Juan, Villefranche, Beaulieu, Eze-sur-Mer, Cap-d'Ail, Monaco, Roquebrune, Menton or Cannes. A one-way ticket costs 48 ff, roundtrip costs 87 ff. The bus also services the Monaco and Menton Airports.

The Auto Nice Transport (ANT), tel: 93.96.31.51, will bring you from the airport to Nice's Place Massena or harbour for 13.50 ff during the day and 16 ff at night.

ANT will also take you from the airport to the railway station. To go from the train to the airport, contact Transport Benavent, tel: 93.88.14.02, departures on arrival of every train.

Finally, to go from the Nice Airport to the Marseille Airport, there is Phocéens Cars, tel: 93.85.66.61. They serve Nice to Aix-en-Provence for 86 ff.

DOMESTIC TRAVEL

Some of the towns in Provence linked by train are Nîmes, Arles, Avignon, Arles, Tarascon, Marseille, Aix-en-Provence, St-Auban, Digne, Toulon, Hyères, Cannes, Nice, Menton and Breil. Stations are generally conveniently located, with taxis and buses waiting for the weary or heavily laden.

Nonetheless, tourists wishing to get into the many wonderful nooks of Provence will need another means of transportation. Bikes are popular, but that both limits the area you can cover and the time you have to spend on important Provençal activities like site exploring and café sitting. It is best either to bring a car or rent one once you've arrived. (For more information about renting, see "Private Transport" below.)

WATER TRAVEL

Unlike the rest of France, Provence is not dissected by many waterways. The three main bodies of water that can be used for transportation and on which some cruises do exist are the Rhône River, the Durance River and the Mediterranean Sea.

Travel by water is both necessary and pleasant to visit any of the many offshore islands. You will find services ferry and boat companies easy to locate and reasonable.

–For the Ile de Bendor, boats leave from Bandol at least 25 times a day. Tel: 94.29.44.34.

–For the Iles d'Hyères, boats leave for Porquerolles from Tour Fondue between six and 19 times a day depending on the season. Boats to Port-Cros and Le Levant from the Port d'Hyères leave slightly less frequently – up to four times a day. The times vary according to the month, so get a schedule from SNCF, the tourist board or tel: 94.58.21.81. You can also take an all-day cruise to the islands from the Toulon Port on the Quai Stalingrad, tel: 94.93.07.56. Finally, in the summer, you can also sail to Porquerolles from Le Lavandou with one trip a day leaving at about 9am and returning in the evening. More frequent boats leave from the same port to Port-Cros and Levant in season.

You can also rent sailboats and yachts along the Mediterranean and the mouth of the Rhône. For example:

Eagle Yacht Charters, 150 Main St, Port Washington, NY 11050
US. Tel: (516) 883-3033. Luxury crewed yachts for charter from any port in the Mediterranean. Bare boats from major Rivieran ports.

Or join a cruise along the Rhône or Camargue in a boat without living accommodation:

Tiki III, Capitaine Edmond Aupy, Rue de Petit-Rhône, 13640 Les-Stes-Maries-de-la-Mer, France. Tel: 90.97.85.89.
Cié de Navigation Mixte, 3, Rue du Président-Carnot, 69002 Lyon, France. Tel: 78.42.22.70; telex: 310000.

PUBLIC TRANSPORT

Buses run between most towns. They won't all be fast or air-conditioned, but they will get you where you are going.
In Aix: Phone the bus station (tel: 42.27.17.91) or try Compagnie Autocars de Provence, 10 Av. de-Lattre-de-Tassigny, 13100 Aix-en-Provence (tel: 42.23.14.26; telex: 420579), whose regular service to many places in Provence may be what you need.
In Marseille: Either phone the bus station (tel: 91.08.16.40) or contact ARTC, 300, Av. du Prado, 13008 Marseille (tel: 91.76.55.35).
In Nice: Phone the Gare Routière (tel: 93.85.61.81), for departures from Nice.

Elsewhere, enquire at the local Syndicat d'Initiative for information.

PRIVATE TRANSPORT

RENTING A CAR

Many visitors to Provence will find renting a car indispensable, and there are many combinations of car/air, car/train and car/hotel arrangements available. Among the major rental agencies represented are Hertz, Avis, Budget and Europcar. Of these, as ever, Hertz is the leader.

No one should have much trouble driving in Provence, once they get used to the daring way local motorists pass one another. A few rules to remember:

1. Drive on the right.
2. Seat belts are mandatory in the front seat.
3. No child under six may ever be in the front seat.
4. Old village streets can be *extremely* narrow – at times, impassable for larger cars. Enter with care.
5. Parking is often difficult in the larger towns and cities. Be sure to examine any signs or kerb markings. Frequently, you are expected to drive right up onto the pavement to park. If you park at a metered spot, you must deposit the amount of money appropriate for the amount of time you believe you will be there, then display the yellow ticket you've received from the meter.

TAXIS

Taxis can be convenient methods of travel in the cities, but don't expect to use many in the countryside.
In Aix
Tel: 42.26.29.30; Tel: 42.27.71 (day) or 42.26.29.30 (night); Tel: 42.21.61.61.
In Marseille
Tel: 91.02.20.20; Tel: 91.06.15.15; Tel: 91.49.91.00; Tel: 91.49.20.20; or Tel: 91.05.80.80.

BICYCLES

Bicycles are a favourite means of getting around all over Provence, by the locals and tourists alike. If you don't want to bring your own, they are also quite easy to rent – either from the train station or a local sporting shop. Those visiting steeper areas or who are less athletically ambitious may want to rent mopeds instead. They are inexpensive, easy to rent and popular with the locals.

ON FOOT

Some distances in spread-out Provence may seem daunting; however, once in a town or city, most sites are within easy walking distance from one another. Also, in certain parts of Provence, particularly the northern regions, the land is liberally crossed with wonderful walking and hiking trails.

HITCHHIKING

Hitchhiking is never the best means of travel, especially for women. If you are determined to thumb it, however, you probably will find it relatively safe and easy.

COMPLAINTS

Complaining is a way of life all over France, so don't expect to have your own complaints greatly honoured. If you really are aggrieved, take it to your local tourist board and be polite but persistent and firm. This goes for your concierge if you have problems with a bill.

WHERE TO STAY

HOTELS

There is no shortage of great places to stay all over Provence, although accommodation quickly fills up during the high season. Anywhere on the Côte and in touristic meccas such as Aix-en-Provence, Arles and Avignon during their festivals, it is important to make reservations well in advance or be willing to stay slightly out of the centre of things.

One extremely easy way to make reservations from your home base is to go with one of the hotel chains that service the bigger towns and cities in Provence. Best among these is the ACCOR chain, with their Sofitel hotels (first class, ultra-deluxe all the way) and Novotel hotels (medium-priced and civilised). Most importantly, reservations can be made worldwide, tel: 1-800-221-4542.

In the smaller towns, hotels will mostly be family-run operations, their service pleasant and their rooms clean and decent. All hotels are classed on a range from ☆ to ☆☆☆☆. Those who want a respectable hotel within a modest budget will find ☆☆ hotels fine. Keep in mind that, except in the larger hotels, you will probably be expected to take demi-pension (continental breakfast) and its cost will be extra.

Following is a selective list of recommended hotels in Provence, arranged by *département*. For more information, you should contact the tourist office of your destination.

VAUCLUSE

If you are determined to stay next door to the Palais in AVIGNON, consider the Cité des Papes. Be sure to request one of the quiet back rooms from where you get a palatial view.
Cité des Papes, 1, Rue Jean Vilar, 84000 Avignon, Tel: 90.86.22.45. 65 rooms, 190–290 ff.

Also in the old town of AVIGNON is the Hôtel de la Garlande. Situated by the Place St-Didier, it is comfortable, straightforward and reasonable.
Hôtel de la Garlande, 20, Rue Galante, 84000 Avignon. Tel: 90.85.08.85.

☆-star, but surprisingly adequate for its category, is the Mignon in AVIGNON. It's not luxurious, but its location is central and its bathrooms sparkle.
Le Mignon, 12, Rue Joseph Vernet, 84000 Avignon. Tel: 90.82.17.30. 15 rooms.

For inexpensive, barebones (but decent) accommodation in AVIGNON, try the Squash Club where bed and breakfast are offered for under 60 ff.
The Squash Club, 32, Boulevard Limbert, 84000 Avignon. Tel: 90.85.27.78.

The Ferme Jamet stands on the Ile de Barthelasse some 3 miles (5 km) from the centre of AVIGNON. Simple, calm and pleasant, its five bungalows and two apartments share a tennis court, pool and fields of cereals, pine trees and vineyards.
Ferme Jamet, Chambre d'Hôtes, Ile de Barthelasse. Tel: 90.86.16.74.

Not only does Hostellerie Le Beffroi in CAROMB have comfortable and affordable rooms with picturesque beamed ceilings, but it also boasts a first-class restaurant much admired by the locals.
Hostellerie Le Beffroi, 84330 Caromb. Tel: 90.62.45.63. 10 rooms.

The *mas* is the traditional Provençal farmhouse, and a stay in one during your visit is *à propos*. Built in the 12th century, the Mas de la Tour now sports a swimming pool and is convenient to the CAVAILLON region.
Mas de la Tour, 84400 Gargas. Tel: 90.74.12.10. 12 rooms, 145-295 ff.

La Mayanelle in GORDES is a very nice little hotel with a wonderful view of the Luberon.
La Mayanelle, 84220 Gordes. Tel: 90.72.00.28. 10 rooms. Closed: 2 January–1 March.

A good starting point for hiking excursions around the GORGES DE NESQUE is the Hameau de la Lauze. Hidden among the woods, this *ferme-auberge* (farm-inn) is the essence of rusticity. It's charming, but don't expect silk sheets.
Le Hameau-de-la-Lauze, Villes-sur-Auzon. Tel: 90.61.83.23.

When visiting Avignon, you may want to stay in nearby VILLENEUVE. La Magnaneraie is a quiet and convenient hotel with an excellent restaurant, a swimming pool and tennis courts. It's not priced to fit into everyone's budget.
Hôtel-Restaurant La Magnaneraie, 37, Rue Camp de Bataille, 30400 Villeneuve-les-Avignon. Tel: 90.25.11.11. 20 rooms, 280-500 ff.

BOUCHES-DU-RHONE

Guests at Le Globe in AIX-EN-PROVENCE generally consider themselves quite lucky to have found such nice rooms at reasonable prices.
Le Globe, 74, cours Sextius, 13100 Aix-en-Provence. Tel: 42.26.03.58. 45 rooms.

For really inexpensive accommodation in AIX contact the Hôtel des Quatre Dauphins in the Quartier Mazarin.
Hôtel des Quatre Dauphins, 54, Rue Roux Alpheron, 13100 Aix-en-Provence. Tel: 42.38.16.39. 78 ff (and up) for doubles, showers 7 ff.

ARLES makes the perfect centre for any visit to the western part of the Bouches-du-Rhône or the Camargue. And, within Arles, you won't find a more conveniently located, pleasantly run and well-priced hotel than Le Cloître. Set on a narrow street just a stone's throw away from the Roman theatre, you will feel Provençal without any discomfort.
Hôtel Le Cloître, 18, Rue du Cloître, 13200 Arles. Tel: 90.96.29.50. 33 rooms, 135-220 ff. Closed: 15 November–15 March.

Very popular and rather more pricey in ARLES is **Hôtel D'Arlatan**, 26, Rue du Sauvage, 13200 Arles. Tel: 90.93.56.66; telex: 441 203. 46 rooms, 170–465 ff.

The Golf Club de Mouriès is a real find. Set in an old château, with a swimming pool and nice views, it is also very inexpensive. Mouriès is just to the east of ARLES and a good back up when the latter town becomes overbooked.
Golf Club de Mouriès, 13890 Mouriès. Tel: 90.47.59.95.

Although CASSIS is somewhat more modest than Cannes, it still falls into the Côte d'Azur summer frenzy. Book well in advance, and you will find waterfront Hôtel Liautaud a reasonably priced and accommodating place to stay. Whatever you do, however, don't wind up at the less salubrious, less well managed and less keenly priced Cassitel across the street.
Hôtel Liautaud, 2, Rue Victor-Hugo, 13260 Cassis. Tel: 42.01.75.37; telex: 441 287. 32 rooms, 180–240 ff.

Very inexpensive and centrally located in CASSIS is **Pension Maguy**, Avenue du Revestel, Cassis 13260. Tel: 42.01.75.37.

Those who prefer to stay outside Arles and have a little more cash to spend on such added luxuries as a lovely veranda and pool will enjoy the Peireiro in FONTVIEILLE.
Hôtel Peireiro, Avenue des Baux, 13990 Fontvieille. Tel: 90.97.76.10. 32 rooms, 250–370 ff.

There is no place better to stay in MARSEILLE than the Sofitel Vieux-Port. Set right on the old harbour with wonderful porches that look out over the yachts and fishing boats, it is within reasonable walking distance of many of Marseille's major attractions and possesses a nice little pool and several restaurants for afternoons when you're too hot to go anywhere. Its unparalleled elegance and sophistication offers the visitor a great haven from the less-lovely aspects of this big city and, at the same time, it manages to retain a real French flavour.
Sofitel Vieux-Port, 36, Boulevard Charles Livon, 13001 Marseille. Tel: 91.52.90.19; telex: 401 270. 130 rooms, 570–925 ff.

CAMARGUE

Visitors to the Camargue may well want to make Arles their home base. Conveniently located on N.453, just east of Arles and *en route* to the Camargue, is the friendly L'Auberge de la Fenière. This attractively converted farmhouse has the added plus of a restaurant that serves both traditional French and Camarguais specialities.
L'Auberge de la Fenière, RN 453, Raphèle, Arles. Tel: 90.98.47.44; telex: 441 237. 25 rooms, 210–488 ff w/o pension, 230–370 ff w/pension.

ALPES-DE-HAUTE-PROVENCE

Located just over 1 mile (2 km) outside of ANNOT, the Hôtel Honnoraty is charming, cheap and friendly – and its restaurant serves very good food.
Hôtel Honnoraty, Les Scaffarels, 14240 Annot. Tel: 92.83.22.03.

A modest, reasonably priced hotel in CASTELLANE is **Ma Petite Auberge**, Avenue F. Mistral, 04120 Castellane. Tel: 92.83.62.06. 18 rooms.

Eight miles (14 km) from Sisteron in CHATEAU-ARNOUX is a comfortable and inexpensive hotel, with a very good restaurant, called **Le Barrasson**, 04160 Château-Arnoux. Tel: 92.64.17.12.

A nice place to make your base for sporting activities in the COLMARS region is Le Chamois. Pleasant and inexpensive, it also has its own restaurant.
Hôtel le Chamois, 04370 Colmars-les-Alpes. Tel: 92.83.41.92 or 92.83.43.29. 26 rooms.

The place to stay in DIGNE for those with discerning taste and large wallets is Le Grand Paris. Its restaurant is excellent.
Hôtel Le Grand Paris, 19, Boulevard Thiers, 04000 Digne. Tel: 92.31.11.15; telex: 430 605. 30 rooms.

Visitors to DIGNE travelling on a more modest budget will find the Coin Fleuri a pleasant hotel and restaurant.
Hôtel Coin Fleuri, 9, Boulevard Victor Hugo, 04000 Digne. Tel: 92.31.04.51. Has 15 rooms, 105–215 ff.

Set on the main square of FORCALQUIER is a medium-priced, authentic 17th-century coaching inn called **Hostellerie des Deux Lions**, 11, Place du

Bourguet, 04300 Forcalquier. Tel: 92.75.25.30. Has a good restaurant.

Also in FORCALQUIER is the Auberge Charembeau. Its claim to fame is its excellent, home-cooking restaurant, which is open only to those staying in the hotel.
Auberge Charembeau, N 100, 04300 Forcalquier. Tel: 92.75.05.69. 11 rooms.

The Francois 1èr is a basic, friendly hotel in MANOSQUE.
Francois 1èr, 18, Rue Guilhempierre, 04100 Manosque. Tel: 92.72.07.99. 25 rooms, 85–180 ff.

Le Colombier in ST-ANDRE-LES-ALPES offers a swimming pool and restaurant without being too pricey. The downside is that it is located about a mile outside of town.
Hôtel le Colombier, Route d'Allos, 04170 St-Andre-les-Alpes. Tel: 92.89.07.11. 24 rooms, 145–235 ff.

Actually in the town of ST-ANDRE is the cute Grand Hôtel du Parc, whose added bonus is its good restaurant.
Grand Hôtel du Parc, Place Eglise, 04170 St-Andre-les-Alpes. Tel: 92.89.00.03.

Just over a mile south of ST-ETIENNE-LES-ORGUES is the very prettily situated St-Clair, with a restaurant and swimming pool. Unfortunately, you do have to pay more here.
Hôtel St-Clair, Chemin du Serre, St-Etienne-les-Orgues. Tel: 92.76.07.09. 28 rooms, 130–200 ff.

The Tivoli in SISTERON is a pleasant, medium-priced hotel with a well-above-average restaurant.
Hôtel Tivoli, Place du Tivoli, 04200 Sisteron. Tel: 92.61.15.16. 19 rooms.

Ski buffs should head on up to the winter sports hotel of Pyjama in SUPER-SAUZE. Geared for the snow seeker, it is also open in July and August.
Hôtel Pyjama, Super-Sauze. Tel: 92.81.12.00. 10 rooms. Open: 15 December–Easter and July–August.

VAR

The Auberge de la Tour, near the church in AUPS, is a comfortable spot with the bonus of a restaurant that serves local specialities.
Auberge de la Tour, 83630 Aups. Tel: 94.70.00.30. 24 rooms.

In BANDOL the sea-perched Ker-Mocotte was once home to the Provençal-born movie star Raimu. It's restaurant is open only to guests staying at the hotel.
Hôtel Ker-Mocotte, Rue Raimu, 83150 Bandol. Tel: 94.29.46.53 telex 400 383. 19 rooms, 210–285 ff.

Very reasonable in cost is BANDOL's **Hôtel Bel Ombra**, Rue La Fontaine, 83150 Bandol. Tel: 94.29.40.90.

The Pont d'Or is a pleasant, inexpensive hotel in BARJOLS.
Hôtel Pont d'Or, Route St-Maximin, 83670 Barjols. Tel: 94.77.05.23. 15 rooms.

Those appalled by the high prices of the Côte will be relieved by those at BORMES-LES-MIMOSAS' **Hôtel Belle-Vue**, Rue Jean Aicard, 83230 Bormes-les-Mimosas. Tel: 94.71.15.17 or 94.71.15.15. 14 rooms.

An ordinary but reliable hotel outside the old town of DRAGUIGNAN is the **Hôtel Parc**, 21, Boulevard de la Liberté, 83300 Draguignan. Tel: 94.68.53.84. Has 20 rooms, 170–260 ff.

The Château in ENTRECASTEAUX is quite expensive but certainly the most unusual place to stay in the area. It offers a sort of bed-and-breakfast – but only in three rooms. The setting is, to put it mildly, sumptuous.
Château d'Entrecasteaux, Entrecasteaux. Tel: 94.04.43.95.

In the heart of the Massif de L'Esterel by FREJUS is the luxurious coaching inn Auberge des Adrets. Placed in a fabulously isolated location, it boasts a tennis court, swimming pool and nearby golf course and riding.
Auberge des Adrets, RN 7, 83600 Fréjus. Tel: 94.40.36.24. 10 rooms.

Pleasant, centrally located and easily affordable in HYERES is the **Hôtel Du Portalet**, 4, Rue Limans, 83400 Hyères. Tel: 94.65.39.40. 18 rooms.

More expensive but by the beach at HYERES-PLAGE is the Pins d'Argent with a swimming pool and restaurant.
Hôtel Pins d'Argent, Hyères-Plage. Tel: 94.57.63.60. 20 rooms, 240–450 ff.

Those wishing to escape the craze of St-Tropez should head up to the pretty, tiny La Claire Fontaine, set on the small main square of LA GARDE-FREINET in the Massif des Maures.
Hôtel La Claire Fontaine, Place Vieille, 83310 La Garde-Freinet. Tel: 94.43.67.41. Has 8 rooms, 150–180 ff.

A simple but comfortable hotel in SALERNES is **Hostellerie Allegre**, 83690 Salernes. Tel: 94.70.60.30. 26 rooms.

Located right next to the Saracen tower in SANARY is the inexpensive little **Hôtel Tour**, Quai General de Gaulle, 83110 Sanary-sur-Mer. Tel: 94.74.10.10. 27 rooms.

The Hôtel des Deux Rocs, set in a lovely spot above the town of SEILLANS, is a converted 18th-century manor house, complete with terrace and fountain outside. The price is medium-high, and it has a restaurant.
Hôtel des Deux Rocs, 83440 Seillans. Tel: 94.76.87.32. 15 rooms.

A centrally located hotel in TOULON of medium-range price is the **Hôtel Amirauté**, 4, Rue Adolphe-Guiol, 83000 Toulon. Tel: 94.22.19.67. 64 rooms.

Less expensive and close to the TOULON tourist office is the **Hôtel Europe**, 7 bis, Rue Chabannes, 83000 Toulon. Tel: 94.92.37.44. 29 rooms.

Probably the most expensive and palatial place in the Var, the Bastide de Tourtour (in TOURTOUR) is strictly for wealthy visitors with a penchant for the pretentious.
Bastide de Tourtour, 83690 Tourtour. Tel: 94.70.57.30; telex: 970 827. Has 26 rooms, 330–1,000 ff.

ALPES-MARITIMES

In AURON, Las Donnas is open in the summer and the winter for moderate comfort and moderate rates.
Las Donnas, Rue Marie Madeleine, 06660 Auron. Tel: 93.23.00.03; telex: 470 300. 48 rooms, 165–280 ff.

Au Logis du Puei is a very nice, quiet hotel with 16 rooms and an excellent restaurant in BOLLENE-VESUBIE.
Au Logis du Puei, 06450 Bollène-Vésubie. Tel: 93.03.01.05. 16 rooms, 150–260 ff w/o pension, 565–685 ff w/pension.

A stay at the Château de la Chevre d'Or in EZE may well be one you will never forget. Its four-star restaurant offers a view of the coast from Nice to Monaco, and its hotel combines a warm provincial elegance with Riviera élan.
Château de la Chevre d'Or, Rue de Barri, 06360 Eze. Tel: 93.41.12.12; telex: 970839. 8 rooms, 900–1,700 ff. Open: March–November.

The Domaine du Foulon is a pleasant enough hotel beside a park about 2½ miles (4 km) outside GREOLIERES.
Domaine du Foulon, Route de Gourdon, 06620 Gréolières. Tel: 93.59.95.02. 13 rooms, 170 ff.

The Hôtel Les Grands Prés is a nice quiet hotel in LEVENS.
Hôtel Les Grands Prés, 06670 Levens. Tel: 93.91.70.35.

Very inexpensive and not exactly deluxe – but acceptable – in ST-ETIENNE-DE-TINEE is the **Pinatelle**, 06660 St-Etienne-de-Tinée. Tel: 93.02.40.36. 14 rooms.

The **Auberge de la Roya** is a comfortable, cheap hotel run by very nice people in the SAORGE. Tel: 93.04.50.19.

COTE D'AZUR

It's hard to find a bargain in high season on the Côte d'Azur, but Postillon in the old town of ANTIBES is pretty close to one. It's pleasant enough – but not by the sea.
Relais du Postillon, 8, Rue Championnet, 06600 Antibes. Tel: 93.34.20.77. 14 rooms, 150–240 ff.

Not too bad – for the Côte – in price and set across from the port in BEAULIEU-SUR-MER is the ☆☆☆-star Frisia. Best are the rooms with terraces.
Hôtel Frisia, 2, Boulevard du Général Leclerc, 06310 Beaulieu-sur-Mer. Tel: 93.01.01.04. 35 rooms, 180–490 ff. Closed: November.

A good half of the visitors to CANNES each year come either for a convention or something like the International Film Festival and don't intend to tighten their belts. But, no matter what, if you want to stay in Cannes, you're going to have to spend a little. So why not spend it at the Sofitel? Standing on the tip of the Croisette, its beautifully appointed rooms that look out over the sea seem right out of a fine country home. Swimmers enjoy the beachfront across the street and the rooftop pool. Everyone enjoys the excellent and courteous service. This is a first-class business and luxury hotel all the way.
Sofitel-Mediteranée, 2, Boulevard Jean Hibert, 06400 Cannes. Tel: 93.38.87.87; telex: 470728. 152 rooms, 350–1,090 ff.

The Auberge Les Santons in MENTON has a great view of the Riviera.
Auberge Les Santons, Colline de l'Annonciade, 06500 Menton. Tel: 93.35.94.10.

NICE has scores and scores of places to stay within all different price ranges. One small and very inexpensive spot is the ☆-star Hôtel Rialto. Only two of its eight rooms have their own showers, but it does have the distinction of being only two blocks from the ocean. Also, most rooms have kitchenettes.
Hôtel Rialto, 55, Rue de la Buffa, 06000 Nice. Tel: 93.88.15.04. 105–180 ff (all doubles), no credit cards.

You can't go wrong with the Sofitel Splendide in NICE, close to the city centre and only 440 yards (400 metres) from the sea.
Sofitel Splendide, 50 Boulevard Victor Hugo, 06048 Nice. Tel: 93.87.63.64; telex: 46.09.38. 130 rooms.

So you want to play with the big boys? Well, then there's only one place for you – Le Grand Hôtel du Cap Ferrat on ST-JEAN-CAP-FERRAT. Exclusive is this hotel's middle name. Its first name is expensive.
Le Grand Hôtel du Cap Ferrat, Boulevard Général de Gaulle, 06290 St-Jean-Cap-Ferrat. Tel: 93.76.00.21; telex: 470184. 66 rooms, 990–2830 ff. Open: July–October.

Another place to throw some money around is the Byblos in ST-TROPEZ.
Hôtel Byblos, Avenue Paul Signac, 83990 St-Tropez. Tel: 94.97.00.04; telex: 470235. 107 rooms, 860–1960 ff. Open: March–November.

CAMPGROUNDS

Camping is one of the most popular ways of visiting rural Provence, for the French and foreigners alike. It is certainly the least expensive type of accommodation for those without resident relatives or friends. The Bouches-du-Rhône has, by itself, nine campsites with ☆☆☆☆ stars (campsites are rated just like hotels), 26 campsites with ☆☆☆ stars, 57 campsites with ☆☆ stars and three campsites with ☆ star. Even along the chic shores of the Côte d'Azur, numerous decent and affordable campgrounds – for recreational vehicles and/or tents – can be found.

For a complete list, check with the tourist office of the area in which you are planning to stay or purchase a *Guide to Camping in France* by Michelin. It's not widely available outside France, but try your local French bookstore.

In the Bouches-du-Rhône
Chambre Syndicale de l'Hôtellerie de Plein Air, Camping Marina-Plage, 13127 Vitrolles. Tel: 42.89.31.46; or Comité Départemental du Tourisme, 6, Rue du Jeune Anacharsis, 13001 Marseille. Tel: 91.54.92.66.

In the Alpes-Maritimes/Côte d'Azur
Centre Départemental de Camping, Office du Tourisme de Nice-Parking, Parking Ferber, 06200 Nice. Tel: 93.21.05.21.

GITES RURAUX

The *Gîtes Ruraux* (farmhouse accommodations) are a great way to dig right into Provençal life. Basically, the *gîtes* are traditionally arranged houses or flats situated nearby a farm or village. They can be rented for anything from a weekend to several weeks and are a great alternative from the hotel circuit for those (especially with children) looking for a peaceful respite in the countryside.

To rent a *gîte*, you first buy a catalogue appropriate to the region where you would like to stay from that *département's* Reservations Service. Inside you will find descriptions, photos, prices and phone numbers for a large selection of farmhouses. When you've found one that appeals to you, you can either contact the owners directly or go through the departmental Reservation Service. The latter will also offer advice and additional information.

In the Vaucluse
Gîtes de France, Chambre Départementale de Tourisme, La Balance, Place Campana - BP 147, 84008 Avignon Cedex. Tel: 90.85.45.00. Catalogue: free.

In the Bouches-du-Rhône
Gîtes de France, Domaine du Vergon, 13370 Mallemort. Tel: 90.59.18.05. Catalogue: 20 ff.

In the Alpes-de-Hautes-Provence
Gîtes de France, Maison du Tourisme, Rond-Pont du 11 Novembre, 04000 Digne. Tel: 92.31.52.39. Catalogue: 30 ff.

In the Var
Gîtes de France, Conseil Général, 83600 Draguignan Cedex-B.P. 215. Tel: 94.67.10.40. Catalogue: 30 ff.

In the Alpes-Maritimes
Gîtes de France, 55, Promenade des Anglais, 06000 Nice. Tel: 93.44.39.39. Catalogue: 30 ff.

YOUTH HOSTELS

There are plenty of youth hostels throughout Provence. If you join AYH in the United States, they will give you a full listing.

In the Bouches-du-Rhône
Service Centre de Vacances, Direction Départementale Jeunesse, 20, Av. de Corinthe, 13006 Marseille. Tel: 91.78.44.88; or Centre d'Information Jeunesse, 4, Rue de la Visitation, 13004 Marseille. Tel: 91.49.91.55.

On the Côte d'Azur
Centre Information Jeunesse, Esplanade des Victoires, 06300 Nice. Tel: 93.80.93.93.

UNIVERSITY RESIDENCES

To arrange a stay in a university residence, you should go through the "Crous" university extension service.

In Aix-en-Provence
Crous, Cité "Les Gazelles", Avenue Jules-Ferry, 13621 Aix Cedex 1. Tel: 42.26.33.75.

In Marseille
Crous, 38, Rue du 141 e-RIA, 13331 Marseille Cedex 3. Tel: 91.95.90.06

FOOD DIGEST

WHAT TO EAT

In place of France's traditional rich cream sauces (although you can always find these, if it's what you want), Provençal cuisine uses only the freshest of pungent herbs, garlic and olive oils to enhance meals of the most delicate local meat, fish or shellfish served with the most succulent, sun-bursting tomatoes, eggplant or asparagus. It is hard to imagine anyone, no matter what the taste or diet, going hungry in Provence.

Among the specialities of Provence are:

gigot de mouton	leg of lamb, especially that of Sisteron
daube	beef braised with spices and red wine
gardianne	a strictly Camarguais bull stew with olives
pieds-paquets	tripe stuffed with garlic, onion, etc.
anything *farci*	stuffed meat, fish, fowl or vegetable
ratatouille	vegetable stew
bourride	white fish served in a clear soup with *aioli*
soupe au pistou	a herby vegetable soup with beans
soupe de poisson	an all-liquid fish soup, served with a spicy mayonnaise called *rouille* and croutons
poutargue	mullet eggs grated in oil – "white caviar"
anchoïade	crushed anchovies and olive oil on bread
brandade	crushed cod with olive oil
tapenade	cream of black olive, served on bread
aïoli	garlic-mayonnaise condiment
goat cheese	
bouillabaisse	a gaggle of different rockfish such as *rascasse*, *fielas* (eel) and *St-Pierre* served whole in a bath of *soupe de poisson*

Perhaps the most celebrated aspect of the cuisine is its use of the *herbes de Provence* – wildly aromatic herbs like basil and oregano, grown all over the dry and sun-parched Provençal countryside.

Of course, where exactly you are in Provence will determine the local speciality. Along the coast it may include fish and crustaceans, while lamb and *farcis* are found inland. The Camargue has its own, exceedingly exotic cuisine, as seems appropriate for the cowboy life.

The Vaucluse and Bouches-du-Rhône are particularly famous for their abundance of luscious produce – the melons of Cavaillon, cherries of Malaucène, strawberries of Carpentras, peaches of Cabannes, olives of the Crau Plain, etc. – but, everywhere, the fruit and vegetables will be fresh.

Various areas are known for their desserts:

From Carpentras:	*berlingots* (a sort of caramel candy); and chocolate truffles.
From Aix:	*calissons* (an almond-paste confectionery).
In Marseille:	*navettes* (a sweet biscuit shaped as a boat); and *pompes à l'huile*.
From Puyricard:	chocolate.
From Allauch:	*suce-miel* (honey-based candy).
The Alpes-Maritimes:	honey cakes and candy.
From all over:	*nougat* (a sugar-paste candy with nuts); and *fougasses* (bread with nuts, olives or cheese).

WHERE TO EAT

Look for basic, family-run restaurants with limited, fresh-daily menus, and you can't go wrong. Markets are another great place for eating. Bring a knife and wander.

VAUCLUSE

For contemporary bistro-style food with a Provençal touch, try La Fourchette II in AVIGNON. They don't take credit cards, but the menu of the day is always affordable.
La Fourchette II, 17, Rue Racine, 84000 Avignon. Tel: 90.85.20.93.

Tante Jeanne's address is a well-guarded, unpublished secret near Apt and not far from the hamlet of BUOUX (*see article on cuisine for more clues*) with no telephone. Nonetheless, true fans of Provençal cuisine should consider the search a small price to pay for having Sunday lunch here. The meal begins at 11.30am precisely.
Tante Jeanne, on an unmarked dirt road, near Buoux. No telephone.

You must be sure to make reservations ahead of time at Le Beffroi in CAROMB, for the locals are committed fans of its delicious menu.
Hostellerie Le Beffroi, 84330 Caromb. Tel: 90.62.45.63.

For a delicious traditional Provençal meal in a cozy, restored *mas*, stop at Mas de Cure Bourse in L'ISLE-SUR-LA-SORGUE.
Le Mas de Cure Bourse, Route de Caumont, RD25, 84800 L'Isle-sur-la-Sorgue. Tel: 90.38.16.58.

Les Lavandes restaurant in MONIEUX has a spectacular view, and its charm is enhanced by the many riders who come into the mountain to dine here, tying their horses in the square adjacent to the restaurant.
Les Lavandes, Place Leon Doux, 84390 Monieux. Tel: 90.64.05.08.

The little Saule Pleureur restaurant in MONTEUX well deserves the reputation that keeps guests coming out of their way to dine here.
Le Saule Pleureur, Le Pont des Vaches, 84170 Monteux. Tel: 90.62.01.35.

You can sit on the tranquil dining terrace of Le Mesclun in SEGURET and point to the vineyard whose wine you would like to have with your meal. In addition, the portions are lavish and the desserts simply irresistible.
Le Mesclun, Rue des Poternes, 84110 Séguret. Tel: 90.46.93.43.

Lou Barri is a charming tea room in SEGURET with a very pretty view.

A bit off the beaten track but worth a stop is the Restaurant de France just northeast in SERIGNAN near Orange.
Restaurant de France, 12, cours Joel-Estève, Serignan. Tel: 94.70.06.83.

Even if you don't choose to stay at the Hostellerie la Magnaneraie in VILLENEUVE-LES-AVIGNON, you can still enjoy a sumptuous feast on its well-shaded pleasant porch.
Hostellerie la Magnaneraie, Rue de Camp de Bataille, 30400 Villeneuve-les-Avignon. Tel: 90.25.11.11.

BOUCHES-DU-RHONE

For regional specialities AIX-EN-PROVENCE-style try **Le Félibre**, hidden away on one of the old town's myriad back streets. To begin, choose the sampler of local appetisers.

More Provençal cuisine can be sampled at AIX-EN-PROVENCE's **Côte d'Aix**.

A good old stand-by in AIX-EN-PROVENCE is Le Madeleine. It's not exotic, but you can be confident that you will eat well.
Le Madeleine, Place de Verdun, 13100 Aix-en-Provence.

An excellent and slightly more atmospheric spot for traditional French food in AIX-EN-PROVENCE is **Le Cave de Mon Oncle Tam**.

Le Temps de L'Heure in AIX-EN-PROVENCE doesn't serve dinner or any hard alcohol and the service can be unbelievable slow, but it's still an extremely popular place for tea and dessert, a cold beer or glass of wine. Its *tarte maman* (apple tart) is not to be beaten.
Le Temps de l'Heure, Rue Vauvenargues, 13100 Aix-en-Provence.

For a late-night *boulangerie* (bakery) in AIX, there is Le Gibassier. Insomniacs will appreciate their 2am–1pm and 2–8pm hours.
Le Gibassier, 46, Rue Espariat, 13100 Aix-en-Provence. Tel: 42.27.53.54.

The perenially popular Vaccarès in ARLES counts among its illustrious clientele writers Michel Tournier, Yvan Audouard and Daniel Boulanger as well as fashion designer Christian Lacroix. Menu at 145–180 ff.
Le Vaccarès, Place Forum, 13200 Arles. Tel: 90.96.06.17.

Another well-liked ARLESIENNE restaurant that specialises in Camargue cuisine is Lou Gardianoun. 75–150 ff.
Lou Gardianoun, Rue Noguier, 13200 Arles. Tel: 90.93.66.28.

Rice is what makes the Camargue go round – and, in turn, ARLES. For some delicious examples in a very interesting place, stop by the **Rizerie du Petit Manusclat**, Le Sambuc, 13200 Arles. Tel: 90.98.90.29.

If you're dying for fresh salad, American health-food style, and don't mind putting up with an uppity owner and slightly inappropriate prices try Vitamine in ARLES.
Vitamine, Rue Dr Fanton, 13200 Arles. Tel: 90.93.77.36.

In CASSIS, most of the harbour-front restaurants can be counted on to serve excellent fresh fish. Outstanding among them, however, is the delectable Chez Gilbert. And, for the quality of the meal, the price is not outrageous.
Chez Gilbert, Quai Baux, 13260 Cassis. Tel: 42.01.71.36.

The well-heeled enjoy La Presqu'île's fabulous view of the cliffs in CASSIS. Its menu, also, is pretty special, although some might say pretentious. Expensive.
La Presqu'île, Direction les Calanques, 13260 Cassis. Tel: 42.01.03.77.

La Regalido in FONTVIEILLE is not only a culinary but an aesthetic treat. The restaurant is set in a restored olive-oil mill brightly decorated with flowers. The menu is fairly costly, but they do take American Express, Diners Club and Visa.
La Regalido, 13990 Fontvieille. Tel: 90.97.60.22.

Les Arsenaulx in MARSEILLE not only has an excellent, elegant and inspired menu, it is imaginatively located within an ancient stone arsenal adjoined to a bookstore and tea house. Much of the fare is fresh fish, but don't expect any old-fashioned homecooking. It's not wildly expensive (120–150 ff) and it's a real find. Ultra-cool as only Marseille can do it – hip without being at all adolescent.
Les Arsenaulx, Restaurant-Salon de Thé, 25, cours d'Estienne d'Orves, 13001 Marseille. Tel: 91.54.77.06.

Only 14 restaurants belong to the "Guild of the Bouillabaisse Marseillaise" – 12 of which are in Marseille. Always phone ahead of time since the dish takes hours to prepare. One guild member along the Vieux Port in MARSEILLE is Le Miramar. It's served elegantly here – for a hefty price.
Le Miramar, 12, Quai du Port, 13002 Marseille. Tel: 91.91.10.40.

Another guild member, located by the edge of the MARSEILLAISE seaside, is **Chez Fonfon**, 140 Vallon des Auffes, 13007 Marseille. Tel: 91.52.14.38.

A third guild member just up the road in the Vallon des Auffres of MARSEILLE is L'Epuisette. The establishment's fresh fish, and *bourride*, served on an open terrace, make for a nice alternative to *bouillabaisse*.
L'Epuisette, Vallon des Auffes, 13007 Marseille. Tel: 91.52.17.82

Each table and chair at Les Thes-Tard in MARSEILLE bears a unique design *à* Picasso, and the clientele is generally equally artsy. The food is earthy in a very French way, and on Sunday mornings you can enjoy a musical brunch American or Viennese style for 50 ff. A la carte menu 45–70 ff, set menus at 35 and 40 ff. Open: 11am–2am. Closed: Tuesday.
Les Thes-Tard, 2, Rue Vian, 13001 Marseille. Tel: 91.42.29.74

For excellent Vietnamese food in MARSEILLE, served with traditional grace and at attractive prices, try Restaurant Le Saigon. Best are the innovative dishes that blend Oriental and Provençal styles. Open: noon–2pm and 7–11pm. Closed: Monday.
Restaurant Le Saigon, 8, cours Jean-Ballard, 13001 Marseille. Tel: 91.33.21.72.

O'Stop is an all-night restaurant in MARSEILLE with a charming, rustic atmosphere. 50–80 ff. Open: 3pm–6am.
O'Stop, Place de l'Opéra, 13001 Marseille. Tel: 91.33.85.34.

The Oustau de Baumanière in LES BAUX is one of France's France's very best restaurants. It is also priced accordingly. Here, you won't find any hearty stews, although the basis is still strictly Provençal, with an emphasis on fresh local produce.
Oustau de Baumanière, 13520 Maussane-les-Alpilles. Tel: 90.97.33.07.

Le Bistro du Paradou is a genuine country café in PARADOU, with one daily meal, one price and a real homespun atmosphere. Unsurprisingly, the cuisine is Provençal.
Le Bistro du Paradou, Avenue de la Vallée des Baux, 13125 La Paradou. Tel: 97.32.70.

CAMARGUE

La Fenière is in a pretty farmhouse conveniently located right outside Arles in RAPHELE. It offers both Camarguais and traditional French cuisine. 120–170 ff.
La Fenière, RN 453, Raphèle, Arles. Tel: 90.98.47.44.

Vegetarians are in for a treat at the Hôtel Le St-Gilloir in ST-GILLES, where the menu offers some tasty vegetable-based local dishes, especially during asparagus season.
Hôtel Le St-Gilloir, St-Gilles.

If heading south along the road to SALINS-DE-GIRAUD, make a stop at the **Boduc Lighthouse** to try Juju's fresh fish.

A good local restaurant in AIGUES-MORTES is **La Camargue**, Rue Republique, 30220 Aigues-Mortes. Tel: 66.53.86.88.

ALPES-DE-HAUTE-PROVENCE

Although it is closed most of June, Sunday nights and all of Monday, La Mangeoire is a popular place in BARCELONETTE.
La Mangeoire, Place 4-Vents, 04400 Barcelonette. Tel: 92.81.01.61.

Le Grand Paris in DIGNE serves pricey regional dishes that have been critically acclaimed.
Le Grand Paris, 19, Boulevard Thiers, 04000 Digne. Tel: 92.31.11.15.

If something special is what you're after, reserve a table for an all-out meal at the beautiful Hostellerie de la Furste just outside Manosque in FURSTE.
Hostellerie de la Furste, Route D4, La Furste. Tel: 92.72.05.95.

The Auberge Charembeau in FORCALQUIER offers good homecooking to the guests in its hotel.

Auberge Charembeau, 04300 Forcalquier. Tel: 92.75.05.69.

A good, inexpensive restaurant in MANOSQUE is **André**, 21 bis, Place Terreau, 04100 Manosque. Tel: 92.72.03.09.

For a pleasant, medium-high priced meal in MOUSTIERS-STE-MARIE reserve a table at **Les Santons**, Place de l'Eglise, 04360 Moustiers-Ste-Marie. Tel: 92.74.66.48.

For high-class country-style cooking in a beautiful setting, try the Auberge de Reillanne in REILLANNE.
Auberge de Reillanne, 04110 Reillanne. Tel: 92.76.45.95.

VAR

Restaurants in the Massif d'Esterel are generally more expensive than their northern brethren, but Auberge de la Rade in AGAY is both good and quite reasonable.
Auberge de la Rade, Bord de Mer, 83700 Agay. Tel: 94.82.00.37.

At Parc in BANDOL you can enjoy medium-priced fish specialities.
Parc, Corniche Bonaparte, 83150 Bandol. Tel: 94.29.52.10.

Lou Calen in COTIGNAC is deservedly one of the most popular quality restaurants in the north central Var. It is set within an equally well-liked hotel, and both are a good find.
Lou Calen, 1, cours Gambetta, 83570 Cotignac. Tel: 94.04.60.40.

Les Deux Cochers in DRAGUIGNAN has a pleasant terrace.
Les Deux Cochers, Boulevard G-Peri, 83300 Draguignan. Tel: 94.68.13.97.

A real find in FOX-AMPHOUX is the **Auberge du Vieux Fox**, Place de l'Eglise, Fox-Amphoux. Tel: 94.80.71.69.

The Bello Visto in GASSIN is aptly named – drinks and/or meals served on its terrace are accompanied by a fabulous view. The establishment also has 9 rooms for overnight guests.
Bello Visto, Auberge la Verdoyante, 83990 Gassin. Tel: 94.56.16.23.

Also in GASSIN is a small *auberge* with good quality meals (but no rooms).
Auberge la Verdoyante, 83990 Gassin. Tel: 94.56.16.23.

For quality dining in HYERES try **Le Delphin**, 7, Rue Roux-Seigneuret, 83400 Hyères. Tel: 94.65.04.27.

La Faucado in LA GARDE-FREINET has the special benefit of a beautiful open-air terrace.
La Faucado, Route Nationale, 83310 La Garde-Freinet. Tel: 94.43.60.41.

Also in LA GARDE-FREINET is a trendy restaurant/bar called **Le Lézard**, 7, Place du Marché, 83310 La Garde-Freinet. Tel: 94.43.62.73.

You will find a good sampling of TOULON's many North-African residents' cuisine at Pascal in the old town.
Pascal, Square L.-Varane, 83000 Toulon. Tel: 94.92.79.60.

Au Sourd is a good standby in TOULON, unless you are there in July when it is, rather inconveniently, closed.
Au Sourd, 10, Rue Molière, 8300 Toulon. Tel: 94.92.28.52.

You'll be assured of a civilised, respectably priced dinner at Chez Nous in ST-MAXIMIN.
Chez Nous, Boulevard Jean Juarès, 83470 St-Maximin-la-Ste-Baume. Tel: 94.78.02.57.

ALPES-MARITIMES

The Château de la Chèvre d'Or in EZE has a ☆☆☆☆-star restaurant with a view from Nice to Monaco.
Château de la Chèvre d'Or, Place Felix-Fauré, 06450 St-Martin Vésubie. Tel: 93.03.21.28.

Les Plantanes, "Chez Mario," offers the specialties of the Saorge Valley: Raviolis!
Les Plantanes, 17, route Nationale, 06540 Fontan. Tel: 93.04.53.06.

Roger Vergé, the chef at Moulin de Mougins (in MOUGINS), is the most influential of all the great chefs of the Côte d'Azur. A visit to his restaurant is to be cherished. In summer, reserve at least two or three weeks ahead of time and, whenever you go, bring a good 600 ff.
Moulin de Mougins, Quartier Notre-Dame de Vie, 06250 Mougins. Tel: 93.75.78.24.

Also in MOUGINS and stressful to the wallet is Le Relais à Mougins. In summer, reserve at least ten days ahead of time.
Le Relais à Mougins, Place de la Mairie, 06250 Mougins. Tel: 93.90.03.47.

Next door also in MOUGINS, for much less frightening prices, is Le Feu-Follet. Reserve a few days ahead of time.

Le Feu-Follet, Place de la Mairie, 06250 Mougins. Tel: 93.90.15.78.

For a well-prepared, unpretentious dining experience at reasonable prices try the Auberge de la Madone in PEILLON.
Auberge de la Madone, Peillon Village, L'Escarène, 06440 Peillon. Tel: 03.79.91.17.

Try Chez Henri for a nice, mid-priced restaurant in ST-PAUL-DE-VENCE. No reservations.
Chez Henri, Place du Village, 06570 St-Paul-de-Vence. Tel: 93.32.82.75.

COTE D'AZUR

A great and fairly reasonably priced meal can be had along the ancient sea wall of ANTIBES at **Les Vieux Murs**, Avenue Admiral-de-Grasse, 06600 Antibes. Tel: 93.34.06.73.

The excellent *soupe de poisson* at Le Portofino along the port of BEAULIEU-SUR-MER is a bargain at 60 ff. Be sure to order the day before since it is made fresh for you.
Le Portofino, Beaulieu Port, 06310 Beaulieu-sur-Mer. Tel: 93.01.16.30.

The other fish restaurant in BEAULIEU is Le Chicorée. It has no fixed menu, since all depends on the day's catch.
Le Chicorée, 5, Rue du Lieutenant Colonelli, 06310 Beaulieu-sur-Mer. Tel: 93.01.01.27.

For the best flame-oven-cooked pizza in CANNES, maybe anywhere, head to Le Pizza on the Croisette. The tiny, tight-shirted proprietors, with classically "Italian" temperaments they do nothing to hide, also offer a wide variety of Italian favourites that anyone can afford.
Le Pizza, Quai St-Pierre, 06400 Cannes.

For a bit of Provençal along the Riviera at medium-high prices (200–500 ff) join the local regulars at La Mère Besson in CANNES. They don't serve lunch in July and August and are always closed on Sunday.
La Mère Besson, 13, Rue des Frères Pradignac, 06400 Cannes. Tel: 93.39.59.24.

Everyone with well-shined shoes knows the exclusive Restaurant Bacon in CAP D'ANTIBES. Reserve a week ahead and ask for a table with a view. (Keep in mind that they only take Diners Club and Visa, and lunch alone starts at 250 ff.)
Restaurant Bacon, Boulevard Bacon, 06600 Cap d'Antibes. Tel: 93.61.50.02.

"La Mère Barale", *restauratrice extraordinaire*, is probably one of the best-known personalities in NICE. Anything out of the kitchen at Chez Barale tastes fabulous, but specialities include *estocaficada* and Provençal ravioli *(ralhola)*. They serve dinner only, for a menu under 200 ff that includes wine. Closed: Sunday, Monday and all of August.
La Barale, 39, Rue Beaumont, 06300 Nice. Tel: 93.89.17.94.

La Merenda is a strictly unpretentious bistro with terrific regional food. It makes for a nice change after the touristic fanfare found in many places along the Côte. Closed: Saturday night, Sunday, Monday and all of February and August. No phone.
La Merenda, 4, Rue Terasse, 06000 Nice.

La Saleya is a funky brasserie, set in a great place to hang out – NICE's flower market.
La Saleya, 06300 Nice. Tel: 93.62.29.62.

The food at Vien Dong in ST-TROPEZ is very good, but that's only half the attraction. Its owner is the other half – a former Mr Universe.
Vien Dong, St-Tropez. Tel: 94.97.09.78.

DRINKING NOTES

Provençal vineyards are rich and varied, producing reds, whites and rosés of decent to superb quality.

The most famous wines made in Provence come from the *département* of the Vaucluse and are more properly classified under the "Côtes du Rhône." Among these, Châteauneuf-du-Pape is the most celebrated. Full-bodied with a fruity flavour, the reds, in particular, are considered among the top French wines. Also excellent are the slightly lesser-bodied reds of Gigondas, the Vaqueyras and Séguret wines, and sweet muscat wine from Beaune de Venise.

The "Côtes de Provence" comprises the region between Aix and Nice, of which the most famous are the Bandols. Aix-en-Provence and the Baux-de-Provence vineyards have their own *appellation*. These wineries are situated between Mount Ste-Victoire and the Rhône in the Arles district. The best of their wines are the delicious whites from Cassis and Palette.

In Aigues-Mortes of the Camargue, the local wine is Listel. It is remarkable for being a *vin de sable*, which means, literally, that it is made from vines that grow directly out of the sand. The white is called *gris de gris* and the red is called *rubis*.

Provence also is the home of the famed *pastis*, which is an anise-flavoured aperitif.

In the wine-growing region of the Vaucluse, you will find numerous *caves* open for wine-tasting. For example:
La Domaine les Palliers, 84190 Gigondas. Tel: 90.65.85.07. Prop. M. Roux.

A couple from the Bouches-du-Rhône are:
Château d'Estoublon-Mogador, 13990 Fontvieille. Tel: 90.54.64.00. Prop. M. Lombrage père & fils. Open for on-site sale of their wines and a chat with

the long-time owner who speaks French and English and is very talkative.

Château Revelette, 13490 Jouques. Tel: 42.63.75.43. Prop. Peter Fischer. Open: Monday, Wednesday and Saturday for tasting, a tour of their cave and on-site sale of their young wines. The chateau is located just south of the Durance.

THINGS TO DO

CITY

Every major city and many of the larger towns in Provence will have at least one museum worth visiting, at least one historical site and maybe also a prehistorical site, a summer music and/or theatre festival, a colourful market, a main boulevard lined with cafés for people watching, and an assortment of good restaurants. Unless you are looking for Monets and Davids (which you *can* find in some cities), you will find no shortage of interesting distractions.

COUNTRY

Many of the smaller towns dotting the countryside sport their own historical monuments, a number of local festivals and possibly a cultural one as well, a clutch of cafés and restaurants, and a vibrant market. In addition, the possibilities for sporting activities is enormous: walking, hiking, biking, skiing (in winter), sailing (along the coast), water skiing and swimming (by the lakes or the coast), fishing, hang gliding, riding, etc.

TOUR OPERATORS

As with travel packages, you'd do well to check with your local travel agent or the appropriate French Tourist Office for information concerning tour operators to the destination of your particular interest. However, listed are some tour groups operating at the time this book went to press:

ON LAND

The Alternative Travel Group, 1–3 George Street, Oxford, OX1 2A2, England. Tel: 0865-251195/6; telex: 83147. "Walking through History" tours in Provence.

Air France Holidays, 69 Boston Manor Road, Brentford Mddx, England. Tel: 081-568 6981.

French & International Travel, 383 George Street, Sydney, NSW 2000, Australia. Tel: (02) 290-2523.

De Lux Vacations, 53, Rue Grignan, 13006 Marseille. Tel: 91.33.12.46; telex: 430415 F.

Provence - Voyages, 3, Boulevard Raspail BP 127, 84007 Avignon Cedex. Tel: 90.86.58.42; telex: 432803 F. Travel agent.

Arles Voyages, 14, Boulevard G. Clemenceau, 13200 Arles. Tel: 90.96.88.73; telex: 401609. Travel agent.

Art Trek, Box 807-F, Bolina, CA 94924, United States. Tel: (415) 868-1836. 18-day painting seminars in the south of France.

Daily-Thorp Travel, 315 W. 57th Street, New York, NY 10019. Tel: (212) 307-1555. Music festival tour to Paris, Aix-en-Provence, Orange, Nice.

The French Experience, 171 Madison Avenue, Ste. 1505. New York, NY 10016. Tel: (212) 683-2445. Self-drive tours to Provence-Riviera.

Serenissima Travel, 381 Park Avenue South, Ste. 914, New York, NY 10016. Tel: (212) 953-7720, (1-800) 358-3330. Deluxe tours in Provence.

Esplanade Tours, 581 Boylston Street, Boston, MA 02116, United States. Tel: (617) 266-7465 or (1-800) 343-7184. Escorted motorcoach tours to Provence & the Côte d'Azur.

ON WATER

Caminav, Base Fluviale de Carnon, 34280 La Grande Motte. Tel: 67.68.01.90. Barge rental on the Petit-Rhône and Camargue.

Navig France, 172 Boulevard Berthier, 75017 Paris. Tel: 1.46.22.10.86; telex: 642502. Deluxe houseboat and barge holidays in the Camargue.

ON A BICYCLE

Fédération Francaise de Cyclotourisme, 8, Rue Jean-Marie-Jego, 75013 Paris. Tel: (1) 45.80.30.21.

ON THE RIVIERA

Kuoni Travel, 3, Boulevard Victor Hugo, 06000 Nice. Tel: 93.16.08.01.44; telex: 460946. Individuals and groups, all over Provence, upper-scale accommodation, tours, cruises, vacation rentals.

Novatours, 14, Avenue de Madrid, 06400 Cannes. Tel: 93.69.47.47; telex: 470934. Multilingual staff, seminars, conventions, holidays in the mountains, exclusive agent to Cure Center of Cannes.

Tourazur-Villegiatour, 7, Promenade des Anglais, 06000 Nice. Tel: 93.37.80.08; telex: 970579. Individuals, groups, vacation rentals on Riviera and Alps.

TOUR GUIDES

You will find that the Offices de Tourisme of most major touristic cities offer excellent sightseeing tours, although they do tend to be in French. If not, they can always help you to find an organisation that does and in the language of your choice. Otherwise, you can check with one of the local travel agents or one of the tour operators listed directly above.

Following are a few, limited names:

RIVER TOURS ON THE RHONE

Tiki III, Capitaine Edmond Aupy, Rue du Petit-Rhône, 13640 Les-Stes-Maries-de-la-Mer. Tel: 90.97.85.89. Boat rides along the Rhône and Camargue.

Europaboat, Route des Stes-Maries-de-la-Mer, 13200 Arles. Tel: 90.93.74.34. Morning and all-day cruises from Arles to Avignon, to Aigues-Mortes, along the Petit-Rhône, "arlesienne" and "camarguaise." During the summer. 100-person capacity.

Mireio, Grands Bateaux de Provence, Allée de l'Oulle, 84000 Avignon. Tel: 90.85.62.25. Air-conditioned with a capacity for 250 people. All year round except 11 January–12 February. Dinner cruises, dancing cruises, lunch cruises to Arles and Roquemaure, etc.

L'Arlène, Agna Viva, 60, Avenue Leclerc, 13200 Arles. Tel: 78158.36.34; telex: 330661. Year-round cruises in a liner with 50 cabins.

Le Cygne, Rue Fourrier, 30300 Beaucaire. Tel: 66.59.45.08. Seven-hour trips from Avignon down the Rhône through Arles to Aigues-Mortes. Discounts for senior citizens.

Cié de Navigation Mixte, 3, av. du Président-Carnot, 69002 Lyon. Tel: 78.42.22.70; telex: 310000. Boat rides along the Rhône and Camargue.

IN THE CAMARGUE

Mas Sauvage, Connaissance de la Camargue Sauvage, Le Paty de la Trinité, 13200 Arles. Tel: 90.97.11.45. Half- to whole-day guided trips in jeep or car or horseback, gypsy evenings, open all year round.

SIGHTSEEING ON THE RIVIERA

Santa Azur, 11, Avenue Jean-Medecin, 06000 Nice. Tel: 93.85.46.81; telex: 461029. Sightseeing by motorcoach.

CTM, 5, square Merimée, 06400 Cannes. Tel: 93.39.79.40; telex: 470810. Sightseeing by motorcoach.

Palais Lascaris, 15, Rue Droite, 06300 Nice. Tel: 93.62.05.54. Guided tours with cultural or historical themes.

Information desk at the train station, 33, Avenue Malaussena, 06000 Nice. Tel: 93.88.28.56. Guided tours by train to Provençal villages.

CULTURE PLUS

MUSEUMS

A complete list of museums worth visiting in Provence and along the Côte d'Azur would be endless. It is best to look directly within the text of this book for descriptions and recommendations. However, here are a few names and times:

PAYS D'ARLES

Musée Réattu in Arles. Open: October–March 10am–12.30pm and 2–5.45pm; April–May 9.30am–12.30pm and 2–7pm; June–September 9.30am–7pm.

All other museums in Arles. Open: November–February 9am–noon and 2–4.30pm; March 9am–12.30pm and 2–6pm; April 9am–12.30pm and 2–6.30pm; May 9am–12.30 and 2–7pm; June–9 August 8.30am–7.30pm.

Château Barbentane in Barbentane. Outside season: open only on Sunday. In season: closed Wednesday July–September, open every day 10am–noon and 2–6pm at 30-minute intervals.

Cathédral d'Images in Les Baux. Open: 18 March–11 November 10am–7pm. (After October closes at 6pm and on Tuesday.)

Musée des Alpilles in St-Rémy. Open: April–June and September–October 10am–noon and 2–6pm; July–August 10am–noon and 2–7pm, also Friday and Saturday 9–11pm, free. November–December 10am–noon and 2–5pm. Closed: January–March

AIX

Musée Granet. Open: 10am–noon and 2–6pm. Closed: Tuesday and holidays.

Musée du Vieil Aix. Open: 10am–noon and 2.30–6pm. Closed: Monday.

Museum d'Histoire Naturelle. Open: 10am–noon and 2–6pm. Closed: Sunday.

Musée Paul-Arbaud. Open: 2–5pm. Closed: Sunday and holidays.

Pavillon Vendome. Open: 10am–noon and 2–6pm. Closed: Tuesday.

Atelier Cézanne. Open: 10am–noon and 2.30–6pm. Closed: Tuesday and holidays.

Fondation Vasarely. Open: 9am–noon and 2–6pm. Closed: Tuesday. (A bus leaves from the Boulevard de la République every half hour.)

MARSEILLE

All museums are closed Tuesdays, and Wednesday mornings. Open: 10am–noon and 2–6.30pm. Admission: free on Sunday mornings, except as noted.

Musée d'Histoire de Marseille. Open: noon–7pm. Closed: Sunday and Monday. Métro: Vieux-Port. Admission: free Wednesday afternoon.
Musée de la Marine. Closed: Sunday. Métro: Vieux Port-Hôtel de Ville.
Musée Cantini. Métro: Vieux Port-Hôtel de Ville.
Musée de Vieux Marseille. Métro: Vieux Port-Hôtel de Ville.
Docks Romains. Métro: Vieux Port-Hôtel de Ville.
Musée d'Archéologie (Borély). Open: 9.30am–12.15pm and 1–5.30pm. Bus: #44 (enter Avenue Clot-Bey) or #19 (enter Parc Borély).
Musée des Beaux-Arts. Métro: Longchamp-Cinq Avenue.
Musée d'Histoire Naturelle. Métro: Longchamp-Cinq Avenue.
Musée Grobet-Labadie. Métro: Longchamp-Cing Avenue.
Musée de Château-Gombert (Museum of Popular Arts and Traditions of Marseille Area). Open: Saturday, Sunday and Monday. Métro: Frais-Vallon, then bus #5

One very special exhibition hall is the **Maison de l'Artisanat et des Métiers d'Art** (21, cours Honoré d'Estienne d'Orves, 13001 Marseille. Tel: 91.54.80.54). The objective of this institution, which represents working artisans throughout the Provence-Alpes-Côte d'Azur region, is to serve as a marketing outlet for and information centre about regional artists and craftspeople.

CAMARGUE

La Palissade, outside Salin de Giraud. On the Route de la Plage de Piemancon. Open: 1 September–15 June Monday–Friday 9am–5pm, weekends by appointment; 16 June–31 August daily, tours 9.30am–2.30pm.
Musée Camarguais, Mas du Pont de Rousty. Open: 1 October–31 March daily except Tuesday 10am–5pm; 1 April–30 September 9am–6pm.
Réserve Nationale de Camargue, La Capelière, Route d.36B. Open: Monday–Saturday 9am–noon and 2–5pm. Admission: free.

ALPES-DE-HAUTE-PROVENCE

Natural History Museum in Riez. Open: April–Oct am and pm, only am the rest of the year. Closed January to mid-February.
Musée Municipal in Digne. Closed: Monday. Admission: free on Sunday.
Alexandra David-Neel Foundation in Digne. Visits by guided tour only. Open: July–September 10.30am, 2pm, 3.30–5pm; October–June 10.30am , 2–4pm.
Le Château Sauvan, 04300 Mane. Tel: 92.75.05.64.

Musée de la Faïence in Moustiers-Ste-Marie. Open: April–October. Closed: Tuesday.
Musée de la Moto in Entrevaux. Open: July and August only.
Musée de la Vallée (Museum of Regional History) in Barcelonette. Open: afternoons only.

TOULON

Naval Museum. Open: 10am–noon and 1.30–6pm. Closed: Tuesday out of season.
Musée de Toulon. Open: 12.30am–7pm.
Musée du Vieux Toulon. Open: 2–6pm. Closed: Sunday.

ELSEWHERE IN THE VAR

Municipal Museum in Hyères. Open: Monday–Friday 10am–noon and 3–6pm, Saturday and Sunday pm only. Closed: Tuesday.
Musée du Freinet in La Garde-Freinet. Open: Monday–Friday 10am–noon. Closed: Saturday and Sunday.
Musée des Arts et Traditions Populaires de Moyenne Provence in Draguinan. Open: 9am–noon and 2.30–6pm. Closed: Sunday morning and Monday.

ALPES-MARITIMES

Musée Fernand Leger in Biot. Open: April–September 10am–noon and 2–6pm; October–March 10am–noon and 2–5pm. Closed: Tuesday.
Musée Picasso (La Guerre et la Paix) in Vallauris. Open: 1 October–31 March 2–5pm; 1 April–30 September 10am–6pm. Closed: Tuesday.
Musée Municipal in Vallauris. Closed: Tuesday.
Musée de l'Automobile in Mougins. Closed: 15 November–15 December.
Musée d'Art et d'Histoire de Provence in Grasse. Closed: November and on weekends.
Villa Fragonard in Grasse. Closed: Saturday and November.
Fondation Maeght in St-Paul-de-Vence. Open: every day.
Chapelle du Rosaire de Matisse in Vence. Open: Tuesday, Thursday and by appointment. Closed: 1 November to mid–December and holidays.

NICE

Musée Nationale Marc Chagall. Open: 1 July–30 September 10am–7pm; 1 October–30 June 10am–12.30pm and 2–5.30pm. Closed: Tuesday.
Musée d'Art Naïf Anatole Jakovsky. Closed: Tuesday and holidays.
Museum d'Histoire Naturelle. Closed: Tuesday, from mid-August to mid-September, and holidays.
Musée Matisse et d'Archéologie. Closed: Sunday

morning, Monday and November.
Musée Massena. Closed: Monday, November and holidays.

ELSEWHERE ON THE COTE

Musée d'Archéologie in Antibes. Closed: November; and Tuesday, except in summer.
Musée Grimaldi Picasso in Antibes. Closed: Tuesday; November; and holidays.
Musée Renoir du Souvenir in Cagnes-sur-Mer. Closed: Tuesday; 15 October–15 November; and holidays.
Musée Cocteau in Menton. Closed: Monday, Tuesday; and holidays
Musée Ephrussi de Rothschild, 062330 St-Jean-Cap-Ferrat. Tel: 93.01.33.09.

GALLERIES

The countryside of Provence has become somewhat of a mecca for painters and artist colonies, and art galleries exhibiting local painters abound. This is particularly true in the Lubéron area of the Vaucluse, the Pays d'Arles and lower Alpes-Maritimes. Larger cities have galleries with changing exhibitions. For names, addresses and current expositions check with the local tourist office.

CONCERTS

Although Provence boasts no well-renowned resident symphonies, music lovers will find no shortage of concerts – classical, traditional or jazz – during the summer here, due to the abundance of cultural festivals. Off-season, pickings become somewhat reduced except in the more cultural cities, such as Marseille, Avignon, Aix-en-Provence, Nice and Monaco, or in abbeys and churches. Following are some random suggestions:

VAUCLUSE

Avignon: Year-round musical soirées at Minit Conservatory.

Concerts by the Orchestre Lyrique de la Region d'Avignon Provence.

Festival of opera, music, etc., second week in July through first week in August (tel: 48.74.59.88).
Lourmarin: Musique d'Eté au Château de Lourmarin, second week in July through August (tel: 90.68.15.23)
Gordes: Festival of jazz and classical music (plus theatre), end July through first week in August

BOUCHES-DU-RHONE

Aix-en-Provence: Festival in July, then varied series from January–April, April–May, and in June.
Arles: All types of music from classical to flamenco during the Festival d'Arles (contact: Hôtel de Ville/ 13637 Arles/tel: 90.93.34.06).
Marseille: The Orchestre Philharmonique de Marseille and sporadic and seasonal concerts at Château Borély (tel: 91.72.41.27); Château Gombert, Cathédrale de la Major (tel: 91.55.04.36); Théâtre aux Etoiles (mostly popular – tel: 91.33.47.97); Centre de la Vieille-Charité (some very prestigious orchestral visitors); Abbaye St-Victor (tel: 91.33.25.86); Port Frioul (tel: 91.91.55.56).
St-Rémy: Free organ concerts at the Collegiale St-Martin, every Saturday at 5.30pm from June–September.

The Music Conservatory, from mid-July to end-August, "L'Argelier," Route d'Avignon (tel: 90.92.08.10).

Concerts at Fondation Armand Panigel, Petite Route des Jardins (tel: 90.92.07.92).
Salon-de-Provence: Classical and jazz concerts at the Château de l'Emperi during the summer.
Tarascon: Year-round classical concerts and organ music at the Collegiale or the auditorium.

VAR

Eguilles: Sporadic classical concerts at the church.
Entrecasteaux: Sporadic classical and jazz performances at the châteaux (tel: 94.04.43.95).
St-Maximin: Sporadic orchestral concerts at the Royal Convent.
Toulon: July music festival.
Le Val: Festival d'Eté à la Campagne with choral music and operettas, during July and August (tel: 94.69.06.15).

ALPES-DE-HAUTE-PROVENCE

Rousset: Sporadic classical concerts at the church.
Sisteron: "Nuits de la Citadelle" music festival during July and August.

ALONG THE COTE D'AZUR

Cannes: Sacred music festival in December.

Les Nuits Musicales du Suquet, second and third weeks in July, classical concerts held at Notre-Dame d'Esperance in the old town. (For information contact: Billetterie du Palais du Festival, La Croisette, 06400 Cannes, tel: 93.3944.44.)
Juan-les-Pins: The Antibes Jazz Festival, three weeks in July. (For information contact: Maison du Tourisme, 11 Place de Gaulle, Antibes 06600, tel: 93.39.44.44.)
Menton: Music festival with performances outside the St-Michel church in the old town. (For information: Palais de l'Europe, 06500 Menton, tel: 93.33.82.22.)
Monaco: Symphony, particularly October–December.

DANCE

Many of the summer festivals incorporate classical and folkloric ballet companies into their schedules, while some actually focus on dance, such as the "Danse à Aix – Festival International" held end-June through the first two weeks in July. Avignon's summer festival has some excellent offerings, while Arles's summer festival attracts some of the most original, ethnic dance groups. The summer Offenbach Festival in Carpentras lets ballet take up half its agenda and, year round, Monte Carlo is quite proud of its local troupe.

OPERA

Established 200 years ago, the Marseille Opera was traditionally known as the European proving ground for divas. It was said that if a singer had a success there, he or she had "arrived," partly because of the opera house's prestige and partly because of the Marseillaise audience's reputation for being ruthless critics.

This is no longer quite as true – although supposedly the Marseillaise audience continues to make most singers apprehensive – but opera fans will still enjoy taking in one of their productions. For information, contact:

Opéra de Marseille, 1, Place Reyer, 13001 Marseille. Tel: 91.55.14.99.

During July and August, Orange boasts all-out opera productions during the "Chorégies d'Orange," poetically staged amidst Roman ruins. Slightly more modest, but still enjoyable, are the operas that form part of the Aix-en-Provence festival and the Offenbach festival in Carpentras.

THEATRE

As is true all over France, practically every town in Provence has its own theatre. Almost all productions, are going to be in French, but even non-French-speaking audience members may find them enjoyable.

Marseille has some of the most interesting theatres, including the celebrated La Criée (30, Quai de Rive-Neuve, 13007 Marseille, tel: 91.54.74.54) run by actor and director Marcel Maréchal. Fans of experimental theatre will find much to interest them in this city, and classicists will be thrilled to find full productions of Ancient Greek plays given in the original language during the summer "Festival des Iles."

Avignon and Aix-en-Provence offer worthwhile theatre all year round, and occasionally in English. In the summer, however, both cities really take off – as do half the towns in Provence – with special, high-quality productions as part of their summer festivals.

CINEMA

Avignon has begun including a French-American Film workshop as part of their summer festival but, without question, the biggest cinema event to take place in Provence every year is the Cannes Film Festival in May. For information, contact:

Le Palais des Festivals, Esplanade Président Georges Pompidou, La Croisette, f-06400 Cannes. Tel: 93.39.01.01; telex: FESTIFI 461 670F.

As it happens, nearby Nice has been steadily cementing some very well-equipped film studios over the past years. If you would like to use their facilities, contact one of the following:

Studio de la Victorine Côte d'Azur, 16, Av. Edouard Grinda, 06200 Nice. Tel: 93.72.54.54; telex: 970056.

LTM, 104, Boulevard St-Denis, 92400 Courbevoie. Tel: 47.88.44.50; telex: 630277.

LTM Corp. of America, 437 W. 16th Street, New York, NY 10011. Tel: (212) 243-9288.

LTM Corp. of America, 1160 N. Las Palmas, Hollywood, CA 90038. Tel: (213) 460-6166; telex: 677693.

Those more interested in seeing films than making them will find movie houses in most of the larger towns and all the cities. The schedules are geared for local French audiences, of course, but anglophones will probably find that about half the films shown are American or English productions with only the subtitles in French.

NIGHTLIFE

CAFES & BARS

A good portion of social activity in Provence centres around the cafés. Much like pubs in England, cafés can be found in even the tiniest towns, filled with locals, a *bière*, a *ballon de rouge* or a *café* in hand, midday, after work, after dinner.

It would be, quite simply, impossible to list all the good cafés in Provence. Superficially, they are all more or less alike: small tables spilling out onto the street, an interior bar, serving mostly beer, wine, soda and coffee with a light menu of sandwiches, pizza and such. Of course, to the insider, each is infinitely different, frequented only by certain members of the town, except for the occasional unaware tourist.

Glance the clientele and patrons before sitting down. If they look like what you are looking for and you get a good feeling, you can't really go wrong.

Avignon: You'll find scores of open-air bars and cafés near the Palais des Papes, although, keep in mind that they will become less pricey the further you go from the palace.

Aix-en-Provence: Similarly, the large cafés right on the Place de la Libération in Aix-en-Provence will be outrageously expensive and rather touristy, but further up the Cours Mirabeau prices become more reasonable. Les Deux Garcons café is the local favourite for "cruising." (53, Cours Mirabeau. Open: 6am–1am).

Arles: Fans of the *cocarde* should visit Le Tambourin on the Place du Forum where the local toreadors hang out.

Marseille: Hipsters should head for the cafés along the Cours Julien, while a still-artsy but slightly older crowd may prefer one of those along the Place Thiars.

St-Tropez: Is famous for its "café society." Most famous of all is its Café Senequier (Quai Jean Juarès, open 8am–midnight), where you get a good chance to yacht- as well as people-watch Another lively favourite, with perhaps more appeal, is the Café des Arts (Place des Lices. Open: 8.30am–4am in July and August. Closes at 8.30pm September–June).

In general, cafés take the place of bars in Provence, but larger towns, major hotels and the cities do have fully-fledged bars. Just remember, that mixed drinks are not the local speciality. You might be wise to explain exactly what it is you want in your drink.

DISCOS

A surprising number of towns have discos, although they won't all be quite like Les Bains in Paris.

Pays d'Arles/Camargue: Most people in the Pays d'Arles either travel up to Avignon, to Montpellier (during the school year) or to Les-Stes-Maries-de-la-Mer to go dancing. Particularly "in" (as of 1992) is **Le 13ème** in Les-Stes-Maries-de-la-Mer, Place des Gitans, tel: 90.97.88.79. Open: 10.30pm–4am.

Arles, Barbentane, Châteaurenard, Eygalières, Maussane, Orgon and St-Rémy-de-Provence all also have at least one discothèque, although their quality isn't guaranteed.

Cassis: Cassis's disco is called **Big Ben** and is located on the Place Clemenceau (tel: 42.01.93.79). It's a perfectly harmless sort of place – the kind of disco where everyone waits for someone else to start the dancing.

Cannes: In Cannes, the best-known disco is **Galaxy**, set above the municipal casino. Despite the fairly steep entrance fee, the interior is of the too-worn velvet type and the clientele seems to match. Locals and those in the know prefer **Le Blitz** at 22, Rue Masse.

Antibes: Further down the beach, **La Siesta** has acquired a certain popularity, despite their bizarre Polynesian theme – with flaming torches and interior concrete lily ponds – and extremely outdated music. It can be found (if anyone cares to) along the Route du Bord de Mer in Antibes (tel: 93.33.01.18).

Nice: Has three particularly fashionable discos: **La Camargue** on the Cours Saleya (usually private), **Chez Ecossais** at 6, Rue Halévy, and **L'Aventure** at 12, Rue Chauvain.

NIGHTCLUBS & CABARETS

You will find a number of nice nightclubs and cabarets in the larger cities of Provence, particularly ones that have jazz. The **Piano Bar** at Hôtel Mercurie in Avignon is a time-honoured spot for enjoying a drink to the sound of casual piano music. One favourite among jazz lovers in Aix-en-Provence is **Le Jazz Club**. In Arles, jazz aficionados patronise **Pub le 37*2** (19, Place Honore Clair, tel: 90.96.11.44).

Marseille has any number of good jazz bars. Behind the Théatre de la Criée is the **Golden Jazz**

Club (40, Rue Plan Fourmiguier, Quai de Rive-Neuve, tel: 91.54.36.36). The towns along the Côte d'Azur also have their fair share. Best to check with your local Syndicat d'Initiative.

GAMBLING

There is no shortage of casinos along the coast of Provence, and Monte Carlo's is probably the world's most famous. To enter them you generally must have an ID card or passport and pay 50–55 ff. (Don't be confused by the widespread chain of supermarkets called "Casino.")

ANTIBES

La Siesta, Route du Bord de Mer, Pont la Brague. Tel: 93.33.01.18. Boule, baccarat, roulette, *chemin de fer*, blackjack. Its adjacent nightclub spills out onto the beach and is decorated in a Polynesian themes – complete with torches.

BEAULIEU

Casino, 8, Avenue Blundell Maple. Tel: 93.01.00.39. Roulette, baccarat, blackjack, 30/40, *chemin de fer*, *banque à tout va*.

CANNES

Palm Beach, Place Franklin D. Roosevelt. Tel: 93.43.91.12. Boule, roulette, 30/40, blackjack, *chemin de fer*. Open: 1 June–31 October. Gambling begins at 5pm.. Also, dinner dances at the "Iron Mask" with terrace orchestras, etc., and a private nightclub called "Jack Pot". Lunch grill called the "Commodore" with a swimming pool.
Municipal Casino, La Croisette. Tel: 93.38.12.11. Boule, roulette, *chemin de fer*, blackjack, 30/40, *banque à tout va*. Open: 1 November–1 May. Gambling begins at 4pm. Year-round (and slightly sleazy) nightclub, "Galaxy."

MENTON

Casino de Menton, Avenue Felix Fauré. Tel: 92.10.16.16. Boule, roulette, blackjack.

NICE

Casino Club, 6, Rue Sacha Guitry. Tel: 93.80.55.70. Boule. Entrance is only 10 ff, no passport required.
Casino Ruhl, 1, Promenade des Anglais. Tel: 93.87.95.87.

MONACO

Monte-Carlo Casino, Place du Casino. Baccarat, roulette, *chemin de fer*, 30/40. Slot machines. Entrance fee of 50 ff, but free for American roulette, blackjack and craps.
Mandelieu-la-Napoule Loews, Boulevard Henri-Clews. American roulette, blackjack, craps.

BOUCHES-DU-RHONE

AIX-EN-PROVENCE

2 bis, Avenue Napoleon-Bonaparte. Tel: 42.26.30.33. Boule, baccarat, roulette, 30/40, blackjack.

CARRY-LE-ROUET

Tel: 42.45.01.58. Boule, baccarat, roulette.

CASSIS

La Rostagne, Avenue Docteur Leriche. Tel: 42.01.78.32. Boule, baccarat, roulette, 30/40, blackjack. Open: all year except May, 3pm–2am.

LA CIOTAT

Avenue Wilson. Tel: 42.83.40.63. Boule. Open: all year round.

SHOPPING

WHAT TO BUY

You will have no trouble finding wonderful gifts and mementos to take home from Provence, without having to spend a lot of money. The best things to buy are local crafts or products, for which the Provençals are so famous. Among these are: *santons* (little figures from clay or sometimes dough used together to create a nativity scene); pottery; faïence (fine ceramics decorated with opaque glazes); brightly coloured fabrics (the Souleiado stores carry the most famous *tissus*, but they are also the most costly); hand-woven baskets; and bath and toiletries made with local herbs like lavender. Or you might just want to pick up a bottle of lavender essence, a little bag of fresh Herbes de Provence or a bottle of delicately green olive oil.

Some towns are inextricably linked with particular products, like the wonderful handblown glass from Biot, the perfumes from Grasse or soap from Marseille. The same goes for certain regions: honey with the Alpes-Maritimes and Alpes-de-Haute-Provence; olive oil and *santons* with the Bouches-du-Rhone; leather products with the Camargue and

Alpes-de-Haute-Provence. And, if you're on the Riviera – well, there really is no better place to buy a bikini.

SHOPPING AREAS

Practically every town will give you an opportunity to shop – if not in a local *atelier* or store, at the market place. But, for some specific suggestions, see below:

VAUCLUSE

AVIGNON

A well-liked leathermaker in Avignon offering accessories, bags and belts in leather, is Vincenette Ranchet-Leron. **Zenaide**, 4, Rue Pavot, 84000 Avignon.

Souleiado, 5, Rue Joseph Vernet, 84000 Avignon. Basically *the* shop for Provençal memorabilia. Located all over Provence and in many cities internationally as well, these stores are most famed for their beautiful fabrics done in bright traditional designs. You can buy the fabric in bolts or already decorating any number of different gift items such as address books or aprons. They also stock a few other random pieces of Provençal handicraft. Their stock is top of the line and you pay quite a lot for it, but to ease the pain they take several credit cards (AE, DC, EC and Visa) and ship internationally.

BONNIEUX

La Bouquière, 84480 Bonnieux. For wooden and educational toys, stop at the atelier of Xavier De Tugny.

GORDES

Souleiado shop for Provençal cloths, etc. Place du Monument.

LE THOR

Louis Ortiz, Chemin Croix de Tallet, Le Thor. Tel: 90.33.91.30. It may seem silly to buy rush caning for chair bottoms while on vacation but, just the same, you might consider bringing some back from Louis Ortiz's workshop, to be attached once you get home. Done in any number of different patterns, it will last forever.

MALAUCENE

Emmanuelle Chouvion, Cours des Isnards, 84340 Malaucène. Deep blue tints, inspired by the colour of lavender, are among the themes chosen for the contemporary tone of Emmanuelle Chouvion's smooth and polished pottery.

VAISON-LA-ROMAINE

This town is becoming a good place to purchase local pottery and paintings as the reputation for its artist community steadily grows.

Souleiado shop for Provençal cloths. 2, Cours J.H. Fabré, 84110 Vaison-la-Romaine.

VALREAS

Stop in at Revoul to buy truffles. **Revoul**, 84500 Valrèas. Tel: 90.35.01.26.

BOUCHES-DU-RHONE

AIX-EN-PROVENCE

Atelier Fouque, 65, Cours Gambetta, 13100 Aix-en-Provence. Tel: 42.26.33.38. All visitors to Aix should make a point of stopping at the atelier Fouque. Here you can both purchase and watch the manufacture of *santons* by one of the craft, Paul Fouque, and his talented daughter, Mireille.

A La Reine Jeanne, 32, Cours Mirabeau, 13100 Aix-en-Provence. Tel: 42.26.02.33. *Calissons*, the local almond-paste sweet, make a nice souvenir to take home. You can find them all over the place but the confectionary store A La Reine Jeanne is centrally located, takes AE, DC and Visa and will ship internationally.

The old town boasts quite a few chic (and not inexpensive) shops, including designer clothing, baby things and antiques.

Souleiado shop for Provençal cloth, etc. Place des Tanneurs, 13100 Aix-en-Provence.

ARLES

L'Art du Bois, 29, Rue des Porcelets, 13200 Arles. Tel: 90.96.77.17. Jesus Sendon is a young man whose painstakingly handcarved woodworking carries on the traditions of the *style Provençale*. Smaller gift items such as music stands and pencil boxes can be purchased directly at his atelier while larger pieces must be comissioned. Not inexpensive but very hard to find elsewhere.

Souleiado shop, 4, Boulevard des Lices, 13200 Arles. Tel: (90) 96.37.55.

Atelier Jannin, 1, Rue du Bastion. 13200 Arles. Juliette Jannin does lovely contemporary designs on cotton.

AUBAGNE

Poterie Provençale, Avenue des Goums, 13400 Aubagne. Tel: (42.03.05.59. French visitors flock here to purchase their bargain-priced, hand-made cookware. Although not elegant, their quick pans, casseroles, etc., are the original sturdy and rustic Provençal dayware.

FONTVIEILLE

Li Mestierau, Val Parisot, 13990 Fontvieille. Tel: 90.97.77.95 Here you will find beautiful painted silk garments and earthy handwoven wools. Clothing, upholsteries and murals can be bought on site or commissioned. A visit to the group workshop, set up commune-style, is itself worth the trip.
Atelier Monleau, Route d'Arles, 13990 Fontvieille. Tel: 90.97.70.63. For some beautiful handmade and handpainted faïence, porcelain and ceramics visit the atelier of the multi-prizewinning Chantal Monleau and family.

LES BAUX-DE-PROVENCE

Like any good tourist centre, Les Baux has lots of shops. The quality is on the whole quite good although the prices are higher than you will find elsewhere.

MARSEILLE

Atelier Carbonel, 47, Rue Neuve Ste-Catherine, 13007 Marseille. Tel: 91.54.26.58. Carbonel's Boutique du Santon is indisputedly one of the most famous and has the added attraction of a large workshop and small "museum."
Souleiado shop for Provençal cloths, etc. 101, Rue Paradis.

MEYRARGUES

Atelier Devouassoux, Legrand Vallat, 13650 Meyrargues. Tel: 42.57.51.10. Huguette Devouassoux is famous for making *santons* in faïence rather than just plain red clay.

ST-REMY-DE-PROVENCE

Known as the centre for the herb market, St-Rémy has many *herboristeries*. Most wonderful, with a colourful variety of bath items (made with local herbs and packaged aesthetically) and a complete selection of freshly dried herbs is: **L'Herbier de Provence**, 34, Boulevard Victor-Hugo, St-Rémy-de-Provence. Tel: 90.92.11.96. Prop. Anne & Vincent du Rothazard.
Souleiado shop, 2, Avenue de la Résistance.

ST-ZACHARIE

La Petite Foux, RN 560, 83640 St-Zacharie. Tel: 92.72.96.02. This roadside workshop/store of the talented young Voelkel offers unique hand-turned pottery to the public. The potter's personal favourites are his blossoming, Oriental-inspired vases and bowls.

ALPES-DE-HAUTE-PROVENCE

DIGNE

Maison de la Lavande, 58, Boulevard Gassendi, 04000 Digne-les-Bains. Tel: 92.31.33.94. Lavender shop for soap, essence, etc.

FORCALQUIER

Atelier Le Point Bleu, 22, Rue Violette, 04300 Forcalquier. At the Atelier Le Point Bleu of Michele Paganucci, you will find painting on silk, scarves, cushions and more.

MOUSTIERS-STE-MARIE

This town is famous for elegant, handpainted faïence, known as "Moustier-ware." Probably the best (and most expensive) pottery in this style can be found at the Atelier Segriès. They take the major credit cards and are prepared to ship internationally.
Atelier Segriès, 04630 Moustiers-Ste-Marie. Tel: 92.74.66.69.

RIEZ

G. Ravel, Allées Louis Gardiol, 04500 Riez. A good, old-fashioned store for lavender honey.

VAR

CUGNES-LES-PINS

Quartier du Puits, Chemin de la Pujeade, 13780 Cuges-les-Pins. Alain Carbone handmakes his leather saddles.

DRAGUINAN

Philippe Rabussier is anpther well-respected leathermaker, who fashions such objects as handbags, belts and wallets. Lot. Les Faisses #13, 83300 Draguignan.

RAMATUELLE

Souleiado store, Rue des Sarassins.

ALPES-MARITIMES

BIOT

La Verrerie de Biot, Chemin des Combes, 06410 Biot. Tel: 93.65.03.00. No visitor to the Côte or Alpes-Maritimes should miss a visit to the Verrerie de Biot. Next to the large atelier where they give demonstrations of glassblowing methods is an even larger shop (they will ship overseas, it is costly but it means you don't have to pay the tax). They take American Express and Visa and are open on Sundays.

OPIO

If you don't know *what* you would like to bring home, except that you want it to be Provençal, stop at the **Huilerie de la Brague**. They stock all sorts of different local specialities, from olive oil to olivewood carvings to Provençal cloths.

VALLAURIS

Famous for its pottery since Picasso developed an atelier there, Vallauris is now a sink of trashy tourist shops. There's only one true potter left: Roger Collet, Montée Ste-Anne, 06406 Vallauris.

MARKETS

If there is one thing that symbolises Provence, it is the open-air markets. After cafés, they are the centre of social activity for the locals, as well as where the Provençal cook obtains the multitude of fresh ingredients that make up the delicious cuisine. Every town has its own – be it tiny or enormous.

Vendors of locally produced vegetables, fruit, herbs, cheese, sausage, meat, honey, flowers, soaps, lavender essence and fresh-baked breads set up their stands at daybreak. In addition, some towns have a *marché au brocante* (literally, junk market) which offers everything from priceless antiques to brand-new items like underwear to honest-to-god junk. If you want to get to know Provence, you have to go to one of their markets.

Here's a list of where and when:

Apt: lively market – Saturday.
Avignon: diverse – daily; produce and flea market – Saturday and Sunday morning; antiques 30 August–2 September.
Bollène: lively market – Monday.
Cavaillon: produce and diverse – Monday.
Châteauneuf-du-Pape: diverse – Friday.
Gordes: diverse – Tuesday.
Isle-sur-Sorgue: antiquities and flea – Saturday and Sunday.
Orange: produce and diverse – mid-April to mid-October.
Roussillon: diverse – Wednesday.
Vaison-la-Romaine: lively market – Tuesday.

BOUCHES-DU-RHONE

Aix-en-Provence: diverse – daily; largest diverse (Old Town) – Saturday; flowers – Monday, Wednesday, Friday, Sunday.
Arles: produce and diverse (Place Lamartine) – Wednesday; produce and diverse (Boulevard des Lices) – Saturday 7am–12.30pm; sheep – 3 and 20 May.
Barbentane: diverse – Wednesday; goats – 19 and 20 May.
Cassis: diverse (Parking du Marché) – Wednesday and Friday.
Châteaurenard: produce and diverse – Sunday 7am–12.30pm.
Fontvieille: produce and diverse (Place de l'Eglise) – Monday and Friday 8am–noon.
Jouques: diverse (Parking Municipal) – Wednesday afternoon and Sunday.
La Ciotat: diverse (Place Evariste Gras) – daily.
Maillane: diverse (Place de l'Eglise) – Thursday.
Marseille: diverse – daily; flower and fish – daily except Sunday.
St-Etienne-du-Grès: fruits and vegetables (Place du Marché) – daily 6am–8pm.
Stes-Maries-de-la-Mer: diverse (Place de la Mairie) – Monday and Thursday.
St-Rémy: produce and diverse – Wednesday and Saturday 7am–1pm; animals – 25 April and 28 October; wine and Provençal – end July.
Salon: diverse (Cours Gimon) – Wednesday; diverse (Canourges) – Friday; diverse (Quartier des Bressons) – Saturday; animals – 6 and 22 May, 29 September and 10 November.
Tarascon: produce and diverse (Place des Halles) – Tuesday 7am–1pm; sheep and asses – first week September and first week October.

ALPES-DE-HAUTE-PROVENCE

Barcelonette: diverse – Wednesday and Saturday.
Castellane: diverse – Wednesday and Saturday.
Colmars-les-Alps: diverse – Tuesday.
Digne: diverse – Wednesday, Thursday, Saturday.
Forcalquier: diverse – Monday.
Manosque: diverse – Saturday.
Moustiers: diverse – Friday.
Sisteron: lively market – Wednesday and Saturday.

VAR

Bormes-les-Mimosas: diverse – Wednesday.
Colobrières: diverse – Sunday.
Draguinan: vegetable – daily; produce and diverse – Wednesday and Saturday.
Fayence: diverse – Thursday and Saturday.
Hyères: major diverse (Place Lefevre) – third Thursday in month; flea market (Place Massillon) – most mornings.
Le Lavandou: diverse – Thursday.
Salernes: diverse – Sunday.
Sanary: diverse – Wednesday.
Seillans: major diverse – first Sunday in July; flower fair – mid-July.
Toulon: flower and vegetable (Cours Lafayette) – every morning; fish (Place de la Poissonerie) – every morning.

ALPES-MARITIMES

Grasse: diverse – Tuesday.
Vence: diverse – Tuesday and Friday.

St-Etienne-de-Tinée: diverse – Friday and Sunday.
St-Jeannet: diverse – Thursday.
Sospel: diverse – Thursday.
Tende: diverse – Wednesday.
Vallauris: diverse – Wednesday and Sunday.

COTE D'AZUR

Antibes: diverse – daily except Monday and Friday.
Cannes: diverse – daily except Monday; grand diverse – Saturday and Sunday.
Menton: diverse – daily; grand diverse – Saturday.
Nice: lively market – daily except Monday.
St-Tropez: lively market – Tuesday and Saturday.

SPORTS

PARTICIPANT

Provence provides plenty of opportunities for the sporting-minded to stretch their legs. Cycling, hiking, rock climbing, swimming, sailing, windsurfing, tennis, golf, riding, skiing and the local favourite of *boules* are some of the most popular athletic activities in the region. Also, if you'd prefer to sit back and watch someone else get sweaty, Provence has some very special offerings: bullfighting, motor rallies, yacht races, horseracing and the ubiquitous *boules*.

You will find that hiking is a predominant activity for the whole region, which is criss-crossed with Grands Randonnées. The GRs are national hiking paths along which you can walk from Nice to Holland or Menton to Spain. Maps by Les Cartes Didier Richard outline these routes and mark relay stations and spots for rock climbing. Of particular interest might be maps #1 Alpes-de-Provence, #9 Haut-Pays Niçois, #19 En Haute Provence and #26 Au Pays d'Azur.

Below are some suggestions as to what you can do where and how to get further information:

VAUCLUSE

The Vaucluse is particularly tempting to the hiker because of the wonderful trails surrounding Mt-Ventoux. You can get information about trails, from the **Comité National des Sentiers de Grande Randonnée**, 8 Avenue Marceau, 75008 Paris (tel: 47.23.62.32) or local tourist boards.

Riding is another big draw in the Vaucluse. For information about holidays geared towards the equestrian, hacking, trekking, riding excursions, etc, plus a free yearly folder with prices and programmes, contact: Association Nationale pour le Tourisme Equestre: **ANTE - Provence**, 28, Place Roger-Salengro, 84300 Cavaillon. Tel: 90.78.04.49.

When in Avignon, you might try the olympic-sized pool at Ile de Barthelasse, the Squash Racket Center at 32, Boulevard Limbert, or the Golf Club at Chateaublanc.

BOUCHES-DU-RHONE

Air sports such as flying, parachuting, gliding, hang gliding and ballooning, earthy sports such as cycling, walking and riding (particularly around Aix, Les Alpilles and in the Camargue); and watersports like swimming, sailing and waterskiing (on the coast) are some of the favourites in the Bouches-du-Rhône.

For information about all the different schools, clubs, courses, etc., contact:
Direction Départementale de la Jeunesse et des Sports, 20, Avenue de Corinthe, 13006 Marseille. Tel: 91.78.44.88.

Despite the heat of the sun, and probably due to the flatness of much of the land, **cyclists** love the Bouches-du-Rhône. For information, contact:
Comité Départemental de Cyclisme, Maison du Quartier des Chutes Lavie, 10, Boulevard Anatole-France, 13004 Marseille. Tel: 91.64.58.79; or
Comité Départemental de Cyclotourisme, 15, Lotissment la Trevaresse, 13540 Puyricard. Tel: 42.92.13.41.

Although this is probably the least favourable of the five Provençal *départements* for **hiking**, there are some worthwhile inclines around Aix-en-Provence and Les Alpilles. Contact:
Comité Départemental de Montagne (Alpinism), M. Gorgeon, 5, Impasse du Figuier, 13114 Puyloubier. Tel: 42.61.48.49; or
Comité Départemental de Randonnée Pedestre (Hiking), 16, Rue de la Rotonde, 13001 Marseille. Tel: 42.21.03.53.

The following towns have **municipal swimming pools**, which can be a relief during long hot days: Aix, Arles, Aubagne, Châteaurenard, Fos, Gemenos, Istres, La Ciotat, Marseille, Martigues, Maussane, Rognac, La Roque d'Antheron, St-Martin de Crau, St-Rémy-de-Provence, Salon-de-Provence and Trets.

The following towns have **marinas**: Berre, Carro, Carry, Cassis, La Ciotat, La Couronne, Fos, Istres, Marignane, Marseille, Martigues, Port de Bouc, Port St-Lous, St-Chamas, Stes-Maries-de-la-Mer and Sausset-les-Pins.

For further information about sailing and windsurfing, contact:

Comité Départemental de Voile et Planche à Voile, SNE Mourièpiane, L'Estaque, 13106 Marseille. Tel: 91.03.73.03.

You will find a nine-hole **golf** course in Bouc Bel Air and 18-hole courses either just completed or almost completed in Les Milles, Cabriès, Allauch, Fuveau and La Valentine. For more information about local golf courses, contact:
Délégation Régionale de Golf, 82, Rue Breteil, 13008 Marseille. Tel: 91.37.01.23.

For information about the more-than-80 **diving centres** in the Bouches-du-Rhône, contact:
Comité Départemental de Sports Sous-Marins, 38, Av. des Roches, 13007 Marseille. Tel: 91.52.55.20.

One nice place to fish is the little lake outside St-Rémy-de-Provence. For other locations, contact:
Fédération Départ. des Assoc. Agrées de Peche, 30, Boulevard de la République, 13100 Aix.
Tel: 42.26.59.15.

ALPES-DE-HAUTE-PROVENCE

As might be expected, the Alpes-de-Haute-Provence has many attractions for the alpinist. Rock climbing, hiking and mountain biking are key activities here, with all levels of difficulty to be found. Also, the rivers that snake between the gorges and the lakes that dot the mountains mean kayaking, rafting, sailing, fishing and all other types of fresh water sports. Finally, in the winter, there are some resorts that offer alpine skiing.

For information about renting kayaks and/or pedalos, details of *Grand Randonnées* (the national hiking paths) routes and rock faces for climbing, maps, etc., in the area of the Grand Canyon de Verdon, contact:
Syndicat Initiative, Rue Nationale, 04120 Castellane. Tel: 92.83.61.14.

Colmars is more or less a centre for sporting activities. Situated by the Verdon and Lance rivers and Allos, Encombrette and Lignin lakes and at the base of several mountain peaks, you can kayak (classed at a level of III and IV), windsurf and sail, hike over 150 miles of marked paths through forests and along mountain slopes, fish for trout and, from November to April, there is alpine or cross-country skiing at either Le Seignus d'Allos (4500–8200 ft, with 24 miles of trails and 9 chairlifts) or La Foux d'Allos/Prloup (6000–8700 ft, with 72 miles of trails, 17 chairlifts, towlifts and gondolas). The town also has three municipal tennis courts, a municipal swimming pool, a sledding hill and a year-round judo school.

For information, contact:
Office Municipal des Sports et des Loisirs, Mairie de Colmars, 04370 Colmars-les-Alpes.
Tel: 92.83.41.92.

Not far from Colmars, deeper into the mountains in the Vallée de l'Ubaye is another resort that beckons the athletic. Annot offers a municipal pool, four tennis courts, more *randonnées pedestres*, fishing, kayaking, cross-country skiing (with a special cross-country ski school) and an equestrian centre (in season). For information, contact:
Syndicat d'Initiative, 04240 Annot. Tel: 92.83.23.03.

The two largest ski resorts in the area are Pra-Loup and Super-Sauze. For information on them, you can contact: **Syndicat d'Initiative**, Place Sept Portes, 04400 Barcelonette. Tel: 92.81.04.71.

VAR

Aside from cycling, hiking and riding, most of the Var's sportive activity takes place by the coast. There you will find all the usual water-related sports: suntanning, girl- and boy-watching, sailing, boating, swimming, parasailing, etc.

Two towns in upper Var have guides available and organise outings for walks and hiking – Val d'Entraunes and Beuil-Valberg. For information, contact:
Syndicat d'Initiative, Val d'Entraunes. Tel: 94.05.51.04; or
Bureau des Guides. Tel: 94.02.42.34.

ALPES-MARITIMES

Hiking is also a big thing in the Alpes-Maritimes, and you can get a great little map called *The Hiker's Handbook* from the Comité Regional du Tourisme Riviera-Côte d'Azur (55, Promenade des Anglais, 06000 Nice, tel: 93.44.50.59). This will tell you everything you need to know – hostels and relay stations, hiking routes, park rules, and names and numbers for climbing schools, guides, regional clubs and assocations. It also contains a good general map.

There are fully equipped **climbing** schools in Cabris, Gourdon, St-Jeannet, La Loubière, St-Dalmas-de-Valdeblore, Le Boréon (by St-Martin-Vésubie), Roquebillière, Belvedere, Valberg, St-Etienne-de-Tinée, Auron, Isola 2000 and Tende. The Mercantour National Park offers a cornucopia of possibilities for outdoor enthusiasts with climbing ascents that range from difficult to very difficult.

The best time for **mountaineering** is from June–October. Always be on your guard for violent late-afternoon storms and remember that bad weather here comes from the east. For mountain weather reports tel: 93.71.81.21.

Some addresses that might be of use:
Regional Council for Hiking Enthusiasts, 2, Rue Deloye, 06000 Nice. Written enquiries only. If within France, please include return postage.
French Alpine Club (FAC), 15, Avenue Jean Medecin, 06000 Nice.

During winter, **skiing** becomes the most popular alpine activity in this department. Twelve smaller resorts exist, each ranging from about 4,500–6,500 ft (1,380–1,980 metres): L'Audibergue (tel: 93.60.31.51); Beuil-Les-Launes (tel: 93.02.30.05); Le Boréon (tel: 93.02.20.08); La Colmiane-Valdeblore (tel: 93.02.84.59); Esteng-D'Entraunes (tel: 93.02.84.59); La Gordoloasque-Belvedere (tel: 93.05.51.36); Gréolières-Les-Neiges (tel: 93.59.70.12); Peira-Cava (tel: 93.93.57.12); Roubion-Les-Buisses (tel: 93.02.00.48); St-Auban (tel: 93.60.43.20); St-Dalmas-Les-Selvage (tel: 93.02.41.02); Tende-Caramagne (tel: 93.04.60.91); Turini-Camp-D'Argent (tel: 93.03.01.02); Val-Pelens-St-Martin-D'Entraunes (tel: 93.05.51.04).

In addition, there are three major resorts – Auron, Isola 2000 and Valberg – within easy driving distance from Nice. For information, contact:

Office du Tourisme/Auron, 06660 St-Etienne-de-Tinée. Tel: 93.23.02.66; telex: 470300. 25 ski lifts, 44 trails, 75 miles of slope. 5,300–8,200 ft (1,600–2,500 metres). Several winter packages are available from the tourist office. Also, summer packages including hiking, tennis courts, riding, fencing and judo.

Office du Tourisme/Isola 2000, 06420 Isola 2000. Tel: 93.23.15.15; telex: 461666 F. 22 ski lifts, 40 trails, 70 miles of slope. 6,000–8,600 ft (1,830–2,620 metres). Ice driving school and snow bike. Several winter packages available. Also, summer packages including tennis lessons, horseback riding lessons, hiking and swimming pool.

Office du Tourisme/Valberg, 06470 Valberg. Tel: 93.02.52.54; telex: 461002 F. 21 ski lifts, 55 trails, 50 miles of slope. 5,300–8,000 ft (1,620–2,440 metres). Also, 30 miles of trail for cross-country skiing. One winter package of six nights "full ski" from January–March.

There are five 18-hole **golf** courses in the Alpes-Maritimes, with average rates of 180 ff on weekdays and 250 ff on weekends:

Golf Bastide du Roy, 06410 Biot. Tel: 93.65.08.48.

Monte-Carlo Golf Club, Mont-Agel, 06320 La Turbie. Tel: 93.41.09.11.

Country Club de Cannes Mougins, 175, Route d'Antibes, 06250 Mougins. Tel: 93.75.79.13.

Golf Club Cannes Mandelieu, RN 98, 06210 Mandelieu-La-Napoule. Tel: 93.49.55.39.

Golf de Valbonne, La Bégude, route D.4, 06560 Valbonne. Tel: 93.42.00.08.

And, of course, there are lots of **stables**. Costs average about 50 ff per hour. For information:

ANTE Provence-Côte d'Azur, Mas de la Jumenterie, Route de St-Cézaire, 06460 St-Vallier-de-Thiey. Tel: 93.42.62.98; or

Comité Hippique Départemental, Stand Municipal des Sports, Parc Layet, 06700 St-Laurent-du-Var. Tel: 93.20.99.64.

COTE D'AZUR

If the French are known for being a fairly unathletic nation, the residents of the Côte d'Azur are an exception. After all, they have great beaches at their feet and sunny skies over their heads all year round. You can find all sorts of water-connected activity that might interest you, plus tennis courts galore, racquet clubs and miniature golf (full golf courses are located inland – *see under Alpes-Maritimes*).

Some places to go or contact:

Aerodrome International Cannes-Mandelieu, 269, Avenue Francis Tonner, 06150 Cannes-La Bocca. Tel: 93.47.11.00. For **aviation**.

Bowling de Nice Acropolis, 5, Esplanade Kennedy, 06300 Nice. Tel: 93.55.33.11. Bowling alley with 24 tracks. Open: all year from 11am–2am with a bar and cocktail lounge.

California Bowling, Route de Grasse, 06600 Antibes. Tel: 93.33.23.95. Bowling alley with a roller-skate disco. Open: all year.

Comité Départemental, C/o F. Parcillie, Le Jura A, 12, Boulevard Henri Sappia, 06000 Nice. Tel: 93.51.22.56. For **cycling**.

Comité Départemental, Fort Carré, 06600 Antibes. Tel: 93.95.05.65. For **diving**. In general, many hotels, beaches and sports shops rent diving equipment. The best place to dive is along the Esterel.

Ecole Francaise de Vol Libre, La Colmiane. Tel: 93.02.83.50; or

Fédération Francaise Vol Libre, 54 bis, Rue de la Buffa, 06000 Nice. Tel: 93.88.62.89. For **hang gliding**. (The first is actually a school.) The wind currents are great all over, but the best locations are considered to be near Gourdon and Fayence.

Association de Polo de Cannes-Mandelieu. Tel: 93.36.81.93. For **polo**. (For riding *see under Alpes-Maritimes.*)

Fédération Départementale des Sociétés de Pêche, 20, Boulevard Victor Hugo, 06000 Nice. Tel: 93.03.24.09. For **fishing**. You can rent boats at most marinas to go fishing in the Mediterranean, but locals prefer going into the mountains after trout. Get a permit or get arrested.

For **sailing**. There is no shortage of places from where to rent or buy sailboats. Almost every city on the Mediterranean has a port with boats for hire. To find out the going prices drop in on the yacht clubs or pubs. Many of the clubs also teach sailing.

Lique Provence Côte d'Azur, 20, Rue des Palmiers, 06600 Antibes. Tel: 93.34.78.07.

Aquasport, Port Gallice, 06160 Juan-les-Pins. Tel: 93.61.20.01.

Camper & Nicholson's Yacht Services, Port Canto, 06407 Cannes. Tel: 93.43.16.75.
South of France Boating, Marina Baie des Anges, Villeneuve-Loubet. Tel: 93.20.55.16.
Yacht Club Beaulieu/St-Jean, Quai White Church, Pourt de Plaisance, 06310 Beaulieu. Tel: 93.01.14.44.
Yacht Club de Monaco, Cale de Halagé, Quai Albert lèr, Monte Carlo. Tel: 93.30.63.63.

For **squash**. All three clubs listed here are open all year.
Squash Club, Les Terriers, Route de Grasse, Sortie Autoroute, 06600 Antibes. Tel: 93.33.35.33.
Squash Club Vauban, Avenue du Maréchal Vauban, 06300 Nice. Tel: 93.26.09.78.
Squash Lou Pistou, Avenue du Zoo, 06700 St-Laurent-du-Var. Tel: 93.31.05.44.

For **tennis**. The Sophia Country Club is the newest and brightest club on the Côte, located among 31 acres of woodland far from the smelly and noisy autoroutes. Yannick Noah's coach, Patrice Hagelauer, is director here, and tennis holidays, junior tennis competitions, private lessons and training for players and coaches are available.
Ligue Côte d'Azur, 5, Avenue Suzanne Lenglen, 06000 Nice. Tel: 93.53.92.98.
Sophia Country Club, Biot-Sophia Antipolis, 06560 Valbonne. Tel: 93.65.26.65.
Tennis Ecole de Villeneuve-Loubet, Route de Grasse, 06270 Villeneuve-Loubet. Tel: 93.20.60.09.

Finally, **windsurfers** are rented at many sports shops and almost every beach along the Côte. Rental costs range from 35–50 ff per hour, and lessons go for 75–90 ff per hour. A spin on water skis costs about 70 ff for six minutes.

SPECTATOR

Bullfighting: Full-fledged bullfights can be seen at various times during the warm months at the arenas in Arles, Nîmes, Les-Stes-Maries-de-la-Mer and Fréjus. At the *grand corrida* in Arles on 15 August, they don't kill the bull, but during Arles's Easter *feria*, Easter Saturday, Sunday and Monday, they do.

If bullfighting seems like more than you can stomach, you might want to take in one of the *cocardes* instead. In the *cocardes*, about 20 men dressed in white run around the ring badgering a series of bulls until they lose their tempers and charge. In the meantime, the men try to collect points by pulling strings and rosettes off their horns. The Society of Prevention of Cruelty to Animals might not love it, but there really is nothing bloody about it (unless one of the men doesn't move fast enough). And it's very much like what a baseball game is to an American – the Provençal folk arrive in droves on Saturday afternoons to cheer for the favourites, eat peanuts and drink beer.

More bull-connected sports and events can be seen in other smaller towns in Provence, particularly those in or near the Camargue, such as the mid-June bull races in Les-Stes-Maries-de-la-Mer or *courses camarguais* in Tarascon.

Boules: Any time you see four or more men gathering around an open space, chances are good that they are going to embark on the Provençal sport of *boules*. Basically, *boules* is somewhat like bowling – you try to knock certain balls with other balls. You can join in and play yourself, but don't unless you know what you're doing. Provençals can be as fervent about their *boules* games as some people are about soccer.

Auto rallies are a particularly popular spectator sport along the Côte d'Azur. The most famous are, of course, the Grand Prix in Monaco at the end of May and the Monte-Carlo Car Rally in January. Other rallies are the Rally of the Roses in Antibes, beginning in October, the Golden Snail old cars rally in Antibes, mid-May, and the vintage car rally in Toulon, in May. For information about racing in the Bouches-du-Rhône, contact:
Lique Regionale Provence, Course de Sport Automobile et Circuits Automobiles, 7, Boulevard Jean-Jaurès, 13100 Aix. Tel: 42.23.33.73.

Tennis lovers around the world direct their attention to Monaco every third week in April for the International Open Tennis Championship.

Bridge: If you are a bridge fan, you might want to head towards St-Tropez during April for the International Bridge Festival. A bridge tournament is held during mid-May in Antibes.

Horses: During May in Cannes you can see both an International Jumping show and the Spring Cup Regatta.

For horseracing addicts, the Cagnes-sur-Mer race track is open from 15 December–16 March and from 28 June–30 August, Tuesday, Friday and Sunday. This is subject to change from year to year, for information tel: 93.20.30.30.

Special Information

CHILDREN

In some ways, Provence is an ideal spot to bring children. It is inexpensive, very safe and spacious. And, although a lot of the things that make Provence so special – the sensuousness, the smells, the great food, the fascinating folklore – may be less appealing to your average toddler or adolescent, they will probably find the bright festivals and ethnic museums more tolerable than they would La Scala or the Louvre.

Most of the larger towns have a *Centre Jeunesse* or youth centres. There, they should be able to advise you of special events or attractions for younger people, although they tend to cater to adolescents and teenagers rather than kids.

A lot of families enjoy the many opportunities to camp in Provence. This is a particularly inexpensive way to accommodate an entire family, and kids quickly make friends with the kids of other campers. Another affordable way to house a brood is by renting one of the *gîtes ruraux*. If you want to stick to hotels, you will find that most have some sort of reduction for families – particularly if they stay in an extra bed in your room.

Little children generally enjoy amusement parks, and in Provence there are a few:

OK Corral, Cuges-les-Pins. Tel: 42.73.80.05. Amusement park along an (American) Western theme with a Wild West show. Open: March–October.

Eldorado City, Ensues-la-Redonne. Tel: 42.79.86.90. Another amusement park with a Western theme. Open: 1 June–15 September.

Marineland, Carrefour RN7/Route de Biot, Antibes. Tel: 933.33.49.49. Marine zoo with dolphins, killer whales and sea elephant shows. Admission and show costs 60 ff. Open: all year. Also aquatic park with 12 giant water toboggans and a wave pool. 68 ff. Open: June–September.

Parc de Loisirs de Barbossi, Domaine de Barbossi, RN7/Route de Fréjus. Tel: 93.49.64.74. Wild West train, electric cars, motorbikes for kids, puppet theatre, pony club, tennis, bowling alley, mini-zoo.

DISABLED

France has established a handful of national organisations that assist disabled visitors. For information about accommodation, means of transport, recreational activities and access to historical monuments, museums, theatres, etc., contact:

Comité National Francais de Liaison pour la Réadaptation des Handicapes, Service accessibilité, 38, Boulevard Raspail, 75007 Paris. Tel: (1) 45.48.90.13; or

La Maison de la France, 8, Avenue de l'Opéra, 75001 Paris. Tel: (1) 42.96.10.23.

Handicapped tourists can also check with the departmental tourist offices and should contact the regional tourist board of the Côte d'Azur for the Riviera-Côte d'Azur Hotel Guide's list of rooms with easy access. A couple of places on or near the Riviera specialise in the disabled guest:

Arc Vacances Association, Le Delta, Avenue des Mouettes, 06700 St-Laurent-du-Var. Tel: 93.07.20.84.

Le Beau Soleil, 12, Boulevard J-Crouet, 06310 Grasse. Tel: 903.36.01.70. Centre for the disabled and the mentally handicapped. Open: all year round with a capacity of 100–150 people.

STUDENTS

The university in Aix is one of the very best in France, and other universities and academic institutions exist in Marseille and along the Côte d'Azur. For more information:

IN THE UK

French Embassy, Service des Echanges Extra-Universitaires, 22 Wilton Crescent, London SW1.

Central Bureau Seymour, Mews House, Seymour Place, London W1H 9PE. Tel: 071-486 5101.

YMCA and YWCA, 2 Weymouth Street, London W1. Tel: 071-636 9722.

IN THE US

American Youth Hostels, Spring Street, New York, NY 10012. Tel: (212) 431-7100.

Council on International Exchange, 205 East 42nd Street, New York, NY 10017. Tel: (212) 661-1414.

FACETS, 989 Sixth Avenue, New York, NY 10017. Tel: (212) 475-4343.

French Embassy, 4101 Reservoir Road, NW, Washington, DC 20007. Tel: (202) 944-6000.

French Cultural Attaché, French Embassy, 40 West 57th Street, 21st floor, New York, NY 10019. Open: 2–5pm, weekdays.

IN CANADA

French Embassy, 464 Wibrod, Ottawa Ontario 6M8 KIN. Tel: (613) 238-5715. Cultural attachés in Edmonton, Montreal, Moncton, Quebec, Toronto, Vancouver and Winnipeg

Comité d'Accueil Canada France (CACF - OTU), 1183, Rue Union, Montreal H3B 3C3. Tel: (514)

875-6172; telex: (055) 62015; or 17 St Joseph, Suite 311, Toronto M4Y 1JB. Tel: (416) 962-0370.

IN AUSTRALIA

French Cultural Attaché, French Embassy, 6 Darwin Avenue, Yarralumia, ACT 2600. Tel: (062) 705111.

Some academic institutions located on the Côte d'Azur that offer language courses:

College International de Cannes, 1, Avenue du Docteur Pascal, 06400 Cannes. Tel: 93.47.39.29; telex: 214 235 ATT College INT. Courses all year round, minimum age 16 yrs. Accommodations, full board or day school.

Alliance Française, Centre Regional d'Examens de l'Allliance Francaise de Paris pour le Côte d'Azur, 1, Rue Vernier, 06000 Nice. Tel: 93.87.42.11. All year round, min 16 years. Optional accommodation.

Centre International d'Etudes Françaises, Université de Nice, Faculte des Lettres, Boulevard E.-Herriot, 06000 Nice. Tel: 93.86.66.43 Two summer sessions, minimum age 18 years.

Ecole du Château, Château Notre-Dame-des-Fleurs, 06140 Vence. Tel: 93.58.28.50; or

FIVE, 1 bis, Rue Molière, 06000 Nice. Tel: 93.52.42.82; telex: 460000. Courses from February–December, minimum age 18 years.

Institut de Français, 23, Avenue Général Leclerc, 06230 Villefranche-sur-Mer. Tel: 93.01.88.44; telex: 970989. Four- or eight-week programmes all year round, minimum age 21 years.

Centre International de Formation Musicale, Conservatoire National de Region, 24, Boulevard de Cimiex, 06000 Nice. Tel: 93.81.01.23. Courses 7–22 July, 24 July–8 August, minimum age 16 years. French language, music, dancing, plastic arts.

A number of tour operators have designed trips especially geared for the student or with education themes. Some names:

Centre des Echanges Internationaux, 104, Rue de Vaugirard, 75006 Paris. Tel: (1) 45.49.26.25; telex: 210311. Non-profit association offering sporting and cultural holidays with stays in the French Riviera (15–30 years).

Club Riviera Langues International, Château Laval, 14, Route de la Badine, 06600 Antibes. Tel: 93.74.36.08. All levels language courses in Antibes/Juan-les-Pins (min. 18 yrs) with half-board or homestays.

Institut de Langue Française de la Côte d'Azur, 11, Boulevard Matignon, 83400 Hyères. Tel: 94.65.03.31; telex: 404705. Elementary to advance courses with or without homestay (min. 18 yrs).

International Council for Cultural Exchange, Inc., 1559 Rockville Picke, Ste. 306-21, Rockville, MD 20852. Tel: (301) 983-9479. Language, culture, advance music and art programmes. Special language programmes for executives, managers and others. High school students to senior citizens.

American Leadership Study Groups, Airport Drive, Worcester, MA 01602. Tel: (617) 757-6369. All-inclusive educational travel programmes with a structured learning emphasis. Adult and special interest groups. French studies at Nice, one- and two-week travel-oriented programmes.

COURSES

Provence is famous for its many artistic schools and colonies. **The Centre National d'Information et de Documentation sur les Métiers d'Art (CNIDMA)** regularly updates a catalogue (90 ff) that lists all available programmes in France, including prices. Some operate only during the summer.

CNIDMA, Musée des Arts Décoratifs, 107, Rue de Rivoli, 75001 Paris.

Below are some suggestions for studying arts, crafts and music in the Vaucluse:

Pottery: Michel Delsarte, Chemin des Chinaïes, Plan de Saumane, 84800 Isle-Sur-La-Sorgue. Tel: 90 20 33 17.

Gabriel Vorburger, Mas Mabelly, 84360 Puget-Sur-Durance. Tel: 90 68 32 08.

Lace-making, spinning, dying and weaving: Mya Lagas-Muller, Le Rocasson, Chemin de Combemiane, 84400 Apt. Tel: 90 74 32 83.

Drawing, painting and sculpture: Bernard Pfriem, Foreign Study/Lacoste, Cleveland Institute of Art, 11141 East Boulevard, Cleveland, Ohio 44106. Tel: (216) 421-4322.

Hélène Marion, Atelier du Beffroi, Les 4 coins, 84330 Caromb.

Marion Lamy, Grande Rue, Cabrières d'Avignon, 84220 Gordes. Tel: 90 76 92 53.

Daniel Robert, Les Compagnons du XXè siècle, Mas de la Rouvière, 84210 Venasque.

Monique Labarthe (Ceramics, Sculpture), 75, Rue Antoine de Très, 84240 La Tour d'Aigues. Tel: 90 77 50 13.

Fine arts restoration: Michel Hebrard, 3, Rue du Puit de la Tarasque, 84000 Avignon. Tel: 90 82 02 68.

Basket-making and caning: Jean Claude Mangematin, Atelier du vannier, 84730 Cabrières d'Aigues.

Photography: Jean-Pierre and Claudine Sudre, 84710 Lacoste. Tel: 90 75 82 70.

Jean Raffegean, Le Jas, Roussillon, 84220 Gordes. Tel: 90 75 62 83.

Ironwork: Gérard Besset, Prats des VAllats, 84240 Grambois. Tel: 90 77 93 84.

Music: Deller Academy of Early Music – Summer course in Lacoste. 61, Oxenturn Road, Wye, Ashford, Kent, England.

PHOTOGRAPHY

Just as painters once flocked to the south with their easels, photographers now find themselves seduced by the magical light and colours of Provence. There's no shortage of photo opportunities in the region, but you should keep in mind a couple of points:

1. By noon the sun is very strong. Unless you don't mind washed-out pictures, you should try to shoot early in the morning or at dusk. You may also want to buy a polarised filter before leaving home.
2. Because of the heat, film left in your car can be damaged. It is best to leave anything you aren't immediately using back at your hotel. Otherwise, try to get a protective and insulated pouch.
3. Unless you absolutely can't wait, you would do well to hold off until you get home to have your film developed. Especially in the smaller towns, the quality of development is not guaranteed and it will generally be quite pricey.
4. Similarly, film purchased in the small towns will be expensive and may be old.

The Pays d'Arles boasts two major events for photographers. The first is the Rencontres Internationales de la Photographie in Arles. The Rencontres include a photographic festival during the first week of July with soirées of photographic spectacles projected upon a giant screen in the Théâtre Antique; a selection of photographic exhibitions set up in varied spots around the town during the first two weeks of July; and over 40 different courses and lectures given by recognised professional photographers during the first three weeks of July. During this period, Arles overflows with photographers from around the world, from the most famous to the most humble.

For more information, contact:
Rencontres Internationales de la Photographie, 16 Rue des Arènes, BP 90, 13632 Arles Cedex.
Tel: 90.96.76.06.

The second event takes place from mid-September into early October. During this time, the renowned Maine Photographic Workshops take up residence in Fontvieille to offer a series of intensive week-long classes. Unlike the Rencontres, the Workshops concentrate on personalised interaction with the instructors and hands-on learning. They also take care of your lodging, etc., in the best of styles. Of course, you do pay for this.

For more information, contact:
Maine Photographic Workshops, Rockport, Maine 04856, United States. Tel: 207-236-8581.

PILGRIMAGES

There are a number of pilgrimages made to spots in Provence throughout the year. The most famous of these is, of course, the gypsy pilgrimage to Les Stes-Maries-de-la-Mer 24–25 May. During this period, gypsies from all over Europe make their way to the seaside town in pilgrimage to their patron saint, Sarah. A large mass is held followed by a procession to the sea carrying the statue of Saint Sarah – plus there is a great deal of dancing and singing and general partying. Some others:

Allauch: Notre-Dame-du-Château – 8 September.
Beaucet: Eglise St-Gens May and September.
Boulbon: St-Marcelin (Procession of the Bottles) – 1 June.
Marseille: Notre-Dame-de-la-Garde – 14 and 15 August.
Marseille: Notre-Dame-de-la-Calline – 8 September.
St-Etienne-du-Grès: Notre-Dame-du Château – end of May.
Ste-Etienne-les-Orgues: Notre-Dame-de-Lure – 8 September.
Stes-Maries-de-la-Mer: Saints Mary Jacoby and Salome – Sunday by 22 October.
Stes-Maries-de-la-Mer: invention of the hunt – Sunday by 4 December.

LANGUAGE

French, of course, is the official language of Provence, although the area does have its own regional dialect of "Provençal." Apart from encounters with older people deep in the country and the names of restaurants, Provençal won't impinge much on the average tourist and, if you speak at least some French, you shouldn't find communication impossible (although their lilting accent may take a while to get used to). The dialect is not an easy one to pick up, even if you speak other romance languages. One tip: *"Lou"* means "the" in Provençal and is used in the names of many establishments.

Some common French terms:

hello	*bonjour*
good bye	*au revoir*
please	*s'il vous plaît*

thank you	*merci*
thank you very much	*merci beaucoup*
you're welcome	*de rien*
Where is/are...?	*Où est/sont...?*
What is it?	*Qu'est-ce que c'est?*
How much?	*Combien?*
Do you have...?	*Avez-vous...?*
When?	*Quand?*
What time is it?	*Quelle heure est-il?*
with/without	*avec/sans*
the airport	*l'aeroport*
the plane	*l'avion*
the train station	*la gare*
the train	*le train*
the bus	*le car*
the car	*la voiture*
the subway	*le métro*
the bank	*la banque*
the exchange	*le change*
customs	*la douane*
the post office	*le bureau de poste*
the bathroom	*les toilettes*
room where the bath is	*la salle des bains*
the police station	*la gendarmerie*
Help!	*Au secours!*
the hospital	*l'hôpital*
the doctor	*le medecin*
the nurse	*l'infirmière*
the dentist	*le/la dentiste*
I am sick.	*Je suis malade*
the drugstore	*la droguerie/la pharmacie*
shampoo	*le shampooing*
soap	*le savon*
shaving cream	*la creme à raser*
toothpaste	*le dentifrice*
tampon	*le tampon (hygienique)*
I want to dial a number	*Je veux composer un numero*
...to call collect	*...téléphoner en PCV*
...make a person-to-person call	*...téléphoner avec préavis*
breakfast	*le petit déjeuner*
lunch	*le déjeuner*
dinner	*le dîner*
a cup	*une tasse*
a glass	*une verre*
a plate	*une assiette*
a bowl	*un bol*
a fork	*une fourchette*
a knife	*un couteau*
a spoon	*une cuillère*
a napkin	*une serviette*
the bill	*l'addition*
the waiter	*le garçon*

the waitress	*la serveuse*
I would like a coffee	*Je veux du café*
coffee with milk	*du café au lait*
tea	*du thé*
wine	*du vin*
beer	*une bière*
mineral water	*de l'eau minerale*
juice	*du jus*
ice cubes	*les glaçons*
butter	*la beurre*
salt	*le sel*
pepper	*le poivre*
sugar	*le sucre*
mustard	*la moutarde*
jam	*la confiture*
oil	*l'huile*
vinegar	*le vinaigre*
bread	*le pain*
eggs	*les oeufs*
vegetables	*les légumes*
salad	*la salade*
meat	*la viande*
beef	*le boeuf*
rib steak	*l'entrecôte*
lamb	*l'agneau*
leg of lamb	*le gigot d'agneau*
cold cuts	*la charcuterie*
ham	*le jambon*
pork	*le porc*
sausage	*la saucisse*
rabbit	*le lapin*
veal	*le veau*
chicken	*le poulet*
fish	*le poisson*
trout	*la truite*
salmon	*le saumon*
red mullet	*le rouget*
shellfish	*les fruits de mer*
lobster	*l'homard*
shrimp	*les crevettes*
dessert	*le dessert*
cheese	*le fromage*
fruit	*les fruits*
pastry	*la patisserie*
ice cream	*la glace*
one	*une*
two	*deux*
three	*trois*
four	*quatre*
five	*cinq*
six	*six*
seven	*sept*
eight	*huit*
nine	*neuf*
ten	*dix*

twenty	*vingt*
thirty	*trente*
forty	*quarante*
fifty	*cinquante*
sixty	*soixante*
seventy	*soixante-dix*
eighty	*quatre-vingt(s)*
ninety	*quatre-vingt-dix*
one hundred	*cent*
one thousand	*mille*

USEFUL ADDRESSES

GOVERNMENT TOURIST OFFICES

There are French Government Tourist Offices around the world. They should be a main source of information.

Argentina: c/o Chambre de Commerce et d'Industrie Franco-Argentine, Bartolome Mitre 559, 8th floor, 1342 Buenos Aires. Tel: 331.24.94.
Australia: Kindersley House, 33 Bligh Street, Sydney NSW 2000. Tel: 231.52.44.
Austria: Hilton Centre 259C/2 Landstrasser, Hauptstrasse 2 A, Postfach 11, 1033 Vienna. Tel: 222.75.70.62.
Belgium: 21, Avenue de la Toison d'Or, 1060 Brussels. Tel: 25.13.73.89.
Brazil: Avenida Paulista 509, 10 andar, CEP 01311 Sao Paulo. Tel: 11.251.25.90.
Canada: 1981 MacGill College,Tour Esso, Suite 490, Montreal Quebec H3 A2 W9, tel: (514) 288-4264; 1 Dundas Street West Suite 2405, Box 8, Toronto Ontario M 5G1Z3, tel: (416) 593.4723.
Denmark: Frederiksberggade 28, 1st floor, DK 1459 Copenhagen K. Tel: 1.11.40.76.
Great Britain: 178 Piccadilly, London WIV 0AL. Tel: 071-493 6594.
Greece: Air France, 4, Rue Karageorgi-Servias, 105-62 Athens. Tel: 1.32.30.501.
Holland: Prinsengracht 670, 1017 KX Amsterdam. Tel: 20.27.33.18.
Hong Kong: Alexandra House, 21st Floor, Chater Road, Hong Kong. Tel: 5.22.31.31.
Italy: Via Sant Andrea 5, 20121 Milan, tel: 2-70.02.68; c/o Air France, Via Vittorio Veneto 93, 00187 Rome.
Japan: Landic #2 Akasaka Building, 10-9 Akasaka 2 Chome, Minotaku Tokyo 107. Tel: 3-582.6965.
Mexico: Paseo de la Reforma, 404 Piso, 06600 Mexico City. Tel: 55.46.91.40.
Portugal: Avenida da Liberdade 244 A, 12000 Lisbon. Tel: 35.54.80.59.
Spain: Granvia 59, 28013 Madrid. Tel: 1.241.88.08; Granvia Corte Catalanes 656, 08010 Barcelona. Tel: 93.302.05.82.
Switzerland: 84, Rue du Rhône, CP 970, CH 1211 Geneva 3. Tel: 22.21.27.49; Bahnhofstrasse 16, Postfach 4979, 8001 Zurich. Tel: 1.211.30.85.
United States: 610 Fifth Avenue, New York, NY 10020-2452, tel: (212) 315-0888; 645 North Michigan Avenue, Chicago, IL 60611, tel: (312) 337-6301; World Trade Center No. 103, 2050 Stemmons Freeway, Dallas, TX 75258, tel: (214) 742-7011; 9401 Wilshire Boulevard, Beverly Hills, CA 90212, tel: (213) 272-2661; 1 Hallidie Plaza, Suite 250, San Francisco, CA 94102, tel: (415) 986-4174.
Venezuela: c/o Air France, Edif. Parque Cristal, Torre Oeste Piso 3, Av. Fransisco de Miranda, Los Palos Grandes, Caracas. Tel: 283.20.44.
West Germany: Westenstrasse 47, Postfach 100-128, 6000 Frankfort/Main 1, tel: 69.75.60.83; Berliner Allee 26, 4000 Dusseldorf, tel: 211-80.375.

TOURIST INFORMATION

Every town in Provence has either an Office du Tourisme or a Syndicat d'Initiative. They can tell you anything you need to know. In the cities and more touristic towns, they often speak a number of languages.

BOUCHES-DU-RHONE

Esplanade Charles de Gaulle, 13200 Arles. Tel: 90.96.29.35; telex: 440096.
Mairie et Syndicat d'Initiative, 13570 Barbentane. Tel: 90.95.50.39.
Hôtel de Manville, Les-Baux-de-Provence, 13520 Maussane-les-Alpilles. Tel: 90.97.34.39.
Mairie et Syndicat d'Initiative, 13990 Fontvieille. Tel: 90.97.70.01.
Rue Lamartine, 13910 Maillane. Tel: 90.95.74.06.
Place Jean Jaurès, 13210 St-Rémy-de-Provence. Tel: 90.92.05.22.
59, Rue des Halles, B.P. #9, 13150 Tarascon. Tel: 90.91.03.52.
56, Cours Gimon, 13300 Salon-de-Provence. Tel: 90.56.27.60.
Place du Général-de-Gaulle, 13100 Aix-en-Provence. Tel: 42.26.02.93.
4, La Canebière, 13001 Marseille. Tel: 91.54.91.11; telex: 430.402.
Place P. Baragnon, 13260 Cassis. Tel: 42.01.71.17; telex: OFTOUCA 441287.

CAMARGUE

Avenue Van Gogh, 13460 Les-Stes-Maries-de-la-Mer. Tel: 90.47.82.55.

ALPES-DE-HAUTE-PROVENCE

Hôtel de Ville, 04200 Sisteron. Tel: 92.61.12.03.
Place du Bourguet, 04300 Forcalquier. Tel: 92.75.10.02.
Place Dr P. Joubert, 04100 Manosque. Tel: 92.72.16.00.
Le Rond-Point, 04000 Digne. Tel: 92.31.42.73.
Syndicat d'Initiative, Moustiers 04360. Tel: 92.74.67.84.
Rue Nationale, 04120 Castellane. Tel: 92.83.61.14.
Rue Principale, 04170 St-Andre-les-Alpes. Tel: 92.89.02.46.
Place Revelly, 04240 Annot. Tel: 92.83.21.44. Syndicat d'Initiative, Old Forge, 04320 Entrevaux.
Syndicat d'Initiative, 04370 Colmars-les-Alpes. Tel: 92.83.41.92.
Place Sept Portes, 04400 Barcelonette. Tel: 92.81.04.71.

VAR

8, Avenue Colbert, 83000 Toulon. Tel: 94.22.08.22.
Allées Vivien, 83150 Bandol. Tel: 94.29.44.34.
Jardins de la Ville, 83110 Sanary-sur-Mer. Tel: 94.74.01.04.
Avenue Belgique, 83400 Hyères. Tel 94.65.18.55.
Rue Jean Aicard, 83230 Bormes-les-Mimosas. Tel: 94.71.15.17.
Place de la Mairie, 83310 La Garde-Freinet. Tel: 94.43.67.41.
Place Republique, 83310 Cogolin. Tel: 94.54.63.18.
Place Hôtel de Ville, 83470 St-Maximin-la-Ste-Baume. Tel: 94.78.02.47.
Boulevard Grisolle, 83670 Barjols. Tel: 94.77.20.01.
Place Mairie, 83630 Aups. Tel: 94.70.00.80.
Rue Victor Hugo, 83690 Salernes. Tel: 94.70.69.02.
9, Boulevard Clemenceau, 83300 Draguignan. Tel: 94.68.63.30.
Le Valat, 83440 Seillans. Tel: 94.76.85.91.

ALPES-MARITIMES

La Ruade, 06660 Auron. Tel: 93.23.02.66.
Place de la Chapelle, 06410 Biot. Tel: 93.65.05.85.
Mairie, 06360 Eze. Tel: 93.41.03.03.
Mairie, 06620 Gourdon. Tel: 93.42.54.83.
Place de la Foux, 06130 Grasse. Tel: 93.36.03.56.
Mairie, 06620 Gréolières. Tel: 93.59.95.16.
Maison d'Isola, 06420 Isola 2000. Tel: 93.23.15.15.
Avenue Mallet, 06250 Mougins. Tel: 93.75.87.67.
L'Escarène, 06440 Peira Cava - Turini Camp d'Argent. Tel: 93.91.57.22.
BP 7, 06260 Puget-Théniers. Tel: 93.05.02.81.
Place Felix-Fauré, 06450 St-Martin Vésubie. Tel: 93.03.21.28.
2, Rue Grande, 06570 St-Paul-de-Vence. Tel: 93.32.86.95.
Mairie, 06470 Sauze. Tel: 93.05.52.62.
Centre Administratif, 06470 Valberg. Tel: 93.02.52.77.
Boulevard Gambetta, 06560 Valbonne. Tel: 93.42.04.16.
Square 8 mai 1945, 06220 Vallauris. Tel: 93.63.82.58.
Place du Grand-Jardin, 06140 Vence. Tel: 93.58.06.38.

COTE D'AZUR

Maison du Tourisme, 06600 Antibes. Tel: 93.33.95.64.
Palais des Festivals, Esplanade Georges Pompidou, 06400 Cannes. Tel: 93.39.01.01.
51, Boulevard Guillaumont, 06160 Juan-les-Pins. Tel: 93.61.04.98.
Avenue de Cannes, 06210 Mandelieu. Tel: 93.49.14.34.
Palais de l'Europe, 06500 Menton. Tel: 93.57.57.00.
2A, Boulevard des Moulins, 98000 Monaco. Tel: 93.30.87.01; telex: 469760.
5, Avenue Gustave V, 06000 Nice. Tel: 93.87.60.60.
Avenue Dénis-Semeria, 06230 St-Jean-Cap-Ferrat. Tel: 93.01.36.86.
23, Avenue Général Leclerc, 83990 St-Tropez. Tel: 94.97.41.21; telex: 461440.

CONSULATES

Foreign consulates located in Marseille include:

Algeria	tel: 91.53.28.99
Argentina	91.71.47.48
Austria	91.53.02.08
Belgium	91.33.25.26
Brazil	91.54.25.51
Canada	91.37.19.37
Chile	91.71.33.74
Colombia	91.37.55.60
Denmark	91.90.80.23
Egypt	91.71.57.57
Finland	91.91.91.62
Great Britain	91.53.43.32
Greece	91.77.54.01
Holland	91.71.47.84
Indonesia	91.71.34.35
Israel	91.77.39.90
Italy	91.47.14.60
Japan	91.71.61.67
Lebanon	91.71.50.60
Libya	91.71.67.02
Mali	91.50.35.65
Mauritania	91.91.91.91
Morocco	91.50.02.96
Panama	91.53.29.24
Peru	91.91.90.55
Portugal	91.76.12.82
Spain	91.37.60.07
Sweden	91.76.30.14
Switzerland	91.53.36.65
Syria	91.54.73.00
South Africa	91.52.81.69
Thailand	91.90.68.52
Tunisia	91.50.28.68

Turkey	91.76.44.40
United States	91.54.92.00
Uruguay	91.76.46.72
USSR	91.77.15.25
Venezuela	91.79.29.03
West Germany	91.77.60.90

FURTHER READING

BELLES LETTERS

Fischer, M.F.K., *Two Towns in Provence*. New York: Vintage Books, 1983.
Fitzgerald, F. Scott, *Tender is the Night*. New York: Scribner's
Giono, Jean, *Regain*.
Hennessy, James Pope, *Aspects of Provence*.
Scipion, Marcel, *Le Clos du Roi*.
Tomkins, Calvin, *Living Well is the Best Revenge*. New York: E.P. Dutton
Wylie, Lawrence, *Village in the Vaucluse*. Third ed., New York: Harper & Row, 1974.

CUISINE

Escudier, Jean-Noel, *La Veritable Cuisine Provençale et Niçoise*. France: Solar, 1982.
Granoux, Monique, *Recueil de la Gastronomie Provençale*. France: Delta 2000/Editions SAEP, 1983.
Vergé, Roger, *Cuisine of the Sun*. edited and adapted by Caroline Conran, New York: MacMillan, 1979.

OTHER INSIGHT GUIDES

Among the other Insight Guides covering this area of France are:

Insight Guide: Côte d'Azur is a companion volume to this book, covering in detail the history, culture and modern-day life of the Riviera.

Insight Pocket Guide: Côte d'Azur provides an ideal introduction to the coast, setting out a range of detailed itineraries.

ART/PHOTO CREDITS

Photography by	**CATHERINE KARNOW except for:**
Page 104, 189	**Ruth Aebi**
8/9	**Douglas Corrance**
206	**Gavin Lewis**
50/51, 52	**National Gallery of Art**
109, 111	**Rex Features**
40, 46, 87, 200	**Anne Roston**
Maps	**Berndtson & Berndtson**
Illustrations	**Klaus Geisler**
Visual Consultant	**V. Barl**

INDEX

A

Abbaye de Silvacane 154
Abbaye de Thoronet 229–230
Abbé de Montmajour 167
Abbé de Senanque 127
agneau de Sisteron 81
Agnes Mortes 187–189
agriculture 54, 59
Aix-en-Provence 46, 147
Albigensians 42
Alexandra David-Néel Foundation 202
Allée des Sarcophages 162
Alpes-de-Haute-Provence 29
Alpes-Maritimes 29
Alpilles 167
Amphitheatre, Arles 165
Amphitheatre, Nimes 115
Ancient Theatre, Orange 117
Annot 206–208
Ansouis 132
Antibes 274
Apt 134
Arc de Triomphe 116
Arles 35, 36, 159, 166
Auberge de la Colombe d'Or 244
Auberge des Adrets 225
Aups 228
Aureille 159
Auron 238
Avignon 71, 104

B

Bandol 217
Baptème des Côtes du Rhône 110
Barbentane 170
Barcelonette 210
Barjols 72, 227
Baroncelli Museum 186
Baux family 39, 40
Beaumes de Venise 112
Bendor 217
Berre Lagoon 60
Beuil 238
Biot 88, 243
Bormes-les Mimosas 220
Bosco, Henri 131
bouillabaisse 78
Brangwyn, Sir Frank 117
Breil 250–251
Brougham, Lord 54, 56

C

Cadarache 194, 201
Camargue 181
Camus, Albert 131
Cannes 270–271
Cannes Film Festival 272–273
Cap d'Antibes 274
Caromb 123
Carpentras 71, 112
Cascade de Courmes 247
Cassis 145, 146
Castellane 205
Cathedral Church of Notre-Dame-du-Bourg 203
Cathédral de Notre-Dame de Nazareth 120
Cathédral St-Saveur 152
Cathédral Ste-Anne 135
Cathédrale d'Images 175
Cathédrale Nôtre-Dame-de-Puy 242
Cavaillon 132, 134
Centre Bourse 141
Cézanne, Paul 53, 150, 153
Chagall 244, 246
Chanel, Coco 57, 278
Chapel of Notre-Dame-de-Lure 198
Chapel of St-Sebastian 238
Chapelle Bleue 249
Chapelle de Notre-Dame du Grosseau 123
Chapelle du Rosoure 245
Chartreuse de la Verne 223
Chartreuse de Val de Bénédiction 111
Chateau de L'Emperi 157
Chateau Grimaldi 274
Château d'If 140
Château de Barben 156
Château de Sauvan 200
Château de Vauvenargues 155
Château du Roy René 170
Châteauneuf-du-Pape 111
Christmas 65–67, 79, 85
Church of Notre-Dame-de-la-Mare 186
Church of St Donat 197
citadel, Sisteron 195
Clement V, Pope 43
Coaraze 248–249
Cocteau, Jean 249, 274
Collegial Ste-Marthe 169
Collobrières 223
Colmars 209–210
Contes 248
cork production 221
Corniche Sublime 204
Corot 176
Corso de la Lavande 202
Cours Mirabeau 148
Cours Saleya 266
Courthézon 112
Côte d'Azur 56, 259–276
Côtes du Ventoux 126
craft 85
crèche 85
cuisine 77

D

Daudet, Alphonse 169
de Craponne, Adam 47
de Lisle, Rouget 49

de Sévigné, Madame 120
dialect 52, 53, 65
Digne-les-Bains 202, 203
Draguignan 230
Dufy 176
Durance River 194, 196

E

Eglise de Notre Dame 167
Eglise de St Vincent 176
Eglise St Didier 107
Eglise St Michel 158
Eglise St Trophime 162
Entraignes-sur-Sorgue 114
Entrecasteaux 229
Entrevaux 208–209, 237
Ernst, Max 231
Esterel 221
Exposition Universelle des Vins et Spiritueux 218
Eze 253

F

Fabré, Jean-Henri 124
faïence 87
festivals 71, 108, 109
Festival d'Arles 165
Félibridge 52, 53
fish 59
foire aux santons 85
Fontaine-de-Vaucluse 135
Fontvielle 87, 168
Forcalquier 199–200
Fort St André 110
Fos-sur-Mer 60
Foundation Maeght 244
Foundation Vasarely 153
Fragonard Museum 242
French Riviera 259–276
Fréjus 223

G

Galeries de la Charité 141
gardiens 184, 186
Garriga, Joseph Gran 165
Gauguin, Paul 54, 167
Gigondas 118
Giono, Jean 130, 198, 199, 201, 204
Glanum 73, 173, 174
Gordes 128, 133
Gorge du Cians 239
Gorges de la Mescla 239
Gorges de Nesque 125, 126
Gorges du Loup 247
Gourdon 247–248
Grand Canyon de Verdon 29, 203, 204, 205
Grasse 240–241, 254
Grès d'Annot 208
Grignan 120
Grimaldi family 268
gypsies 71, 186

H – J

handicraft 85
herbs 172
High Chapel of St Michel 186
Hitchcock, Alfred 253
Hôtel de Manville 175
Hyères 218–219
Iles d'Hyères 219
Iles d'Or 219
Iles des Lerins 274
industry 55
Isola 2000 238
James, Henry 114
Jardins des Vestiges 141
Jean Cocteau Museum 274
Juan-les-Pins 274

L

La Brique 250
La Ciotat 146
La Garde-Freinet 221, 222
La Turbie 252
Lac de Castillon 206
Lacoste 129
Lacroix, Christian 166
language, (Provençal) 36, 41, 67
lavender 201, 202, 254
Le Grau de Roi 189
Le Hameau de la Lauze 126
Le Luc 225
Le Pen, Jean-Marie 61
Les Antiques 173, 174
Les Baux 175, 176
Les Dentelles de Montmirail 119
Les Mées 196
Les Terres Marines 121
Les-Stes-Maries-de-la-Mer 36, 71, 185
Levant 220
Léger, Fernand 243, 244
Ligurians 33
Listel 189
Lourmaria 131
Luberon mountains 48, 129

M

Maeght, Airné 244
Maillane 172
Maison Carrée 115
Maison de Tartarin 169
Malaucène 122
Manosque 200–201
Marseille 35, 42, 48, 49, 139, 144
Martigues 49 176
Mary Magdalene 71
Mas du Pont du Rousty 182
Massif de L'Estérel 224
Massif des Maures 29, 221
Matisse, Henri 245
Maures 222
McGarvie-Munn, Ian 229
Menton 274
Mercantour, river 236
Méjanes 184
mining 54
mistral (wind) 29, 121

Mistral, Frédéric 39, 41, 53, 108, 163, 164, 172, 187, 204
Monaco 267–270
Monieux 125
Montagne de Lure 197
Moulin de Lettre 168
Mount Ventoux 28, 122, 124
Mouriès 159
Moustiers-Ste-Marie 87, 203, 204
Moyenne Corniche 253
Mt-Chauve 248
Museum Arlaten 53, 163
Museum of Perfume 241
Museum of Sound and Mechanical Instruments 224
Museum of Underwater Archaeology 224
Musée Arbaud 87, 149
Musée Boumian 187
Musée Calvet 108
Musée Cantini 87
Musée d'Art Païen 163
Musée de L'Emperi 157
Musée de l'Histoire de Marseille 86, 141
Musée de la Faïence 87, 204
Musée de la Moto 209
Musée de Toulon 217
Musée de Vieux-Marseille 141
Musée des Alpilles 173
Musée des Beaux-Arts 113
Musée des Docks Romains 141
Musée du Vieil Aix 85, 151
Musée Extraordinaire 132
Musée F.-Benoit d'Art Chrètien 163
Musée Granet 150
Musée Historique du Centre Var 225
Musée Judeo-Comtadin 134
Musée Lapidaire 104
Musée Municipal, Orange 117
Musée Municipal, Vallauris 242
Musée National Fernand Léger 243
Musée National Picasso 242
Musée Nationale de Ceramique 88
Musée Naval 217
Musée Réattu 164

N – O

Napoleon Bonaparte 52
national anthem 49
Nice 266–267
Nietzsche, Friedrich 253
Nimes 114
Nostradamus 47, 157, 158, 173
Notre-Dame-des-Doms, Cathedral of 107
Noves 171
ochre 54, 55, 134
Offenbach Festival 113
Old Quarter 154
olive oil 158, 159, 250
Orange 116

P

Palais des Papes 105
Parc National du Mercantour 236
Parlement de Provence 46
pastis 129
Pavillon Vendôme 152, 153
Peillon 249
perfume 254
Perrier water 115
Petit Train des Pignes 237
Petrarch 113, 122, 135, 172
Pénitents 196
Picasso 88, 155, 164, 242, 243, 244, 274
Pierre-Ecrite 196
Piène 251
Place d'Albertas 150
Plage de Piemenson 184
Point Sublime 205
Pont d'Avignon 71
Pont St Benezet 105
Pope Clement V 43
Porquerolles 219, 220
Port-Camargue 189
Port-Cros 220
Port Grimaud 274
pottery 86, 87, 88
Priory of Ganagobie 197
Priory of Salagon 200
Provençal dialect 47
Provençal Fronde 48
Provençal Renaissance 41, 47
Provençal, language 36, 41

Q – R

Quartier Ancien 150
Quartier Mazarin 149, 150
Quartier Panier 141
ratatouille 79
Rencontres Internationales de la Photographie 167
René of Anjou 44
Renoir 176
Revolution 49
Riez 201
Riviera 259–276
Roussillon 131
Roya Valley 250
Rubens 242

S

St Baume 71
St Blaise 33
St Didier 108
St-Etienne-de-Tinée 238
St Laurent 158
St-Martin-Vésuloie 240
St-Maximin-la-Ste Baume 226
St Paul 245
St-Paul-de-Vence 243–244
St Raphael 223–224
St Rémy de Provence 172, 174
St Siffrein Cathedral 113
St Tropez 271–274
St Trophime 36
St Victor's Basilica 142
Ste Maxime 224
Salernes 228
Salle Henri Rousseau 164
Salon-de-Provence 156
Sanary-sur-Mer 218
santons 85, 87
Saorge 250
Saracens 37, 221
Sarrians 112

Sault museum 33
Sault 125
Seillans 231
Senanque 128
Serignan 124
Serre-Ponçon 196
Sert, Jose-Luis 244
Séguret 118
Signal de Lure 198
Silvacane 155
Sisteron 194–195
Smollett, Tobias 56
Sophia Antipolis 262, 276
Sospel 251
soupe de poisson 77
synagogue 112

T

Tarascon 71, 169
Tavel wines 114
Tende 249–250
Théâtre Antique 165
Toulon 215–216
Tourette-sur-Loup 246
tourism 54, 56, 57, 61
Tourtour 229
Tower of Philip the Fair 110
Tricolore 49
Trophée des Alpes 252
troubadours 39, 40, 41, 175

U – Z

Utelle 240
Waldensians 47, 48
wine 111, 112, 114, 118, 155, 217
Zola, Emile 53